DESIGN OF
REINFORCED CONCRETE STRUCTURES

Design of Reinforced Concrete Structures

(Fourth Edition)

P. Dayaratnam
Formerly Professor and head of Civil Engineering
Formerly Dean of Research and Development
Indian Institute of Technology, Kanpur
Formerly Vice Chancellor, J.N.T. University, Hyderabad

Oxford & IBH Publishing Co. Pvt. Ltd.
New Delhi

DESIGN OF REINFORCED CONCRETE STRUCTURES

Oxford & IBH Publishing Company Pvt. Ltd.
113-B Shahpur Jat
Asian Village Side
New Delhi 110 049, India

Fax: (011) 4151 7559
Email: oxford@oxford-ibh.in

Last Reprint 2017

ISBN 978-81-204-1419-8

Printed at Chaman Enterprises, New Delhi.

To
My Alma Mater
Harries Higher Elementary School
Sattenapalli, Andhra Pradesh, India

Preface to the Fourth Edition

Design methods and quality control in concrete structures are changing as the technology is improving. The codes on design and construction practices also change depending on the trends in technology. IS: 456 is the mother code for design and construction of plain and reinforced concrete structures and it was originally developed in 1953 and revised in 1957, 1964, and 1978. The book on Design of Reinforced Concrete Structures was written based on the IS: 456-1978. The code is now revised in 1999 and the highlights of the revision are:

☐ Emphasis on design criteria based on the durability considerations of the structures

☐ A change in the acceptability criteria of concrete is made,

☐ Minimum strength of concrete to be used in reinforced concrete is increased to M20 from M15,

☐ Working Stress Design (WSD) was given equal weightage as that of Limit State Design in 1978. However, in the present revision, it is given less importance and the chapter is moved to appendix.

☐ Some other minor changes in the specifications are made.

The book needs a suitable revision consequent to the revision of the code of practice and also because of the improved technological changes in practice during the last one-decade since the third edition was published. The important revisions in the book are:

☐ The section on the durability of concrete structures is expanded incorporating latest trends in durability considerations and the present code of practice on plain and reinforced concrete,

☐ A section on Quality control in concrete structures is added incorporating the concrete acceptability criteria based on the probability of failure and on IS: 456-1999 draft,

☐ A brief note on non-destructive testing is added to the section on quality control,

☐ As the Ultimate Strength Design (USD) of RC Structures is phased out in almost all the codes of practice, so the chapter on the Ultimate Strength Design is reduced to minimum level. The chapter is not deleted as the concepts in the strength design are worth learning for a structural Engineering student.

☐ Working Stress Design is the earliest method of design used in many countries. However, it is either phased out or being phased out in many countries. Several chapters with many examples were included in the earlier edition. But the present version has retained only typical examples of importance.

☐ The materials and the production costs also change from place to place and from time to time. The basic unit costs used in the earlier edition have undergone considerable change. Extrapolated unit costs are used at the moment just to give an indication of the relative cost of the items such as concrete, steel and formwork. The cost calculations must be taken as guidelines but not in absolute terms.

The book must be useful universally even though it may be code dependent, therefore the three methods of design used by many countries are included with more emphasis on limit state design method. The examples are illustrated as lucidly as possible with drawings.

P. Dayaratnam

Preface to the First Edition

The invention of Portland cement in the middle of the nineteenth century introduced remarkable changes in the building industry. The use of cement concrete and reinforced concrete structures has made several projects not only feasible but relatively economical and durable. The combination of cement concrete and steel has resulted in achieving architectural imaginations of master builders, establishing long lasting variety of structures in short duration, and enabled skilled and less skilled workers to produce strong and durable buildings. The art of structural design became a common engineering practice during the last sixty years. The design of reinforced concrete structure is being taught to all Civil Engineering and Architectural Students in all parts of the world. The methodology design has been undergoing modifications during the last fifty years. Working stress design, Ultimate strength design and Limit states design are the three methods commonly practiced.

Working Stress Design (WSD) is probably the earliest codified method of design of reinforced concrete structures, and it is based on a criterion that the actual stresses developed in the materials under the action of working loads must be limited to a set of allowable values. The method also constraints that the deformations of the structures or elements must be within the acceptable values. An elastic-linear structural analysis is considered to be a basis in determination of stresses in the materials. The book is divided into two parts, the first part deals with superstructures and the second deals with foundations. Chapter one gives a brief review of the important characteristics of the materials used in the reinforced concrete. Chapters two to six present the working stress design of variety of structures.

Limit State Design (LSD) is relatively a recent method and is developed during late sixties and brought into practice during seventies. The concepts on structural safety and serviceability are rationalised better, consequently certain amount of sophistication is introduced. The structures are designed to provide adequate strength, serviceability and durability. Partial safety factors applied to loads and materials provide the required safety and serviceability of the structure. The design forces and deformations are arrived at by the limit or the elastic analysis depending on the type of the limit state. Chapters seven to ten present the design of structural components by the Limit State Design.

Ultimate Strength Design (USD) is primarily based on strength concept. The working loads are enhanced by multiplying them with load factors to give a hypothetical loads called ultimate loads. Then the design forces on the members are obtained by an elastic structural analysis under the action of the ultimate loads. The members are proportioned such that the strength of the number is not less than the ultimate design force. In a way, though not strictly, the ultimate strength design may be considered as the strength limit state of the limit state design. In view of the introduction of the limit state design, and an overlap between the two methods, this method is not discussed in as much details as the other methods. Chapter twelve presents design of beams by ultimate strength approach. There is so much common in the design of columns by limit state and ultimate strength design methods, therefore the design of columns by this method is not presented. Chapter eleven presents design of water tanks. There are many components such a beams, columns, slabs, shells, balconies, staircases and foundations in water tanks. This chapter therefore presents a project oriented design rather than element oriented approach. In addition, the design requirements of water retaining structures are illustrated through a set of examples. This chapter demonstrates the overall design of a structure and provides good training in synthesis.

The part two of the book consisting of chapters thirteen to seventeen presents design of reinforced concrete foundations. The working stress design of shallow, deep and machine foundations is presented in three separate chapters. The limit state design of shallow foundations and the working stress design of retaining walls are presented in two chapters.

The book presents a large number of illustrative examples with reinforcement details. The examples not only illustrate the method of design but give an insight to the economical designs. The students should be trained to produce a structurally sound design and further they should be made aware of the influence of the design parameters on the economy. In many cases, the cost comparisons of solutions with variations in the sensitive design variable be considered as a useful tool. The relative costs of materials are likely to differ with time and place; further the value system attached to the cost of materials, and spatial requirements and constraints may present another problem in economics of design. While it is impossible to establish a comparative study in absolute terms, but the impact of the variations of the design parameters on the cost as presented in this book helps the students to provide economical solutions. The book also provides solutions of a set of problems by the three methods along with the relatives costs. The columns in actual practice are invariably subjected to combined axial force and bending, the design of which involves trial and correction iterations. A direct and systematic approach of the design of columns under combined action by working stress and limit states design is presented. There are a large number of types of floor slabs in construction. Direct design moment coefficients to the design of flat slabs, waffle slabs and circular slabs with different boundary conditions are listed in the book. The design of slabs by yield the theory which results into an economical design is presented in detail applying to many types of slabs such as simple, flat, waffle and circular slabs. There are many variations and boundary conditions to the foundations of walls, columns and shafts. This book takes up most of the variations in the foundations and illustrates the design through many examples.

One of the strengths of the book is that the examples given are the solutions to either full or part of structures which are already constructed. This is an outcome of several lectures given to students and practicing engineers, and design already implemented. Various Indian Codes of Practice are extensively used in this book so as to enable the designer to be in tune with the accepted practices. Many thanks are due to Indian Standard Institute, New Delhi. The preparation of the manuscript was assisted by many persons at IIT Kanpur along with the financial support of the Quality Improvement Programme for which the author is very thankful.

P. DAYARATNAM

Notations

A = area of cross-section
A_c = area of concrete
A_g = gross area
A_s = area of steel or reinforcement
A_{sc} = area of compression steel
A_{st} = area of tension steel
a = size of pile
B = breadth or width
B_f = width of footing
B_p = width of plate
b = width, width of beam sections
b_f = width of flange
b_w = width of web
C = total compressive force on concrete, or shear strength of soil
C_a = adhesion strength of soil
C_r = load reduction factor for long columns
c = cohesion
D = diameter of bars
D_d = relative density of the densest soil
D_f = depth of foundation
DL = dead load or fixed load
d = effective depth of reinforcement from the compression fibre, or deflection of subgrade
d_f = settlement of footing
d_p = settlement of plate in bearing test on soil
E = Young's modulus of material
EL = earthquake load
e = eccentricity of load or M/P or void ratio or elongation
F = force, load factor
F_d = load factor applied to dead loads
F_e = load factor applied to seismic load
F_l = load factor applied to live load
F_w = load factor applied to wind load
FS = factor of safety
f_{ck} = characteristic strength of concrete
f_y = yield or proof strength of steel
G = specific gravity

h = depth, height

h_o = maximum depth

I = moment of inertia of a section

jd = leverarm

K = stiffness, moment resistance coefficient, coefficient of later earth pressure

k = subgrade modulus

L = span

LL = live load

M = bending moment

MPa = mega Pascals = $10^6 N/m^2$

$M20$ = grade of concrete having a characteristic strength of 20 N/mm^2

N_c, N_q, N_r = bearing capacity factors

P = axial force

P_h = horizontal component of soil pressure

P_v = vertical component of soil pressure

p = pressure or perimeter

p_o = pressure at depth h_o

Q_u = ultimate bearing capacity of the soil or first moment of area

q = bearing pressure or permanent surcharge pressure

R = radius

r = radius or radius of gyration, h/h_o

S = surface area, shear resistance in soil

s = slenderness ratio, spacing reinforcement bars

T = temperature

t = thickness

V = transverse shear force

V_e = effective shear force

W = concentrated load

WL = wind load

w = unit weight or distributed load

Z = section modulus

z = depth of soil

$\in / \varepsilon$ = strain

γ = density of material, partial safety factor

μ = Poisson's ratio or coefficient of friction

η = efficiency factor

ϕ = angle of internal friction

σ = stress

σ_{acb} = allowable bending compressive stress

σ_{acc} = allowable axial compressive stress

σ_{at} = allowable tensile stress

σ_{bt} = allowable bending tension

σ_c = compressive stress

σ_{cb} = bending compressive stress
σ_{cc} = axial compressive stress
σ_{co} = compressive stress in composite steel
σ_e = effective normal stress in soil
σ_{sc} = compressive stress in steel

Contents

PART I

RCC
SUPERSTRUCTURES

Chapter 1

Properties of Materials used in Concrete

1.1 Introduction

Lime soorkee concrete remained in use for many centuries until cement concrete came into use in the later part of the 19th century with the advent of portland cement. The main ingredients of concrete are: aggregate and cement. Aggregate forms the bulk volume and filler of the concrete whereas cement acts as the binder. In addition, water which is used to combine the materials plays a very important role in binding. These three basic materials mixed in different proportions result into concrete of variables strengths. Some additives and admixtures are added to give certain desired properties to the concrete. This chapter is devoted to a review of the important characteristics of the ingredients of concrete.

1.2 Definitions in Aggregates

Crushed or uncrushed materials derived from the natural sources such as rocks, gravel, boulders and sand for production of concrete are called *aggregates.* There are two main groups in aggregates. *Fine aggregates* having particle size less than 5 mm, and 90 to 100 per cent of which must be able to pass through 4.75 mm sieve. Natural sand, finely crushed stone, and crushed gravel are treated as fine aggregates. *Coarse aggregate,* on the other hand, is the stone material most of which is retained on 4.75 mm sieve. Uncrushed gravel or stone which is the result of natural disintegration and crushed gravel or stone are the usual coarse aggregates. Aggregates, whether fine or coarse, must be hard, durable, clean and free from coal, mica, iron pyrites, shale, clay, sea shells, alkalies, and other organic materials. Unless otherwise specified, aggregate usually refers to that from stone and which is used in concrete constructions. There are other types like *broken brick-aggregate, cinder aggregate, light weight, slag and heavy weight aggregate,* etc. which are used in some concrete special constructions. Coarse and fine aggregates together are used in concrete whereas only fine aggregate with either cement or lime is used in mortar for plastering and jointing.

Coarse aggregates are classified into two main groups: (i) single-size aggregate and (ii) graded aggregate. *Single-size aggregate* is based on a nominal size specification. It contains about 85 to 100 per cent of the material which passes through that specified size of the sieve and zero to 25% of which is retained in the next lower sieve. Table 1.1 gives different single-size aggregates. A graded aggregate contains more than one single-size aggregate and its requirements are also given in Table 1.1. About 90 to 100 per cent of the aggregate passes through the sieve whose size is same as that of the aggregate and further 30 to 70 per cent must pass through the next lower size of the sieve. Another type called gap graded aggregate is also used in concrete at times.

Table 1.1 Size Classification of Coarse Aggregate

IS sieve size (mm)	Percentage passing single-size aggregate of nominal size (mm)						Percentage passing of graded aggregate of nominal size (mm)			
	63	40	20	16	12.5	10	40	20	16	12.5
80	100	–	–	–	--	–	100	–	–	–
63	85-100	100	–	–	–	–	–	–	–	–
40	0-30	85-100	100	–	–	–	95-100	100	–	–
20	0-5	0-20	85-100	100	–	–	30-70	95-100	100	100
16	–	–	–	85-100	100	–	–	–	90-100	–
12.5	–	–	–	–	85-100	100	–	–	–	90-100
10	0-5	0-5	0-2	0-30	0-45	85-100	10-35	25-55	30-70	40-85
4.75	–	–	0-5	0-5	0-10	0-20	0-5	0-10	0-10	0 10
2.36	–	–	–	–	–	0-5	–	–	–	–

The classification of fine aggregate specified by the Indian Standards is given in Table 1.2. Grade I is the coarsest and grade IV is the finest of the fine aggregate. For 1st grade, about 90 to 100 per cent of the material must pass through 4.75 mm sieve and about 60 to 80 per cent must pass through the next standard sieve, namely 2.36 mm. The first three graded zones are usually acceptable for reinforced or prestressed concrete constructions. However, the 4th grade zone fine aggregate is rather too fine and will decrease workability of the concrete.

Table 1.2 Classification of Fine Aggregate

IS sieve size	Percentage passing for grading zones			
	I	II	III	IV
10 mm	100	100	100	100
4.75 mm	90-100	90-100	90-100	95-100
2.36 mm	60-95	75-100	85-100	95-100
1.18 mm	30-70	55-90	75-100	90-100
600 micron	15-34	35-59	60-79	80-100
300 micron	5-20	8-30	12-40	15-50
150 micron	0-10	0-10	0-10	0-15

*For crushed sands, the permissible limit passing through 150-micron sieve is increased to 20%.

All-in-aggregate: Aggregate containing both coarse and fine aggregates is called *all-in-aggregate*. For a good concrete mix, it is desirable to mix appropriate quantities of the aggregates to develop good compaction and required strength. Sometimes when all-in-aggregate is available, it can be used in the concrete by an addition of appropriate amount of single-sized aggregate. Typical grading of all-in-aggregate is given in Table 1.3.

1.3 Physical Requirements of Aggregate

Aggregate shape can be either round, angular or a combination of the both but should not

Table 1.3 All-in-aggregate Grading

IS sieve	Percentage passing for nominal size of	
	40 mm	20 mm
80 mm	100	–
40 mm	95–100	100
20 mm	45–75	95–100
4.75 mm	25–45	30–50
600 micron	8–30	10–35
150 micron	0–6	0–6

Table 1.4 Percentage of Permissible Maximum Limit of Deleterious Materials in Aggregates

Substance	Fine aggregate		Coarse aggregate	
	Uncrushed	Crushed	Uncrushed	Crushed
Coal and lignite	1.0	1.0	1.0	1.0
Clay lumps	1.0	1.0	1.0	1.0
Soft particles	0	0	3.0	0
Material passing 75 micron IS sieve	3.0	15.0	3.0	1.0
Shale	1.0	–	0	0
Sum of the materials	5.0	–	5.0	5.0

be flaky. It should not contain deleterious materials. The maximum limit of deleterious materials in an aggregate is given in Table 1.4.

Size of coarse aggregate is specified by a sieve size. A 40 mm aggregate normally means that 100 per cent of the aggregate should pass through 63 mm IS sieve (next higher of 40 mm IS-sieve), 85 to 100 per cent pass through 40 mm IS sieve and about 0 to 30 per cent may pass through the next lower size, i.e. 20 mm sieve. *Very large* aggregate is of 80 mm to 150 mm size, *large* aggregate is 40 mm to 80 mm, *medium size* aggregate is 20 mm to 40 mm, and *small* coarse aggregate is 4.75 mm to 20 mm.

The important physical characteristics of aggregates are: (1) specific gravity, (2) bulk density, (3) moisture content, (4) absorption value, (5) aggregate crushing strength, (6) abrasion value, (7) flakiness index, (8) elongation index, (9) presence of deleterious materials, (10) soundness, and (11) potential reactivity.

Specific gravity of a material is the ratio of its weight to the weight of the equivalent volume of distilled water. The specific gravity of a material in any form either in large or small aggregate is same as that of the base stone from which it is derived. It varies from 2.6 to 2.8.

Bulk density is the weight of the aggregate per unit volume. *Void ratio* is the ratio of the volume of the voids to the bulk volume of the aggregate and is given by

$$\text{Void ratio} = \frac{\text{bulk volume} - \text{volume of solid aggregate}}{\text{bulk volume}} \tag{1.1}$$

Void ratio in single-size aggregate is around 0.42 to 0.46 and in graded aggregates it is around 0.38 to 0.44. Less is the void ratio better is the grading of the aggregate.

The void ratio can easily be determined from the amount of water needed to fill the voids of a measured volume of the dry aggregate. The bulk density of the coarse aggregate which depends on the void ratio varies from 0.54 to 0.62 of the specific weight of the material. The bulk density of highly well packed aggregate can, however, be as high as 0.7 times the specific weight. Its actual range is 1500 kg/m^3 to 1900 kg/m^3. High density aggregate may have a bulk density in the range of 1800 kg/m^3 to 2100 kg/m^3.

Moisture content and absorption value: Aggregate when exposed to moisture collects moisture on its surface and absorbs certain portion of the moisture through its pores. The total moisture content with the aggregate is divided into two parts. One is the surface moisture which is the difference in weight of the moist and the surface dry aggregate and is called *moisture content*

$$\% \text{ moisture content} = 100 \left(\frac{\text{weight of moist aggregate} - \text{weight of surface dry aggregate}}{\text{weight of surface dry aggregate}} \right) \quad (1.2)$$

The other portion of water *absorbed* in the pores of the aggregate is equal to the difference in weight of the air-dry and bone-dry aggregate.

$$\% \text{ absorption value} = \left(\frac{\text{weight of the absorbed water content}}{\text{weight of the dry aggregate}} \right) 100 \quad (1.3)$$

Moisture content increases the bulk volume of fine aggregate upto a point whereas the absorbed water content does not change the volume of the aggregate. The increase in volume of the sand which is caused by a moisture film pushing the particles apart is called *bulking of sand*. The volume of sand increases with increase in the moisture content upto a certain limit. but further increase in the moisture content reduces the bulk volume of the sand. The maximum bulking of the sand depends on the type of sand and it is about 30 per cent with normal occurrence around 8 per cent of the moisture content. The ratio of the volume of moist sand to that of dry sand is called *bulking factor*. The bulking factor varies from 1.0 to 1.32 depending on the moisture content and the type of sand. Measurement of fine aggregate by volume in concrete making is not desirable because of the possibility of bulking and uncertain filling or compaction condition.

$$\text{Bulking factor} = \frac{\text{volume of the moist aggregate at optimum moisture}}{\text{volume of the dry sand}} \quad (1.4)$$

Strength of aggregate: The strength of an aggregate depends on the nature of the base rock. size and shape of the aggregate and also the impurities it contains. Aggregate *crushing. crushing value* and *impact value* are the usual indices used to specify the strength quality of an aggregate. The aggregate crushing value is given by

$$\text{Aggregate crushing value} = \frac{100 \, W_2}{W_1} \quad (1.5)$$

where W_1 = weight of surface dry sample of aggregate sample passing through 12.5 mm IS sieve and retained on 10 mm IS sieve.

W_2 = weight of the material passing through 2.36 mm IS sieve after the application of a specified load in a specified manner on the sample.

There is a simple procedure employing simple apparatus for the determination of the crushing strength of an aggregate. The aggregate crushing value is considered to be an index of the strength of the aggregate. Higher the crushing value, lower is the strength of the aggregate. The crushing value more than 40 indicates a weak aggregate. The following maximum limits of crushing value are recommended:

For plain concrete	: 45
For reinforced concrete for buildings	: 40
For concrete for roads and prestressed and high performance concrete	: 30

The aggregate strength can also be determined by the crushing strength of a rock sample cut to a specific size. A minimum of 80 MN/m^2 (MPa) as core crushing strength of rock is normally acceptable for good concrete aggregate.

$$\text{Aggregate impact value} = \frac{100 \; W_2}{W_1} \qquad (1.6)$$

where W_1 = weight of oven-dry sample of aggregate passing through 12.5 mm IS sieve but retained on 10 mm IS sieve.

W_2 = weight of the fractured aggregate finer than 2.36 mm IS sieve formed after 15 impacts of a special hammer falling through 380 mm on the sample.

An impact value of upto 30 indicates good quality; however, a value upto 45 is acceptable for concrete. Aggregate crushing value or impact value for concrete should not be less than 45 and for wearing surfaces such as roadways and runways it should not be less than 30. Table 1.5 gives typical properties of some aggregates.

Table 1.5 Physical Characteristics of Coarse Aggregate

Name	Core crushing strength N/mm^2	Aggregate values		Abrasion value ± 3
		Crushing	Impact	
Granite	150 to 220	15 to 30	12 to 25	18
Basalt	180 to 250	10 to 25	10 to 25	17
Limestone	120 to 180	20 to 40	20 to 40	15
Quartzite	300 to 350	15 to 30	15 to 80	19
Flint	150 to 250	15 to 30	15 to 30	19

The aggregate strength is also indicated by a load required for ten per cent fines. Oven-dry sample similar to that used in the impact test is subjected to a specific penetration and the load for ten per cent fines is determined.

$$\text{Load required for 10 per cent fines} = \frac{14 \; W}{p + 4} \qquad (1.7)$$

where W = load in ten thousand newtons producing the specified penetration and 7.5 to 12.5 percentage of fines passing through 2.36 mm IS sieve.

p = mean percentage of fines from two tests at W load (should be between 7.5 and 12.5)

Abrasion value of aggregate: Aggregates used in roads and runways are subjected to constant wear. In addition to good strength, the aggregate must have good wear resistance. There are special abrasion apparatus such as Los-Angeles, Deval machines designed to test the wear of an aggregate. Abrasion values can be expressed as

$$\text{Abrasion value} = \frac{100 \, (W_1 - W_2)}{W_1} \qquad (1.8)$$

where W_1 = original weight of the dry aggregate within certain particle size.
    W_2 = weight of aggregate retained after sieving through 1.7 mm sieve and after abrasion test of the sample.

    Equation (1.8) gives a simple relation which is modified depending on the type of test. Wear resistance of a material is inversely proportional to its abrasion value and some ranges of these values are given in Table 1.5.

Flakiness Index: The shape of the aggregate for use in the concrete should not be flaky, that is, thin and oblong. The flakiness of an aggregate is indicated by sieving the aggregate through a set of sieves having oblong openings. *Flakiness index* is the weight of aggregate passing through a set of special sieves expressed as a percentage of the total weight of the sample.

Elongation Index: It is the percentage weight of the sample material retained on various length gauges.

Bond of aggregate: Bonding of an aggregate with cement in the concrete depends on the shape, size, texture and grading of the aggregate. The angularity number *is the index* that gives the interlocking of aggregate in the concrete. Interlocking of the aggregate is inversely proportional to the flakiness index or the elongation index.

Soundness of an aggregate: Aggregates when exposed to changes in temperature and weather tend to disintegrate into smaller particles. More disintegration implies less soundness. The soundness of an aggregate is measured by subjecting it to alternate wetting in saturated solution of sodium or magnesium sulphate and drying it in an oven through a set of cycles. The reduction in the particle size of the aggregate which is obtained through a sieve analysis indicates the soundness or unsoundness of the aggregate. The soundness is more or less inversely proportional to the moisture absorption.

Alkali aggregate reaction: The alkalies present in cement react with silica of the aggregate in the presence of moisture resulting into a gel. The silica-alkali gel increases the volume of concrete which is undesirable. Alkali-aggregate reaction is measured by mechanical and chemical tests.

    Heavy weight and *light weight* aggregates are used for special concretes.

1.4 Introduction of Portland Cement

Cement is the main bonding material used in building construction. Natural, portland and aluminous cements are the three main natural groups. Compounds of lime are the main elements having the basic property of cementing. Portland cement is obtained by mixing of calcareous (*lime stone* or *chalk*) and argillaceous (clay or shale containing *silica* and *alumina*) materials, burning the mixture at clinkering temperatures and grinding the clinker along with some gypsum into a fine powder.

Dry process: The crushed raw materials in appropriate proportions are ground into fine powder in a grinding mill. The dry powder is blended into a fine mixture in a blending silo with the help of compressed air. The dry mixture is sieved and fed into a rotating granulator along with appropriate quantity of water and then air dried to form into granules. The granules also called pellets are baked by hot air and fed into a rotating kiln where the material undergoes some chemical reactions when exposed to high temperature (1300 C to 1500 C). The clinker formed in the kiln is cooled and ground with gypsum in ball mills. The powder ground to a satisfactory level (about 10^{11} particles per newton) is separated as cement.

Wet process: Finely broken raw materials are mixed with water in separate wash mills to form fine mixtures. The mixtures are screened, mixed in appropriate proportions and the resulting slurry is pumped into storage tanks. The slurry is kept under constant stirring against any sedimentation. The slurry having proper proportions is fed into a rotary kiln. Pulverized coal which gets burned in kiln is blown in from the other end of the kiln. The slurry as it moves down undergoes chemical changes and forms into clinker. The clinker is then let through cooling chambers and afterwards pulverized in ball-mills along with gypsum. The powder ground to a satisfactory level is separated as cement.

1.5 Principal Compounds of Portland Cement

The principal compounds of cement are given in the Table 1.6. Basic chemical compounds in cement are CaO, SiO_2, Al_2O_3, Fe_2O_3, MgO, K_2O, Na_2O, and SO_3. The calcium oxide gets fused with silica, alumina and iron oxide resulting into tricalcium silicate ($3CaO.SiO_2 = C_3S$), dicalcium silicate ($2CaO.SiO_2 = C_2S$), tricalcium aluminate ($3CaO.Al_2O_3 = C_3A$) and tetracalcium aluminoferrite ($4CaO, Al_2O_3.Fe_2O_3 = C_4AF$). These are the main compounds of the final portland cement. First C_3A and C_4AF are formed in the kiln and then C_2S forms. After saturation C_2S turns into C_3S. Tricalcium silicate (C_3S) is mostly responsible for early strength whereas the dicalcium silicate (C_2S) contributes to the strength at later stage. The tricalcium aluminate contributes towards the strength developed in the first 48 hours.

Table 1.6 Principal Compounds of Ordinary Portland Cement

Compound	Composition		Approximate
Tricalcium silicate	$3CaO.SiO_2$	= (C_3S)	45-58
Dicalcium silicate	$2CaO.SiO_2$	= (C_2S)	15-32
Tricalcium aluminate	$3CaO.Al_2O_3$	= (C_3A)	6-13
Tetracalcium aluminoferrite	$4CaO.Al_2O_3.Fe_2O_3$	= (C_4AF)	6-12
Minor compounds	(MgO, K_2O, Na_2O)		2-3

Table 1.7 Chemical Requirements of Ordinary, Rapid-hardening and Portland Pozzolana Cements

	Substance	Percentage limits
1.	Ratio of lime to silica, alumina and iron oxide $$\frac{CaO - 0.7\ SO_3}{2.8\ SiO_2 + 1.2\ Al_2O_3 + 0.65\ Fe_2O_3}$$	Not more than 1.02 and not less than 0.66
2.	Ratio of alumina to that of iron oxide	Not less than 0.66
3.	Insoluble residue	Not more than 2.0
4.	Magnesia	Not more than 6.0
5.	Total sulphur content, calculated as sulphuric anhydride (SO_3)	Not more than 2.75
6.	Total loss on ignition	Not more than 5.0

1.6 Physical Requirements of Portland Cement

Fineness: The number of the particles of cement powder in a unit weight reflects the fineness of a cement. It is also indicated by the surface area of the particles per unit weight. The development of the strength of the cement is proportional to the fineness of the cement. Table 1.8 gives the fineness requirements. Soundness, setting time and compressive strength are the other physical tests normally required. Table 1.8 also gives the acceptable limits of these physical tests.

Setting of cement: Cement paste changes from fluid state to hard state. This process is called *setting of cement*, and is measured by standard test on cement paste. The premature hardening of the cement paste is called *false set*. The hardening process caused by the reaction of pure C_3A with water is called *flash set*. It is prevented by the addition of gypsum to the cement.

Initial set of a cement indicates the beginning of noticeable stiffening of the cement paste and the *final set* indicates the beginning of the hardening of the paste. The normal consistency, initial and final sets are determined by vicat needle or by Gillmore needle tests. The limits of initial and final setting times are indicated in Table 1.8. As the cement comes in contact with water, dissolution and reaction of the constituents begins generating high heat. After few minutes the reactions and heat generation slows down, while cement builds up hydrated products. This first stage lasts about four to five hours. Again reactions get accelerated after few hours resulting in the product of calcium silicate gel. During this period the cement paste gets hardened and improves in strength.

Typical hydration of cement are:

$$3CaO.Al_2O_3 + 6H_2O \rightarrow 3CaO \cdot Al_2O_3 \cdot 6H_2O \tag{a}$$

(The very first reduction if gypsum is not present).

$$3CaO.Al_2O_3 + 3(CaSO_4.2H_2O) + 25H_2O \rightarrow 3CaO \cdot Al_2O_3\ 3CaSO_4 \cdot 31H_2O \tag{b}$$

(Calcium hydro sulfoaluminate)

$$2(3\ CaO.SiO_2) + 6H_2O \rightarrow 3CaO \cdot 2SiO_2 \cdot 3H_2O + 3Ca(OH)_2 \tag{c}$$

$$3(2\ CaO.SiO_2) + 6H_2O \rightarrow 3CaO \cdot 2SiO_2 \cdot 3H_2O + 3Ca(OH)_2 \tag{d}$$

$$4CaO.Al_2O_3.Fe_2O_3 + 2Ca(OH)_2 + 10H_2O \rightarrow 3CaO \cdot Al_2O_3 \cdot 6H_2O + 3CaO \cdot Fe_2O_3 \cdot 6H_2O \tag{e}$$

Normal consistency: The state of standard condition of cement to obtain a specified consistency of cement paste is called normal consistency. It is that condition for which a penetration of 33 to 35 mm by a standard 10 mm needle is obtained in 30 seconds. The water content for normal consistency varies from 20 to 26 per cent. The normal consistency of cement paste is used in standard physical tests of cement. Specific gravity of ordinary portland cement varies from 3.12 to 3.16. The percentage of voids in cement is around 40 and bulk density is about 18000 to 20000 N/m^3. Usually cement is supplied in bags of 50 kg wt.

Soundness of cement: The soundness of a cement is indicated by the rate of expansion of the cement paste during hardening process. The soundness of cement is inversely proportional to the expansion of the cement paste. Expansion of cement paste can be due to slow hydration or reaction of the compounds such as free lime, magnesia and calcium sulphate. Hydration of the free lime is delayed because of the cement film around the free lime. Expansion of cement paste takes place as the free lime gets hydrated. Fine grinding of the raw materials or aeration of the cement decreases the unsoundness of cements.

Strength of cement: Strength of cement is determined from the compressive test on hardened cement-sand mortar specimens. Specimens of the cement mortar have to be cast, cured and tested as per standard specifications. Table 1.8 gives the acceptable strength limits.

1.7 Special Cements

Ordinary portland cement is known as cement for a layman. It is commonly used in cement mortar, plain cement concrete (PCC), reinforced cement concrete (RCC), prestressed cement concrete (PSC), light weight cement concrete, etc. At the moment the ordinary portland cement (OPC) comes in the three grades based in the strength quality. The grades are: 33 grade, 43 grade and 53 grade. The number say 33 refers to the strength of cement mortar standard cube in MPa. There are special situations in which ordinary portland cement is not very efficient where cements having special properties such as quick setting, slow setting, sulphate resistance cements are used.

Rapid-hardening cement or high-early strength cement: The strength developed by rapid-hardening cements in about 72 hours is almost equal to that of the ordinary cement developed in 28 days. The factors that influence the early strength are chemical composition, degree of chemical combination of the raw materials, blending, grinding and burning of the raw materials and the fineness of cement.

Portland-pozzolana cement: A pozzolana is a siliceous material which reacts with lime in the presence of moisture producing calcium silicate. Cement produced by grinding Portland cement clinker and pozzolana together with addition of gypsum is called portland-pozzolana cement. It can also be obtained by a uniform blending of the portland cement with fine pozzolana. The percentage of pozzolana in such cements varies from 10 to 30 per cent and it is usually around 20 per cent. Pumicite and shales, volcanic ashes, fly ash and diatomaceious earth have pozzolanic properties. Portland-pozzolana cement produces less heat of hydration and offers greater resistance to the attack of aggressive waters. It develops the strength at a lower rate in the same tune of ordinary cement with extended period of

Table 1.8 Physical Requirements of Portland Cements

Tests	Ordinary	Rapid hardening	Low heat	Pozzo-lana
1. Fineness test				
(a) After sieving on 90-micron IS sieve the residue by weight not to exceed, (per cent)	10	5	–	5
or				
(b) Specific surface m^2/kg by air permeability methods not less than	225	325	320	300
2. Soundness test				
(a) Expansion unaerated cement by Le Chatelier Method not to exceed (mm)	10	10	10	10
(b) Expansion of aerated sample by being spread out to a depth of 75 mm at relative humidity of 50 to 80 per cent for a period of 7 days, shall not be more than (mm)	5	5	5	5
(c) All cements having magnesia content more than three per cent should be tested by autoclave test. The expansion of unaerated cements on autoclave test, should not have more than (per cent)	0.8	0.8	0.8	0.8
3. Setting time				
Initial setting time (minutes) not less than	30	30	60	30
Final setting time (minutes) not more than	600	600	600	600
4. Compressive strength (MPa) Minimum average compressive strength of atleast three mortar cubes (area of face = 50 cm^2) composed of 1:3 cement and standard sand by weight, and (0.25 P^* + 3) per cent (of combined weight of cement and sand) water and prepared, stored and tested after the periods mentioned below:				
(a) $24 \pm \frac{1}{2}$ hours, not less than	–	16	–	–
(b) 72 ± 1 hour, not less than	16	27.5	10	–
(c) 168 ± 2 hours not less than	22	–	16	22
(d) 672 ± 4 hours (336 hr ± 4 hr for pozzolana)	–	–	35	31
($^*P$ = percentage of water required to produce a paste of normal consistency)				
5. Heat of hydration in calories per gram weight				
(a) After 7 days	–	–	65	–
(b) After 28 days	–	–	75	–

moist curing. It is particularly useful in marine and hydraulic structures and also for mass concrete construction. The chemical and physical requirements of the cement are listed in Tables 1.7 and 1.8 respectively.

Portland blast-furnace-slag cement: Hot slag from blast-furnace is granulated with cold water and then mixed with portland cement clinker in about 1 : 1 to 1 : 3 proportions. The mixture along with some gypsum is ground to the same fineness as that of portland cement. This cement yields about the same strength as that of the ordinary portland cement and can be used in place of ordinary cement. However, one has to be careful about the chemical composition of slag.

Slag cement: Slag cements are produced by grinding blast-furnace slag with hydrated lime. The hot slag is granulated with cold water, dried, ground and mixed with high calcium hydrated lime. The amount of lime may vary from 25 to 45 per cent depending on the nature of the slag. The mixture is then pulverised into a fine powder. The specific gravity of the slag cement is around 2.8 as compared to 3.14 ordinary portland cement. The strength of the slag cement is usually lower than that of the ordinary cement. Therefore, slag cement is used in masonry and other less important constructions.

High-alumina cement: This cement is made by grinding a compound of bauxite and limestone. The main compounds are calcium-aluminate, hydrates and hydrated colloidal alumina. The cement attains strength at early stage which at 72 hr of curing it is about the same as that of portland cement at 28 days. Most of the heat of hydration takes place in about 24 hours, resulting in high heat of hydration. This cement can be used for underwater construction. As it is difficult to control heat of hydration in mass concreting, its use in mass concrete works in dry weather is not desirable.

Expansive (or expanding) cement: Expansive agents such as gypsum, bauxite, calcite, etc. are burnt to form clinker. It is then ground with Portland cement clinker to form cement. The tendency of shrinking of cement during drying is reduced in expansive cements. In the presence of water, calcium aluminate and calcium sulphate which are derived from the agents react to form calcium sulphoaluminate hydrate resulting in expansion of the paste. A stabilizing agent such as blast-furnace slag reacts with excess calcium sulphate and stabilizes the expansion. This cement usually contains 5 to 14 per cent of expansive agent, 10 per cent of stabilizer and the rest being portland cement clinker.

White cement: Chalk or limestone that is used as a raw material needs to be free from impurities and very little of oxides of iron and manganese should be permitted to make white cement. Oil is used as a fuel in place of coal ash. Grinding of the clinker is done in a special mill so as to avoid contamination by iron oxide. White cement is used for architectural purposes, mosaic floors, fixing glazed tiles, etc.

Natural cement: Cement rock containing clayey limestone upto 25 per cent is calcined and grounded to obtain natural cement. It is easier to manufacture natural cement which is inferior to Portland cement.

Hydraulic cement: It is finely ground material which on addition of appropriate quantity of water is capable of hardening both underwater and in air by chemical interaction of its constituents. Most cements come under this classification.

Masonry cement: It is a product obtained by intergrinding of mixture of portland cement clinker with inert materials such as limestone, conglomerates, and dolomite, and gypsum and air-entraining agent in suitable proportions. This has slow-hardening and high workability and water retentivity properties. It is suitable for masonry construction.

Oil-well cement: Hydraulic cement suitable for resisting high pressure and temperature in sealing water and gas pockets and setting castings during the drilling and repair of oil-wells is called oil-well cement. This cement contains reduced content of tricalcium aluminate and a retarding additive to suit the requirements.

Hydrophobic cement: Ordinary portland cement clinker when ground with an additive which will impart a water repelling property is called hydrophobic cement. The water repelling property is eliminated when the cement is mixed with water. This cement can be stored for long periods in wet and highly humid climates.

Low heat portland cement: Cement containing about 40% of C_2S and low percentages of C_3S and C_3A generates low heat of hydration.

Sulphate resistance cement: Should contain not more than 5% of C_3A to resist sulphate action.

1.8 Cement Concrete Admixtures

Type of admixtures: Some properties of cement such as setting time, colour, impermeability can be modified by the addition of suitable compounds called *admixtures*. The main purposes of admixtures are:

    (a) Improve workability.
    (b) Accelerate setting time.
    (c) Reduce setting time.
    (d) Aid curing.
    (e) Decrease permeability
    (f) Improve wear resistance.
    (g) Improve durability.
    (h) Reduce shrinkage.
    (i) Reduce weight.
    (j) Reduce bleeding.
    (k) Reduce heat of hydration.
    (l) Impart colour.

The admixtures are usually finely ground powders having certain chemical or physical reactions with cement thus producing the desired action. These admixtures are added to concrete or mortar at the time of actual use or sometimes are premixed with cement. The quantity of admixture is usually limited to one to three per cent so as to minimize secondary effects. The chloride content in admixtures should be limited to that specified by the standards.

Workability admixtures: Powdered hydrated lime, diatomaceous earth, bentonite and fly ash can be used as admixtures to improve workability. The workability of concrete increases to a limited extent by the addition of these powdered materials. However, beyond certain proportions the strength of concrete will decrease, while the shrinkage will increase.

Some patented admixtures which are added in small quantities are available in the market. Calcium chloride, stearate, some *oily compounds* which function as wetting or dispersing agents are used as admixtures. Air in the form of minute bubbles in the concrete improves the workability. A *foaming agent* or a chemical which produces *gas bubbles* on reaction with cement is used as an air entrained agent. *Natural resins, sulphonated soaps* and oils *are the common* basic material for this purpose. Aluminium and zinc powders, hydrogen peroxide are some of the elements used to produce gas to increase workability. Air entrained concrete is usually lighter and less strong compared with ordinary concrete.

Accelerators: Admixtures which cause early setting and hardening are called accelerators. Powdered calcium chloride is one of the commonly used accelerators. Calcium chloride of one to two per cent of cement content is likely to reduce the initial setting time by 10 to 15 minutes. Some soluble carbonates and silicates also help in nearly setting time and their effect on hardening after third day is lessened. The heat of hydration is likely to be more in the first two days. Too much of accelerator can cause too early setting, thus poor workability and consequently low ultimate strength. Some of the accelerators such as sodium silicate may result in poor strength and durability. Chlorides must be avoided in reinforced concrete.

Retarders: Retarders are admixtures which prolong the setting and hardening time. Where laying of concrete can take more time because of distance or other problems, retarding agents are used. Gypsum is one of the commonly used retarders. Indiscriminate use of retarders, can, however, effect the ultimate strength.

Air entraining agent: Air entraining agent is similar to the water-reducing agent. Neutralized vinsol resin when added to cement disperses the cement particles and entraps air. Thus the workability of cement is obtained with less water. Some of the trade names of admixtures are: cemix (mortar plasticizer), impero (water proofing compound), snowsol (stabilizing solution), surfacit (concrete surface hardening liquid), hardcrete (cement water proofing and hardening liquid), and colorum (decorative waterproof cement paint).

Super-plasticizer: Low water-cement ratios of the order 0.3 to 0.45 are used in high strength concretes. Super-plasticizer, an admixture is added to such concretes to improve workability. High workability of 150 to 200 mm slump are used in high strength and high performance concretes.

1.9 Water Used in Cement Concrete

Water used in concrete mixing and curing should be free from solids, acids, alkalies, organic materials and salts. Potable and *clean bathing water* is normally but not always acceptable for concrete mixing. The permissible limits of solids in water are given in Table 1.9.

1.10 Reinforcement Used in RCC

The following reinforcement material can be used in RCC:

Table 1.9 Permissible Limits of Solids in Water (pH value not less than 6)

	Material	Maximum limit
1.	Suspended	2000 mg/litre
2.	Organic	200 mg/litre
3.	Inorganic	3000 mg/litre
4.	Sulphates as SO$_4$	500 mg/litre
5.	Chlorides	2000 mg/litre for PCC[*a] 1000 mg/litre for RCC[*b]
	pH value of water should not be less than 6	

a PCC = plain cement concrete
b RCC = reinforced cement concrete

(a) Mild and medium tensile steel bars (MS).
(b) Mild, medium tensile and high yield strength deformed bars (HYSD or HYD).
(c) Cold twisted deformed bars of steel (CTD).
(d) Rolled steel from structural steel in composite construction.

The bars must be free from dust, loose rust, oil, paint, loose mill scales, etc. before placing in position and laying concrete. Usually 5 mm to 16 mm bars are used in slab and shell construction, 10 mm to 28 mm bars are used in beam and column* construction. In some cases, bars upto 40 mm are used in beams and columns.

Minimum requirements for reinforced bars are given in Table 1.10.

Table 1.10 Minimum Requirements of Reinforcement Bars (stress in MPa). The Quality is Associated with Steels Produced in India

No.	Property	Mild Steel Grade I	Mild Steel Grade II	Deformed bars Medium tensile	Deformed bars Mild steel	Deformed bars Medium tensile	CTD, (HYSD) Cold twisted (high yield)	Hard drawn
1.	Ultimate tensile stress	410	370	570	410	550	490	560
2.	Yield/proof stress							
	Dia* ≤ 40 mm	250	230	350	250	350	415	500
	Dia > 40 mm	230	210	320	230	320	415	450
3.	% elongation (on guage 8D)*							
	Dia < 10 mm	20	20	17	23	20	14.5	7.5
	Dia ≥ 10 mm	23	23	20	23	20	14.5	7.5
4.	Maximum tolerances							
	In length: Length not specified	–	– 25 mm; + 75 mm					
	Min length specified	–	+ 75 mm					
	Max. length specified	–	– 50 mm					
	In weight							
	Dia ≤ 8 mm	–	± 4%					
	Dia > 8 mm	–	± 2.5%					
	In diameter							
	Upto 25 mm	–	± 0.5%					

*D is diameter

1.11 Pozzolana

A pozzolana is a siliceous material which when finely divided reacts with lime in moist condition producing calcium silicate. Cements are produced using pozzolana material such as pumicite, diatomaceous clay, certain shales and clays, fly ash, etc. In such cements, the percentage of pozzolana may vary from 10 to 30 per cent.

Table 1.11 Different Cement Concrete Mixes

No.	Type of concrete	Grade
1.	Lean cement concrete (LCC)	M5 to M7.5
2.	Plain cement concrete (PCC)	M10 to M25
3.	Reinforced cement concrete (RCC)	M20 to M35
4.	High strength concrete (prestressed concrete) (PSC)	M35 to M60
5.	High performance concrete	M45 to M120

Following are some of the properties of pozzolanic cements when compared with ordinary cements:
- (a) Curing time is 20 to 30 per cent more.
- (b) Heat of hydration is less.
- (c) Gains strength in about 40 to 50 days time which is equivalent to 28 days portland cement.
- (d) Less reactive with salts and sulphates.
- (e) Reduced permeability in lean mixes.
- (f) Improve workability.
- (g) Reduced bleeding.
- (h) Increased shrinkage.
- (i) May have lower strength if used in excess quantity.

1.12 Grades of Concrete

Basically five major types of concretes are used in construction, which are:
- (a) Lean cement concrete (LCC)
- (b) Plain cement concrete (PCC)
- (c) Reinforced cement concrete (RCC)

Table 1.12 Nominal Mix Proportions by Volume (See Table 1.22 for Better Proportions)

Grade	Nominal mix
Lean concrete M5	1 : 5 : 10
M7.5	1 : 4 : 8
Plain concrete M10	1 : 3 : 6
M15	1 : 2 : 4
M25	$1 : 1\frac{1}{2} : 3$

(d) High strength concrete (HSC or PSC)

(e) High performance concrete (HPC)

The grade of the concrete is indicated by the strength of 150 mm cube tested after 28 days of water curing in surface dry-saturated condition. For example, grade M10 means it is a concrete whose 150 mm cube strength is 10 N/mm^2. Table 1.11 gives normal ranges of concrete used in different construction. There are some concretes which are proportioned nominally instead of designing the mix. Such concretes are used in small jobs or in lean concretes. Table 1.12 gives the nominal mixes which are measured by volume. The aggregate used should be all in one type.

1.13 Strength of Concrete

The strength of concrete is normally associated with the hardened concrete and it depends on several factors. The strength of the concrete increases with:

1. Decrease in water-cement ratio
2. Increase with fineness of cement
3. Increase with curing time
4. Age of concrete
5. Strength of the aggregate
6. Increase in size of aggregate
7. Grading, texture, shape and size of aggregate
8. Temperature of water at curing

Approximate relation of strength of concrete with water-cement ratio is given by

$$\text{Strength} = K\left(\frac{C}{W+C+A}\right)^2 \tag{1.9}$$

where A, C and W are absolute volumes of air, cement and water, and K is a constant.

Approximate strength in relation with the age of concrete with respect to that of 28 days curing is given in Table 1.13. The designer can give an appropriate correction to the strength of concrete when he is certain about the time of actual application of the load.

The strength of the concrete is the compressive strength of the 150 mm cube tested after 28 days of curing in water in saturated surface dry condition. There are other strengths such as prism, flexure, tension, split, shear, bond, etc. which are generally inter related with the compressive cube strength.

The tensile strength in bending which is called strength in flexure of concrete can be

Table 1.13 Ratio of Strengths of Concrete at Different Ages with Respect to 28 days Strength

Period after casting	Age factor
7 days	0.65 to 0.7
28 days	1.0
3 months	1.10 to 1.15
6 months	1.15 to 1.20
12 months	1.20 to 1.25

obtained from the cube strength and it is given by

$$\text{Modulus of rupture} = f_{cr} = 0.7 \sqrt{f_{ck}} \qquad (1.10)$$

$$\text{Tensile strength} \quad = f_{ct} = 0.35 \sqrt{f_{ck}} \qquad (1.11)$$

where f_{ck} = crushing strength of 150 mm cube and the units are in N/mm^2 only.

Modulus of rupture or flexural strength is usually used in determining cracking strength of concrete in bending members. Table 1.14 gives the value of tensile stresses in bending, diagonal tension, etc. Ultimate bond stresses of the concrete for 0.25 mm slip and for practically no slip are also given in Table 1.14.

Table 1.14 Modulus of Rupture and Bond Strengths (stresses in N/mm^2)

	Concrete (N/mm^2) = f_{ck} =	10	15	20	25	30	35	40	45
(1)	Modulus rupture	2.2	2.7	3.1	3.5	3.8	4.1	4.4	4.7
(2)	Bond stress for 0.25 mm slip								
	Plain bars	1.4	1.9	2.4	2.9	3.3	3.4	3.5	3.5
	Deformed bars	2.4	3.5	4.4	5.2	5.8	6.3	6.5	6.6
(3)	Bond stress for no slip in :								
	Plain bars	1.2	1.5	1.7	2.0	2.2	2.5	2.7	2.8
	Deformed bars	1.7	1.9	2.1	2.5	2.8	3.1	3.4	3.5

Different characteristic strengths of concrete are determined by the following tests.

(a) *Compressive strength:* 150 mm concrete cube is cured in water for 28 days and is tested in the moist saturated condition in compression. This forms the basis for *characteristic strength of the concrete.* The compressive strength of the concrete in any element of a structure varies with slenderness ratio, type of stress distribution and age of concrete. Table 1.15 gives the compressive strength of prisms with respect to that of the 150 mm cube.

The strength of a prism specimen decreases with increase in height to side ratio and stabilizes when this ratio is about 4.

The strength of the concrete determined through the cube specimens varies with the size of the cube. The strength of the specimen increases with decrease in its size. Table 1.16 gives approximate distribution of strength of concrete based on the size of the cube.

The cube strength is estimated to be more than that of a cylinder of similar size. Its relation is:

$$f_{cy} = (0.70 \text{ to } 0.80) f_{cu} \qquad (1.12)$$

where

f_{cy} = cylinder strength having height to diameter ratio equal to two, and

f_{cu} = strength of concrete cube having the same size as that of the diameter of the cylinder.

The lower value applies to smaller sizes. The relation for 150 mm cube can be expected as:

$$f_{cy} = 0.75 \, f_{cu} \qquad (1.13)$$

The strength of concrete is also influenced by the moisture content in the concrete at the time of testing. Moisture content in concrete provides lubrication effect and reduces the strength when compared with a dry sample.

Table 1.15 Strength of Prisms vs 150 mm Cube Strength

$\left(\dfrac{\text{Height}}{\text{Side}}\right)$	0.5	1.0	2.0	3.0	4.0	5 and more
Relative strength	1.5	1.0	0.8	0.72	0.67	0.66

Table 1.16 Strength of Concrete Cubes with Respect to that of 150 mm Cube

Cube size (in mm) =	100	150	200	300
Relative strength with 150 mm cube	1.05	1.0	0.95	0.87

Strength of a dry sample = (1.1 to 1.20) strength of the saturated sample   (1.14)

The stress strain relation of concrete in compression may be taken as

$$\sigma_c = E_c \varepsilon \left(1 - \frac{\varepsilon}{2\varepsilon_0} \right) \tag{1.15}$$

where σ_c = compressive strain, ε = compressive strain

ε_0 = ultimate strain (about 0.0035 to 0.0045)

E_c = modulus of elasticity of concrete.

The modulus of elasticity of the concrete E_c (in N/mm^2) can be taken as

$$E_c = 5000 \sqrt{f_{ck}} \text{ to } 5700 \sqrt{f_{ck}} \tag{1.16}$$

The compressive strength capacity of the concrete in bending compression is higher than that under axial compression and its relation can be given as

$$f_{cb} = k_b f_{ca} \tag{1.17}$$

where f_{cb} and f_{ca} are the compressive strengths of concrete in bending and in axial force respectively; k_b is a constant which is taken from 1.2 to 1.25.

Flexural strength: A plain concrete when subjected to pure bending force fails by flexure. The maximum tensile stress resisted by the concrete in flexure is called *flexural strength*. Usually small beams cured for 28 days are tested in moist condition under two point loading. The flexural strength is also called *modulus* of rupture and it is computed by

$$f_{cr} = \frac{M_{cr}}{Z} \tag{1.18}$$

where f_{cr} = modulus of rupture (cracking tension)

M_{cr} = pure BM on the section at failure or cracking

Z = modulus of the section

The relation between the strength of cube and modulus of rupture is given earlier. The value of modulus of rupture depends marginally on size of the beam, its length and moist condition, besides age, curing etc.

Splitting strength: Tensile strength of concrete is obtained indirectly by *split cylinder* test which is also called *Brazilian* test. A 150×300 mm cylinder is subjected to bearing compression on the cylindrical faces causing splitting. The horizontal split tensile stress is computed as

$$f_{cs} = \frac{P}{\pi r l} \qquad (1.19)$$

where r and l are radius and length of the cylinder and P is the axial load at which splitting takes place. Approximate relation between modulus of rupture, splitting strength and direct tensile strength are:

$$f_{cs} = 0.66 \, f_{cr} \qquad (1.20)$$

$$f_{ct} = 0.5 \, f_{cr} \qquad (1.21)$$

where f_{ct} = direct tension strength.

Rate of loading on strength: Recorded strength of a concrete specimen depends on the rate of loading to some extent. The strength shows an increase with increase in the rate of loading. The normal rate of loading recommended is about $2 N/mm^2/min$. This means that the rate of loading on 15 cm cube is about 45000 N/min. Very slow of rate of loading introduce creep in concrete.

1.14 Durability of Concrete Structures

The main factors, which influence the durability and life of a structure, are listed and explained briefly later.

Factors Effecting Durability

1. Physical and mechanical factors: This consists of Internal and external factors.

(a) *Internal factors:* Compaction, Porosity and permeability of concrete, Surface finish, Duration of curing of the concrete; Cover to the reinforcement and Surface cracking.

(b) *External factors:* Abrasion and erosion (wear & tear), Exposure to wetting and drying (moisture) and Freezing and thawing.

2. Chemical Factors: Chemical composition of Aggregate, Cement and water used, Exposure to Acid action, Sulphate attack, Internal Aggregate alkaline reaction, Other chemical aggression actions.

3. *Biological and Environmental Factors:* Biological growth, such as moss, algae.

4. *Non-structural factors:* Shape and size of structural elements, Drainage, Joints (inter-connections, construction and expansion joints), Inserts, Bearings, Railings, Anchorage and Fixtures. There is a certain interdependence and combination of the above factors which result in deterioration of the concrete.

Physical Factors

The main physical effects that influence the durability are:

Compaction and Moisture transport: Good workable concrete leading to dense and well-compacted concrete minimizes the transport of moisture within. The porosity or permeability of the concrete is a direct result of poor compaction. Transportation of moisture through concrete generates a number of physical and chemical reactions resulting in deterioration of the concrete. Diffusion of air, carbon dioxide and chloride ions etc. is assisted by the transport of the moisture. The carbonation leading to increase in volume results in cracking of the concrete. Similarly the presence of moisture and air in concrete will lead to chemical reactions in concrete and corrosion of reinforcement, resulting in cracking and spalling of the concrete.

Surface Finish: Water in contact with the concrete surface is transported through capillary action even in well-compacted concrete. It is therefore necessary to prevent retention and accumulation of moisture at the surface of the concrete. Uneven and rough concrete surfaces retain water and allow water/moisture transportation leading to slow degradation of the concrete. The cracked and micro-crack surfaces of concrete permit the movement of water freely into the concrete. An impermeable concrete surface is the most ideal finish but even such surfaces will develop micro cracking in course of time due to external temperature and internal chemical changes of the concrete. It is therefore necessary to take adequate precautions to maintain an even and uncracked concrete surface.

Wetting and drying: Some concretes are exposed to frequent wetting and drying. The water carries dissolved salts such as sulphates, chlorides, carbonates etc. into the pores of the concrete under wet condition. The moisture in the concrete evaporates during drying leaving dissolved salts in crystallized state. The accumulation of such crystals increases the volume leading to cracking of the concrete. Further the concentrated crystallized chemicals act on the aggregates and reinforcement causing carbonation and corrosion. Efflorescence of crystallized salts at the surface of concrete is also a result of wetting and drying.

Freezing and thawing: Freezing and thawing of water occurs in some extreme climates. The water or moisture transported into the pores of the concrete when exposed to freezing and thawing causes volume changes in addition to the acceleration of chemical degradation. This leads to cracking of the concrete and deterioration.

Chemical Factors

A number of chemical reactions takes place due to internal structure or external exposure of the concrete. The moisture transport phenomenon plays a key role in the deterioration of the concrete. Some of the important chemical attacks are briefly mentioned.

Carbonation, Acid formation and attack: Moisture along with environmental gases, liquids and solids may lead to acid formations which react with hydrates of cement. The attack will break the chemical bonds in the calcium silicates and other hydrates thus destroying the very strength of hardened concrete. The extent of deterioration depends on the density and porosity of the concrete, cement and aggregate properties and the environmental conditions.

Sulphate attack: The sulphate ions from soil or water or even from cements react with calcium hydroxide resulting into sulphates in the concrete. The increase in the volume due to formation of sulphates within the pores of the concrete result in cracking and further degradation of the concrete. The presence of aluminates in cement has a tendency to expand in volume on reaction with sulphates.

Alkali and chemical attacks: Concrete is normally saturated with lime therefore has an alkali environment. The sodium and potassium ions in the alkali solutions attack the silica of the aggregate. The rate of attack depends on the active silica in the aggregate and dusty particles, porosity of the concrete and moisture transport. Concrete surface when exposed to acidic environment, the acids transported by moisture react with aggregates resulting in increase of volume and consequent cracking.

Biological and Environmental Factors

The continuous presence of moisture on the surface of concrete promotes biological growth such as moss, algae and small plants. The penetration of such products and roots of plants into the concrete results in cracking of the concrete.

Corrosion of Reinforcement

Steel reinforcement when exposed to moisture along with oxygen; will result in formation of oxides of iron. Similarly the chloride ions accelerate the formation of oxides of iron. Normally the alkali environment in the concrete with pH value in the range of 12 provides good protection against formation of iron oxide. The diffusion of carbon dioxide into concrete leads to formation of calcium carbonate on reaction with calcium hydroxide. Therefore the alkali environment, the pH value decreases to 9 and decrease of the protection to the reinforcement. The dissolution of positively charged iron ions and the combination of the released electrons with moisture and oxygen results into hydroxyl ions. The ion activates the formation of ferric oxide. The formation of ferric oxide on the surface of the reinforcement results into increase in volume. The rusted volume can be as much as 50 to 200 per cent of the iron content.

Stress Corrosion in Pre-tensioned Steel

A chemical process in corroded stressed steel wires or bars will result in sudden splitting of the wires into a brittle failure. This phenomenon is present in stressed steels of prestressed concrete constructions.

Influence of Non-structural Components

The discontinuities, construction and expansion joints, corrosive and even non-corrosive fixtures and inserts are exposed to moisture, wear and tear, shrinkage and expansion etc. influence the durability of the concrete.

Design Considerations

The choice of materials, design of components and detailing has to be done based on the exposure conditions. All structures can broadly be classified into five groups based on the durability design considerations. These exposure conditions are:

1. Protected (Mild) environment,
2. Moderate exposure,
3. Severe exposure,
4. Very severe exposure and
5. Extreme exposure.

Mild exposure: Interiors of dwellings, offices, commercial complexes and workshops of non-aggressive environment and other structures protected against weathering, wetting and drying are classified in this group.

Moderate exposure· Foundations buried under non-aggressive soils and ground water, structures exposed to rain and not under frequent wetting and drying, outside non-aggressive environment, high humidity halls, running water (canals, dams weirs etc.) bath rooms and water tanks come under this class.

Severe exposure: Structures exposed to frequent wetting and drying of ordinary water, foundations in aggressive soils, coastal regions, occasional freezing (not annual), partially submerged under water and subjected to constant vibrations come under this classification.

Very severe exposure: Structures exposed to sea water spray, extreme freezing, chemical fumes, colour dying halls, partially submerged sea water (all harbour structures in contact with sea water), septic tanks come under this classification.

Extreme exposure: Concrete roads; structures under constant abrasion in wet and dry conditions, direct contact with aggressive chemicals, floors of chemical plants are classified in this category.

Design Parameters

Design parameters for durability considerations are listed below:

(a) *Materials:* Quality and Quantity of aggregate, Type and Quantity of cement, Water Quality, Water-Cement ratio, Admixtures, Concrete mix and its workability, Reinforcement, Inserts and Fixtures.

(b) *Detailing and Workmanship:* Surface finish, shape and texture, Cover to the reinforcement and precautions, Finishing of points, Drainage of water, Curing of concrete, Accessibility for maintenance and Quality assurance scheme.

(c) *Maintenance:* Maintenance scheme should have Inspection regulations, systematic maintenance scheme and its implementation, Immediate attention to damages and repairs, and Prevention of accumulation of water.

Design Recommendations

Structural Concrete: The constituents of concrete, namely aggregate, cement, water and admixtures must satisfy the standard requirements. Further there are limits on qualities and qualities of the basic materials to suit the exposure requirements. Similarly the mix proportions, methods of mixing, laying, consolidation, finishing, formwork and curing are important. Testing methods and quality assurance must be given adequate importance.

Cements and Water: Portland cements of grades 33, 43 and 53 are suitable for all exposure conditions. Portland composite (pozzolana) cements can also be used depending on the type of construction conditions. 33-grade cement is normally slower in hardening when compared with the 53 grade. A minimum quantity of cement content, especially in reinforced concrete is required to provide adequate alkaline environment to protect the steel from corrosion. Similarly maximum limits is suggested to minimize shrinkage and heat of hydration effects. Shrinkage of concrete increase with water content hence a reduction in water-cement ratio enables higher cement content. Water used in concrete making should conform to the standards and should not contain oil, organic matter, acids etc. The water-cement ratio needs to be as low as possible not only for strength consideration but also to minimize the porosity of the concrete. The maximum water-cement ratio admissible for different exposure conditions is given in Table 1.17.

Admixtures: Any admixture added to cement or concrete in small quantity to improve the properties such as workability, setting time or to decrease the permeability should not exceed 5 per cent and normally in the range of 2 per cent or as recommended by the manufacture. Admixtures or additives should not contain any chlorides in any form in reinforced or prestressed concrete constructions.

Strength of concrete: The strength of concrete need not necessarily indicate its durability but yet it is considered as a desirable index for durability. Concrete must have a minimum strength for durability considerations. It indirectly reflects the quality of production and porosity. Minimum acceptance grades of concretes under different exposure conditions are listed in Table 1.17.

Reinforcement: Prevention of any corrosion of reinforcement should be the main aim. Diffusion of air into the concrete, and penetration of chlorides and sulphates into concrete be minimized. Some suggestions on cement and water contents were already explained and the limits are indicated in the Table 1.17 to protect the reinforcement from corrosion. The other suggestions and design parameters are:

(a) Uninterrupted period of curing to avoid micro-cracking,
(b) Blended cements with slag or fly ash or silica fumes can be used,
(c) Limit the total sodium oxide to a maximum of 0.6%,
(d) Use Low water-cement ratio and
(e) Provide adequate cover to the reinforcement as given in Table 1.18.

The reinforcement is an important and highly sensitive element in RCC and PSC constructions. Besides designing the reinforcement for strength, adequate precautions must be taken to protect it against corrosion. The most important factors in this regard are: the compactness of the concrete and the cover to the reinforcement, proper placing of reinforcement and cover is the most often neglected factor. Table 1.18 suggests the minimum covers to be provided to the reinforcement. Similarly the clear spacing of the bars and cables be adequate enough to achieve good and sound concrete around the bars. A special protective

Table 1.17 Recommended Limits of Materials for Durability

(Quantities by weight unless otherwise stated)

Material/Property	Exposure Classification			
	Mild (1)	Moderate (2)	Severe (3)	Very Severe & extreme (4 & 5)
1. **Aggregate**				
Grading	Normal	Normal	Good	Good
Crushing value (Maximum)	40	40	30	30
Aggregate density (Minimum in kg/cum)	2000	2200	2200	2400
% of Silts/clay (less than 0.15 mm)			(Maximum %)	
In Coarse	2	2	1	1
In Sand	4	3	2	2
Sulphates Admissible (max %)	1	1	0.5–1.0	0.5–1.0
2. **Cement (OPC)**				
Minimum (kg/cum)				
PCC	220	250	260	300
RCC	300	300	350	400
PSC	350	350	400	400
Maximum content (kg/cum) (For Minimum shrinkage)		← 550 (unless special considerations are given) →		
3. Max. Water-Cement ratio				
PCC	0.60	0.60	0.60	0.45
RCC	0.55	0.50	0.45	0.45-0.40
PSC	0.50	0.50	0.45	0.45-0.40
		(Better avoid PSC in highly aggressive environment)		
4. Minimum concrete grade				
PCC	M10	M15	M20	M20, M25
RCC	M20	M25	M25	M35, M40
PSC	M35	M40	M45	M45
5. Water quality for mixing minimum pH value	4.5	6.0	6.0	6.5
6. Maximum chloride content: Maximum acid soluble Chloride content as % of chloride ion by mass of concrete				
PCC		← 1.00% →		
RCC		← 0.15% →		
PSC		← 0.10% →		

treatment to the concrete surface or to the reinforcement be provided in aggressive environment as of extreme exposure.

Inserts: Steel or galvanized iron or stainless steel inserts be chosen depending on the exposure conditions. Wrongly placed or wrong inserts exposed to moisture or retention of moisture will cause surface damages which will lead to deterioration of the structure. Table 1.18 recommends desirable type of inserts.

Table 1.18 Minimum Cover to Reinforcement and Spacing etc.

Material/Property	Exposure Classification			
	Mild (1)	Moderate (2)	Severe (3)	Very Severe & extreme (4 & 5)
1. Minimum cover (in mm) to reinforcement bars in RCC (The cover specification applies to main/secondary reinforcement)				
Slab	15	15	20	25 to 35
Walls	20	20	25	25 to 35
Beams	25	30	45	50 to 60
Columns	40	45	50	50 to 70

(1) A wearing or maintenance protective coat to be provided to the concrete exposed to the extreme exposure (Reinforcement be coated with protective coating)

(2) Cover not less than the maximum size of bar

2. Minimum clear spacing of reinforcement: Diameter of bar or maximum size of aggregate + 5 to 10 mm

3. Min. cover (in mm) to prestressed concrete wires or cables				
Slabs	20	25	25	25-30
Beams	25	30	45	45-75
4. Inserts projecting beyond the concrete surface : made of	Steel	GI	GI	Stainless steel

5. Minimum clear spacing = 1.5 to 2 times the diameter of the cables/ducts in PSC ducts.

1.15 Shrinkage, Creep and Thermal Expansion

The shrinkage of concrete is directly proportional to the amount of water present in the concrete at the time of mixing and to some extent to the amount of cement content and type of aggregate. Approximate shrinkage strain (ε_{so}) in concrete is:

$$\varepsilon_{so} = 0.0003 \qquad (1.22)$$

Half of the shrinkage takes place in the first one month and seventy-five per cent in the first few months after commencement of drying.

Creep is the time dependent strain under sustained loading. It is a function of water content, cement, type of aggregate, age of loading.

Ultimate creep strain is obtained by

$$\varepsilon_{cc} = C_c \varepsilon_{ce} \qquad (1.23)$$

where ε_{cc} = ultimate creep strain, C_c = creep coefficient, and ε_{ce} = initial elastic strain.

The total strain is equal to the sum of creep strain and elastic strain. The creep coefficients are given in Table 1.19.

Table 1.19 Creep Coefficients (Excluding Elastic Strain)

Age of concrete at loading	7 days	28 days	1 year
Creep coefficient	2.2	1.6	1.1

Thermal Expansion of concrete depends on the type of aggregate. It is in the range of 0.6×10^{-5} to 1.2×10^{-5}. For granite stone aggregate its value in 0.8×10^{-5} per degree of Celsius.

1.16 Workability of Concrete

Workability of concrete is a property which indicates the ease in placing and compacting the concrete. It is measured by any of the following tests:

  (a) Slump test.
  (b) Compaction factor.
  (c) Vee-bee (or V-B) consistometer.
  (d) Flow test.
  (c) Remoulding test.
  (f) Ball penetration or kelley ball test.

The workability of concrete depends on several factors such as:

  (a) Proportional to water-cement ratio.
  (b) Proportional to size of aggregate.
  (c) Proportional to grading of the aggregate.
  (d) Inversely proportional to aggregate cement ratio.
  (e) Directly proportional to irregular shape of the aggregate.
  (f) Decreases with flakeness of the aggregate.
  (g) Proportional to the air in the mix.
  (h) Inversely proportional to time of transit.

Recommended *workabilities* for different conditions are given in Table 1.20. Minimum desirable aggregate to cement ratio for good *workability* of concrete is given in Table 1.21.

1.17 Segregation and Bleeding of Concrete

Segregation is the property of the ingredients to separate from each other while placing the concrete. However, in order to obtain good compaction, concrete must have cohesiveness. Good workability, therefore, does not segregate.

Segregation can be avoided by proper grading of aggregates and appropriate vibration.

Table 1.20 Recommended Workabilities of Concrete

	Placing condition	Quality	Compaction factor	Slump (in mm)	Vee-bee time (in seconds)
			Recommended workability		
1.	Shallow sections with vibration	Very low	0.75–0.8	0–10	20–10
2.	Lightly reinforced with vibration	Low	0.8–0.85	10–30	10–5
3.	Lightly reinforced without vibration or heavy reinforced with vibration	Medium	0.85–0.92	25–75	5–2
4.	Heavily reinforced without vibration	High	0.92–1.0	75–175	–
5.	Pumping concrete	Very high	–	160–200	–

Table 1.21 Minimum Aggregate-cement Ratio for Workability (Angular Granite)

Water cement	High	Medium	Low
		Workability	
0.35	2.1-2.3	2.2-2.4	2.5-3.0
0.40	2.6-3.0	2.7-3.2	3.0-4.0
0.45	3.1-3.3	3.3-3.8	3.5-5.0
0.50	3.5-4.0	3.8-4.2	4.5-6.0
0.60	4.4-4.7	4.8-5.2	5.5-6.5
0.70	5.3-5.6	5.7-6.2	6.0-7.5
0.80	6.0-6.2	6.3-6.6	7.0-8.0

Throwing concrete instead of placing properly and over vibration will cause segregation even in well graded mix. The property of fresh concrete in which the water in the mix tends to rise to the surface while placing and compacting is called *bleeding*. This also causes over-settlement of the concrete. Bleeding is usually reflected through the quality of the cement. High calcium aluminate or addition of calcium chloride or high alkaline content in concrete and rise in temperature are likely to cause bleeding. Lean mixes bleed more when compared to the rich ones. Bleeding can be minimized by addition of finer aggregate or pozzolana or aluminium powder.

1.18 Concrete Mix Proportions

As far as possible the concrete mix proportions should be designed using reliable methods. However, in small jobs, nominal mix proportions suggested in Table 1.22 can be adopted.

Concrete Mix Design: There are several methods used in the design of concrete mix. Even though each method has an approach, there is certain amount of trial that is needed in all the methods. A useful method is described here. Information needed for the mix design is:

1. Grade of concrete.
2. Workability.
3. Type of supervision and level of quality control.
4. Supply of cement and aggregate to determine their properties.
5. Exposure conditions.

Table 1.22 Recommended Proportions of Nominal Mix (Also see Table 1.12)

Grade of concrete	Total quantity of dry aggregate for 50 kg of cement		Water in litre
	in kg	in litre (approximate)	
M10	480	300	34
M15	350	220	32
M20	250	160	30
M25	160	100	27

Note: The ratio of coarse to fine aggregate is generally 1 : 2. However, it can be adjusted with $1 : 1\frac{1}{2}$ to 1 : 3 to provide good workability and strength. For a graded aggregate the ratios can be $1 : 1\frac{1}{2}$, 1:2 and 1 : 3 for 10 mm, 20 mm and 40 mm aggregate respectively.

(a) *Laboratory cube strength:* The strength of concrete cube prepared in the field will be less than that prepared and cured under controlled condition in the laboratory. Usually the strength of the field cured cube is likely to be 10 to 20 per cent less than that of the laboratory one depending on the type of supervision. Therefore design strength of the laboratory prepared cube be computed. Table 1.23 suggests recommended strengths of laboratory prepared cubes for different mix designs,

(b) *Water cement ratio:* Having decided upon the strength of the cube to be prepared, select the appropriate water cement ratio from Table 1.24 which gives the water-cement ratio for use of 20 mm size angular (crushed) aggregate. Marginal adjustment in water cement ratio has to be made for variation in the aggregate size. Let the water cement ratio be *w*. The quantity of water can be reduced by addition of water reducing agents.

(c) *Determine the following quantities*

1. Check whether the cement conforms to the specification.

Table 1.23 Strength of Concrete Mixes for Laboratory Design

Mix (N/mm^2)	Strength of laboratory cured concrete cube			
	Type of supervision			
	Excellent	Good	Average	Poor
15	16.5	18	19	20
20	22	24	25	26
25	27.5	29	31	32
30	33	35	38	39
40	45	47	50	
45	50	53		

Note: IS code requirements of strengths of laboratory cubes are slightly different from those suggested in this table.

Table 1.24 Approximate Water-cement Ratio(*w*) for OPC grade 33

Mix (N/mm^2)	10	15	20	25	30	35	40	45	50	55
Water/cement	1.0	0.8	0.72	0.63	0.56	0.52	0.44	0.40	0.36	0.34

2. Compute: (i) specific gravity,
               (ii) bulk density, and
               (iii) fineness modulus of coarse and fine aggregates.

(d) *Workability:* The ratio of the total aggregate with cement is selected, based on the workability requirement, as given in Table 1.21. Let the ratio be G.

(e) *Proportioning the coarse and fine aggregate:* Having obtained the ratio of the aggregate to the cement, the ratio between the fine and the coarse aggregates need to be determined. This can be determined either based on void ratio or fineness moduli. Normal limits of fineness moduli of aggregates are given in Table 1.25.

Table 1.25 Normal Range of Fineness Moduli (FM)

	Maximum size of aggregate	Range of (FM)
(1)	Very fine aggregate (not recommended for concrete)	1.0 to 2.0
(2)	Fine aggregate	2.0 to 3.5
(3)	Coarse aggregate of maximum size	
	20 mm	5.8 to 6.4
	25 mm	6.0 to 7.0
	40 mm	6.5 to 7.5
	60 mm	7.0 to 8.0
(4)	Mixed aggregate of maximum size	
	20 mm	4.7 to 5.2
	25 mm	5.0 to 5.6
	40 mm	5.3 to 6.0
	60 mm	5.5 to 6.2

The ratio of fine aggregate to the combined aggregate can be obtained as

$$R = \frac{F_a - F_o}{F_a - F_s} \qquad (1.24)$$

where R = ratio of fine to combine aggregate; F_a, F_s and F_o are the fineness moduli of coarse, fine and combined aggregates respectively.

(f) *Quantities of material for one cubic metre of concrete:* The total solid volume of all the ingredients when added should give one cubic metre of concrete. An allowance of 1 to 4 per cent is given for air content.

$$\frac{1}{1000}\left(W + \frac{C}{s_c} + \frac{A_a}{s_s} + \frac{A_s}{s_s}\right) = (1 - \text{ratio of air content}) \qquad (1.25)$$

$$= (1 - a)$$

where a = ratio of the air content

and W, C, A_a, A_s are the weights of water, cement, and coarse and fine aggregates respectively: s_c, s_a and s_s are the specific gravities of cement, coarse and fine aggregates. The value of $s_c = 3.15$.

Equation (1.25) can be used to obtain weight of cement after getting values of the different ratios from the respective tables:

$$W = wC \tag{1.26}$$

$$A_g = (A_a + A_s) = GC \tag{1.27}$$

(G is taken from Table 1.21. based on workability.)

The relation between A_a and A_s is obtained from Eq. (1.24), and is

$$R = \frac{A_s}{A_a + A_s} = \frac{F_a - F_c}{F_a - F_s} \tag{1.28}$$

$$\text{or} \qquad A_a = \frac{(1 - R) A_s}{R} \tag{1.29}$$

From Eqs. (1.27) and (1.28)

$$A_s = R(A_a + A_s) = RGC \tag{1.30}$$

$$A_a - (1 - R) GC \tag{1.31}$$

Substituting Eqs. (1.26), (1.30) and (1.31) in Eq. (1.25) gives

$$C\left[w + \frac{1}{s_c} + \frac{(1 - R)\ G}{s_a} + \frac{RG}{s_s}\right] = 1000\ (1 - a) \tag{1.32}$$

s_s, s_a and s_s are obtained from experimental results and w and G are obtained from the tables. R can be obtained after selecting a desired modulus of the mixed aggregate.

The value of C is solved from Eq. (1.32) and then A_s and A_a from the respective equations.

EXAMPLE 1.1 Design of a concrete mix.

Data: Characteristic strength $f_{ck} = 40$ N/mm^2

Workability is medium (0.85 to 0.92 compaction factor or 25 to 60 mm slump).

Two different coarse aggregates with maximum sizes as 30 mm, and 20 mm are to be used along with the fine aggregate. The properties are:

For coarse aggregate:

$$F_{a2} = 7.2,\ F_{a1} = 6.5,\ s_{a1} = s_{a2} = 2.7$$

For fine aggregate:

$$F_s = 2.8,\ s_s = 2.6$$

Good supervision is provided.

Design procedure

(a)* *Strength requirement and water cement ratio:* Laboratory strength of the cube can be obtained Table 1.23 and it is

$$f_{ck}^* = 47 \quad \text{for } f_{ck} = 40 \quad \text{N/mm}^2$$

From Table 1.24 the water cement ratios are:

$$w = 0.40 \ \textit{for } f = 45$$

$$w = 0.36 \text{ for } f = 50$$

From interpolation for $f_{ck}^* = 47$ the value of w is

$$w = 0.40 - \frac{(0.04) \times 2}{5} = 0.384$$

(b) *Workability and aggregate proportions:* There are three types of aggregates. The total of the aggregate with respect to cement for medium workability with $W/C = 0.384$ can be obtained from Table 1.21.

The value G from Table 1.21 are:

$$G_1 = 2.2 \text{ to } 2.4 \text{ for } w = 0.35$$
$$G_2 = 2.7 \text{ to } 3.1 \text{ for } w = 0.40$$

Use $G_1 = 2.3$ for $w = 0.35$ and $G_2 = 2.9$ for $w = 0.40$

For $w = 0.384$

$$G = 2.3 + \frac{(2.9 - 2.3)}{(0.4 - 0.35)}(0.384 - 0.35)$$

$$= 2.3 + \frac{0.6}{0.06}(0.034) = 2.71, \text{ use } G = 2.75$$

Total aggregate quantity $A_g = GC = 2.75C$

Select the combined fineness moduli from Table 1.25.

Let F_{o2} = combined FM of 30 and 20 mm aggregate = 5.6
F_{o1} = combined FM of 20 mm and fine aggregate = 5.0

From Eq. (1.24), we have

$$R_1 = \frac{F_{a1} - F_{o1}}{F_{a1} - F_s} = \frac{6.5 - 5.0}{6.5 - 2.8} = \frac{15}{37}$$

$$R_2 = \frac{F_{a2} - F_{c2}}{F_{a2} - F_{o1}} = \frac{7.2 - 5.6}{7.2 - 5} = \frac{16}{22}$$

$$R_1 = \frac{A_s}{A_s + A_{a1}} = \frac{15}{37}$$

$$\text{or } 37A_s = 15(A_s + A_{a1})$$

$$\text{or } A_s = \frac{15A_{a1}}{22}$$

$$R_2 = \frac{A_s + A_{a1}}{A_s + A_{a1} + A_{a2}} = \frac{16}{22}$$

$$A_s + A_{a1} = \frac{8}{11}(A_s + A_{a1} + A_{a2}) = \frac{8}{11}A_g = \frac{8}{11}(2.75\,C)$$

$$A_s = \frac{15}{37}(A_s + A_{a1}) = \frac{15}{37}\left(\frac{8}{11}\right)(2.75)\,C = (0.295)(2.75)\,C = 0.811\,C$$

$$A_{a1} + A_{a2} = A_g - A_s = 2.75\,C - 0.811\,C = 1.939\,C$$

From Eq. (1.25) and assuming two per cent air content

$$\left(w + \frac{C}{s_c} + \frac{A_{a2} + A_{a1}}{s_a} + \frac{A_s}{s_s}\right) = 980$$

Substituting the respective values in the above equation, we get

$$0.384\,C + \frac{C}{3.14} + \frac{1.939\,C}{2.8} + \frac{0.811\,C}{2.6} = 980$$

34

$$C = \frac{980}{1.707} = 574 \text{ kg}$$

$$A_s = 0.811(574) = 465 \text{ kg}$$

$$A_{a1} = \frac{22}{15} A_s = 682 \text{ kg}$$

$$A_{a2} = 1.939\, C - A_{a1} = 431 \text{ kg}$$

The proportions are: Cement : sand : (20 mm + 30 mm aggregate)

574 : 465 : (682 + 431) or 1 : 0.81 : (1.19 + 0.75) or

1 : 0.81 : 1.96 and W/C = 0.384.

Note: The cement content required is too high, therefore use plasticizer/super plastizer and redesign the mix.

1.19 Curing of Concrete

Moist curing: Exposed surfaces of green concrete after about 6 to 10 hours of casting must be kept under damp or wet condition by continuous sprinkling of water, or covered with jute sacks which are kept under wet condition. The concrete can also be kept under ponding of water. In any case, the surface should be wet at least for minimum of 7 days or as specified by the engineer. A canvas or polythene sheet when placed over the wet surface helps in reducing loss of water.

Steam curing: Steam curing with proper control can also be done. The concrete must be allowed to set and harden for about 3 hours, then the concrete be covered under tarpaulin or put in steam chambers. Steam at about 80 to 100°C be allowed over the concrete gradually raising temperature of the chamber upto 70 to 90°C in about 2 hours time. Steam curing for about 8 to 10 hours should result into an equivalent 7 days strength of water cured concrete. Steam curing of 18 to 24 hours should give a strength equal to 28 days water cured cubes. The steam be cut off slowly in about an hour's time and the concrete allowed to cool in the chamber for another hour. Improperly done steam curing can cause cooking and spalling of the surface concrete.

1.20 Sampling of Concrete for Quality Control

Normally accepted specimens to test the strength of the concrete are 150 mm cubes. Beam specimens of $150 \times 150 \times 700$ mm are used to determine the modulus of rupture. A minimum of six cubes (3 for 7 days strength and 3 for 28 days strength) are required for a sample. The sampling is done at random and the total number of samples for a mix design depends on the total quantity of the concrete. Table 1.26 gives the minimum number of samples desired for a volume of the concrete and each sample consisting of six cubes.

Table 1.26 Number of Test Sample

Concrete in m^3	1 to 5	6 to 15	16 to 30	31 to 50	51 +
No. of samples	1	2	3	4	4+ (one for 50 m^3 or part)

In addition, sampling should be done whenever there is a change in the material lot or in the mix proportions. The specimens be cured one day less than the required period in water and sent to laboratory for saturation with water for a day and for testing on the specified date.

1.21 Quality Assurance in Concrete Structures

The quality assurance of concrete structures in a country like India varies from reliable level to uncertain level. In the final analysis it is the quality of the structure to be assured and not just the quality of the materials as dominantly known today. Components that control the quality of a structure are:

1. Quality of the basic materials such as aggregate, cement, water, Admixtures, etc.
2. The relative quantities of basic materials in concrete and its production,
3. Scaffolding, formwork, Construction and fabrication of the structure,
4. Transportation, placing, compaction and finishing of the concrete,
5. Tolerance limit, specifications in quantities and qualities, detailing,
6. Curing of Concrete,
7. Durability design considerations and maintenance requirements.

At the moment, the quality control specifications lean heavily towards items one and two of the list because of the historical and technological background. The idea that the quality control in the basic materials and the production of the concrete ensures the quality of the final structure is a necessary condition but not sufficient. There are a number of inter links between the quality of basic materials and the final structure. Between the cup and the lip, there could be a number of slips.

Once the designer specifies a characteristic strength of concrete, it is the responsibility of the field engineer to ensure that the quality of the concrete produced must meet the characteristic value. The quality assurance is obtained through tests on concrete samples. Test specimen for compressive strength is 150 mm concrete cube. Each sample consists of a minimum of six concrete cubes of which three cubes are tested on 7 days of curing and the remaining on 28 days of curing. Flexure tests are undertaken wherever tensile strength is specified. The characteristic strength and test statistics are connected by:

$$f_m = f_{ck} + ks$$

where $f_m =$ mean strength of samples of a large population,

      $f_{ck} =$ characteristic strength,

      $k =$ an index related to the probability of acceptance criteria,

      $=$ 1.65, in case of large population test samples in which not more than five per cent of the sample strengths are allowed to fall below the specified value, 1.41, in case of limited four consecutive samples with the same probability

      $s =$ standard deviation of the data or equivalent acceptable data; in the absence of data, the following coefficient of variation can be adapted: 25% for M15 concrete, 20% for M20 to M25 concrete, 15% for M25 to M40 concrete, and 12% for M40 to M60 concrete. The Indian standard code of practice suggests slightly different acceptability criteria and values.

Codal Provisions (IS : 456-1999 draft) on Quality Control

The clauses in the codes of practice on the quality control of structures are primarily on materials and hardened concrete and not on the final structure. Besides, the pre-assigned checks on the input materials, the main clauses on the quality control and acceptability of concrete as per the proposed code IS : 456-1999 (final draft) are:

For acceptability of concrete, the quality of concrete must satisfy the following:

$$f_m \geq f_{ck} + 0.825 \ s; \text{ or}$$

$$f_m \geq f_{ck} + A, \text{ and}$$

$$f_j \geq f_{ck} - B$$

where f_{ck} = characteristic strength,

f_m = mean strength of four consecutive samples,

f_j = strength of a sample,

A = 3 MPa for M15 concrete and 4 MPa for M20 grade concrete and above

B = 3 MPa for M15 concrete and 4 MPa for M20 grade concrete and above

The concrete is liable to be rejected if it is porous or honeycombed, improper construction joints and tolerances on size of the members. It can also be rejected for improper placement of reinforcement and inadequate cover, and not following the specifications.

Concrete Cube Test

The test specimen mentioned in the acceptance criteria is 150 mm concrete cube. Further, the following conditions be incorporated:

1. Sample is random,
2. The mould of the cube is well-finished steel or cast iron,
3. The cube is poured and vibrated under careful control,
4. Cured under good moist condition,
5. Tested on 28th day of curing the sample. Very often cubes at 7 days cured cubes are also tested for quick reference. Tested under surface dry condition.

The concrete CUBE TEST has become a demi-god in quality assurance of concrete structures. It is a hope and an imagination that the concrete in the structure has the same qualities of the sample! The code further states the following on the Inspection and Testing of the structure:

Care should be taken to see that:
   a) Design and details are being capable of being executed,
   b) Clear instructions on the inspection standards,
   c) Clear instructions on the permissible deviations,
   d) 'Elements critical to workmanship, structural performance, durability and appearance are identified'
   e) There is a system to verify that the quality is satisfactory in the individual parts specially the critical ones.

Deficiencies of Acceptable Criteria in Cube Test

A concrete sample normally consists of 3 standard cube specimens to be tested on 7th day, and another 3 specimens to be tested on 28th day of curing. Standard specifies, how the samples to be collected, cast, cured and tested. The construction in India is mostly through site mixed concrete. Consequently, there are some inherent deficiencies in the standard cube test criteria. Some of the weak points in the acceptance criteria are:

1. The batch of concrete from which the samples are collected may be prepared with care even though the samples are supposed to be chosen randomly,
2. Filling, compaction or vibration of samples is different from placing and vibration of the concrete in the structure,
3. The formwork and shuttering to the concrete in structure is not as grout leakproof as compared with the concrete cube moulds.
4. The time difference in transportation and laying of the concrete in position could be different from that of the preparation of cube specimens,
5. The detailing of reinforcement that might cause honeycombing or porous concrete is not reflected in making of the cubes,
6. The curing of the concrete in structure is not as systematic as that in cubes.
7. Testing of 3 specimens as a sample and taking an average strength of the sample is not reliable value.
8. There is a likelihood of less quality control in real concrete structure, detailing of reinforcement, formwork and curing as the actual structure is not subjected to testing.

Non-destructive Testing

Methods of non-destructive testing are developed during the last four decades. Reliable equipment for such non-destructive testing is manufactured only in recent times. Further, a number of types of non-destructive tests are available. Ultrasonic non-destructive testing has been accepted as a reliable method of testing of welding in all-important steel structures. At the moment, rebound hammer test, ultrasonic test, and core-cutting sampling test are adapted. The core cutting is adapted to a limited extent because of practical difficulties associated with core drilling, specimen preparation and drilling in thin elements, etc. Rebound hammer test indicates the hardness of the concrete surface from which the strength can be extrapolated. The test can be carried out quite extensively and a reliable statistical approach can be applied for acceptance of the test. Some of the strong points in quality control through rebound hammer test are:

1. A large number of readings can be taken in short duration,
2. Large number of locations on the structure can be tested,
3. A statistical approach with confidence can be applied,
4. Actual structure is tested, therefore the test result reflects the totality of the final product indicating the concrete mix, its consolidation, surface finish, formwork and curing,
5. The builder/contractor or even the supervisor will be careful in concreting and curing.

An indiscriminate application of the non-destructive testing without proper correlation factors can lead to less dependable results. Therefore, it is necessary to specify the method of testing.

specimen preparation, instrument calibration, correction factors and acceptable criteria. A number of precautions to be taken and correction factors to be applied to the rebound hammer test. These special aspects are:

1. Surface texture of the concrete,
2. The size of the element under test,
3. The thickness of the element,
4. Surface moisture content in the concrete at the time of testing,
5. Level of maturity (age) of the concrete at the time of testing with respect to 28 days strength,
6. Maintenance of the test hammer,
7. Stability of the supporting base from which the testing is undertaken.

A rationalized statistical approach to obtain the strength of the concrete through test hammer is more dependable when compared to the result of standard cube test. A computer aided program and a method of testing is available with the authors of the book.

Chapter 2

Working Stress Design of Beams

2.1 Introduction

In working stress design (WSD), structures are analysed by elastic-linear structural theory and the stress resultants acting on the members are computed. Each member is designed to resist the total forces acting on it, especially at the critical sections. The stress resultants are as follows:

1. *Axial force* (either tension or compression): A member subjected to primary tension is called a *tie*, and a member subjected to primary compression is called a *strut*. When the strut is in the upright position and may be subjected to bending, it is called a *column*.

2. *Shear force:* Total force acting transverse to the axis of a member is called *transverse shear* or simply *shear force*. The shear force is dominant in beams and present in columns or ties if they are subjected to combined axial force and bending.

3. *Flexure or bending moment:* A slender member subjected to bending moment is called *beam* and the bending moment is invariably coupled with transverse shear force.

4. *Torsion:* Torsion is caused by cumulative moment effect of shear stresses about the axis of a member. Torsion is not commonly present in most building structural elements. Sometimes, the beams are subjected to torsion, but ties and columns are subjected to torsion very rarely.

After having analysed for the stress resultants, one should proportion the member such that the actual stress caused in the member should be limited to an allowable stress. The main design criterion of the working stress design can be stated as:

$$\sigma_i(M, GE, GS, L) \le \sigma_{ai}(M, GE, GS, L, LC) \tag{2.1}$$

where $\sigma_i(M, GE, GS, L)$ = actual stress in the member which is a function of:

M, the material properties.

GE, the geometry of the element and is usually connected by area, second moment of the area and the shape of the cross-section, effective length, etc.

GS, the geometry of the structure such as interconnection of the members, general shape of the structure, etc.

L, the loads acting on the structure. The loads can broadly be classified as permanent or fixed loads also called as dead loads; live loads which are either movable or moving without acceleration: dynamic loads which have the effect of the acceleration and they are wind, seismic, blast, etc. and lastly the environmental loads. Under each of the above types one can have several sub-classifications.

$\sigma_{ai}(M, GE, GS, L, LC)$ = allowable stress, specified by the code. The stress could be normal or shear stress and it is a function of the quantities listed above. The allowable stress is taken equal to yield or ultimate or buckling stress divided by a factor of safety. It is also taken as a function of load combinations (LC). Alternatively, one can say that the factor of safety is reduced in case of combined load condition and it is because the occurrence of load combination may be rather small.

The allowable stress can be expressed as

$$\sigma_{ai} = \frac{f_k}{\alpha} \qquad (2.2)$$

where f_k = limiting strength which may be equal to yield stress, or crushing strength depending on situation or buckling stress, or modulus of flexure, etc.

α = factor of safety (Table 2.1)

Table 2.1 is only a guideline and invariably the allowable stresses are listed by the code of practice. In addition to the allowable stress criterion, the deflection limitations are also imposed in the design. This may be expressed as

$$v_i \leq v_{ai} \qquad (2.3)$$

where v_i = actual deflection in i th load combination

v_{ai} = allowable deflection

Sometimes, even the strains are to be limited.

Table 2.1 Guidelines to Factors of Safety

Condition	Limit strength	Factor of safety
1. Tension in reinforcement	Yield or proof stress	
a) Direct tension	"	1.60
b) Tension caused by bending	"	1.66
2. Compression in reinforcement	Buckling or compatible strain-stress	2.00
3. Compression in concrete		
a) Direct compression in short members	Crushing strength	4.00
b) Bending compression	"	3.00
c) Direct compression in long members	Correction due to slenderness	4.00
4. Tension in concrete		
a) Direct tension	Modulus of rupture	3.00
b) Bending tension	"	2.00
c) Diagonal tension (due to shear)	"	3.00

Note: The factor of safety is reduced by 75 to 80 per cent of the values listed above in case of combination of live load with wind or earthquake loads.

2.2 Moment Capacity of Rectangular Sections

The following assumptions are made in the calculation of the moment capacity of RCC beams:

1. Plane sections before bending remain plane even after bending. (The members are assumed to be slender, that is, the span to depth ratio is more than 5.)
2. The members are prismatic and the axis of the beam is always normal to the transverse plane even after bending. (This is true in all slender members in which shear deformation is negligible.)
3. A perfect bond exists between the concrete and the reinforcement. (No slip takes place between reinforcement bars and the concrete.)

4. Tensile resistance of the concrete is neglected and the section on the tension face of the neutral axis is assumed to be cracked.
5. Linear superposition of stresses irrespective of load sequences will hold good. The material behaviour is linear and elastic.

Consider a rectangular section as shown in Fig 2.1a. The notations are:

b = width, h = overall depth = D
d = depth of steel from extreme compression fiber, it is known as effective depth
nd = neutral axis depth from the extreme compression fibre
σ_{ab} = compressive stress in bending
σ_{st} = tensile stress in steel
C = total compression on the section
T = total tension on the section from the reinforcement
ϵ_c = compressive strain in concrete, ϵ_s = tensile strain in steel

Figure 2.1b indicates the strains. The neutral axis of the section based on the compatibility of strains can be obtained from similar triangles of the strains in the figure.

$$\text{This is } \frac{nd}{d-nd} = \frac{\epsilon_c}{\epsilon_s} \tag{2.4}$$

$$\text{or } nd = \frac{\epsilon_c d}{\epsilon_c + \epsilon_s} \tag{2.5}$$

The strains in the materials can be expressed as

$$\epsilon_c = \frac{\sigma_{cb}}{E_c} \tag{2.6}$$

$$\epsilon_c = \frac{\sigma_s}{E_s} \tag{2.7}$$

where E_c and E_s are the moduli of elasticity of concrete and steel respectively.
The neutral axis distance can be obtained from the above equation as

$$nd = \frac{\sigma_{cb}d}{\sigma_{cb} + \dfrac{E_c \sigma_s}{E_s}} = \frac{d}{1 + \dfrac{\sigma_s}{m\,\sigma_{cb}}}$$

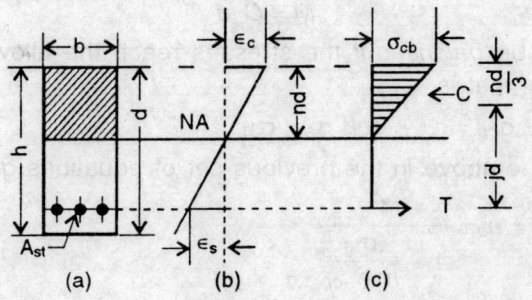

Fig. 2.1 Working Strain and Stress Distributions.

$$\text{or } n = \cfrac{1}{1 + \cfrac{\sigma_s}{m\sigma_{cb}}} \tag{2.8}$$

in which $m = E_s/E_c$ = modular ratio
The value of m is usually taken as

$$m = \frac{280}{3\sigma_{cb}} \tag{2.9}$$

and rounded of to the nearest integer.

$$n = \frac{1}{1 + 3\sigma_s/280} \text{ (approximate)} \tag{2.10}$$

The total compressive force acting on the concrete is obtained from the triangular area of stress as shown in Fig. 2.1c and is given by

$$C = \frac{1}{2} b\sigma_{cb} \, nd = \frac{nbd\sigma_{cb}}{2} \tag{2.11}$$

The total tensile force on the section is

$$T = A_{st}\sigma_s \tag{2.12}$$

where A_{st} = area of the tension reinforcement.

Assuming no axial force acting on the section, the equilibrium of the forces gives

$$T = C$$

$$\text{or } A_{st}\sigma_s = \frac{nbd\sigma_{cb}}{2} \tag{2.13}$$

$$\text{or } A_{st} = \frac{nbd\sigma_{cb}}{2\sigma_s} \tag{2.14}$$

$$\text{or } n = \frac{2A_{st}\sigma_s}{bd\sigma_{cb}} \tag{2.15}$$

In case the sectional properties are specified, the neutral axis is defined by Eq. (2.15). However, if the stresses are fixed, the area of the tension reinforcement is obtained from the condition as given in Eq. (2.14).

The bending moment on the section is given by

$$M = C\,jd \tag{2.16}$$

A section is said to be *balanced* if the stresses reach the allowable values in concrete and steel simultaneously, that is

$$\sigma_{cb} = \sigma_{acb} \text{ and } \sigma_s = \sigma_{st} \tag{2.17}$$

The substitution of the above in the previous set of equations gives

$$n = n_b = \cfrac{1}{1 + \cfrac{\sigma_{st}}{m\,\sigma_{acb}}} \tag{2.18}$$

$$C = 0.5\, n_b\, bd\, \sigma_{acb} \tag{2.19}$$

$$A_{stb} = 0.5n_b\, bd\, \frac{\sigma_{acb}}{\sigma_{st}} \qquad (2.20)$$

$$j = \left(1 - \frac{n_b}{3}\right) \qquad (2.21)$$

$$M_b = Cjd \qquad (2.22)$$

$$0.5n\,(1 - n_b/3)\, bd^2\, \sigma_{acb} = Kbd^2\, \sigma_{acb} = Rbd^2 \qquad (2.23)$$

$$\frac{A_{stb}}{bd} = 0.5n_b\, \frac{\sigma_{acb}}{\sigma_{st}} = p_0\, \frac{\sigma_{acb}}{\sigma_{st}} \qquad (2.24)$$

where $p_0 = 0.5n_b$ $\qquad\qquad\qquad\qquad\qquad\qquad\qquad\qquad$ (2.25)

Basically there are two types of problems. They are: First, the analysis of stresses for a given set of sectional properties which is called an analysis problem; secondly, the design problem in which the allowable stresses are preassigned and one is expected to design the section. Most of the problems in this book fall under design. The allowable stresses and other design coefficients are listed in Table 2.2.

Table 2.2 Allowable Stresses in Bending and Design Coefficients (Stresses are in MPa or N/mm^2).

Concrete	σ_{acb}	m	Values of R (MPa) stress in steel	
			140	230
M15	5.0	19	0.865	0.650
M20	7.0	13	1.211	0.910
M25	8.5	11	1.470	1.105
M30	10.0	9	1.730	1.300
M35	11.5	8	1.990	1.495
		n_b	0.400	0.289
		j	0.867	0.904
		K	0.173	0.130
		p_o	0.200	0.145

2.3 Design of Under- and Over-reinforced Sections

The size width and depth of a section of a beam are often fixed by architectural considerations. In such circumstances, one may not be able to design a balanced section. A section in which the actual stress in the reinforcement reaches the full allowable value before the extreme compression fibre of concrete is subjected to the full allowable stress is called an *under-reinforced section*. In such sections, the area of reinforcement is lower than the corresponding area of the balanced section. The depth of neutral axis will be less than the corresponding value of the balanced section. On the other hand, a beam section in which the extreme concrete compression fibre is subjected to full stress even before the stress in the reinforcement attains the full permissible value is called *over-reinforced section*. As the name indicates, the area of the reinforcement provided is more than that of a balanced section. The concrete in under-reinforced section and the reinforcement in the over-reinforced section are under-utilized. One may tend to conclude that such sections are not likely to be economical. The total cost of the beam depends not only on the actual material used but also on the surface area of the beam. The cost of formwork is also important in the overall cost. It may be that the balanced sections are likely to be of lower cost but one cannot completely conclude that way because the relative cost of steel is very high when compared with that of the concrete and the influence of the cost of the formwork.

The depth of a section in under-reinforced section would be more than that of a balanced section. The depth of neutral axis is obtained from the equilibrium Eq. (2.13) in which

$$\sigma_s = \sigma_{st}$$

Equation (2.14) gives

$$\sigma_{cb} = \frac{2A_{st}\,\sigma_{st}}{nbd} \tag{2.26}$$

Equation (2.8) gives

$$n = \frac{1}{1 + \dfrac{\sigma_{st}}{m\,\sigma_{cb}}} \tag{2.27}$$

The moment equilibrium gives

$$A_{st}\left(d - \frac{nd}{3}\right)\sigma_{st} = M \tag{2.28}$$

$$\text{or}\quad A_{st}\left(1 - \frac{n}{3}\right) = \frac{M}{d\sigma_{st}} \tag{2.29}$$

The unknown values are A_{st}, n and σ_{cb} and they can be solved from Eqs. (2.26), (2.27) and (2.28).

Equations (2.26) and (2.27) give

$$n = \frac{1}{1 - \dfrac{nbd}{2m\,A_{st}}} = \frac{2m\,A_{st}}{2m\,A_{st} - nbd}$$

$$\text{or}\quad (bd)n^2 - (2m\,A_{st})n - 2m\,A_{st} = 0 \tag{2.30}$$

section is *under-reinforced* if

$$n < n_b \tag{2.31}$$

It is *over-reinforced* if

$$n > n_b \tag{2.32}$$

When the area of steel is given, Eq. (2.30) will give the value of n from which one can judge the nature of the section.

2.4 Doubly Reinforced Section

Sometimes, the depth of the section is restricted to less than that of a balanced section because of architectural or other considerations. In such cases additional reinforcement is provided in the compression and tension zones so as to improve the moment capacity. The compressive reinforcement is given by

$$A_{sc} = \frac{M - M_r}{(d - d_c)\sigma_{sc}} \qquad (2.33)$$

where d_c = depth of compressive reinforcement from the extreme compression fibre

σ_{sc} = allowable compressive stress in the compression reinforcement and it is given by

$$\sigma_{sc} = 1.5m\left(\frac{nd - d_c}{nd}\right)\sigma_{acb} \qquad (2.34)$$

The corresponding tension reinforcement is

$$A_{st2} = \frac{M - M_r}{(d - d_c)\,\sigma_{st}} \qquad (2.35)$$

The total area of the tension reinforcement is

$$A_{st} = A_{stb} + A_{st2} \qquad (2.36)$$

Figure 2.2(a) indicates the notation of doubly-reinforced section. Figure 2.2(b) shows the relative location of the neutral axis (NA) in different types of beams.

2.5 Design for Transverse Shear Force

The transverse shear force on a section causes diagonal tension in the concrete. The maximum shear stress in an uncracked section occurs at the centroid of the section whereas in a cracked section it remains constant from neutral axis till the effective depth of the steel. In an uncracked section, the average shear stress is given by

$$\tau_V = \frac{VQ}{Ib} \qquad (2.37)$$

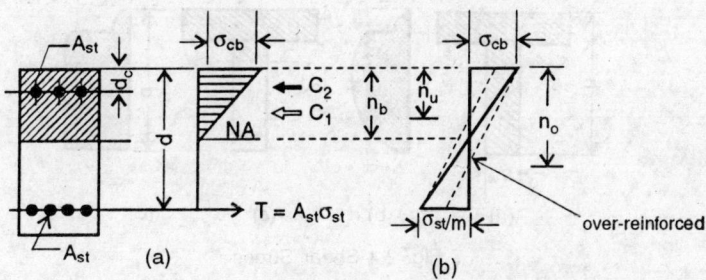

Fig. 2.2(a) Typical Stress-Strain Relations;
(b) Comparison of NA in Different Types of Sections

where τ = shear stress
V = shear force
Q = first moment of area of the section from free edge to the point under consideration
I = moment of inertia
b = width of the section at the point under consideration

Typical variation of shear stress in rectangular and *T* sections is shown in Figs. 2.3b and 2.4b respectively. However, the picture gets modified in the case of cracked reinforced concrete section. The maximum shear stress in cracked reinforced sections can be expressed as

$$\tau = \frac{V}{b_w jd} \tag{2.38}$$

where b_w = width of web
jd = lever arm distance

In design problems, one is more concerned with the average shear stress and should limit it within the allowable stress. The average shear stress also called *nominal shear stress* in cracked sections is given by

$$\tau_v = \frac{V}{b_w d} \tag{2.39}$$

Figures 2.3d and 2.4d illustrate the actual and average shear stress distributions. The capacity

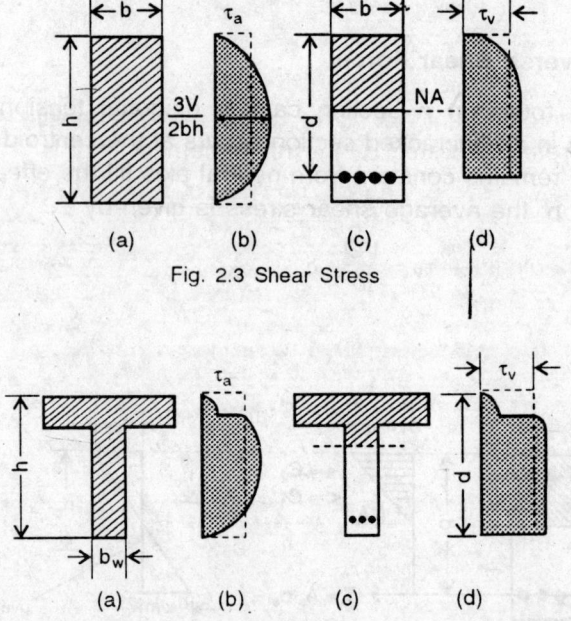

(a)　　(b)　　(c)　　(d)

Fig. 2.3 Shear Stress

(a)　　(b)　　(c)　　(d)

Fig. 2.4 Shear Stress

of a material against shear is controlled by the tension capacity of the material. In a pure shear stress state the principal tensile stress is same as the shear stress

$$\sigma_t = \tau \tag{2.40}$$

and it acts at 45° with the plane of shear. In working stress design, the shear stress is directly connected by the allowable shear stress. If the shear stress exceeds an allowable value, then shear reinforcement in form of stirrups is to be provided. The total tensile force in excess of the capacity of the concrete must be resisted by the transverse reinforcement. The shear reinforcement is usually provided as vertical stirrups or sometimes as inclined bars.

Considering all these aspects, the design of the transverse reinforcement as per IS:456 is given below.

Case 1 For $V \leq 0.5\, V_a$ No shear reinforcement is really needed. (2.41)

where $V_a = \tau_c\, bd$, $\tau_c =$ shear stress resisted by concrete.

Case 2 For $0.5\, V_a \leq V \leq V_a$ (2.42)

Provide only nominal transverse reinforcement at spacing given below.

$$s_v \leq \frac{A_{sv}\, f_y}{0.4\, b_w} \tag{2.43}$$

where f_y = characteristic strength of the shear reinforcement and limited to 415 MPa, 0.4 is in MPa (N/mm^2) and A_{sv} = area of stirrup steel.

The spacing is also restricted by

$$s_v \leq 0.75\, d \tag{2.44}$$

$$\leq 450 \ \text{mm} \tag{2.45}$$

Note: The minimum shear reinforcement resists a shear stress of 0.4 N/mm^2 and the nominal reinforcement is provided for this level.

Case 3 For $V_a \leq V \leq V_{c\,max}$ (2.46)

The transverse reinforcement be so designed that the tensile stress in excess of the allowable on the concrete is resisted by the reinforcement, that is

$$s_v \leq \frac{A_{sv}\, \sigma_{sv}\, d}{(V - V_a)} = \frac{A_{sv} \sigma_{sv}}{b_s\, (\tau - \tau_c)} \tag{2.47}$$

and S_v is subject to the limits as in case 2.

Case 4 For $V \geq V_{c\,max}$

Redesign the size of the web such that V is less than $V_{c\,max}$ and then provide the reinforcement.

The allowable and maximum shear forces are given by

$$V_a = \tau_c\, bd \tag{2.48}$$

$$V_{c\,max} = \tau_{c\,max}\, bd \tag{2.49}$$

where the values of τ_c and $\tau_{c\,max}$ are given in Table 2.3.

2.6 Development Lengths

Main bars in bending members: The tensile force in reinforcement is transferred through bond between steel and concrete. The stress in the reinforcement can be permitted provided that there is enough bond length beyond the point where such a stress is allowed. The length required to develop the designed strength of the bar is called development length and it is obtained by equating the bond strength to the tensile strength as given below:

$$\pi \phi \tau_{bd}\, L_d = \frac{\pi \phi^2}{4}\, \sigma_s$$

$$\text{or} \ \ L_d = \frac{\phi \sigma_s}{4 \tau_{bd}} \tag{2.50}$$

Table 2.3 Allowable Shear Stress in Concrete (τ_c). and Maximum Shear Stress $\tau_{c\,max}$ (in MPa)

$\dfrac{100\,A_s}{bd}$	(τ_c)				
	M15	M20	M25	M30	M35
0.25	0.22	0.22	0.23	0.23	0.23
0.50	0.29	0.30	0.31	0.31	0.31
0.75	0.34	0.35	0.36	0.37	0.37
1.00	0.37	0.39	0.40	0.41	0.42
1.25	0.40	0.42	0.44	0.45	0.45
1.50	0.42	0.45	0.46	0.48	0.49
2.0	0.44	0.49	0.51	0.53	0.54
2.5	0.44	0.51	0.55	0.57	0.58
3.0	0.44	0.51	0.57	0.60	0.62
$\tau_{c\,max}$	1.60	1.80	1.90	2.20	2.30

A_s = area of the tension steel which continues at least one effective depth beyond the critical shear plane.

where ϕ = diameter of the bar

σ_s = actual stress in the bar

τ_{bd} = design bond stress given in Table 2.6

When the ends of bars are provided with bends or hooks, an additional bond capacity is obtained. Allow a length of 4ϕ for every 45° bent subject to a maximum of 16ϕ, and allow 16ϕ for U hooks.

Development length in stirrups: The stirrups be bent at 90° around a bar of its own diameter and then continued beyond the end of the curve at least *eight times the diameter* of the stirrup. Similarly, the other lengths are given in Table 2.4.

Table 2.4 Minimum Required Development Length in Stirrups Beyond the Curve (in mm)

Diameter (mm)	Curve of the bent		
	90°	135°	180°
6	50	40	25
8	65	50	35
10	80	60	40
12	95	70	50
16	130	95	65

Table 2.5 gives the full development lengths of different bars for ready reference. It includes the allowance for U-hooks in plain bars.

Development length in ties is suggested to be taken twice as much as that required in bending members as per the code. In the opinion of the author this is much on the conservative side. It should be taken 1.25 times that of the length derived in the bending members. The reinforcement bar in bending tension or indirect tension is subjected to tensile stress only.

Table 2.5 Full Development Length of Bars in which full Permissible Stress in Steel is Allowed and U-hooks in Plain Bars are Accounted (in mm)

Diameter of bar in mm (ϕ)	Grades of concrete					
	15		20		25	
	Plain	Deform	Plain	Deform	Plain	Deform
	42 ϕ	51 ϕ	23 ϕ	46 ϕ	29 ϕ	41 ϕ
6	252	306	168	276	138	246
8	336	408	224	368	184	328
10	420	510	280	460	230	410
12	504	612	336	552	276	492
14	588	714	392	644	322	574
16	672	816	448	736	368	656
18	756	918	504	826	414	738
20	840	1020	560	920	460	820
22	942	1122	616	1012	506	902
25	1050	1275	700	1150	575	1025
28	1176	1428	784	1288	644	1148
32	1344	1632	896	1472	736	1312
36	1512	1836	1008	1656	828	1476

Table 2.6 Allowable Stress in Concrete (all stresses are in MPa)

Type of stress	Grade of concrete					
	M10	M15	M20	M25	M30	M35
1. Compressive						
a) Bending (σ_{bc})	3	5	7	8.5	10	11.5
b) Direct (σ_{cc})	2.5	4	5	6	8	9.0
2. Bond (τ_{bd})						
a) Plain bars (tension)	–	0.6	0.8	0.9	1.0	1.1
b) Deformed bars (tension)	–	0.84	1.12	1.26	1.4	1.54
c) Compression	1.25 times that in tension					

2.7 Permissible Stresses and other Minimum Requirements

The permissible stresses in concrete listed in Table 2.6 can be *increased* in certain conditions as given below:

a) 25 per cent in handling and transportation.

b) 33 per cent in combined load condition of *wind* and *dead* load. Further increase may be affected as given in IS 875 for different wind zones.

c) 33 per cent in combined load conditions of *dead load and earthquake.*

d) The compressive stresses may be increased in case the design loads will act later than 28 days of curing (age). The percentage increase various from 10% at 3 months age to 20% at the age of one year.

e) For slender beams where the span is greater than 30 times the width of the compression flange, the permissible stresses are (1.75 - L/40b) σ_{cb} where L = span, *b* = width of the flange.

f) For columns in which the ratio of effective height (L) to the least side (b) is more than 12, compressive stress be reduced by a reduction factor ($1.25 - L/48b$).

g) For *slender beams* in which $Wh/L > 0.5$, apply the reduction factor given in (*h*), where W = total load, h = overall depth, L = span.

h) In case of structure subjected to several *repeated loads* of magnitude close to the working load, the permissible stresses be *decreased* by about *ten per cent*.

The limiting strains in concrete are:
Ultimate compressive strain without reinforcement = 0.0035
Ultimate compressive strain with reinforcement   = 0.005
Cracking tension strain   = 0.00015

Design for lateral stability and deflection: Concrete beams in general are stable, specially when supporting a floor or a continuous medium. However, isolated beams have the tendency of lateral buckling. In such isolated beams, the stability is ensured by designing the width (*b*) of the beam such that

$$b \ge L/60 \tag{2.51}$$

$$\text{or} \quad b^2 \ge Ld/100 \tag{2.52}$$

The *permissible stresses* in steel reinforcement are given in Table 2.7

Table 2.7 Permissible Stresses in Steel Reinforcement (MPa)

	Type of stress	Mild steel (IS : 432)	HYSD or CTD (Fe 415)
1.	Tension		
	a) Bars upto ϕ 20 mm	140	230
	b) Bars over ϕ 20 mm	130	230
2.	Compressive stress in column bars	130	190
3.	Compressive stresses in $\quad$ (1.5 m) (stress in concrete at that level) beam and slab bars where compression in concrete accounted		
4.	Compressive stress in which the compression in concrete is neglected.		
	upto ϕ 20 mm	140	190
	over ϕ 20 mm	130	190

Deflection: A chapter on serviceability limit state gives a detailed discussion on the deflections.

Deflection in reinforced concrete structures is usually small because of large moment of inertia. Recommended span to depth ratios of beams are listed in Table 2.9 for good stiffness. *Permissible effective spans:* The effective spans of beam can be taken as given here.

a) Effective span = distance between centre to centre of supports or clear distance
   (of simply   between supports + effective depth of the beams whichever is
   supported)   smaller

Table 2.8 Recommended Effective Span to Depth Ratios

Type of beam or slab	Effective span/depth
Cantilever beams	7 to 14
Simply supported beams	18 to 25
Continuous beams	26 to 30
Cantilever slabs	8 to 15
One-way simply supported slabs	20 to 30
Two-way simply supported slabs	30 to 35
Continuous two-way slabs	30 to 40

Table 2.9 Area of Steel for Different Bars (mm^2)

Size in mm	No. of bars						
	1	2	3	4	5	6	7
5	19	39	58	78	98	117	137
6	28	56	84	113	141	169	197
8	50	100	150	201	251	301	351
10	78	157	235	314	392	471	549
12	113	226	339	452	565	678	791
14	153	307	461	615	769	923	1077
16	201	403	604	806	1008	1209	1411
18	254	508	763	1017	1272	1526	1781
20	314	623	942	1256	1570	1884	2199
22	380	760	1140	1520	1900	2280	2660
25	490	980	1472	1963	2454	2944	3435
28	615	1230	1845	2460	3075	3690	4305
32	804	1608	2412	3216	4021	4825	5629
40	1256	2513	3769	5026	6283	7539	8796

b) Effective span (continuous) = same as that of simply supported beams, in case the width of supported is less than clear span/12

= clear span + $\frac{1}{2}$ (width of support or depth of the member whichever is smaller), in case of one end is continuous

= clear span, in case both ends are continuous

c) Effective span (for roller or rocker bearings) = distance between the centre to centre of the bearings

Cover and spacing requirements for steel: Minimum cover to steel reinforcement should not be less than:

a) 25 mm or twice the diameter of the bar at the *ends* of beams and slabs.

b) 25 mm or diameter of the bar for longitudinal reinforcement in *beams*.

c) 15 mm or diameter of the bar in case of reinforcement in protected *slabs*.

d) 40 mm or the diameter of the bar in case of main reinforcement in *columns* of least dimension *more than 20 cm.* 25 mm for *columns* with least dimension *less than 20 cm* and diameter of bar less than 12 mm.

Table 2.10 Perimeter for Different Bars (cm)

Bar size in mm	Number of bars							
	1	2	3	4	5	6	7	8
5	1.57	3.14	4.71	6.26	7.85	9.42	10.99	12.56
6	1.88	3.76	5.65	7.53	9.42	11.30	13.18	15.02
8	2.51	5.02	7.53	10.05	12.56	15.07	17.59	20.10
10	3.14	6.28	9.42	12.56	15.71	18.85	21.99	25.13
12	3.77	7.54	11.31	15.06	18.85	22.62	26.38	30.16
14	4.52	9.04	13.57	18.09	22.62	27.14	31.16	36.19
16	5.02	10.05	15.08	20.10	25.13	30.16	35.18	40.21
18	5.65	11.31	16.96	22.66	28.27	33.93	39.58	45.24
20	6.28	12.56	18.84	25.13	31.41	37.69	43.98	50.20
22	6.91	13.82	20.73	27.64	34.56	41.47	48.38	55.29
25	7.85	15.70	23.56	31.41	39.27	47.12	54.97	62.83
28	8.79	17.59	26.38	35.18	43.98	52.77	61.52	70.65
32	10.05	20.10	30.15	40.21	50.26	60.31	70.37	80.52
40	12.56	25.13	37.69	50.26	62.83	75.39	87.96	100.52

e) 15 mm or diameter of the bar in any *other reinforcement*.

f) 50 mm to 75 mm in case of concrete members subjected to spray of *sea water*. (see durability requirement also)

Further increase in thickness in special cases:

g) Increase the thickness of cover by 15 mm to 40 mm in case the concrete surface is exposed to action of harmful chemicals, acid vapour, saline atmosphere, etc.

h) Increase the thickness of cover by 10 to 25 mm to resist fire for about 30 to 100 minutes.

i) Special bitumin coating may be provided over the concrete in case it is exposed to sever chemical actions.

Spacing of reinforcement be restricted to:

a) Clear spacing of the *main bars* should be greater than diameter of the large bar or 5 mm more than the nominal maximum size of the coarse aggregate.

b) The above value may be reduced to 2/3 of the nominal maximum size of coarse aggregate in case *needle vibrators* are used, provided sufficient space is left between the groups for the needle to go in.

c) The clear vertical spacing between two rows of bars should not be less than 15 mm or maximum size of coarse aggregate or maximum size of the bar.

d) Spacing of the *main bars in slabs* shall not be more than three times the *effective depth* of the slab or *450 mm*.

e) Spacing of the distribution reinforcement in slabs shall not be more than *five times* the effective *depth* or *450 mm*.

f) vertical spacing of distribution reinforcement in slabs and beams should not be more than 750 mm.

2.8 Design Examples of Rectangular Sections

EXAMPLE 2.1 *Design of balanced section:* Design of a simply supported beam of effective span 6 m subjected to the following loads and specifications:

Superimposed load	$w_s = 9.5$ kN/m
Live load	$w_l = 15.0$ kN/m
Grade of concrete	$f_{ck} = 20.0$ MPa $= 20(10^6)$ N/m^2
Grade of reinforcement	$f_y = 415$ MPa
Span	$L = 6$ m

$\sigma_{st} = 230$ MPa, $\sigma_{sv} = 140$ MPa; $\sigma_{cb} = 7$ MPa

Let the depth of the beam be assumed in the range of $L/15$ for the purpose of calculation of the self-weight.

Let $\qquad\qquad\qquad h = 0.80$ m and $b = 0.25$ m

Self weight $= w_g = (0.80)(0.25)(25) = 5.0$ kN/m

$\qquad\qquad w_d = w_g + w_s = 14.5$ kN/m

The maximum bending moment will occur at mid-span and is equal to

$$M_w = \frac{w_d L^2}{8} + \frac{w_1 L^2}{8} = (14.5 + 15)\frac{36}{8} = 132.75 \text{ kNm}$$

Design a balanced section by equating the balanced moment capacity equal to the active moment, i.e.

$\qquad\qquad\qquad M_r = M_w$, which leads to

$\qquad\qquad Kbd^2 \sigma_{cb} = 132.75(1000)$ Nm

From Table 2.2 $\qquad K = 0.130, p_0 = 0.175$

$\qquad$ Let $\quad b = 0.25$ m,

$\qquad$ then $d^2 = \dfrac{132750}{0.130(0.25)(7)(10^6)}$

$\qquad$ or $\qquad d = 0.764$ m

$\qquad$ then $\quad h = 0.764 + 0.036 = 0.8$ m

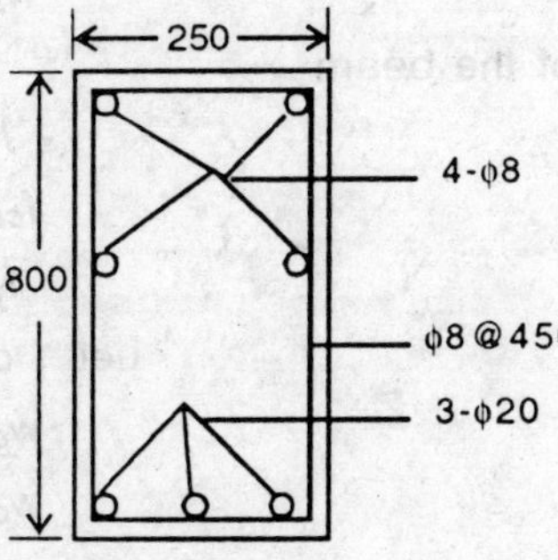

$$A_{stb} = p_0 bd \frac{\sigma_{cb}}{\sigma_{st}}$$

$$= 0.145\,(250)(764)(7)/230 = 842 \text{ mm}^2$$

Fig. 2.5 Example 2.1

Weight of the beam assumed checks well with actual one.
Provide 3 numbers of ϕ 20 of which one bar can be cranked.

$$A_{st} = 942 \text{ mm}^2$$

Provide 2 numbers of 8 mm bars at top as hanger bars.

Design for shear: Maximum design shear force which occurs at distance d from the centre of support is

$$V = (14.5 + 15)(3 - 0.74) = 66 \text{ kN}$$

Nominal shear stress

$$= \tau_v = \frac{V}{bd} = \frac{66\,000}{250(746)} = 0.35 \text{ N/mm}^2$$

The shear capacity of M20 concrete for 0.33 reinforcement (from Table 2.3) is $= \tau_c = 0.25$ N/mm^2

Let two-legged 8 mm bars be selected as stirrups, then the area of stirrups = 100 mm^2 $= A_{sv}$

Design shear force $= V_s = V - \tau_c bd = 66000 - 44750 = 18520N$

The spacing of the stirrups

$$s_v = \frac{A_{sv}\,\sigma_{sv}d}{V_s} = \frac{(100)(230)(764)}{18250} = 963 \text{ mm}$$

Spacing based on the minimum shear reinforcement is

$$s_v = \frac{A_{sv}f_y}{0.4\,b} = \frac{100(415)}{(0.4)\,(250)} = 450 \text{ mm}$$

or, Maximum spacing = 0.75 d or 450 mm

Provide 8 mm two-legged stirrups at 450 mm c/c.

EXAMPLE 2.2 *Design of under-reinforced section:* Design of a simply supported beam of span 6 m subjected to the following specifications:

$$w_s = 10 \text{ kN/m and } w_l = 15 \text{ kN/m}$$

Size of the beam

$$h = 0.80 \text{ m}, \ b = 0.3 \text{ m and}$$
$$f_{ck} = 20 \text{ MPa}; \ \sigma_{cb} = 7 \text{ MPa}$$
$$f_y = 415 \text{ MPa}; \ \sigma_{st} = 230 \text{ MPa}$$

Let $\quad d = h - 0.04 = 0.76 \text{ m}$
$$w_g = (0.3)\,(0.8)\,(25) = 6 \text{ kN/m}$$
$$w_d = w_g + w_s = 16\text{kN/m}$$
$$w_a = (w_d + w_l) = 31\text{kN/m}$$

$$M = \frac{w_a L^2}{8} = \frac{31\,(36)}{8} = 139.50\,\text{kNm}$$

Balanced moment capacity of the section is

$$M_{rb} = Kbd^2\,\sigma_{cb} = 0.130\,(0.3)\,(0.76)^2\,\frac{(7)\,(10^6)}{1000} = 157.68\text{kNm}$$

The balanced moment capacity of the section is higher than the external moment, the section can be designed as an under-reinforced one.

Moment capacity of the section is

$$M_r = \sigma_{st}\,A_{st}\,jd = 230\,A_{st}\,(0.904)\,(0.76)\,(10^6) \quad \text{Nm} = 158.20\,A_{st}$$

Equating the resisting capacity to the moment, we have

$$A_{st} = \frac{139.500}{158.20} = 880 \text{ mm}^2$$

Provide 3 numbers of 20 mm dia bars at the bottom and 2 numbers of 8 mm dia bars at the top as hanger bars.

$$\text{Area of concrete} = A_c = 0.3 (0.8) \qquad = 0.24 \text{ m}^2$$

$$\text{Total area of steel} = A_{st} + A_{sc} = 942 + 100 = 1042 \text{ mm}^2$$

Check for shear stress: Maximum shear force at a distance d from the centre of the

$$\text{support is } V = w_a (3 - 0.76) \qquad = 68\ 900 \text{ N}$$

$$\text{Nominal shear stress} = \frac{69\ 900}{300\ (760)} \qquad = 0.30 \text{ N/mm}^2$$

$$\text{Concrete shear capacity} = \tau_c \qquad = 0.27 \text{ N/mm}^2$$

The shear stress exceeds the allowable value marginally, so the minimum shear reinforcement governs the design. Provide 8 mm bars at 450 mm.

EXAMPLE 2.3 A portico of size 6 by 3 m is supported by two beams fixed into a wall and spaced at 4 m apart. Design the slab and beams.

$$\text{Data and assumptions:} \quad w_l = 1.5 \text{ kN/m}^2$$

$$f_{ck} = 15 \text{ MPa}; \ \sigma_{cb} = 5 \text{ MPa}$$

$$f_y = 250 \text{ MPa}; \ \sigma_{st} = 140 \text{ MPa}$$

The design parameters from Table 2.2 for $f_y = 250$ MPa are:

$$n = 0.40, \ K = 0.173, \ p_0 = 0.20$$

600 mm high half brick (115 mm) wall is provided as a parapet along the edge of the portico.

Design of the slab: The slab is supported by two beams spaced at 4 m apart. Therefore, it can be designed as one way a slab with overhang of 1 m on either side which means that there is a cantilever of 1 m for the slab.

$$a = 1 \text{ m}, L = 4 \text{ m}$$

Figure 2.6 illustrates the cross section of the portico. The slab is analysed and designed for a width of 1 m.

Assume thickness of slab = t = span/25 = 4/25 = 0.16 m
Weight of the slab $= w_g = t\gamma_c = 0.16(25) = 4.0 \text{ kN/m}^2$
Let the finish load $= w_s = 1.0 \text{ kN/m}^2$
$$w_d = w_g + w_s = 5 \text{ kN/m}^2$$
$$w_a = (w_d + w_l) = (6.5) \text{ kN/m}^2$$
Weight of parapet $\quad W = (0.115) (0.6) (20) = 1.38 \text{ kN/m}$

The negative BM on the slab due to self weight and also due to the concentrated load of the parapet

$$M_{a1} = \frac{w_a a^2}{2} + Wa = \frac{6.5}{2} + 1.38 = 4.63 \text{ kN/m}$$

in which a is taken as the effective cantilever span and equal to 1 m

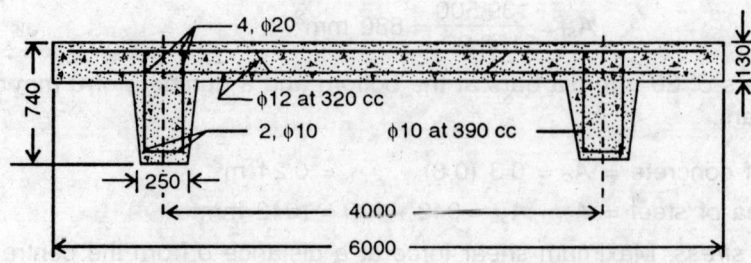

Fig. 2.6 Beam, Example 2.3 (dimensions in mm).

(actual effective span is $a = (1 - 0.115/2)$).

The positive BM which occurs at mid-span is

$$M_{a2} = \frac{w_a L^2}{8} - M_{a1} = \frac{6.5\,(16)}{8} - 4.63 = 8.37 \text{ kNm/m}$$

The section should be designed for the maximum of the above two. Equating the moment to the balanced moment capacity, we have

$$K b d^2 \sigma_{cb} = M_{a2}$$

$$0.173\,(1)\,d^2\,(5)(10^6) = 8370 \text{ Nm/m}$$

$$d = \sqrt{\left(\frac{8370}{0.173(5)(10^6)}\right)} = 0.098 \text{ m}$$

Let the overall depth of the slab be $h = 0.13$ m, which is less than the assumed for weight calculations. Revise the calculation with actual self-weight

$$w_g = 0.13(25) = 3.3 \text{ kN/m}^2$$

$$w_s = 1.0 \text{ kN/m}^2$$

$$w_l = 1.5 \text{ kN/m}^2$$

$$w_a = (w_g + w_s + w_l) = 5.8 \text{ kN/m}^2$$

$$M_{a2} = \frac{w_a L^2}{8} - \left(\frac{w_a a^2}{2} + W a\right)$$

$$= 5.8(2) - (5.8/2 + 1.38) = 7.29 \text{ kNm/m}$$

$$d = \sqrt{\frac{M_{a2}}{Kb\,\sigma_{cb}}} = \sqrt{\frac{7.29(1000)}{0.173(1)(5)(10^6)}} = 0.09 \text{ m}$$

$$A_{st} = p_0 b d \frac{\sigma_{cb}}{\sigma_{st}} = 0.2(1000)(90) \frac{5}{140} = 643 \text{ mm}^2/\text{m}$$

Provide 12 mm dia bars at 160 mm c/c at the bottom and crank alternate bars at 1 m from the beam. Provide 0.2 per cent distribution reinforcement

$$A_s = \frac{0.2}{100}\,(1000)(110) = 220 \text{ mm}^2/\text{m}$$

8 mm bars at 200 mm c/c.

The negative bending moment is less than half of the positive BM; therefore, the reinforcement which is cranked from the bottom would provide adequate resistance if extended at the top in the overhang portion. The details of the reinforcement are shown in Fig. 2.6.

Design of the beam: Each beam supports half of the slab; therefore, the load of the slab is superimposed on the beam. The beam has a cantilever span of 3 m supporting 3 m wide slab. The beam section is assumed to be below the slab; therefore, it is designed as a rectangular section. The design of T-beam will be discussed later. Let the average size of the beam for the purpose of self-weight computations be 250×400 mm. The beam would taper towards the free end. It is also assumed that the portico supports a facia of 600 mm high brick parapet wall. The weight of the wall should be accounted for in the bending moment computations.

Self-weight of the beam $= w_g = 0.25(0.4)(25) = 2.5$ kN/m.

Load from the slab including the finishing coat/3 m width
$$W_s = 5.8(3)(3) = 52.2 \text{ kN}$$

Weight of parapet at free end
$$W_{p1} = 3(1.38) = 4.14 \text{ kN}$$

Weight of the parapet at the side edge
$$W_{p2} = 3(1.38) = 4.14 \text{ kN}$$

The design moment at the fixed end of the beam is
$$M_a = W_s \left(\frac{L}{2}\right) + W_{p1} (L) + W_{p2} \left(\frac{L}{2}\right) + w_g \left(\frac{L^2}{2}\right)$$
$$= 52.2(1.5) + 4.14(3. + 1.5) + (2.5)(4.5) = 107.46 \text{ kNm}$$

Equating the balanced moment capacity of the beam to the external moment, we have
$$Kbd^2 \sigma_{cb} = M_a = 107\,460 \text{ Nm}$$

$$d = \sqrt{\frac{107\,460}{0.173(0.25)(5)(10^6)}} = 0.703 \text{ m}$$

$$A_{st} = \frac{p_o \, bd\sigma_{cb}}{\sigma_{st}} = 0.2(250)(703)(5)/140 = 1250 \text{ mm}^2$$

Provide 4 numbers of 20 mm, then
$$A_{st} \text{ (provided)} = 1256 \text{ mm}^2$$

$$\frac{100 \, A_{st}}{bd} = \frac{125\,600}{250(703)} = 0.72$$

Overall depth at fixed end
$$h = d + 10 + 25 = 740 \text{ mm}$$

Depth at free end = 250 mm
The average depth = (250 + 740)/2 = 495 mm

The self-weight is approximately equal to that assumed and is 9 kN.
Design for shear: Maximum shear force which occurs at the support is
$$V = W_s + W_g + W_{p1} + W_{p2} = 52.2 + 9.0 + 4.14 + 4.14 = 69.5 \text{ kN}$$

Nominal shear stress is

$$\tau_v = \frac{V}{bd} = \frac{69\,500}{250(703)} = 0.4 \text{ N/mm}^2$$

The shear capacity of M15 concrete from Table 2.3
(for $100\, A_{st}/bd = 0.7$) is

$$\tau_c = 0.33 \text{ N/mm}^2$$

Stirrups have to be designed.
Provide two-legged 10 mm diameter MS bars, then the spacing of the stirrups is

$$S_v = \frac{A_{sv}\,\sigma_{sv}}{b\,(\tau_v - \tau_c)} = \frac{157(140)}{250(0.4 - 0.33)} = 1340 \text{ mm}$$

Maximum spacing of the stirrups is

$$S_{max} = \frac{A_{sv}\,f_{sv}}{0.4b} = \frac{157(250)}{(0.4(250))} = 392.5 \text{ mm or,}$$

$$= 0.75d = 391 \text{ mm}$$

Final design details: Size of the web of the beam at free end

$$b = 250\,\text{mm}, \quad h_w = 250 - 130 = 120 \text{ mm}$$

Size of the beam at fixed end

$$b = 250 \text{ mm}, \quad h_w = 740 - 130 = 610 \text{ mm}, \quad d = 700 \text{ mm},$$

where h_w = depth of the web alone.

Reinforcement (MS bars)
4 numbers of 20 mm dia at top near the fixed end, and
2 numbers of 20 mm dia bars curtailed at 1.5 m from support.
Two-legged 10 mm dia stirrups at 390 mm c/c
(number of stirrups = 8)
2 numbers of 10 mm bars at bottom stirrup support as bars.

EXAMPLE 2.4 *Doubly reinforced section:* Design of a simply supported beam of span 6 m subjected to the following loads and specifications:

$$\text{Superimposed load} = w_s = 9.5 \text{ kN/m}$$
$$\text{Live load} \qquad = w_l = 15.0 \text{ kN/m}$$
$$f_{ck} = 20 \text{ MPa}, \qquad f_y = 415 \quad \text{MPa}$$
$$b = 0.20 \text{ m}, \qquad h = 0.5 \text{ m}$$
$$A_c = (0.2)(0.5) = 0.10 \text{ m}^2$$

Self-weight of the beam = $w_g = (0.2)(0.5)(25) = 2.5$ kN/m

The maximum moment which occurs at the mid span is

$$M_a = (w_g + w_s + w_l)\,\frac{L^2}{8}$$

$$= (2.5 + 9.5 + 15)\,\frac{36}{8} = 27\left(\frac{36}{8}\right) = 121.5 \text{ kNm}$$

The overall depth of the section is 500 mm and after allowing for a cover of 25 mm, the effective depth can be taken as

$$d = 500 - 25 - 25 = 450 \text{ mm}$$

The design coefficients for a balanced section taken from Table 2.2 are:

$$n = 0.289, \ K = 0.130, \ p_0 = 0.135$$

The balanced moment capacity of the section is

$$M_r = Kbd^2 \sigma_{cb} = 0.130(0.2)(0.45)^2 (7)(10^6) = 36 \ 855 \text{ Nm}$$

The moment is more than the balanced moment capacity; therefore, the section is to be designed as a doubly reinforced section.

The moment to be resisted by the compression reinforcement is

$$M = M_a - M_r = 121 \ 500 - 36 \ 855 = 84 \ 645 \text{ Nm}$$

The centre of compression steel from the extreme compression fibre is taken as 55 mm as two rows of bars are likely to be placed at top

$$d_c = 55 \text{ mm}$$

The stress at the level of the compression steel is

$$\sigma_c = \sigma_{cb} \left(1 - \frac{d_c}{nd} \right)$$

$$= 7 \left(1 - \frac{55}{0.289 \ (450)} \right) = 4.04 \text{ N/mm}^2$$

The stress in the compression steel is

$$\sigma_{sc} = 1.5 \ m \ \sigma_c = 1.5(13)(4.04) = 79 \text{ N/mm}^2$$

This value is less than 190 MPa; therefore use only

$$\sigma_{sc} = 79$$

$$d - d_c = 450 - 55 = 395 \text{ mm} = 0.395 \text{ m}$$

The moment capacity of the compression reinforcement is

$$M_{rc} = A_{sc} \ \sigma_{sc} \ (d - d_c)$$

Equating the moment capacity of the compression reinforcement to that of the unbalanced, we have

$$A_{sc} \ \sigma_{sc} \ (d - d_c) = M = 84 \ 645 \text{ Nm}$$

$$A_{sc} = \frac{84645}{0.395 \ (79)} = 2713 \text{ mm}^2$$

The total tensile reinforcement is

$$A_{stb} = p_c \ bd \frac{\sigma_{cb}}{\sigma_{st}}$$

$$= (0.145)(200)(450) \left(\frac{7}{230} \right) = 370 \text{ mm}^2$$

$$A_{st2} = A_{sc} \frac{\sigma_{sc}}{\sigma_{st}} = 2710 \left(\frac{79}{230} \right) = 932 \text{ mm}^2$$

60

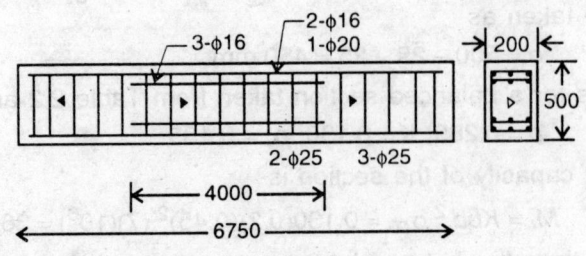

Fig. 2.7 Example 2.4

$$A_{st} = A_{stb} + A_{st2} = 1302 \text{ mm}^2$$

Provide b = 200 mm, *h* = 500 mm
  d = 450 mm, d_c = 55 mm

5 numbers of 16 mm bars and 1 number of 20 mm bar at top. Four number of 25 mm bars and 1 number of 20 mm bars at bottom

$$A_{sc} = 2768 \text{ mm}^2, \ A_{st} = 1369 \text{ mm}^2$$
$$A_s = A_{st} + A_{sc} = 4137 \text{ mm}^2$$

2.9 Moment Capacity of a Triangular Section

Let *b* = width of the section at depth *d* from apex.
  d = depth of the reinforcement from apex.

Figure 2.8 gives general dimensions and notations. Consider a thin strip at depth *x* from apex, then the width and stress at that depth are:

$$b_y = \frac{bx}{d}$$

$$\sigma_y = \frac{(nd - x)\sigma_{cb}}{nd}$$

This distance *x* is selected from top for convenience and it is equal to (*nd* − *y*). The compressive force on an infinitesimal area is

$$dC = b_y \, \sigma_y \, dx = \frac{bx(nd - x) \, \sigma_{cb}}{nd \cdot d}$$

The total compressive force on the section is

$$C = \int dC = \int_0^{nd} \frac{bx(nd - x)}{nd \, d} \sigma_{cb} \, dx$$

$$= \frac{\sigma cb}{nd \, d} \left[\frac{x^2 nd}{2} - \frac{x^3}{3} \right]_0^{nd} = \frac{n^2 bd}{6} \sigma_{cb} \tag{2.53}$$

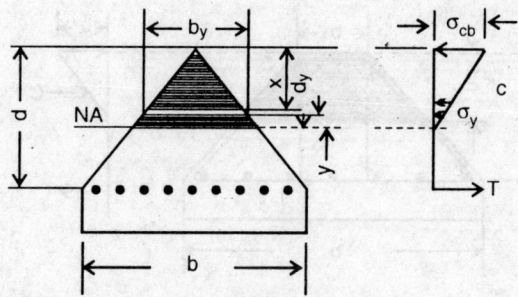

Fig. 2.8 Triangular Section.

The equilibrium of forces in the horizontal direction gives

$$A_{st}\,\sigma_{st} = C = \frac{n^2 bd\,\sigma_{cb}}{6}$$

$$\text{or} \quad A_{st} = \frac{n^2 bd\,\sigma_{cb}}{6\sigma_{st}}$$

Moment of the compressive force about the centre of steel gives

$$M_r = \int_0^{nd} (d - x)\,dC$$

$$= \frac{b\sigma_{cb}}{nd^2} \int_0^{nd} [x\,(nd - x)(d - x)]\,dx$$

$$= \frac{b\sigma_{cb}}{nd^2} \left[\frac{nd^2\,x^2}{2} - \frac{(nd + d)\,x^3}{3} + \frac{x^4}{4} \right]_0^{nd}$$

$$= \frac{n^2\,(2 - n)\,bd^2\,\sigma_{cb}}{12} = R_1 bd^2 \tag{2.54}$$

where $R_1 = n^2\,(2 - n)\,\sigma_{cb}/12 = K_1\sigma_{cb}$

Table 2.11 gives the coefficients for triangular section.

Table 2.11 Balanced Moment Coefficients for Triangular Section

σ_{st} =	140 MPa			σ_{st} =	230 MPa		
K_1 =	0.0211				0.0119		
σ_{cb} =	5	7	8.5		5	7	8.5
R_1	0.106	0.148	0.179		0.06	0.083	0.101

2.10 Moment Capacity of a Trapezoidal Section

The moment capacity of a trapezoidal section can be computed from the sum of the moments of the middle rectangular and the triangular sections as shown in Fig. 2.9 and it is

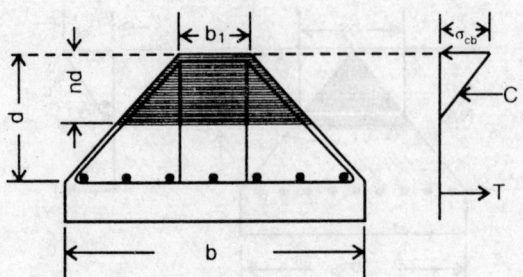

Fig. 2.9 Trapezoidal Section.

$$M_r = (Kb_1 + K_1(b - b_1)) \, d^2 \, \sigma_{cb} \tag{2.55}$$

$$= (Rb_1 + R_1(b - b_1)) \, d^2 \tag{2.56}$$

in which the first term corresponds to the moment capacity of the rectangular portion of the section with width as b_1 and the second portion corresponds to the triangular portion with base width $b - b_1$.

b_1 = width of the extreme compression fibre

b = width of the section at the level of steel

$$R = \frac{n}{2}\left(1 - \frac{n}{3}\right)\sigma_{cb}$$

$$R_1 = \frac{n^2 \, (2 - n)}{12}\sigma_{cb}$$

The area of the reinforcement can be obtained by equating the tension in steel to the total compressive force:

$$A_{st}\,\sigma_{st} = C_1 + C_2$$

$$= \frac{nb_1 \, d\sigma_{cb}}{2} + \frac{n^2 \, (b - b_1) \, d\sigma_{cb}}{6} = \frac{nd}{2}\left(b_1 + \frac{n(b - b_1)}{3}\right)\sigma_{cb}$$

$$A_{st} = \frac{nd}{2}\left(b_1 + \frac{n(b - b_1)}{3}\right)\frac{\sigma_{cb}}{\sigma_{st}} \tag{2.57}$$

2.11 Working Moment Capacity of a Circular Section

Consider a circular cross-section of radius R and the depth of steel at distance d from the extreme compression fibre. Consider a thin element at a distance y from the neutral axis (see Fig. 2.10). We have $b_y = 2R \sin \theta$

The neutral axis distance is given by

$$nd = (R - R \cos \phi) = R\,(1 - \cos \phi) \tag{2.58}$$

where ϕ = semi-central subtended angle.

The distance y is measured from NA and its relation with the subtended angle is

$$y = R \cos \theta - R \cos \phi = R(\cos \theta - \cos \phi) \tag{2.59}$$

$$dy = - R \sin \theta \, d\theta$$

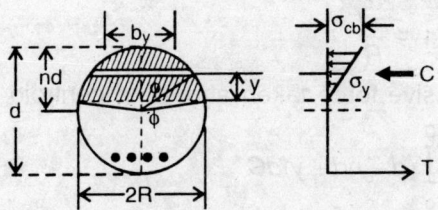

Fig. 2.10 Circular Cross-section.

The value of the compressive stress at a distance y is

$$\sigma_y = \frac{y}{nd}\,\sigma_{cb}$$

The compressive force on an element of thickness dy at distance y from the neutral axis is

$$dC = (b_y\,dy)\,\sigma_y$$

$$= 2R\sin\theta(-R\sin\theta\,d\theta)\frac{y\sigma_{cb}}{nd}$$

$$= -\frac{2R^3\sin^2\theta\,(\cos\theta - \cos\phi)}{R(1-\cos\phi)}\,\sigma_{cb}\,d\theta$$

$$= \frac{-2R^2\,\sigma_{cb}}{(1-\cos\phi)}\,(\sin^2\theta\cos\theta - \sin^2\theta\cos\phi)\,d\theta$$

The total compressive force on the section is

$$C = -\frac{(2R^2\sigma_{cb})}{1-\cos\phi}\int_{\phi}^{0}(\sin^2\theta\cos\theta - \sin^2\theta\cos\phi)\,d\theta$$

$$= \frac{2R^2\,\sigma_{cb}}{1-\cos\phi}\left[\frac{\sin^3\theta}{3} - \frac{\cos\phi}{2}\left(\theta - \frac{\sin 2\theta}{2}\right)\right]_{0}^{\phi}$$

$$= \left[\frac{2R^2\,\sigma_{cb}}{1-\cos\phi}\left[\frac{\sin^3\phi}{3} - \frac{\cos\phi}{2}\left(\phi - \frac{\sin 2\phi}{2}\right)\right]\right]$$

$$= \frac{R^2\sigma_{cb}}{3(1-\cos\phi)}\left[2\sin^3\phi + \frac{3\cos\phi}{2}(\sin 2\phi - 2\phi)\right] \tag{2.60}$$

The value of ϕ can be obtained from the compatibility condition in which n is

$$n = \frac{1}{1 + \dfrac{\sigma_{st}}{m\sigma_{cb}}}$$

which is equal to
$$n = \frac{R}{d}(1 - \cos \phi)$$

therefore
$$\cos \phi = 1 - \frac{nd}{R} \qquad (2.61)$$

The moment of the compressive force taken about the centroid of steel is

$$M_r = \int_{\phi}^{0} (d - nd + y)dC$$

The value of d can be expressed as a function of the radius of the section and cover distance. Let d_1 = distance of the extreme tension fibre to the centre of the steel, then

$$d = 2R - d_1$$

and
$$d - nd + y = d(1 - n) + R(\cos \theta - \cos \phi)$$
$$= (2R - d_1)(1 - n) + R(\cos \theta - \cos \phi)$$
$$= R(2 - 2n + \cos \theta - \cos \phi) - d_1(1 - n)$$

$$M_r = \int_{\phi}^{0} (d - nd + y)dC$$

$$= \int_{\phi}^{0} [R(2 - 2n + \cos \theta - \cos \phi) - d_1(1 - n)]$$

$$= \left[\frac{-2R^2 \sigma_{cb}}{1 - \cos \phi}(\sin^2 \theta \cos \theta - \sin^2 \theta \cos \phi) \right] d\theta$$

$$= \frac{2R^2 \sigma_{cb}}{1 - \cos \phi} \int_{0}^{\phi} [R(2 - 2n) - d_1(1 - n)](\sin^2 \theta \cos \theta - \sin^2 \theta \cos \phi) \, d\theta$$

$$+ \int_{0}^{\phi} R \sin^2 \theta (\cos \theta - \cos \phi)^2 \, d\theta$$

$$= \frac{R^2 \sigma_{cb}}{3(1 - \cos \phi)} [R(2 - 2n) - d_1(1 - n)] \left[2 \sin^2 \phi + \frac{3 \cos \phi}{2} \cdot (\sin 2\phi - 2\phi) \right]$$

$$+ \frac{2R^3 \sigma_{cb}}{1 - \cos \phi} \int_{0}^{\phi} (\sin^2 \theta \cos^2 \theta - 2 \sin^2 \theta \cos \theta \cos \phi + \sin^2 \theta \cos^2 \phi) \, d\theta$$

the second expression on RHS is

$$= \frac{2R^3 \sigma_{cb}}{1 - \cos \phi} \left[\frac{1}{32}(4\phi - \sin 4\theta) - \frac{2 \cos \phi}{3}(\sin^3 \theta) + \frac{\cos^2 \phi}{4}(2\theta - \sin 2\theta) \right]_{0}^{\phi}$$

$$= \frac{R^3 \sigma_{cb}}{16(1 - \cos \phi)} [4\phi - \sin 4\phi - \frac{64}{3}\cos \phi \sin^3 \phi + 8\cos^2 \phi (2\phi - \sin 2\phi)]$$

therefore $M_r = \dfrac{R^2}{3(1 - \cos \phi)} \sigma_{cb}\Big[\{R(2 - 2n) - d_{1(1 - n)}\}$

$$\left\{2\sin^3 \phi + \frac{3}{2} \cos \phi (\sin 2\phi - 2\phi)\right\}$$

$$+ \frac{3R}{16}\left\{4\phi - \sin 4\phi - \frac{64}{3} \cos \phi \sin^3 \phi + 8\cos^2 \phi (2\phi - \sin 2\phi)\right\}\Big] \qquad (2.62)$$

2.12 Examples of Design of Non-rectangular Section

EXAMPLE 2.6 *Design of a triangular section:* A cantilever beam which supports a portico is made of an inverted triangular section. The effective span of the beam is 3 m and the live and dead loads from the slab over it are 3 kN/m and 1 kN/m respectively. Design the cross-section.

Design data. Let self-weight be

$$w_g = 2 \text{ kN/m}$$

Weight of slab $= 3 \text{ kN/m}$
Dead load $= w_d = 5 \text{ kN/m}$
Live load $= w_l = 1 \text{ kN/m}$
Total load $w = 6 \text{ kN/m}$ and Span $L = 3$ m

Use M15 concrete with mild steel bars, then the compatible neutral axis distance is $n = 0.404$
  Design of section: Let the width of the base be selected approximately in the range of $L/10$, then

$$b = 0.3 \text{ m}$$

The working moment capacity of the section is

$$M_r = R_1\, bd^2$$

where $\qquad R_1 = \dfrac{n^2 (2 - n)}{12} \sigma_{cb} = \dfrac{0.4^2(2 - 0.404)(5)}{12} = 0.106 \text{ N/mm}^2$

The maximum bending movement on the beam is

$$M = \frac{wL^2}{2} = \frac{6(9)}{2} = 27 \text{ kNm} = 27\,000 \text{ Nm}$$

Equating the moment capacity to the external bending movement, we have

$$0.106(10^6)(0.3)d^2 = M = 27000$$

$$\text{or } d = \sqrt{\frac{27000}{0.106(0.3)(10^6)}} = 0.91 \text{ m}$$

Use the overall depth of the section as

$$h = d + 0.04 = 0.95 \text{ m}$$

Let the depth of the section be decreased to 150 mm free.
Area of cross-section at the fixed end $= 0.5(0.3)(0.95) = 0.1425 \text{ m}^2$
Area of cross-section at the free end $= 0.5(0.3)(0.15) = 0.0225 \text{ m}^2$

Approximate average weight of the beam $= \dfrac{(0.1425 + 0.0255)(24)}{2} = 1.98$ kN/m

Assumed weight is close to the actual one.

The area of steel required is

$$A_{st} = \frac{n^2 b d\sigma_{cb}}{6\sigma_{st}} = \frac{(0.4)^2(0.3)(0.91)(5)(10^6)}{(6)(140)} = 265 \text{ mm}^2$$

Minimum percentage of steel be about 0.15 per cent, then minimum steel required is

$$A_{st} = \frac{0.15}{100} \frac{(0.3)(0.95)(10^6)}{2} = 214 \text{ mm}^2$$

Moreover, the minimum steel is also subject to

$$A_{st} = \frac{0.85 \, A_c}{f_y} = \frac{0.85(300)(910)}{250(2)} = 464 \text{ mm}^2$$

Provide six numbers of 10 mm bars, at top. Figure 2.11 illustrates the cross-section at support.

Design for shear. Maximum shear force at a distance *d* from the support is

$$V = w(L) = 6(3) = 18 \text{ kN}$$

The nominal shear stress is obtained by dividing the shear force by the area of the effective concrete.

$$A_c = \frac{bd}{2} = 150 \,(910) \text{ mm}^2$$

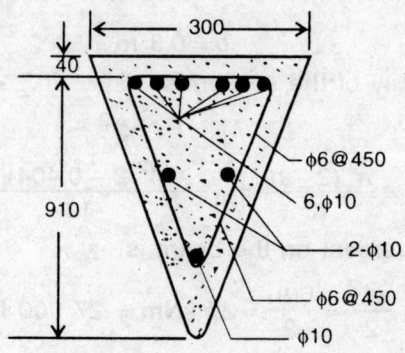

300

40

910

φ6@450

6,φ10

2-φ10

φ6@450

φ10

Fig. 2.11 Example 2.6.

The nominal shear stress is

$$\tau_v = \frac{V}{A_c} = \frac{18000}{150(910)} = 0.13 \text{ MPa}$$

Provide two-legged 6 mm stirrups at 450 mm c/c as nominal.

EXAMPLE 2.7 *Design of a trapezoidal section:* A cantilever beam which supports a portico slab is made of trapezoidal section. The effective span of the beam is 3 m. The live and dead load from the portico on to the beam are 1 kN/m and 3 kN/m. Design the trapezoidal section.

Design data: The data are same as that of Example 2.6 except for the shape of the section.

Let $w_g = 2$ kN/m

$w_a = 3$ kN/m

Given $w_d = 5$ kN/m

$\underline{w_l = 1 \text{ kN/m}}$

6 kN/m

Design of section: Maximum bending moment on the beam occurring at the support is

$$M = \frac{wL^2}{2} = \frac{6(9)}{2} = 27 \text{ kNm} = 27000 \text{ Nm}$$

The resisting moment capacity of the section is (M15 concrete)

$$M_r = Rb_1 d^2 + R_1(b - b_1)d^2$$

in which

$$R_1 = \frac{n}{2}\left(1 - \frac{n}{3}\right)\sigma_{cb} = \frac{0.4}{2}\left(1 - \frac{0.4}{3}\right)(5) = 0.865 \text{ MPa}$$

$$R_1 = \frac{n^2(2-n)}{12}\sigma_{cb} = 0.106 \text{ MPa}$$

Let $b = 0.300$ m and $b_1 = 0.150$ m

Then the resisting moment is

$$M_r = (0.865(0.15) + 0.106\ (0.15))\ 10^6\ d^2 = 0.1473\ (10^6)\ d^2 \text{ Nm}$$

Equating the moment capacity of the external moment, we have

$$0.1473\ (10^6)d^2 = M = 27000 \text{ Nm or } d = 0.43 \text{ m}$$

Use $h = d + 0.04 = 0.47$ m

Let the depth of the beam at free end be reduced to 150 mm. The area of the section at support is

$$\frac{0.47\ (0.3 + 0.15)}{2} = 0.106 \text{ m}^2$$

The area of the section at free end is

$$0.15\frac{(0.3 + 0.15)}{2} = 0.034 \text{ m}^2$$

Approximate weight of the beam is

$$w_g = \frac{(0.106 + 0.034)25}{2} = 1.72 \text{ kN/m}$$

The assumed weight is 2 kN/m, which is slightly higher than the actual value, hence no revision is needed. The area of the reinforcement needed is

$$A_{st} = \frac{nd}{3}\left(b_1 + \frac{n(b-b_1)}{3}\right)\frac{\sigma_{cb}}{\sigma_{st}}$$

$$= (0.2)(0.43)\left(1.15 + \frac{(0.4)(0.15)}{3}\right)\frac{5(10^6)}{140} = 528 \text{ mm}^2$$

Minimum area of the reinforcement needed is

$$A_{sm} = \frac{0.15(0.3 + 0.15)(0.47)10^6}{100} \frac{1}{2} = 581 \text{ mm}^2$$

and it is also controlled by

$$A_{sm} = 0.85 \frac{(0.3 + 0.15)(0.43)10^6}{2(140)} = 587 \text{ mm}^2$$

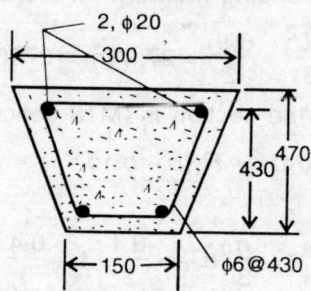

Fig. 2.12 Example 2.7.

Provide two numbers of 20 mm bars at top. Figure 2.12 illustrates the sectional details.

EXAMPLE 2.8 *Design of a trapezoidal section RCC beam:* A simple supported beam of trapezoidal cross-section with 300 m and 150 mm widths at top and bottom fibre is subjected to a working moment of 27000 Nm. Design the section using M15 concrete and mild steel reinforcement.

Design of the section. The resisting moment capacity of a trapezoidal section is

$$M_r = Rb_1 d^2 + R_1 (b - b_1)d^2$$

in which

$$R = \frac{n}{2}\left(1 - \frac{n}{3}\right)\sigma_{cb} = 0.865 \text{ MPa}$$

$$R_1 = \frac{n^2 (2 - n)}{12} \sigma_{cb} = 0.106 \text{ MPa}$$

The width of the section at the extreme compression fibre is

$$b_1 = 0.30 \text{ m}$$

The width of the section at the tensile reinforcement level is

$$b = 0.15 \text{ m}$$

Therefore, the working moment capacity of the section is

$$M_r = Rb_1 d^2 + R_1(b - b_1)d^2$$

$$= (0.865(0.3) + 0.106(-0.15))10^6 d^2 = 0.246(10^6) d^2 \text{Nm}$$

The working moment capacity must be more than the moment acting on the section. This leads to

$$M_r = 0.246(10^6)d^2 \geq M = 27000$$

or $\quad d \geq \sqrt{\dfrac{27000}{0.246(10^6)}} = 0.331$ m

The area of reinforcement is

$$A_{st} = \frac{nd}{2}\left(b_1 + \frac{n(b-b_1)}{3}\right)\frac{\sigma_{cb}}{\sigma_{st}}$$

$$= (0.2)(0.331)\left(0.3 + \frac{(0.4)(-0.15)}{3}\right)\frac{5}{140}$$

$$= 6.68(10^{-4})\ m^2 = 668\ mm^2$$

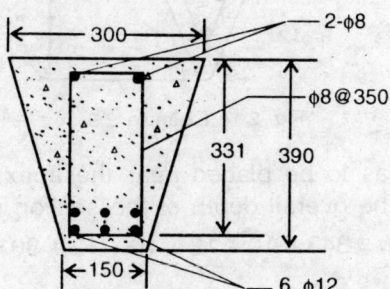

Fig. 2.13 Example 2.8.

Provide 6 number of 12 mm bars and $h = 390$ mm. Figure 2.13 illustrates the section.

EXAMPLE 2.9 *Design of a triangular sectioned RCC beam:* A simply supported beam should have a triangular cross section with width at the top of the section as 300 mm. The section is to be designed to resist a working moment of 27000 Nm with M15 concrete and mild steel.

Design of the section: The cross-section of the beam is an inverted triangle with reinforcement at the bottom vertex point. The resisting moment capacity of a trapezoidal section can be used in the computation with the width of the section at the reinforcement set equal to zero. The various parameters are (vide previous example):

$$R = 0.865\ MPa, \quad R_1 = 0.106\ MPa$$

$$b_1 = 0.30\ m, \quad\quad b = 0$$

The substitution of different values in the expression of working moment capacity gives

$$M_r = (0.865(0.3) - 0.106(0.3))10^6\ d^2 = 0.2298(10^6)d^2$$

$Mr > M$ gives

$$d > \sqrt{\frac{27000}{0.2298(10)^6}} = 0.343\ m$$

The area of the reinforcement needed is

$$A_{st} = \frac{nd}{2}\left(d_1 + \frac{n(b-b_1)}{3}\right)\frac{\sigma_{cb}}{\sigma_{st}}$$

$$= \frac{(0.4)(0.343)}{2}\left(0.3 - 0.4\frac{(0.3)}{3}\right)\frac{5(10^4)}{140} = 642 \text{ mm}^2$$

Provide 6 numbers of 12 mm bars.

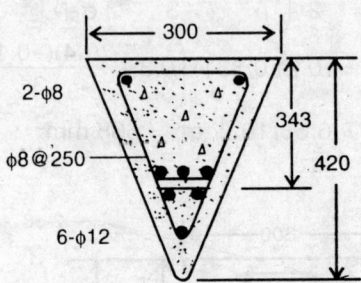

Fig. 2.14 Example 2.9.

Since the reinforcement has to be placed near the apex, it is arranged in a triangular manner as shown Fig. 2.14. The overall depth of the section is

$$h = 343 + 10 + 12 + 10 + 12 + 30 = 420 \text{ mm}$$

2.13 Stress Analysis of a Section

One may be interested in the determination of stresses in a given RCC cross-section under a particular load or in the determination of moment capacity of the given section. Such a problem is normally treated as an analysis problem as the design variables are fixed. Consider a rectangular section in which the sectional details (b, d, h, A_{st}) and bending moments are given.

Under these conditions, one would like to compute the stresses in concrete and steel and check whether these stresses are within the limits. The plane section before bending will remain plane after bending leading to a compatible relation in the strains in steel and concrete. Figure 2.15a gives the notations of a given cross section, and Fig. 2.15b gives the strain variation across the depth. The compatibility of the strains is:

$$\frac{\varepsilon}{x} = \frac{\varepsilon_s}{d-x} \tag{2.63}$$

where ε_c = actual compressive strain in the extreme concrete fibre
  ε = actual strain in the reinforcement
  x = neutral axis distance

In a design problem, we preassign the full allowable stresses. However, in an analysis problem, these strains or stresses are not fixed ahead; therefore, the problem formation is different. Assuming a linear stress-strain relation both in steel and concrete, the strain compatibility equation can be rearranged as

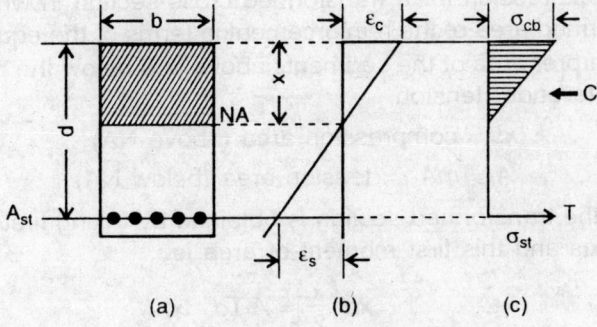

Fig. 2.15 Strain, Stress Variations.

$$\frac{\sigma_{cb}}{E_c x} = \frac{\sigma_{st}}{E_s(d - x)} \qquad (2.64)$$

$$\text{or} \quad \sigma_{st} = \frac{E_s \sigma_{cb}}{E_c} \frac{(d - x)}{x} \qquad (2.65)$$

where σ_{cb} = bending compression in the extreme fibre of the concrete

$\qquad \sigma_{st}$ = tensile stress in the steel

(For convenience σ_{cb} and σ_{st} are referred as the actual stresses but not the full allowable values.)

Let $\dfrac{E_s}{E_c} = m =$ modular ratio, then the compatibility equation (2.65) can be rewritten as:

$$\sigma_{st} = m \frac{(d - x) \, \sigma_{cb}}{x} \qquad (2.66)$$

The total compressive and tensile forces on the cross-section are

$$C = \frac{bx \, \sigma_{bc}}{2} \text{ and} \qquad (2.67)$$

$$T = A_{st} \, \sigma_{st} \qquad (2.68)$$

The equilibrium of forces in the horizontal direction gives

$$\frac{bx \, \sigma_{cb}}{2} = A_{st} \, \sigma_{st}$$

$$\text{or} \quad \sigma_{st} = \frac{bx \, \sigma_{cb}}{2 A_{st}} \qquad (2.69)$$

Eqs. (2.66) and (2.69) result in

$$\frac{m \, (d - x) \, \sigma_{cb}}{x} = \frac{bx \, \sigma_{cb}}{2 A_{st}} \text{ or}$$

$$0.5bx^2 = A_{st} \, m \, (d - x) \qquad (2.70)$$

This above equation establishes the neutral axis (NA) distance. It can also be considered as the first moment area expression for a transformed cross-section in which

$mA_{st} = A_t$ = transformed area of the reinforcement in terms of the equivalent concrete area. Therefore, one can interpret area of the segments above and below the neutral axis neglecting the area of the concrete under tension.

$$xb = \text{compression area (above NA)}$$

$$A_t = mA_{st} = \text{tension area (below NA)}$$

The centroidal axis of the transformed section is obtained by taking first moment of the areas about the centroidal axis and this first moment of area is

$$xb\left(\frac{x}{2}\right) = A_t(d - x)$$

$$\text{or} \quad 0.5\,bx^2 = A_t(d - x) \tag{2.71}$$

Equations (2.70) and (2.71) are exactly the same. Therefore, the equilibrium condition in the horizontal direction will result into first moment of area equation of the transformed cross-section. The neutral axis of the section is exactly same as the centroidal axis of the transformed cross-section neglecting the area of concrete in tension. The neutral axis distance can be obtained from the equilibrium equation as

$$x^2 + \frac{2 A_t}{b}x - \frac{2 A_t d}{b} = 0$$

Let $\dfrac{A_t}{b} = a_t$ = transformed area of the reinforcement per unit width

$$\frac{x}{d} = n$$

$$\frac{A_t}{bd} = \frac{mA_{st}}{bd} = mp_t$$

Then the above equation can be expressed as

$$n^2 + 2mp_t m - 2mp_t = 0$$

The solution of the above quadratic equation is

$$n = -mp_t + \sqrt{m^2 p_t^2 + 2mp_t)}$$

$$\text{or} \quad n = mp_t(\sqrt{1 + 2/mp_t} - 1) \tag{2.72}$$

The neutral axis distance (the centroidal axis distance of the transformed section) is given by the above expression.

The compatibility expression (2.66) can be rewritten as

$$\sigma_{st} = \frac{m(1 - n)}{n}\sigma_{cb} \tag{2.73}$$

The moment of the compressive force about the centre of the steel gives

$$M = C\left(d - \frac{x}{3}\right) = \frac{bx}{2}\left(d - \frac{x}{3}\right)\sigma_{cb}$$

$$= bd^2 n\left(1 - \frac{n}{3}\right)\sigma_{cb} \tag{2.74}$$

Alternatively, it can also be expressed as

$$M = T\left(d - \frac{x}{3}\right) = A_{st}\left(d = \frac{x}{3}\right)\sigma_{st}$$

$$= A_{st}\, d\left(1 - \frac{n}{3}\right)\sigma_{st} \tag{2.75}$$

The analysis problem can be bifurcated into: First, in which the bending moment is given and stresses are to be found out; second, the limits on the stresses are assigned and the limiting moment is to be found.

Case 1: Given the bending moment to calculate the stresses. Equations (2.74) and (2.75) establish the relation between the bending moment and the stresses. The maximum compressive stress in concrete and tensile stress in steel are given by

$$\sigma_{cb} = \frac{2M}{bd^2 n\,(1 - n/3)} \tag{2.76}$$

$$\sigma_{st} = \frac{M}{A_{st}d\,(1 - n/3)} \tag{2.77}$$

If both the above stresses are within the allowable values, then the section is adequate to resist the bending moment. However, if any one of the two exceed the allowable value, then the section is not upto the desired level.

Case 2: Given the allowable stresses, to find the limiting bending resistance.

The limiting bending resistance of a section is that which is obtained by limiting the stresses within the allowable values. There is a particular relation between the actual stresses which is based on the compatibility of strains. Even if one of the two actual stresses reaches the limit, resistance of the section is governed by the particular stress. In case the actual stress in the reinforcement reaches the full allowable value earlier than the maximum compression stress in the concrete, the resistance is governed by the reinforcement. Such a section in which the steel controls the moment capacity is called the *under-reinforced* section. The condition is

$$\sigma_{st} = \sigma_{ast} \text{ and } \sigma_{cb} < \sigma_{acb}$$

where σ_{ast} = allowable tensile stress in steel

σ_{acb} = allowable bending compression in concrete

Then the resisting moment of the section is

$$M_r = A_{st}\,\sigma_{ast}\, d(1 - n/3) \tag{2.78}$$

Similarly, a section can be said to be *over-reinforced* if the actual extreme compressive stress in the concrete reaches its limit first. Then the working moment capacity of the section is given by

$$M_r = n(1 - n/3)\,\frac{bd^2}{z}\,\sigma_{acb} = Kbd^2\,\sigma_{acb} \tag{2.79}$$

One can arrive at the controlling resisting moment by first testing for the controlling stress. First assign $\sigma_{cb} = \sigma_{acb}$, then check if $\sigma_{st} = m(1 - n)\,\sigma_{cb} < \sigma_{ast}$, the resisting moment in which case is given by Eq. (2.79), otherwise Eq. (2.78) controls the capacity.

The sensitivity of the neutral axis and the working moment capacity can be obtained from

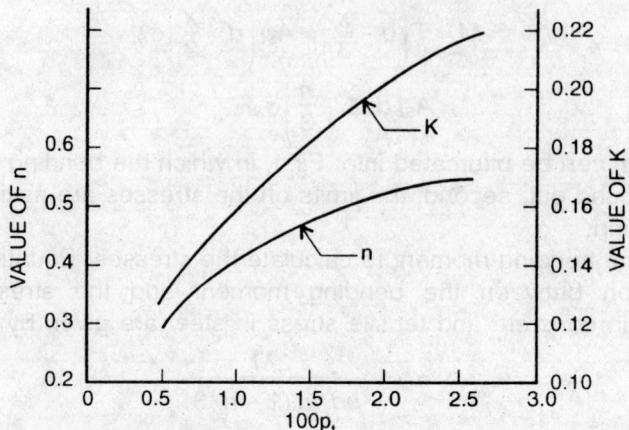

Fig. 2.16 Values of Neutral Axis Distance w.r.t. Percentage of Steel.

the earlier expression. Figure 2.16 illustrates the variations of the neutral axis and the working moment coefficient with respect to the percentage of tension reinforcement.

2.14 Doubly Reinforced Concrete Section

Consider a section in which the compression reinforcement is also provided. The strain compatibility in the section when subjected to pure bending is shown in Fig. 2.17b. The compatible strains are:

$$\frac{\in_c}{x} = \frac{\in_{st}}{d - x} \tag{2.80}$$

The strain in compression reinforcement is

$$\in_{sc} = \frac{\in_c}{x}(x - d_c) \tag{2.81}$$

Equation (2.80) can be expressed in terms of stresses as

$$\sigma_{st} = m\frac{(d - x)}{x}\sigma_{cb}$$

The steel in the compression area is constrained by the concrete against lateral expansion; therefore, the stress in the compression steel is taken 1.5 times the compatible stress

$$\sigma_{sc} = E_s \in_{sc}(1.5) = 1.5\,E_s\frac{(x - d_c)\in_c}{x}$$

$$= 1.5m\frac{(x - d_c)}{x}\sigma_{cb} \tag{2.82}$$

The compressive forces on the concrete section and in the compression reinforcement, and the tensile force in the tensile reinforcement are:

$$C_1 = \frac{xb}{2}\sigma_{cb} \tag{2.83}$$

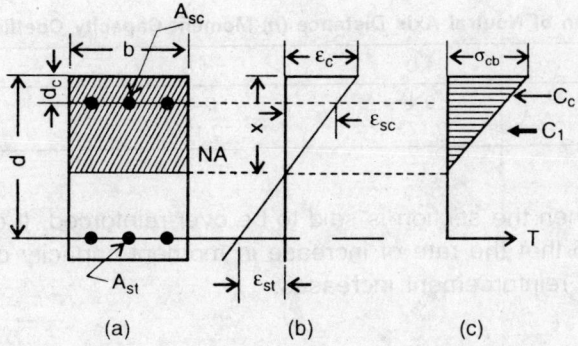

Fig. 2.17 Strain and Stress Variations in Doubly Reinforced Section.

$$C_2 = A_{sc}\,\sigma_{sc} = 1.5m\,\frac{(x - d_c)}{x}\,A_{sc}\,\sigma_{cb} \tag{2.84}$$

$$T = A_{st}\,\sigma_{st} = m\,\frac{(d - x)}{x}\,A_{st}\,\sigma_{cb}$$

The equilibrium of forces on the section gives

$$C_1 + C_2 = T$$

or

$$\frac{xb}{2}\,\sigma_{cb} + 1.5m\,\frac{(x - d_c)}{x}\,A_{st}\,\sigma_{cb} = m\,\frac{(d - x)}{x}\,A_{st}\,\sigma_{cb}$$

or

$$\frac{x^2 b}{2} + 1.5m\,(x - d)\,A_{st} = m(d - x)\,A_{st} \tag{2.85}$$

The above equation can also be interpreted as the first moment of area of the transformed cracked cross-section with 1.5 weightage to the compressive reinforcement. The distance of the neutral axis is given by this expression, which can be rearranged as

$$n^2 + 3m\,(n - d_c/d)\,p_c - 2m\,(1 - n)\,p_t = 0$$
$$n^2 + m\,(3p_c + 2p_t)n - m\,(3p_c\,d_c/d + 2p_t) = 0 \tag{2.86}$$

Taking moment of the forces about the centre of tension steel gives

$$M = C_1\left(d - \frac{x}{3}\right) + C_2\,(d - d_c)$$

$$= \frac{xb}{2}\left(d - \frac{x}{3}\right)\sigma_{cb} + 1.5m\,A_{sc}\,(d - d_c)\,\sigma_{cb} \tag{2.87}$$

There could be two types of analysis problems. Given the bending moment acting on the section, the actual stresses in the concrete and steel are to be evaluated; and the other problem is, given the allowable stresses, one has to calculate the working moment capacity of the section. As discussed in the earlier article, the actual stresses in the concrete and steel need not reach the allowable stresses at the same time. The stress in compression reinforcement may never reach the full allowable value. A section in which the actual stress in the tension reinforcement reaches the allowable value before the compressive stress on the concrete reaches its limit is called an *under-reinforced* cross-section. If the reverse of the

Table 2.12 Variation of Neutral Axis Distance (n) Moment Capacity Coefficient (A'), for m = 13

100 p_t	0.5	1.0	1.5	2.0	2.5
n	0.3	0.4	0.46	0.51	0.54
K	0.135	0.173	0.195	0.212	0.221

process takes place then the section is said to be over-reinforced. It can be seen from Table 2.12 or from Fig. 2.16 that the rate of increase in moment capacity coefficient decreases as the percentage of the reinforcement increases.

2.15 Analysis Examples

EXAMPLE 2.10 A simply supported beam of effective span 6 m has the following cross-sectional details: b = 250 mm, h = 780 mm, d = 740 mm, A_{st} = 603 mm^2 (that is, 3 nos of ϕ16). Determine the stresses caused in the beam by a UDL of 24.2 kN/m superimposed load

Data provided:

$$b = 250 \text{ mm}, \ d = 740 \text{ mm}, \ A_{st} = 603 \text{ mm}^2.$$

$$w_s = 24.2 \text{ kN/m}, \ L = 6 \text{ m}, \ f_{ck} = 20 \text{ MPa}$$

$$f_y = 415 \text{ MPa}, \ m = 13 \text{ m}, \ \sigma_{acb} = 7 \text{ MPa}, \ \sigma_{ast} = 230 \text{ MPa}.$$

Analysis: The self-weight of the beam is

$$w_g = 0.25(0.78)(25) = 4.9 \text{ kN/m}$$

$$w_t = w_s + w_g \qquad = 29.1 \text{ kN/m}$$

The maximum bending moment caused by the total load is

$$M = \frac{w_t L^2}{8} = \frac{29.1(36)}{8} = 130.95 \text{ kNm}$$

The transformed area of the tension reinforcement is

$$A_t = m \, A_{st} = 13 \, (603) = 7839 \text{ mm}^2$$

The centroidal axis of the cross-section which is same as the neutral axis of the section is obtained from the first moment of the transformed cracked cross-section area about NA and it is given by

$$0.5 \, bx^2 = A_t \, (d - x)$$

That is, $0.5(250)x^2 = 7839 \, (740 - x)$ or $x^2 + 62.712x - 46406 = 0$

$$\text{or} \quad x = \frac{-62.712 + \sqrt{62.712^2 + 4(46406)}}{2} = 186 \text{ mm}$$

The resisting moment capacity of the section is obtained by taking either moment of the compression force about the CG of the steel or moment of the tension force in the steel about the CG of the compression force. It is given by

$$M_r = C \left(d - \frac{x}{3} \right) = \frac{bx}{2} \left(d - \frac{x}{3} \right) \sigma_{cb}$$

$$\text{or } M_r = T\left(d - \frac{x}{3}\right) = A_{st}\,\sigma_{st}\left(d - \frac{x}{3}\right)$$

Equating M_r to M, we can solve for σ_{cb} and σ_{st} which are:

$$\sigma_{cb} = \frac{2M}{bx\left(d - \dfrac{x}{3}\right)} = \frac{2(130.95)(10^6)}{250(186)(740 - 62)} = 8.3 \text{ MPa}$$

$$\sigma_{st} = \frac{M}{A_{st}\left(d - \dfrac{x}{3}\right)} = \frac{(130.95)(10^6)}{603(740 - 62)} = 293 \text{ MPa}$$

The actual stresses caused are far higher than the allowable stresses.

Note: Vide Example 2.1 in which the beam was designed as a balanced section with the area of the tension reinforcement as 804 mm² whereas in present case reinforcement of only 603 mm² is provided.

EXAMPLE 2.11 Calculate the moment capacity of a rectangular section having the following data:

$$b = 250 \text{ mm}, \qquad h = 780 \text{ mm}, \qquad d = 740 \text{ mm}, \ m = 13$$
$$A_{st} = 603 \text{ mm}^2, \ \sigma_{ast} = 230 \text{ MPa}, \ \sigma_{acb} = 7 \text{ MPa}.$$

Analysis. The transformed area of the reinforcement is

$$A_t = mA_{st} = 13(603) = 7839 \text{ mm}^2$$

The location of the neutral axis is obtained from the first moment of area of the transformed cracked section about NA and it is

$$0.5\ bx^2 = A_t\ (d - x)$$

$$\text{or} \quad 0.5(250)\ x^2 + 7839\ x - 7839(740) = 0$$

$$\text{or} \quad x^2 + 62.712\ x - 46406 = 0$$

$$\text{or} \quad x = \frac{-62.712 + \sqrt{(62.712^2 + 4(46406))}}{2} = 186 \text{ mm}$$

Assuming that the concrete has reached its full stress, the corresponding stress in the reinforcement can be obtained from the compatibility of the strains, and it is

$$\sigma_{st} = m\left(\frac{d - x}{x}\right)\sigma_{acb}$$

$$= \frac{13(740 - 186)(7)}{186} = 271 \text{ MPa} > \sigma_{ast}$$

The actual stress in reinforcement exceeds the allowable value if the limit in compression in concrete is reached earlier. Therefore, the stress in the reinforcement governs the moment capacity. The moment capacity of the section is given by

$$M_r = T\left(d - \frac{x}{3}\right) = A_{st}\,\sigma_{ast}\left(d - \frac{x}{3}\right)$$

$$= 603\ (230)\ (740 - 62) = 94.04\ (10^4) \text{ Nmm}$$

EXAMPLE 2.12 A balanced RCC section is modified to reduce the working moment capacity by reducing the area of the tension reinforcement to half. Determine the moment capacity of the modified section with respect to the balanced section. The design data of the balanced section is:

$$f_{ck} = 20 \text{ MPa}, \quad f_y = 415 \text{ MPa}, \quad \sigma_{ast} = 230 \text{ MPa}$$

$$n_b = 0.289, \quad K_b = \frac{n_b}{2}\left(1 - \frac{n_b}{3}\right) = 0.1305; \quad j = 0.904$$

$$p_0 = \frac{n_b \, \sigma_{acb}}{2\sigma_{ast}} = \frac{0.289(7)}{2(230)} = 0.0044$$

Analysis: The balanced working moment capacity of the section is

$$M_{rb} = A_{st} \, \sigma_{ast} \, jd = p_u \, (j) \, bd^2 \, \sigma_{ast} = 0.004 \, bd^2 \, \sigma_{ast}$$

The area of the tension reinforcement in the balanced section is

$$A_{stb} = p_0 bd = 0.0044 \, bd$$

The area of the tension reinforcement in the modified section is

$$A_{st} = 0.5 \, A_{stb} = 0.0022 \, bd$$

The first moment of the transformed area of the cracked section which gives the neutral axis distance is

$$\frac{bx^2}{2} = m \, A_{st} \, (d - x)$$

or $\quad bx^2 + 2m \, A_{st} \, x - 2mA_{st}d = 0$

or $\quad n^2 + 2mp_t \, n - 2mp_t = 0$

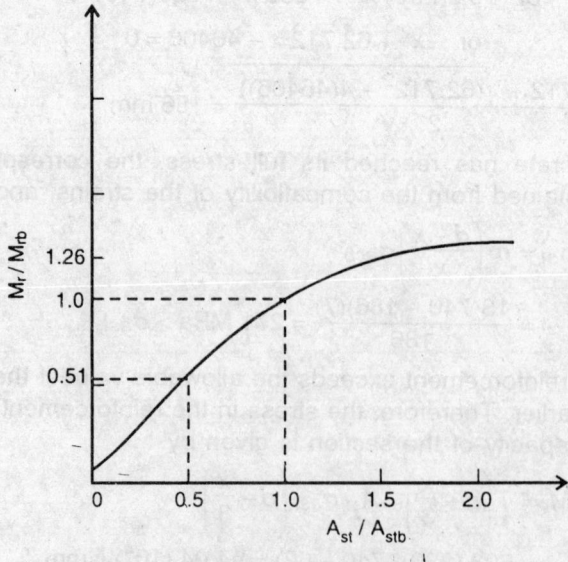

Fig. 2.18 Variation of Bending Moment with Increase in Tension Reinforcement.

where $n = x/d$.

For $m = 13$ and $p_t = 0.0022$, we have the neutral axis distance equation as

$$n^2 + 0.0572n - 0.0572 = 0$$

$$\text{or} \quad n = 0.212$$

The working moment capacity of the section is governed by the tension stress in the reinforcement and is

$$M_r = A_{st}jd\,\sigma_{ast} = \frac{A_s}{bd}\,(j)\,bd^2\,\sigma_{ast}$$

$$= 0.0022\left(1 - \frac{0.212}{3}\right)bd^2\,\sigma_{ast} = 0.00204\,bd^2\,\sigma_{ast}$$

The ratio of the moment capacity of the modified section to that of the balanced section is

$$\frac{M_r}{M_{rt}} = \frac{0.00204}{0.0044} = 0.51$$

It may be noted that if the area of the tension reinforcement is decreased by fifty per cent, the corresponding moment capacity is reduced by 49 per cent.

PROBLEMS

2.1   Determine the neutral axis of a cross-section for compatible stresses of 8 and 280 N/mm² in concrete in compression and steel in tension respectively. Assume the modular ratio as 13.

2.2   Determine the balanced working moment capacity of square cross-section with distance of the reinforcement at 0.9 time size of the cross-section. Assume a modular ratio of 18. The moment capacity should be expressed in σ_{cb}, σ_{st} and size of section.

2.3   Determine the working moment capacity of a diamond-shaped cross-section with its side a and included angle 90°. The depth of reinforcement from the apex is 80% of the overall depth as shown in the Fig. 1.

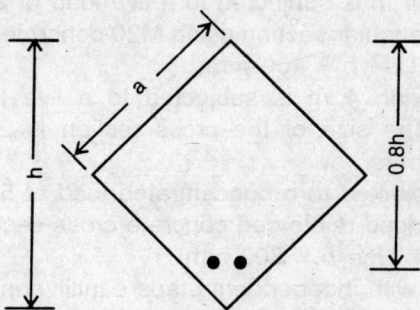

Fig. 1 Problem 2.3.

2.4   A square cross-section has a notch cut out in the compression flange as shown in Fig. 2. The size of the notch is 1/3rd of the size of the section. Derive the working moment

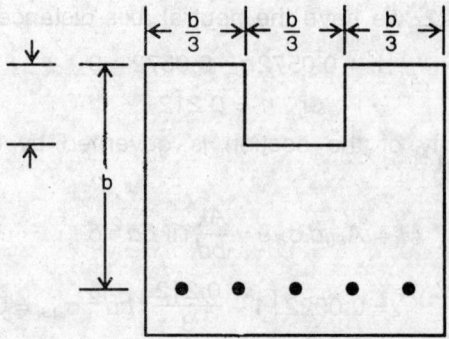

Fig. 2 Problem 2.4.

capacity of the section for the following data : σ_{cb} = 5 N/mm^2, σ_{st} = 230 N/mm^2 and m = 13.

2.5 Derive the working moment capacity of a circular cross-section with reinforcement at a distance of 0.8 times the diameter from the top compression fibre. Calculate the percentage of reinforcement needed for a balanced cross-section. Assume σ_{cb} = 7N/mm^2.σ_{st} = 230 N/mm^2, and m = 13.

2.6 A simply supported beam of span 6 m is subjected to a live load of 16 kN/m. Design a balanced reinforced concrete cross-section beam with width equal to 60% of the effective depth. Sketch the reinforcement details. Use σ_{cb} 7 N/mm^2, σ_{st} = 230 N/mm^2 and m = 13.

2.7 A simply supported beam of 6 m effective span is subjected to a live load of 16 kN/m. The size of the beam is 400 by 400 mm. Design the reinforcement details using M25 concrete and HYSD-Fe415 bars both for tension and shear reinforcement.

2.8 A simply supported beam of effective span 6 m is subjected to a live load of 16 kN/m. The size of the beam is 250 by 400 mm. Design the reinforcement details both for bending and shear using M20 concrete and HYSD-Fe415 bars.

2.9 A cantiliver beam span 4 m is subjected to a live load of 20 kN/m. Design a balanced reinforced concrete rectangular section using M20 concrete and HYSD-Fe415 bars both for bending and shear. Use b = 400 mm.

2.10 A cantiliver beam of span 4 m is subjected to a live load of 20 kN/m. Design a reinforcement detail if the size of the cross-section is 200 by 500 mm using M25 concrete HYSD-Fe415 bars.

2.11 A cantiliver beam is subjected to a concentrated load of 5 kN at the free end of the 4 m span. Design a balanced reinforced concrete cross-section using M25 concrete and mild steel reinforcement. Use b = 300 mm.

2.12 A staircase is provided with independent steps cantilivering from an upright reinforced concrete wall. The length of the step is 1.2 m and it is designed for a load of 1.5 kN placed on 1.1 m from the fixed support. The width of the step is 300 mm. Design a balanced reinforced concrete cross-section using M20 concrete and HYSD-Fe415 bars.

2.13 A staircase slab spans between two walls separated by a distance of 4.8 m. The width of the staircase slab is 1.2 m and it has two landings of 1 m at each end and having a length of 1.2 m each. Design the slab for a live load of 4 kN/m² using M20 concrete HYSD-Fe415 bars. The rise and tread of the staircase are 150 and 300 mm respectively.

2.14 A reinforced concrete cross-section of width 400 mm effective depth 800 mm is provided with 4 numbers of 20 mm bars as tension reinforcement. The beam is simply supported with an effective span of 8 m and subjected to uniformly distributed load of 24 kN/m. Determine the stresses in concrete and reinforcement. Assume a modular ratio of 13.

2.15 A reinforced concrete cross-section of 200 mm width and 400 mm effective depth is provided with 4 numbers of 20 mm bars in the tension zone and 3 numbers of 20 mm bars the compression zone at depth 50 mm from the extreme compression fibre. The cross-section is subjected to a bending moment of 150 kNm. Determine the stresses in concrete and reinforcement assuming a modular ratio of 13.

2.16 A double over-hang beam of total length 8 m is subjected to a UDL of 16 kN/m. The over-hang span is 1.6 m on each end. Design the reinforcement concrete beams for uniform depth throughout and using M25 concrete and HYSD-Fe415 bars. The depth of the section should not exceed 400 mm.

2.17 An electrical transmission line pole of 8 m total length is embedded into ground to a depth of 1.5 m. The wind load acting on the pole is transmitted as 3 kN at 0.5 m from the top of the pole. Design a reinforced concrete square cross-section pole using M25 concrete and HYSD-Fe415 bars.

2.18 A boundary wall of 2 m high above ground level is subjected to wind force of 1.5 kN/m². Design a reinforced concrete wall with foundation depth 1 m below ground level. Use M20 concrete and mild steel bars. The booting of the wall need not be designed and the wall is assumed to be stable under the wind load condition.

2.19 A boundary wall of 2.5 m high above ground level is subjected to 1.5 kN/m² wind load. The wall is designed with square pilasters spaced at 2.5 m intervals. Design the wall as well as the pilasters in reinforced concrete using M25 concrete and HYSD-Fe415 bars. Assume that the wall is stable under the wind load condition.

2.20 A precast reinforced concrete pole of 8 m length and having square cross-section of 300 mm size is provided with 4 numbers of 20 mm HYSD-Fe415 bars placed one at each corner of the cross-section. Clear cover to the bars is 25 mm HYSD-Fe415 bars placed one at each corner of the cross section. Clear cover to the bars is 25 mm. The member is lifted from the ground holding it at its middle. Determine the stresses in the reinforcement. Assume a modular ratio of 13.

2.21 Calculate the actual stresses in concrete and reinforcement of a rectangular cross section beam for the following data: $b = 300$ mm, $d = 750$ mm, $A_{st} = 604$ mm², $M = 100$ kNm.

2.22 A rectangular reinforced concrete section is subjected to an external bending moment of 150 kNm. Determine the structural safety of the section based on working stress design for the following data : $b = 400$ mm, $d = 800$ mm, $A_{st} = 804$ mm², $\sigma_{ast} = 230$ N//mm², $\sigma_{acd} = 7$ N/mm².

2.23 Calculate the working moment capacity of a rectangular cross-section reinforced beam for the following data: $b = 200$ mm, $d = 600$ mm, $A_{st} = 603$ mm², $\sigma_{ast} = 230$ N/mm², and $\sigma_{acb} = 7$ N/mm². State whether the section is over-reinforced or under-reinforced?

2.24 Determine the moment capacity of a rectangular cross-section reinforced concrete beam for the following data: $b = 200$ m, $d = 600$ mm, $A_{st} = 1206$ mm^2, $\sigma_{ast} = 230$ N/mm^2, and $\sigma_{acb} = 7$ N/mm^2. State whether the moment capacity of a rectangular section is over-reinforced or under-reinforced?

2.25 The tension reinforcement in a singly reinforced balanced rectangular cross-section is increased by 200% of the balance steel. Determine the enhanced moment capacity of the cross-section as a function of balanced moment capacity, assuming all other properties remaining the same. (M20, HYSD-Fe415)

2.26 The tension reinforcement in a balanced rectangular cross-section beam is reduced to 1/3rd of the balanced reinforcement. Determine the revised moment capacity of the section as a function of balanced moment capacity assuming all other data remaining the same. (M20, HYSD-Fe415)

2.27 Determine the stresses in concrete and steel in doubly reinforced rectangular cross-section beam for the following data:
$b = 400$ mm, $d = 700$ mm, $A_{st} - 1206$ mm^2, $A_{sc} = 603$ mm^2, $m = 13$, and bending moment = 180 kNm. If the allowable tensile stress in steel is 230 N/mm^2, determine whether the cross-section is safe against the working moment.

2.28 Determine the moment capacity of a double reinforced rectangular cross-section beam for the following data:
$b = 300$ mm, $d = 600$ mm, $m = 13$, $A_{st} = 1407$ mm^2,
$A_{sc} = 804$ mm^2, $\sigma_{ast} = 230$ N/mm^2, $\sigma_{acb} = 7$ N/mm^2.

Chapter 3

WSD of Design of T-Beams

3.1 Introduction

T-beams are commonly used in building and bridge construction without special effort. The beam slab construction in RCC or prestressed concrete often results into a T-beam section. I-beams are not common in RCC construction as only one of the flanges is subjected to compressive force and the other is subjected to tensile force. The flange under tension does not serve any structural purpose except housing the reinforcement. The flange under tension is more of a burden on the beam rather than of any help. On the other hand, the T-beam sections provide a large concrete cross-sectional area of the flange to resist the compressive force. T-beams are very advantageous in simply supported spans and the inverted T-beam sections have an advantage in cantilever beams. Whether someone likes it or not, T-section is usually built into the simply supported beam-slab construction. Continuous beam and frame elements do not have that much of benefit of the T-beam flange action. The critical moments occur mostly at the supports and the next critical moments occur near mid-spans. The sections at the support are subjected to negative bending moment which results into compression on the bottom fibre, the slabs are invariably at the top fibres of the beams and are under tension. Consequently, the section at the support is to be treated as a rectangular cross-section. The sections of the beam near the mid-spans are under positive bending moments. Here the section can be considered as a T-section by including certain portion of the slab as the flange of the section. The thickness of RCC slabs is usually in the range of 100 mm to 250 mm in buildings. The width of the flange to be accounted in the moment resistance of the beam is governed by shear lag and it is:

Width of flange $\leq$ centre to centre distance of the beams or

$$\leq \frac{L}{3} \quad \text{or}$$

$$\leq (b_w + 12t) \quad \text{or}$$

$$\leq \left(\frac{L_0}{6} + b_w + 6t \right)$$

where t = thickness of the flange (same as the slab)
     b_w = width of web
     L_0 = distance between the two consecutive points of contraflexure
     L = span

Moment capacity in case neutral axis in flange: In case the neutral axis of the section falls in the top flange, the section can be treated as rectangular with width equal to the width of flange for the purpose of bending resistance. The area of steel is obtained from the rectangular beam formula. The width of the web is designed to resist the shear force. However, in most beams, the NA is likely to fall in the web of the beam. The next section deals with beams having the NA in the web.

84

3.2 Resisting Moment of T-Beam Sections

Figure 3.1 illustrates the stress distribution on a T-beam section with NA in the web. The compressive force below the flange area is usually neglected as it contributes about ten per cent or even less against bending resistance. For more accurate analysis, the compression in the web below the flange can be included. It is common practice to design beams neglecting the compression in the web.

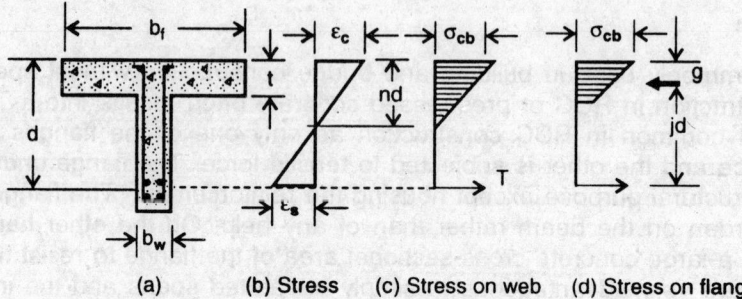

(a)   (b) Stress   (c) Stress on web   (d) Stress on flange

Fig. 3.1 T-Beam Section.

For a balanced section the neutral axis is given by

$$n = \frac{1}{1 + \sigma_{st}/m\sigma_{cb}}$$ (3.1)

The compressive stress at the bottom fibre of the flange is

$$\left(1 - \frac{t}{nd}\right)\sigma_{cb}$$

The total compressive force on the flange is

$$C = \frac{bt}{2}\left(\sigma_{cb} + \left(1 - \frac{t}{nd}\right)\sigma_{cb}\right)$$

$$= bt\sigma_{cb}\left(1 - \frac{t}{2nd}\right)$$ (3.2)

The CG of the compressive force from the top of the compression fibre and the moment lever arm are:

$$g = \frac{t}{3}\frac{3nd - 2t}{2nd - t}$$ (3.3)

$$jd = (d - g)$$ (3.4)

and the resisting moment is

$$M_b = Cjd = bt\sigma_{cb}\left(1 - \frac{t}{2nd}\right)jd$$ (3.5)

Let $t/d = r$, the Eqs. (3.2) to (3.5) can be written as

$$C = bt\left(1 - \frac{r}{2n}\right)\sigma_{cb}$$ (3.6)

$$g = \frac{r}{3}\left(\frac{3n-2r}{2n-r}\right)d \tag{3.7}$$

$$j = 1 - \frac{r}{3}\left(\frac{3n-2r}{2n-r}\right) \tag{3.8}$$

$$M_b = \left(1 - \frac{r}{2n}\right)jr\,\sigma_{cb}\,bd^2 = K_t bd^2\,\sigma_{cb} = R_t bd^2 \tag{3.9}$$

where $R_t = \left(1 - \dfrac{r}{2n}\right)jr\,\sigma_{cb}$ $\qquad$ (3.10)

The area of tension reinforcement is given by

$$A_s = \frac{M_b}{jd\,\sigma_{st}}$$

$$= \left(1 - \frac{r}{2n}\right)r\frac{\sigma_{cb}}{\sigma_{st}}\,bd = p_0\,bd\,\frac{\sigma_{cb}}{\sigma_{st}} \tag{3.11}$$

where $\qquad\qquad p_0 = \left(1 - \dfrac{r}{2n}\right)r$ $\qquad$ (3.12)

For the purpose of design, Eq. (3.5) can be rearranged as

$$d = \frac{t}{2n} + \frac{M}{bt\,j\,\sigma_{cb}} \tag{3.13}$$

The modular ratio can be approximated as

$$m = 280/3\sigma_{cb} \text{ then, the NA distance is}$$

$$n = \frac{1}{1 + 3\sigma_{st}/280} \tag{3.14}$$

The value of j varies from 0.90 to 0.93. It increases as $r = t/d$ decreases. For all practical purposes the value of j can be selected in the range of 0.9 to 0.91. Table 3.1 gives values of n, j and K_t for different slabs and depth ratios.

Table 3.1 Design Coefficients of T-Beam

	$r = 0.15$	0.2	0.25	0.15	0.2	0.25
		$n = 0.4$			$n = 0.289$	
j	0.93	0.91	0.89	0.93	0.92	0.91
K_t	0.113	0.136	0.153	0.103	0.120	0.129

3.3 Design Method in T-Beams

The thickness of the slab which forms a part of T-beam is designed independent of the beam. The normal thickness of the slab in beam and slab construction is 100 mm to 250 mm. Only in the case of heavy floors the thickness of the slab exceeds 250 mm. The main design therefore consists of:

1. Select width of the web (b_w) which may be taken as

$$b_w = c_1 t \tag{3.15}$$

in which t = thickness of the slab and c_1 varies from 1.5 to 3 depending on the shear force on the beam. For roof beams select $c_1 = 1.5$ and for very heavy floors select $c_1 = 3$.

The width of the web is usually governed by the shear force and area required for housing the tension reinforcement.

2. Calculate width of the flange

$b_f \leq$ centre to centre distance of beams

$\leq L/3$

$\leq 12t + b_w$

$\leq L_0/6 + b_w + 6t$

The lowest of the above is chosen.

3. Assume approximate overall depth (h) of the beam. It may be selected in the range of

$$h = \frac{L}{16} \text{ to } \frac{L}{12} \text{ for simply supported floor beams}$$

$$h = \frac{L}{12} \text{ to } \frac{L}{8} \text{ for cantilever beams}$$

$$h = \frac{L}{30} \text{ to } \frac{L}{16} \text{ for continuous beams.}$$

4. Calculate approximate BM due to dead load of the beam.
5. Calculate BM due to superimposed loads.
6. Estimate the value of j from Eq. (3.8) or from Table 3.1 with the assumed value of h.
7. Calculate effective depth (d) of the beam from Eq. (3.13).
8. Compute overall depth ($h = 1.1d$) and compare it with that assumed for the purpose of dead load. Modify the BM due to dead load suitably and recalculate the value of d in case there is reasonable variation between the two values of d.
9. Calculate the actual value of j from Eq. (3.8)
10. Compute the area of steel from Eq. (3.11), and select appropriate bars.
11. Calculate maximum shear force and the nominal shear stress

$$\tau_v = \frac{V}{b_w d} \tag{3.16}$$

Estimate the allowable shear stress for the percentage of reinforcement and design the transverse reinforcement according to the following cases.

Case 1: $\tau_v < 0.5 \tau_c$

No shear reinforcement is normally required.

Case 2: $0.5 \tau_c \leq \tau_v \leq \tau_c$

Nominal shear reinforcement is desired and it is governed by the following maximum spacing:

$$s_v \leq \frac{A_{vy} f_y}{0.4 b_w} \tag{3.17}$$

$$\leq 0.75 d$$

$$\leq 450 \text{ mm}$$

where A_{sv} = total area of the stirrup

Case 3: $\tau_c \leq \tau_v \leq \tau_{c\,max}$

The spacing of the stirrups is given by

$$s = \frac{A_{sv}\,\sigma_{sv}\,d}{V - V_a} = \frac{A_{sv}\,\sigma_{sv}}{b_w\,(\tau_y - \tau_c)} \qquad (3.18)$$

where V_a is the allowable shear force and is given by $V_a = b_w d\,\tau_c$ (3.19)

The spacing is subject to the limits as given in case 2.

Case 4: $\tau_v \geq \tau_{c\,max}$

$\tau_{c\,max}$ is the allowable shear stress in the concrete with stirrups. If the nominal shear stress exceeds this value, the section has to be modified by increasing b_w or and d such that the nominal shear stress is within the limits.

The design of T-beams for shear is very similar to that of rectangular sections.

12. The design check for deflections be made as per the section in Chapter 2.
13. Curtailment and bending of the main reinforcement at about L/5 from support.
14. Check for the development lengths.

3.4 Design Examples

EXAMPLE 3.1. *Design of RCC singly reinforced simply supported beam.* A floor slab 120 mm thick is supported by simply supported beams spaced 4 m apart integral with the slab. The clear span of the beam is 8 m. The beam is designed for the following:

Live load	$= 4$ kN/m^2
Concrete grade f_{ck}	$= 20$ Mpa
Steel grade f_y	$= 415$ Mpa

$t = 120$ mm; $\sigma_{cb} = 7$ MPa, $\sigma_{st} = 230$ MPa, $n = 0.289$

Let the web thickness $b_w = 0.23$ m = thickness of one-brick wall

Effective span L = clear span + 5% of the clear span

$$= 8 + 8/20 = 8.4 \text{ m}$$

The flange width is controlled by the minimum of the following:

Centre to centre distance of beams = 4 m or

$$= \frac{L_0}{6} + b_w + 6t = \frac{8.4}{6} + 0.23 + 6(0.12) = 2.316 \text{ m}$$

$$= b_w + 12t = 0.23 + 12(0.12) = 1.67 \text{ m}$$

$$= L/3 = 8.4/3 = 2.8 \text{ m}$$

Select the flange width as the least of the above, i.e.

$$b_f = 1.67 \text{ m}$$

Let the overall depth of the beam be taken as

$$h = \frac{L}{20} = \frac{8.4}{20} = 0.42 \text{ m}$$

Self-weight of web $= w_g = 2000$ N/m

Weight of slab $= w_s = 0.12(4)(25000) = 12000$ N/m

Live load $= w_1 = 4000(4) = 16000$ N/m

Total load $= w = \overline{30000 \text{ N/m}}$

Let the total load be $= w = 30000$ N/m

$$\text{Let } r = \frac{t}{d} = \frac{0.12}{0.4} = 0.3, \text{ then}$$

$$j = 1 - \frac{r(3n - 2r)}{3(2n - r)} = 0.903$$

The critical bending moment which occurs at the mid-span is

$$M = \frac{wL^2}{8} = \frac{30000(8.4)^2}{8} = 264600 \text{ Nm}$$

The effective depth of steel can be obtained from Eq. (3.13) as

$$d \geq \frac{t}{2n} + \frac{M}{btj\sigma_{cb}} = 0.207 + 0.208 = 0.415 \text{ m}$$

Use $d = 0.42$ m, we have

$$nd = 0.289(0.42) = 0.121 > t = 0.12 \text{ m}$$

The neutral axis falls in the web and the self-weight is slightly more than assumed. The overall depth $h = d + 0.08 = 0.50$ m.

The revised bending moment with actual self-weight $= M = 266\,000$ Nm

The area of the tensile reinforcement is

$$A_{st} = \frac{M}{jd\sigma_{st}} = \frac{266000}{0.903(0.42)(230)} = 3049 \text{ mm}^2$$

Provide 5– $\phi25$ *and* 2 – $\phi20$ bars then the area of steel provided is

$$A_{st} = 3078 \text{ mm}^2$$

$$\frac{100 \, A_{st}}{b_w d} = \frac{307800}{(230) \, (420)} = 3\%.$$

Design for shear: The maximum shear force occurs at a distance d from the support and is

$$V = w\left(\frac{L}{2} - d\right) = 30000 \, (4.2 - 0.42) = 113\,400 \text{ N}$$

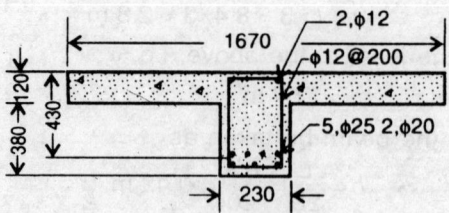

Fig. 3.2 Example 3.1.
(dimensions in mm)

$$\tau_v = \frac{V}{bd} = \frac{113\,400}{230\,(420)} = 1.17 \text{ MPa} = 1.17 \text{ N/mm}^2$$

Assume half of the main bars are continued till the end. The maximum allowable stress is

$$\tau_{c\,max} = 1.8 \text{ N/mm}^2$$

and allowable shear stress for 1.55 per cent steel is

$$\tau_c = 0.47 \text{ N/mm}^2$$

Shear reinforcement must be provided beyond the shear capacity of the concrete. Provide 12 mm MS two-legged stirrups, then the area of the stirrup steel is

$$A_{sv} = 2(113) = 266 \text{ mm}^2$$
$$\sigma_{sv} = 140 \text{ (for MS bars)}$$

The spacing of the stirrups at the maximum shear force is

$$s_v = \frac{A_{sv}\,\sigma_{sv}}{b_w\,(\tau_v - \tau_c)} = \frac{226(140)}{230(1.17 - 0.47)} = 200 \text{ mm}$$

Spacing of the shear reinforcement based on the minimum steel is:

$$s_{v\,max} = \frac{A_{sv}\,f_y}{0.4 b_w} = \frac{226(250)}{04(230)} = 614 \text{ mm or}$$

$$= 0.75 \ d = 210 \text{ mm}$$

Provide the stirrups at 200 mm c/c near the support.

EXAMPLE 3.2 *Design of T-beam for 15 m span*

The design data of the problem are:

$$\text{Clear span} = L_0 = 15 \text{ m}$$

Superimposed load including weight of the slab = 22 kN/m
Live load consists of two concentrated moving loads of 150 kN each spaced 2 m apart.

$$f_{ck} = 25 \text{ MPa}, \qquad f_y = 415 \text{ MPa}$$

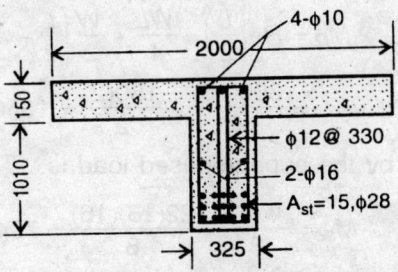

Fig. 3.3 Example 3.2 (dimensions in mm).

Width of top flange $b_f = 2$ m, Thickness of flange $t = 150$ mm

Let effective span $L = L_0 + \dfrac{L_0}{15} = 16$ m

The maximum bending moment due to the moving loads will occur under one of the loads and when the centroid of the two loads is at the same distance of the load from the mid span. This moment is

$$M_{l1} = \frac{2W_1}{L}\left(\frac{L}{2} - \frac{a}{4}\right)\left(\frac{L}{2} - \frac{a}{4}\right)$$

$$= \frac{300}{16}(8 - 0.5)^2 = 1054.688 \text{ kNm}$$

This occurs at a distance $x_1 = \dfrac{L}{2} - \dfrac{a}{4} = 7.5$ m from the support.

Let the self-weight of the web be

$$w_g = 0.3(1)(25) = 7.5 \text{ kN/m}$$

The bending moments caused by self-weight and the superimposed loads at x_1 are

$$M_{g1} = w_g \frac{L}{2}(x_1) - w_g \frac{x_1^2}{2}$$

$$= 7.5(8)(7.5) - \frac{7.5(7.5)^2}{2} = 239.5 \text{ kNm}$$

$$M_{s1} = w_s \frac{L}{2}(x_1) - w_s \frac{x_1^2}{2}$$

$$= 22(7.5)(8 - 3.75) = 701.25 \text{ kNm}$$

Total moment at distance x_1

$$M_{t1} = M_{11} + M_{g1} + M_{s1} = 1995.438 \text{ kNm}$$

There is another possibility of maximum BM occurring at mid-span as the maximum BM due to DL occurs at mid-span and this moment is computed for comparison. When one of the moving load is at mid-span, the reaction at the left end support is

$$R_1 = \frac{W}{2} + \frac{W}{L}\left(\frac{L}{2} - a\right)$$

The bending moment at the mid-span is

$$M_{l2} = R_1\left(\frac{L}{2}\right) = \frac{WL}{4} + \frac{W}{2}\left(\frac{L}{2} - a\right) = \frac{2WL}{4} - \frac{Wa}{2}$$

$$= \frac{300(16)}{4} - \frac{150(2)}{2} = 1050 \text{ kNm}$$

The bending moment caused by the superimposed load is

$$M_{s2} = \frac{w_s L^2}{8} = \frac{22(16)(16)}{8} = 704 \text{ kNm}$$

Assume the size of the web for the purpose of self-weight as $= 0.3 \times 1$ m
Self-weight $w_g = 0.3(1)(25) = 7.5$ kN/m
Bending moment due to self-weight

$$M_{g2} = \frac{w_g L^2}{8} = \frac{7.5(256)}{8} = 240 \text{ kNm}$$

The total BM is

$$M_{t2} = (M_{g2} + M_{s2} + M_{l2}) = (240 + 704 + 1050) = 1994 \text{ kNm}$$

From the two critical moments the larger of the two is taken as the design moment.

$$M_t = M_{t1} = 1995.438 \text{ kNm} \simeq 2000 \text{ kNm}$$

The effective depth of section is

$$d = \frac{t}{2n} + \frac{M}{bt j \sigma_{cb}} = 0.26 + 0.78 = 1.04 \text{ m}$$

The area of the tension reinforcement is obtained as

$$A_{st} = \frac{M}{jd\, \sigma_{st}} = \frac{2000\,000}{0.9\,(1.04)\,(230)} = 9290 \text{ mm}^2$$

Provide 15 numbers of 28 mm bars

$$A_{st} \text{ (provided)} = 15\,(615) = 9225 \text{ mm}^2$$

As the area of steel provided is marginally smaller than required, increase the effective depth proportionately

$$d = \frac{1.04(9290)}{9225} = 1.048 \text{ m}$$

The 15 bars are arranged in three rows of 5 each, with clear spacing between the bars as 35 mm. Therefore, the overall depth of the beam with 35 mm cover is

$$h = d + 14 + 35 + 28 + 35 = 1160 \text{ mm}$$

Depth of the web $\qquad = h - t = h_w = 1010 \text{ mm}$

Assuming the clear spacing of the bars as 30 mm, the minimum width of the web is computed for accommodating the five bars in a row with a cover of 30 mm:

$$b_w \text{ (minimum)} = 5(28) + 4\,(30) + 2\,(30) = 320 \text{ mm}$$

Let $\qquad b_w = 325 \text{ mm}$

$$w_g = 0.325\,(1.01)\,(25) = 8.2 \text{ kN/m}$$

The assumed weight for moment calculation is only slightly less than the actual; therefore, the moment calculation does not need a revision. The maximum shear force is computed on the actual weight.

Design for shear. The maximum shear force occurs at the support when one of the moving loads is next to the support. The left end reaction caused by the moving loads is

$$R_3 = \frac{W(L-a)}{L} = 2W - \frac{W_a}{L}$$

$$= 300 - \frac{150(2)}{16} = 281.25 \text{ kN}$$

The maximum shear force due to self-weight, superimposed load and live loads are:

$$V_2 = \frac{w_g L}{2} = \frac{8.2(16)}{2} = 65.6 \text{ kN}$$

$$V_s = \frac{w_g L}{2} = \frac{22\,(16)}{2} = 176 \text{ kN}$$

Total shear force at support $\qquad V_t = 522.84$ kN

Percentage ratio of the reinforcement is

$$\frac{100\,A_{st}}{b_w\,d} = \frac{922500}{(325)\,(1048)} = 2.8$$

The nominal shear stress is

$$\tau_v = \frac{V}{b_w d} = \frac{520\,450}{525\,(1048)} = 1.52 \text{ MPa}$$

The allowable shear stress of the M 25 concrete for fifty per cent of the bars curtailed

$$\tau_c = 0.46 \text{ MPa}$$

and the maximum allowable shear stress with reinforcement is

$$\tau_{max} = 1.9 \text{ MPa}$$

The nominal shear stress is more than the design shear strength without reinforcement and less than the maximum allowable value with reinforcement, therefore, shear reinforcement should be designed.

Provide four-legged 12 mm diameter HYSD stirrups, then the area of the stirrup steel is

$$A_{sv} = 4\,(113) = 452 \text{ mm}^2$$

The spacing of the reinforcement is

$$A_{sv} = \frac{A_{sv}\,\sigma_{sv}}{b_w\,(\tau_c - \tau_v)} = \frac{452(230)}{325(1.52 - 0.46)} = 300 \text{ mm}$$

The maximum spacing of the stirrups is

$$s_{max} = \frac{A_{sv}\,f_y}{0.4 b_w} = \frac{452\,(415)}{0.4\,(325)} = 1142 \text{ mm or}$$

$$= 0.75\,d$$

Provide 6 numbers of four-legged ϕ12 stirrups in the first 2 m from the supports, 4 numbers in the next 2 m and 7 numbers in the next 4 m upto the mid-span.

EXAMPLE 3.3. *Design of under-reinforced T-beam:* A simply supported beam of effective span 8.4 m is subjected to a live load of 16 kN/m and superimposed load of slab of 11.22 kN/m.

Design the beam with web and flanges given below:

$\qquad b_w = 230$ mm, $\qquad h_w = 480$ mm,
$\qquad b_f = 1670$ mm, $\qquad t = 120$ mm,
Also $\qquad f_{ck} = 20$ MPa, $\qquad f_y = 415$ MPa,
$\qquad \sigma_{cb} = 7$ MPa, $\qquad \sigma_{st} = 230$ MPa.

Overall depth $= h = h_w + t = 600$ mm

(This problem is same as that given in Example 3.1, except that the overall depth is given in this case.)

Design for bending moment. Self-weight of the web

$$w_g = 25((0.23)(0.48) + 1.67(0.12)) = 7.77 \text{ kN/m}$$
$$w_s = 11.22 \text{ kN/m}$$
$$w_l = 16.0 \text{ kN/m}$$
$$\text{Total load } w_t = \overline{35.00 \text{ kN/m}}$$

Maximum bending moment is

$$M = \frac{w_t L^2}{8} = \frac{35.00(8.4)^2}{8} = 309.70 \text{ kNm}$$

Since the width of the web is small, the reinforcement may have to be adjusted in three rows. Let the depth of web below the centre of the reinforcement be 90 mm. The effective depth of steel is

$$d = h - 0.09 = 0.6 - 0.09 = 0.51 \text{ m}$$

The NA in a balanced section would be

$$nd = 0.289 (0.51) = 0.147 \text{ m}$$

Since the neutral axis distance is more than the thickness of the flange, the bending moment capacity of a balanced section is given by :

$$M_b = \left(1 - \frac{t}{2nd}\right) jb \, bt \, \sigma_{cb}$$

where $r = \dfrac{t}{d} = \dfrac{0.12}{0.51} = 0.235$, $n = 0.289$, $j = 0.905$

and $M_b = \left(1 - \dfrac{0.12}{2(0.289)(0.51)}\right)(0.904)(0.51)(1.67)(0.12)(7) = 0.38383 \text{ MNm}$

The balanced moment capacity M_b is larger than the external bending moment; therefore, the depth provided is more than that required for balanced section. The beam will be designed as an under-reinforced section. The area of tension steel required is

$$A_{st} = \frac{M}{jd \, \sigma_{st}} = \frac{309700}{0.904(0.51)(230)} = 2918 \text{ mm}^2$$

Provide 6 numbers of 25 mm bars in two rows, then

$$A_{st} \text{ (Provided)} = 6 (490) = 2940 \text{ mm}^2$$
$$d = 0.51 \text{ m (maximum available } d \text{ is 0.53 m)}$$

Design for shear: The critical section for shear is at a distance d from the effective support, In fact, it can be taken at distance d from the clear end; however, in the absence of the support details, one should take it from the effective support. The maximum shear force is

$$V = w_t \left(\frac{L}{2} - d\right) = 35.12(4.2 - 0.51) = 129.59 \text{ kN}$$

$$\frac{100 \, A_{st}}{b_w \, d} = \frac{294000}{230 (510)} = 2.5\%$$

The allowable shear stress is $= \tau_c = 0.41$
The allowable shear force on the concrete section is

$$V_a = \tau_c \, b_w \, d = 0.41 (230)(510) = 48100 \text{ N}$$

The nominal shear stress would certainly be less than the maximum allowable shear stress allowed with transverse reinforcement; therefore, design stirrups to resist the extra shear force. Provide two-legged 10 mm stirrups, then the area of the stirrups is

$$A_{sv} = 2(78.5) = 157 \text{ mm}^2$$

The spacing of the stirrups is

$$s_v = \frac{A_{sv} \, \sigma_{sv} \, d}{V - V_a} = \frac{157 \, (230)(510)}{81490} = 226 \text{ mm}$$

$$s_{max} = \frac{A_{sv} \, f_y}{0.4 \, b_w} = \frac{157(415)}{0.4(230)} = 708 \text{ mm, or}$$

$$= 0.75d$$

Provide two-legged ϕ 10 stirrups at 225 mm in the end for 1500 mm, then provide at 350 mm. (Total number of stirrups 39.)

EXAMPLE 3.4: *Doubly reinforced T-beam* : A simply supported beam of effective span 8.4 is subjected to the following loads and specifications:

$w_s = 11.52$ kN/m, $w_t = 16$ kN/m,

$b_f = 1670$ mm, $t = 120$ mm,

$b_w = 230$ mm, $h_w = 300$ mm,

$f_{ck} = 20$ MPa, $f_y = 415$ MPa,

$\sigma_{cb} = 7$ MPa, $\sigma_{st} = 230$ MPa.

$h = h_w + t = 300 + 120 = 420$ mm

Design the beam. This problem is same as Example 3.1 or 3.3, except that the depth specified is different from the other two. It is desirable to check the deflection in this problem.

Design for bending moment: The self weight of the beam is

$$w_g = 25(0.23(0.3) + 1.67(0.12)) = 6.48 \text{ kN/m}$$

$$w_s = 11.52 \text{ kN/m}$$

$$w_l = 16.0 \text{ kN/m}$$

$$\overline{w_t = 34.00 \text{ kN/m}}$$

The maximum bending moment caused by the load is

$$M = \frac{w_t \, L^2}{8} = \frac{34(8.4)^2}{8} = 299.88 \text{ kNm}$$

For a balanced T-beam section, the neutral axis distance is

$$nd = 0.289 \, d$$

Since the width of the web is small, so the depth of CG of steel above the bottom fibre can be taken as 90 mm; then

$$d = h - 90 = 0.33 \text{ mm and } nd = 0.289 \, d = 0.095 \text{ m}$$

The distance of the neutral axis is less than the thickness of the flange. Hence the section is to be treated as rectangular section and, therefore, the balanced moment resistance of the section is

$$M_b = \frac{n}{2}\left(1 - \frac{n}{3}\right)bd^2\,\sigma_{cb} = 0.289\,(0.904)\,(1.67)(0.33)^2(7) = 0.16623 \text{ MNm}$$

The actual bending moment is more than the balanced capacity; therefore, the beam be designed as a doubly reinforced one. Let the depth of the compression steel from top fibre be

$$d_c = 60 \text{ mm}$$

Then the compatible stress in steel is

$$\sigma_{sc} = (1.5\,m)\,\frac{(nd - d_c)\,\sigma_{cb}}{nd} = 1.5\,(13)\,\frac{(0.095 - 0.06)(7)}{0.095} = 50 \text{ MPa}$$

The bending moment to be resisted by the compression reinforcement is

$$M_c = M - M_b = 0.29988 - 0.16623 = 0.13365 \text{ MNm}$$

The area of the compression reinforcement required is

$$A_{sc} = \frac{M_c}{(d - d_c)\,\sigma_{sc}} = \frac{133650}{0.27\,(50)} = 9900 \text{ mm}^2$$

The area of the tension steel is computed together for balanced and for the doubly reinforced one and it is

$$A_{st} = \frac{M_b}{jd\,\sigma_{st}} + \frac{M_c}{(d - d_c)\,\sigma_{st}}$$

$$= \frac{166230}{0.904(0.33)(230)} + \frac{133650}{0.27\,(230)} = 2423 + 2152 = 4575 \text{ mm}^2$$

Provide 19 numbers of ϕ 25 and 2 numbers of ϕ 20 bars in the top flange, and 9 numbers of ϕ 25 and 1 numbers of ϕ 20 bars at the bottom.

$$A_{sc} \text{ (provided)} = 9938 \text{ mm}^2$$
$$A_{st} \text{ (provided)} = 4611 \text{ mm}^2$$

The area of the compression reinforcement is high as the allowable compatible stress in steel is rather row. This example requires a revision as d_c assumed will be far less than actually available.

Design for shear: The critical shear force plane is at a distance d from the effective support and it is

$$V = w_t\left(\frac{L}{2} - d\right) = 34000\,(4.2 - 0.33) = 131580 \text{ N}$$

Nominal shear stress in the concrete is

$$\tau_v = \frac{V}{b_w d} = \frac{131\,580}{230(330)} = 1.73 \text{ MPa}$$

The maximum allowable shear stress in the concrete with transverse reinforcement is

$$\tau_{cmax} = 1.8 \text{ MPa}$$

As the nominal shear stress is less than the maximum allowable, the section does not need any modification but transverse reinforcement is required

$$\frac{100\,A_{st}}{b_w d} = \frac{461100}{230(330)} = 6\%$$

The allowable shear stress in concrete for 3% or more reinforcement is

$$\tau_c = 0.51 \text{ MPa}$$

The allowable shear force on the concrete is

$$V_a = b_w \, d \, \tau_c = 230(330)(0.51) = 38710 \text{ N}$$

Let two-legged 12 mm bars be used for stirrups, then the area of the stirrup is

$$A_{sv} = 2(113) = 226 \text{ mm}^2$$

The spacing of the transverse reinforcement is

$$s_v = \frac{A_{sv} \, \sigma_{sv} \, d}{V - V_a} = \frac{226(230)(330)}{131580 - 32710} = 175 \text{ mm}$$

The allowable maximum spacing of the stirrups is

$$s_{max} = 0.75 \, d = 247 \text{ mm}$$

Provide φ 12 stirrups at 175 mm for the extreme 1.75 m on each end then at 240 mm.

Deflection calculations: In the present case the beam is built into the slab, therefore deflections of the beam slab system is to be computed. However, such a computation is beyond the scope of this book. An upper bound in deflection can be calculated by treating the beam as independent. The moment of inertia of the cracked section by including the effect of the reinforcement is computed about the neutral axis.

$$nd = 0.289(0.33) = 0.095 \text{ m}$$

$$I_{cr} = \frac{b(nd)^3}{3} + mA_{sc} (nd - d_c)^2 + mA_{st} (d - nd)^2$$

$$= \frac{2(0.095)^3}{3} + 13(0.009938)(0.035)^2 + 13(0.004611)(0.33 - 0.06)^2$$

$$= (57.2 + 15.8 + 437) \, 10^{-5} = 510(10^{-5}) \text{ m}^4$$

The maximum deflections which occur at the mid-span of the beam due to dead and live load are:

$$v_d = \frac{5w_d \, L^4}{384 \, EI_{cr}} = \frac{5(18)(8.4)^4}{384(25491)(0.0051)} = 8.9 \text{ mm}$$

$$v_l = \frac{5(16)(8.4)^4}{384(25491)(0.0051)} = 8 \text{ mm}$$

The maximum deflection including creep with creep coefficient of 2.1 is

$$v_t = c_c \, v_d + v_l = 2.1(8.9) + 8 = 26.7 \text{ mm}$$

The allowable total deflection is

$$v_a = \frac{L}{250} = \frac{8400}{250} = 33.6 \text{ mm}$$

The deflection is within the limits.

Note: The effect of the tension reinforcement dominates in the moment of inertia of the

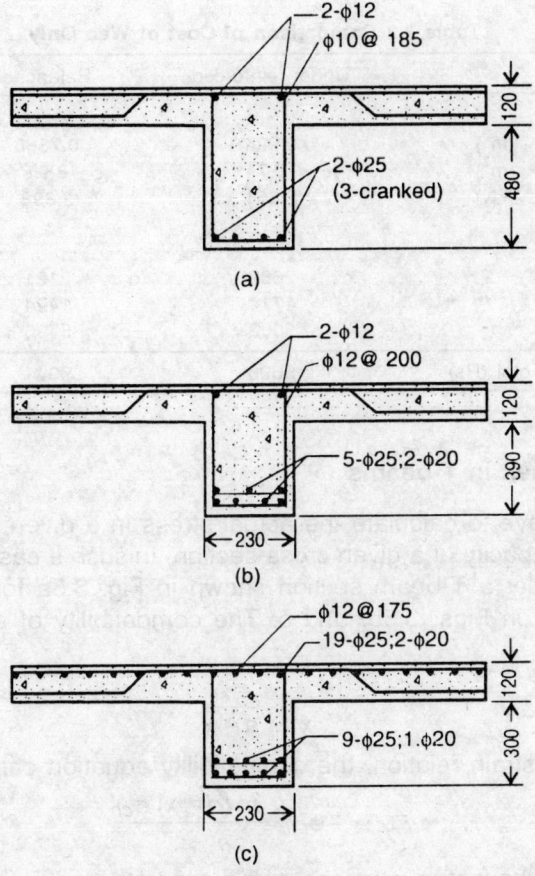

Fig. 3.4. Example 3.5.

cracked section. 80 per cent of the inertia is contributed by the tension reinforcement. Therefore, by neglecting the effect of the reinforcement, one is likely to underestimate the moment of inertia. Figure 3.4 illustrates details of the under-reinforced, balanced and over-reinforced sections. (This is an uneconomical design.)

EXAMPLE 3.5. *Comparison of costs of under-reinforced, balanced and doubly reinforced beams:* In a beam slab construction, the thickness of the slab is 120 mm and the effective span of the beam is 8.4 m. The beam is designed as under reinforced, balanced and doubly reinforced (Examples 3.3, 3.1 and 3.3 respectively). The unit cost of the materials including supplying, laying etc. are:

$$M20 \text{ concrete} = Rs. \ 2700/m^3$$
$$Formwork = Rs. \ 160/m^2$$
$$Steel = Rs. \ 2000/kN$$

The cost of the web and the reinforcement are computed in Table 3.2. The total length of the beam is taken as 9 m.

Table 3.2 Comparison of Cost of Web Only

Item		Under-reinforced	Balanced	Doubly reinforced
Materials				
Concrete	(m^3)	0.994	0.7866	0.621
Formwork	(m^2)	10.71	8.91	7.47
Reinforcement	(kN)	2.651	2.898	10.803
Cost (in Rs.)				
Concrete (Rs. 2700/m^3)		2682	2121	1680
Formwork (Rs. 160/m^2)		1714	1424	1196
Steel (Rs. 2000/kN)		5302	5796	21606
Total (Rs)		9698	9341	24472

3.5 Analysis of Stresses in T-beams

Sometimes one may have to calculate the actual stress in a given cross section or evaluate the resisting moment capacity of a given cross-section. In such a case the problem is basically one of analysis. Consider a T-beam section shown in Fig. 3.5a for which strain and stress distributions are shown in Figs. 3.5b, and c. The compatibility of strains can be expressed as:

$$\frac{\epsilon_c}{x} = \frac{\epsilon_{st}}{d-x} \qquad (3.20)$$

Using the linear stress-strain relation, the compatibility equation can be rewritten as

$$\sigma_{st} = \frac{m\,(d-x)\,\sigma_{cb}}{x} \qquad (3.21)$$

where σ_{cb} and σ_{st} are the *actual stresses* in the concrete and steel but not necessarily the full allowable stresses. In the earlier sections, these are referred as full allowable stresses. In this section the allowable stresses are indicated by *subscript a* for proper identification.

If the neutral axis falls in the flange portion, the problem reduces to a rectangular sectioned beam which was discussed earlier. T-beam section becomes relevant when the neutral axis falls in the web portion. The total top flange and a small portion of the web comes into action. The level of the stress and force in the web is usually small when compared with the corresponding values in the flange. The moment contribution of the force in the web is much

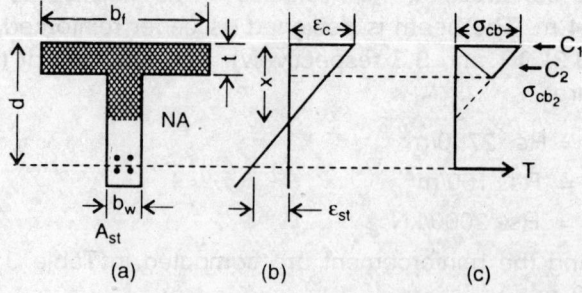

Fig. 3.5. Strain and Stress Variation.

less than the corresponding values from the flange. Therefore, the contribution of the stress in the web is neglected.

The compressive stress at the bottom of the compression flange is

$$\sigma_{cb2} = \frac{(x - t)\sigma_{cb}}{x} \tag{3.22}$$

The total compressive force on the flange can be expressed as two triangular stress distribution one due to σ_{cb} and the other due to σ_{cb2}. These total forcess are:

$$C_1 = \frac{b_f t \sigma_{cb}}{2}$$

$$C_2 = \frac{b_f t \sigma_{cb2}}{2} = \frac{b_f t(x = t)\,\sigma_{cb}}{2x} \tag{3.24}$$

The total tension force in the reinforcement is

$$T = A_{st}\,\sigma_{st} = A_{st}\,\frac{m(d - x)\,\sigma_{cb}}{x} \tag{3.25}$$

Neglecting the contribution of web in compression and also the tensile stress in the concrete, the equilibrium of the forces on the cross-section are:

$$C_1 + C_2 = T$$

or $$\frac{b_f t\,\sigma_{cb}}{2} + \frac{b_f t(x - t)\sigma_{cb}}{2x} = A_{st}\,\frac{m\,(d - x)\,\sigma_{cb}}{x}$$

or $$\frac{b_f t}{2}\left(1 + \frac{x - t}{x}\right) = A_{st}\,\frac{m\,(d - x)}{x}$$

or $$b_f t\left(x - \frac{t}{2}\right) = A_{st}\,m\,(d - x) \tag{3.26}$$

This equation is same as the equation corresponding to first moment of area of the transformed cross-section neglecting the web contribution. Hence the neutral axis of the section will correspond to the centroidal axis of the transformed cracked cross-section. Therefore, the neutral axis can be obtained from the first moment of area of the section. The equation can be rearranged as

$$x = \frac{mA_{st}\,d + 0.5 b_f t^2}{mA_{st} + b_f t} \tag{3.27}$$

The moment of the compression forces about the centroid of the reinforcement is:

$$M = C_1\left(d - \frac{t}{3}\right) + C_2\left(d - \frac{2t}{3}\right)$$

$$= \frac{b_f t \sigma_{cb}}{2}\left(d - \frac{t}{3}\right) + \frac{b_f t(x - t)\,\sigma_{cb}}{2x}\left(d - \frac{2t}{3}\right)$$

$$= \frac{b_f t}{2}\left(d - \frac{t}{3} + \frac{(x - t)}{x}\left(d - \frac{2t}{3}\right)\right)\sigma_{cb}$$

$$= \frac{b_f t}{2x}\left(xd - \frac{xt}{3} + (x - t)\left(d - \frac{2t}{3}\right)\right)\sigma_{cb}$$

$$= \frac{b_f t}{2x}\left(2xd - xt - dt + \frac{2t^2}{3}\right)\sigma_{cb}$$

$$= \frac{b_f t}{2}\left((2d - t) - \frac{t}{x}\left(d - \frac{2t}{3}\right)\right)\sigma_{cb} \tag{3.28}$$

If the bending moment on the section is given, one can compute the actual stress in the concrete from the above equation, or if the stresses in the materials are limited, then the moment capacity can be obtained from the same equation. In the latter problem one should ensure which one of the two actual stresses reach the corresponding allowable value and the limit to the lower bound. In case the stress in concrete reaches its limit earlier, then the moment capacity is given by Eq. (3.28) with

$$\sigma_{cb} = \sigma_{acb}$$

Otherwise compute the actual stress in concrete corresponding to the full allowable tension in steel. That is

$$\sigma_{cb} = \frac{x\,\sigma_{ast}}{m(d - x)}$$

EXAMPLE 3.6 *Analysis for stresses:* A T-beam section with the following details is subjected to a bending moment of 264.6 kNm. (Fig. 3.6)

$$t = 0.12 \text{ m},\ b_f = 1.67 \text{ m},\ b_w = 0.23 \text{ m},\ h_w = 0.38 \text{ m}$$
$$A_{st} = 3430 \text{ mm}^2 = (7 - \phi\ 25),\ m = 13;\ d = 0.42 \text{ m}$$

Determine the stresses in the concrete and the reinforcement.

Analysis: Let x = neutral axis distance from the top fibre; then the stresses in the reinforcement and in the bottom fibre of the flange are:

$$\sigma_{st} = \frac{m(d - x)\sigma_{cb}}{x} = \frac{13(0.42 - x)\sigma_{cb}}{x}$$

$$\sigma_{cb_2} = \frac{(x - 1)\sigma_{cb}}{x} = \frac{(x - 0.12)\sigma_{cb}}{x}$$

It is assumed that the neutral axis falls in the web. Then the total compressive force in the top flange and tensile force in the reinforcement are

$$C = \frac{b_f t\sigma_{cb}}{2} + \frac{b_f t\sigma_{cb_2}}{2}$$

$$= \frac{b_f t}{2}\left(1 + \frac{x - t}{x}\right)\sigma_{cb}$$

$$= (0.1002)\left(1 + \frac{x - 0.12}{x}\right)\sigma_{cb}$$

$$T = A_{st}\sigma_{st} = 13A_{st}\frac{(0.42 - x)\sigma_{cb}}{x}$$

$$= 0.04453\frac{(0.42 - x)\sigma_{cb}}{x}$$

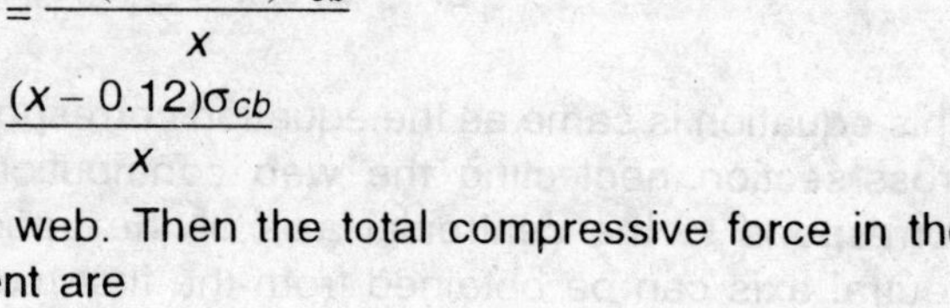

Fig. 3.6 Example 3.6.

Neglect the compressive force in the web and then equate the total compression to tensile force:

$$C = T$$

or $\quad 0.1002 \left(1 + \dfrac{x - 0.12}{2}\right) = 0.04453 \dfrac{(0.42 - x)}{x}$

$$x + x - 0.12 = 0.445 \, (0.42 - x)$$

or $\quad 2.445x = 0.3069$

or, so $\quad x = 0.126$ m

Since x is greater t the neutral axis falls in the web. The error due to neglecting the compression in web will be very small as x is close to t. The bending of the compressive force about the reinforcement is

$$M_r = \frac{b_f t \, \sigma_{cb}}{2} \left(d - \frac{1}{3}\right) + \frac{b_f t}{2} \left(\frac{x - t}{x}\right) \sigma_{cb} \left(d - \frac{2t}{3}\right)$$

$$= 1.67(0.06) \, (0.42 - 0.04) + \frac{(0.126 - 0.12)}{0.126} \, (0.42 - 0.08) \, \sigma_{cb}$$

$$= 0.396 \, \sigma_{cb}$$

Equate the external bending moment to the moment resistance M_r, then we have

$$M_r = M$$

or $\quad 0.0396 \, \sigma_{cb} = 0.2646$ MNm

or $\quad \sigma_{cb} = \dfrac{0.2646}{0.0396} = 6.6$ MPa

The corresponding tensile stress in the reinforcement is

$$\sigma_{st} = \frac{m(d - x) \, \sigma_{cb}}{x}$$

$$\frac{13(0.42 - 0.126) \, (6.6)}{0.126} = 200.2 \text{ MPa}$$

Note: This example is same as Example 3.1 except that the tensile reinforcement is taken as 3430 mm^2 instead of 3078 mm^2. This example may be considered as slightly over-reinforced section.

EXAMPLE 3.7 Moment capacity of an under-reinforced section: Determine the working moment capacity of a T-beam section with the following details:

$$b_f = 1.67 \text{ m}, \ t = 0.12 \text{ m}, \ d = 0.51 \text{ m}$$
$$A_{st} = 2940 \text{ mm}^2, \quad \sigma_{acb} = 230 \text{ MPa}$$

Analysis: The stresses in the reinforcement and the bottom fibre of the flange are:

$$\sigma_{st} = \frac{m \, (d - x)}{x} \, \sigma_{cb} = \frac{13(0.51 - x)}{x} \, \sigma_{cb}$$

$$\sigma_{cb_2} = \frac{(x - t)}{x} \, \sigma_{cb} = \frac{(x - 0.12)}{x} \, \sigma_{cb}$$

It is assumed that the neutral axis falls in the web and the compressive force from the web the flange is negligible, then the total force in the flange is

$$C = \frac{b_f t}{2} \sigma_{cb} + \frac{b_f t}{2} \sigma_{cb_2}$$

$$= \frac{b_f t}{2} \left(1 + \frac{x - t}{x}\right) \sigma_{cb} = \frac{0.1002(2x - 1)}{x} \sigma_{cb}$$

The total tensile force in the reinforcement is

$$T = A_{st} \sigma_{st} = 13(0.00294) \frac{(0.51 - x)}{x} \sigma_{cb}$$

$$= 0.03822 \frac{(0.51 - x)}{x} \sigma_{cb}$$

Equating the compression to the tensile force, we have

$$C = T$$

or $\quad 0.1002(2x - 0.12) = 0.03822(0.51 - x)$

or $\quad x = \dfrac{0.03152}{0.23862} = 0.132$ m

Since x is greater than t, the neutral axis falls in the web. Now it is necessary to determine which of the two stresses reaches the full value, first. Let

Let $\sigma_{cb} = \sigma_{acb} = 7$ MPa

Then the tensile stress is steel is

$$\sigma_{st} = \frac{m(d - x)}{x} \sigma_{acb}$$

$$= \frac{13(0.51 - 0.132)(7)}{0.132} = 260 \text{ MPa}$$

The actual stress in the steel is 260 MPa as against 230 MPa allowable. This implies that the tensile stress in the reinforcement reaches its full stress earlier than the corresponding compressive stress in concrete. The corresponding compressive stress in the extreme fibre of concrete is

$$\sigma_{cb} = \frac{x \sigma_{ast}}{m(d - x)} = \frac{0.132(230)}{13(0.51 - 0.132)} = 6.18 \text{ MPa}$$

The working moment capacity of the section is obtained by taking moment of the compressive force about the centre of the reinforcement and it is

$$M_r = \frac{b_f t}{2} \sigma_{cb} \left(d - \frac{t}{3}\right) + \frac{b_f t}{2} \frac{(x - t)}{x} \sigma_{cb} \left(d - \frac{2t}{3}\right)$$

$$= 0.1002 \left[(0.51 - 0.04) + \frac{(0.132 - 0.12)}{0.132}(0.51 - 0.08)\right] 6.18$$

$$= 0.3152 \text{ MNm}$$

Note: See Example 3.6 in which a beam is designed to resist a moment of 0.30975 MNm. The area of the tension reinforcement needed was computed as 2918 mm² and provided as 2940 mm². The moment capacity of the same section is computed in this example and it is found to be 0.3152 MNm.

EXAMPLE 3.8 *Analysis of doubly reinforced T-beam:* Determine the moment capacity of a T-beam section for the following data: (Fig. 3.7)

$$b_f = 1.64 \text{ m}, \qquad b_w = 0.23 \text{ m}, \qquad t = 0.12 \text{ m},$$
$$h_w = 0.3 \text{ m} \qquad d = 0.33 \text{ m}, \qquad d_c = 0.06 \text{ m},$$
$$m = 13, \qquad \sigma_{acb} = 7 \text{ MPa}, \qquad \sigma_{ast} = 230 \text{ MPa}.$$
$$A_{st} = 4611 \text{ mm}^2; \qquad A_{sc} = 9938 \text{ mm}^2,$$

Analysis: Let x = neutral axis distance from the top fibre, then the stresses in the tension and compression reinforcement, and in the bottom of the compression flange are:

$$\sigma_{st} = \frac{m(d-x)}{x}\,\sigma_{cb} = \frac{13(0.33-x)}{x}\,\sigma_{cb}$$

$$\sigma_{sc} = 1.5\,\frac{m(x-d_c)}{x}\,\sigma_{cb} = \frac{19.5(x-0.06)}{x}\,\sigma_{cb}$$

$$\sigma_{cb2} = \frac{(x-t)}{x}\,\sigma_{cb} = \frac{(x-0.12)}{x}\sigma_{cb}$$

It is assumed that the neutral axis falls in the web and the contribution of the web under compression is negligible. Then the total compressive and tensile forces are:

$$C_1 = \frac{b_f t}{2}\,\sigma_{cb} = 0.1002\,\sigma_{cb}$$

$$C_2 = \frac{b_f t}{2}\,\sigma_{cb2} = 0.1002\,\frac{(x-0.12)}{x}\,\sigma_{cb}$$

$$C_3 = A_{sc}\,\sigma_{sc} = 0.009938\,(19.5)\,\frac{(x-0.06)}{x}\,\sigma_{cb}$$

$$= 0.1937\,\frac{(x-0.06)}{x}\,\sigma_{cb}$$

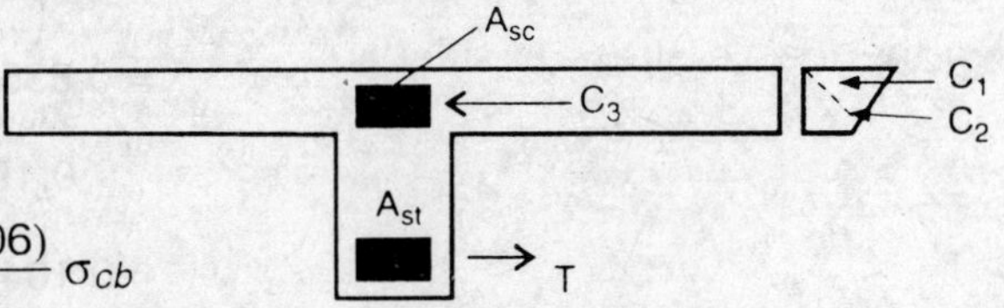

$$T = A_{st}\,\sigma_{st} = 0.004611(13)\,\frac{(0.33-x)}{x}\,\sigma_{cb}$$

$$= 0.05994\,\frac{(0.33-x)}{x}\,\sigma_{cb}$$

Fig. 3.7 Example 3.8.

Equating the total compressive force to the tensile force, we have

$$C_1 + C_2 + C_3 = T$$

or $\quad 0.10002(2x-0.12) + 0.1937(x-0.06) = 0.05994(0.33-x)$

$$2x = 0.12 + 1.934(x-0.06) = 0.5982(0.33-x)$$

$$x = 0.07 \text{ m}$$

The neutral axis distance is less than the thickness of the slab; therefore, revise the compressive forces are:

$$C_1 = \frac{b_f x\,\sigma_{cb}}{2} = 0.835\,x\sigma_{cb}; \quad C_2 = 0$$

$$C_3 = 0.1934\,\frac{(x-0.06)}{x}\,\sigma_{cb}$$

Now equating the total compressive force to the tensile force, we have

$$0.835x + 0.1934 \frac{(x - 0.06)}{x} = 0.05994 \frac{(0.33 - x)}{x}$$

or $\quad x^2 + 0.2316(x - 0.06) - 0.0717(0.33 - x) = 0$

$$x^2 + 0.3033x - 0.0376 = 0$$

$$x = 0.094 \text{ m}$$

It is necessary to determine which of the actual stress reaches the corresponding allowable value earlier. Let

$$\sigma_{cb} = \sigma_{acb} = 7 \text{ MPa}$$

Then the tensile stress in steel is

$$\sigma_{st} = \frac{13(0.3 - 0.094)(7)}{0.094} = 229 \text{ MPa}$$

Since $\sigma_{st} < \sigma_{ast}$, the compressive stress governs the design and the beam is over-reinforced. The working moment capacity of the section is obtained by taking moments about the tension reinforcement and it is

$$M_r = C_1 \left(d - \frac{x}{3} \right) + C_3(d - d_c)$$

$$= 0.835 \times (0.33 - 0.31)(7)$$

$$+ 0.1934 \frac{(0.094 - 0.06)(7)}{0.094} (0.33 - 0.06)$$

$$= 0.1643 + 0.1322 = 0.2965 \text{ MNm}$$

Note: See Example 3.4 in which the beam is designed for a moment of 0.29988 MNm whereas from the reverse process of analysis, we find that the moment capacity is 0.2965 MNm. The small difference is due to the rounding off.

PROBLEMS

3.1 A simply supported reinforced concrete beam is subjected to a working bending moment of 700 kNm and a shear force of 300 kN. Design a T-beam cross-section to resist the above forces using the following data: Maximum width of flange = 1000 mm, thickness of flange = 150 mm, σ_{st} = 230 N/mm^2, σ_{cb} = 5 N/mm^2 and σ_{st} = 230 N/mm^2.

3.2 Derive an expression for working moment capacity of reinforced concrete T-section in terms of width and thickness of the flange, effective depth and allowable stresses in concrete and steel. Also calculate the percentage of tension steel with respect to the total web area.

3.3 A reinforced concrete T-beam with the properties given below is subjected to a working moment of 500 kNm. Determine a maximum stresses in the concrete and steel. b_f = 900 mm, t = 120 mm, d = 600 mm, A_{st} = 4900 mm^2, m = 13 and b_w = 300 mm.

3.4 A reinforced concrete T-beam cross-section with the details given below is subjected to an external working bending moment of 320 kNm. The maximum allowable stresses in concrete and reinforcement are 7 and 230 N/mm^2 respectively. Determine whether the cross-section is structually safe in working stress design for the moment given above. Given $b_f = 1.6$ m, $t = 0.12$ m, $d = 0.51$ m, and $A_{st} = 2918$ mm^2, b_w 250 mm.

3.5 A doubly reinforced T-beam section with the dimensions given below is subjected to a maximum working bending moment of 300 kNm. Determine the stresses in the concrete and reinforcement respectively. $b_f = 1600$ mm, $t = 120$ mm, $b_w = 250$ mm, $d = 330$ mm, $A_{st} = 4900$ mm^2, $A_{sc} = 9938$ mm^2, $d_c = 60$ mm and $m = 13$.

3.6 Determine the working moment capacity of a singly reinforced concrete T-section for the following details: $b_f = 1000$ mm, $t = 120$ mm, $b_w = 250$mm, $d = 500$ mm, and $A_{st} = 1256$ mm^2, $\sigma_{acb} = 7$ N/mm^2, $\sigma_{ast} = 230$ N/mm^2.

3.7 Determine the working moment capacity of reinforced concrete T-section for the following details: $b_f = 1200$ mm, $t = 150$ mm, $b_w = 400$ mm, $d = 600$ mm $A_{st} = 4900$ mm^2 the maximum allowable stresses in concrete and steel are 7 and 230 N/mm^2.

3.8 Determine the moment capacity of a doubly reinforced T-section for the following details: $b_f = 1500$ mm, $t = 150$ mm, $b_w = 400$ mm, $d = 500$ mm, $A_{st} = 4900$ mm^2, $A_{sc} = 7350$ mm^2, $\sigma_{acb} = 7$ N/mm^2, $\sigma_{ast} = 230$ N/mm^2, $m = 13$ and $d_c = 60$ mm.

3.9 A simply supported beam of an effective span of 10 m is subjected to a uniformly distributed live load of 30 kN/m, the thickness and width of the flange are restricted to 120 and 1200 mm respectively. Design a T-beam section by working stress design with M 20 concrete and HYSD-Fe415 Bars. Use allowable shear stress is 0.4 N/mm^2.

3.10 A simply supported beam of an effective span 8 m is subjected to an uniformly distributed live load of 23 kN/m and a moving load of 150 kN. Design a singly reinforced T-beam by working stress design using M20 concrete and HYSD-Fe415 bars. The width and the thickness of the flange are restricted to 1000 and 200 respectively.

3.11 A simply supported beam of an effective span 8 metres is subjected to uniformly distributed live load of 30 kN/m. The architect has specified the following dimensions for the beam: $b_f = 1000$ mm, $t = 160$ mm, $b_w = 300$ mm and $d = 600$ mm. Design a reinforcement details by working stress design using M20 concrete and HYSD-Fe415 bars.

3.12 A simply supported beam of an effective span 12 metres is subjected to uniformly distributed live load of 20 kN/m. Design a reinforced concrete T-section by working stress design subjected to the following limitations: $b_f = 1500$ mm, $t = 300$ mm, $h = 800$ mm, $b_w = 400$ mm, $\sigma_{acb} = 7$N/mm^2 and $\sigma_{ast} = \sigma_{sv} = 230$ N/mm^2.

Chapter 4

WSD of Design of RCC Slabs

4.1 Introduction

Reinforced cement concrete (RCC) slabs are most commonly used in floors and roofs of buildings. In RCC slabs, the thickness is small when compared with the other two dimensions namely, spans. Further, the loads acting and the reactions on the slab are treated as functions of the x and y coordinates, whereas in the case of beams, the shear and bending moment forces are functions of the length only. When the ratio of the length to the width of a slab is more than two, most of the load is carried by the short span (also called width). In such a case the slab is called *one-way slab*. One-way slabs are analysed and designed like beams spanning across the width. A slab having length to width ratio less than two is called *two-way slab.* Here the load is carried in two directions; however, more is carried in the shorter span when compared with that of the long span. The thickness of RCC slabs range from 75 mm to 300 mm. The recommended thickness of slabs in different constructions is given in Table 4.1.

Table 4.1 Recommended RCC Slab Thickness

	Type	L/t
1.	One-way slabs	18 to 25
2.	Cantilever slabs	7 to 12
3.	Two-way and flat slabs	20 to 35

The effective span of a slab can be taken as:
Effective span = clear span + effective depth in simply supported slabs and
= clear span in fixed slabs

Minimum steel and cover requirements: To minimize the shrinkage and temperature effects, and the consequent cracking, one must provide a minimum reinforcement in the slabs. Minimum cover to the reinforcement should also be provided so as to make steel effective and avoid damage to it which might be caused by the environment forces such as humidity, moisture, temperature, etc.

The minimum percentage of reinforcement in slabs

= 0.15% of gross area of the cross-section in case of mild steel, and
= 012% of gross area of the section for high yield strength deformed bars or welded fabric.

Maximum area of reinforcement is limited to four per cent of the cross-section. *Diameter of the reinforcement bar* should not be more than one-eighth of the thickness of the slab. *Spacing of the main reinforcement* should not be more than two times the thickness of slab and it should be more than five times the diameter of the reinforcement. *Minimum cover to reinforcement* should be more than 15 mm or diameter of the bar.

4.2 Design of One-way Slab

The maximum bending and shear forces are computed using the beam analysis for one metre width of the slab. The resisting bending moment per metre width (M_r) of a balanced design slab can be expressed as (see Chapter 2):

$$M_r = Kbd^2 \sigma_{cb} \tag{4.1}$$

where $K = \dfrac{nj}{2}; j = 1 - \dfrac{n}{3}$

The effective depth of a balanced section is calculated as

$$d \geq \sqrt{\left(\frac{M}{Kb\sigma_{cb}}\right)} \tag{4.2}$$

where M = external bending moment on unit width of the slab.

The area of steel for balanced or under-reinforced cross-section is given by

$$A_s = \frac{M}{\sigma_{st} jd} \tag{4.3}$$

in which j can be obtained as $j = 1 - n/3$ where $n = 0.40$ and 0.289 for mild steel and HYSD bars respectively.

Table 4.2 gives the moment capacities of different slabs in which the overall thickness (h) is in the range of 90 mm to 210 mm. The effective depth in these slabs is assumed to be ($h - 20$) mm for slabs up to 140 mm thick and ($h - 25$) mm for slabs from 150 to 210 mm thick. The area of steel required is also given in Table 4.2.

4.3 Design of Two-way Slab

In case of slabs having aspect ratio (long span/short span) less than two, the load is distributed between the two spans and the bending moment is a function of both the spans. Such slabs are designed as two-way slabs having reinforcement in both the directions. Table 4.3 gives bending moment coefficients for two-way slab subject to the following conditions:

a) The corners of the slabs are prevented from lifting. (Adequate reinforcement against torsion is provided at the corners.)

b) The slab in each direction is divided into one middle and two edge strips. The middle strip is taken as of 0.75 width and each edge strip as of 0.125 width.

c) The maximum bending moment in the middle strip per unit width is given by

$$M_x = \alpha_x wL_x^2 \tag{4.4}$$

$$M_y = \alpha_y wL_x^2 \tag{4.5}$$

where M_x and M_y are the BM's per unit width, α_x and α_y are the moment coefficients given in Table 4.3, L_x and L_y are the effective spans in x and y directions, and w is the intensity of load (L_x is the short span).

d) The bending moment in the edge strip can be taken equal to 30% of that of the middle strip.

e) The supports are treated as relatively rigid.

Main reinforcement: The variation of bending moment on the plane of the middle line varies with maximum at the middle point and decreases in a sinusoidal manner. Let

Table 4.2 Working Moment Capacities of 1000 mm Width Slabs and Area of Steel Required (M is Nm/m; A_s = Area of Steel in mm²/m and h is in mm, Cover 20 mm to 25 mm)

σ_{ct}	σ_{st}	140		190		230	
	h	A_s	M_r	A_s	M_r	A_s	M_r
	90	505	4282.60	307	3630.90	213	3087.00
	95	541	4916.25	329	4168.12	229	3543.73
	100	577	5593.60	351	4742.40	244	4032.00
	105	613	6314.65	373	5353.72	259	4551.75
	110	650	7079.40	395	6002.10	274	5103.00
	115	686	7887.85	417	6687.52	289	5685.75
	120	722	8740.00	439	7410.00	305	6300.00
	125	758	9635.85	461	8169.52	320	6945.75
	130	794	10575.40	483	8966.10	335	7623.00
	135	830	11558.64	504	9799.72	350	8331.75
5.0	140	866	12585.60	526	10670.40	366	9072.00
	145	902	13636.24	548	11578.12	381	9843.75
	150	938	14770.59	570	12522.90	396	10647.00
	155	974	15928.65	592	13504.72	411	11481.75
	160	974	15928.65	592	13504.72	411	11481.75
	165	1010	17130.39	614	14523.60	427	12348.00
	170	1046	18375.84	636	15579.52	442	13245.74
	175	1083	19664.99	658	16672.50	457	14175.00
	180	1119	20997.84	680	17802.52	472	15135.74
	185	1153	22374.39	702	18969.60	487	16128.00
	190	1191	23794.64	724	20173.72	503	17151.74
	195	1227	25258.59	746	21414.89	518	18206.99
	200	1263	26766.23	768	22693.12	533	19293.74
	205	1299	28317.59	790	24008.40	548	20411.99
	210	1335	29912.64	812	25360.72	564	21561.74
	90	851	7227.50	517	6115.20	331	4777.50
	95	911	8296.87	554	7020.00	354	5484.37
	100	972	9440.00	590	7987.20	378	6240.00
	105	1033	10626.87	627	9016.80	402	7044.37
	110	1094	11947.50	664	10108.80	425	7897.50
	115	1154	13311.87	701	11263.20	449	8799.37
	120	1215	14750.00	738	12480.00	473	9750.00
	125	1276	16261.87	775	13759.20	496	10749.37
	130	1337	17847.50	812	15100.79	520	11797.50
	135	1397	19506.87	849	16504.79	543	12894.37
	140	1458	21239.99	886	17971.19	567	14040.00
	145	1519	23046.87	923	19499.99	591	15234.38
	150	1580	24927.50	959	21091.19	614	16477.50
	155	1641	26881.87	996	22744.79	638	17769.37
	160	1641	26881.87	996	22744.79	638	17769.37
	165	1701	28909.99	1033	24460.79	662	19110.00
	170	1762	31011.87	1070	26239.19	685	20499.37
	175	1823	33187.50	1107	28079.99	709	21937.50
	180	1884	35436.87	1144	29983.19	733	23424.37
	185	1944	37759.99	1181	31948.79	756	24960.00
	190	2005	40156.87	1218	33976.79	780	26544.37
	195	2066	42627.50	1255	36067.19	803	28177.50
	200	2127	45171.87	1292	38219.99	827	28859.57
	205	2187	47789.99	1328	40435.19	851	31590.00
	210	2240	50481.87	1365	42712.79	874	33369.37

M_x = bending moment on x-plane (same as BM along x-axis)

This moment is a function of both x and y axes. Even though the subscript is only x, the bending moment is really a function of both the directions. The critical moment on the critical strip depends on the boundary conditions at $x = 0$ and L_x. The depth of the section and the reinforcement requirement are based on the critical moment on the central strip. The bending moment M_x at $x = 0.5 L_x$ for $y = 0$ to L_y varies close to a sinusoidal variation. Consequently, the spacing requirement of the reinforcement for M_x along y axis varies with maximum at the edges (that is, $y = 0, L_y$) and minimum at central unit strip. To minimize the problems of detailing of the reinforcement, the slab in each direction is divided into three strips. The middle 3/4 portion is called the middle strip and the two extreme edges of 0.125 L width are called edge strips. A uniform spacing of the reinforcement is recommended in each of the strips. In small slabs, the width of the edge strip is too small to be considered as a unit for the purpose of separate spacing of the reinforcement and so the spacing of the reinforcement can be same throughout the slab. The main reinforcement in each direction is subject to some minimum limits as discussed earlier. The bending moment coefficients recommended by the codes of practice do not necessarily correspond to the peak moments but to some statically equivalent averaging values. The main reinforcement in the edge strips can be equal to half of that in the middle strip or the minimum reinforcement.

Torsion reinforcement: The transverse shear force when combined with twisting moment on the slabs gives rise to Kirchoffs shear force. The tendency of this shear force is to lift the corner of the slabs from the supports. If the corners of the slabs are prevented from lifting up (which is common in most slabs) then a torsional moment is developed at each of the corners of the slab. The prevention of the lifting up of the corners can be due to either the walls over the slab or continuity of the slab over rigid or beam supports. Torsional reinforcement must be provided at the corners of the slab. The torsional moment depends on the aspect ratio, however, for convenience of design a reinforcement which is equal to three-fourths of the main reinforcement of the short span is recommended both at the top and bottom and in each direction at each of the corners. The torsional reinforcement must be placed in the edge for a distance equal to 1/5 of short span in both directions and both faces of the slab. Table 4.3a gives the moment coefficients in slabs with restrained corners. Table 4.3b gives the moment coefficients of simply supported slabs with warping corners. No torsional reinforcement is needed in case of simply supported ends with warping corners.

4.4 Design for Shear

Shear stress is not normally very critical in slabs, however, one should ensure that the nominal shear stress is less then the allowable shear stress. The allowable shear stress in solid slabs is given by:

$$\tau_{cs} = k_s \tau_c \tag{4.6}$$

where k_s = modified shear stress coefficient given in Table 4.5

The shear failure is likely to occur at a distance d from the face of the direct contact support if the resting face of the slab is different from the loading face. Otherwise, the failure is at the face of contact of the support.

Table 4.3 (a) Bending Moment Coefficient for Rectangular Panels Supported on Four Sides with Provision for Torsion at Corners (based on IS 456:1978)

Case	Type of panel and moments considered	Short span moment coefficients (α_x)								Long span coefficients (α_y) for all values of L_y/L_x
		$L_y/L_x = 1.0$	1.1	1.2	1.3	1.4	1.5	1.75	2.0	
1	2	3	4	5	6	7	8	9	10	11
1.	Interior panels									
	Negative moment at continuous edge	0.032	0.037	0.043	0.047	0.051	0.053	0.060	0.065	0.032
	Positive moment at mid-span	0.024	0.028	0.032	0.036	0.039	0.041	0.045	0.049	0.024
2.	One short edge discontinuous									
	Negative moment at continuous edge	0.037	0.043	0.048	0.051	0.055	0.057	0.064	0.068	0.037
	Positive moment at mid-span	0.028	0.032	0.036	0.039	0.041	0.044	0.048	0.052	0.028
3.	One long edge discontinuous									
	Negative moment at continuous edge	0.037	0.044	0.052	0.057	0.063	0.067	0.077	0.085	0.037
	Positive moment at mid-span	0.028	0.033	0.039	0.044	0.047	0.051	0.059	0.065	0.028
4.	Two adjacent edges discontinuous									
	Negative moment at continuous edge	0.047	0.053	0.060	0.065	0.071	0.075	0.084	0.091	0.047
	Positive moment at mid-span	0.035	0.040	0.045	0.049	0.053	0.056	0.063	0.069	0.035
5.	Two short edges discontinuous									
	Negative moment at continuous edges	0.045	0.049	0.052	0.056	0.059	0.060	0.065	0.069	–
	Positive moment at mid-span	0.035	0.037	0.040	0.043	0.044	0.045	0.049	0.052	0.035
6.	Two long edges discontinuous									
	Negative moment at continuous edges	–	–	–	–	–	–	–	–	0.045
	Positive moment at mid-span	0.035	0.043	0.051	0.057	0.063	0.068	0.080	0.088	0.035
7.	Three edges discontinuous (One long edge continuous)									
	Negative moment at continuous edge	0.057	0.064	0.071	0.076	0.080	0.084	0.091	0.097	–
	Positive moment at mid-span	0.043	0.048	0.053	0.057	0.060	0.064	0.069	0.073	0.053
8.	Three edges discontinuous (One short edge discontinuous)									
	Negative moment at continuous edge	–	–	–	–	–	–	–	–	0.057
	Positive moment at mid-span	0.043	0.051	0.059	0.065	0.071	0.076	0.087	0.096	0.043
9.	Four edges discontinuous									
	Positive moment at mid-span	0.056	0.064	0.072	0.079	0.085	0.089	0.100	0.107	0.056

Table 4.3 (b) Bending Moment Coefficients in Simply Supported Slabs with no Torsion Reinforcement at the Corners

Moment	Aspect ratio									
	1.0	1.10	1.20	1.30	1.40	1.50	1.75	2.00	2.5	3.0
M_x	0.062	0.074	0.084	0.093	0.099	0.104	0.113	0.118	0.122	0.124
M_y	0.062	0.061	0.059	0.055	0.051	0.046	0.037	0.029	0.020	0.014

Table 4.4 Area in mm^2 of Bars Per Metre Width

Spacing of bars in mm	Diameter of the bars							
	6	8	10	12	14	16	18	20
90	314	559	873	1257	1710	2234	2827	3491
95	298	529	827	1191	1620	2116	2679	3307
100	283	503	785	1131	1539	2011	2545	3142
105	269	479	748	1077	1466	1915	2424	2992
110	257	457	714	1028	1399	1828	2313	2856
115	246	437	683	983	1339	1748	2213	2732
120	236	419	654	942	1283	1676	2121	2618
125	226	402	628	905	1232	1608	2036	2513
130	217	387	604	870	1184	1547	1957	2417
135	209	372	582	838	1140	1489	1885	2327
140	202	359	561	808	1100	1436	1818	2244
145	195	347	542	780	1062	1387	1755	2167
150	188	335	524	754	1026	1340	1696	2094
155	182	324	507	730	993	1297	1642	2027
160	177	314	491	707	962	1257	1590	1963
165	171	305	476	685	933	1219	1542	1904
170	166	296	462	665	906	1183	1497	1848
175	162	287	449	646	880	1149	1454	1795
180	157	279	436	628	855	1117	1414	1745
185	153	272	425	611	832	1087	1376	1698
190	149	265	413	595	810	1058	1339	1653
195	145	258	403	580	789	1031	1305	1611
200	141	251	393	565	770	1005	1272	1571
205	138	245	383	552	751	981	1241	1532
210	135	239	374	539	733	957	1212	1496
215	132	234	365	526	716	935	1184	1461
220	129	228	357	514	700	914	1157	1428
225	126	223	349	503	684	894	1131	1396
230	123	219	341	492	669	874	1106	1366
235	120	214	334	481	655	856	1083	1337
240	118	209	327	471	641	838	1060	1309
245	115	205	321	462	628	821	1039	1282
250	113	201	314	452	616	804	1018	1257
255	111	197	308	444	604	788	998	1232
260	109	193	302	435	592	773	979	1208
265	107	190	296	427	581	759	960	1186
270	105	186	291	419	570	745	942	1164
275	103	183	286	411	560	731	925	1142
280	101	180	280	404	550	718	909	1122
285	099	176	276	397	540	705	893	1102

Table 4.5 Modified Shear Stress Coefficient (k_s)

Slab thickness (mm)	150 or less	175	200	225	250	275	300 or more
k_s	1.3	1.25	1.20	1.15	1.10	1.05	1.00

The nominal shear stress is calculated as:

$$\tau_v = \frac{V}{bd} \tag{4.7}$$

where V is the shear force per unit width and b is the unit width.

The effective depth of a slab based on the shear strength criterion is given by

$$d = \frac{V}{b\tau_c} \tag{4.8}$$

In case of beams, one can provide shear reinforcement if the nominal shear stress exceeds the allowable value, whereas in solid slabs, such a transverse reinforcement is not economical. Often the depth of the concrete slab is usually adjusted so as to meet the shear force requirement.

4.5 Design Examples

EXAMPLE 4.1 *Design of one-way slab:* A simply supported slab of clear spans 4 m by 10 m is subjected to live load of 2.5 kN/m². Design the slab.

Design data

$$\text{Long span } L_y = 10 \text{ m, Short span } L_x = 4 \text{ m}$$
$$\text{Live load } \quad w_l = 2.5 \text{ kN/m}^2, \ f_{ck} = 20 \text{ MPa, } f_y = 415 \text{ MPa}$$

The allowable stresses and design coefficients for balanced section are :

$$\sigma_{cb} = 7 \text{ MPa}, \qquad \sigma_{st} = 230 \text{ MPa}$$
$$n = 0.289, \qquad j = 0.904$$
$$K = 0.13 \qquad p_0 = 0.145$$

Design for bending: Let the effective spans be 4.1 and 10.1 m; the aspect ratio of the slab is:

$$\frac{L_y}{L_x} = \frac{10.1}{4.1} = 2.46$$

The slab has to be designed as a one-way slab as the aspect ratio is more than 2. For the purpose of computing the bending moment, one has to have the self-weight of the slab. Let the thickness of the slab be in the range of $L_{xo}/24$ say $t = 175$ mm.

The self-weight of the slab

$$w_g = \gamma t = 25(0.175) = 4.375 \text{ kN/m}^2$$

Let the effective short span of the slab be

$$L = L_x = L_{xo} + d = 4.1 \text{ m}$$

The total working load on the slab is

$$w_t = w_g + w_l = 6.875 \text{ kN/m}^2$$

The maximum bending moment which occurs at the middle of the short span is

$$M = \frac{w_t L^2}{8} = \frac{6875(4.1)^2}{8} = 14446 \text{ Nm/m}$$

Equating the moment capacity to the external bending moment on the slab, we have

$$d = \sqrt{\frac{M}{Kb\sigma_{cb}}} = \sqrt{\frac{0.014446}{0.13(1)(7)}} = 0.13 \text{ m; Let } d = 0.15 \text{ m}$$

The overall thickness of the slab

$$t = d + 0.02 = 0.17 \text{ m}$$

This value is slightly less than the assumed.
Area of the tension reinforcement is

$$A_{st} = \frac{M}{jd\sigma_{st}} = \frac{14446}{0.904(0.15)(230)} = 463 \text{ mm}^2$$

Use 12 mm bar at 240 mm spacing, then the tension reinforcement provided is

$$A_{st} \text{ (provided)} = \frac{1000(113)}{240} = 470 \text{ mm}^2/\text{m}$$

$$\frac{100 \, A_{st}}{bd} = \frac{47000}{1000(150)} = 0.31$$

Design for shear: The critical shear plane is at a distance d from the support and the shear force on this plane is

$$V = w_t \left(\frac{L}{2} - d \right) = 6875(2.05 - 0.15) = 13063 \text{ N}$$

The nominal shear stress in the slab is

$$\tau_v = \frac{V}{bd} = \frac{0.013063}{(0.15)} = 0.09 \text{ MPa}$$

The allowable shear stress for 0.3% tension reinforcement is 0.23 (1.25) MPa, which is far higher than the nominal shear stress. Provide minimum 0.12% distribution reinforcement in the other direction.

$$\text{Minimum steel} = \frac{0.12(170)(1000)}{100} = 210 \text{ mm}^2/\text{m}$$

Provide 8 mm HYSD bars at 240 mm spacing as distribution steel.

EXAMPLE 4.2 *Design of simply supported two-way slab with restrained corners:* A simply supported slab with clear spans of 3.5 m and 4.5 m is subjected to a live load of 3 kN/m^2. Design the slab with the following data:

Clear spans are $L_{xo} = 3.5$ m, $L_{yo} = 4.5$ m
$w_l = 3$ kN/m^2, $f_{ck} = 20$ MPa, $f_y = 415$ MPa

The allowable stresses and the design coefficients for a balanced section are:

$$\sigma_{cb} = 7 \text{ MPa}, \quad \sigma_{st} = 230 \text{ MPa}$$

$$n = 0.289, \ j = 0.904, \ K = 0.13$$

Corners of the slab are constrained against lifting.

Design for bending: Let the thickness of the slab be assumed as 150 mm for the purpose of computing self-weight, then

Self-weight $w_g = 25(0.15) = 3.75 \text{ kN/m}^2$

Total load on the slab is

$$w_t = w_g + w_l = 6.750 \text{ kN/m}^2$$

The aspect ratio of the slab is

$$r = \frac{L_y}{L_x} = \frac{4.65}{3.65} = 1.27$$

Since the aspect ratio is less than 2, the slab must be designed as a two-way slab. The bending moment coefficients can be obtained from Table 4.3a. The slab must be treated as discontinuous on four edges and the moment coefficients must be interpolated from the values given for aspect ratio of $r = 1.2$ and 1.3. The coefficients from Table 4.3a (where the corners of the slab are assumed to be prevented from lifting up) are:

$$\alpha_x = 0.072 \text{ and } \alpha_y = 0.056 \text{ for } r = 1.2$$

$$\alpha_x = 0.079 \text{ and } \alpha_y = 0.056 \text{ for } r = 1.3$$

The interpolation of the coefficients for $r = 1.27$ are:

$$\alpha_x = 0.072 + \frac{0.079 - 0.072)}{(1.3 - 1.2)} = (1.27 - 1.2) = 0.0769 \text{ and}$$

$$\alpha_y = 0.056$$

The maximum bending moments on the short and long spans which occur at the mid-point are:

$$M_x = \alpha_x w L_x^2 = 0.0769(6750)(3.65)^2 = 6915 \text{ Nm/m}$$

$$M_y = \alpha_y w L_x^2 = 0.056(6750)(3.65)^2 = 5036 \text{ Nm/m}$$

Equating the moment capacity to the larger of the above two moments, we have

$$d = \sqrt{\frac{0.006915}{0.13(7)}} = 0.087 \text{ m}$$

Let $h = 105$ mm

The thickness of the slab is less than that assumed so the self-weight bending moment decreases.

The self-weight $= 25000(0.105) = 2625 \text{ N/m}^2$

Total load $\quad w_t = 2625 + 3000 = 5625 \text{ N/m}^2$

Use $w_t = 5700 \text{ N/m}^2$ and $L_x = 3.6$ m

The bending moments are recomputed as

$$M_x = 0.0769(5700)(3.6)^2 = 5680 \text{ Nm/m}$$

$$M_y = 0.0560(5700)(3.6)^2 = 4136 \text{ Nm/m}$$

$$d = \sqrt{\frac{0.005680}{0.914}} = 0.079$$

Use $d = 0.08$ m

Area of the tension reinforcement in the short span direction is

$$A_{sx} = \frac{M_x}{jd\sigma_{st}} = \frac{5680}{0.904(0.08)(230)} = 341 \text{ mm}^2/\text{m}$$

Provide 10 mm bars at 200 mm spacing in the x-direction.
The actual reinforcement provided is

$$A_{st} \text{ (provided)} = \frac{1000}{200} (78.5) = 392 \text{ mm}^2/\text{m}$$

The total depth of the slab with 15 mm clear cover to the reinforcement is
$$h = d + 0.5 \, \phi + 15 \text{ mm} = 80 + 5 + 15 = 100 \text{ mm}$$

The percentage of reinforcement in the short span direction is

$$p = \frac{100 A_{sx}}{bd} = \frac{39200}{1000(80)} = 0.48$$

The effective depth available for steel in the y-direction is
$$d_y = 80 - 10 = 70 \text{ mm}$$

The area of the tension reinforcement in the y-direction is

$$A_{sy} = \frac{M_y}{jd_y\sigma_{st}} = \frac{4136}{0.904(0.07)(230)} = 284 \text{ mm}^2/\text{m}$$

Minimum reinforcement required in the slab is taken as 0.12% and is

$$A_{sm} = 0.0012 \, b \, h = 0.0012(1000)(100) = 120 \text{ mm}^2/\text{m}$$

Provide 8 mm bars at 175 mm spacing in the y-direction, then the actual **area of** steel provided is

$$A_{sy} \text{ (provided)} = \frac{50(1000)}{175} = 285 \text{ mm}^2/\text{m}$$

Check for shear stress: The load distribution in the short span direction can be approximated as

$$w_x = \frac{L_y^4 w_t}{L_x^4 + L_y^4} = 0.725 \, w_t$$

The maximum shear force which occur, at a distance d from the support is
$$V = w_x (0.5 L_x - d) = 0.725(5700)(1.8 - 0.08) = 7108 \text{ N/m}$$

The nominal shear stress on the section is

$$\tau_v = \frac{V}{bd} = \frac{7108(10^{-6})}{1(0.08)} = 0.09 \text{ MPa}$$

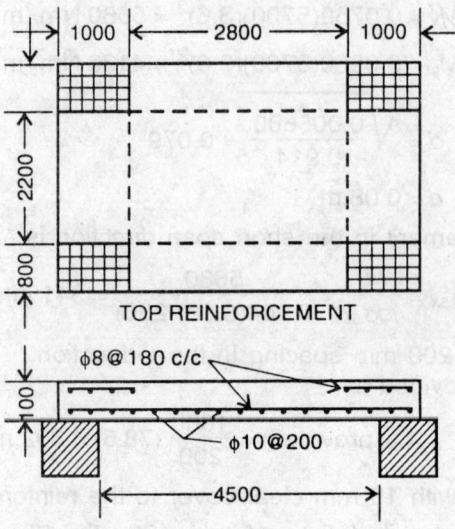

1000 | 2800 | 1000

2200

800

TOP REINFORCEMENT

$\phi 8 @ 180$ c/c

100

$\phi 10 @ 200$

4500

Fig. 4.1. Example 4.2.

The nominal shear stress is far lower than the allowable values; therefore, the section is safe against shear failure.

Design for torsion at corners: The corners of the slab were assumed to be constrained, so torsional reinforcement must be provided at top and bottom of the slab and it should be extended upto 1/5 of the span from the ends. The area of the reinforcement in each of the four layers should not be less than seventy-five per cent of the reinforcement at the middle of the short span. Therefore, the torsional reinforcement is

$$A_t = 0.75 \, A_{sx} = 0.75(341) = 268 \text{ mm}^2$$

Provide 8 mm bars at 180 mm spacing at top in both directions from edge to $L/5$ distance.

Reinforcement details: The slab in each direction is divided into one middle strip of 0.75 L and two edge strips of each 0.125 L on each edge. The reinforcement computed is that for the middle strip. The reinforcement in the edge strip is usually governed by that required for torsion on each face. The spacing of the reinforcement at the bottom face in the edge strips can be at 1.33 times that of the short span reinforcement. In small span slabs, edge span width is so small that it is inconvenient to change the spacing of the reinforcement, therefore, provide the same spacing of the reinforcement in the entire length of the slab. Figure 4.1 shows the reinforcement details. Also note that the assumed effective spans are slightly longer than the actual ones. The effective span is $L_{xo} + d = 3.58$ m against 3.6 m assumed. The bending moments computed are slightly more than the actual ones because of the assumed longer spans. The design is on the safer side.

EXAMPLE 4.3 *Design of a simply supported two-way slab with warping corners:* A simply supported slab with clear spans of 3.5 m and 4.5 m is subjected to a live load of 3 kN/m². Design the slab with the corners unconstrained.

Clear spans are: 3.5 m and 4.5 m

$$w_l = 3 \text{ kN/m}^2, \qquad f_{ck} = 20 \text{ MPa}, \qquad f_y = 415 \text{ MPa}$$

The allowable stresses and the design coefficients for a balanced section are:

$$\sigma_{cb} = 7 \text{ MPa}, \qquad \sigma = 230 \text{ MPa}$$

$$n = 0.289, \qquad j = 0.904, \qquad K = 0.13$$

Design for bending: Let the thickness of the slab be assumed as 150 mm for the purpose of computing weight and the effective spans.

$$\text{Self-weight } w_g = 2500(0.15) = 3750 \text{ N/m}^2$$

The total load on the slab is

$$w_t = w_g + w_l = 6750 \text{ N/m}^2$$

The effective spans be assumed as

$$L_x = L_{xo} + t = 3.65 \text{ m}$$
$$L_y = L_{yo} + t = 4.65 \text{ m}$$

The aspect ratio of the slab is

$$r = \frac{L_y}{L_x} = \frac{4.65}{3.65} = 1.27$$

Since the aspect ratio of the slab is less than 2, the slab must be designed as a two-way slab. The bending moment coefficients for two-way slab with warping corners can be obtained using Table 4.4 and are

$$\alpha_x = 0.084, \text{ and } \alpha_y = 0.059 \text{ for } r = 1.2$$
$$\alpha_x = 0.093, \text{ and } \alpha_y = 0.055 \text{ for } r = 1.3$$

From interpolation we have the coefficients as:

$$\alpha_x = 0.084 + \frac{(0.093 - 0.084)}{(1.3 - 1.2)}(1.27 - 1.2) = 0.0889$$

$$\alpha_y = 0.059 - \frac{(0.059 - 0.055)}{(1.3 - 1.2)}(1.27 - 1.2) = 0.0562$$

The maximum bending moments which occur at the mid-span on the middle strips are:

$$M_x = \alpha_x w_t L_x^2 = 0.0889(6750)(3.65)^2 = 7995 \text{ Nm/m}$$

$$M_y = \alpha_y w_t L_x^2 = 0.0562(6750)(3.65)^2 = 5054 \text{ Nm/m}$$

Equation the moment capacity to the larger of the above two moments, we have

$$K\sigma_{cb} bd^2 = 7995 \text{ Nm/m}$$

$$\text{or } d = \sqrt{\frac{0.007995}{(0.13)(7)}} = 0.094 \text{ m}$$

Use the overall depth of the slab as

$$h = 0.094 + 0.021 = 0.115 \text{ m}$$

The assumed thickness of the slab was 0.15 m as against 0.115 m actual. Therefore, the actual loads and bending moments are recomputed:

$$w_g = 25000(0.115) = 2875 \text{ N/m}^2$$
$$w_t = w_g + w_l = 5875 \text{ N/m}^2$$

Use $w_t = 5900 \ N/m^2$ and $L_x = 3.6$ m

$$M_x = 0.0889(5900) \ (3.6)^2 = 6798 \ Nm/m$$

$$M_y = 0.0562(5900) \ (3.6)^2 = 4297 \ Nm/m$$

$$d = \sqrt{\frac{0.006798}{0.13 \ (7)}} = 0.086 \ m$$

Use= 0.09 m and $h = 0.115$ m

The area of the tension reinforcement in the short span direction is

$$A_{sx} = \frac{M_x}{jd \ \sigma_{st}} = \frac{6798}{0.904 \ (0.09) \ (230)} = 363 \ mm^2/m$$

Provide 10 mm bars at 215 mm spacing, then the actual reinforcement provided is

$$A_{sx} \ (\text{provided}) = 78.5 \ \frac{(1000)}{215} = 365 \ mm^2/m$$

$$p = \frac{100 \ A_{sx}}{bd} = \frac{36500}{1000(90)} = 0.4\%$$

The effective depth available for the reinforcement in the y-direction is

$$d_y = d - \phi = 0.08 \ m$$

The area of the reinforcement in the y-direction is

$$A_{sy} = \frac{M_y}{jd_y \ \sigma_{st}} = \frac{4297}{0.904 \ (0.08) \ (230)} = 256 \ mm^2/m$$

The minimum reinforcement needed is

$$A_{sm} = 0.0012 \ bh = 0.0012(1000)(115) = 138 \ mm^2/m$$

Provide 8 mm bars at 195 mm spacing.
The actual area of steel provided is 256 mm²/m

Check for shear stress: The load distribution along the short span direction can be taken as

$$w_n = \frac{L_y^4 \ w_t}{L_x^4 + L_y^4} = 0.727 \ w_t = 4290 \ N/m$$

The maximum shear force occurs at a distance d from the support and it is

$$V = w_x \ (0.5 \ L_x - d) = 4290(1.8 - 0.09) = 7336 \ N/m$$

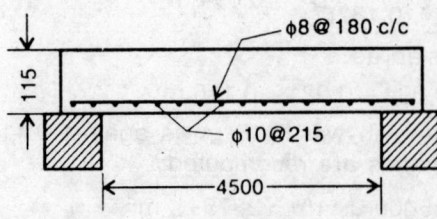

Fig. 4.2 Example 4.3.

The nominal shear stress is

$$\tau_v = \frac{V}{bd} = \frac{7336}{1000(90)} = 0.08 \text{ MPa}$$

The allowable shear stress is far higher than the nominal shear stress; therefore, the slab is safe in shear. Figure 4.2 shows the design details.

EXAMPLE 4.4 *Design of two-way slab with two adjacent sides continuous:* A floor slab of a room clear size 4 by 5 m is continuous along the two adjacent supports and discontinuous at the other two supports. The slab is subjected to a live load of 3 kN/m². Design the slab with the following design data:

$$\text{Clear spans: } L_{xo} = 4 \text{ m, } L_{yo} = 5 \text{ m}$$

$$w_l = 3000 \text{ N/m}^2, f_{ck} = 15 \text{ MPa, } f_y = 415 \text{ MPa}$$

The allowable stresses and design coefficients for a balanced section are:

$$\sigma_{cb} = 5 \text{ MPa, } \sigma_{st} = 230 \text{ MPa, } = K = 0.13$$

$$n = 0.289, j = 0.904$$

Design for bending: Let the thickness of the slab be taken as 0.15 m for the purpose of computation of the self-weight.

$$\text{Self-weight} = 25000(0.15) = 3750 \text{ N/m}^2$$

Total load on the slab is $\qquad w_t = w_g + w_l = 6750 \text{ N/m}^2$

The effective spans are taken as $\qquad L_x = 4 + 0.15 = 4.15 \text{ m}$

$$L_y = 5 + 0.15 = 5.15 \text{ m}$$

The aspect ratio of the slab is $\qquad r = \dfrac{L_y}{L_x} = \dfrac{5.15}{4.15} = 1.24$

The bending moment coefficient from Table 4.3 for two adjacent edges as discontinuous are: *Negative BM coefficients* (at the continuous supports)

$$\alpha_x = 0.060, \text{ and } \alpha_y = 0.047 \text{ for } r = 1.2$$

$$\text{and } \alpha_x = 0.065, \text{ and } \alpha_y = 0.047 \text{ for } r = 1.3$$

The interpolated BM coefficients for $r = 1.24$ are:

$$\alpha_{xn} = 0.06 + \frac{(0.065 - 0.060)(1.24 - 1.20)}{1.3 - 1.2} = 0.062$$

$$\alpha_{yn} = 0.047$$

The critical moments are at the continuous support points and at the mid-span. The negative bending moments occur at the continuity points.

The subscript n refers to negative BM and similarly the subscript p refers to the positive BM.

Positive BM coefficients are (at mid-span):

$$\alpha_x = 0.045 \text{ and } \alpha_y = 0.035 \text{ for } r = 1.2$$

$$\alpha_x = 0.049 \text{ and } \alpha_y = 0.035 \text{ for } r = 1.3$$

The interpolated BM coefficients for $r = 1.24$ are:

$$\alpha_{xp} = 0.045 + \frac{(0.049 - 0.045)(1.24 - 1.20)}{1.3 - 1.2} = 0.0466$$

$$\alpha_{yp} = 0.035$$

The thickness of the slab is governed by the absolute maximum bending moment in the entire slab, and it occurs at the support on the short span. It is

$$M_{xn} = \alpha_{xn} \, w \, L_x^2 = 0.062(6750)(4.15)^2 = 7208 \text{ Nm/m}$$

The effective depth of the slab is obtained by equating the moment capacity to the maximum BM and is:

$$d = \sqrt{\frac{0.007208}{1\,(0.65)}} = 0.105 \text{ m}$$

Use $h = 0.105 + 0.02 = 0.125$ m.

As the assumed value is more than the actual one, the self weight of the slab and the effective spans are modified:

$$w_g = 25000(0.125) = 3125 \text{ N/m}^2$$

$$w_t = w_g + w_l = 6125 \text{ N/m}^2$$

Use $w_t = 6200 \text{ N/m}^2$ and $L_x = 4.105$ m.

The negative and positive BMs are :

$$M_{xn} = 0.062\,(6200)(4.105)^2 = 6478 \text{ Nm/m}$$

$$M_{xp} = 0.0466\,(6200)(4.105)^2 = 4869 \text{ Nm/m}$$

$$M_{yn} = 0.047\,(6200)(4.105)^2 = 4910 \text{ Nm/m}$$

$$M_{yp} = 0.035\,(6200)(4.105)^2 = 3657 \text{ Nm/m}$$

The maximum effective depth needed is

$$d = \sqrt{\frac{0.006478}{0.65}} = 0.10 \text{ m}$$

Use $d = 0.10$ m and $h = 0.120$ m

The assumed depth was 0.125 m; therefore, the computed bending moments are on the safer side, yet very close. The tension reinforcements at different locations are:

$$A_{sxn} = \frac{M_{xn}}{jd\sigma_{st}} = \frac{6478}{0.904(0.1)(230)} = 311 \text{ mm}^2/\text{m}$$

$$A_{sxp} = \frac{4869}{0.904(0.1)(230)} = 234 \text{ mm}^2/\text{m}$$

The effective depth of steel in long span direction

$$d_y = 0.1 - 0.01 = 0.09 \text{ m}$$

The area of the reinforcements are:

$$A_{syn} = \frac{4910}{0.904(0.09)(230)} = 269 \text{ mm}^2/\text{m}$$

$$A_{syp} = \frac{3657}{0.904(0.09)(230)} = 196 \text{ mm}^2/\text{m}$$

Use 8 mm diameter bars in both the directions, then the spacing of the bars at the bottom face of the slab are: (s_x = spacing of the *x* directional reinforcement)

$$s_x = 210 \text{ mm} \quad \text{and} \quad s_y = 250 \text{ mm}$$

Crank alternate bars in both directions and then provide extra bars at top near the support as listed below:

φ 10 bars at 350 mm c/c in *x* direction
φ 10 bars at 260 mm c/c in *y* direction

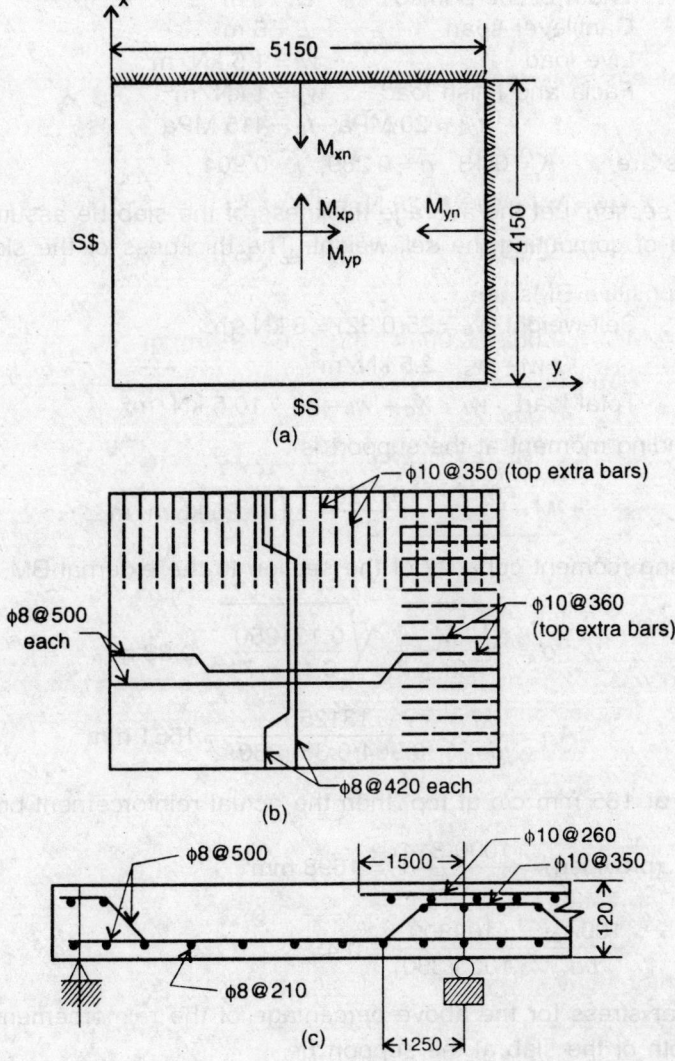

Fig 4.3 Example 4.4.

The minimum reinforcement is

$$A_{sm} = \frac{0.12(120)(1000)}{100} = 144 \text{ mm}^2$$

This value is less than that provided. There is no need to check for shear stress in this type of slabs as the shear stress will be only nominal and would not require revision or shear reinforcement. Figure 4.3 illustrates the reinforcement detail.

EXAMPLE 4.5 *Cantilever slab:* A portico slab has a free cantilever of 5 m with width 6 m. Design the slab with HYSD steel and M20 concrete.

Width of the portico $B = 6$ m
Cantilever span $L = 5$ m
Live load $w_l = 1.5 \text{ kN/m}^2$
Facia and finish load $w_s = 1 \text{ kN/m}^2$
$f_{ck} = 20$ MPa, $f_y = 415$ MPa

Design coefficients are: $K = 0.13$, $n = 0.289$, $j = 0.904$

Design of the section: Let the average thickness of the slab be assumed as $L/15 = 0.32$ m for the purpose of computing the self-weight. The thickness of the slab at free edge be used 0.15 m only.

$$\text{Self-weight } w_g = 25(0.32) = 8 \text{ kN/m}^2$$
$$w_l + w_s = 2.5 \text{ kN/m}^2$$
$$\text{Total load } \quad w_t = w_g + w_l + w_s = 10.5 \text{ kN/m}^2$$

The maximum bending moment at the support is

$$M = \frac{w_t L^2}{2} = \frac{10.5(25)}{2} = 131.25 \text{ kNm/m}$$

Equating the working moment capacity of the section to the external BM, we obtain

$$d = \sqrt{\frac{M}{K\sigma_{cb}}} = \sqrt{\frac{0.131250}{0.13 \times 7}} = 0.38 \text{ m}$$

$$A_{st} = \frac{M}{j d \sigma_{st}} = \frac{131250}{0.904(0.38)(230)} = 1661 \text{ mm}^2$$

Provide ϕ *20* bars at 185 mm c/c at top, then the actual reinforcement provided is

$$A_{st} \text{ (provided)} = \frac{1000(314)}{185} = 1698 \text{ mm}^2$$

$$\frac{100 A_{st}}{bd} = \frac{169800}{1000(380)} = 0.45$$

The allowable shear stress for the above percentage of the reinforcement is $\tau_c = 0.29$ MPa. Let the overall depth of the slab at the support be

$$h = d + 0.5 \, \phi + 0.025 = 0.415 \text{ m}$$

Let overall depth at the free end = 0.15 m
The average depth = (0.415 + 0.15) = 0.283 m

Since the assumed average thickness is more than the actual value, therefore the design is on the safer side. The distribution reinforcement be provided at 0.12%, that is

$$A_{sx} = \frac{0.12(283)(1000)}{100} = 340 \text{ mm}^2/\text{m}$$

Provide φ 12 *bars* at 275 mm c/c along the width of the slab.
Actual self-weight is

$$w_g = 25(0.283) = 7.1 \text{ kN/m}^2$$
$$w_t = 7.1 + 2.5 = 9.6 \text{ kN/m}^2$$

Check for shear: The critical section for shear force is at from the support and the shear force at this section is

$$V = w_t(L) = 9.6(5) = 48 \text{ kN/m}$$

The nominal shear stress is

$$\tau_v = \frac{V}{bd} = \frac{48000}{1000(380)} = 0.126 \text{ MPa}$$

The allowable shear stress is higher than the nominal shear stress, so the section is safe in shear.

Check for deflection: Moment of inertia of the cracked section including the effect of the reinforcement is:

$$I_{cr} = mA_{st}(d-nd)^2 + \frac{b(nd)^2}{3} = 0.0018 + 0.00044 = 0.00224 \text{ m}^4$$

The gross moment of inertia of the section is

$$I = \frac{bh^3}{12} = \frac{1(0.415)^2}{12} = 0.00596 \text{ m}^4$$

The effective sectional moment of inertia can be taken as

$$I_e = 0.5(I + I_{cr}) = 0.0032 \text{ m}^4$$

The Young's modulus of the concrete is

$$E_c = 5700\sqrt{20} = 25491 \text{ MPa}$$

The deflection caused by the permanent load alone is

$$v_d = \frac{w_d L^4}{8E_c I_c} = \frac{0.0081(625)}{8(25491)(0.0032)} = 0.008 \text{ m}$$

Let the creep coefficient be $c_c = 2.1$, then the total deflection is

$$v_t = c_c v_d + v_1 = 0.019 \text{ m}$$

Allowable deflection including the creep effect is:

$$v_a = \frac{L}{250} = \frac{5}{250} = 0.02 \text{ m}$$

The deflection is within the limit.

Development length: Development length is important in all structures, and more so in the cantilever beams. The development length is given by

$$L_d = \frac{\phi \sigma_s}{4\,\tau_{bd}} = \frac{20(230)}{4(1.12)} = 958 \text{ mm}$$

All the tension bars should have a development length of 958 mm, therefore, care must be taken to provide this embedded length at the support. Curtail alternate bars at 3 m from support.

This design does not include the design of the fixed support either for bending or for stability. In case of cantilever slabs or beams, careful attention should be paid to the design of the support and development length.

4.6 Flat Slabs

Beam and slab construction has the advantage of providing intermediate beam supports to the slabs, thus reducing the effective span of the slabs. But the beams require larger depths, leading to more heights of buildings. In some situations, specially in warehousing, etc. it is desirable to have larger clear ceiling heights. Flat slabs are the slabs which rest directly on the columns without beams, and such a construction provides a larger clear ceiling height for the same total height of the building. In addition, the formwork requirement is also reduced when compared with the beam and slab construction. Flat slabs are invariably two-way slabs and rest on several columns. Sometimes the top of the columns are widened so as to provide wider base to support the slab. Such widened portions are called *column heads*. There is a limit up to which one can treat the widened portion as a part of the column. The width must be limited to the portion within 90° of the segment as shown in Figs. 4.4b and c. Any projection beyond the column head should be treated as thickening of the slab at the column head. Such thickened portions of the slab, are called *drops*. The drops are sometimes known as *capitals* of the columns. In each panel, the slab is divided into columns and middle strips in each direction. The width of the *column strip* is equal to half of the column spacing and it is placed half on either side of the column line. In case of unequal spans, it can be taken equal to half of the average. In addition, it should also be restricted to 0.5 times the column spacing in any direction. The *middle* strip is the one bounded by the column strips and its width is equal to the spacing of the columns minus the width of the column strip. The width is usually equal to or greater than half of the spacing of the columns.

Let b_{ci} = width of column strip in the i th column row

b_{mi} = width of the middle strip

L_{xi} = spacing of the columns in the x-direction in the i th panel

L_{yi} = spacing of the columns in the y-direction in the i th panel

Then widths of the column and middle strips spacings in x direction are given by

$$b_{ci} = 0.25\,(L_{yi-1} + L_{yi}) < 0.25\,(L_{xi} + L_{xi+1}) \tag{4.9}$$

$$b_{mi} = L_{yi} - 0.25\,(L_{yi} + L_{yi+1}) \tag{4.10}$$

Similarly, the widths of the strip in the y-direction. In most situations, the spacing of the columns in one direction is same in all the panels; therefore, the calculation of the widths of the strips is normally a very simple exercise.

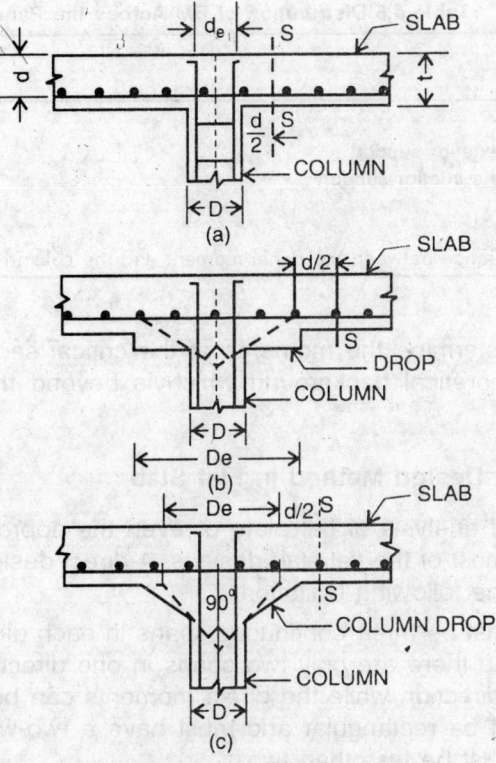

SS=Critical Section for Shear

Fig. 4.4 Column Heads.

The slabs can be analysed as equivalent frames having idealized continuous wall supports along the column lines. The stiffness of the column is divided by the panel width and considered as the stiffness of the vertical element per unit width of the frame. The analysis has to be done in both directions independently as two sets of independent frames. Such an idealization introduces undue bending moments into the middle strip. Therefore, the moment and shear force computed by this method must be proportioned with higher weightage to the column strip when compared with that of the middle strip. The analysis is to be carried by loading only three-fourths of the total live load in each panel and full dead load. However, in case of mat foundation slabs, full load coming from the column should be taken as the load and also no reduction should be given to liquid loads. The frames should be analysed for two load conditions, namely, all panels loaded and alternative panels loaded. The critical section for design of moment is the section at the face of the column or the drop. The critical section for shear force design is peripheral line around the column at a distance 0.5 d from the face of the column or the face of the drop. Usually the thickness of the drop is taken large enough to eliminate the shear failure of the section around the periphery of the column. The section and the reinforcement must be designed to withstand the weighted proportioned moment as given in Table 4.6.

A more accurate method of analysis would be to treat the slab as an elastic plate supported

Table 4.6 Distribution of BM Across the Panel

Strip and its boundary conditions	Per cent of the total BM
1. *Column strip*	
a) Negative BM at exterior support	100
b) Negative BM at the interior support	75
c) Positive BM	60
2. *Middle strip*	
The difference between the panel moment and the column strip moment.	

on columns and then determine the moments at the critical sections. Such analysis would require considerable theoretical background which is beyond the scope of undergraduate studies.

4.7 Approximate Direct Design Method in Flat Slab

The rigorous methods of analysis of flat plate or even the approximate frame analysis may not be really needed in most of the flat slab designs. A direct design moment can be obtained for flat slabs satisfying the following limitations:

1. At least there must be three continuous spans in each direction. This is true in many flat slabs. In case there are only two spans in one direction, one can apply a frame analysis in that direction while the direct moments can be used lengthwise.
2. The panels must be rectangular and must have a two-way action. The aspect ratio in each panel must be less than two.
3. The end span should not be larger than the interior spans.
4. The ratio of the successive span lengths should be within 0.75 to 1.33. In most of the buildings fall in this zone except those which have middle corridor.
5. A cantilever projection of about one third of the exterior span can be permitted with appropriate modification in the bending moments.
6. The design live load should not be more than three times the dead load. This is a major constraint and when ignored it may effect the magnitude of the positive bending moment.

The sum of the magnitudes of the positive and average of the negative bending moments is equal to

$$M_0 = \frac{WL_0}{8} \tag{4.11}$$

Consider a free body diagram of a half panel of a flat slab. Let M_n and M_p be the negative and positive bending moments at the face of the column and at the mid span respectively. If load acting on the clear panel be W, then the equilibrium of moments of the free body diagram gives

$$|M_n| + |M_p| = \frac{W}{2}\left(\frac{L_0}{2} - \frac{L_0}{4}\right) = \frac{WL_0}{8}$$

where M_0 = sum of the magnitudes of the positive and average of the negative bending moments, that is

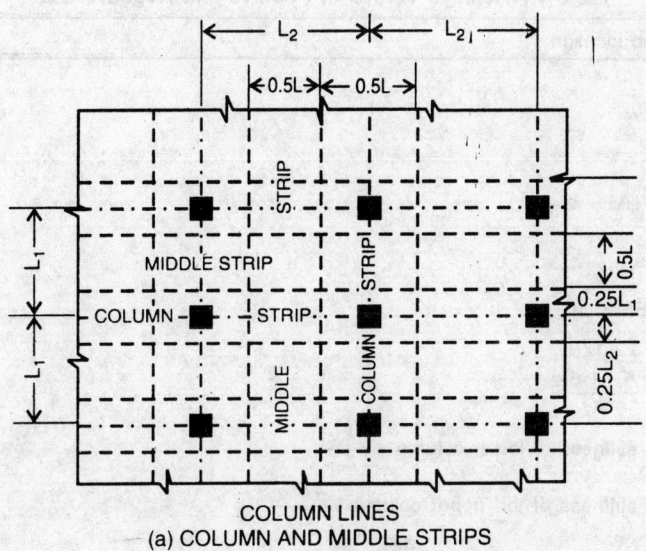

(a) COLUMN AND MIDDLE STRIPS

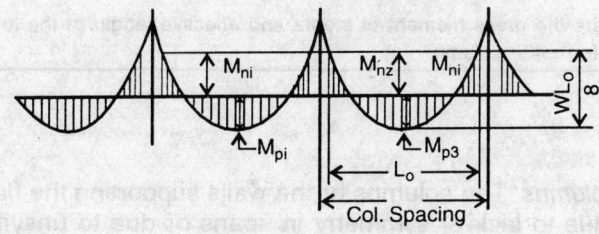

(b) Typical Bending Moment

Fig. 4.5. Flat Slabs

$$M_0 = M_3 + 0.5 (M_1 + M_2) \tag{4.12}$$

M_3 = positive bending movement

M_1 and M_2 = magnitudes of negative bending moments at the face of the columns.

L_0 = clear span extending from face to face of the columns or capitals. It should be equal to the larger of the two unequal adjacent spans and > 0.65 spacing of the columns.

W = Design load on the clear span.

The relative magnitudes of negative and positive bending moments can be calculated using the ratios given in Table 4.7.

The magnitude of the bending moment can be obtained as

$$M_i = c_i M_0 \tag{4.13}$$

where the moment coefficients, c_i are listed in Table 4.7.

The distribution of any one of the moments at a section between the column and middle strips should be as per the coefficients given in Table 4.6.

Table 4.7 Relative Values of Positive and Negative BM

Bending moment and location	c_i
1. *Interior spans*	
a) Negative BM = M_{ni}	0.65
b) Positive BM = M_{pi}	0.35
2. *End span*	
a) Exterior negative BM = M_{n1}	$\dfrac{0.65}{\alpha}$
b) Positive BM = M_{p3}	$0.63 - \dfrac{0.28}{\alpha}$
c) Interior negative BM = M_{n2}	$0.75 - \dfrac{0.1}{\alpha}$

where $\alpha = 1 + 1/\alpha_c$

$$\alpha_c = \frac{K_{c1} + K_{c2}}{K_2}$$

K_{c1} = stiffness of lower column = $\dfrac{4EI_{c1}}{L_{c1}}$.

K_{c2} = stiffness of the upper column = $\dfrac{4EI_{c2}}{L_{c2}}$

K_s = stiffness of the slab = $\dfrac{4EI_s}{L}$

I_{c1}, L_{c1} are the gross moment of inertia and effective length of the lower column,

Similarly I_{c2}, L_{c2}, for upper column.

Design moments on columns: The columns or the walls supporting the flat slabs are subjected to bending moments due to lack of symmetry in spans or due to unsymmetrical load on the spans or due to rigid joints between the column and slab. The bending moment on the column which is either above or below the slab is approximated as.

$$M_c = 0.08 \frac{((w_d + 0.5w_1) \, L_0^2 - w_d L_{01}^2) \, L_2}{\alpha} \tag{4.14}$$

where L_{01} = clear shorter span and α is the same as the value given in Table 4.7.

To avoid possible damages to columns due to lateral and unsymmetrical loads on the slabs, the columns are to be designed with at least the minimum relative stiffnesses. The minimum value of α_c is given in Table 4.8.

Design for shear: The slabs are likely to fail by diagonal tension around the columns rather than along a section parallel to the column line. The critical section for shear is a peripheral plane at a distance $0.5\,d$ from the face of the column or the column head or drop. In case there is an opening near this zone, appropriate deduction in the peripheral length should be made proportional to the distance of the critical section to that of the opening from the centre of the column.

Let B = size of the opening

    $d_1 = 0.5 \, (a + d)$ distance of the opening from centre of the column

        = distance of the critical section from the centre of the column

   a = size of the square column

Table 4.8 Minimum Permissible Relative Stiffness of Column in Flat Slabs

$\dfrac{\text{Live load}}{\text{Dead load}}$	$\dfrac{L_s}{L_1}$	α_c (minimum)
0.5	0.5 to 2	0
1.0	0.5 to 1	0.7
	1.25	0.8
	2.0	1.2
2.0	0.5	1.3
	1.0	1.6
	1.25	1.9
	2.0	4.9
3.0	0.5	1.8
	1.0	2.3
	1.25	2.8
	2.0	13.0

Then the ineffective width to be deducted from the length of the critical section is given by

$$b_c = \frac{B(0.5)(a+d)}{d_1} \qquad (4.15)$$

the length of the critical section for shear is

$$b_0 = 4(a+d) - b_d \qquad (4.16)$$

Typical notations for subscripts refer to:

c = column strip
m = middle strip
n = negative BM
p = positive BM
1 = exterior critical section usually at the face of the exterior columns or drops
2 = interior column face of the exterior panel
3 = middle point in the exterior panel
i = interior panels, further
M_{n1} = negative BM at section 1 on the panel
M_{nc1} = negative BM at section 1 on the column strip
M_{pi} = positive BM at the interior panel
M_{pci} = positive BM at the interior panel on the column strip

Similarly, the other notations and Figure 4.5 illustrates the notations.

4.8 Design Examples of Flat Slabs

EXAMPLE 4.6 *Flat roof slab without column heads:* A roof slab is supported on columns spaced 5 metres apart in both the directions. The size of the square columns is 440 mm and the live load on the roof is 1500 N/m². The load of the waterproof treatment course on the slab is 2000 N/m². Design a flat slab without drops or column heads. Height of the column above the mat foundation is 6 m.

Design data:

>Spacing of the columns $L_1 = L_2 = L = 5$ m
>Size of the column $a = 0.44$ m
>Live load $w_l = 1.5$ kN/m^2
>Superimposed load $w_s = 2$ kN/m^2
>$f_{ck} = 15$ MPa, $f_y = 415$ MPa

The working stress design coefficients are:

>$n = 0.289$, $j = 0.904$, $K = 0.13$, $p = 0.145$

Design of the section for moment: The design is done by using direct design method. For the purpose of estimating the self-weight of the slab, let the thickness of the slab be assumed in the range of $L/20$ for the slabs without column heads.

>Let the thickness of the slab $= t = 0.24$ m
>Self-weight $= 0.24(25) = 6.0$ kN/m^2
>Total dead load $= w_d = w_g + w_s = 8$ kN/m^2

Clear spacing between the columns is

$$L_0 = L - a = 5 - 0.44 = 4.56 \text{ m}$$

The total design load in a panel is

$$W = w_t L L_0 = (w_d + w_l) L L_0 = 9.5(5)(4.56) = 216.6 \text{ kN}$$

Sum of the magnitudes of bending moments in a panel

$$M_0 = \frac{WL_0}{8} = \frac{216.6(4.56)}{8} = 123.462 \text{ kNm} = 0.123462 \text{ MNm}$$

The negative bending moment acting on the column strip in the exterior columns is likely to govern the depth of the slab. The shear force is also likely to be the governing force for the depth. Therefore, the depth of the section is calculated first based on the bending moment and then shear force.

>Width of the column strip $b = 2.5$ m
>Magnitude of the negative BM at the face of the columns in the interior panels is

$$M_{ni} = 0.65 M_0 = 0.082503 \text{ MNm}$$

The column is resting on a mat foundation; therefore, the base of it can be treated as fixed. The end walls of the hall restrain the free lateral movement of the roof slab, therefore, the top of the column can be assumed as fixed in position and rotation. Hence the effective height of the column is

$$L_c = 0.65 \ H = 0.65 \ (6) = 3.9 \text{ m}$$

The moment of inertia of the column is

$$I_c = \frac{a^4}{12} = \frac{(0.44)^4}{12}$$

The relative stiffness of the column is

$$K_c = \left(\frac{I_c}{L_c} \right) = \frac{0.44^4}{12(3.9)} = 8(10^{-4})$$

The relative stiffness of the slab panel (with $t = 0.22$ m) is

$$K_s = \frac{I_s}{L_1} = \frac{5(0.22)^3}{12\,(5)} = 8.87\,(10^{-4})$$

$$\alpha_c = \frac{K_c}{K_s} = \frac{8}{8.87} = 0.9$$

The ratio of the live load to dead load is

$$\frac{w_l}{w_d} = \frac{1.5}{8} = 0.1875$$

The minimum value of α_c for the above ratio and for an aspect ratio of 1 can be obtained from Table 4.8 and is zero; therefore, the relative stiffness of the column is adequate.

$$\alpha = 1 + 1/\alpha_c = 2.111$$

The exterior panel negative, and positive bending moments and the interior panel negative bending moment coefficients are taken from Table 4.7 and are:

$$c_1 = \frac{0.65}{\alpha} = 0.308$$

$$c_3 = 0.63 - \frac{0.28}{\alpha} = 0.497$$

$$c_2 = 0.75 - \frac{0.1}{\alpha} = 0.702$$

The corresponding bending moments are:

$$M_{n1} = c_1 M_0 = 0.0380 \text{ MNm}$$

$$M_{p3} = c_3 M_0 = 0.0613 \text{ MNm}$$

$$M_{n2} = c_2 M_0 = 0.08667 \text{ MNm}$$

in which M_{n1} and M_{p3} are the negative and positive bending moments in the exterior panel. The bending moments on the *column strip* of 2.5 m width at different locations are computed using Table 4.6 and these moments are:
Negative BM at the face of the exterior column is

$$M_{nc1} = M_{n1} = 0.038 \text{ MNm}$$

The positive BM in the exterior panel is

$$M_{pc3} = 0.6\ M_{ps} = 0.03673 \text{ MNm}$$

The negative BM in the interior column face of the exterior panel is

$$M_{nc2} = 0.75\ M_{n2} = 0.06500 \text{ MNm}$$

The negative BM in the interior panels is

$$M_{nci} = 0.75\ M_{ni} = 0.06188 \text{ MNm}$$

The values listed are only the magnitudes, and the depth of the slab is controlled by the largest of the bending moments. The minimum effective depth of the slab is given by

$$d = \sqrt{\frac{M_{nc2}}{Kb\sigma_{cb}}} = \sqrt{\frac{0.065}{0.13(2.5)(5)}} = 0.2 \text{ m}$$

Use d = 0.2 m *and D* = 0.23 m

The overall depth of the section will be 0.23 m and it checks closely with that assumed for the purpose of self-weight computation. Therefore, no revision in load need to be done.

Design of section for shear. In case of beams, one can provide shear reinforcement if the nominal shear stress exceeds the allowable value whereas in case of slabs the shear reinforcement is to be avoided and, therefore, one has to provide sufficient depth of the slab to limit the nominal stress within the allowable value. It is, therefore, desirable that the adequacy of the depth of the section should be checked at the earliest. The percentage of tension reinforcement in a balanced section is

$$p = 100 \, p_0 \, \frac{\sigma_{cb}}{\sigma_{st}} = 14.5 \, \frac{(5)}{415} = 0.17$$

The minimum reinforcement should also be limited by 0.12% or

$$p_{min} = \frac{0.85(100)}{f_y} = 0.205$$

The allowable shear stress for diagonal tension failure is

$$\tau_c = 0.16 \, \sqrt{f_{ck}} = 0.62 \text{ MPa}$$

The critical shear plane is the peripheral plane at a distance 0.5 *d* from the face of the column. The length of the critical section is

$$b_0 = 4(a + d) = 4(0.44 + 0.20) = 2.56 \text{ m}$$

The shear force on the plane is

$$V = w_t \, (L_1 L_2 - (a + d)^2) = 9.5(25 - 0.64^2) = 233.6 \text{ kN}$$

The nominal shear stress is

$$\tau_y = \frac{V}{b_0 d} = \frac{0.2336}{2.50(0.20)} = 0.46 \text{ MPa}$$

The nominal shear stress is less than the allowable value; therefore, there is no need for transverse reinforcement or thickening of the slab.

Design of reinforcement
Column strip: The negative BM at exterior support is $M = M_{nc1}$.
The area of the reinforcement at top near the column line is

$$A_{st1} = \frac{M_{nc1}}{jd\sigma_{st}} = \frac{38000}{0.904(0.20)(230)} = 914 \text{ mm}^2$$

Provide 9 numbers of 12 mm bars in the 2.5 m width of the column strip at top near the column line.
The positive BM at mid span of exterior panel is $\cong M = M_{pc3}$.
The area of the tension steel at the bottom of the mid span is

$$A_{st2} = \frac{M_{pc3}}{jd\sigma_{st}} = \frac{36730}{0.904(0.2)(230)} = 883 \text{ mm}^2$$

Provide 8 numbers of ϕ 12 bars in the column strip at the bottom face of the mid-span.
The negative bending at the interior column of end panel is $M_{nc2} = 0.065$ MNm.

The area of the tension reinforcement is

$$A_{st2} = \frac{65000}{0.904(0.2)(230)} = 1563 \text{ mm}^2$$

Provide 14 numbers of 12 mm bars at top over the column line.
The positive BM in the column strip of the *interior panel* is

$$M_{pci} = (0.35)(0.75)\, M_0 = 0.0324 \text{ MNm}$$

The area of the tension reinforcement is

$$A_{sti} = \frac{32400}{0.904(0.2)(230)} = 779 \text{ mm}^2$$

Provide 7 numbers of φ *12 bars at the bottom.*

Design of middle strip: The bending moment in the middle strip is obtained by subtracting the BM on the column strip from the total BM on the panel. One can calculate the number of bars required by simple proportion of the bending moments subject to the minimum reinforcement.

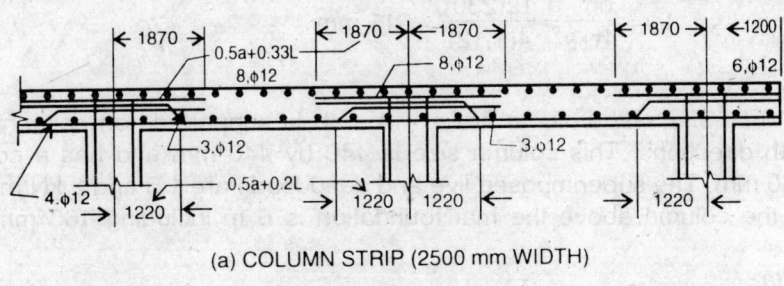

(a) COLUMN STRIP (2500 mm WIDTH)

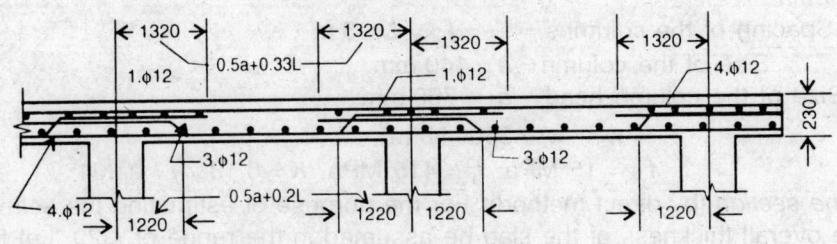

(b) MIDDLE STIP (2500 mm WIDTH)

Fig. 4.6 Reinforcement Details in Flat Slab, Example 4.6.

Table 4.9 Reinforcement Details Example 4.6 (interior panels)

No. of bars and length			Location	Strip
(1)	5,5000	straight	At the mid span at bottom face	Column
	3,7070	cranked	From bottom	
(2)	5,3740	extra	Over the exterior column line and at top face	
(3)	8,3740	extra	Over the interior column line and at top face	
(4)	4,5000	straight	At the mid span and at bottom face	Middle
	3,6520	cranked	From bottom	
(5)	4,2640	extra	Over the exterior column line and at top face	
(6)	4,2640	extra	Over the interior column line and at top face	

1. The negative BM at the exterior support is zero; however, a nominal reinforcement is provided at top. The nominal reinforcement is

$$= \frac{0.12bD}{100} = \frac{0.12(2500)(230)}{100} = 690 \text{ mm}^2$$

Provide 7 numbers of ϕ 12 bars at top in the middle strip at the edge.

2. The positive BM in the exterior panel is

$$= (0.35)(0.25)(M_0) = 0.0108 \text{ MNm}$$

The area of reinforcement needed for this moment is less than the minimum required so, provide 7 numbers of ϕ 12 at the bottom.

It can be seen that the bending moments in the middle strip are only nominal and the minimum tension reinforcement governs the design. The minimum reinforcement is 7 numbers of 12 mm bars in 2.5 m width. The reinforcement details are given in Fig. 4.6. The curtailment and cranking of the bars is recommended as per the normal practice.

Development length is given by

$$L_d = \frac{\phi \sigma_s}{4\tau_{bd}} = \frac{12(230)}{4(1.12)} = 615 \text{ mm}$$

EXAMPLE 4.7 *Flat slab with column heads* : A slab is supported on columns spaced at 5 m apart in both directions. This column size is 440 by 440 mm and has a column head of size 760 by 760 mm. The superimposed live and dead leads are 1.5 and 2 kN/m² respectively. The height of the column above the mat foundation is 6 m including 160 mm high column head.

Design data

Spacing of the columns $= L_1 = L_2 = L = 5$ m

Size of the column $= a = 440$ mm

Size of the column head $= a_1 = 760$ mm

$$w_s + w_l = 3500 \text{ N/m}^2$$

$$f_{ck} = 15 \text{ MPa}, \ f_y = 415 \text{ MPa}, \ K = 0.13, \ j = 0.904$$

Design of the section (by direct method): For the purpose of estimating the self-weight of the slab, let the overall thickness of the slab be assumed in the range of $L/20$. Let the thickness of the slab $t = 0.24$ m

Self-weight $w_g = 0.24(25) = 6 \text{ kN/m}^2$

The total dead load $\quad w_d = 6 + 2 = 8 \text{ kN/m}^2$

The total design load $\quad w_t = 8 + 1.5 = 9.5 \text{ kN/m}^2$

Clear spacing between the column heads is

$$L_0 = L - a_1 = 5 - 0.76 = 4.24 \text{ m}$$

The total design load in a panel is

$$W = w_t L L_0 = 9.5(5)(4.24) = 201.4 \text{ kN}$$

The sum of the magnitudes of the positive and negative bending moments in a panel is

$$M_0 = \frac{W L_0}{8} = \frac{201.4(4.24)}{8} = 106.742 \text{ kNm}$$

The depth of the section is normally governed by the negative bending moment on the column strip, therefore, the depth of the slab is computed first based on the critical BMs in column strip.

Width of the column strip = $b = 2.5$ m

The magnitude of the negative BM at the face of the column head in the interior panel is

$$M_{nl} = 0.65\ M_0 = 0.65(106.742) = 69.382\ \text{kNm}$$

The column can be treated as fixed as base and constrained against rotation at the base of the column head, then the effective height of the column is

$$L_c = 0.65\ (\text{height of column} - \text{height of the column head})$$

$$= 0.65(6 - 0.16) = 3.796\ \text{m}$$

The moment of inertia of the column is

$$I_c = \frac{a^4}{12} = \frac{0.44^4}{12}$$

The relative stiffness of the column is

$$K_c = \frac{I_c}{L_c} = \frac{0.44^4}{12(3.796)} = 8.2(10^{-4})$$

The approximate relative stiffness of the slab is computed by assuming the thickness of the slab as 0.22 m (on the safer side instead of 0.24 m which was used for self-weight purpose).

$$K_s = \frac{I_s}{L} = \frac{5(0.22)^4}{12(5)} = 8.87(10^{-4})$$

$$\alpha_c = \frac{K_c}{K_s} = \frac{8.2}{8.87} = 0.9$$

The ratio of the live load to the dead load is

$$= \frac{w_l}{w_d} = \frac{1.5}{8} = 0.1875$$

The relative stiffness of the column is adequate as the minimum acceptable value of α_c for the above ratio and an aspect ratio of one is zero as seen from Table 4.8.

$$\alpha = 1 + \frac{1}{\alpha_c} = 2.111$$

The exterior negative and positive BMs and the interior bending moment coefficients are taken from Table 4.7, and are:

$$c_1 = \frac{0.65}{\alpha} = 0.308$$

$$c_3 = 0.63 - \frac{0.28}{\alpha} = 0.492$$

$$c_2 = 0.75 - \frac{0.1}{\alpha} = 0.702$$

The corresponding bending moments are:

$$M_{ni} = c_1 M_0 = 0.0329 \text{ MNm}$$
$$M_{p3} = c_3 M_0 = 0.0525 \text{ MNm}$$
$$M_{n2} = c_2 M_0 = 0.0749 \text{ MNm}$$

The bending moments on the column strip at different locations are computed using Table 4.6. These moments are:

Negative BM at the exterior face of the exterior panel

$$M_{nc1} = M_{n1} = 0.0329 \text{ MNm}$$
$$M_{pc3} = (0.6) M_{p3} = 0.0315 \text{ MNm}$$

The negative BM in the interior column face of the interior panel:

$$M_{nc2} = 0.75 \, M_{n2} = 0.0562 \text{ MNm}$$

The negative BM in the interior panels

$$M_{nc1} = 0.75 \, M_{ni} = 0.0520 \text{ MNm}$$

The values listed are only the magnitudes of the moments and the thickness of the slab is governed by the largest magnitude of the BMs. The effective depth of the slab is given by

$$d = \sqrt{\frac{M_{nc2}}{bK\sigma_{cb}}} = \sqrt{\frac{0.0562}{0.65(2.5)}} = 0.186 \text{ m}$$

Using $d = 0.19$ m, then the overall thickness of the slab is

$$t = 0.19 + 0.03 = 0.22 \text{ m}$$

The thickness of the slab was assumed as 0.24 m for the purpose of self-weight: therefore, the moments computed are slightly on the safer side.

The actual weight $w_g = 0.22(25) = 5.5 \text{ kN/m}^2$

$$w_t = 5.5 + 2.0 + 1.5 = 9 \text{ kN/m}^2$$

Design for shear: The allowable shear stress for diagonal tension is

$$\tau_c = 0.16 \sqrt{f_{ck}} = 0.62 \text{ MPa}$$

The critical shear plane is the peripheral plane at a distance 0.5 d from the face of the column head. The total length of the critical shear plane section is

$$b_0 = 4(a_1 + b) = 4(0.76 + 0.19) = 3.8 \text{ m}$$

The total shear force on this plane is

$$V = w_t (L_1 L_2 - (a_1 + d)^2) = 9(25 - 0.95^2) = 216.88 \text{ kN}$$

The nominal shear stress is

$$\tau_v = \frac{V}{b_0 d} = \frac{216880}{3800(190)} = 0.3 \text{ MPa}$$

The nominal shear stress is within the allowable value.

Design of reinforcement

Column strip: The reinforcement is calculated first in the column strip and then in the middle strip. The negative BM at the exterior support is M_{nc1} and the area of the reinforcement which is to be laid at the top near the column is:

$$A_{st1} = \frac{M_{nc1}}{jd\sigma_{st}} = \frac{32900}{0.904(0.19)(230)} = 833 \text{ mm}^2$$

Provide 8 numbers of 12 mm bars in the 2.5 m width of the column strip near the column line. A_{st} (provided) = 904 mm^2

$$\frac{100\, A_{st}}{bd} = \frac{90\,400}{2500(190)} = 0.19$$

The positive BM at mid span is $M_{pc3} = 0.0315$ MNm and the area of the reinforcement needed is

$$A_{st3} = \frac{M_{pc3}}{jd\,\sigma_{st}} = \frac{31500}{0.904(0.19)(23)} = 798 \text{ mm}^2$$

Provide 7 numbers of 12 mm bars. Then the actual steel provided is $A_{st} = 791$ mm^2 and the percentage reinforcement is 0.17.

The negative bending moment at the interior of the end panel is $M_{nc2} = 0.0562$ MNm, and the area of the reinforcement required is

$$A_{st2} = \frac{56200}{0.904(0.19)(230)} = 1423 \text{ mm}^2$$

Provide 13 numbers of 12 mm bars at top. A_{st} (provided) = 1469 mm^2.

Design of middle strip: The bending moments in the middle strip are obtained by subtracting the bending moments of the column strip from the total bending moments in the panel at the corresponding section. The minimum reinforcement at any of the section can be taken as 0.12% of the area of the concrete and it is

$$A_{smin} = \frac{0.12\, bD}{100} - \frac{0.12(2500)(230)}{100} = 690 \text{ mm}^2$$

Therefore, the minimum number of 12 mm bars at any given section is 7. The bending moment on the middle strip is two-thirds of that of the column strip in the positive BM zone and it is one-third in the negative BM zone. It can be observed that the amount of the reinforcement needed is governed by the minimum requirement rather than by the bending moments. Figure 4.7 illustrates the reinforcement details. The development length of the bars is

$$L_d = \frac{\phi\sigma_s}{4\tau_{bd}} = \frac{12(230)}{4(1.12)} = 616 \text{ mm}$$

Comment about the desirability of column heads or drops: The provision of column heads or drops reduces the effective span and consequently the bending moment. Any reduction in the cost of concrete due to smaller thickness of the slab may partly be compensated by the addition of the concrete at the column heads and extra formwork. The length of the cranked bars or top bars is increased in the case of slabs with column heads, so some of the gain in the decrease of the area of the reinforcement is reduced. The drops or column heads become almost unavoidable if the nominal shear stress exceeds the allowable value. In general such a situation arises when: (1) the spacing of the columns is large, say more than 6 m; (2) the size of the column is very small when compared with the panel size, say less than $L/15$ and (3) the total load intensity on the slab is high, say more than 40 kN/m^2. Otherwise,

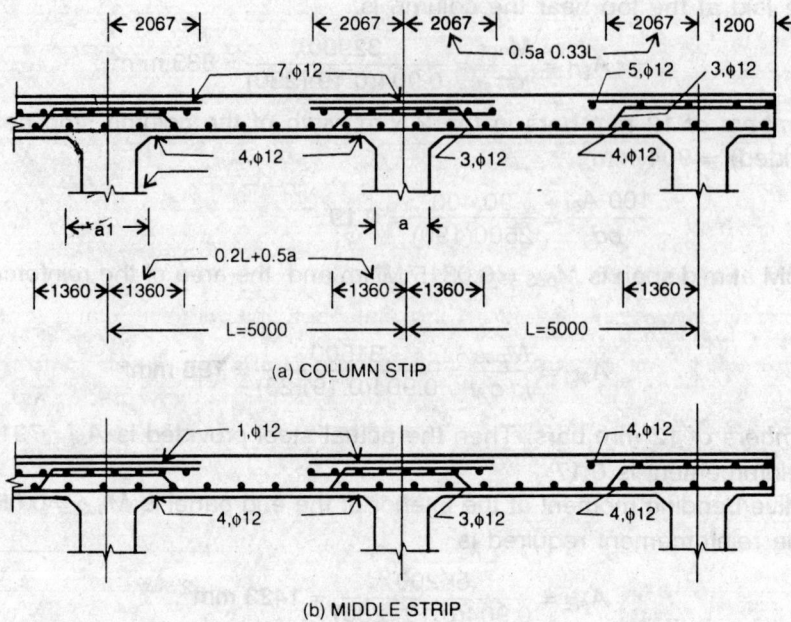

Fig. 4.7 Reinforcement Details of Flat Slab of Example 4.7.

it is desirable to design the flat slabs without column heads because of not only economic aspect but also esthetic and functional aspects.

4.9 Grid Floors

An assembly of intersecting beams placed close to each other and interconnected to a slab is called *grid* or *waffle* floor. When two sets of beams intersect at right angle to each other, such a gird is called *orthogrid*. In case these beams are placed parallel to the diagonals, it is called a *diagrid*. Grid slabs or floors are found to be economical and practical for clear spans of 8 to 25 m. Large unobstructed covered space can be obtained with flat floor surface without taking away much height of the floor by the beams. The spacing of the beams in each direction can be different depending on the architectural considerations. A minimum grid spacing of one metre and a maximum spacing of two metres appears to be reasonable. In case the beams are unequally spaced in both the directions, one of the spacings can be less than even one metre. A span can be divided into 5 to 12 grids so as to present a pleasant appearance. About 6 to 8 grids in each direction will result into a minimum bending moment per unit width. Therefore, it is advisable to provide about 6 to 8 grids in each direction.

The grid floor acts as an orthotropic slab and floor slab be designed as an orthotropic plate. There are other theories which can also be applied; however, for convenience only the orthotropic plate theory is presented in this book. The bending moments per unit width of an orthotropic plate can be expressed as

$$M_x = (c_x + c_y \nu e) \, w \, L_x^2 \qquad (4.17)$$

$$M_y = \frac{(c_y + \nu c_x)\, w L_y^2}{r\sqrt{e}} \tag{4.18}$$

where c_x and c_y = coefficients of the bending moments
ν = Poissons ratio

$$e = \sqrt{\frac{D_x}{D_y}} \tag{4.19}$$

D_x and D_y = area moments of inertia of the plate about x and y planes respectively
r = aspect ratio = L_y/L_x
M_x and M_y = Bending moments on the plate per unit width in x and y planes.

The bending moment coefficients have been obtained by the orthotropic plate theory and are available in the literature. One of the earliest investigators who presented such a plate theory was S. Timoshinko. His work deserves a special mention in the area of elastic theory and applications. The moment coefficients are listed in Table 4.10 for ready reference. The design

Table 4.10 Bending Moment Coefficients in Orthotropic Slab

$r\sqrt{e}$ =	1	1.2	1.4	1.6	1.8	2.0	2.5	4
c_y	0.037	0.034	0.030	0.026	0.021	0.017	0.01	0.002
c_x	0.037	0.052	0.067	0.079	0.088	0.096	0.11	0.123

procedure consists of selection of grid spacing and a beam size, then the relative inertia in both the directions is calculated. Next, compute the bending moments and provide appropriate reinforcement. The procedure is illustrated through an example.

EXAMPLE 4.8 *Design of grid floor:* A hall of dimensions 12.2 m by 18.3 m is to be covered by a flat roof. Design a grid floor.

    Design data

$L = L_x = 12.2$ m, $L_y = 18.3$ m,
$f_{ck} = 15$ MPa, $f_y = 415$ MPa,
$\sigma_{cb} = 5$ MPa, $\sigma_{st} = 230$ MPa.

Roof finish load = $w_s = 2$ kN/m^2

Live load   = $w_l = 1.5$ kN/m^2

Selection of grid and grid properties: Let the floor be divided into 8 grids in the short span direction and 12 in the long span direction. This makes the spacing of the grids in both the directions same.

Aspect ratio $= r = \dfrac{18.3}{12.2} = 1.5$

Grid spacing $= a = \dfrac{12.2}{8} = \dfrac{18.3}{12} = 1.525$ m

The thickness of the slab is usually selected nominal and in the range of $a/15$ subject to a minimum of about 60 mm. Similarly, the depth of the web of the grid beam can be selected in the range of $L/25$ to $L/15$ depending on the load intensity on the floor.

Let the following dimensions be selected.

$$\text{Thickness of the slab} = t = 100 \text{ mm}$$
$$\text{Thickness of the web} = b_w = 150 \text{ mm}$$
$$\text{Depth of the web} \quad = h_w = 600 \text{ mm}$$

The grid slab can be considered as a T-beam section in each direction with flange thickness same as the thickness of the slab and the width of the flange equal to the grid spacing. The flange width should be limited to $15t + b_w$ in this case. The properties of one grid beam, the beam being a T-section, are computed first:

$$b_f = a = 1525 \text{ mm}$$

The centroidal axis of the T-section is obtained by taking moment of the area about the top fibre. The centroidal axis distance from the top fibre is

$$y = \frac{b_w h_w(t + 0.5h_w) + b_f t (0.5t)}{b_w h_w + b_f t}$$

$$= \frac{150(600)(400) + 1525(100)(50)}{242500} = 180 \text{ mm} = 0.18 \text{ m}$$

The gross moment of inertia of the section is

$$I = \frac{1.525(0.1)^3}{12} + 1.525(0.1)(0.18 - 0.05)^2$$

$$+ \frac{0.15(0.6)^3}{12} + 0.15(0.6)(0.4 - 0.18)^2 = 0.0476 \text{ m}^2$$

The grid beams are exactly same in each direction so the properties of the grid slab in either direction are same

$$(D_x = D_y) \frac{1}{a} = \frac{0.0476}{1.525} = 0.0312 \text{ m}^3$$

$$e = \sqrt{\frac{D_x}{D_y}} = 1$$

The Poissons ratio for concrete can be taken equal to 0.1.

Design of the section:

Roof finish load $\hspace{6cm} w_s = 2 \text{ kN/m}^2$

Slab-weight $\hspace{6cm} w_{gs} = 0.1(25) = 2.5 \text{ kN/m}^2$

Weight of the web $\quad w_g = \dfrac{(1.525 + 1.375)(0.15)(0.6)(25)}{(1.525)(1.525)} = 2.8 \text{ kN/m}^2$

Live load $\hspace{7cm} w_l = 1.5 \text{ kN/m}^2$

$$\text{Total load} = 8.8 \text{ kN/m}^2$$

Use $\hspace{3cm} w_t = 9 \text{ kN/m}^2$

The moment coefficients for $\sqrt{e} = 1.5$ are taken from the Table 4.12 and they are

$$c_y = 0.028, \quad c_x = 0.073$$

The corresponding bending moments are :

$$M_y = \frac{(c_y + c_x \, ve) w_t L_y^2}{n\sqrt{e}}$$

$$= \frac{(0.028 + 0.0073)(9)(18.3)^2}{1.5} = 70.89 \text{ kNm/m}$$

$$M_x = (0.073 + 0.0028)(9)(18.2)^2 = 101.27 \text{ kNm/m}$$

The bending moment per a unit of one grid spacing would equal to the product of the grid spacing and the bending moment per unit width. The effective depth of the reinforcement from the top fibre can be taken as

$$d = h - 0.05 = 0.7 - 0.05 = 0.65 \text{ m}$$

The area of the tension reinforcement needed per each beam spanning in the short span (same as x-axis direction) is

$$A_{stx} = \frac{aM_x}{jd\sigma_{st}} = \frac{1.525(101\,270)}{0.9(0.65)(230)} = 1150 \text{ mm}^2$$

Provide 4 numbers of 20 mm bars in each beam at the bottom.

The area of the reinforcement provided is

$$A_{st} = 4(314) = 1256 \text{ mm}^2$$

The percentage of the reinforcement in the T-section is

$$p = \frac{100\,A_{st}}{A_c} = \frac{125600}{242500} = 0.5\%$$

The bending moment in the grid along the long span direction is about three-fourths of that in the short span direction.

Provide three numbers of 20 mm bars in the grid in the long span direction.

Check for shear stress

The load in the middle strip level is almost transferred to the short span. Therefore, the total load at the middle grid beam spanning in the short span is

$$W = w_t a L_x$$

The shear force on a grid beam at a distance d from the end support is

$$V = \frac{W}{2} - w_t a d$$

$$= w_t a \left(\frac{L_x}{2} - d \right) = 9(1.525)(6.1 - 0.65) = 74.8 \text{ kN}$$

The nominal shear stress in the beam is

$$\tau_v = \frac{V}{b_w d} = \frac{74\,800}{150(650)} = 0.77 \text{ MPa}$$

The percentage of reinforcement in the web is

$$p_t = \frac{100\,A_{st}}{b_w d} = \frac{125600}{150\,(650)} = 1.3\%$$

The allowable shear stress in M15 concrete for 1.2% reinforcement is 0.34 MPa. The allowable shear force on the section is

$$V_a = b_w d\tau_c = 152(650)(0.34) = 33\,150 \text{ N}$$

The design shear force on the beam is

$$V_s = V - V_a = 74\,800 - 33\,150 = 41\,650 \text{ N}$$

Select two-legged 8 mm HYSD stirrups, then the area of the stirrup is

$$A_{sv} = 100 \text{ mm}^2$$

The spacing of the stirrups is

$$s_v = \frac{A_{sv}\,\sigma_{sv}d}{V_s} = \frac{100(230)(650)}{41650} = 359 \text{ mm}$$

The maximum spacing of the stirrups is

$$s_{max} = \frac{A_{sy}\,f_y}{(0.4)b_w} = \frac{100(415)}{0.4(150)} = 691 \text{ mm}$$

Provide two-legged 8 mm HYSD stirrups at 350 mm spacing in the short span beams and 450 mm spacing in long span beams.

The deflections in the slab and particularly in grid floor are negligibly small; therefore, there is a no-need to check for the deflection limits.

Design of slab: The slab over the web of the grids considered as continuous in both directions, square in plan and supported by the grid beams at the spacing *a*. The load on the slab including its weight is

$$w = w_{ga} + w_s + w_t = 6.0 \text{ kN/m}^2$$

The maximum negative bending moment in the slab is

$$m = 0.06 \, wa^2 = 0.06(6000)(1.525)^2 = 837 \text{ Nm/m}$$

The flexural stress in the plain concrete slab is

$$\sigma_{cr} = \frac{m}{z} = \frac{837(6)}{(100)^2} = 0.5 \text{ MPa}$$

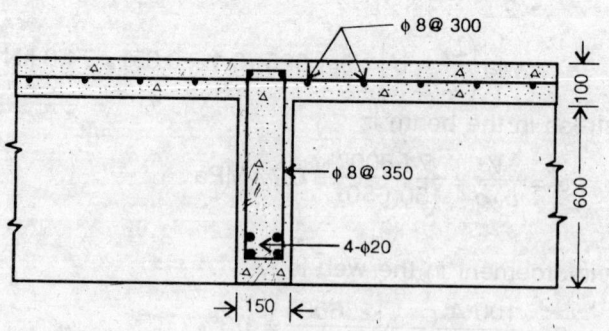

Fig. 4.8 Grid Beam along the Short Span. Example 4.8.

The allowable fracture stress in the concrete is

$$\sigma_{acr} = \frac{0.7\sqrt{f_{ck}}}{2.5} = 1.08 \text{ MPa}$$

The flexural tension is only half of the allowable stress; therefore, provide only nominal reinforcement in the slab. About 0.12% reinforcement in each direction must be adequate in the slab.

$$A_s = \frac{0.12(100)(1000)}{100} = 120 \text{ mm}^2/\text{m}$$

Provide 8 mm HYSD bars at 300 mm spacing in the middle level of the slab in each direction. Figure 4.9 illustrates the typical reinforcement details of the grid slab.

4.10 Circular Slabs with Simple Boundary Conditions

Circular slabs are commonly used in well foundations and water tanks. A slab having a ratio of the diameter to the thickness more than 10 can be treated as a thin plate and the bending analysis applied to the thin plates can be extended to the reinforced concrete slabs. The Poissons ratio of the concrete in the slabs can be taken as zero to be on the safe side as the slab develops cracks on the tension face. The bending moments and deflection at critical points can be expressed as:

$$M_R = c_R wR^2 \tag{4.20}$$

$$M_\theta = c_\theta wR^2 \tag{4.21}$$

$$v = c_V \frac{wR^4}{EI_e} \tag{4.22}$$

where M_R = bending moment per unit width on the radial plane
M_θ = bending moment per unit width on the circumferential plane
R = radius of the slab in plan
v = deflection
c_R and c_θ are the bending moment coefficients
c_V = deflection coefficient

M_R and M_θ are also called *radial* and *circumferential* bending moments respectively. They are functions of the radius in case of symmetrically loaded slabs. Most common boundary conditions of the slabs are the simply supported and fixed ended on the peripheries. The boundaries of the isolated foundation slabs are treated as free. The bending moment coefficients for simple slabs are listed in Table 4.11.

The section of the slab is designed like that of a rectangular slab. The reinforcement can be placed in the radial and circumferential directions. The spacing of radial bars decreases towards the centre. There will be considerable crowding of bars at the centre of the slab. The radial bars are curtailed towards the centre while keeping the amount of reinforcement to the desired plus that required for the development length. The reinforcement can also be placed in a rectangular grid as in the case of the rectangular slabs.

EXAMPLE 4.9 *Design of simply supported slab:* A circular slab of effective radius 3 m is subjected to a uniformly distributed load of 5 kN/m², an is simply supported at the boundary.

Table 4.11 Bending Moment and Deflection Coefficients in Circular Slabs

Support conditions	Location	c_R	c_θ	c_v
Simply supported	centre	$\dfrac{3}{16}$	$\dfrac{3}{16}$	$\dfrac{8}{64}$
	support	0	$\dfrac{2}{16}$	0
Fixed ended	centre	$\dfrac{1}{16}$	$\dfrac{1}{16}$	$\dfrac{1}{64}$
	support	$-\dfrac{1}{8}$	0	0

Design the slab with M20 concrete and HYSD bars.

Design data $R = 3$ m, $w_t = 5$ kN/m^2, $f_{ck} = 20$ MPa, $f_y = 415$ MPa, $K = 0.13$, $\sigma_{st} = 230$ MPa.

Design of the section: Assume the thickness of the slab around $R/12$ for the purpose of computing the self-weight. Let $t = 0.20$ m.

Self-weight $w_g = 0.20 \ (25) = 5$ kN/m^2

$$w_l = 5 \text{ kN/m}^2$$

Total load $\overline{w_t = 10 \text{ kN/m}^2}$

The maximum bending moment on the slab is at the centre and is given by

$$M_{max} = M_R = M_\theta = c_R \ w_t \ R^2 = c_\theta \ w_t \ R^2$$

(where $c_R = c_\theta = 3/16$ from Table 4.11)

$$= \frac{3}{16} \ (10)(9) = 16.875 \ \text{kNm/m}$$

Equating the bending moment to the moment capacity, we have

$$M_r = M_{max} \text{ leads to, } Kbd^2 \ \sigma_{cb} = 0.016875 \text{ MNm/m}$$

$$d = \sqrt{\frac{0.016875}{0.13 \ (1)(7)}} = 0.136 \text{ m}$$

Provide $d = 0.15$ m, and $t = 0.17$m, then

$$A_{st} = \frac{M}{jd \ \sigma_{st}} = \frac{16875}{0.904 \ (0.15)(230)} = 541 \text{ mm}^2/\text{m}$$

Provide 12 mm bars at 200 mm spacing.

Figure 4.9 illustrate the reinforcement details in two alternative ways.

The peripheral length $= 2\pi R = 18.86$ m

Number of the bars required at the edge assuming double the spacing of that at the centre is

$$N = \frac{2\pi R}{2 \ (0.2)} = 47$$

Therefore, start with 48 bars at the periphery and curtail one-third at a distance 1.5 mm from

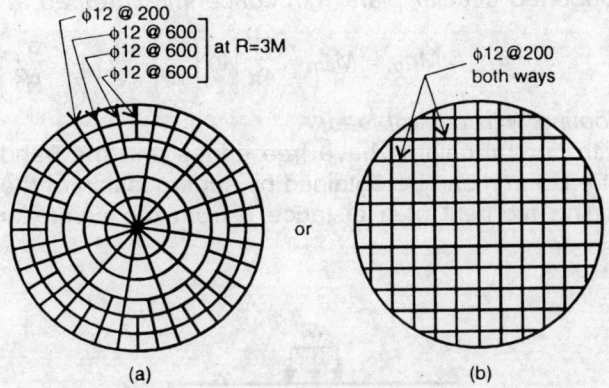

Fig. 4.9 Layout of Reinforcement and Bars in Circular Slabs, Example 4.9.

edge, then one-third of the bars at a distance 2.5 m from the edge. The number of bars at radius 0.5 m from the centre are 16. The spacing of these bars is

$$s = \frac{2\pi r}{16} = \frac{2(3.14)(0.5)}{16} = 0.2 \text{ m}$$

The circumferential bars are kept at spacing of 0.2 m throughout at the bottom face of the slab. Alternatively, one can provide 12 mm both ways in a rectangular grid as shown in Fig. 4.9b.

4.11 Circular Slab with different Load and Boundary Conditions

The analysis of circular slabs with different load and boundary conditions is beyond the scope of this book. However, the maximum bending moments which are important in the design of slabs can be taken from standard textbooks on theory of plates. One of the useful books that gives several useful results is *Theory of Plates and Shells* by Timoshenko and Woinowsky-Krieger. Some bending moment coefficients are listed here for ready reference. The subscript m refers to the maximum value.

Case 1 Simply supported slab with central load W:

$$M_{Rm} = \frac{W}{4\pi} \left(\ln\left(\frac{R}{a}\right) - \frac{1}{4}\left(1 - \frac{a^2}{R^2}\right) \right) \tag{4.23}$$

$$M_{\theta m} = \frac{W}{4\pi} \left(\ln\left(\frac{R}{a}\right) + \frac{1}{4}\left(3 - \frac{a^2}{R^2}\right) \right) \tag{4.24}$$

where a = radius of the area over which the load is placed.

Both are at the face of the load.

Case 2 Fixed end plate with central load W:

$$M_{Rm} = -\frac{W}{4\pi} \text{ at support} \tag{4.25}$$

$$M_{\theta m} = \frac{W}{4\pi} \ln\left(\frac{R}{a}\right) \tag{4.26}$$

146

Case 3 Simply supported circular plate with concentric line load at radius a

$$M_{Rm} = M_{\theta m} = \frac{W}{4\pi}\left(\ln\left(\frac{R}{a}\right) + \frac{1}{2}\left(1 - \frac{a^2}{R^2}\right)\right) \text{ at } r = a \qquad (4.27)$$

Case 4 Circular footing with central load:

Circular footing with central column have free edges and the bending moment which is critical at the face of the column can be obtained by superposition of the two plates as shown in Fig. 4.10. The bending moment at a distance a from the centre of an inverted slab as shown in Fig. 4.10b is

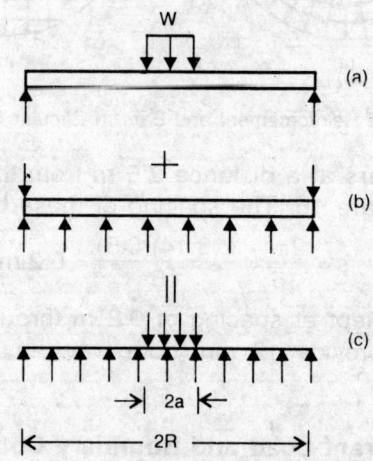

Fig. 4.10 Superposition of Two Plate Problems.

$$M_{R2} = \frac{-3w}{16}(R^2 - a^2) \qquad (4.28)$$

where $W = \pi R^2 w$

Therefore, Eq. (4.28) gives

$$M_{R2} = \frac{-3W}{16\pi}\left(1 - \frac{a^2}{R^2}\right) \qquad (4.29)$$

Superimposing Eqs. (4.29) and (4.23), we have

$$M_R = \frac{W}{4\pi}\left(\ln\left(\frac{R}{a}\right) - \frac{1}{4}\left(1 - \frac{a^2}{R^2}\right)\right) - \frac{3W}{16\pi}\left(1 - \frac{a^2}{R^2}\right)$$

$$= \frac{W}{4\pi}\left(\ln\left(\frac{R}{a}\right) - \frac{1}{4} + \frac{a^2}{4R^2} - \frac{3}{4} + \frac{3}{4}\frac{a^2}{R^2}\right)$$

$$= \frac{W}{4\pi}\left(\ln\left(\frac{R}{a}\right) - 1 + \frac{a^2}{R^2}\right) \qquad (4.30)$$

$$M_\theta = \frac{W}{4\pi}\left(\ln\left(\frac{R}{a}\right) + \frac{3}{4} - \frac{a^2}{4R^2} - \frac{3}{4} + \frac{a^2}{4R^2}\right)$$

$$= \frac{W}{4\pi}\ln\frac{R}{a} \tag{4.31}$$

The critical bending moments in the circular footing can be expressed from Eqs. (4.30) and (4.31) as

$$M_R = c_R W \tag{4.32}$$
$$M_\theta = c_\theta W \tag{4.33}$$
$$\text{where}\quad c_R = (1n\,(R/a) - 1 + a^2/R^2)/4\pi \tag{4.34}$$
$$c_\theta = (1n\,(R/a))/4\pi \tag{4.35}$$

For ready reference, these moment coefficients are listed in Table 4.12.

Table 4.12 BM Coefficients of Circular Footings

	R/a				
	3	5	6	7	8
c_R	0.036	0.052	0.065	0.077	0.087
c_θ	0.110	0.0128	0.143	0.155	0.165

EXAMPLE 4.10 *Design of circular footing:* A column of 500 mm diameter transfer a load of 1200 kN to a foundation. The safe net bearing capacity of the soil is 65 kN/m². Design a circular footing.

Design data

$$\text{Superimposed load} = W_s = 1200 \text{ kN}$$
$$\text{Radius of the column} = a = 250 \text{ mm}$$
$$\text{Safe net bearing capacity} = p_a = 65 \text{ kN/m}^2$$
$$\text{Use}\qquad f_{ck} = 15 \text{ MPa}, \ f_y = 250 \text{ MPa}$$
$$K = 0.173, \qquad j = 0.867$$

Design of foundation

$$\text{Superimposed load} = W_s = 1200 \text{ kN}$$
$$\text{Let the difference in the load of the slab and the soil} = \underline{120 \text{ kN}}$$
$$\text{Total net load} = W = \underline{1320 \text{ kN}}$$

$$\text{Bearing area needed} = A_f = \frac{W}{p_a} = \frac{1320}{65} = 20.308 \text{ m}^2$$

$$\text{Radius of the footing} = \sqrt{\frac{A_f}{\pi}} = 2.54 \text{ m}$$

Use $R = 2.6$ m, then $R/a = 2.6/0.25 = 10\,4$

Design of the section: The maximum bending moments occur at $r = a$ and they are

$$M_R = \frac{W_s}{4\pi}\left(\ln\frac{R}{a} - 1 + \frac{a^2}{R^2}\right)$$

$$= \frac{1200}{4\pi} (\ln 10.4 - 1 + 1/10.4^2) = 129.1 \text{ kNm/m}$$

$$M_\theta = \frac{W_s}{4\pi} \ln \frac{R}{a} = 223.6 \text{ kNm/m}$$

Equating the maximum moment to the balanced moment capacity, we have

$$K bd^2 \sigma_{cb} = M_\theta = 0.2236 \text{ MNm/m}$$

$$\text{or } d = \sqrt{\frac{0.02236}{0.173(5)}} = 0.51 \text{ m}$$

Check for shear stress: The critical shear stress plane is at a distance 0.5 *d* from the face of the column. The peripheral distance of the critical shear plane is

$$b_0 = \pi(2a + d) = \pi(0.5 + 0.51) = 3.172 \text{ m}$$

Shear force on this plane is

$$V = \frac{W_s}{\pi R^2} (\pi R^2 - \pi(a + 0.5d)^2) = W_s \left(1 - \frac{(a + 0.5d)^2}{R^2} \right)$$

$$= W_s \left(1 - \frac{(2a + d)^2}{4R^2} \right)$$

$$= 1200 \left(1 - \frac{(1.01)^2}{4 \times 2.6^2} \right) = 1154 \text{ kN}$$

The nominal shear stress is

$$\tau_v = \frac{V}{b_0 d} = \frac{1.154}{3.172(0.51)} = 0.71 \text{ MPa}$$

The allowable shear stress for diagonal tension is

$$\tau_c = 0.16\sqrt{f_{ck}} = 0.62 \text{ MPa}$$

Since the shear stress is more than the admissible value, increase the thickness of the slab to 0.58 m.

$$d = 0.58$$

$$b_0 = \pi(a + d) = 3.39 \text{ m}$$

$$V = W_s \left(1 - 0.25 \frac{(1.08)^2}{2.6^2} \right) = 1148 \text{ kN}$$

$$\tau_v = \frac{V}{b_0 d} = \frac{1.148}{3.39(0.58)} = 0.58 \text{ MPa}$$

This value is less than the admissible shear stress; therefore, the thickness is acceptable.

$$A_{s\theta} = \frac{M_\theta}{jd\, \sigma_{st}} = \frac{223\,600}{0.867(0.58)(140)} = 3176 \text{ mm}^2/\text{m}$$

$$A_{sr} = \frac{M_R}{jd\, \sigma_{st}} = \frac{129\,100}{0.867\,(0.57)(140)} = 1865 \text{ mm}^2/\text{m}$$

Design of reinforcement: The area of the circumferential steel is

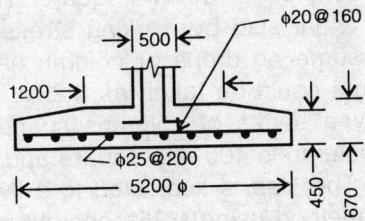

Fig. 4.11 Sectional Details of Circular Slab Footing of Example 4.10.

Provide 25 mm bars at 200 mm spacing as circumferential and 20 mm bars at 160 mm as radial reinforcement. The reinforcement detail is illustrated in Fig. 4.11.

PROBLEMS
(Use IS : 456)

4.1 A rectangular slab is simply supported on 4 edges with effective spans 3 m × 8 m. It is subjected to a live load of 4 kN/m². Design a reinforced concrete slab using M20 concrete and HYSD-Fe415 bars by working stress method.

4.2 A long slab of effective spans 3 m × 8 m simply supported along 3 m edges and has continuity at 8 m long supports. The slab is subjected to a live load of 4 kN/m². Design a RCC slab using M20 concrete and mild steel bars by working stress method. Sketch the reinforcement and structural details.

4.3 A verandah slab of 10 m length cantilevers from a fixed wall to 2 cantilever span. The load on the slab is 4 kN/m². Design a RCC slab with M25 concrete and HYSD-Fe415 bars by working stress method. Sketch the reinforcement details and indicate the minimum embedment length.

4.4 A slab is simply supported on 4 edges with effective span of 4.5 m by 5.5 m and subjected to a live load of 3 kN/m². Design a reinforced concrete slab assuming the corners of the slab are allowed to warp and using M20 concrete and HYSD-Fe415 bars. Design the slab by working stress method giving reinforcement details both in plan and in cross-section.

4.5 A simply supported slab of effective spans 4.5 m × 5.5 m is subjected to a live load of 3kN/m². Design a RCC slab assuming the corners of the slab are fixed against warping. Design the slab with M20 concrete and HYSD-Fe415 bars by working stress method. Sketch the reinforcement details.

4.6 A 5m square slab simply supported at 2 adjacent edges and having continuity connection at the other two adjacent supports is subjected to a live load of 3 kN/m². Design a reinforced concrete slab by working stress method using M25 concrete and HYSD-Fe415 bars. Sketch the reinforcement details.

4.7 A rectangular slab of 4m × 6m is simply supported on two opposite short walls and continues on the other two long walls and it is subjected to a live load of 3 kN/m². Design a reinforced concrete slab by working stress method using M20 concrete and HYSD-Fe415 bars.

4.8 A flat slab is supported by a set of columns spaced at 4.5 m intervals in two orthogonal directions. The size of the column is 400 mm square. The slab is subjected to a live load of 2.5 kN/m². Design a flat slab by working stress method using M20 concrete and HYSD-Fe415 bars. Assume no drops or column heads in the construction. The effective height of the column could be taken as 3.2 m.

4.9 A flat slab is supported by a series of columns spaced at 6 m in two orthogonal directions. The size of the column is 400 mm square and provided with a column head of size of 600 mm square. The slab is subjected to a live load of 5 kN/m². Design a flat slab by working stress method using M25 concrete and HYSD-Fe415 bars.

4.10 A flat slab with drop panels having square drop panels of 1 m² is supported by columns spaced at 6 m in two perpendicular directions. The size of the column is 500 mm square. The slab is subjected to a live load of 60 kN/m². Design a reinforced concrete flat slab by working stress method using M20 concrete and HYSD-Fe415 bars. Assume an effective height of the column as 4m. Sketch reinforcement details

4.11 A rectangular floor slab simply supported on 4 edges having effective span of 10 m by 14 m and is subjected to a live load of 3.0 kN/m². The slab is designed as a waffle floor with grid spacing of 1 m. Design the waffle floor by working stress method using M20 concrete and HYSD-Fe415 bars.

4.12 A floor slab circular in plan and having radius of 4 m is subject to a live load of 3 kN/m². Design a reinforced concrete slab by working stress method using M20 concrete and mild steel reinforcement. Sketch the reinforcement details. Given the maximum radial and circumferential bending moments on the slab equal to $(3/16) wR^2$ where the w is intensity of the load and R is the radius of the slab.

4.13 A column carrying a load of 1000 kN is supported by a circular footing resting on a soil having a net safe bearing capacity of 100 kN/m². Assume the column diameter as 500 mm. Design a RCC circular footing by working stress method using M25 concrete and mild steel reinforcement. The maximum radial and circumferential bending moments are given by

$$M_R = \frac{W_s}{4\pi}\left(\ln\frac{R}{a} - 1 + \frac{a^2}{R^2}\right)$$

$$M_\theta = \frac{W_s}{4\pi}\left(\ln\frac{R}{a}\right)$$

where W_s = total superimposed load

R = outer radius of the slab

a = radius of circular column

4.14 A water tank with circular cylindrical supporting shaft of mean radius 6 m is resting on circular raft foundation. The total load carried by the shaft is 20 MN. The thickness of the shaft wall is 150 mm. The safe bearing capacity of the soil is 70 kN/m². Design a RCC raft by working stress method using M20 concrete and HYSD-Fe415 bars. Sketch the reinforcement details (use a bending moment coefficient given in the book).

Chapter 5

Working Stress Design of RCC Compression Members

5.1 Introduction

An upright member subjected to dominant compressive force with or without bending moment and transferring the load to ground is called a *column*. If the same member is oriented in an arbitrary direction, it is normally referred as a *strut*. Usually struts are under pure compression when compared with the columns. Columns are the common elements in most buildings and are usually subjected to combined axial compression and bending. The columns in industrial buildings and framed structures are not only subjected to high compression but also reasonably large bending and shear forces. A very short column having height to side ratio less than four is called a *stub column*. Poles and even posts which are upright and subjected to some compression but dominant bending moment are often classified as beams. Similarly, *pilaster* is an upright element sandwiched between two walls belongs to the class of beams. Large size upright members used in bridges and similar construction are called *piers*. Piers are basically under high compression with some bending moment. *Wells* which support piers or columns or bridges, etc. are also compression members subjected to combined compression, bending and shear. There is also another class of compression members which is called a *pedestal*, and is provided at the base of the columns over the foundations or of an independent free end stub columns. Pedestals are usually designed as plain masonry elements and its height to side ratio is less than three.

This chapter deals with the design of columns. The columns may be broadly classified as:

1. Tied columns.
2. Spiral columns.
3. Composite columns.
4. In-filled columns.

A *tied column* is the one in which the longitudinal bars (also called the main reinforcement) are held in position by separate ties spaced at equal intervals along the length. Tied columns can be of any shape, some, typical ones are shown in Fig. 5.1 A *spiral column* is usually circular in cross-section (can also be of other shapes) and the longitudinal bars are wrapped by a closely spaced spiral. A *composite column* consists of steel or cast iron structural member encased in concrete. Nominal or main reinforcement bars positioned with ties or spiral are placed around the structural member. Figures 5.2 and 5.3 illustrate the spiral and composite columns. A composite column can also be called as *encased column*. *Concrete filled* column is the one having a steel pipe filled with plain concrete. It can also be provided with some reinforcement, though this is not usually done.

The behaviour of a column depends on its type and boundary conditions. The axial deformation or the lateral deflections of the columns are negligible when compared with the

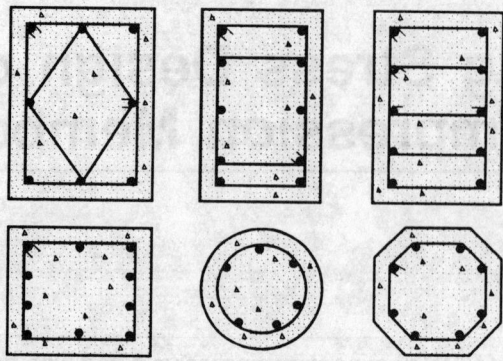

Fig. 5.1 Cross-sections of Tied Columns.

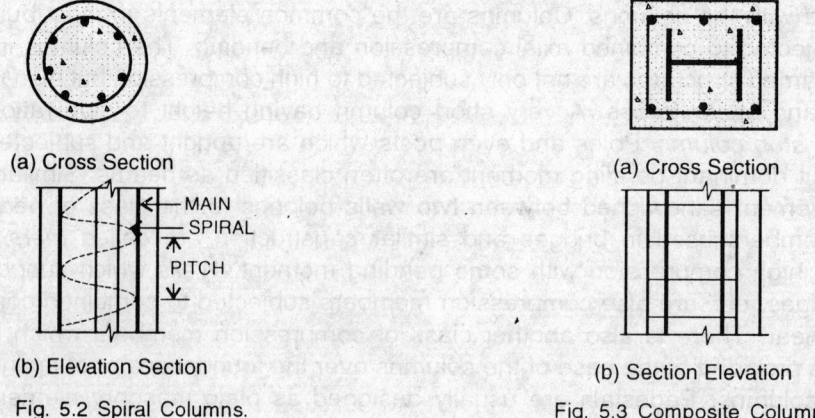

(a) Cross Section

MAIN
SPIRAL
PITCH

(b) Elevation Section

Fig. 5.2 Spiral Columns.

(a) Cross Section

(b) Section Elevation

Fig. 5.3 Composite Columns.

deflections in the beams. However, the drift of the column in tall buildings is comparable with other deformations. Buckling of the concrete columns is not so frequent when compared with the problem in metal columns as the size of the masonry columns is large enough to minimize buckling. However, the slenderness of the column must be taken into account in the design. A column is classified as a *short column* if the ratio of the effective length to the least side is less than 12; otherwise, it is called a *long column*. This ratio in any case should be restricted to 60. As the columns are under primary compression, the failure is usually sudden with a warning shown through the spalling of the cover. Spirally reinforced column shows more ductility when compared with the tied columns.

5.2 Permissible Stresses and other Limitations

The columns are so designed such that the actual stress in the materials of the column is limited to an allowable or permissible stress. In the original elastic design, a strain compatibility between the materials was assumed, but later it was found to be not only uneconomical but untrue. Hence the allowable stresses recommended by the codes of practice need not necessarily satisfy the compatibility of strains. The allowable axial compressive stress is taken about eighty per cent of that in bending or even less. The permissible stresses are given in

Table 5.1. The allowable compressive stress in concrete may be approximately expressed as

$$\sigma_{cc} = \frac{f_{ck}}{4} \tag{5.1}$$

Most columns are subjected to axial force and some bending moment. Some columns are idealized as axially loaded, and even such columns should be designed for a *minimum eccentricity*. This eccentricity may be arising out of the support conditions, or load conditions or construction tolerances. The minimum eccentricity normally recommended is

$$e_0 \geq 20 \text{ mm} \tag{5.2}$$

$$\geq \frac{\text{Lateral dimension}}{30} + \frac{L}{500}$$

One need not make a special effort to incorporate this limit in the design as the allowable stresses in the materials are selected so as to account for the minimum eccentricity.

Minimum reinforcement requirements: All reinforced columns must have a minimum percentage of reinforcement to resist temperature, shrinkage, creep, accidental loads, and to cope up with the normal tolerances such as a minimum eccentricity allowed in the construction. Similarly, a maximum amount of steel that can be placed without causing problem in construction and to maintain the bonded action of the member is also to be adhered. The *percentage limits of the reinforcement* are:

$$0.8 \leq p \leq 6 \tag{5.3}$$

where p = percentage of the longitudinal reinforcement

In case the reinforcement is not taken into account in load carrying capacity of the column, then the minimum percentage of steel can be reduced to the normal shrinkage reinforcement, and is given by

$$p_n \geq 0.15 \tag{5.4}$$

where p_n is the nominal percentage of the reinforcement. The *minimum* amount of reinforcement is further controlled by the minimum diameter and number of the main bars. The main bars in the column refer to the longitudinal bars. The limits are

$$\phi \geq 12 \text{ mm} \tag{5.5}$$

$$N \geq 4, \text{ for rectangular section} \tag{5.6}$$

$$N \geq 6, \text{ for circular or other multiple face shapes}$$

where ϕ = diameter of the main bar and
N = number of the main bars

The maximum spacing of the longitudinal bars should also be limited to 300 mm. The longitudinal bars are held in position by ties or spiral in the transverse direction. All bars at the corners must be supported in two directions to avoid any lateral bucking of the bars. The allowable stress in the longitudinal bars is higher than the corresponding strain compatible stress in concrete. In addition, the bars are to be treated as slender columns with semi-rigid lateral continuous support. As the longitudinal bar resists the load, it has a tendency to buckle; but this lateral buckling is constrained by the concrete. For large forces in the bars, the lateral expansion due to Poissons effect in steel will induce splitting tendency in the concrete and also the lateral buckling tendency of the bars which in turn introduce spalling of the concrete.

Table 5.1 Permissible (allowable) Stresses (in MPa)

f_{ck}	Concrete					
	10	15	20	25	30	35
σ_{cc} or $\sigma_{acc}*$	2.5	4	5	6	8	10

	Reinforcement	
Type of steel	Mild or medium	HYSD
σ_{sc} or σ_{asc}	130	190

	Composite-metal core	
Type of metal	Cast iron	Steel
σ_{so}	70	125

Infilled shell

$$\sigma_{sc} = 120 - 0.0034(L/r_s)^2$$
(L/r_s) = slenderness ratio of the shell
$f_y \geq 240$

Notes: (i) The permissible stresses under wind or earthquake load condition can be increased by $33\frac{1}{3}\%$, and are modified under combined bending and axial force.

(ii) Wherever there is a difference between the actual stress and allowable stress, the allowable values are indicated by an additional subscript *a*. Example σ_{cc} = actual direct compressive stress also treated as allowable value when there is no confusion. Otherwise, the corresponding allowable stress in indicated by σ_{acc}.

Both these tendencies are prevented by adequate *transverse reinforcement* and proper cover to the main bars. The transverse reinforcement can be in the form of either ties or spiral. The diameter and the spacing of the transverse reinforcement must be limited to a minimum so as to hold the longitudinal bars in position or to contain the concrete against bursting.

The diameter of the ties is controlled by

$$\phi_v \geq \phi/4 \tag{5.7}$$
$$\geq 5 \text{ mm}$$

where ϕ_v = diameter of the tie.

The spacing of the ties must be limited to the least of the following:

$$S_v \leq \text{ least lateral dimension of the column}$$
$$\leq 16\phi \tag{5.8}$$
$$\leq 48\phi_v$$

where S_v = spacing of the ties.

Similarly, the pitch of a helical reinforcement must not exceed certain limits.

$$\text{pitch} \leq 75 \text{ mm} \tag{5.9}$$
$$\leq \frac{\text{core diameter of the column}}{6}$$

In addition to holding the longitudinal bars in position, the spiral reinforcement also contains concrete in the core: therefore, the pitch is usually limited to a small value. However, it should be large enough to permit placing of the concrete and interlocking of the cover concrete with that of the core. Hence, the pitch should not be less than certain minimum limit, and it is given by

$$\text{pitch} \geq 30 \text{ mm} \tag{5.10}$$

$$\geq 3 \, \phi_v$$

$$\geq \text{size of the aggregate}$$

The ties should preferably be loops; however, open ties or direct two bar tie can be provided if there is at least one outer loop which contains all the main bars. In any given row of any face if there are more than two longitudinal bars, intermediate bars need not be tied in two directions provided that its spacing is less than 75 mm. In case the spacing of the main bars exceeds 75 mm, each of the main bars must be restrained in two perpendicular directions by appropriate loops or links. A loop means a closed and a link means open interconnection.

A minimum cover to the main reinforcement should be ensured so as to protect the steel from corrosion and support the main bars from buckling. Corrosion of the reinforcement not only damages the steel but increases the volume of the bar. Consequently, the concrete is split open and further damage and failure of the column takes place. All columns exposed to weather must be adequately protected. The recommended covers to the main reinforcement are given in Table 5.2.

Table 5.2 Minimum Cover to the Main Reinforcement

	Exposure condition	Clear cover* (mm)
1.	Indoor and protected (except for 12 mm bars)	40
2.	Outdoor and exposed to normal environment	40
3.	Outdoor and exposed to saline atmosphere:	
	for bars upto 25 mm	40
	for bars over 25 mm	50
4.	Exposed to chemical fumes	40 to 50
5.	In sea water	60 to 75
6.	One hour protection for exposure of 400°C	50 mm

* In no case cover to the bars should be less than the size of the bar. The minimum cover to the ties or spiral should not be less than 25 mm. See durability criteria also.

5.3 Effective Height of Columns

The classification of the columns as long or short and the correction factor for long columns are based on the slenderness ratio of the column. The slenderness ratio of a column is defined as the ratio of the effective height to the radius of gyration of the section. That is

$$k = \frac{L_c}{r} \tag{5.11}$$

where L_c = effective height and r = radius of gyration.

The buckling of the column is initiated in the plane about which the slenderness ratio is largest. This factor is very important in steel columns. In case of reinforced concrete or masonry columns, the slenderness ratio is usually less than 50 and, therefore, the columns are likely to get crushed rather than buckled. However, a reduction in the load carrying capacity of the column be made in slender columns. For this purpose another factor which may be called *slenderness factor sometimes called as slenderness ratio* is defined. It can be expressed as

156

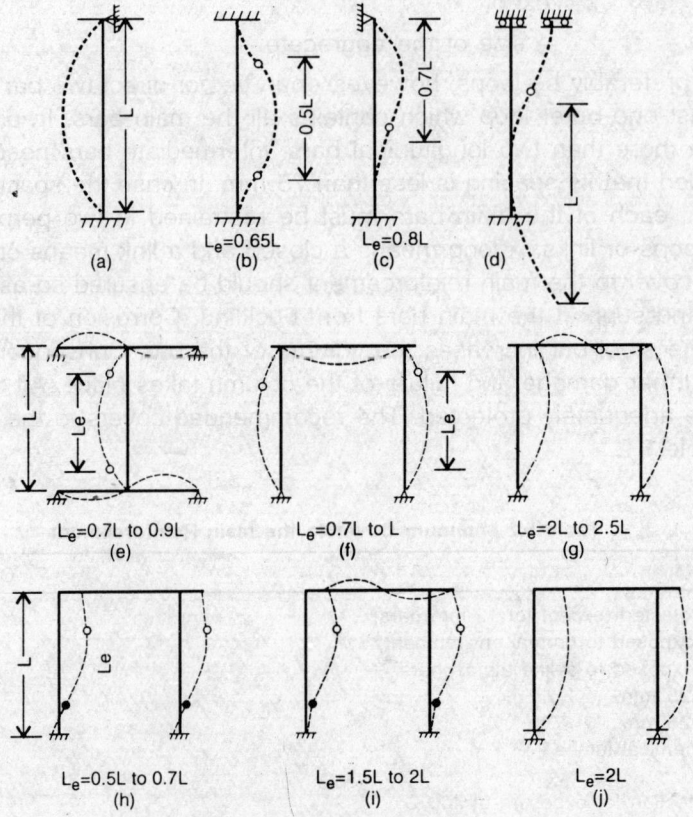

Fig. 5.4. Effective Column Heights.

$$g = \frac{L_{eb}}{b} \text{ or } \frac{L_{eD}}{D} \tag{5.12}$$

where g = slenderness factor (sometimes referred as slenderness ratio)
   b = width
and D = depth of the section
   L_{eb} and L_{eD} = effective column height about width and depth planes

The effective height of a column is the distance between the two consecutive points of contraflexure or zero BMs. In case of columns simply supported at both ends, the effective height of the column is same as the distance between the support points. For convenience of design, the idealized effective heights of the column are given in Table 5.3 and are illustrated in Fig. 5.4.

The radius of gyration for rectangular section columns is

$$r = \frac{b}{\sqrt{12}} = 0.29b \tag{5.13}$$

Table 5.3 Effective Length of Columns

Boundary conditions	Effective lengths	
	Theoretical	Recommended
1. Simply supported at both the ends	L	L
2. Fixed at both the ends	0.5 L	0.65 L
3. Fixed at one end and hinged at the other	0.7 L	0.8 L
4. Fixed at one end and constrained against rotation at the other end	L	1.2 L
5. Fixed at one end and free at the other end	2 L	2 L
6. Columns in portal frames with fixed bases and having lateral drift (approximate)		1.5 L
7. Columns in portal frames with hinged bases and having lateral drift (approximate)		2 L

The slenderness ratio of the rectangular sectioned column is

$$k = \frac{L_e}{r} = \frac{L_e}{0.29b} = \frac{3.45\, L_e}{b}$$

The slenderness ratio divided by the slenderness factor gives

$$\frac{k}{g} = 3.45 \tag{5.14}$$

Similar ratio for circular columns is 4. A column having a slenderness ratio more than 45 is treated as a long column and the corresponding slenderness factor is selected 12. The maximum slenderness factor for RCC columns should be restricted to 60.

In case of columns subjected to only axial force, a clear bifurcation point exists at the time of buckling. However, most columns are subjected to at least a nominal bending moment in addition to the axial force. The bending moment causes deflection and the effect of the axial force on the deflection is the amplification problem. The $P - \Delta$ effect or drift problem in columns increases deflections and causes large deformations even before buckling. Figure 5.5 illustrates the drift problem. Large deformations must be avoided, so that the safety of the column is controlled by the permissible deformations than the buckling load.

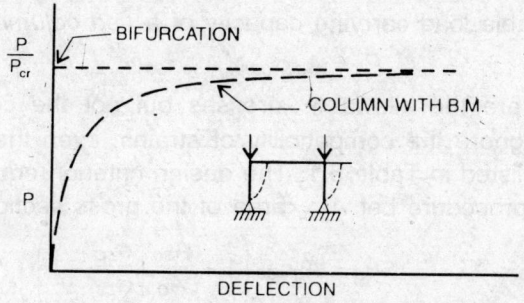

Fig. 5.5 Drift Effect on Columns.

5.4 Design of Columns by Elastic and Compatible Strain Theory

The plane sections of columns before loading will continue to remain plane after loading. This assumption leads to compatibility of strains in the concrete and the reinforcement. The stress in the reinforcement is therefore equal to modular ratio times the stress in the concrete at the level of reinforcement. Hence, the allowable load carrying capacity of a column based on the strain compatibility is given by

$$P_s = A_c \, \sigma_{cc} + A_{sc} \, \sigma_{sc}$$

$$= (A_c + A_{sc}m) \, \sigma_{cc} \tag{5.15}$$

where A_c = area of the concrete

A_{sc} = area of the compression reinforcement = A_s

σ_{cc} = allowable compressive stress in concrete

σ_{sc} = compressive stress in steel = $m\sigma_{cc}$

However, the reinforcement bars are restrained against later expansion by the concrete, therefore, the reinforcement bars are treated as members in plain strain problem rather than plain stress problem. The approximate compatible stress in the steel is moderated by the restrained Poissons effect. Hence, the stress in steel is taken as

$$\sigma_{sc} = 1.5 \, m\sigma_{cc} \tag{5.16}$$

The load carrying capacity of the column is therefore expressed as

$$P_{eo} = (A_c + 1.5m \, A_{sc}) \, \sigma_{cc} \tag{5.17}$$

The load carrying capacity based on the idealized elastic theory has been found to be uneconomical as the level of stress in steel is far below the allowable capacity. Consider the case of M20 concrete along with HYSD bars. The allowable stress in concrete is 5 MPa and the corresponding admissible stress in the reinforcement is only 97.5 MPa whereas the proof stress of the steel is 415 MPa. The expensive steel is very much underutilized and, consequently, the elastic theory is found to be uneconomical. The compatible elastic theory of columns is therefore not applied in the design.

5.5 WSD of Short Columns

In view of the fact that the elastic design is very much on the conservative side, an equilibrium approach which violates the compatibility of strains has been recommended in the working stress design. The allowable load carrying capacity of a *tied column* section is given by

$$P_a = A_c \, \sigma_{cc} + A_{sc}\sigma_{sc} \tag{5.18}$$

where the σ_{cc} and σ_{sc} are the allowable stresses but not the compatible stresses. The equilibrium model does ignore the compatibility of strains, even then it is a safe load. The permissible stresses are listed in Table 5.1. The design criterion equation can be rearranged to suit the direct design procedure Let A_g = area of the gross section, then

$$P_a = A_g\sigma_{cc}\left(1 + \frac{A_{sc}}{A_g}\left(\frac{\sigma_{sc}}{\sigma_{cc}} - 1\right)\right) \tag{5.19}$$

Let the area of the steel be expressed as a ratio of the gross section. That is

$$p = \frac{A_{sc}}{A_g} \qquad (5.20)$$

The resisting capacity should be more than the external load; thus, the design criterion along with Eq. (5.19) can be rearranged as

$$A_g \geq \frac{P}{(1 + p(q-1))\sigma_{cc}} \qquad (5.21)$$

where $q = \sigma_{sc}/\sigma_{cc}$

Equation (5.21) gives the gross cross-section of the column for any given percentage of the reinforcement and for a given set of materials. The design criterion along with Eq. (5.18) can be rearranged as

$$A_c \geq \frac{P}{\sigma_{cc}} - A_{sc}\frac{\sigma_{sc}}{\sigma_{cc}} \qquad (5.22)$$

Equations (5.21) and (5.22) can be considered as the direct design expressions.

The ratio of the strength of the steel to that of concrete is in the range of 20 to 40 whereas the corresponding cost ratio is in the range of 60 to 90. It indicates that the cost of the column increase with the increase in the area of the steel.

The load carrying capacity of a helically reinforced column is about 5% more than that of a tied column. Therefore, it can be expressed as

$$P_{as} = 1.05\,(A_c\sigma_{cc} + A_{sc}\sigma_{sc}) \qquad (5.23)$$

The specifications regarding spacing of the main and transverse reinforcement and cover should be adhered to.

5.6 WSD Criterion for Long Columns

The columns are treated as long if the slenderness factor is more than 12, that is

$$\frac{L_e}{b} \geq 12 \qquad (5.24)$$

In such cases, the load carrying capacity of the column is reduced and the load reduction factor is given by

$$C_r = 1.25 - \frac{L_e}{48b} \qquad (5.25)$$

where $L_e/b = $ larger of the slenderness factors.

The load carrying capacity of the column is given by

$$P_{al} = C_r\,(A_c\sigma_{cc} + A_{sc}\sigma_{sc}) \qquad (5.26)$$

The design procedure of the design of the long columns is same as that of short columns except for the reduction factor.

5.7 Design Examples of Tied Columns

EXAMPLE 5.1 *Design of short tied column:* A column resting on an independent footing supports a flat slab. The total load on the column is 590 kN and the height of the slab above the footing is 4.5 m. Design the column with M20 concrete and HYSD bars.

Design of the Section: The allowable stresses in the concrete and steel are:

$$\sigma_{cc} = 5 \text{ MPa}, \ \sigma_{sc} = 190 \text{ MPa}$$

Assume a short column with 1% reinforcement in ($A_{sc} = 0.01 \ A_g$). Then the allowable load carrying capacity of the column is

$$P_a = A_c \sigma_{cc} + A_{sc} \ \sigma_{sc} = 5(A_g - A_{sc}) + 190 \ A_{sc}$$

$$= 5 \ A_g + 185 \ A_{sc} = 5A_g + 185 \ \frac{(A_g)}{100} = 6.85 \ A_g$$

Assuming the self-weight of the column as 10 kN, the gross load on the column is

$$P = 590 + 10 = 600 \text{ kN}$$

Equating the working load capacity of column to the external load, we have

$$P = P_a = 6.85 \ A_g$$

$$A_g = \frac{P}{6.85} = \frac{600 \ 000}{6.85} = 87 \ 591 \text{ mm}^2$$

Assuming a square section, the column size should be equal to

$$b = D = \sqrt{A_g} = 296 \text{ mm and}$$

$$A_{sc} = pA_g = 0.01(87 \ 591) = 876 \text{ mm}^2$$

Use 4 numbers of 16 mm bars, then the actual area of the steel provided is

$$A_{sc} = 4(201) = 804 \text{ mm}^2$$

Then the gross area of the section is given by

$$A_g = \frac{P - 190 \ A_{sc}}{5} + A_{sc}$$

$$= \frac{600 \ 00 - 190(804)}{5} + 804 = 90 \ 252 \text{ mm}^2$$

or $b = 300.4 \text{ mm}$

The weight of the column is

$$W_g = (0.3004)^2 (4.5)(25) = 10.1 \text{ kN}$$

So the assumed weight is reasonable.

$$\frac{100 \ A_{sc}}{A_g} = \frac{80400}{90252} = 0.9\%$$

The column can be treated as fixed at one end and hinged at the other end. Then the effective length of the column can be taken as

$$L_e = 0.8 \ L = 0.8(4.5) = 3.6 \text{ m}$$

The slenderness factor of the column is

$$\frac{L_e}{b} = \frac{3.6}{0.3004} = 12$$

The column can be treated as a short one and hence its capacity is same as computed. Select the size of the column as $b = D = 305 \text{ mm}$.

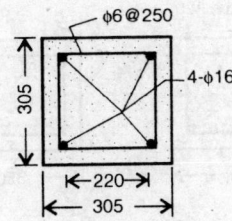

Fig. 5.6 Cross-section of Column in Example 5.1.

Design of the ties: The diameter of the tie can be selected as one-fourth the diameter of the main bars subject to a minimum of 5 mm.
Let the diameter of the tie = 6 mm then the spacing of the ties should be limited to

$$s \leq b = 305 \text{ mm}$$
$$\leq 16\phi = 16(16) = 256 \text{ mm}$$
$$\leq 48\phi = 48(6) = 288 \text{ mm}$$

Provide ϕ 6 ties at 250 mm spacing.

EXAMPLE 5.2 *Design of short column:* A column of effective height of 3.6 m is subjected to an axial load of 600 kN including the self-weight, and the size of the column is fixed at 350 mm × 300 mm. Design the column using M20 concrete and HYSD bars.

Design of main reinforcement: The allowable stresses in the material are:

$$\sigma_{cc} = 5 \text{ MPa and } \sigma_{sc} = 190 \text{ MPa}$$

The slenderness factor of the column is

$$g = \frac{L_e}{b} = \frac{3.6}{0.3} = 12$$

The column can be treated as a short one as the slenderness factor is equal to 12.

$$A_g = bD = 300(350) = 105\,000 \text{ mm}^2$$

The load carrying capacity of the column is

$$P_a = A_c\sigma_{cc} + A_{sc}\,\sigma_{sc} = (A_g - A_{sc})\,5 + A_{sc}(190)$$
$$= (105\,000)5 + 185\,A_{sc}$$

Equating the working load capacity of the column to the external load, we have

$$(105\,000)5 + 185\,A_{sc} = 600\,000 \text{ N}$$

$$\text{or} \quad A_{sc} = \frac{600\,000 - 525\,000}{185} = 405 \text{ mm}^2$$

The minimum percentage of steel required is 0.8 per cent and it is

$$A_{scm} = 0.008\,A_g = 0.008(105\,000) = 840 \text{ mm}^2$$

Provide 6 numbers of ϕ 14 bars, then

$$A_{sc} = 6(154) = 924 \text{ mm}^2$$

Select 6 mm ties, then the spacing of the ties is given by

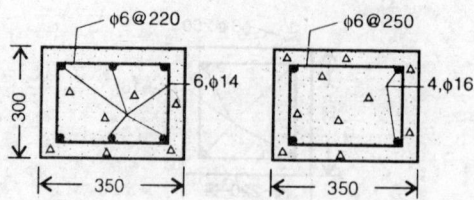

Fig. 5.7 Two Alternate Cross-sections of Column in Example 5.2.

$$s \le b = 300 \text{ mm}$$
$$\le 16\phi = 16(14) = 224 \text{ mm}$$
$$\le 48\phi_v = 48(6) = 288 \text{ mm}$$

Provide 6 mm ties 220 mm spacing.

Note: Problem of Examples 5.1 and 5.2 is the same, though the solutions are slightly different. In Example 5.2 the area of the concrete provided is more than that of Example 5.1, and the area of the compression reinforcement also works out to be more because of the minimum requirements. By selecting a smaller diameter for the main bars, the number of ties also increased because of the maximum spacing limitation. If the diameter of the main bar is selected as ϕ 16 instead of ϕ 14, then the 4 numbers of ϕ 16 are not adequate, and 5 numbers of ϕ 16 is equal to 1005 mm which is about 20% more than the minimum required. One can apply a slightly different approach and provide only 4 numbers of 16 mm bars, which is explained here.

The capacity of 4 numbers of ϕ 16 bars is

$$P_{sc} = A_{sc} \, \sigma_{sc} = 804(190) = 152 \, 760 \text{ N}$$

Load to be carried by the concrete is

$$P - P_{sc} = 600 \, 000 - 152 \, 760 = 447 \, 240 \text{ N}$$

The minimum area of the concrete needed to resist the load is

$$A_{cm} = \frac{P - P_{sc}}{\sigma_{cc}} = \frac{447 \, 240}{5} = 89 \, 448 \text{ mm}^2$$

The corresponding gross area required is

$$A_{gm} = A_{cm} + A_{sc} = 90 \, 252 \text{ mm}^2$$

The percentage of the reinforcement on the minimum gross area required is

$$p = \frac{100 \, A_{sc}}{A_{gm}} = \frac{80 \, 400}{90 \, 252} = 0.89$$

Minimum required area of steel is provided, however, the gross area of the section actually provided is $300(350) = 150 \, 000 \text{ mm}^2$ which is more than the minimum gross area required. One need not really pay a penalty by having provided more area of the concrete than minimum required. This design is also acceptable. That is *provide 4 numbers* of ϕ 16 as main bars and ϕ 6 ties at 250 mm spacing.

EXAMPLE 5.3 *Design of long tied column*: A column is subjected to an axial load of 700 kN and has an effective height of 5.275 m. Design the column with M 20 concrete and HYSD bars.

$$L_e = 5.275 \text{ m}, \ f_{ck} = 20 \text{ MPa}$$

$$\sigma_{cc} = 5 \text{ MPa}, \ \sigma_{sc} = 190 \text{ MPa}$$

$$f_y = 415 \text{ MPa}, \ \sigma_{sc}/\sigma_{cc} = 38$$

Design of the section: Assuming about 1% steel and the column as short one, the load carrying capacity of the section is

$$P_a = A_c \sigma_{cc} + A_{sc} \sigma_{sc}$$

$$= A_g \sigma_{cc} \left(1 + \frac{A_{sc}}{A_g} \left(\frac{\sigma_{sc}}{\sigma_{cc}} - 1 \right) \right)$$

$$= A_g(5)(1 + 0.01(37)) = 1.37(5) \ A_g$$

Let $P_a = P$, then

$$A_g = \frac{P}{1.37(5)} = \frac{700\,000}{1.37(5)} = 102\,183 \text{ mm}^2$$

Assuming a square cross-section, the size of the column is

$$b = D = \sqrt{A_g} = 320 \text{ mm}$$

The slenderness factor for the column is

$$g = \frac{L_e}{b} = \frac{5.275}{0.32} = 16.5$$

The slenderness factor is larger than 12, so the column should be treated as a long one and an appropriate reduction factor should be applied. It is better to select an appropriate number of reinforcement bars and then determine the size of the column. Since the size of the column is likely to be around 320 mm, we can use a slenderness factor of 16.

Approximate area of the compression reinforcement is

$$A_{sc} \text{ (approx)} = 0.01 \ (A_g) = 0.01(102\,183) = 1022 \text{ mm}^2$$

Provide 4 numbers of 18 mm bar, then the area of the compression steel provided is

$$A_{sc} = 4(254) = 1016 \text{ mm}^2$$

The allowable load carrying reduction factor is

$$C_r = 1.25 - \frac{g}{48} = 1.25 - \frac{16}{48} = 0.917$$

The equivalent short column load is

$$P_0 = \frac{P}{C_r} = \frac{700\,000}{0.917} = 763\,360 \text{ kN}$$

The load to be carried by the concrete section is

$$P_c = P_0 - A_{sc}\sigma_{sc} = 763\,360 - 1016(190) = 570\,320 \text{ N}$$

The minimum required area of the concrete is

$$A_c = \frac{P_c}{\sigma_{cc}} = \frac{570\,320}{5} = 114\,064 \text{ mm}^2$$

$$A_{sc} \qquad\qquad = 1016 \text{ mm}^2$$

$$A_g = A_c + A_{sc} \qquad = 15080 \text{ mm}^2$$

The size of the square cross section is

$$b = D = \sqrt{A_g} = \sqrt{115\ 080} = 339 \text{ mm}$$

Use $b = D = 340$ mm, then

$$g = \frac{L_e}{b} = \frac{5275}{340} = 15.5$$

The value of g assumed is 16 against an actual value of 15.5. Therefore, the design is just right.

Design of transverse reinforcement: Select 6 mm dia ties which is more than $\phi/4 = 18/4 = 4.5$ mm. The spacing of the ties is

$$s \le b = 340 \text{ mm}$$

$$\le 16\phi = 16(18) = 288 \text{ mm}$$

$$\le 48\ \phi_v = 48\ (6) = 288 \text{ mm}$$

Provide 6 mm ties at 280 mm spacing. Figure 5.8 illustrates the section.

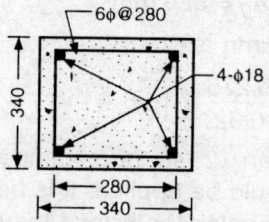

Fig. 5.8 Cross-section of Column in Example 5.3.

EXAMPLE 5.4 *Design of column with maximum per cent of steel:* A column is subjected to a load of 700 kN including the self weight. The effect length of the column is 5.275 m. Design the smallest possible size column with M 20 concrete and HYSD bars.

$$L_e = 5.275 \text{ m}, \quad \sigma_{sc} = 190 \text{ MPa}, \quad \sigma_c = 5 \text{ MPa}$$

The smallest size column is possible with the largest percentage of the reinforcement. So let $p = 6\%$.

Design of the section: The column will be a long one, therefore, one can design the section by trial and correction. First determine the size of the section for a short column with 6% steel. The allowable load on the short column is

$$P = A_c\,\sigma_{cc} + A_{sc}\,\sigma_{sc}$$

$$= A_g\,\sigma_{cc}\left(1 + \frac{A_{sc}}{A_g}\left(\frac{\sigma_{sc}}{\sigma_{cc}} - 1\right)\right)$$

$$= A_g\,(5)(1 + 0.06(37)) = 16.1\ A_g$$

The approximate size of the column assuming it as a square short one is

$$a = \sqrt{A_g} = \sqrt{\frac{P}{16.1}} = \sqrt{\frac{700\ 000}{16.1}} = 209 \text{ mm}$$

The slenderness factor of the column is

$$\frac{L_e}{a} = \frac{5275}{209} = 25.2$$

Slenderness factor of the column is 25.2 which is more than 12. This leads to the design a long column. Since the load will be enhanced, the size of the column required will also increase and consequently the slenderness factor will decrease. Therefore, assume the slenderness factor as 23. The load reduction factor is

$$C_r = 1.25 - \frac{L_e}{48b} = 1.25 - \frac{23}{48} = 0.77$$

The corresponding equivalent short column load is

$$P_0 = \frac{P}{C_r} = \frac{700\,000}{0.77} = 909\,090 \text{ N}$$

The gross area of the column for 6% steel is then given by

$$A_g = \frac{P_0}{16.1} = 56\,465 \text{ mm}^2$$

The size of the square column is

$$a = (b = D) = \sqrt{A_g} = 238 \text{ mm}$$

The slenderness factor of the column is

$$g = \frac{L_e}{b} = \frac{5275}{238} = 22$$

The assumed value is 23 against an approximate value of 22 which is slightly on the safer side. The actual size can be rounded off after selecting the area of the steel. The maximum allowable area of the reinforcement is

$$A_{sc} = pA_g = 0.06(238)(238) = 3399 \text{ mm}^2$$

The actual area of the reinforcement must be selected less than the above value so as not to exceed the 6% reinforcement. Select 8 numbers of 22 mm bars, then

$$A_{sc} = 8(380) = 3040 \text{ mm}^2$$

The load carrying capacity of the steel is

$$P_s = A_{sc}\sigma_{sc} = 3040(190) = 577\,600 \text{ N}$$

The actual slenderness factor of the column can be assumed as 22, so the reduction factor and the corresponding equivalent short column load are:

$$C_r = 1.25 - \frac{22}{48} = 0.7916$$

$$P_0 = \frac{P}{C_r} = \frac{700\,000}{0.7916} = 884210 \text{ N}$$

The load to be carried by the concrete is

$$P_c = P_0 - P_s = 306\,610$$

$$A_c = \frac{P_c}{\sigma_{cc}} = 61\,322 \text{ mm}^2$$

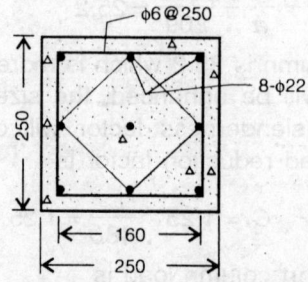

Fig. 5.9 Cross-section of Column in Example 5.4.

$$A_g = A_c + A_{sc} = 64362 \text{ mm}^2$$
$$b = D = \sqrt{A_g} = 253 \text{ mm}$$

The slenderness assumed is slightly more than the actual.

Provide 6 mm ties at 280 mm in two pairs as shown in Fig 5.9 and

$$p = 304\,000 / 62\,500 = 4.86$$

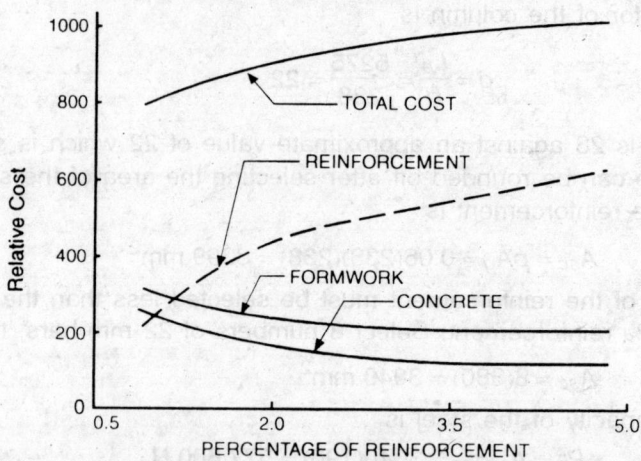

Fig. 5.10 Relative Cost Variation in RCC Columns.

Figure 5.10 illustrates the influence of variation of the percentage of the reinforcement on the cost of different components of the column.

5.8 Design Aid Tables for Axially Loaded Columns

It could be seen from previous examples that the design is done by trials. Even though the design method is simple, it can be further simplified with the aid of design aids. Table 5.4 gives the recommended reinforcement ties for a given area of the concrete. It also gives the load capacities of the reinforcement and the concrete for different characteristic strengths of steel and concrete. This table helps in arriving at the desired load capacity by selecting appropriate reinforcement and the area of the concrete. The use of the table is illustrated through examples.

EXAMPLE 5.5 Design a column of 3.6 m effective length to carry a total load of 600 kN (including the self-weight) with M20 concrete and HYSD bars.

Solution. We can assume about 50 to 70 per cent of the load to be carried by the concrete. So the approximate load to be carried by the concrete is about 400 kN to 500 kN. Move down in column 10 (of Table 5.4) which corresponds to M20 concrete and select the zone of the capacity in the range of 400 to 500 kN. In that zone, the sum of the capacities of the concrete and the reinforcement must make upto 600 kN. This happens at

$$A_g = 900 \text{ cm}^2 \text{ and } A_{sc} = 4 \text{ Numbers } \phi\ 16.$$

The corresponding capacities from the Table 5.5 are

$$P_c = 446 \text{ kN}$$
$$P_s = 152 \text{ kN}$$
$$\text{Total} = 598 \text{ kN}$$

The total capacity is 598 kN as against 600 kN load. Therefore, modify the concrete area marginally. If a square column is selected, size of the column should be

$$b = D = \sqrt{A_g} = 300 \text{ mm}$$

$$\text{Slenderness factor} = \frac{3.60}{0.3} = 12$$

$$\text{Use } b = D = 305 \text{ mm}$$

4 numbers of ϕ 16 main bars, and ϕ 6 ties at 250 mm spacing. This is exactly same as the solution given in Example 5.1.

5.9 Composite Columns

A composite column consists of steel or cast iron core with reinforcement around the core and all put together in concrete. Here again an equilibrium model is selected which does not satisfy the compatibility of strains and such a model can be called a *plastic model* as the stresses in the materials do not have the elastic compatibility of the strains. The working load carrying capacity of the column consists of three parts. The working load carrying capacity of each of the items can be expressed as

$$P_c = A_c \sigma_{cc} \tag{5.26}$$
$$P_s = A_{sc} \sigma_{sc} \tag{5.27}$$
$$P_{s0} = A_{s0} \sigma_{s0} \tag{5.28}$$

where P_{cs} P_s and P_{∞} are the working load capacities of the concrete, the reinforcement steel and the composite core respectively.

A_{s0} and σ_{s0} are the area of the cross-section of the composite core and allowable compressive stress in the material of the core respectively.

The allowable stresses in the core material are given in Table 5.1. The total working load capacity of a short column is given by

$$P_a = P_c + P_s + P_{s0}$$

or $\qquad P_a = A_c\sigma_{cc} = A_{sc}\sigma_{sc} + A_{s0}\sigma_{s0}$ (5.29)

The allowable stress on the steel reinforcement and on the core composite are more than the corresponding compatible stresses. Therefore, there is a tendency of spalling of the concrete or buckling of the reinforcement. Such a tendency can be eliminated by providing proper bonding and transverse reinforcement. The minimum amount of the transverse reinforcement as prescribed for the normal RCC column must be used in the composite columns. In addition, the area of the core must be limited to a *maximum of* 20% of the gross area of the cross section, sometimes even 20% core with 5% of the reinforcement becomes rather too large for good workability of placing the concrete. There should be a good working and bonding space between the core and the spiral or the ties. A minimum 75 mm *clearance* between the core and the helical reinforcement or 50 mm *clearance* between the core and the ties must be maintained. A reduction correction factor as discussed earlier should be applied to in the case of design of long columns. Composite columns are normally provided for large loads and where the size restrictions are severe.

EXAMPLE 5.6 *Design of a composite column:* A column of effective length 5.275 m is subjected to a load of 690 kN. Design a composite reinforced concrete column using structural steel core and reinforcement bars, and M20 concrete.

Effective length $L_e = 5.275$ m

Add extra 10 kN towards self-weight
Total design load $P = 690 + 10 = 700$ kN
The allowable stresses in MPa are:

$\sigma_{cc} = 5$, $\sigma_{sc} = 130$, and $\sigma_{s0} = 125$, $f_{ck} = 20$, $f_y = 230$

Design of the section: Assume the column as a short one and about 40% of the load is carried by the core, then the approximate area of the structural steel section is about $280000/125 = 2240$ mm^2. Select ISMB 150 as the core composite section, then

$$A_{sc} = 1900 \text{ mm}^2$$

Use only about 0.8% reinforcement, then the load carrying capacity of the section is

$$P_a = A_c\sigma_{cc} + A_{sc}\sigma_{sc} + A_{s0}\sigma_{s0}$$
$$= (A_g - A_{sc} - A_{s0})\,\sigma_{cc} + 0.008\,A_g\,\sigma_{sc} + 1900(125)$$
$$= (0.992\,A_g - 1900)(5) + 0.008\,A_g\,(130) + 237\,500$$
$$= 6\,A_g + 228\,000 \text{ N.}$$

The working load carrying capacity of the column can be equated to the design load, then

$$6A_g + 228\,000 = P = 700\,000 \text{ N}$$

or $\qquad A_g = \dfrac{700\,000 - 228\,000}{6} = 78\,667 \text{ mm}^2$

Assuming a square cross-section, the size of the column is

$$b = D = \sqrt{A_g} = 280.5 \text{ mm}$$

Table 5.4 Capacity Calculations for Column Sections

No.	Gross area of section (cm²)	Longitudinal reinforcement				Ties (in mm)		Concrete capacity (kN)		
		Nos	Dia (mm)	Capacity in kN for allowable stress in N/mm²		Dia	Spacing	M15	M20	M25
				130	190					
1	2	3	4	5	6	7	8	9	10	11
1	200	4	12	58.7	85.8	5	125	78	97	118
2	250					5	150	98	122	147
3	300					5	150	118	147	177
4	350							138	172	207
5	400							158	198	237
6	450							178	222	267
7	500					5	175	198	247	297
8	550					5	175	218	272	327
9	600	4	14	80	116.9	5	200	236	296	355
10	650							256	321	385
11	700							276	346	415
12	750							296	371	445
13	800	4	16	104	152	6	250	316	396	475
14	850							336	421	505
15	900							356	446	535
16	1000							396	496	595
17	1100	4	18	132	193	6	275	436	545	654
18	1200							476	595	714
19	1300	4	20	163	232	6	300	515	644	772
20	1400							555	694	832
21	1500							595	744	892
22	1600	8	16	209	305	6	250	693	792	950
23	1700							673	842	1010
24	1800							713	892	1070
25	1900							753	942	1130
26	2000	8	18	264	386	8	275	740	991	1188
27	2200							872	1091	1308
28	2400							950	1191	1428
29	2600	8	20	326	476	8	300	1028	1287	1544
30	2800							1108	1387	1664
31	3000							1188	1487	1784
32	3300	4, 4	20, 25	418	610	8	300	1306	1633	1959
33	3600							1426	1783	2139
34	3900							1546	1933	2319
35	4200	12	20	489	714	8	300	1664	2000	2497
36	4600	12	20	489	714	8	300	1824	2280	2737
37	5000	12	22	592	866	8	350	1984	2480	2977

Slenderness factor $= \dfrac{L_e}{b} = \dfrac{5.275}{0.2805} = 18.8$

Since the column is a long one, a load carrying reduction factor has to be applied. One should also ensure that there is enough clearance and cover space to the reinforcement. There should be a minimum of 50 mm clearance between the composite core and the ties, and about 40 mm cover to the main reinforcement. This requires minimum sizes as

$$b_{min} = 80 + 2(50 + 40) = 260 \text{ mm}$$
$$D_{min} = 150 + 2(50 + 40) = 340 \text{ mm}$$

Whereas the size required is 280.5 mm. Therefore, reduce the core area by selecting a lower structural shape. Select ISMB 125, the area of which is

$$A_{so} = 1660 \text{ mm}^2$$

and the minimum size of the square column is

$$D_{min} = 125 + 2(50 + 40) = 315 \text{ mm}$$

Use $b = D = 320$ mm then slenderness factor

$$= \frac{5.275}{0.32} = 16.48$$

$$C_r = 1.25 - \frac{16.48}{48} = 0.906$$

The corresponding equivalent short column load is

$$P_0 = \frac{P}{C_r} = \frac{700\,000}{0.906} = 772\,200 \text{ N}$$

$$A_g = 320(320) = 102\,400 \text{ mm}^2$$

Assuming approximately 2% reinforcement, the area of the concrete required is

$$A_c = A_g - A_{sc} - A_{so} = A_g - 0.02\,A_g - A_{so}$$
$$= 0.98\,A_g - A_{so}$$
$$= 0.98(102400) - 1660 = 98692 \text{ mm}^2$$

The working load carrying capacity of the section is

$$P_a = A_c\,\sigma_{sc} + A_{sc}\sigma_{cc} + A_{so}\,\sigma_{so}$$
$$= 98\,692(5) + A_{sc}\,(130) + 1660(125)$$
$$= 130\,A_{sc} + 700\,960 \text{ N}$$

Equating the load capacity of the design load, we have

$$A_{sc} = \frac{P_0 - 700\,960}{130} = 548 \text{ mm}^2$$

Select 4 numbers of ϕ 14 bars, then the actual area of the reinforcement provided is

$$A_{sc} = 4(153) = 615 \text{ mm}^2$$

and the percentage of the reinforcement is

$$p = \frac{100\,A_{sc}}{A_g} = \frac{61500}{102400} = 0.6\%$$

A reduction of 2% was given in the gross area towards the reinforcement, therefore the design is slightly on the safer side.

Transverse reinforcement: Assume 6 mm bars as ties, then the maximum spacing of the tie is limited to

$$s \le b = 320 \text{ mm}$$
$$s \le 16\phi = 16(14) = 224$$
$$s \le 48\phi_v = 48(6) = 288$$

Figure 5.11 illustrates the section.

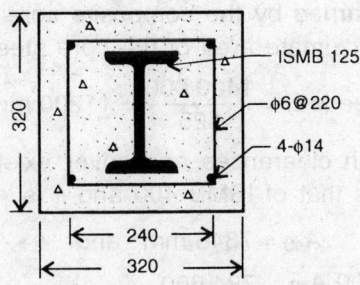

Fig. 5.11 Cross-section of Column in Examples 5.6.

Note: The problem of this example is same as that of the earlier Example 5.3. One can see that there is not much of a gain in providing the composite column for small loads. The total percentage of the steel in this case is 2.2 which is not high.

EXAMPLE 5.7 *Design of large composite column:* The load on an industrial building column is 2050 kN and the column is supporting cladding in one direction and is hinged at top by bracing and fixed at the base. The total height of the column is 12 m. Design a column such that the width and depth are restricted to 350 and 680 mm.

Distance between supports $L = 12$ m
Superimposed load $W_s = 2050$ kN
Concrete M20, HYSD steel and MS core are used.
The allowable stresses in MPa are:

$$\sigma_{cc} = 5, \ \sigma_{sc} = 190, \ \sigma_{s0} = 125, \ f_y = 415$$
$$b_{max} = 350 \text{ mm}, \ D_{max} = 600 \text{ mm}$$

Design of the section: Since one end of the column is fixed and the other end is hinged the effective length of the column w.r.t. depth is 0.7 L = 0.7(12) = 8.4 m (the buckling is prevented in the other direction). The maximum slenderness factor

$$g = \frac{L_e}{D_{max}} = \frac{8.4}{0.6} = 14$$

The load reduction factor is

$$C_r = 1.25 - \frac{14}{48} = 0.958$$

Let the self-weight of the column be about 150 kN, then the design load on the column is

$$P = W_s + 150 = 2200 \text{ kN}$$

Equivalent short column load is

$$P_0 = \frac{P}{C_r} = \frac{2200}{0.958} = 2296 \text{ kN}$$

Use

$$P_0 = 2300 \text{ kN}$$

Maximum load carried by the gross concrete area, assuming 300 by 600 mm section, is

$$P_{c0} = A_g \sigma_{cc} = (300)(600)(5) = 900\,000 \text{ N} = 900 \text{ kN}$$

This load is far less than the load to be carried by the section. Assume a composite section where the approximate load carried by the composite core is 2300 kN, less 900 kN, which is equal to 1400 kN. The approximate area of the core steel is

$$A_{s0} \text{ (needed)} = \frac{1400\,000}{125} = 11200 \text{ mm}^2$$

Use ISMB 400 so that enough clearances and cover exists in 300 by 600 mm size. The actual area of the core steel is that of ISMB 400 and it is

$$A_{s0} = 7846 \text{ mm}^2 \text{ and}$$

$$\frac{100\,A_{s0}}{A_g} = \frac{784\,600}{180\,000} = 4.36\%$$

Assume approximate percentage of the compression reinforcement as about 2, the load carrying capacity of the section is

$$P_a = (A_g - A_{sc} - A_{s0})\,\sigma_{cc} + A_{sc}\,\sigma_{sc} + A_{s0}\,\sigma_{s0}$$
$$= 0.98\,A_g\,\sigma_{cc} + A_{sc}\,\sigma_{sc} + A_{s0}\,(\sigma_{s0} - \sigma_{cc})$$
$$= 0.98(180\,000)(5) + 7846(125 - 5) + A_{sc}\,\sigma_{sc}$$
$$= 1823520 + A_{sc}\,\sigma_{sc}$$

Equating the capacity of the column to the equivalent short column load, we have

$$P_a = 1823\,520 + A_{sc}\,\sigma_{sc} = P_0 = 2300\,000 \text{ N}$$

or

$$A_{sc} = \frac{2300\,000 - 1823\,520}{190} = 2508 \text{ mm}^2$$

Provide 8 numbers of 20 mm bars, then the actual area of the reinforcement is

$$A_{sc} = 8(314) = 2512 \text{ mm}^2$$

The percentage of the reinforcement is

$$p_{sc} = \frac{251\,200}{180\,000} = 1.4\%$$

This is less than the assumed one and therefore, is on the safer side. The total percentage of the steel including the core is

$$p = 4.36 + 1.40 = 5.76\%$$

Design of transverse reinforcement: Select 6 mm diameter ties, then the spacing of the ties is limited to

$$s \leq b = 300 \text{ mm}$$
$$\leq 16\phi = 16(20) = 320 \text{ mm}$$
$$\leq 48\,\phi_v = 288 \text{ mm}$$

Provide 6 mm *ties* at 20 mm spacing.

The minimum depth of the column based on the minimum clearance and cover is

$$D_{min} = 400 + 2(50 + 40) = 580 \text{ mm}$$

$$b_{min} = 140 + 2(50 + 40) = 320 \text{ mm, so provide 320 mm width.}$$

Figure 5.12 illustrates the section.

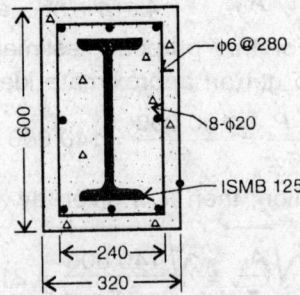

Fig. 5.12 Cross-section of Column in Example 5.7.

5.10 Infilled Columns

Steel pipes filled with concrete and used as columns are called *infilled* columns or concrete filled columns. Concrete when confined in a tube is subjected to triaxial state of stress. The load carrying capacity of the column in filled can be expressed as

$$P_a = A_c \, \sigma_{cc} \left(\frac{1 - 0.000025L^3}{r_c^2} \right) + A_{sc} \sigma_{sc} \tag{5.30}$$

where L = unsupported length of the column

A_{sc} = area of section of the shell of the tube

r_c = radius of gyration of the concrete core

$$\sigma_{sc} = 120 - 0.0034 \frac{L^2}{r_s^2} \, \text{N/mm}^2$$

r_s = radius of gyration of the shell

The infilled pipes are used for architectural purpose in buildings and also for small loads. A confined concrete normally should have higher strength when compared with the unconfined concrete used in the tied columns etc. However, the codes of practice recommend a lower allowable stress which depends on the slenderness ratio of the concrete fill is used. No additional reduction need to be made for the slenderness factor applied to the long tied columns.

EXAMPLE 5.12 *Design of infilled columns:* The design data are:

$$L_e = 5.275 \text{ m}$$
$$\sigma_{cc} = 5 \text{ MPa,} \quad P = 700 \text{ kN}$$

Let　　　a = radius of the concrete

t = thickness of the shell

Design of the section: The radius of gyration of the concrete core and steel shell are :

$$r_c = \sqrt{\frac{I_c}{A_c}} = \sqrt{\frac{\pi a^4}{4\pi a^2}} = \frac{a}{2}$$

$$r_s = \sqrt{\frac{I_s}{A_g}} = \sqrt{\frac{\pi((a+t)^4 - a^4)}{4\pi((a+t)^2 - a^2)}} = \frac{1}{2}\sqrt{((a+t)^2 + a^2)}$$

The design is carried by trial. For this purpose, estimate the gross area of concrete by neglecting the area of the shell to get an approximate idea. Then

$$A_c = \frac{P}{\sigma_{cc}} = \frac{100\,000}{5} = 140\,000 \text{ mm}^2$$

Let a_c = radius of such a section, then it is given by

$$a_c = \sqrt{\frac{A_c}{\pi}} = \sqrt{\frac{140\,000}{\pi}} = 210 \text{ mm or diameter} = 420 \text{ mm.}$$

Let the internal diameter of the pipe be 350 mm, and the thickness of the shell of the pipe be 4 mm, then the properties of the section are:

$$r_c = \frac{a}{2} = \frac{175}{2}$$

$$r_s = 0.5\sqrt{(179^2 + 175^2)} = 125 \text{ mm}$$

The corresponding allowable stresses are

$$\sigma_{acc} = 5\left(1 - 0.000025\left(\frac{5275}{8745}\right)^2\right) = 4.55 \text{ MPa}$$

$$\sigma_{asc} = 120 - 0.0034\left(\frac{L}{r_s}\right)^2$$

$$= 120 - 0.0034\left(\frac{5275}{125}\right)^2 = 114 \text{ MPa}$$

The working load carrying capacity of the column is

$$P_a = A_c\,\sigma_{acc} + A_{sc}\,\sigma_{asc}$$

$$= \pi a^2\,\sigma_{acc} + \pi(2a + t)t\,\sigma_{asc}$$

$$= \pi(175)^2\,(4.55) + \pi\,(354)(4)(114) = 944\,880 \text{ N}$$

The steel pipes are normally rolled in standard diameter with three to four thicknesses in each size.

5.11 Spirally Reinforced Columns

Circular cross-sectioned columns can have either independent ties or a continuous spiral tie or a continuous helical tie which is called a spiral to hold the main bars in position. A column in which a helical tie is provided to hold the main bars is called a *spirally reinforced* column.

The helical tie which is called a spiral, holds not only the longitudinal bars but also the concrete in partial confinement. The capacity of the column is enhanced by the circumferential action of the spiral and, in addition, the ductility of the column is improved. The total compressive strain capacity in the concrete confined by the spiral reinforcement is four to six times that corresponding to the tied columns. A load deformation behaviour of tied, spiral and infilled columns have a similar deformation character upto a point close to the maximum load, but beyond this point there is a substantial difference in the behaviour. The tied columns get split and fall off when the load reaches the ultimate load. Whereas in the case of a spirally reinforced column, the concrete cover over the spiral will fall off slowly, thus lowering the resistance upto a point. The helical reinforcement contains the concrete inside the helical area and absorbs the load through hoop action. The load carrying capacity of the column increases marginally as the crushing strength of confined concrete is improved. The composite column behaves similar to the tie column and fails as soon as the spalling of the cover takes place. The infilled concrete column behaviour is very similar to that of a spiral column. There is an improved crushing strength of the concrete and ductility in the behaviour. Figure 5.13 illustrates the behaviour of columns.

The load carrying capacity of a spiral column can be expressed as

$$P_a = 1.05 \, (A_c \, \sigma_{cc} + A_{sc} \, \sigma_{sc}) \tag{5.31}$$

It can be rearranged with respect to the gross area of the concrete as

$$P_a = 105 \, ((A_g - A_{sc}) \, \sigma_{cc} + A_{sc} \, \sigma_{sc}) \tag{5.32}$$

$$= 1.05 \, A_g \, \sigma_{cc} \, (1 + p(q - 1))$$

where $p = A_{sc}/A_c$ and $q = \sigma_{sc}/\sigma_{cc}$

For a given set of materials and a percentage of the main reinforcement, the gross area of cross section of the column can be obtained by equating the capacity to the load.

At least there must be six number of the longitudinal bars with minimum diameter as 12 mm. The diameter of the spiral reinforcement must be more than or equal to one-fourth the diameter of the bar or 5 mm.

Upper and lower bounds to the pitch of the spiral reinforcement ensure a partial confinement to the concrete and at the same time allow workability and interlocking of the cover with core. The pitch limitations are given by

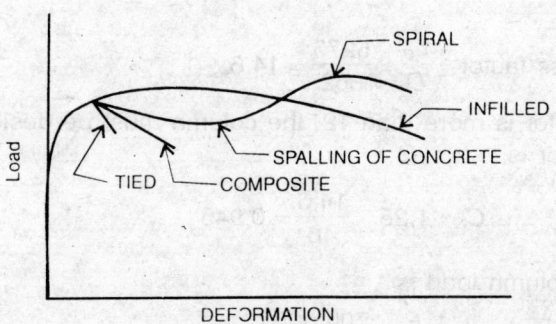

Fig. 5.13 Typical Load Transverse Deformation Behaviour of Columns.

$$s \leq \frac{\text{core diameter of the column}}{6}$$

$$\leq 75 \text{ mm}$$

$$s \geq 30 \text{ mm}$$

$$\geq 1.5 \, \phi$$

$$\geq \text{ maximum size of the aggregate}$$

$$\geq 3\phi_v$$

The minimum cover and reinforcement and other limitations are already stated.

EXAMPLE 5.8 *Design of spiral column with minimum reinforcement:* A column of effective length 5.275 m is subjected to a total load of 700 kN. Design a spiral column with M20 concrete and HYSD bars.

$$L_e = 5.275 \text{ m and } P = 700 \text{ kN}$$

$$f_{ck} = 20 \text{ MPa} \quad \sigma_{cc} = 5 \text{ MPa}$$

$$f_y = 415 \text{ MPa} \quad \sigma_{sc} = 190 \text{ MPa}$$

Design of the section : Let the percentage of the reinforcement be about 0.8, the load carrying capacity of the column assuming it as a short one is

$$P_a = 1.05 \, (A_c \, \sigma_{cc} + A_{sc} \, \sigma_{sc})$$

$$= 1.05 \, ((A_g - A_{sc})(5) + A_{sc} \; 190)$$

$$= 1.05 \, A_g \left(5 + (190 - 5) \frac{A_{sc}}{A_g} \right)$$

$$= 1.05 \, A_g \, (5 + 185(0.008)) = 6.804 \, A_g$$

where A_{sc}/A_g is assumed as 0.008.
Equating the load on the column to the load capacity, we have

$$A_g = \frac{P}{6.804} = \frac{700\,000}{6.804} = 102\,880 \text{ mm}^2$$

The approximate diameter of the column is

$$D = \sqrt{\frac{4A_g}{\pi}} = 362 \text{ mm.}$$

$$\text{Slenderness factor} = \frac{L_e}{D} = \frac{5275}{362} = 14.6$$

As slenderness factor is more than 12, the column must be designed as a long one and the load reduction factor is

$$C_r = 1.25 - \frac{14.6}{48} = 0.945$$

The equivalent short column load is

$$P_0 = \frac{P}{C_r} = \frac{700\,000}{0.945} = 740\,740$$

The area of the reinforcement needed is

$$A_{sc} = 0.008 \ A_g = 823 \ \text{mm}^2$$

Provide 6 numbers of 14 mm bars, then the actual area of the reinforcement is

$$A_{sc} = 6(154) = 924 \ \text{mm}^2$$

The load carrying capacity of the steel is

$$P_s = 1.05 \ A_{sc} \ \sigma_{sc} = 1.05(924)(190) = 184 \ 333 \ \text{N}$$

The load to be carried by the concrete is

$$P_c = P_0 - P_s = 740 \ 740 - 114 \ 333 = 556 \ 407 \ \text{N}$$

$$\text{Area of the concrete} = A_c = \frac{P_c}{1.05(5)} = 105 \ 981 \ \text{mm}^2$$

The gross area of the section is

$$A_g = A_c + A_{sc} = 106 \ 905 \ \text{mm}^2$$

The diameter of the column is

$$D = \sqrt{\frac{4 \ A_g}{\pi}} = 369 \ \text{mm}$$

Provide $D = 370$ mm and 6 numbers of $\phi 14$ bars.
This value is more than that assumed for slenderness factor.
Transverse reinforcement: The outer diameter enclosing the main bars is

$$D_c = D - 2(\text{cover}) = 370 - 80 = 290 \ \text{mm}$$

Select 5 mm diameter spiral reinforcement, then the pitch of the spiral is limited by

$$s \leq \frac{D_c}{6} = \frac{290}{6} = 48 \ \text{mm}$$

$$\leq 75 \ \text{mm and}$$

$$\geq 1.5\phi = 1.5(14) = 21 \ \text{mm}$$

$$\geq 3 \ \phi_v = 15 \ \text{mm}$$

$$\geq 30 \ \text{mm}$$

Provide 5 mm bar spiral with a pitch of 45 mm and percentage of steel is

$$p = \frac{100 \ A_{sc}}{A_g} = 0.86$$

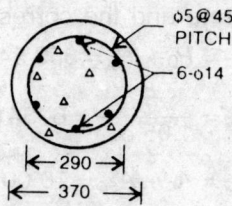

Fig. 5.14 Cross-section of Column in Example 5.8.

EXAMPLE 5.9 *Design of spiral column with maximum percentage of the reinforcement:* A column of effective length 5.275 mm is subjected to axial load of 700 kN. Design a spirally reinforced column with M20 concrete and HYSD bars. Keep the size of the column to the minimum possible.

$$L_e = 5.275 \text{ m} \qquad P = 700 \text{ kN}$$

$$\sigma_{cc} = 5 \text{ MPa}, \qquad \sigma_{sc} = 190 \text{ MPa}$$

Design of the section: The size of the column is to be kept as low as possible; therefore, select about 6% reinforcement. The load carrying capacity is

$$P_a = 1.05 \, (A_c \sigma_{cc} + A_{sc} \, \sigma_{sc})$$

$$= 1.05 \, A_g \, \sigma_{cc} \, (1 + p(q-1)) = 16.9 \, A_g$$

where $p = 0.06$ and $q = \sigma_{sc}/\sigma_{cc} = 38$

Let the column be assumed as a short one and then equating the load capacity to the load gives

$$A_g = \frac{P}{16.9} = \frac{700\,000}{16.9} = 41\,420 \text{ mm}^2$$

The approximate diameter of the column is

$$D = \sqrt{\frac{4 \, A_g}{\pi}} = 229 \text{ mm}$$

The slenderness factor = g $\qquad = L_e/D = 23$

Use a slenderness factor of 22, then the corresponding load reduction factor is

$$C_r = 1.25 - \frac{22}{48} = 0.7916$$

The equivalent short column load is

$$P_0 = \frac{P}{C_r} = 884\,210 \text{ N}$$

The area of the reinforcement required for 6% reinforcement is

$$A_{sc} = 0.06 \, A_g = 2480 \text{ mm}^2$$

Provide 7 numbers of 22 mm bars. The area of the reinforcement provided and the corresponding load capacity are:

$$A_{sc} = 7(380) = 2660 \text{ mm}^2$$

$$P_s = 1.05 \, A_{sc} \, \sigma_{sc} = 530\,670 \text{ N}$$

The load to be carried by the concrete and the corresponding concrete areas are:

$$P_c = P_0 - P_s = 353\,540 \text{ N and}$$

$$A_c = \frac{P_c}{1.05 \, \sigma_{cc}} = 67341 \text{ mm}^2$$

$$A_g = A_c + A_{sc} = 70\,001 \text{ mm}^2$$

The diameter of the column is

$$D = \sqrt{\frac{4A_g}{\pi}} = 298 \text{ mm}$$

$$\frac{L_e}{D} = 17.7$$

Since the slenderness factor assumed was 22 as against a value of 17.7, the size of the column can be decreased. Let

$$D = 280 \text{ mm}$$

$$\frac{L_e}{D} = \frac{5275}{210} = 18.8$$

and $C_r = 0.8583$

The equivalent short column load would be

$$P_0 = \frac{P}{C_r} = 815\,533 \text{ N}$$

Loads to be carried by the concrete and the corresponding concrete areas are:

$$P_c = P_a - P_s = 284\,863 \text{ N}$$

$$A_c = \frac{P_c}{1.05\,\sigma_{cc}} = 54\,259 \text{ mm}^2$$

$$A_g = A_c + A_{sc} = 56\,919 \text{ mm}^2$$

$$D = \sqrt{\frac{4A_g}{\pi}} = 269 \text{ mm}$$

Use $D = 275 \text{ mm}$

$$p = \frac{100\,A_{sc}}{A_g} = \frac{266\,000}{59\,395} = 4.48$$

Design of spiral reinforcement: The core diameter is

$$D_c = D - 80 = 195 \text{ mm}$$

Select 5 mm spiral wire, then the pitch of the spiral is controlled by

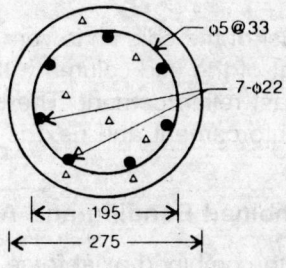

Fig. 5.15 Cross-section of Column in Example 5.9.

$$s \le \frac{D_c}{6} = 33 \text{ mm}$$
$$\le 75 \text{ mm}$$
$$\ge 1.5 \, \phi = 33 \text{ mm}$$
$$\ge 3\phi, = 18 \text{ mm}$$
$$\ge 30 \text{ mm}$$

Provide 5 mm bar spiral with a pitch of 33 mm (Fig 5.15).

EXAMPLE 5.10. *Cost comparison of the tied and spiral columns:* A column of effective length 5.275 m is subjected to an axial load of 700 kN. Compare the cost of tied, composite, spiral and infilled columns with different percentages of steel.

The unit costs of materials including supplying and placing the materials are:

Cost of M20 concrete	= Rs. 1500/m^3
Cost of HYSD bars	= Rs 2000/kN
Cost of formwork-rectangular section	= Rs 160/m^2
Cost of form work-circular section	= Rs 200/m^2
Cost of mild steel	= Rs. 1800/kN

Cost comparisons: The details along with costs are given in Table 5.5.

Table 5.5 Comparison of different Designs of Columns

Details	RCC — Tied		Composite		RCC	Spiral	Infilled
Percentage steel	0.7	2.0	4.86	MS, 2.6	0.86	4.48	4.6
D (mm)	340	300	250	320	370	275	350
A_{sc} (mm^2)	1016	1885	3040	2676	924	2600	4448
A_c (m^2)	0.1145	0.0881	0.0595	0.0997	0.1066	0.567	0.0962
Weight of main bars (N)	418	776	1251	1101	380	1095	1830
Weight of transverse bars (N)	52	52	100	52	114	167	0
Cost in Rs.							
Concrete	906	697	470	789	843	449	769
Steel	1140	1656	2702	2086	1206	2524	3294
Formwork	1246	1100	916	1173	1227	912	0
Total	3239	3453	3729	4043	3276	3885	4055
Relative cost w.r.t. tied column with minimum steel	1	1.05	1.13	1.23	1.01	1.18	1.23

It may be observed that the cost of the column is very much influenced by the percentage of the steel. The lowest cost is that of the tied column with minimum reinforcement. The next lowest is the spiral column with least reinforcement. The highest cost is that of a tied column with highest percentage of the reinforcement and next is that of the infilled column.

5.12 Columns Subjected to Combined Bending and Axial Force – Uncracked Section

Columns are invariably subjected to combined axial force and bending moment. The bending moment might be caused by the eccentricity of an external load, or by bending moment from the action of a transverse load, or due to the end moments coming from the interconnected members. There are very few columns in practice which are subjected to pure axial loads. Most members in practice act as plane frame members either because of symmetry in load or structure. In such cases, the columns are subjected to uniaxial bending moment in addition to the axial force. There are other members which are subjected to biaxial bending moment. The columns subjected to combined bending and axial force are broadly classified into two groups. They are:

a) Uncracked sections and
b) Cracked sections.

If the bending moment is not large when compared to the axial force, the tensile stress caused by the moment will be compensated by the compressive stress caused by the axial force, and thus resulting into no tension or small tension in the concrete. If the net tensile stress is less than the cracking stress, then the section is treated as uncracked section. If the net tensile stress caused in the concrete exceeds the cracking stress, the cracked portion of the section should be appropriately accounted in the design. Such a section is called a cracked section.

The uncracked sections are analysed by elastic theory assuming transformed cross-section. The reinforcement in the compression zone is restrained against lateral expansion, so the equivalent transformed section is taken equal to 1.5 times modular ratio times the actual area of steel.

The transformed tensile and compressive reinforcement areas are:

$$A_{ts} = mA_{st} \tag{5.33}$$

$$A_{tc} = 1.5mA_{sc} \tag{5.34}$$

where A_{ts} and A_{tc} are the transformed areas of the tension and compression steels. There are two sub-groups in the uncracked sections. Case one is when the entire section is under compression and the second case is that a part of the section is under net tension, yet uncracked. Theoretically one should solve these two problems with different criteria, but they are combined into one and governed by a condition corresponding to the no tension condition. The axial and bending stresses are computed as:

$$\sigma_{cc} = \frac{P}{A_c + 1.5mA_{sc}} \tag{5.35}$$

$$\sigma_{cb} = \pm \frac{M}{Z_t} \tag{5.36}$$

where σ_{cc} = actual compressive stress in concrete due to axial load

σ_{cb} = actual bending stress in concrete due to bending

Z_t = modulus of the section and it is = $\dfrac{2I_t}{D}$

I_t = transformed moment of inertia of the column

The design criterion of the column is

$$\frac{\sigma_{cc}}{\sigma_{acc}} + \frac{\sigma_{cb}}{\sigma_{acb}} \leq \alpha \tag{5.37}$$

where $\alpha = 1$ for $DL + LL$ combination and $= 1.333$ for Wind or earthquake load condition. For the net tension developed in the section to be within the allowable tensile stress, the following inequalities should be satisfied.

$$(\sigma_{cb} - \sigma_{cc}) \leq 0.25\,(\sigma_{cb} + \sigma_{cc}) \tag{5.38}$$

$$\leq 0.6\,f_{acr} \tag{5.39}$$

where σ_{acc} = allowable compressive stress in concrete

σ_{acb} = allowable bending compressive stress in concrete

f_{acr} = allowable modulus of rupture

Note: σ_{cc} is to be treated as the actual stress and in the limit it tends to the allowable value. In the earlier portion of the chapter, this was treated as the allowable stress. Actually the allowable compressive stress in concrete is denoted by σ_{acc}, which includes an additional subscript *a*. This extra subscript was omitted in the earlier discussion for convenience and also to be consistent with the notations adopted in the code. $0.6f_{acr}$ can be equated to 0.75 times the modulus of rupture of the concrete at 7 days curing.

The inequality (5.38) can be rearranged as

$$\sigma_{cb} \leq 1.67\,\sigma_{cc} \tag{5.40}$$

In general the area of the compression reinforcement is placed equally at each of the bending faces, then the moment of inertia of the section can be taken as

$$I_t = \frac{bD^3}{12} + \frac{(m-1)A_{sc}D_s^2}{4} \tag{5.41}$$

where D_s = distance between the centroids of the compression and nominal tension face reinforcements.

In a normal design problem, one has to solve for the design variables such as *b*, *D*, and A_{sc} for given values of *P* and *M*. However, such a direct solution in this problem is rather difficult, so the problem is solved by pre-assigning a section and testing the adequacy of it for the given loads.

The transformed properties of the section can be modified as:

$$A_t = A_c + 1.5mA_{sc} = A_g\,(1 + (1.5m - 1)p) \tag{5.42}$$

$$I_t = \frac{bD^3}{12} + \frac{(m-1)A_{sc}D_s^2}{4} = A_g\frac{D^2}{12} + A_g\,(m-1)\,p\,\frac{D_s^2}{4}$$

$$= \frac{A_g}{12}\,(1 + 3(m-1)pc^2)D^2 \tag{5.43}$$

where $$c = \frac{D_s}{D},\ p = \frac{A_{sc}}{A_g}$$

Let $$e = \frac{M}{P} \tag{5.44}$$

Then the interaction formula is modified to suit the design problem:

$$\frac{\sigma_{cc}}{\sigma_{acc}} + \frac{\sigma_{cb}}{\sigma_{acb}} \leq \alpha$$

$$\frac{P}{A_g}\left(\frac{1}{(1+(1.5m-1)p)\sigma_{acc}}+\frac{6\beta}{(1+3(m-1)pc^2)\sigma_{acb}}\right)\le\alpha \qquad (5.45)$$

or $$A_g\le\frac{P}{\alpha}\left(\frac{1}{(1+(1.5m-1)p)\sigma_{acc}}+\frac{6\beta}{(1+3)(m-1)pc^2)\sigma_{acb}}\right) \qquad (5.46)$$

where $\beta=\dfrac{e}{D}$

Inequality (5.46) will give the gross area of the section for a given percentage of steel and depth of the section.

EXAMPLE 5.11 *Design of short column for combined axial force and bending:* A column of effective length of 3.6 m is subjected to an axial force of 700 kN and bending moment of 81.9 kNm in live load condition. Design the column with M 20 concrete and HYSD bars.

$L_e = 3.6$ m, $\qquad P = 700$ kN, $\qquad M = 81.9$ kNm,
$f_{ck} = 20$ MPa, $\qquad f_y = 415$ MPa,
$\sigma_{acc} = 5$ MPa, $\qquad \sigma_{acb} = 7$ MPa $\qquad m = 13$.

Design of the section: The effective eccentricity of the load is

$$e=\frac{M}{P}=\frac{81.9}{700}=0.117\text{ m}$$

Assume that the column is a short one and the percentage of the reinforcement is $p = 0.008$.

As a first trial assume:

$$c=\frac{D_s}{D}=\frac{D-100}{D}=0.82$$

$$\beta=\frac{e}{D}=0.2$$

Since the loads are due to the live load conditions, $\alpha = 1.0$.
The first and the second terms in the inequality (5.46) are:

$$f_1=(1+(1.5m-1)p)\sigma_{acc} \qquad (5.47)$$
$$=(1+(1.5(13)-1)0.008)(5)=574$$
$$f_2=(1+3(m-1)pc^2)\sigma_{acb} \qquad (5.48)$$
$$=(1+3(12)0.008(0.82)^2)(7)=8.356$$
$$A_g=\frac{P}{\alpha}\left(\frac{1}{f_1}+\frac{6\beta}{f_2}\right)=\frac{700000}{1}\left(\frac{1}{5.74}+\frac{6(0.2)}{8.356}\right)$$
$$=700\,000\left(\frac{1}{5.74}+\frac{6(0.2)}{8.356}\right)=224\,478\text{ mm}^2$$

Let $\qquad D = 600$ mm, then

$$b=\frac{A_g}{D}=371\text{ mm}$$

The actual design non-dimensional coefficients are:

$$\beta = \frac{e}{D} = \frac{117}{600} = 0.195$$

$$c = \frac{D_s}{D} = \frac{D-100}{D} = \frac{500}{600} = 0.833$$

These values are very close to the assumed variables. Therefore, finalize the actual area of the reinforcement and the slenderness factor. Let

Let $b = 360$ then $\qquad \dfrac{L_e}{b} = \dfrac{3600}{360} = 10$

The column is a short one so no reduction factor need to be applied. The approximate area of the reinforcement is

$$A_{sc} = pA_g = 0.008(600)(371) = 1776 \text{ mm}^2$$

Use 6 numbers of 20 mm bars, then

$$A_{sc} = 6(314) = 1884 \text{ mm}^2$$

(Three bars on each face of the depth of the section)

Let $\qquad\qquad D = 600 \text{ mm} \qquad b = 360 \text{ mm},$

$$A_g = bD = 216\,000$$

$$A_c = A_g - A_{sc} = 204\,116 \text{ mm}^2$$

Cover to the main bars = 40 mm, then

$$D_s = D - 2(40) - \phi = 500 \text{ mm}$$

Then the properties of the section are:

$$A_t = A_g + (1.5m - 1)A_{sc} = 240\,854 \text{ mm}^2$$

$$I_t = \frac{bD^3}{12} + (m - 1)\frac{A_{sc}}{4} d_s^2 = 0.7893\,(10^{10}) \text{ mm}^4$$

$$\sigma_{cc} = \frac{P}{A_t} = \frac{700\,000}{240\,854} = 2.91 \text{ MPa}$$

$$\sigma_{cb} = \frac{M}{Z_t} = \frac{81900.000(300)}{0.7893\,(10^{10})} = 3.11 \text{ MPa}$$

The interaction formula gives

$$\frac{\sigma_{cc}}{\sigma_{acb}} + \frac{\sigma_{cb}}{\sigma_{acb}} = \frac{2.93}{5} + \frac{3.11}{7} = 0.581 + 0.444 = 1.025 > 1$$

The net tension is

$$\sigma_{cb} - \sigma_{cc} = 0.18 \text{ MPa (small)}$$

Since the value of the interaction formula is violated marginally, *select the following* details

$$b = 365 \text{ mm}, \ D = 600 \text{ mm}$$

$$A_{sc} = 1384 \text{ mm}^2.$$

Design of transverse reinforcement
Select 6 mm tie, then the spacing is governed by

$$s \leq b = 365 \text{ mm}$$
$$\leq 16\phi = 320 \text{ mm}$$
$$\leq 48\phi_v = 288 \text{ mm}$$

Provide 6 mm ties at 280 mm spacing.

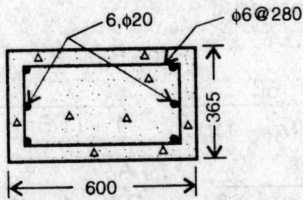

Fig. 5.16 Cross-section of Column in Example 5.11.

Direct design approach: In most problems of combined axial force and bending, the width of the column can be assumed and the depth of the column can be solved for a given percentage of reinforcement.

Inequality 5.47 can be expressed as

$$D \leq \frac{P}{\alpha b}\left(\frac{1}{f_1} + \frac{6e}{Df_2}\right) \tag{5.49}$$

where f_1 and f_2 are given in Eqs. (5.48) and (5.49).

In the limit, inequality (5.50) can be rearranged as

$$D = \frac{w}{f_1} + \frac{6ew}{Df_2} \tag{5.50}$$

where $w = P/b\alpha$ = modified load per unit width.

The same can be expressed as

$$D^2 - \frac{wD}{f_1} - \frac{6\,ew}{f_2} = 0 \tag{5.51}$$

$$D = \frac{w}{2f_1}\left(1 + \sqrt{\left(1 + \frac{24\,ef_1f_1}{wf_2}\right)}\right) \tag{5.52}$$

All the quantities on the right hand side of Eq. (5.53) are either known or can be assumed, therefore, the depth of the column can be computed.

5.13 Columns Subjected to Combined Bending and Axial Force—Cracking Limits

Some columns are subjected to large bending moments combined with small forces. The maximum tensile stress in a section can be obtained as

$$\sigma_{ct} = \frac{M}{z_t} - \frac{P}{A_t}$$

or $\qquad \sigma_{ct} = \sigma_{cb} - \sigma_{cc}$

where σ_{ct} = tensile stress

$$\sigma_{cb} = \frac{6Pe}{A_g D (1 + 3(m-1)pc^2}$$

$$\sigma_{cc} = \frac{P}{A_g (1 + (1.5m + 1)p)}$$

Let f_{cr} = allowable cracking tensile stress of the concrete
The section is then said to be cracked if

$$\sigma_{ct} \geq f_{cr}$$

$$\frac{P}{A_g} \left(\frac{6\beta}{1 + 3(m-1)pc^2} - \frac{1}{1 + (1.5m-1)p} \right) \geq f_{cr} \tag{5.53}$$

$$\frac{6\beta}{1 + 3(m-1)pc^2} \geq \frac{f_{cr} A_g}{P} + \frac{1}{1 + (1.5m-1)p}$$

The cracking will take place if

$$\frac{e}{D} = \beta \geq \frac{1 + 3(m-1)pc^2}{6(1 + (1.5\ m-1)p)} \left(1 + \frac{f_{cr} A_g (1 + (1.5m-1)p)}{P} \right) \tag{5.54}$$

One can see that the eccentricity ratio for a cracked section is not a simple function. However, the approximate range of the e/D for which the section is likely to crack is about 0.22 to 0.36.

There may be situations where a column has to be designed as an uncracked section irrespective of the eccentricity. Such situations may arise in water retaining and similar structures. The depth of the column is then controlled by the tensile stress in the concrete and it may be that the overall depth has to be as much as 4 to 5 times the eccentricity of the load. The governing equation for the design of an uncracked section with large eccentricity can be expressed as

$$\sigma_{ct} \leq \sigma_{act} = \frac{f_{cr}}{\text{factor of safety}} = 0.6(0.7)\sqrt{f_{ck}}$$

$$\leq 0.42\sqrt{f_{ck}} = \sigma_{act} \tag{5.55}$$

This will lead to

$$\frac{P}{A_g} \left(\frac{6\beta}{1 + 3(m-1)pc^2} - \frac{1}{1 + (1.5m-1)p} \right) \leq \sigma_{act} \tag{5.56}$$

The value of A_g can be obtained by trails and it is given by

$$A_g \geq \frac{P}{\sigma_{act}} \left(\frac{6\beta}{1 + 3(m-1)pc^2} - \frac{1}{1 + (1.5m-1)p} \right) \tag{5.57}$$

5.14 Design of Cracked Columns

The total reinforcement in cracked columns can be divided into tension and compression steels. Let

A_{sc} = compression reinforcement
A_{st} = tension reinforcement

nd = neutral axis distance from the compression face
d = distance of the tension steel from the extreme compression face
P = axial force
M = bending moment and
d_c = distance of the compression steel from compression fibre.

Figure 5.17 illustrates the cross section, and Fig. 5.17b indicates the external forces and the corresponding resisting forces. Tensile strength of the concrete is neglected.

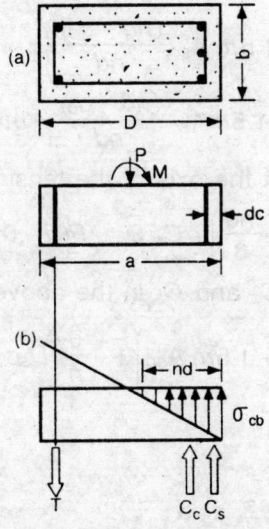

Fig. 5.17 Notations in Cracked Column.

Let C_c = compressive force resisted by the concrete
C_s = compressive force resisted by the reinforcement
T = tensile resistance of the reinforcement
σ_{acb} = compressive bending stress in the extreme fibre

Then the compressive stress in the steel can be expressed by the compatibility consideration and it is

$$\sigma_c = 1.5m \frac{(nd - d_c)}{nd} \tag{5.58}$$

where $1.5m$ = enhanced modular ratio.
The corresponding tensile stress in the steel is

$$\sigma_{st} = m \frac{(d - nd)\sigma_{cb}}{nd} \tag{5.59}$$

The total compressive forces in the concrete and steel, and the tensile force in the steel are:

$$C_c = \frac{bnd\,\sigma_{cb}}{2} \tag{5.60}$$

$$C_s = A_{sc} \frac{(1.5m)(nd - d_c)}{nd}$$

$$T = A_{st}\sigma_{st} \tag{5.61}$$

The value of C_s is slightly over-estimated as no deduction is allowed for the area of the concrete section in the compression zone. One can use $(m - 1)$ in place of m. However, in view that the problem is a plane strain one, this approximation is safe.

The equilibrium of forces gives

$$C_c + C_s - T = P, \text{ or}$$

$$\left(\frac{nbd}{2} + 1.5m\,A_{sc} \frac{(nd - d_c)}{nd} \right)\sigma_{cb} - A_{st}\,\sigma_{st} = P \tag{5.62}$$

$$\text{or } \sigma_{cb}\left(\frac{nbd}{2} + 1.5mA_{sc} \frac{(nd - d_c)}{nd} - mA_{st} \frac{(d - nd)}{nd} \right) = P \tag{5.63}$$

The moment of forces taken about the axis of the tension steel gives

$$C_c\left(d - \frac{nd}{3} \right) + C_s D_s - P(d - 0.5D) = M \tag{5.64}$$

The substitution of the values of C_c and C_s in the above gives

$$n\left(1 - \frac{u}{3} \right)\frac{bd^2}{2}\,\sigma_{cb} + 1.5m\,A_{sc}\left(1 - \frac{u}{n} \right)D_s\sigma_{cb} - P\left(1 - \frac{D}{2d} \right)d = M \tag{5.65}$$

where $u = \dfrac{d_c}{d}$

Let $\quad e = \dfrac{M}{P}$, then Eq. (5.65) gives

$$0.5n\left(1 - \frac{n}{3} \right)bd^2\sigma_{cb} + 1.5m\,A_{sc}\left(1 - \frac{u}{n} \right)D_s\,\sigma_{cb} - P\left(e + d - \frac{D}{2} \right) = 0 \tag{5.66}$$

Equations (5.63) and (5.66) are the two governing conditions of a cracked column. In case of design problems, the known quantities are: $P, M, m, \sigma_{cb} = \sigma_{acb}$ or $\sigma_{st} = \sigma_{ast}$. The design parameters are:

$$A_{st} . A_{sc}, b \text{ and } D$$

In addition, the values d_c, D_s and d are associated with the reinforcement details which can be worked out from D, A_{sc} and A_{st}.

The value of n can be computed from the compatibility of the strains for full allowable stresses. There are four main design parameters of which only two can be evaluated from the governing conditions. Therefore, the designer must make some assumptions. Usually one can assume b and A_{sc}/A_{st} to solve the problem. In addition, the values of d_c, D_s and d or D must be assumed so as to solve the problems. The ranges of these values are:

$d_c = 0.15d$ for small columns,

$\quad = 0.1d$ for larger columns where $d > 600$ mm

$D_s = d - d_c$ and is in the range of

$\quad = 0.85d$ for small columns,

$\quad = 0.9d$ for larger columns

$D = d + d_c$ and is in the range of
 = 1.15d for small columns,
 = 1.1d for larger columns

In the design problem, one is given the allowable stresses and the external loads to determine the size of the column and the reinforcement details. As discussed earlier, there are some design details such as cover, etc. which depend on the design parameters. Such details have to be assumed to start with to solve the problem. For the columns of depth less than 600 mm, the following can be assumed for all practical purposes:

$$d_c = 0.15 \ d, \tag{5.67}$$
$$D_s = 0.85 \ d \text{ and} \tag{5.68}$$
$$D = 1.15 \ d \tag{5.69}$$

And for $d > 600$ mm, the following can be assumed:

$$d_c = 0.1 \ d \tag{5.70}$$
$$D_s = 0.9 \ d \text{ and} \tag{5.71}$$
$$D = 1.1 \ d \tag{5.72}$$

One of the major confusing questions to be answered is whether the column should be designed for full stress condition, that is, the actual stresses in steel and concrete reaching the full allowable levels.

$$\sigma_{st} = \sigma_{ast} \text{ and } \sigma_{cb} = \sigma_{acb}$$

From the compatibility of strains, the balanced neutral axis is given by

$$\frac{nd}{\sigma_{acb}} = \frac{m(d - nd)}{\sigma_{ast}} \tag{5.73}$$

Irrespective of the M and P, if one forces design of a balanced section, then there is a danger of its coming out with uneconomical depths, sometimes even with unrealistic depths. This full stress design is reasonable in the pure moment case but not in the columns where a large compression area is required to resist P. It is, therefore, recommended that one should design the column with full compressive stress in concrete and then check for the acceptability of the stress in the steel. Very often the tensile stress in the reinforcement is far less than the allowable value. The question that arises is then how to solve for the neutral axis. Equations (5.66) and (5.69) which govern the equilibrium of forces should be taken as the governing conditions from which the neutral axis can be obtained by iteration. The two equations can be rearranged as:

$$\left(\frac{nbd}{2} + 1.5mq \ A_{sc} \left(1 - \frac{u}{n} \right) - mA_{st} \left(\frac{1}{n} - 1 \right) \right) \sigma_{acb} = P \tag{5.74}$$

$$\left(\frac{n}{2} \left(1 - \frac{n}{3} \right) \frac{bd^2}{2} + 1.5mq \ A_{st} \left(1 - \frac{u}{n} \right) D_s \right) \sigma_{acb} - P \left(e + d - \frac{D}{2} \right) = 0 \tag{5.75}$$

where $\quad q = \dfrac{A_{sc}}{A_{st}}$

Let $p_t = A_{st}/bd$, then the above equation can be written as

$$bd\left(\frac{n}{2} + 1.5mqp_t\left(1 - \frac{u}{n}\right) - mp_t\left(\frac{1}{n} - 1\right)\right)\sigma_{acb} = P \tag{5.77}$$

and $$bd^2\left(\frac{n}{2}\left(1 - \frac{n}{3}\right) + 1.5mqp_t\left(1 - \frac{u}{n}\right)\frac{D_s}{d}\right)\sigma_{acb} - Pd\left(\frac{e}{d} + 1 - \frac{D}{2d}\right) = 0$$

or $$bd\left(\frac{n}{2}\left(1 - \frac{n}{3}\right) + 1.5mqp_t\left(1 - \frac{u}{n}\right)\frac{D_s}{d}\right)\sigma_{acb} - P\left(\frac{e}{d} + 1 - \frac{D}{2d}\right) = 0 \tag{5.78}$$

One must assume a percentage of the tension reinforcement, a ratio of the compression reinforcement to that of the tension and relative values of d_c, D_s and D with respect to d, and then the problem can be solved by iteration. It is rather difficult to give a close-form solution to this design problem, and it is explained through illustrative examples.

EXAMPLE 5.12 *Design of cracked short column subjected to reversible bending moments:* A column of effective and also total length of 3.6 m is subjected to an axial force of 660 kN and a bending moment of 350 kNm. The direction of the bending moment is reversible under the normal live load condition. Design the column with M 20 concrete and HYSD bars.

$$L = L_e = 3.6 \text{ m}, \ f_{ck} = 20 \text{ MPa}, \ f_y = 415 \text{ MPa}$$

$$\sigma_{acb} = 7 \text{ MPa}, \ \sigma_{ast} = 230 \text{ MPa}, \ m = 13$$

Superimposed load	$= 660$ kN
Reversible BM	$M = 350$ kNm
Let the self-weight of the column	$= 40$ kN
Total axial force	$P = 700$ kN
Let	$d_c = 0.15d, \ D_s = 0.85d, \ D = 1.15d$

Since the moment–axial force ratio is 0.5 m, which is large, the column can be assumed as cracked. The bending moment can act from either direction, so the reinforcement on either face must be same, that is

$$A_{sc} = A_{st}$$

or $$q = \frac{A_{sc}}{A_{st}} = 1$$

Let $b = 500$ mm, then $\dfrac{L_e}{b} < 12$

Let $\dfrac{100 \, A_{st}}{bd} = 1\%$ or $p_s = 0.01$

$$e = \frac{M}{P} = \frac{350}{700} = 0.5 \text{ m} = 500 \text{ mm}$$

Let $u = \dfrac{d_c}{d} = 0.15$

Design of the section: The substitution of various quantities in the governing conditions. namely, Eqs. (5.77) and (5.78) gives

$$bd\left(0.5n + 1.5(13)(1)(0.01)\left(1 - \frac{0.15}{n}\right) - (13)(0.01)\left(\frac{1}{n} - 1\right)\right)(7) = 700\ 000\ N \text{ and} \qquad \text{(a)}$$

$$bd\left(\frac{n}{2}\left(1 - \frac{n}{3}\right) + 1.5(13)(1)(0.01)\left(1 - \frac{0.15}{n}\right)(0.85)\right)(7) - 700\ 000\left(\frac{500}{d} + 1 - 0.57\right) = 0 \qquad \text{(b)}$$

The above two equations can be simplified for $b = 500$ mm as

$$d(n^2 + 0.65\ n - 0.3185) = 400\ n \qquad \text{(c)}$$

and $$d^2(3n^2 - n^3) + 0.9945n - 0.149) - 516nd - 600\ 000\ n = 0 \qquad \text{(d)}$$

The section of the above two nonlinear simultaneous equations can be obtained by trail,

Try $n = 0.6$, then the Eqs. (c) and (d) give

$$d = 556 \text{ mm}$$

and $1.3117\ d^2 - 309.6d - 360\ 000 = 0$

or $$d = 655 \text{ mm}$$

Try $n = 0.55$, then the Eqs. (c) and (d) give

$$d = 644 \text{ mm}$$

and $1.1391d^2 - 263.8d - 330\ 000 = 0$

or $$d = 677 \text{ mm}$$

Try $n = 0.54$, then the Eqs. (c) and (d) give

$$d = 667 \text{ mm}$$

and $1.105d^2 - 278.6d - 324\ 000 = 0$

or $$d = 674 \text{ mm}$$

The values of d is converging towards 674 mm, therefore use

$$d = 680 \text{ mm and } n = 0.545$$

The corresponding areas of the reinforcement are:

$$A_{st} = A_{sc} = 0.01(bd) = 3400 \text{ mm}^2$$

Provide 7 numbers of 25 mm bars on each face, then

$$A_{st} = A_{sc} = 3430 \text{ mm}^2$$

$$p_t = \frac{A_{st}}{bd} = 0.01$$

All the 7 bars can be provided in a row, so the other design parameters can be estimated for a cover of 40 mm, and they are

$$d_c = 40 + 13 = 53 \text{ mm}, D_s = d - d_c = 627 \text{ mm}$$

$$D = d + \phi/2 + 40 = 735 \text{ mm}$$

Check for the capacity of the column

Use $$D = 735 \text{ mm}, d_c = 55 \text{ mm}, nd = 0.545(680) = 370 \text{ mm}$$

The capacity of the column should be evaluated by substituting the actual design details in the governing equations. The axial load capacity of the column is

$$P_a = \left(\frac{nbd}{2} + 1.5m\, A_{sc} \frac{(nd - d_c)}{nd} - mA_{st} \frac{(d - nd)}{nd}\right)\sigma_{acb}$$

$$= (92650 + 56943 - 37359)(7) = 785635 \text{ N}$$

The allowable moment capacity of the section is

$$M_a = \left(\frac{nbd}{2}\left(d - \frac{nd}{3}\right) + 1.5m\, A_{sc} \frac{(nd - d_c)}{nd} D_s\right)\sigma_{acb}$$

$$= P(d - 0.5D)$$

$$= (51\ 491\ 566 + 37\ 297\ 426)(7) - 245\ 510\ 937$$

$$= 376\ 012\ 711 \text{ Nmm} = 376.0 \text{ kNm}$$

The working axial force and moment capacity of the column are 785 kN and 376 kNm against the corresponding required values of 700 kN and 350 kNm. The column is safe against the loads. *Provide four-legged* 8 mm bars at 320 mm spacing. Figure 5.18 illustrates the sectional details.

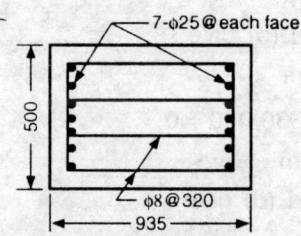

Fig. 5.18 Cross-section of Column in **Example 5.12**.

EXAMPLE 5.13 *Design of cracked long column:* A column supported at 3.6 m apart with hinge and roller is loaded with 660 kN and 350 kNm of axial force and bending moment respectively. Design the column with width not exceeding 200 mm.

$$L_e = L = 3.6 \text{ m}$$

Let $f_{ck} = 20$ MPa, $f_y = 415$ MPa, then

$$\sigma_{acb} = 7 \text{ MPa}, \sigma_{ast} = 230 \text{ MPa}, m = 13$$

Superimposed load = 660 kN

Let the self-weight be = 40 kN

Total axial load P = 700 kN

Bending moment $M = 350$ kN

Width of the column b = 200 mm

$$\frac{M}{P} = \frac{350}{700} = 0.5 \text{ m} = 500 \text{ mm}$$

Since the eccentricity is large, the column can be assumed to have a cracked section.

Assume $A_{st} = A_{sc} =$ about 3%, that is

$$p_t = p_c = \frac{A_{st}}{bd} = \frac{1.1\, A_{st}}{bD} = 0.03(1.1) = 0.033$$

Use $p_t = 0.033$; $q = \dfrac{A_{sc}}{A_{st}} = 1$

$$u = \frac{d_c}{d} = 0.15, \; c = \frac{D_s}{d} = 0.85, \; D = 1.15d$$

Design of the section. Slenderness factor of the column

$$\frac{L_e}{b} = \frac{3.6}{0.2} = 18$$

The column is to be treated as long one and the load reduction coefficient is

$$C_r = 1.25 - \frac{L_e}{48b} = 0.875$$

The corresponding short column equivalent load

$$P_o = \frac{P}{C_r} = 800,000 \text{ N}$$

$$e = \frac{M}{P_o} = \frac{350}{800} = 0.4375 \text{ m} = 438 \text{ mm}$$

The governing condition for axial force limiting the compressive stress in concrete to the allowable value is taken from Eq. (5.77) and it is

$$bd\left(0.5n + 1.5m\,qp_t\left(1 - \frac{u}{n}\right) - mp_t\left(\frac{1}{n} - 1\right)\right)\sigma_{acb} = P_0 \qquad \text{(a)}$$

The moment equilibrium gives (Eq. 5.78)

$$bd\left(\frac{n}{2}\left(1 - \frac{n}{3}\right) + 1.5mqp_t\left(1 - \frac{u}{n}\right)c\right)\sigma_{acb} - P_0\left(\frac{e}{d} + 1 - \frac{D}{2d}\right) = 0 \qquad \text{(b)}$$

The substitution of the different assumed design variables in the above two expressions gives

$$200d\left(0.5n + 1.5(13)(1)(0.333)\left(1 - \frac{0.15}{n}\right) - 13(0.033)\left(\frac{1}{n} - 1\right)\right)(7) = 800\,000 \text{ N}$$

and

$$200d\left(\frac{n}{2}\left(1 - \frac{n}{3}\right) + 1.5(13)(1)(0.033)\left(1 - \frac{0.15}{n}\right)(0.85)(7)\right) - 800\,000\left(\frac{438}{d} + 1 - 0.57\right) = 0$$

These two equations reduce to:

$$d(n^2 + 2.145n - 1.051) = 1143n \qquad \text{(c)}$$

and $\qquad d^2(3n^2 - n^3 + 3.282n - 0.492) - 1474\,nd - 1\,501\,714n = 0 \qquad$ (d)

The solution of the above two nonlinear simultaneous equations can be obtained by iteration.

Try $n = 0.71$

from Eq. c, $d = 931$ mm

and $2.993\,d^2 - 1047d - 1066\,217 = 0$ from Eq. d

or $d = 797$ mm

Try $n = 0.72$

$$d = 813 \text{ mm}$$

and $3.053d^2 - 1061d - 1081\,234 = 0$

or $d = 794$ mm

The value of d is converging towards 810 and of n to 0.725; therefore, use

$$n = 0.725 \quad \text{and} \quad d = 810 \text{ mm, then}$$

$$A_{st} = A_{sc} = 0.033bd = 5544 \text{ mm}^2$$

Provide nine numbers of 28 mm bars on each face, then the actual steel provided is

$$A_{st} = A_{sc} = 9(615) = 5535 \text{ mm}^2$$

The width of the column is only 200 mm; therefore, the nine bars must be arranged in three rows and provide 40 mm clear spacing between the rows, then

$$nd = 587 \text{ mm}$$

$$d_c = \text{cover} + 1.5 \phi + \text{clearance between the bars}$$

$$= 40 + 42 + 40 = 122 \text{ mm}$$

Check for the capacity of the column. Use $d_c = 125$ mm on each face.

$$D = d + d_c = 810 + 125 = 935 \text{ mm}$$

$$D_s = d - d_c = 81 - 125 = 685 \text{ mm}$$

$$\frac{d_c}{d} = 0.15, \quad \frac{D_s}{d} = 0.85, \quad \frac{D}{d} = 1.15$$

The compatible stresses in the reinforcement for full compressive stress in the concrete are :

$$\sigma_{sc} = 1.5m \frac{(nd - d_c)\,\sigma_{acb}}{nd} = 107.43 \text{ MPa}$$

$$\sigma_{st} = m \frac{(d - nd)\,\sigma_{acb}}{nd} = 34.57 \text{ MPa}$$

The total compressive and tensile forces in the concrete and in the reinforcement are:

$$C_c = \frac{(nbd - A_{sc})\,\sigma_{acb}}{2} = 410\,900 \text{ N}$$

$$C_s = A_{sc}\,\sigma_{sc} = 594\,625 \text{ N}$$

$$T = A_{st}\,\sigma_{st} = 191\,345 \text{ N}$$

Net axial allowable force $P_a = C_c + C_s - T = 814\,180$ N

Allowable bending moment on the column

$$M_a = C_c\left(d - \frac{nd}{3}\right) + C_s\,(D_s) - P\left(d - \frac{D}{2}\right) = 386(10^6) \text{ Nmm} = 386 \text{ kNm}$$

The allowable axial load and the bending moment are more than the actual forces; therefore, the column is safe. Fig. 5.19 illustrates the column section.

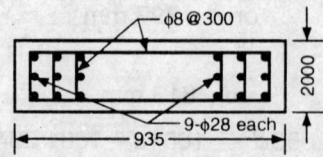

Fig. 5.19 Cross-section of Column in Example 5.13.

5.15 Tensile Stress in Steel Controlling the Design of the Cracked Sections

In some beam columns the bending action is dominant when compared with the axial force. In the examples discussed so far, it can be seen that the stress in tensile reinforcement never reached full stress. The question of the compressive stress in steel reaching the full allowable value may never arise in high yield strength deformed bars. In case of beam columns where the axial force is relatively small, there is a tendency that the stress in steel in tension may reach its allowable limit. The neutral axis distance for a balanced section is given by

$$n = \frac{m\sigma_{acb}}{m\sigma_{acb} + \sigma_{ast}}$$

$$= \frac{1}{1 + \dfrac{\sigma_{ast}}{m\sigma_{acb}}}$$

$$= \frac{1}{1 + \dfrac{3\sigma_{ast}}{280}} \tag{5.79}$$

The values of n for mild steel and HYSD bars are 0.4 and 0.283. If the value of n obtained from the governing equations given in the previous section falls less than the above values, the expressions are not valid since the stress in steel will exceed the allowable value. The forces in the column segments are given in Eqs. (5.60) and (5.61). The force equilibrium equation (5.62) is rewritten here:

$$\left(\frac{nbd}{2} + 1.5mA_{sc} \frac{(nd - d_c)}{nd} \right) \sigma_{cb} - A_{st}\,\sigma_{st} = P \tag{5.62}$$

If the tensile stress in steel controls the design, the compressive stress in the concrete is either less than or at the most equal to, its allowable limit. Therefore, the stress in concrete is expressed in terms of the stress in steel and it is

$$\sigma_{cb} = \frac{nd\sigma_{ast}}{m(d - nd)} \tag{5.80}$$

The substitution of Eq. (5.80) in Eq. (5.62) gives

$$\left(\frac{nbd}{2} + 1.5 \left(A_{sc} \frac{(nd - d_c)}{nd} \right) \frac{nd}{(d - nd)} - A_{st} \right) \sigma_{ast} = P$$

or
$$bd \left(\left(n + 3mqp_t \left(1 - \frac{u}{n} \right) \right) \frac{n}{m(1 - n)} - 2p_t \right) \sigma_{ast} = 2P \tag{5.81}$$

Similarly, the moment equilibrium equation (5.66) can be expressed as

$$bd^2 \left(\frac{n}{2} \left(1 - \frac{n}{3} \right) + 1.5mqp_t \left(1 - \frac{u}{3} \right) c \right) \frac{n\sigma_{ast}}{m(1 - n)} - P \left(e + d - \frac{D}{2} \right) = 0 \tag{5.82}$$

or
$$bd \left(n(3 - n) + 9mqp_t \left(1 - \frac{u}{n} \right) c \right) \frac{n\sigma_{ast}}{m(1 - n)} - 6P \left(\frac{e}{d} + 1 - \frac{D}{2d} \right) = 0 \tag{5.83}$$

Eqs. (5.81) and (5.83) govern the design.

EXAMPLE 5.14 *Design of beam-column* : A column of effective span 3.6 m is subjected to a bending moment of 100 kNm and axial force of 45 kN. Design the beam column with M20 concrete and HYSD bars.

$$\text{Effective length } L_e = 3.6 \text{ m}$$
$$\text{Axial load} = 45 \text{ kN}$$
$$\text{Let self weight} = 5 \text{ kN}$$
$$\text{Total load } P = 50 \text{ kN}$$
$$\text{Bending moment } M = 100 \text{ kNm}$$
$$f_{ck} = 20 \text{ MPa}, f_y = 415 \text{ MPa}$$
$$\sigma_{acb} = 7 \text{ MPa}, \sigma_{ast} = 230 \text{ MPa}$$

Neutral axis distance of a balanced section is given by $n_b = 0.283$

Let $b = 300 =$ mm then

Slenderness factor = 3.6/0.3 = 12. Therefore, the column can be treated as a short one.

$$e = \frac{M}{P} = \frac{100}{50} = 2 \text{ m} = 2000 \text{ mm}$$

The eccentricity is very large, so one may assume that the stress in the tension reinforcement dominates the design.

Design of the section: To solve the two nonlinear simultaneous equations, one must assume certain design parameters. Let

$$u = \frac{d_c}{d} = 0.15, \ p_t = 0.004$$

$$c = \frac{D_s}{d} = 0.85, \ q = 1$$

$$\frac{D}{d} = 1.15, \ m = 13, \ b = 300 \text{ mm}$$

The substitution of these values in the governing equations 5.81 and 5.83 gives

$$(bd)\left(n^2 + 0.156\,(n - 0.15)\right)\left(\frac{1}{13(1 - n)} - 0.008\right)\sigma_{ast} = 2P$$

or
$$(300)d\,(n^2 + 0.156n - 0.023 - 0.104\,(1 - n)) = \frac{26(1 - n)\,P}{\sigma_{ast}}$$

or
$$d(n^2 + 0.26n - 0.127) - 18.84\,(1 - n) = 0 \tag{a}$$

and
$$bd^2\,(3n^2 - n^3 + 0.4(n - 0.15))\,\frac{230}{13(1 - n)} - 6P\,(2000 + 0.43d) = 0$$

or
$$(300)d^2(3n^2 - n^3 + 0.4n - 0.06) - 16956\,(1 - n)(2000 + 0.43d) = 0$$

or
$$d^2(3n^2 - n^3 + 0.4n - 0.06) - 24.3\,(1 - n)d - 113040(1 - n) = 0 \tag{b}$$

Equations (a) and (b) can be solved by iteration.

Try $n = 0.3$, then Eqs. (a) and (b) give

$$d = 322 \text{ mm}$$

$$\text{and } 0.333\, d^2 - 16.31d - '79\,128 = 0$$

$$\text{or } d = 513 \text{ mm}$$

Try $n = 0.26$, then Eqs. (a) and (b) give

$$d = 1700 \text{ mm}$$

and

$$0.229d^2 - 17.24d - 83649 = 0$$

$$\text{or } d = 643 \text{ mm}$$

It may be noticed that there is a reverse tendency in the values of d obtained from Eqs. (a) and (b) by assuming $n = 0.3$ and $n = 0.26$. This implies that the value of n must be between 0.26 and 0.3. The next iteration can be tried with $n = 0.277$.
Try $n = 0.277$, then Eqs. (a) and (b) give

$$d = 626 \text{ mm}$$

$$\text{and } 0.2597\, d^2 - 17.57d - 817\,27 = 0$$

$$\text{or } d = 596 \text{ mm}$$

The value of d is very sensitive at the range of $n = 0.277$. Therefore, use $d = 620$ mm and $n = 0.278$.

It can also be noticed that the value of n is slightly less than $n_b = 0.283$, which ensures that tensile stress in the reinforcement governs the design. Using $n = 0.28$ and $d = 620$ mm we can get the other design variables:

$$A_{st} = p_t bd = 0.004(300)(620) = 744 \text{ mm}^2$$

Try 4 *numbers of* 16 mm bars at each face. Then the distance of the centre of the compression steel from the outer fibre is

$$d_c = 40 + \phi/2 = 48$$

Try

$$d_c = 50 \text{ mm, then}$$

$$D_s = d - d_c = 570 \text{ mm}, \quad D = d + d_c = 670 \text{ mm}$$

$$A_{st} = A_{sc} = 4(201) = 804 \text{ mm}^2$$

Check for capacity of the section: The actual load capacity of the section must be computed using the actual values.

$$nd = 172 \text{ mm and } d - nd = 448 \text{ mm}$$

$$\sigma_{cb} = \frac{nd\sigma_{ast}}{(d - nd)m} = 6.79 \text{ MPa}$$

$$\sigma_{sc} = \frac{(nd - d_c)\,\sigma_{ast}}{(d - nd)} = 62.63 \text{ MPa}$$

$$C_c = (nbd - A_{sc})\,\sigma_{cb}/2 = 175\,182 \text{ N}$$

$$C_s = A_{sc}\,\sigma_{sc} = 50\,354 \text{ N}$$

$$T = A_{st}\,\sigma_{st} = 184\,920 \text{ N}$$

The total allowable axial force on the section is

$$P_a = C_c + C_s - T = 40\,616 \text{ N} < P = 50\,000 \text{ N}$$

The total compressive force resisted by the concrete (C_c) is very close to the total tensile stress resisted by the steel (T) when the neutral axis distance is very close to that of the balanced section. The equations are rather unstable; therefore, the exact solution needs several iterations. In addition, the area of the tensile reinforcement provided is more than the actually needed, which also affects the result. Therefore,

Provide 3 numbers of 16 mm bars and 1 number of 12 mm bar at each face:

$$A_{sc} = A_{st} = 3(201) + 113 = 716 \text{ mm}^2$$

This will not alter the values of d, nd, d_c, C_c etc. Only C_s and T will have to be recomputed.

$$C_s = 716(62.63) = 44\,843 \text{ N}$$

$$T = 716(230) = 164\,680 \text{ N}$$
$$P_a = C_c + C_s - T = 55\,345 \text{ N} > P$$

Now one can see the sensitivity of the expressions; by reducing the amount of the reinforcement, the force condition is satisfied. However, the moment capacity is likely to be reduced; therefore, undue reduction in the tension steel should not be made.

The allowable moment capacity of the section is

$$M_a = C_c\left(d - \frac{nd}{3}\right) + C_s(d - d_c) - P(d - 0.5D)$$

$$= 175\,182\left(620 - \frac{172}{3}\right) + 22\,843(620 - 50) - 50\,000(d - 0.5D)$$

$$= 9856\,907 + 255\,560\,510 - 14\,250\,000$$

$$= 109\,873\,582 \text{ Nmm} = 109.87 \text{ kNm} > M$$

The section is adequate. Provide 6 mm ties at 250 mm spacing.

5.16 Design of Columns Subjected to Wind Load or Seismic Forces

The wind and seismic loads have a tendency of causing overturning of the structure and local bending moment on the columns. The overturning effect will cause additional compressive forces in the columns on the leeward side. The live load on the structure under seismic or wind load condition is likely to be about half of that normally accepted in buildings and it may be close to the full design load in case of warehouses, water tanks and other important structures. The axial forces in the columns under seismic or wind load conditions are usually marginally more than those under the normal load conditions. But more important factor is the bending moment in such load conditions. The duration of the severe seismic load can be anywhere from few seconds to two minutes and that of gust wind is less than five minutes. Therefore, the allowable stresses in the short duration wind and seismic load conditions are taken one-third more than the corresponding values under normal load conditions. The structures should be designed to withstand the severe gust and seismic loads with the enhanced allowable stresses. The probability of simultaneous occurrence of the gust wind and earthquake is considered to be less than one in ten million. One, therefore, need not design the structure for simultaneous occurrence of the two cases. The design methodology of the columns is same as that which was discussed earlier but with the enhanced allowable stresses. The design of uncracked columns under wind or seismic load condition has already been discussed. This section deals with the cracked columns.

EXAMPLE 5.15 *Design of long column subjected to seismic loads:* A column of effective length 3.6 m is subjected to the following critical loads:
1. Combined dead and live load

$$P_1 = 2000 \text{ kN}$$

$$M_1 = 0$$

2. Seismic load condition

$$P_2 = 700 \text{ kN}$$

$$M_2 = 350 \text{ kNm}$$

Design the column with M20 concrete and HYSD bars. The width of the column is not to exceed 200 mm, and its depth is also to be kept the minimum possible.

$$L_e = 3.6 \text{ m}, \quad b = 200 \text{ mm}$$

$$f_{ck} = 20 \text{ MPa}, \quad f_y = 415 \text{ MPa}, \quad \sigma_{asc} = 190 \text{ MPa}$$

$$\sigma_{acb} = 7 \text{ MPa}, \quad m = 13, \quad \sigma_{acc} = 5 \text{ MPa}$$

$$\text{Slenderness factor} = \frac{L_e}{b} = 18 > 12$$

Long column load reduction factor is

$$C_r = 1.25 - \frac{18}{48} = 0.875$$

The corresponding short column equivalent loads are

$$P_{01} = \frac{P_1}{C_r} = 2286 \text{ kN}$$

$$P_{02} = \frac{P_2}{C_r} = 800\,000 \text{ N}$$

Assume a total reinforcement in the column as 6% so as to keep the column size least. That is $p = 0.06$. Assume the percentage of the steel and other design variable as

$$\frac{A_{st}}{bd} = p_t = 0.033, \quad p_c = 0.033$$

$$u = 0.15, \quad c = 0.85, \quad D = 1.15d$$

Design of section. Because of the large bending moment in the seismic load condition, this load condition appears to be critical. However, the axial load under normal load is twice as much as that under the seismic load with lower allowable stress. Therefore, the column need to be designed for both the load conditions.

The allowable axial load on the column for 6% reinforcement is

$$P_a = A_c \sigma_{acc} + A_s \sigma_{asc}$$

$$= (A_g - A_s)\sigma_{acc} + A_g\sigma_{asc}$$

$$= A_g\left((1-p) + p\frac{\sigma_{asc}}{\sigma_{acc}}\right)\sigma_{acc}$$

$$= A_g\left((1-0.06) + 0.06\frac{(190)}{5}\right)5 = 16.1\,A_g$$

Equating the allowable load to the external load, we have

$$P_a = P_{01}$$

$$A_g = \frac{P_{01}}{16.1} = 141\,964 \text{ mm}^2$$

or $D_1 = \dfrac{A_g}{b} = 710$ mm

Design of the section for seismic load condition. Various assumed design variables are substituted in the governing Eqs. (5.77) and (5.78), then we have

$$bd\left(0.5n + 1.5mq\,p_t\left(1 - \frac{u}{n}\right) - mp_t\left(\frac{1}{n} - 1\right)\right)\sigma_{acb}\,(1.333) = P_{02} \qquad \text{(a)}$$

and $$bd\left(\frac{n}{2}\left(1 - \frac{n}{3}\right) + 1.5mq\,p_t\left(1 - \frac{u}{n}\right)c\right)\sigma_{acb}\,(1.333) - P_{02}\left(\frac{e}{d} + 1 - \frac{D}{2d}\right) = 0 \qquad \text{(b)}$$

or $$d(n^2 + 2145n - 1.051) = 858n \text{ and}$$

$$d^2(3n^2 - n^3 + 3.282n - 0.492) - 1105nd - 1126\,235n = 0 \qquad \text{(d)}$$

The two nonlinear simultaneous equations should be solved by iteration.

Try $n = 0.65$, then Eqs. (c) and (d) give

$$d = 727.5 \text{ mm}$$
and $2.634d^2 - 718.5d - 732\,005 = 0$
or $d = 681$ mm

Try $n = 0.73$, then Eqs. (c) and (d) give

$$d = 597 \text{ mm and}$$
and $3.113d^2 - 807d - 822\,151 = 0$
or $d = 660$ mm

Since the tendency of the variation of d has changed for $n = 0.65$ and 0.73, the actual value on n will lie between 0.65 and 0.73.

Try $n = 0.69$, then we have

$$d = 654 \text{ mm}$$
and $2.872d^2 - 762.5d - 777\,102 = 0$
or $d = 670$ mm

The value of d is converging towards 670 mm.
Therefore, select

$$d = 675 \text{ and } n = 0.68$$
$$p_t = p_c = 0.033bd = 4455 \text{ mm}^2$$

Provide 5 *numbers* of 28 mm and 3 numbers of 25 mm bars in each face in three rows. Let the clear spacing between the bars be 40 mm, then the depth of the compression steel from the face is

$$d_c = \text{cover} + \phi + 40 + \phi/2,$$
$$= 40 + 28 + 40 + 14 = 122 \text{ mm}$$

Check for the capacity of the column.

Provide $d_c = 125$ mm, on each face

$$D_s = d - d_c = 675 - 125 = 550 \text{ mm}$$

$$D = d + d_c = 675 + 125 = 800 \text{ mm}$$

$$\frac{d_c}{d} = 0.185, \quad \frac{D_s}{d} = 0.82, \quad \frac{D}{d} = 1.19$$

$$nd = 459 \text{ mm}, \quad A_{st} = A_{sc} = 4545 \text{ mm}^2$$

The overall depth of the section required based on the live load conditions was 710 mm, whereas that for seismic load is 800 mm. Therefore, the seismic load condition governs the design.

The compatible stresses in the reinforcement for full compressive stress in the concrete under seismic load conditions are:

$$\sigma_{sc} = 1.5m\frac{(nd - d_c)}{nd}(1.333)\sigma_{acb} = 132.4 \text{ MPa}$$

$$\sigma_{st} = m\frac{(d - nd)}{nd}(1.333)\sigma_{acb} = 57.08 \text{ MPa}$$

The total compressive and tensile forces in the concrete and in the reinforcement are:

$$C_c = (nbd - A_{sc})\,\sigma_{cb}\,1.333 = 428\,294 \text{ N}$$

$$C_s = 1.333\,A_{sc}\,\sigma_{sc} = 802\,143 \text{ N}$$

$$T = 1.333\,A_{st}\,\sigma_{st} = 345\,813 \text{ N}$$

Total compressive force on the column is

$$P_a = C_c + C_s - T = 884\,614 \text{ N}$$

whereas the load acting on the column under the wind load condition is 800 kN as against an allowable axial load of 884.6 kN. The allowable bending moment on the column is

$$M_a = C_c\left(d - \frac{nd}{3}\right) + C_s\,(D_s) - P_{02}\,(d - 0.5D) = 444.73 \text{ kNm}$$

The allowable bending moment along with the axial force is more than bending moment on the column; therefore, the column is safe against the external loads.

Note: Development lengths of reinforcement bars are often overlooked in the reinforcement details of the columns. The development lengths in compression reinforcement when laps are provided are normally taken care. The bond strength for compression bars in calculating the development length can be taken about 1.25 times the value corresponding to the tensile bars. When the columns are designed as cracked columns, one must be very careful in providing adequate development lengths for the tension bars near the ends of the columns. The compressive stress or the tensile stress in the reinforcement bars is present near the ends of the columns. Therefore, careful attention be given for the detailing of the reinforcement in the columns.

5.17 Design of Uncracked Circular Cross-sectioned Columns Controlled by Compression

The design of the uncracked rectangular columns was already discussed in Section 5.12. The expressions for circular cross-section are developed here on the same lines as done in Section 5.12.

The stresses caused by the axial force and the bending moment are obtained from Eqs. (5.36) and (5.37), that is

$$\sigma_{cc} = \frac{P}{A_c + 1.5mA_s}$$

$$\sigma_{cb} = \pm \frac{M}{Z_t}$$

where

$$Z_t = \frac{2I_t}{D}$$

The area of the reinforcement is assumed to be uniformly spaced at a circle of diameter D_c, then the moment of inertia of the reinforcement is

$$I_s = \frac{8\pi D_c^3 t}{64} = \frac{A_s D_c^2}{8} \tag{5.84}$$

where D_c = diameter of the circle of the reinforcement bars

t = equivalent thickness of steel uniformly spread (see Fig. 5.20)

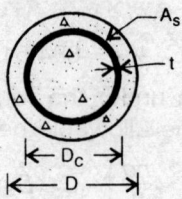

Fig. 5.20 Notations and Idealized Continuous Reinforcement in Circular Column.

$$I_t = I_c + (m-1) I_s = \frac{\pi D^4}{64} + (m-1) \frac{A_s D_c^2}{8} \tag{5.85}$$

The properties of the transformed cross-section can be modified as:

$$A_t = A_g (1 + (1.5m - 1)p)$$

$$I_t = \frac{\pi D^2}{4} \left(\frac{D^2}{16} + \frac{2(m-1)A_s D_c^2}{16 A_g} \right)$$

$$= A_g \left(\frac{D^2}{16} \right) (1 + (m-1) 2pc^2) \tag{5.86}$$

where

$$c = \frac{D_c}{D}$$

The interaction formula which is controlled by the compression failure is given by (refer Eq. 5.37)

$$\frac{P}{A_g}\left(\frac{1}{(1+(1.5m-1)P)\,\sigma_{acc}}+\frac{8\beta}{(1+2(m-1)pc^2)\,\sigma_{acb}}\right)<\alpha \tag{5.87}$$

$$\text{where } \beta=\frac{e}{D}$$

The above equation is rearranged as:

$$\frac{P}{A_g}\left(\frac{1}{f_1}+\frac{8\beta}{f_2}\right)\leq\alpha \tag{5.88}$$

where

$$f_1=(1+(1.5m-1)p)\,\sigma_{acc} \tag{5.89}$$

$$f_2=(1+(m-1)\,2pc^2)\,\sigma_{acb} \tag{5.90}$$

EXAMPLE 5.16 *Design of spiral column subjected to combined forces:* A column of effective length 3.6 m is subjected to an axial force of 700 kN and bending moment of 81.9 kNm in live load condition. Design the column with M20 concrete and HYSD bars.

$$L_e=3.6\text{ m},\quad P=700\text{ kN},\quad M=81.9\text{ kNm}$$

$$e=\frac{M}{P}=\frac{81.9}{700}=0.117\text{ m}$$

$$f_{ck}=20\text{ MPa},\quad f_y=415\text{ MPa},\quad m=13$$

$$\sigma_{acc}=5\text{ MPa},\quad \sigma_{acb}=7\text{ MPa}$$

Let the percentage of the reinforcement be on the lower side say about 0.8%.
Let $p=0.008$.

Design of the selection: Select the following values as a first trial

$$c=\frac{D_c}{D}=0.82,\quad \beta=\frac{e}{D}=0.2$$

The load condition is the live load, so $\alpha=1.0$.
The substitution of the above in Eqs. (5.89) and (5.90) gives

$$f_1=(1+(1.5\,m-1)p)\sigma_{acc}=5.74$$

$$f_2=(1+(m-1)pc^2)\sigma_{acb}=7.45$$

The capacity of the spiral column is 1.05 times the tied column; therefore, Eq. (5.88) can be modified by dividing P by 1.05. So from Eq. (5.88), we have

$$A_g=\frac{P}{1.05\,\alpha}\left(\frac{1}{f_1}+\frac{8\beta}{f_2}\right)$$

$$=\frac{700\,000}{1.05}\left(\frac{1}{5.74}+\frac{8(0.2)}{7.45}\right)=259\,320\text{ mm}^2$$

The gross area of the cross-section is equal to $0.785\,D^2$, so by equating this to the above value, we have

$$D=575\text{ mm}$$

then

$$A_s=0.008\,A_g=2066\text{ mm}^2$$

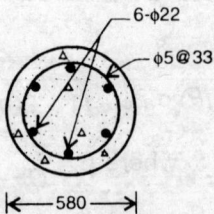

Fig. 5.21 Cross-section of Column in Example 5.16.

Provide 6 *numbers of* 22 *mm* bars, then

$$A_s = 6(380) = 2280 \text{ mm}^2$$

The non-dimensionalized quantities for the above size are:

$$D_c = D - (40 + 40 + 22) = 575 - 102 = 473 \text{ mm}$$

$$c = \frac{D_c}{D} = \frac{473}{575} = 0.822$$

$$\beta = \frac{e}{D} = \frac{117}{575} = 0.200$$

These value check well with the assumed ones.
Check for the capacity of the section.

$$A_g = 259\,540 \text{ mm}^2$$

$$A_s = 2280 \text{ mm}^2$$

$$A_c = 257\,260 \text{ mm}^2$$

$$A_t = A_c + (1.5m - 1)A_s = 299\,440 \text{ mm}^2$$

$$I_t = \frac{\pi D^4}{64} + \frac{(m - 1)}{8} A_s D_c^2 = 5748\,460\,560 \text{ mm}^4$$

$$Z_t = 19994\,645 \text{ mm}^3$$

$$\sigma_{cc} = \frac{P}{A_t} = 2.33$$

$$\sigma_{cb} = \frac{M}{Z_t} = 4.10$$

The interaction formula with the enhanced 1.05 capacity for the spiral column is

$$\frac{1}{1.05}\left(\frac{\sigma_{cc}}{\sigma_{acc}} + \frac{\sigma_{cb}}{\sigma_{acb}}\right) = \frac{1}{1.05}\left(\frac{2.33}{5.0} + \frac{4.10}{7.0}\right) = 1.01$$

Since the value is only 1% more than the allowed value, select the diameter of the column on the higher side. The pitch of the spiral reinforcement of 5 mm diameter bar is controlled by

$$s \leq \frac{D_c}{6} = \frac{473}{6} = 79 \text{ mm}$$

$$\leq 75 \text{ mm}$$

$$\leq 1.5\,\phi = 33 \text{ mm}$$

$$\geq 3\phi_v,\ \text{mm} = 15 \text{ mm}$$

$$> 30 \text{ mm}$$

Final design details are

$$D = 580 \text{ mm}$$

$$A_g = (6 - 22\phi) = 2280 \text{ mm}^2$$

5 mm spiral at 33 mm pitch.

Table 5.7 gives a comparison of design details of tied and spiral column.

Table 5.7 Comparison of Design Details of Tied and Spiral Column $P = 700$ kN, $M = 81.9$ kNm.

Detail	Tied	Spiral
Example No.	5.16	5.29
b, D(mm)	370, 600	ϕ 580
A_s (mm²)	1884	2280
A_s (m²)	0.222	0.265
Relative A_s	1	1.210
Relative A_g	1	1.190

5.18 Cracking Limits and the Failure Modes

A column is said to be cracked if the tensile stress in the concrete exceeds the flexural strength of the concrete, and it is given by

$$\frac{M}{Z_t} - \frac{P}{A_t} \leq f_{cr} \tag{5.91}$$

Substituting the expressions for Z_t and A_t, we have

$$\frac{P}{A_g}\left(\frac{8\beta}{1 + (m-1)pc^2} - \frac{1}{1 + (1.5m - 1)p}\right) \geq f_{cr}$$

or

$$\frac{8\beta}{1 + (m-1)pc^2} \geq \frac{f_{cr} A_g}{P} + \frac{1}{1 + (1.5m - 1)p}$$

Cracking of the concrete will take place for the above condition, which can be further modified as

$$\frac{e}{D} \geq \frac{f_{cr} A_g(1 + (m-1)pc^2)}{8P} + \frac{1 + (m-1)pc^2}{8(1 + (1.5m - 1)p)} \tag{5.92}$$

$$\geq \frac{1 + (m-1)pc^2}{8(1 + (1.5m - 1)p)}\left(\frac{f_{cr} A_g(1 + (1.5m - 1)p)}{P} + 1\right) \tag{5.93}$$

The design of the columns subjected to combined axial force and bending moment is controlled by an interaction formula. Figure 5.22 illustrates the critical points in the design space of a column. Point E in the interaction diagram corresponds to zero moment or zero eccentricity on the column. The corresponding load is equal to the ideal axial load capacity of the column. However, a column must be designed for a minimum eccentricity, no matter

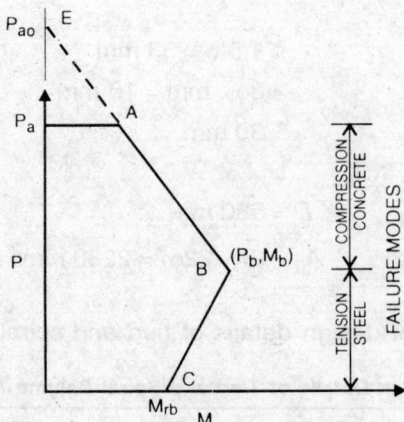

Fig. 5.22 Interaction Curve in the Design of Columns Subjected to Combined Bending and Compression.

whether it is axially loaded or not. Point *A* refers to this axial load with minimum eccentricity for which a column is to be designed. Point *C* refers to the allowable moment capacity of the column under pure moment case as a balanced section. Line *AB* refers to the failure mode of the column compression and line *BC* refers to the failure mode of the column by limiting tension in the reinforcement. Point *B* is the critical point at which the tensile stress in the steel reaches the allowable value under combined axial force and bending. Once the three points *A*, *B* and *C* are established for a column, the capacity of the column can be obtained from the linear interpolation.

PROBLEMS
(Use IS 456)

5.1 A stub column fixed at base and free at the loaded end is subjected to an axial live load of 1 MN. The cantiliver height of the column is 1.2 m. Design a reinforced concrete square column by working stress method using M25 concrete and HYSD-Fe415 bars. Sketch the reinforcement detail indicating main and transverse reinforcement.

5.2 A column fixed at base and hinged at 4 m above the base level is subjected to an axial live load of 5000 kN. Design a reinforced concrete rectangular column with width equal to 400 mm by working stress method using M25 concrete and HYSD-Fe415 bars.

5.3 A column fixed at base and free at top having a height of 3 m is subjected to an axial live load of 700 kN. The column size is 400 × 250 mm. Design the reinforcement details of the column by working stress method using M25 concrete and HYSD-Fe415 bars.

5.4 A column hinge connected at both the ends having a supported height of 5 m is subjected to an axial live load of 500 kN. Design a reinforced concrete square cross sectioned column with minimum possible size by working stress method using M20 concrete and HYSD-Fe415 bars.

5.5 A reinforcement concrete column fixed at a base and hinged at top of height of 4 m, is subjected to an axial live load of 800 kN. Design a reinforced concrete column of square cross section with minimum possible percentage of reinforcement by working stress method using M25 concrete and HYSD-Fe415 bars.

5.6 Design a reinforced rectangular tied column of an effective height 8 m with breadth not exceeding 400 mm and having 3 percentage of reinforcement by working stress method using M25 concrete and HYSD-Fe415 bars. The column is subjected to an axial live load of 1 MN.

5.7 A free standing wall of height of 4 m is subjected to an axial live load of 400 kN/m. Design a reinforced concrete wall by working stress design with 1% steel and using M20 concrete and HYSD-Fe415 bars.

5.8 A column of an effective height 6.5 m is subjected to an axial live load of 2.5 MN. Design a rectangular composite column with ISMB 400 and HYSD-Fe415 bars. The area of ISMB 400 is 7846 mm^2. Permissible stress in the material is 125 N/mm^2. Design the column by working stress method using M25 concrete and HYSD-Fe415 bars and sketch the reinforcement details.

5.9 A column is subjected to an axial live load of 4 MN. The effective height of the column is 6 metres and the size of the column is to be restricted by 550 × 900 mm. Design a composite reinforced concrete column by working stress method using Rolled Steel Beam Section, M30 concrete and HYSD-Fe415 bars. Use 2 nos. of ISMB 400 where the area of one beam section is 7846 mm^2.

5.10 A column is subjected to an axial live load 700 kN. Its effective height is 3m. Design an infilled concrete column by working stress method using M20 concrete and Mild Steel Structural Tube. Assume the thickness of the tube as 3 mm for all outer diameter of the tubes varying from 250 mm to 400 at increments of 50 mm.

5.11 A column fixed at base and free at top having a height of 3 m is subjected to an axial live load of 700 kN. Design a circular spiral reinforced concrete column by working stress method using M20 concrete and HYSD-Fe415 bars. Design two columns—one with minimum and another with maximum permissible reinforcements. Sketch the reinforcement details.

5.12 A column of effective height 5.5 m is subjected to an axial live load of 700 kN. Design a reinforced concrete circular spiral column by working stress method using M20 concrete and HYSD-Fe415 bars subjected to a 350 mm as the maximum diameter of the column.

5.13 A column of effective height of 4 m is subjected to an axial force of 800 kN and bending moment of 100 kNm. Design a rectangular sectioned column by working stress method using M20 concrete and HYSD-Fe415 bars such that the width of the column should not to exceed 400 mm. Use minimum reinforcement.

5.14 A column of 6 m effective length is subjected to an axial live load of 800 kN and bending moment of 70 kNm under wind load condition. Design the column with M20 concrete and HYSD-Fe415 bars by working stress method assuming width of the column not to exceed 400 mm. Also the percentage of reinforcement to be kept within 1%.

5.15 A column of effective length of 5 m is subjected to an axial live load of 900 kN and bending moment of 80 kNm. The size of the column and its reinforcement details are given below. Determine whether the column when made of M20 concrete and HYSD-Fe415 bars can withstand the above load. b = 350 mm, D = 500 mm, A_s = 1708 mm^2, D_s = 420 mm. The reinforcement is equally divided on two opposite faces.

5.16 The reinforced concrete column is subjected to an axial live load of 800 kN. The effective height of the column is 5 m and its size and reinforcement details are:

$b = 425$ mm, $D = 500$ mm, $A_s = 1708$ mm^2 equally divided on two faces. The column is made of M20 concrete and HYSD-Fe415 bars. Investigate what is the maximum bending moment that the column can resist under wind load-condition with the same axial live load.

5.17 A reinforced concrete column of effective height 6 m having $b = 500$ mm and $D = 700$ mm is provided with 4 numbers of 25 mm bars at each opposite faces of the depth of the column. The column is made of M20 concrete and HYSD-Fe415 bars. Investigate whether the column is capable of resisting 1.0 MN axial live load alongwith the bending moment of 200 kNm under wind load condition. Assume the clear cover to the reinforcement as 40 mm.

5.18 A column of effective height equal to 3.2 m is subjected to an axial force of 600 kN and bending moment of 400 kNm. Design the column with M30 concrete and HYSD-Fe415 bars by working stress design. Assume the width of the column as 500 mm. Investigate whether the column is cracked or uncracked section, provide unequal reinforcement at the tension and compression faces.

5.19 A long reinforced concrete column of 7 m effective height is subjected to an axial live load of 600 kN and bending moment of 300 kNm. Design a column with width not exceeding 300 mm using M30 concrete and HYSD-Fe415 bars. Provide unequal reinforcement at the tension and compression faces of the column.

5.20 A column of effective height 8 m is subjected to an axial force of 40 kN and bending moment of 100 kNm. Design a reinforced rectangular sectioned column with width as 300 mm by working stress design using M30 concrete and HYSD-Fe415 bars.

5.21 A reinforced concrete column section is subjected to an axial force of 1000 kN and bending moment of 100 kNm under live load condition. Design an Uncracked concrete column section by working stress design using M20 concrete and Mild Steel Reinforcement. Provide minimum reinforcement.

Chapter 6

WSD of Design of Miscellaneous Structures

6.1 Design of Ties or Tension Members

Some structural elements such as ties in trusses, cylindrical pipes, etc. are subjected to direct tensile force acting on the cross-section. Such members are designed on two basis. In one case, the section can be designed as cracked if it is not exposed to severe environment; otherwise, it should be designed as an uncracked one. In both the cases the tensile force is to be resisted by the reinforcement subject to a limit in the allowable tensile stress in the concrete. The design criteria could be stated as:

$$\sigma_{ast} A_s \geq P \tag{6.1}$$

$$\text{or} \qquad \sigma_{act}(A_c + (m-1)A_c) \geq P \tag{6.2}$$

where P = axial tensile force

A_s = area of the reinforcement

A_c = area of the concrete section

σ_{ast} = allowable tensile stress in steel (σ_{st})

σ_{act} = allowable direct tensile stress in concrete = (σ_{ct})

The allowable stresses normally recommended are given by

$$\sigma_{ast} = \frac{f_y}{\gamma_s} \tag{6.3}$$

$$\sigma_{act} = \frac{f_{cr}}{\gamma_c} \tag{6.4}$$

where γ_s = factor of safety applied to the steel and it is taken about 1.8 in case the surface of the concrete is exposed to ordinary environment and it is 2.7 in case the member is exposed to aggressive atmosphere or to the water face.

γ_c = factor of safety applied to the concrete in tension. It is about 1.8 for ordinary structures and 2.7 = 1.5 (1.8) in case of water retaining structures.

The allowable stresses in different cases are listed in Tables 6.1 and 6.2. Table 6.2 also gives the allowable bending tensile stresses in the concrete. The approximate relation between the allowable bending stress and flexural stress is given by

$$\sigma_{abt} = \frac{f_{cr}}{\gamma_{cb}} \tag{6.5}$$

where γ_{cb} = factor of safety applied to bending tensile strength of the concrete, it is approximately equal to 1.25 for ordinary exposure and 1.8 in case of water retaining walls.

σ_{abt} = allowable bending tensile stress in concrete.

Table 6.1 Allowable Tensile Stress in Steel

	Ordinary exposure		Water-retaining	
	MS	HYSD	MS	HYSD
σ_{ast} (MPa)	140	230	100	150

Table 6.2 Allowable Tensile Stresses in Concrete

Concrete	Ordinary exposure		Water retaining	
	Direct	Bending	Direct	Bending
M15	1.6	2.3	1.1	1.5
M20	1.8	2.5	1.2	1.7
M25	2.0	2.7	1.3	1.8
M30	2.2	3.0	1.5	2.0
M35	2.4	3.3	1.6	2.2

The limitations on the crack width are indirectly controlled by the allowable stress in the concrete. The strain in steel at the allowable stress level is given by

$$\in_{as} = \frac{\sigma_{ast}}{E_s} \qquad (6.6)$$

The allowed strains in HYSD bars are:
  a) in ordinary structures:

$$\in_{as} = \frac{230}{200\,000} = 0.00115 \qquad (6.7)$$

  b) in water retaining structures:

$$\in_{st} = \frac{150}{200\,000} = 0.00075 \qquad (6.8)$$

Applying the compatibility of the strains between steel and concrete, the tensile strains in the concrete at the allowable stress levels of HYSD bars are 0.00115 and 0.00075 in ordinary and water retaining structures respectively. In both the cases, the strains are large. However, the stress in the transformed cross-section is kept below the fracture stress so that the concrete does not crack.

EXAMPLE 6.1 *Design of a tie:* A tie element of a truss of a roof of a workshop is subjected to a tensile force of 500 kN. Design the tie with M 20 concrete and HYSD bars.

$$\text{Tension force} = P = 500 \text{ kN}$$

The truss is in a protected environment, so one can use the following allowable stresses (in MPa):

$$\sigma_{ast} = 230, \ \sigma_{act} = 1.8, \text{ and } m = 13$$

The effect of the self-weight of the element on the stresses is neglected.
  Design of the section. The area of the reinforcement is given by

$$A_s = \frac{P}{\sigma_{ast}} = \frac{500\,000}{230} = 2174 \text{ mm}^2$$

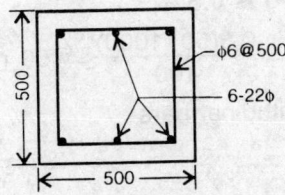

$\phi6@500$

$6-22\phi$

Fig. 6.1 Sectional Details of RCC Tie of Example 6.1.

Provide 6 numbers of 22 mm bars, then the actual area of the reinforcement is

$$A_s = 6(380) = 2280 \text{ mm}^2$$

The area of the concrete section required can be calculated from Eq. (6.2) and it is

$$A_c = \frac{P}{\sigma_{act}} - (m-1) A_s = \frac{500\,000}{1.8} - 12(2280) = 250\,420 \text{ mm}^2$$

Provide 500 by 500 mm square cross-section. Also provide ties of 6 mm at 500 mm spacing.

EXAMPLE 6.2 *Design of a pipe:* A water main of one metre radius is subjected to a pressure of 4 m head of water. Design the pipe with M 20 concrete and HYSD bars.

Radius of the pipe = R = 1 m

Water pressure = 4 m head = 40 kN/m^2 = w

f_{ck} = 20 MPa, f_y = 415 MPa.

Since the pipe carries water, it should be designed as a watertight structure. The allowable stresses in such a structure are:

σ_{act} = 1.2 MPa and σ_{ast} = 150 MPa.

The hoop tension in the pipe is

$$P = wR = 40(1) = \text{kN/m}$$

(The self-weight of the pipe is neglected.)

Design of the section. The pipe is designed for 1 m width. The area of the tension reinforcement required is

$$A_s = \frac{P}{\sigma_{ast}} = \frac{40\,000}{150} = 266 \text{ mm}^2/\text{m}$$

Provide 8 *mm* bars at 150 mm spacing, then

$$A_s = \frac{1000(50)}{150} = 333 \text{ mm}^2/\text{m}$$

The area of the concrete required is given by

$$A_c = \frac{P}{\sigma_{act}} - (m-1) A_s = \frac{40\,000}{1.2} - 12(333) = 30\,940 \text{ mm}^2$$

The thickness of the pipe wall is

$$t = \frac{A_c}{b} = \frac{30\,940}{1000} = 30.94 \text{ mm}$$

212

Provide a minimum thickness of 50 mm so as to have 20 mm cover to the reinforcement. The minimum reinforcement needed is 0.3%.

$$A_{sm} = \frac{0.3(50)(1000)}{100} = 150 \text{ mm}^2$$

Provide 10 numbers of 8 mm longitudinal bars.

6.2 Tie Beams

Tie beams are the members subjected to primary tension and some bending moment. Ties in arch bridges and bowstring girder bridges, and foundation beams in industrial buildings are some of the common tie beams. The tie beam can be designed as cracked or uncracked section. In most cases except in water retaining structures, these are designed as cracked sections. Consider a rectangular section as shown in Fig. 6.2a which is subjected to tension and bending moments as indicated in the Fig. 6.2b. The axial tension and the bending moment can be replaced by an equivalent eccentric force such that

$$e = \frac{M}{P} \tag{6.9}$$

Figure 6.2d indicate the strains caused in the cross section. The strain compatibility gives

$$\frac{\epsilon_c}{nd} = \frac{\epsilon_{st}}{d - nd} \tag{6.10}$$

$$\text{or } \epsilon_{st} = \frac{(1 - n)\epsilon_c}{n} \tag{6.11}$$

$$\text{and } \epsilon_{sc} = \frac{(n - u)\epsilon_c}{n} = \left(1 - \frac{u}{n}\right)\epsilon_c \tag{6.12}$$

where ϵ_{sc} = compressive strain in the compression steel

ϵ_c = compressive strain in the concrete

ϵ_{st} = tensile strain in the steel

$u = d_c/d$

d_c = distance of the steel from the outer fibre

d = distance of the tension reinforcement from the compression face

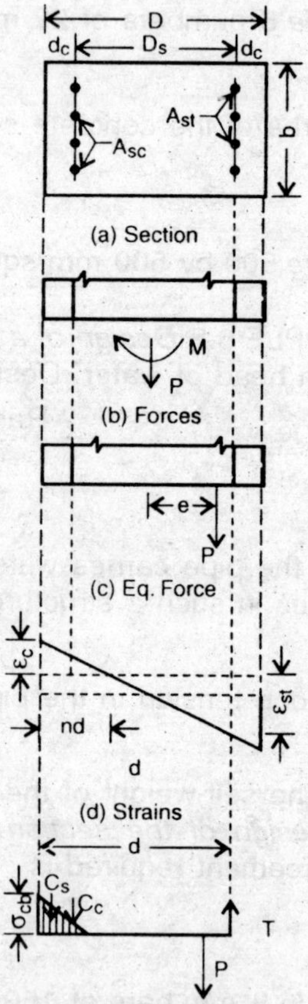

Let σ_{st} = actual tensile stress in the tension steel

σ_{sc} = actual compressive stress in the steel

σ_{cb} = bending compression at the extreme fibre of the concrete

With the stress-strain relations being linear, we have from Eqs. (6.11) and (6.12)

$$\sigma_{st} = \frac{m(1 - n)\sigma_{cb}}{n} \tag{6.13}$$

Fig. 6.2 Notations of Cross-section Strain-stress Distribution of Tied Beam.

$$\sigma_{sc} = \frac{1.5m(n-u)\sigma_{cb}}{n} \qquad (6.14)$$

A factor of 1.5 m is applied to compression reinforcement instead of m, and the bars under compression are in plain strain. Neglecting the effect of the area of the compression reinforcement occupying an area, the resisting forces in concrete and steel are obtained as:

$$C_c = \frac{n\,bd\,\sigma_{cb}}{2} \qquad (6.15)$$

$$C_s = A_{sc}\sigma_{sc} = \frac{1.5m(n-u)A_{sc}\sigma_{cb}}{n} \qquad (6.16)$$

$$T = A_{st}\sigma_{st} = \frac{m(1-n)A_{st}\sigma_{cb}}{n} \qquad (6.17)$$

The equilibrium of the forces gives

$$T - (C_c + C_s) = P$$

or $$\left(\frac{m(1-n)A_{st}}{n} - \frac{1.5m(n-u)A_{sc}}{n} - \frac{nbd}{2}\right)\sigma_{cb} = P \qquad (6.18)$$

or $$bd\left(m\left(\frac{1-n}{n}\right)p_s - \frac{1.5(n-u)}{n}p_c - \frac{n}{2}\right)\sigma_{cb} = P \qquad (6.19)$$

where $$p_s = \frac{A_{st}}{bd} \qquad (6.20)$$

and $$p_c = \frac{A_{sc}}{bd} \qquad (6.21)$$

The equilibrium of bending moments taken about the tension steel gives

$$C_c\left(d - \frac{nd}{3}\right) + C_s D_s = M - \frac{PD_s}{2} \qquad (6.22)$$

or $$\frac{nbd^2}{2}\left(1 - \frac{n}{3}\right)\sigma_{cb} + \frac{1.5A_{sc}m(n-u)}{n}\sigma_{cb}D_s = P(e - 0.5D_s) \qquad (6.23)$$

or $$bd^2\left(K + cp_c\frac{1.5m(n-u)}{n}\right)\sigma_{cb} = Pd(g - 0.5c) \qquad (6.24)$$

where $$K = \frac{n}{2}\left(1 - \frac{n}{3}\right) \qquad (6.25)$$

$$c = \frac{D_s}{d} \qquad (6.26)$$

$$g = \frac{e}{d} \qquad (6.27)$$

Equations (6.19) and (6.24) are the governing equations to solve for n and σ_{cb} in case of an analysis problem. In case of a design problem, one can preassign p_s, p_c, c, u and σ_{cb} and solve for b and d. In case of very small bending moment M and large P, the neutral axis will fall outside the section and it will be uneconomical to adopt the stress distribution shown in the figure. Figure 6.3 gives a typical interaction diagram for a tie beam. Point E refers to the axial load capacity of a tie and it is given by

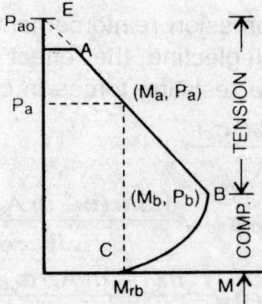

Fig. 6.3 Interaction Diagram for Tied Beam.

$$P_{ao} = A_s \, \sigma_{ast} \tag{6.28}$$

Line *EB* refers to tension failure in which the tensile stress in steel reaches the limit before the compressive capacity of the concrete. The critical pair at point *B* is give by the condition of simultaneous occurrence of the allowable tension and compressive strains in the steel and the concrete. These values are:

$$P_b = T - C_c - C_s$$

$$= A_{st} \, \sigma_{ast} - \frac{n_b}{2} \, bd \, \sigma_{acb} - \frac{(n_b - u)}{n_b} (1.5 \, m - 1) \, A_{sc} \, \sigma_{acb}$$

$$P_b = A_{st} \, \sigma_{ast} - \left(\frac{n_b}{2} + \frac{(n_b - u)}{n_b} (1.5 \, m - 1) \, p_c \right) bd \, \sigma_{acb}$$

$$= bd \, \sigma_{acb} \left(p_s \, q - \left(\frac{n_b}{2} + \frac{(n_b - u)}{n_b} (1.5 \, m - 1) \, p_c \right) \right) \tag{6.29}$$

where $\quad q = \dfrac{\sigma_{ast}}{\sigma_{acb}}$ \hfill (6.30)

$$n_b = \frac{1}{1 + \dfrac{\sigma_{ast}}{m \sigma_{acb}}} \tag{6.31}$$

The moment equilibrium gives

$$C_c \left(1 - \frac{n_b}{3} \right) d + C_s \, D_s = P_b \, (e_b - 0.5 \, D_s) \tag{6.32}$$

$$bd^2 \, \sigma_{acb} \left(K + \frac{(n_b - u)(1.5 \, m - 1)}{n_b} \, p_c c \right) = P_b e_s \tag{6.33}$$

$$e_s = e_b - 0.5 \, D_s \tag{6.34}$$

From Eqs. (6.29) and (6.33), we have

$$e_s = \frac{K + (1 - u/n_b)(1.5 m - 1) p_c c}{0.5 n_b + (1 - u/n_b)(1.5 m - 1) p_c} \tag{6.35}$$

Line CB of Fig. 6.3 represents the compression failure of the concrete. The capacity of the tie beam in tension mode is given by the intersection line diagram of the line EB of the Fig. 6.3. The linear relation gives

$$\frac{P_{ao} - P_a}{P_{ao} - P_b} = \frac{M_a}{M_b} = \frac{P_a e}{P_b e_b} \tag{6.36}$$

$$\text{or } P_a = P_{ao} - (P_{ao} - P_b)\frac{P_a e}{P_b e_b}$$

$$\text{or } P_a\left(1 + \frac{(P_{ao} - p_b)}{P_b}\frac{e}{e_b}\right) = P_{ao}$$

$$\text{or } P_a = \frac{P_{ao}}{1 + \left(\dfrac{P_{ao}}{P_b} - 1\right)\dfrac{e}{e_b}} \tag{6.37}$$

The use of the interaction diagram is illustrated with an example.

EXAMPLE 6.3 *Analysis of tie beam:* A tie beam whose section is shown in Fig. 6.4 is subjected to 800 kN of tension and 200 kNm bending moment. If the beam is made of M20 concrete and HYSD bars, check the capacity of the section to withstand the loads.

$$P = 800 \text{ kN}, \ M = 200 \text{ kNm} = 0.2 \text{ MNm}$$

$$f_{ck} = 20 \text{ MPa}, \ f_y = 415 \text{ MPa}$$

$$\sigma_{acb} = 7 \text{ MPa}, \ \sigma_{ast} = 230 \text{ MPa}$$

$$b = 400 \text{ mm}, \ d = 640 \text{ mm}, \ d_c = 60 \text{ mm}$$

$$D_s = 580 \text{ mm}, \ n_b = 0.283$$

$$A_{st} = 4920 \text{ mm}^2, \ A_{sc} = 2460 \text{ mm}^2$$

$$u = d_c/d = 0.09375 \quad e = M/P = 0.25 \text{ m}$$

Analysis: The critical loads on the section are:

$$P_{ao} = A_s\sigma_{ast} = (4920 + 2460)(230)(10^{-6}) = 1.697 \text{ MN}$$

$$P_b = T - C_c - C_s$$

$$P_b = A_{st}\sigma_{ast} - \frac{n_b b d\,\sigma_{acb}}{2} - \left(1 - \frac{u}{n_b}\right)(1.5m - 1)A_{sc}\sigma_{acb}$$

$$= 0.00492(230) - \frac{0.283}{2}(0.4)(0.64)(7)$$

$$- \left(1 - \frac{0.09375}{0.283}\right)(18.5)(0.00246)(7)$$

$$= 1.1316 - 0.2536 - 0.213 = 0.665 \text{ MN}$$

From Eq. (6.32) we have

$$P_b e_s = C_c\left(1 - \frac{n_b}{3}\right)d + C_s D_s$$

$$= 0.2536(0.904)(0.64) + 0.213(0.58) = 0.2703$$

$$e_s = \frac{0.2703}{P_b} - \frac{0.2703}{0.665} = 0.406 \text{ m}$$

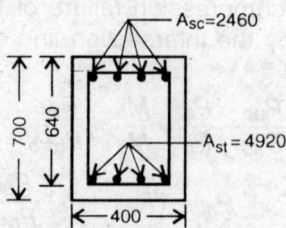

Fig. 6.4 Sectional Details of Tied Beam of Example 6.3.

$$e_b = e_s + 0.5D_s = 0.406 + 0.28 = 0.686 \text{ m}$$

$$\frac{e}{e_b} = \frac{0.25}{0.686} = 0.364$$

$$\frac{P_{ao}}{P_b} = \frac{1.697}{0.665} = 2.55$$

From Eq. (6.37) we have the capacity of the section as

$$P_a = \frac{P_{ao}}{1 + (P_{ao}/P_b - 1)\, e/e_b}$$

$$= \frac{1.697}{1 + 1.55(0.364)} = 1.085 \text{ MN} > P$$

The capacity of the section is adequate.

6.3 Design of Staircases

There are variety of staircases in practice depending on the architectural desires and needs. The staircases in residences are normally subjected to smaller loads whereas the staircases in the public buildings are subjected to overcrowding in peak hours. Table 6.3 gives recommended live loads on the staircases. The loads on the landing could be taken equal to that on the steps but it should be distributed equally in two directions in case of square landings supported on two adjacent sides. The load distribution is usually idealized as uniform over the entire span as done in the case of slabs. The weight of the railings, etc. can be treated as included in the live load if the load is taken spread over the entire width of the staircase. The railing has to be designed subjected to lateral load.

The desired width of the staircase and lengths of the landings are suggested in Table 6.4.

The effective spans of the staircases should be chosen depending on the supports to the landings and the arrangement of the flights.

Table 6.3 Recommended Live Load on Balconies and Staircases

	Locality	Load in kN/m^2
1.	Service stairs for maintenance as in water tanks, catwalks etc.	1.5
2.	Residential buildings	3.0
3.	Office and public buildings, industrial buildings etc.	5.0
4.	Isolated steps *load per step* at free edge (kN)	1.5

	Locality	width (m)
1.	Service stairs for maintenance, catwalks, two-floor residences	1.0–1.2
2.	Apartments, offices, etc.	1.2–1.6
3.	Public houses	1.2–2.0

EXAMPLE 6.4 *Cantilevered steps:* Steps are fixed into a RCC wall cantilevering by 1.2 m from the face of the wall in a residential building. Design RCC steps with M15 concrete and MS bars.

$$\text{Cantilever span } L = 1.2 \text{ m}$$
$$\text{Load per isolated step } W_1 = 1.5 \text{ kN}$$

The working stress design coefficients are:

$$K = 0.173, \quad j = 0.867, \quad \sigma_{cb} = 5 \text{ MPa},$$
$$\sigma_{st} = 140 \text{ MPa}, \quad \text{Tread} = b = 350 \text{ mm}$$

Design of the section: For the purpose of calculating the bending moment, assume the thickness of the step as $L/8 = 0.15$ m.

$$\text{Self-weight} = 1.2 \ (0.35) \ (0.15) \ (25) = 1.6$$
$$\text{Weight of railing} = 0.4$$
$$\text{Live load} = W_l = 1.5$$
$$\text{Design load} = W \underline{= 3.5 \text{ kN}}$$

Maximum BM, which is negative and occurs at the support, is

$$M = \frac{WL}{2} = \frac{3500 \ (1.2)}{2} = 2100 \text{ Nm}$$

Equating the moment to the resisting capacity of the Section, we get

$$Kbd^2\sigma_{cb} = M = 2100 \text{ Nm}$$

$$d = \sqrt{\frac{0.002100}{0.173 \ (0.35) \ (5)}} = 0.083 \text{ m}$$

Maximum shear force $V = 3500$ N
The nominal shear stress for above value d is

$$\tau_v = \frac{0.003500}{0.35 \ (0.078)} = 0.12 \text{ MPa}$$

This value is small and will be within the allowable limit.
Use $d = 0.10$ m and $t = 0.14$ m

$$A_{st} = \frac{M}{jd \, \sigma_{st}} = \frac{2100}{0.867 \ (0.10) \ (140)} = 158 \text{ mm}^2$$

Provide 2 numbers of 10 mm bars at top, then the area of the reinforcement works out to be 157 mm² which is slightly less than desired; therefore, increase the depth accordingly.

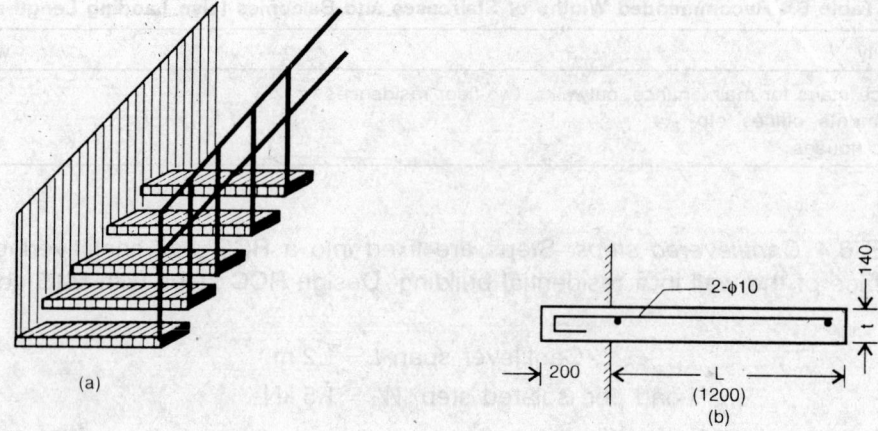

Fig. 6.5 Cantilever Steps Staircase of Example 6.4.

Final design details are:

$$b = 350 \text{ mm}, \ L = 1200 \text{ mm}$$
$$t = 140 \text{ mm}, \ d = 120 \text{ mm}$$

The development length is given by

$$L_d = \frac{\phi \sigma_s}{4 \tau_{bd}} = \frac{\phi(140)}{4(1)} = 35 \ \phi$$

Provide a U-bent at the end of the bar, then the net development length is

$$L_d - 8\phi = 27\phi = 270 \text{ mm}$$

At least 270 mm length must be embedded into the wall for proper anchor.

Note: The cantilevered steps introduce overturning and bending moments into the wall. The wall must be designed for the same.

EXAMPLE 6.5 *Design of staircase slab:* A staircase consists of 14 steps, each of 300 mm tread and 180 mm rise, plus two landings of each 1.25 m length. The width of the staircase is 1.2 m and it is to be used in a public business building. Design the staircase.

    Width of staircase $b = 1.2$ m
    Tread = 0.3 m, Rise = 0.18 m
    Number of steps = 14
    Going length = 14(0.3) = 4.2 m
    Landing length = 1.25 m
    Effective span = going + 1 m for each landing
$$L = 4.2 + 1 + 1 = 6.2 \text{ m}$$
    Live load $w_l = 5$ kN/m^2

        Let $f_{ck} = 15$ MPa and $f_y = 415$ MPa, then
$$\sigma_{cb} = 5 \text{ MPa}, \ \sigma_{st} = 230 \text{ MPa}, \ K = 0.13, \ j = 0.904$$

Design of the section: For the purpose of computing self-weight of the slab, the thickness of the slab is assumed as $L/20 = 0.31$ m.

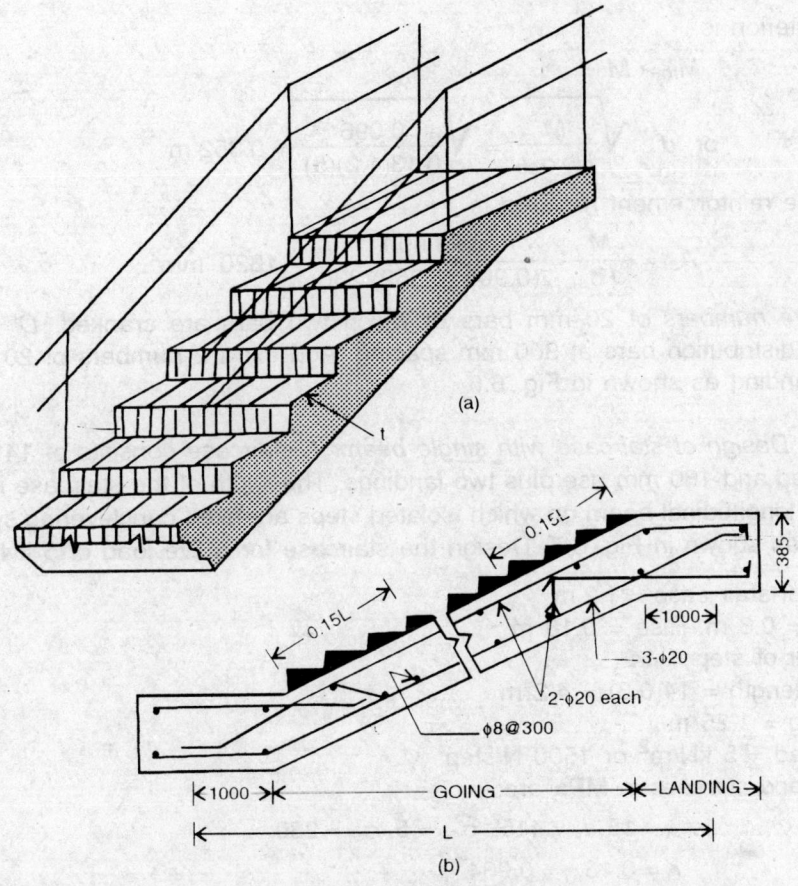

Fig. 6.6 Staircase Slab and Reinforcement Details of Example 6.5.

$$\tan \theta = \frac{180}{300} = 0.6 \text{ or } \theta = 31°$$

Weight of the slab $= \dfrac{1.2(0.31)(25)}{\cos \theta}$ $= 10.8 \text{ kN/m}$

Weight of the steps $= \dfrac{(0.3)\,(0.18)\,(1.2)(25)}{2\,(0.3)} = 2.7 \text{ kN/m}$

Live load $= 1.2(5) = 6.0 \text{ kN/m}$

Extra for railing $= 0.5$

Total load $\underline{w = 20 \text{ kN/m}}$

Maximum bending moment occurs at the mid-span and is

$$M = \frac{wL^2}{8} = \frac{20(6.2)(6.2)}{8} = 96 \text{ kNm}$$

The balanced working moment of the slab section is

$$M_{rb} = Kbd^2 \sigma_{cb}$$

The design criterion is

$$M_{rb} \geq M$$

$$\text{or} \quad d \geq \sqrt{\frac{M}{Kb\,\sigma_{cb}}} = \sqrt{\frac{0.096}{0.13(1.2)(5)}} = 0.352 \text{ m}$$

The area of the reinforcement required is

$$A_{st} = \frac{M}{jd\,\sigma_{st}} = \frac{96\,000}{(0.904)(0.352)(230)} = 1320 \text{ mm}^2$$

Provide five numbers of 20 mm bars of which two bars are cranked. $D = 0.385$. Also provide 8 mm distribution bars at 300 mm spacing. Add extra 2 numbers of 20 mm bars at top near the landing as shown in Fig. 6.6.

EXAMPLE 6.6 *Design of staircase with single beam:* A staircase consists of 14 steps, each of 300 mm tread and 180 mm rise plus two landings. The width of the staircase is 1.2 m and it consists of a longitudinal beam on which isolated steps are fixed cantilevering symmetrically on either side as shown in Fig. 6.7. Design the staircase for a live load of 5 kN/m².

    Width of stair case = 1.2 m
    Tread = 0.3 m, Rise = 0.18 m
    Number of step = 14
    Going length = 14(0.3) = 4.2 m
    Landing = 1.25 m
    Live load = 5 kN/m² or 1500 N/step

The strengths and stresses in MPa are:

$$f_{ck} = 15, \quad f_y = 415, \quad \sigma_{cb} = 5, \quad \sigma_{st} = 230$$

$$K = 0.13, \quad j = 0.904$$

Design of the step: Assume the width of the beam = 300 mm

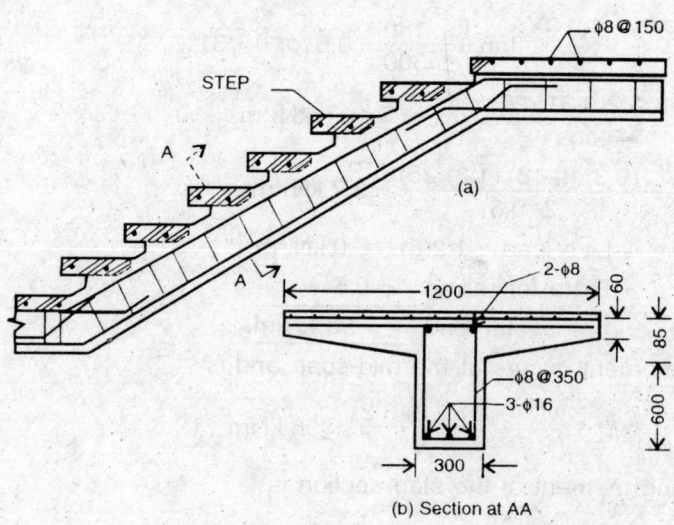

(b) Section at AA

Fig. 6.7 Beam and Step Staircase Details of Example 6.6.

The cantilever span of the step $= \dfrac{1.2 - 0.3}{2} = 0.45$ m

$$L_1 = 0.45$$

Let the thickness of the step be 150 mm for the purpose of computing self-weight.

Self-weight $= 0.45(0.3)(0.15)(25) = 0.5$

Live load on each step $= 1.5$

Total load $= W_1 = 2.2$ kN

Bending moment on each step is

$$M = \dfrac{0.5(0.45)}{2} + 1.5(0.45) = 0.79 \text{ kNm}$$

$$d = \sqrt{\dfrac{M}{Kb\,\sigma_{cb}}} = \sqrt{\dfrac{0.00079}{0.13(0.3)(5)}} = 0.065$$

Use $t = 0.085$ m including surface finishing of 20 mm and cover 15 mm.

$$A_{st} = \dfrac{M}{jd\,\sigma_{st}} = \dfrac{790}{0.904(0.065)(230)} = 59 \text{ mm}^2$$

Provide 2 numbers of 8 mm bars in each step near the top surface. Also provide 8 mm bars at 150 mm spacing in the landing in both the directions.

Design of the beam: The effective span = Going + 2 landing lengths

$$= 4.2 + 2.5 = 6.7 \text{ m}$$

Slope of the beam 180 mm in 300 mm

$$\tan \theta = \dfrac{180}{300} = 0.6$$

$$\theta = 31°$$

Let the depth of the beam $= L/11 = 600$ mm for self weight computations.

Weight of steps $= 1.2(0.085)(25) = 2.6$ kN/m

Self-weight $= \dfrac{0.3(0.60)(25)}{\cos \theta} = 5.3$

Add extra for fillets, etc. $= 0.6$

Live load $= (1.2)(5) = 6.0$

Total load $w = 14.5$ kN/m

Maximum bending moment on the beam is

$$M = \dfrac{wL^2}{8} = \dfrac{14.5(6.7)^2}{8} = 81.36 \text{ kNm}$$

The depth of the beam is

$$d = \sqrt{\dfrac{M}{Kb\,\sigma_{cb}}} = \sqrt{\dfrac{0.08136}{0.13(0.3)(5)}} = 0.65 \text{ m}$$

$$A_s = \dfrac{M}{jd\,\sigma_{st}} = \dfrac{81360}{0.904(0.65)(230)} = 602 \text{ mm}^2$$

222

Provide 3 numbers of 16 mm bars at the bottom

$$A_{st} \text{(provided)} = 3(201) = 603 \text{ mm}^2$$
Bend one of the bars near the support.

Provide 2 numbers of 10 mm bars at top as hanger bars.

$$A_{st} = A_{st} + A_{sc} = 603 + 157 = 760 \text{ mm}^2$$

Let $D = 0.65 + 0.035 = 0.685$ m.

The self-weight assumed checks approximately with the actual load.

Check for shear force. The maximum shear force occurs at the support under the landing and at a distance d from support it is

$$V = 14.5 \left(\frac{6.7}{2} - 0.65 \right) = 39.15 \text{ kN}$$

The nominal shear stress in the beam is

$$\tau_v = \frac{V}{bd} = \frac{0.03915}{0.3(0.65)} = 0.2 \text{ MPa}$$

The stress is less than the allowable value, therefore, only nominal stirrups be provided. *Provide* 8 mm stirrups

$$s = \frac{A_{sv} f_y}{0.4b} = \frac{100(415)}{0.4(300)} = 350 \text{ mm}$$

After deducting the thickness of the step, the depth of the web is

$$h_w = D - t = 685 - 85 = 600 \text{ mm}$$

EXAMPLE 6.7 *Design of folded plate staircase.* A staircase of 1.2 m width as shown in Fig. 6.8a is supported at one face and has a fold of the plate at the other end. There are 14 steps of each 300 mm tread and 180 mm rise. The length of the landing is 1.25 m and the load acting on the staircase is 5 kN/m². Design the staircase.

Design data and Analysis

Number of steps = 14
Length of going $= 14(0.3) = 4.2$ m

Effective span = Going + 2 m of landings

$= 4.2 + 2 = 6.2$ m

Gradient of the staircase is $= \tan\theta = \frac{0.18}{0.30} = 0.6$

or $\theta = 31°$

Half flight height $= H = 14(0.18) = 2.52$ m

The strength and the allowable stresses in MPa of the materials used in the construction are:

$$f_{ck} = 15, \ f_y = 415, \ \sigma_{cb} = 5, \ \sigma_{st} = 230$$
$$K = 0.13, \ j = 0.904$$

Let the thickness of the slab be assumed as 350 mm for the purpose of computing self-weight.

$$\text{Self-weight} = 1.2(0.35)(25)/\cos\theta = 12.2 \text{ kN/m}$$

$$\text{Weight of steps} = \frac{1.2(0.18)(25)}{2} = 2.8$$

$$\text{Live load} = 1.2(5) = 6.0$$

$$\text{Total load} = w = 21.0 \text{ kN/m}$$

There is an inplane shear force at the interface between the two flights slab at the unsupported landing. The free body diagrams of the two successive flights are shown in Fig. 6.8b. The value of the inplane force F can be calculated by taking moments of the forces in the free body diagram of any of the flights as shown in Fig. 6.8. Consider the lower flight, the moment equilibrium about the support gives

$$FH = \frac{wL_o^2}{2} \tag{6.38}$$

$$F = \frac{wL_o^2}{2H} = \frac{21(4.2+2.5)^2}{2(2.52)} = 187 \text{ kN}$$

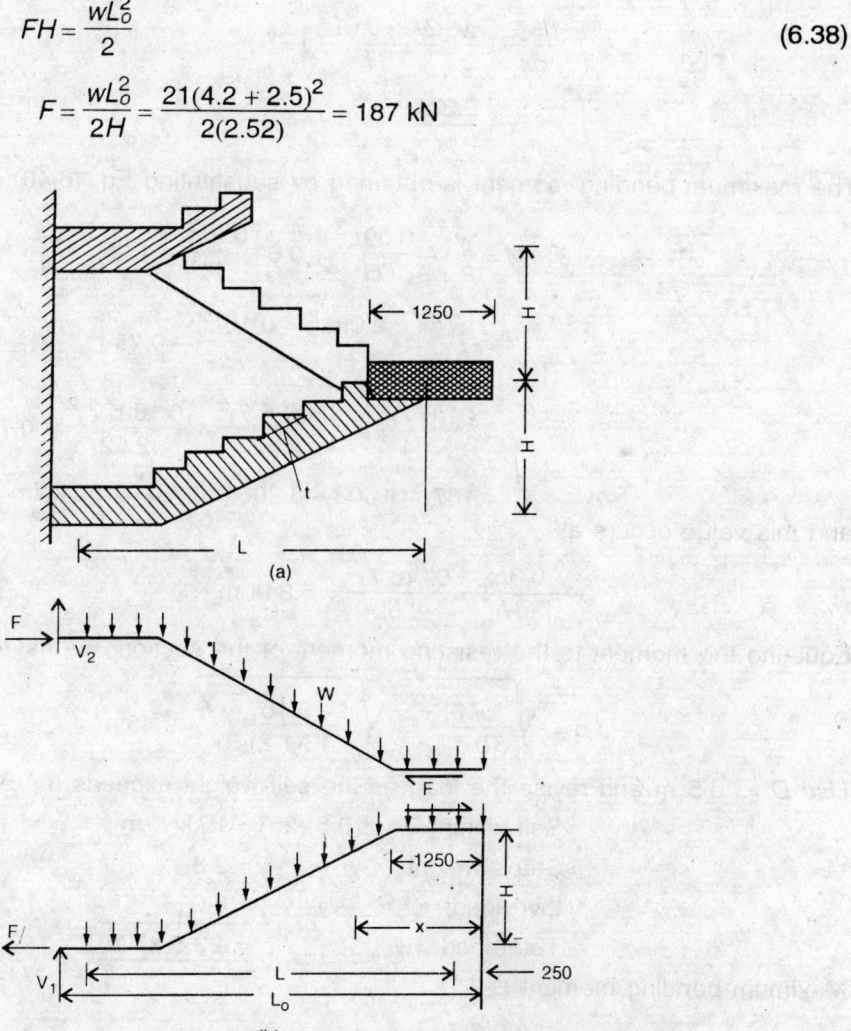

Fig. 6.8 Folded Plate Staircase and Free Body Diagram.

Note: The transverse shear force at the interface is neglected when compared with the inplane shear. The effective support is taken at 1 m from the face of the landing.

The bending moment at a distance x from the face of the landing is

$$M_x = \frac{-wx^2}{2} + F(x - 1.25)\tan\theta$$

$$M_x = \frac{-wx^2}{2} + \frac{wL_0^2}{2H} 0.6(x - 1.25)$$

$$= -\frac{w}{2}\left(x^2 - \frac{0.6L_0^2}{H}(x - 1.25)\right) \tag{6.39}$$

The bending moment is maximum for

$$\frac{dM_x}{dx} = \frac{w}{2}\left(\frac{2x - 0.6L_0^2}{H}\right) = 0$$

$$x = \frac{0.3L_0^2}{H} \tag{6.40}$$

The maximum bending moment is obtained by substituting Eq. (6.40) in Eq. (6.39).

$$M = \frac{w}{2}\left(-\frac{0.09L_0^2}{H} + 0.6\left(\frac{0.3L_0^2}{H} - 1.25\right)\right)\frac{L_0^2}{H}$$

$$= \frac{wL_0^2}{2H}\left(-\frac{0.09L_0^2}{H} + \frac{0.18L_0^2}{H} - 0.75\right)$$

$$= \frac{21(6.7)^2}{5.04}\left(-\frac{0.09(6.7)^2}{2.52} + \frac{0.18(6.7)^2}{2.52} - 0.75\right)$$

$$= 187(-1.603 + 3.206 - 0.75) = 160 \text{ kNm}$$

and this value occurs at

$$x = \frac{0.3L_0^2}{H} = \frac{0.3(6.7)^2}{2.52} = 5.344 \text{ m}$$

Equating the moment to the resisting moment of the section, we have

$$d = \sqrt{\frac{M}{Kb\,\sigma_{cb}}} = \sqrt{\frac{0.16}{0.13(1.2)(5)}} = 0.45 \text{ m}$$

Use D = 0.5 m and revise the load as the self-weight exceeds the assumed value.

Self-weight = 1.2(0.5)(25) =	15 kN/m
Steps weight	= 2.8
Live load	= 6.0
Total load =w	= 23.8 kN/m

Maximum bending moment is

$$M = \frac{23.8(6.7)^2}{5.04}(-1.603 + 3.206 - 0.75) = 181.3 \text{ kNm}$$

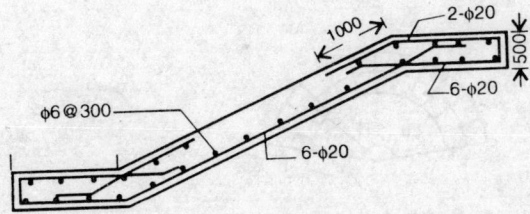

Fig. 6.9 Reinforcement Details of Folded Plate Staircase of Example 6.7.

The value of d and the area of the reinforcement are:

$$d = \sqrt{\frac{M}{Kb\,\sigma_{cb}}} = \sqrt{\frac{0.1813}{0.13(1.2(5)}} = 0.48 \text{ m}$$

$$A_{st} = \frac{M}{jd\,\sigma_{st}} = \frac{181300}{0.904(0.48)(230)} = 1817 \text{ mm}^2$$

Provide 6 numbers of 20 mm bars at the bottom of the slab.

$$A_{st} \text{ (provided)} = 6(314) = 1884 \text{ mm}^2$$

The maximum *negative bending* moment occurs at the starting of the free landing and it is

$$M = \frac{wa^2}{2} = 23.8 \frac{(1.25)^2}{2} = 18.59 \text{ kNm}$$

The area of the reinforcement needed is

$$A_{st} = \frac{M}{jd\,\sigma_{st}} = \frac{18590}{0.904(0.48)(230)} = 186 \text{ mm}^2$$

Shear stress at the fold is $= \dfrac{F}{t(1.25)}$

$$= \frac{wL_0^2}{2Ht(1.25)} = \frac{0.023800(6.7)(6.7)}{2(2.52)(.48)(1.25)} = 0.4 \text{ MPa}$$

Provide two numbers of 20 mm bars at top face and extra 4 numbers of 20 mm bars in the fold. Figure 6.9 illustrates the reinforcement details.

6.4 Design of Helicoidal Staircase

Helicoidal staircase as shown in Fig. 6.10 is a special type of staircase used in buildings of special architecture. The staircase can be analysed as a curved beam or a helicoidal girder in space. The analysis of such girders is not in the scope of this book, however, the resultant forces on the girder have been computed by various investigators. The reader can see the work of AC Scordelis, on Internal Forces in Uniformly Loaded Helicoidal Girders, Journal of ACI, April 1960 for further clarifications. The three critical stress resultants acting on a girder are: the bending moments about two principal planes and torsional moment. The transverse shear and the axial thrust also exist on the girder but they are considered to be not very significant.

226

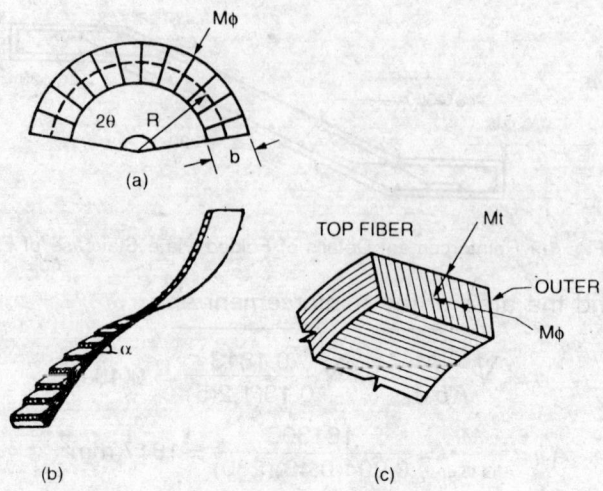

Fig. 6.10 Notations of Helicoidal Staircase.

Out of the three, the bending moment about the radial line and the torsion are dominant. These two quantities by curved beam analysis of girder ends fixed are:

$$M_\phi = wR_c^2 (c \cos \phi - 1) \tag{6.41}$$

$$M_t = wR_c^2 (c \sin \phi - \phi) \tag{6.42}$$

where R_c = radius of the centroidal line of the girder

$\quad\quad = R + e$

$\quad R$ = radius of the centre line of the girder

$\quad e$ = eccentricity of the load action = $b^2/12R$ $\tag{6.43}$

$\quad \phi$ = angle measured from the middle point of the curve towards the support

$\quad M_\phi$ = torsional moment

$\quad M_t$ = circumferential bending moment

$\quad w$ = uniformly distributed load

$$c = \frac{2(q+1)\sin\theta - 2q\theta\cos\theta}{(q+1)\theta - (q-1)\sin\theta\cos\theta} \tag{6.44}$$

$$q = \frac{EI}{GJ} \tag{6.45}$$

$\quad \theta$ = half of the semicentral angle of the curve

$\quad EI$ = flexural rigidity

$\quad GJ$ = torsional rigidity

For all practical purposes E/G is taken as 2 and

$$I = \frac{bh^3}{12}$$

$$J = \frac{b^3 h}{3.5} \text{ (where } b < h \text{) or}$$

$$= \frac{h^3 b}{3.5} \quad \text{for } h < b$$

$$q = \frac{2bh^3(3.5)}{12(b^3 h)} = 0.58 \left(\frac{h}{b}\right)^2 \text{ for } b < h$$

$$= 0.58 \text{ for } h < b.$$

In case of helical slabs, the width b is usually large and the centroid of the load on the slab does not coincide with the centre line of the slab. In such a case, the radius of the centre line of the slab should be replaced by the radius of the centroid of the load. The eccentricity of the load is given by Eq. (6.43). The helicoid is actually a space structure, whereas the analysis made is that of a curve girder; therefore, the bending moment obtained by the curved girder analysis are likely to be on the higher side.

EXAMPLE 6.8 *Design of Helicoidal Staircase:* Two floors separated by 4 m are connected by a helicoidal staircase of 32° slope (one in 1.6 gradient). Design a solid slab 1.2 m wide helicoidal staircase with a working load of 5 kN/m^2. Keep the subtended angle as 120°.

Data and Analysis. The properties of the staircase are:

$$\alpha = 32°, \ \tan \alpha = 0.625$$

$$2\theta = 120° = 2.1 \text{ radians}, \ b = 1.2 \text{ m}$$

$$\text{Total rise } = H = 4 \text{ m} = R(2\theta) \tan \alpha$$

$$\text{or} \quad R = \frac{H}{2 \theta \tan \alpha} = \frac{4(3)}{2\pi(0.625)} = 3.06 \text{ m}$$

$$e = \frac{b^2}{12R} = \frac{1.44}{12(3.06)} = 0.04 \text{ m}$$

$$R_c = R + e = 3.1 \text{ m}$$

Let $f_{ck} = 20$ MPa, $f_y = 415$ MPa

$$\text{then } \sigma_{cb} = 7 \text{ MPa}, \ \sigma_{st} = 230 \text{ MPa and } K = 0.13$$

For the purpose of calculating the self-weight, let the thickness of the slab be assumed as

$$t = \frac{2\theta R}{15} = \frac{(2.1)(3.1)}{(15)} = 0.43 \text{ m}$$

$$\text{Self-weight} = \frac{0.43(1.2)(25)}{\cos \alpha} = 15.2 \text{ kN/m}$$

$$\text{Live load} = 1.2(5) = 6.0$$

$$\text{Total load} = w = \underline{21.2 \text{ kN/m}}$$

$$Use \ w = 22 \text{ kN/m}$$

Since the width is larger than the depth of the slab, the ratio of the flexural rigidity to the torsional rigidity is

$$q = \frac{EI}{GJ} = \frac{2bh^3}{12} \frac{3.5}{(bh^3)} = 0.58$$

$$\text{and} \quad c = \frac{2(q+1)\sin\theta - 2q\,\theta\cos\theta}{(q+1)\,\theta - (q-1)\sin\theta\cos\theta}$$

$$= \frac{2(1.58)(0.866) - 2(0.58)(1.05)(0.5)}{1.58(1.05) + (6.42)(0.866)(0.5)} = 1.16$$

The critical moments are at $\phi = 0$ and $\phi = 60°$, so they are obtained from Eqs. (6.42) and (6.43). For $\phi = 0°$ (at the mid-span)

$$wR_c^2 = 22(3.1)^2 = 211.42$$

$$M_{\phi c} - wR_c^2 \, (c\cos\phi - 1) = 211.42(1.16 - 1) = 33.83 \text{ kNm}$$

$$M_t = wR_c^2 \, (c\sin\phi - \phi) = 0$$

At $\phi = 60°$ (the fixed ends)

$$M_{\phi f} = wR_c^2 \, (c\cos\phi - 1)$$

$$= 211.42 \, (1.16\cos 60 - 1) = -88.8 \text{ kNm}$$

$$M_{tf} = 211.42(1.16\sin 60 - 1.05) = -9.6 \text{ kNm}$$

Design of the section: The maximum bending moment occurs at the supports. In addition, there is torsion at the section; therefore, the equivalent moment at the support will be

$$M_e = M + T\left(\frac{1 + h/b}{1.7}\right)$$

$$= 88.8 + 9.6\left(\frac{1 + 0.43/1.2}{1.7}\right) = 94.67 \text{ kNm}$$

(Negative equivalent bending moment including the effect of the torsion, $M = M_{\phi f}$ and $T = M_{tf}$.)

The bending moment at the supports is almost three times that of at the mid-span; it is therefore, desirable to obtain the depth of the slab by equating the resisting moment to the maximum BM.

$$M_r \geq M_e$$

$$\text{or} \quad 0.13bd^2 \, \sigma_{cb} \geq 0.09647 \text{ MNm}$$

$$\text{or} \quad d \geq \sqrt{\frac{0.09647}{0.13(1.2)(7)}} = 0.3 \text{ m}$$

Then the overall depth of the slab is

$$t = d + 0.05 = 0.35 \text{ m}$$

$$\text{Self-weight} = \frac{0.35(1.2)(25)}{\cos\alpha} = 12.5 \text{ kN}/m$$

$$\text{Total load} = 12.5 + 5 = 17.5 \text{ kN/m}^2$$

The bending moments are revised for the reduced load by proportioning according to the modified load, and are:

$$M_{\phi c} = 33.83(17.5)/22 = 26.9 \text{ kNm}$$

$$M_e = 96.47(17.5)/22 = 76.74 \text{kNm}$$

The reinforcement at mid-span and at the support are :

$$A_{sp} = \frac{M_{\phi c}}{jd\,\sigma_{st}} = \frac{26\,900}{0.904(0.3)(230)} = 431 \text{ mm}^2$$

$$A_{sn} = \frac{M_e}{jd\sigma_{st}} = \frac{76\,740}{0.904(0.3)(230)} = 1230 \text{ mm}^2$$

(The subscripts p and n indicate positive and negative BM reinforcements.)

   Provide the following reinforcement. 5 *numbers of 12 mm* bars at the bottom face through, 6 *numbers of 18 mm* bars at the top near the support for a distance of $R\theta/2 = 3.1\,(1.05)/2$ = 1.62 m from each end of the slab.

   The bond development length is

$$L_d = \frac{\phi\sigma_{st}}{4\tau_{bd}} = \frac{\phi(230)}{4(1.2)(1.6)} = 30\phi = 540 \text{ mm}$$

Figure 6.11 illustrates the reinforcement details.

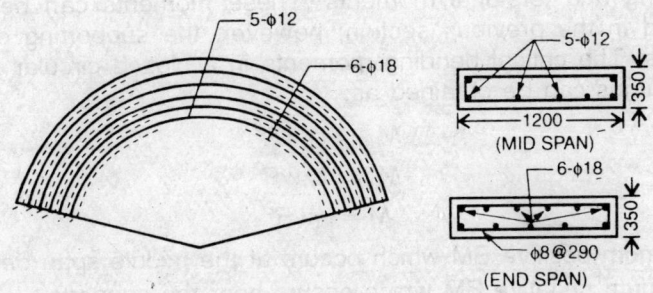

Fig. 6.11 Reinforcement Details of Helicoidal Staircase of Example 6.8.

Design for shear: The maximum vertical force at a distance d from the support is

$$R_o = w(R\theta - d) = 17.5(3.1(1.05) - 0.3) = 51.7 \text{ kN}$$

The transverse shear force at this section is

$$V = R_o \sin\alpha = 51.7 \sin 31 = 26.6 \text{ kN} = 0.0266 \text{ MN}$$

The torsion at this section needs to be converted into an equivalent shear force and the net shear force is

$$V_e = V + \frac{1.6T}{b} = 0.0266 + \frac{1.6(0.0096)}{1.2} = 0.04 \text{ MN}$$

$$\tau_v = \frac{V_e}{bd} = \frac{0.04}{1.2(0.3)} = 0.11 \text{ MPa}$$

Select two-legged 8 mm stirrups, the spacing of which is controlled by

$$A_{sv} \geq \left(\frac{T}{b_1 d_1} + \frac{V}{2.5 d_1}\right)\frac{s_v}{\sigma_{sv}}$$

where $A_{sv} = 2(50) = 100$ mm^2

$b_1 = 1200 - 100 = 1100$ mm

$d_1 = d - d_c = 300 - 50 = 250$ mm

then

$$100 \geq \left(\frac{9600\ 000}{1100(250)} + \frac{26.600}{2.5(250)} \right) \frac{s_v}{230}$$

$$\geq 0.336\ s_v$$

or $s_v = 296$ mm

The spacing is also controlled by:

$$s_{y\ max} \leq \frac{A_{sv}\sigma_{sv}}{b(\tau_v - \tau_c)} = < 0,\ \text{not applicable}$$

Provide two-legged 8 mm stirrups at 290 mm.

6.5 Curved Beams

Curved girders are used to support circular slabs, porticos, water tanks, etc. The beams are subjected to bending and torsional moments. These moments can be obtained from the expressions derived in the previous section; however, the supporting columns will involve boundary conditions. The critical bending moments in a closed circular girder supported by equally spaced columns can be obtained as:

$$M_{\phi p} = c_p wR^2 \tag{6.46}$$

$$M_{\phi n} = c_n wR^2 \tag{6.47}$$

$$M_t = c_t wR^2 \tag{6.48}$$

where $M_{\phi p}$ = maximum positive BM which occurs at the middle span between the columns

$M_{\phi n}$ = maximum negative BM which occurs near the supports

M_t = maximum twisting moment which occurs at quarter points in a segment

c_p, c_n, c_t = moment coefficients given in Table 6.5

The maximum torsion occurs at a location where the bending moment is zero and shear force is not critical. Therefore, the torsional moment does not influence the design very much.

EXAMPLE 6.9 *Design of circular girder:* A girder circular in plan having a mean radius of 5 m is supported by eight columns spaced at equal intervals. The girder supports a circular

Table 6.5 Moment Coefficients in Curved Beams

Number of support column	c_p	c_n	c_t
4	0.110	0.215	0.033
6	0.047	0.093	0.010
8	0.023	0.052	0.004
9	0.019	0.042	0.003
10	0.015	0.034	0.002
12	0.009	0.024	0.001

slab of 0.35 m thick and is subjected to a live load of 5 kN/m². The outer radius of the slab is 5.5 m. Design the circular girder.

Design data and analysis

Diameter of the slab = D = 2(5.5) = 11 m

Thickness of the slab = 0.35 m

Weight of the slab = 0.35(25) = 8.75 kN/m²

     Live load = 5.0

     Total load = 13.75 kN/m²

Total load from the slab = $W = \dfrac{\pi D^2}{4}(13.75) = 1306.5$ kN

Load on the girder = $\dfrac{W}{2\pi R} = \dfrac{1306.5}{3.141(10)} = 41.6$ kN/m

Let the self-weight be = 0.2(25) = 5.0

Total load on the girder = w = 46.6 kN/m

Use w = 48 kN/m.

Let f_{ck} = 20 MPa and f_y = 415 MPa, then

    σ_{cb} = 7 MPa, σ_{st} = 230 MPa and K = 0.13"

The maximum bending moments on the girder with eight supports are computed using the coefficient of Table 6.5, and they are

$$M_{\phi p} = c_p w R^2 = 0.023(48)(25) = 27.6 \text{ kNm}$$

$$M_{\phi n} = c_n w R^2 = 0.052\,(48)(25) = 62.4 \text{ kNm}$$

$$M_t = c_t w R^2 = 0.004(48)(25) = 4.8 \text{ kNm}$$

(The subscripts imply negative or positive BMs; only the magnitudes of moments are given.)

Design of the section: Let the width of the beam = 0.35 m

The resisting moment capacity of the section should be more than the maximum BM on the girder. Therefore,

$$M_r \ge M_{\phi n}$$

$$\text{or } K b d^2 \, \sigma_{cb} \ge 0.0624 \text{ MNm}$$

$$\text{or } d \ge \sqrt{\frac{0.0624}{0.13(0.35)(7)}} = 0.45 \text{ m}$$

Then the overall depth $= h = d + 0.05 = 0.50$ m

The self-weight = 0.35(0.50)(25) = 3.5 kN/m

The self-weight assumed is 5 kN/m whereas the actual weight is 3.5 kN/m. The assumed design load is 48 kN/m as against an actual value of 45.2 kN/m which is only marginally greater. The areas of the reinforcement at the mid span and at the support are:

$$A_{sp} = \frac{M_{\phi p}}{jd\,\sigma_{st}} = \frac{27600}{0.904(0.45)(230)} = 295 \text{ mm}^2$$

$$A_{sn} = \frac{M_{\phi n}}{jd\sigma_{st}} = \frac{62\,400}{0.904(0.45)(230)} = 667 \text{ mm}^2$$

Provide 3 numbers of 12 mm bars at the bottom face and 3 number of 18 mm bars at the top. It is not recommended to curtail or bend the bars in this case, as the spacing of the columns is small and in addition there is torsional moment on the beam.

Design for shear: The total load on each column is

$$W_0 = \frac{2\pi R w}{8} = \frac{5\pi(45.2)}{4} = 178 \text{ kN}$$

The critical section of shear stress is at a distance d from the face of the column. Let the size of the column be 400 mm, then the critical shear force on the section is

$$V = \frac{W_0}{2} - w(0.5a + d) = 89 - 45.2(0.65) = 59.6 \text{ kN}$$

The nominal shear stress on the section is

$$\tau_v = \frac{V}{bd} = \frac{0.0596}{0.35(0.45)} = 0.38 \text{ MPa}$$

The percentage of tensile reinforcement at support is

$$\frac{100 A_{st}}{bd} = \frac{100(3)(254)}{350(450)} = 0.5\%$$

The allowable shear stress in the section for 0.5% tension reinforcement is 0.3 MPa. Only a nominal transverse reinforcement be provided.

Try two-legged 8 mm stirrups.
The maximum spacing allowed is

$$S_{max} = \frac{A_{sv} f_y}{0.4b} = \frac{100(415)}{0.4(350)} = 296 \text{ mm}$$

Provide two-legged 8 mm stirrups at 295 mm spacing throughout. Figure 6.12 illustrates the sectional details.

EXAMPLE 6.10 *Design of semicircular beam:* A semicircular portico slab is supported by a curved beam. The slab thickness is 150 mm and the outer fibre radius is 3.5 m. The slab is wholly supported by the beam whose ends are fixed to heavy columns. Design the beam for a live load of 3 kN/m² on the slab.

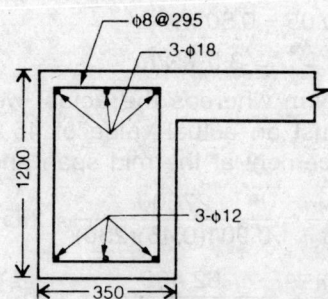

Fig. 6.12 Sectional Details of Curved Beam of Example 6.9.

Design data and analysis.

Let $f_{ck} = 25$ MPa, $f_y = 415$ MPa, then we have

$\sigma_{cb} = 8.5$ MPa, $\sigma_{st} = 230$ MPa, $K = 0.13$, and $j = 0.904$

Live load = 3 kN/m^2
Outer radius of the slab = $R_0 = 3.5$ m
Let the radius of the curved beam = $R = 2.7$ m.

$$
\begin{aligned}
\text{Weight of the slab} &= 0.15(25) &&= 3.75 \text{ kN/m}^2 \\
\text{Live load} &&&= 3 \\
\text{Total} &= w_s &&= 6.75 \text{ kN/m}^2
\end{aligned}
$$

$$\text{Total load from the slab} = \frac{\pi R_0^2 w_s}{2}$$

Load from the slab per metre length of the beam is

$$\frac{\pi R_0^2 w_s}{2 \pi R} = \frac{3.5^2(6.75)}{5.4} = 15.3 \text{ kN/m}$$

Let the size of the beam be assumed as 0.4×0.6 m for the purpose of calculating the self-weight. Assume a parapet well of 0.115 m by 0.6 m is over the slab.

Then the weight of the parapet = 0.115(0.6)(22) = 1.5 kN/m
Self-weight = 0.4(0.6)(25) = 6 kN/m
Total load = 15.3 + 1.5 + 6 = 22.8 kN/m
Use w = 23 kN/m.

The beam is analysed as a curved beam as discussed in Section 6.4 The coefficients associated with the curve beam are:

Semi-central angle = $90° = \pi/2$

$$q = \frac{EI}{GJ} = \frac{2bh^3}{\dfrac{12hb^3}{3.5}} = \frac{3.5h^2}{6b^2} = \frac{3.5}{6}\left(\frac{0.6}{0.4}\right)^2 = 1.3$$

$$c = \frac{2(q+1)\sin\theta - 2q\theta\cos\theta}{(q+1)\theta - (q-1)\sin\theta\cos\theta} = \frac{2}{\pi/2} = \frac{4}{\pi} = 1.27$$

The critical moments at the mid-span and at the support are obtained from Eqs. (6.42) and (6.43):

$$e = \frac{b^2}{12R} = \frac{0.16}{12(2.7)} = \text{very small}$$

Therefore $R_c = R$

$$wR^2 = 23(2.7)(2.7) = 167.7 \text{ kNm}$$

At $\phi = 0$, the mid span section

$$M_{oc} = wR^2 (c\cos\phi - 1) = 167.7(1.27 - 1) = 45.3 \text{ kNm}$$

$$M_{tc} = wR^2 (c\sin\phi - \phi) = 0$$

At $\phi = 90°$, that is at the support, we have

$$M_{of} = wR^2 (c\cos\phi - 1) = -167.7 \text{ kNm}$$

234

$$M_{sf} = wR^2(c \sin \phi - \phi) = 167.7(1.27 - 1.57) = -50.3 \text{ kNm}$$
$$= T$$

Design of the section. The maximum bending and torsional moments occur at the support. The equivalent moment at the support is (the negative BM)

$$M_e = M + T\frac{(1 + h/b)}{1.7} = M_{\phi f} + M_{tf}\frac{(1 + 0.6/0.4)}{1.7}$$

$$= 167.7 + 50.3\frac{(2.5)}{1.7} = 242 \text{ kNm} = 0.242 \text{ MNm}$$

The moment capacity M_r must be more than the external BM; therefore

$$M_r = K\,bd^2\,\sigma_{cb} \geq M_e = 0.242 \text{ MNm}$$

$$\text{or } d = \sqrt{\frac{0.242}{0.13(0.4)(8.5)}} = 0.74 \text{ m}$$

where $K = 0.13$ and b is assumed as 0.4 m.

$$\text{Let } h = d + 0.06 = 0.8 \text{ m.}$$
$$\text{Self-weight } = 0.4(0.8)(25) = 8 \text{ kN/m}$$
$$\text{Total load } = 15.3 + 1.5 + 8 = 24.8 \text{ kN/m}$$

The assumed load = 23 kN/m less than the actual value so the moments are revised.

The bending moments are :

At $\phi = 0$ and at $\phi = 90°$

$$M_{\phi c} = 48.8 \text{ kNm}$$

$$M_{\phi f} = 180.8 \text{ kNm}$$

$$M_{tf} = 54.2 \text{ kNm}$$

$$M_e = 180.8 + 54.2\frac{(1 + 0.8/0.4)}{1.7} = 276.5 \text{ kNm}$$

The value of d is given by

$$d = \sqrt{\frac{M}{Kb\,\sigma_{cb}}} = \sqrt{\frac{0.2765}{0.13(0.4)(8.5)}} = 0.79 \text{ m}$$

$$h = 0.79 + 0.05 = 0.84 \text{ m.}$$

The areas of the bottom and top reinforcement are:

$$A_{sp} = \frac{M_{\phi c}}{jd\,\sigma_{st}} = \frac{48\,800}{0.904(0.79)(230)} = 297 \text{ mm}^2$$

$$A_{sn} = \frac{M_e}{jd\,\sigma_{st}} = \frac{276\,500}{0.904(0.79)(230)} = 1684 \text{ mm}^2$$

Provide 2 numbers of 16 mm bars at the bottom and four numbers of 22 mm bars and one 16 mm bar at the top.

$$A_{sn} = 4(380) + 201 = 1720 \text{ mm}$$

$$\frac{100\,A_{sn}}{bd} = \frac{172\,000}{400(800)} = 0.5\%$$

Design for shear: The permissible shear stress in the concrete for 0.5% tension reinforcement is 0.31 MPa:

$$\tau_c = 0.31 \text{ MPa},$$

Shear capacity of the section is $= V_c = \tau_c bd = 0.098$ MN. The critical shear force occurs at a distance d from the support and it is equal to

$$V = w(R\theta - d) = 24.8(2.7(1.57) - 0.79) = 85.5 \text{ kN} = 0.085 \text{ MN}$$

There is a torsional force at the section which should be converted into equivalent shear:

$$V_e = V + 1.6\frac{T}{b} = 0.085 + 1.6\frac{(0.0542)}{0.4} = 0.302 \text{ MN}$$

The nominal shear stress is

$$\tau_v = \frac{V_e}{bd} = \frac{0.302}{0.4(0.79)} = 0.95 \text{ MPa}$$

The allowable shear stress without web reinforcement is 0.31 MPa, and with web reinforcement is 1.9 MPa. Therefore, shear reinforcement must be provided.

Select two-legged 12 mm stirrups, the area of which is

$$A_{sv} = 2(113) = 226 \text{ mm}^2$$

The spacing of the reinforcement is governed by

$$A_{sv} \geq \left(\frac{T}{b_1 d_1} + \frac{V}{2.5 d_1}\right)\frac{s_v}{\sigma_{sv}}$$

where $b_1 = b - 2 \text{ cover} - \phi = 400 - 50 - 22 = 328$ mm

$$d_1 = d - d_c = 790 - 50 = 740 \text{ mm}$$

Therefore, we have

$$226 \geq \left(\frac{54\,200\,000}{328(740)} + \frac{85\,000}{2.5(740)}\right)\frac{s_v}{230}$$

$$\geq (223 + 46)\frac{s_v}{230} = 1.18 s_v$$

$$s_v \leq \frac{226}{1.18} = 191 \text{ mm}$$

The maximum spacing of the stirrups is also controlled by

$$s_{v\,max} \leq \frac{A_{sv}\sigma_{sv}}{b(\tau_v - \tau_c)} = \frac{226(300)}{400(0.95 - 0.31)} = 203 \text{ mm}$$

Provide two-legged 12 mm closed loop stirrups at 190 mm spacing. Figure 6.13 gives design details.

6.6 Portico Slabs and Frames

Even though the portico slabs or beams are easy to design by conventional method, there are some factors such as counterbalance loads, embedment lengths, etc. which are often ignored. Most portico slabs are designed as cantilevers or overhang slabs, therefore, the sagging effects are seen to be dominant. Architects bring out imaginary shapes of these

236

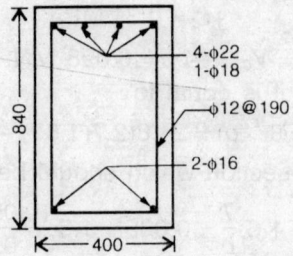

Fig. 6.13 Sectional Details of Semi-circular Curved Beams of Example 6.10.

porticos, which require careful reinforcement detailing. Therefore, this section is devoted to detailed designed of a typical structure.

EXAMPLE 6.11 *Design of self-supporting portico frame-slab:* A self-supporting frame as shown in Fig. 6.14a is provided with a hanging slab at the top of the frame. The bottom slab acts as a foundation slab. Design the slab and column system.

Design data and analysis: No access is provided to the roof slab of the portico; therefore, the live load on the roof can be 750 N/m². However, one should design such proticos for unexpected and uninvited visitors on the roof.

$$\text{Roof live load} = 3 \text{ kN/m}^2$$
$$\text{Roof finish load} = 2 \text{ kN/m}^2$$
$$\overline{\text{Superimposed load} = 5 \text{ kN/m}^2}$$

Let $f_{ck} = 20$ MPa, $f_y = 415$ MPa, then we have
$\sigma_{cb} = 7$ MPa, $\sigma_{st} = 230$ MPa, $K = 0.13$, $j = 0.904$

The design consists of the following:
1. Roof slab.
2. Cantilever beams.
3. Columns.
4. Foundation (stability and bearing pressure considerations).
5. Foundation slab.
6. Foundation beams.

Design of the roof slab. The roof slab is supported by two beams spaced at 4 m apart and having overhangs of 1 m on each side. The slab is to be designed as a one way strip with double overhangs. Let the thickness of the slab be assumed 160 mm for the purpose of computing self-weight.

$$\text{Self-weight} = 0.16(25) = 4 \text{ kN/m}^2$$
$$\text{Superimposed load} = 5 \text{ kN/m}^2$$
$$\overline{\text{Total load} = w = 9 \text{ kN/m}^2}$$
$$\text{Cantilever span} = (a - 0.5b_w) = 1 - 0.2 = 0.2 = 0.8 \text{ m}$$

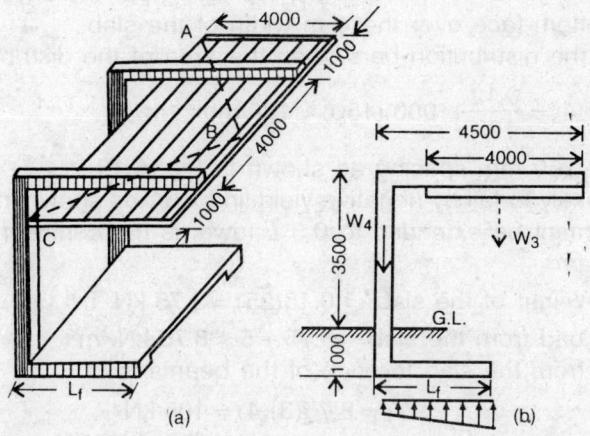

Fig. 6.14 Self-supporting Frame and Slab Portico Example 6.11.

(where b_w = width of the beam and it is taken equal to 0.4 m)

The maximum negative bending moment which occurs at the face of the beam is

$$M_n = \frac{w(a - 0.5w)^2}{2} = \frac{9000(0.64)}{2} = 2880 \text{ Nm/m}$$

The maximum positive bending moment which occurs at the mid-span is

$$M_p = w(a + 0.5\ L)(0.5L - 0.5(a + 0.5L))$$

$$= 9000(3)(2 - 1.5) = 13500 \text{ Nm/m} = 0.0135 \text{ MNm/m}$$

The subscripts n and p refer to the negative and positive bending moments. The depth of the slab is controlled by the positive bending moment and it is given by

$$d = \sqrt{\frac{M_p}{Kb\sigma_{cb}}} = \sqrt{\frac{0.0135}{0.13(1)(7)}} = 0.122 \text{ m}$$

The overall depth of the slab will then be

$$t = d + 0.028 = 0.15 \text{ m}$$

This value is only 10 mm less than the average thickness assumed; therefore, the bending moments computed are just right. The areas of the reinforcement to resist the positive and negative bending moments are:

$$A_{sp} = \frac{M_p}{jd\sigma_{st}} = \frac{13500}{0.904(0.122)(230)} = 532 \text{ mm}^2/\text{m}$$

$$A_{sn} = \frac{M_n}{jd\sigma_{st}} = \frac{2880}{0.904(0.122)(230)} = 105 \text{ mm}^2/\text{m}$$

Minimum reinforcement is

$$A_{sm} = \frac{0.85\ bd}{f_y} = \frac{0.85(1000)(122)}{415} = 250 \text{ mm}^2/\text{m}$$

Provide 14 numbers of 12 mm bars at the top face near the beams and 18 numbers of 12 mm bars at the bottom face over the 4 m width of the slab.

Provide 0.12% of the distribution bars, then the area of the distribution reinforcement is

$$A_s = \frac{0.12}{100}(1000)(150) = 180 \text{ mm}^2/\text{m}$$

Provide 8 mm bars at 250 mm spacing as shown in Fig. 6.15.

Note: The slab is likely to fail by negative yield line *ABC* as shown in Fig. 6.14a. Therefore, the negative reinforcement be extended to 0.3 *L* towards the centre from the beam.

Design of the roof beam

Weight of the slab $= 0.15(25) = 3.75 \text{ kN/m}^2$

Load from the slab $= 3.75 + 5 = 8.75 \text{ kN/m}^2$

Total load transferred from the slab to each of the beams is

$$W_1 = 8.75(3)(4) = 105 \text{ kN}$$

This load is acting at a distance $2 + 0.5 = 2.5$ m from the column face. Let the average size of the beam be assumed as 0.4 m by 0.7 m for the purpose of computing self-weight.

Self-weight $= W_2 = 0.4(0.7)(4.5)(25) = 31.5 kN$

The maximum negative bending moment on the beam which occurs at the column face, is

$$M = W_1(2.5) + W_2(4.5)/2$$
$$= 105(2.5) + 31.5(2.25) = 333.8 \text{ kNm}$$

The beam acts as an inverted T-beam upto a certain point. However, the section at the column face is a rectangular one. Therefore, the effective depth of the beam is given by

$$d = \sqrt{\frac{M}{b_w K \sigma_{cb}}} = \sqrt{\frac{0.3338}{0.4(0.13)(7)}} = 0.96 \text{ m}$$

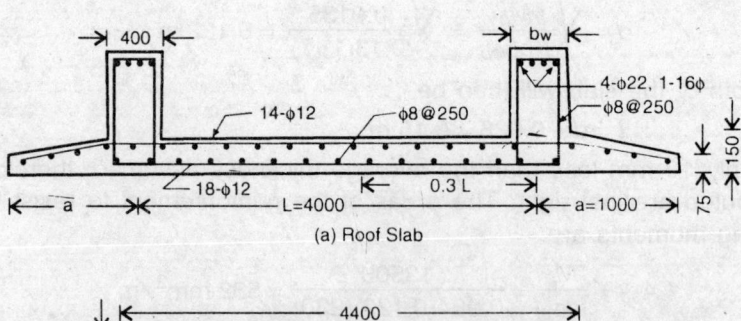

(a) Roof Slab

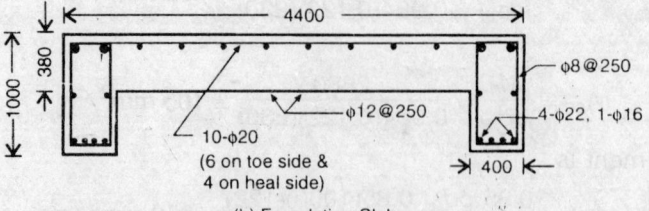

(b) Foundation Slab

Fig. 6.15 Reinforcement Details of Frame of Example 6.11.

where $b_w = 0.4$ m

Then the overall depth is given by

$$h = d + 0.04 = 1.0 \text{ m}$$

The actual bending moment is slightly higher than that computed and the actual self-weight is more than the assumed value. The area of the tension reinforcement is given by

$$A_{st} = \frac{M}{jd\sigma_{st}} = \frac{333\,800}{0.9(0.96)(230)} = 1680 \text{ mm}^2$$

Provide 4 numbers of 22 mm bars and one 16 mm bar in one row. The actual area provided is 1720 mm^2 which is slightly more than that needed. This takes care of the extra bending moments due to increase in the self-weight. The depth of the beam at the outer face of the slab is reduced to 250 mm; therefore, the weight of the beam assumed is satisfactory.

Design of the beam for shear. The maximum shear force on the beam, which occurs at the face of the column, is:

$$V = W_1 + W_2 = 105 + 31.5 = 136.5 \text{ kN}$$

The nominal shear stress is

$$\tau_v = \frac{V}{b_w d} = \frac{0.1365}{0.4(0.96)} = 0.36 \text{ MPa}$$

The percentage of the tension reinforcement is

$$\frac{100\,A_{st}}{b_w d} = \frac{100(1720)}{400(960)} = 0.45\%$$

The allowable shear stress for the above percentage reinforcement is:

$$\tau_c = 0.3 \text{ MPa}$$

Therefore, transverse reinforcement should be provided. Select two-legged 8 mm bars as stirrups, then the spacing is

$$s_v = \frac{A_{sy}\sigma_{sy}}{b_w(\tau_v - \tau_c)} = \frac{100(230)}{400(0.36 - 0.3)} = 958 \text{ mm}$$

and $\quad s_{v\max} = \dfrac{A_{sv}f_y}{0.4b_w} = \dfrac{100(415)}{0.4(400)} = 259 \text{ mm}$

Provide two-legged 8 mm stirrups at 250 mm spacing.

Design of the column. The axial force acting on the column is same as the shear force on the beam, and it is

$$P = V = 136.5 \text{ kN}$$

The column is also subjected to constant bending moment over its length and bending moment is same as that of the maximum on the beam. Therefore,

$$M = 333.8 \text{ kNm}$$

(The actual BM at the centre line of the column will be slightly more.)
The eccentricity on the column would be

$$e = \frac{M}{P} = \frac{333.8}{136.5} = 2.45 \text{ m}$$

The eccentricity is very large; therefore, the tensile stress in the reinforcement may control the design. (See section 5.15 in which the design of column failing in tension was discussed.)

$$\text{Assume} \quad u = \frac{d_c}{d} = 0.15, \quad p = 0.004, \quad c = \frac{D_s}{d} = 0.85$$

Assume the areas of the tensile and compressive reinforcements to be equal, the $q = 1$. Let $b = 400$ mm, same as that of the beam.

Eqs. (5.81) and (5.83) which are the governing equations, are:

$$bd\left(n^2 + 3mp\left(1 - \frac{u}{n}\right)\frac{n}{m(1-n)} - 2p\right)\sigma_{ast} = 2P$$

$$bd\left(n(3-n) + 9mqp\left(1 - \frac{u}{n}\right)c\right)\frac{n\sigma_{ast}}{m(1-n)} - 6P\left(\frac{c}{d} + 1 - \frac{D}{2d}\right) = 0.$$

The values of m, n etc. are:

$m = 13$, $n = 0.283$ and let $D/d = 1.15$ then the substitution of various quantities in Eq. (5.81) gives

$$bd\left(n^2 + 0.156\frac{(n-0.15)}{13(1-n)} - 0.008\right) = \frac{2P}{\sigma_{ast}} = \frac{2(136\,500)}{230}$$

$$\text{or} \quad 400d(n^2 + 0.156n - 0.023 - 0.104(1-n)) = \frac{26(136500)}{230}(1-n)$$

$$\text{or} \quad d(n^2 + 0.26n - 0.127) = 38.3(1-n) \tag{a}$$

Similarly, Eq. (5.83) gives

$$bd^2(3n^2 - n^3 + 0.4(n - 0.15))\frac{230}{13(1-n)} - 6P(2450 + 0.43d) = 0$$

$$\text{or} \quad d^2(3n^2 - n^2 + 0.4n - 0.06) - \frac{78(1-n)P}{230(b)}(2450 + 0.43d) = 0$$

$$\text{or} \quad d^2(3n^2 - n^3 + 0.4n - 0.06) - 115(1-n)(2450 + 0.43d) = 0$$

$$\text{or} \quad d^2(3n^2 - n^3 + 0.4n - 0.06) - 49(1-n)d - 281\,750(1-n) = 0 \tag{b}$$

Equations (a) and (b) are solved by iteration.

Try $n = 0.26$, the above equations give:

$$d = 3450 \text{ mm}$$
$$\text{and } 0.229d^2 - 36.6d - 267\,660 = 0$$
$$\text{or } d = 1165 \text{ mm}$$

Try $n = 0.288$, then we have

$$d = 884 \text{ mm}$$
$$\text{and } 0.28d^2 - 35d - 200\,186 = 0$$
$$\text{or } d = 910 \text{ mm}$$

The values of d are converging towards 900 mm; therefore, try $d = 0.9$ m and $n = 0.3$. (Since n is more than n_c, the tension in the steel does not govern the design.)

$$A_{st} = 0.004bd = 0.004(400)(900) = 1280 \text{ mm}^2$$

Provide 4 numbers of 20 mm bars on each face then

$$A_{st} = A_{sc} = 4(314) = 1242 \text{ mm}^2$$

Use $d = 920$ mm, $d_c = 60$ mm

$$D_s = d - d_c = 860 \text{ mm}, \quad D = d + d_c = 980 \text{ mm}$$

Since $n > 0.283$, the failure is by compression.

Check for the capacity of the column

$$nd = 0.30(920) = 276 \text{ mm}$$

$$d - nd = 644 \text{ mm},$$

$$\sigma_{st} = \frac{(d - nd)m\,\sigma_{acb}}{nd} = \frac{644(13)(7)}{276} = 212 \text{ MPa}$$

$$\sigma_{sc} = \frac{(nd - d_c)\sigma_{st}}{d - nd} = \frac{216(212)}{644} = 71 \text{ MPa}$$

The compressive and tensile forces on the section are:

$$C_c = \frac{nbd\sigma_{acb}}{2} = \frac{0.3(400)(900)7}{2} = 378\,000 \text{ N}$$

$$C_s = A_{sc}\sigma_{sc} = 1242\,(71) = 88\,182 \text{ N}$$

$$T = A_{st}\sigma_{st} = 1242\,(212) = 263\,304 \text{ N}$$

$$P_a = C_c + C_s - T = 202\,878 \text{ N} > P = 136\,500 \text{ N}$$

The axial load capacity is far higher than the load acting riot he columns.
The allowable bending moment is

$$M_a = C_c\left(d - \frac{nd}{3}\right) + C_s(d - d_c) - P(d - 0.5\,D)$$

$$= 378(0.92 - 0.092) + 88.182(0.86) - 136.5\,(0.92 - 0.49)$$

$$= 331 \text{ kNM} < M = 333.8 \text{ kNm}$$

The moment capacity of the section is slightly less than the desired, so modify the **depth** marginally.

Provide 8 mm *ties* at 350 mm spacing.

Design of the foundation. The foundation consists of two beams extended from the columns and a slab between the beams. The overall width of the slab including the **width** of the beam is

$$B = L + b_w = 4.4 \text{ m}$$

where the width of the beam is taken same as that of the columns. The foundation **has to** be designed for stability, bearing pressure and for the strength. A factor of **safety** of 2 is applied to the stability of the structure.

Let the thickness of the foundation slab be = 300 mm
Weight of the slab = 0.3(25) = 7.5 kN/m²
Load from the roof slab and the roof beams

$$W_3 = 2(W_1 + W_2) = 273 \text{ kN}$$

Weight of the column

$$W_4 = 2(4.5)(0.4)(0.98)(25) = 88 \text{ kN}$$

For simplicity of computations, the weight of the roof beam is also placed at the CG of the roof slab. This is to be on the safer side.

The weight of the foundation slab is

$$W_5 = 4.4 \, L_f(7.5) = 33 \, L_f$$

Neglect the weight of the overburden soil for the purpose of stability to be on the safe side, and take moment about the tip of the foundation slab.

The overturning moment is

$$M_1 = W_3(2.25 - L_f) = 273(2.25 - L_f)$$

The stabilizing moment is

$$M_2 = W_4(0.5D + L_f) + \frac{W_5(L_f)}{2}$$

$$= 88(0.49) + (88 + 16.5 \, L_f)L_f$$

For stability, we have

$$M_2 \geq 2M_1$$

$$\text{or} \quad 43.1 + (88 + 16.5L_f)L_f \geq 2(273)(2.25 - L_f)$$

$$\text{or} \quad L_f^2 + 38.4 \, L_f - 72 \geq 0$$

$$\text{or} \quad L_f > 1.83 \text{ m}$$

The minimum length of the foundation required based on the stability consideration is 1.83 m. The length is also governed by the bearing pressure consideration; therefore, one should calculate the length based on the allowable pressure. The maximum bearing pressure on the soil occurs at the toe of the foundation slab and it is given by

$$q = \frac{W}{A_f} + \frac{M_f}{Z_f} \leq q_a$$

where W = the net load on the soil

M_f = BM about on the slab

A_f = area of the foundation slab = 4.4 L_f

Z_f = modulus of the section of the foundation area

$$= \frac{4.4 L_f^2}{6} = 0.73 \, L_f^2$$

Since the safe net bearing capacity is given, only the net load of the foundation slab is considered.

The difference in weight of the concrete foundation and the equivalent replaced soil is

$$W_6 = 4.4 \, L_f (25 - 18) = 31 L_f \, \text{kN}$$

The net load on the foundation is

$$W = W_3 + W_4 + W_6 = 273 + 88 + 31L_f$$

$$= 361 + 31L_f \, \text{kN}$$

The moment about the middle line of the foundation slab is

$$M_f = W_3(2.25 - 0.5L_f) - W_4(0.49 + 0.5 \, L_f)$$

$$= 273(2.25 - 0.5\,L_f) - 88(0.49 + 0.5\,L_f)$$
$$= 571 - 180.5 L_f$$

The maximum bearing pressure on the soil is

$$q_1 = \frac{W}{A_f} + \frac{M_f}{Z_f} = \frac{361 + 31\,L_f}{4.4 L_f} + \frac{571 - 180.5 L_f}{0.73 L_f^2}$$

$$= \frac{(82 L_f + 7 L_f^2) + (782 - 247 L_f)}{L_f^2}$$

$$= \frac{7 L_f^2 - 165 L_f + 782}{L_f^2} \le q_a = 100 \text{ kN/m}^2$$

$$\text{or } 93 L_f^2 + 165 L_f - 782 \ge 0$$

$$\text{or } L_f > 2.2 \text{ m.}$$

The least bearing pressure, which occurs at the heel of the slab, is

$$q_2 = \frac{W}{A_f} - \frac{M_f}{Z_f}$$

$$= \frac{361 + 31(2.2)}{4.4(2.2)} - \frac{571 - 180.5(2.2)}{0.73(2.2)^2}$$

$$= 44.3 - 49.2 = -4.9 \text{ kN/m}^2$$

There is a negative pressure at the heel side, or the tension on the soil which is not **admissible**. Increase the length of the foundation.

Provide $L_f = 2.4$ m then

$$A_f = 4.4(2.4) = 10.56 \text{ m}^2, \quad Z_f = \frac{4.4(2.4)^2}{6} = 4.224 \text{ m}^2$$

Difference in the weight of the foundation slab concrete and the equivalent soil is

$$W_6 = B L_f t\,(\gamma_c - \gamma_s) = 4.4(2.4)(0.3)(25 - 18) = 22 \text{ kN}$$

$$\text{Total load} = W = W_3 + W_4 + W_6 = 273 + 88 + 22 = 383 \text{ kN}$$

$$BM = M_f = W_3(2.25 - 1.2) - W_4(0.49 + 1.2)$$

$$= 273(1.05) - 88(1.69) = 137.93 \text{ kNm}$$

The maximum and minimum pressures on the soil are:

$$q_1 = \frac{W}{A_f} + \frac{M_f}{Z_f} = \frac{383}{10.56} + \frac{137.93}{4224}$$

$$= 36.3 + 32.7 = 69 \text{ kN/m}^2$$

$$q_2 = 63.3 - 32.7 = 3.6 \text{ kN/m}^2$$

Design of the foundation slab. The foundation slab is supported by two beams spaced at 4.0 m apart and is subjected to a maximum soil reaction of 69 kN/m^2. The effective load acting on the slab is the net force less the effect of slab weight.

$$\text{Effective load} = 69 - \frac{22}{4.4(2.4)} = 67 \text{ kN/m}^2$$

This is the pressure at the tip of the slab whereas the pressure at the other end is very small. It is, therefore, desirable to design the 2.4 m length of the slab for the average reaction.

$$W_8 = \text{Effective load} = W_3 + W_4 = 361 \text{ kN}$$

The maximum bending moment which occurs at the mid-span of the slab is

$$M = \frac{W_8 L}{8} = \frac{361(4)}{8} = 180.5 \text{ kNm}$$

The depth of the slab will be given by

$$d = \sqrt{\frac{M}{Kb\,\sigma_{cb}}} = \sqrt{\frac{0.1805}{0.13(2.4)(7)}} = 0.29 \text{ m}$$

Use

$$h = 0.29 + 0.09 = 0.38$$

$$A_{st} = \frac{M}{jd\,\sigma_{st}} = \frac{180\,500}{0.904(0.29)(230)} = 2994 \text{ mm}^2$$

Provide 10 numbers of 20 mm bars of which six be placed on the one metre toe side and four on the heel side.

The maximum shear force is

$$V = \frac{W_8}{2} - w_2 d = \frac{W_s}{2}\left(1 - \frac{2d}{BL_f}\right)$$

$$= 180.5\left(1 - \frac{0.58}{4.4(2.4)}\right) = 170 \text{ kN}$$

The nominal shear stress is

$$\tau_v = \frac{V}{L_f d} = \frac{0.17}{2.4(0.29)} = 0.24 \text{ MPa}$$

The allowable shear stress for 0.5% reinforcement is 0.3 MPa. Therefore, the section is adequate against shear force.

The area of the distribution reinforcement is

$$A_s = \frac{0.12}{100}(370)(1000) = 444 \text{ mm}^2/\text{m}$$

Provide 12 mm bars at 250 mm spacing.

Design of the foundation beam. The effective total load on each foundation beam is $\frac{W_8}{2}$.

The bending moment on the beam is same as that coming from the column.

$$M = 333.8 \text{ kNm}$$

The width of the beam is $= b = 0.4$ m.

The beam acts as an L-section, however, it is advantageous to design it as a rectangular sectioned beam to minimize the reinforcement.

Provide the beam section same as that of the roof beam.

While the bending moment on the foundation beam is same as that of the roof beam, shear force is more because of the weight of the column.

The critical shear force and stress on the section are

$$V = \frac{361}{2} = 180.5 \text{ kN and}$$

$$\tau_v = \frac{V}{bd} = \frac{0.1805}{0.4(0.77)} = 0.57 \text{ MPa}$$

The allowable shear stress is 0.3 MPa, therefore, transverse reinforcement to be provided. Select two-legged 8 mm stirrups, then the spacing of the stirrups is

$$s_v = \frac{A_{sv}\,\sigma_{sv}}{b(\tau_v - \tau_c)} = \frac{100(230)}{400(0.27)} = 212 \text{ mm}$$

$$s_{vmax} = \frac{A_{sv}f_y}{0.4b} = \frac{100(415)}{0.4(400)} = 200 \text{ mm}$$

Provide two-legged 8 mm stirrups at 250 mm spacing

The development length is

$$L_d = \frac{\phi\sigma_{st}}{4\tau_{bd}} = \frac{230\phi}{4(2.1)(1.6)} = 30\phi$$

PROBLEMS

6.1 A tie beam of a bowstring arch bridge is subjected to an axial tensile force of 800 kN. Design the tie beam by working stress method using M20 concrete and HYSD-Fe415 bars. Assume allowable tensile stress in concrete as 1.8 N/mm^2 and modular ratio of 13. If the span of the tie beam is 16 m and assumed to be hinged at both the ends, compute the bending stresses in the tie caused due to self-weight of the beam and the tension force.

6.2 A circular cylindrical penstock is subjected to a pressure of 100 m head of water. The diameter of the pipe is 4 m. Design a reinforced concrete pipe b". working stress method using M25 concrete and HYSD-Fe415 bars. Assume the allowable stress in tension in concrete as 1.6 N/mm^2. The allowable tensile stress in steel is 150 N/mm^2.

6.3 Derive the governing equations for working stress design of a reinforced concrete cross section subjected to an axial tension and bending moment. Assume the area of steel to be equally divided on the two bending faces. Show that the equilibrium equations can be written as

$$bd\left(\frac{m(1-n)p_s}{n} - \frac{1.5\,nm}{n-u} - \frac{n}{2}\right)\sigma_{cb} = P$$

and $$bd^2\left(K + cp_c\,\frac{1.5\,mn}{n-u}\right)\sigma_{cb} = Pd(g - 0.5c)$$

where $K = n(1 - n/3)/2$

$c = D_s/d$

$g = e/d$, D_s = distance between tension and compression by enforcement

$p_s = A_{st}/bd$, $p_c = A_{sc}/bd$

$u = d_c/d$, d_c = distance of compression reinforcement from the compression fibre

6.4 A tie beam of a rectangular cross-section 600 mm by 800 mm is subjected to an axial tension of 700 kN and bending moment of 300 kN. The beam is made of M20 concrete and HYSD-Fe415 bars with an area of steel 9800 mm^2 equally divided on tension and compression faces. The distance of the steel from the outer fibre of concrete is 60 mm. Analyse the beam and determine the stresses in the reinforcement.

6.5 A staircase slab simply supported at 5 m apart has an inclination of 1 in 2.0 and the width of the slab is 1.5m. It is subjected to a live load of 4 kN/m^2. Design the staircase slab by working stress method using M25 concrete and HYSD-Fe415 bars.

6.6 A staircase of a width consists of 1 central beam on which a set of independent steps are fixed overhanging 1 metre on either side of the beam. The size of the step is 350 mm by 150 mm. The slope of the beam supporting the steps is having a span of 4 m in plan. Design the steps and the beam for the following loads: (a) uniformly distributed load of 5 kN/m^2, and (b) 1.5 kN per each step placed at 0.1 m from the end of the step. The two loads do not act together simultaneously. Assume the width of the beam as 400 mm and use M20 concrete and HYSD-Fe415 bars.

6.7 A staircase of 3 m width is made of an independent steps of 350 mm by 150 mm fixed on two longitudinal beams symmetrically placed at 1.6 m apart. In other words, each step is resting as a double overhanged member on the two longitudinal beams. The slope of the beams is in one two with effective span of 4.5 m in the plan. Design the staircase beams using M20 concrete and HYSD-Fe415 bars for the following load conditions:
(a) Uniformly distributed load of 5 kN/m^2; or
(b) Concentrated load of 1.5 kN placed at each of the two free ends of the step.

6.8 A staircase is constructed on a vertical wall from which steps cantiliver to a span of 1.5 m. The rise and tread of the step is 160 mm by 320 mm. Design the staircase using M20 concrete and HYSD-Fe415 bars for the following loads:
(a) Uniformly distributed load of 4 kN/m^2; or
(b) a concentrated load of 1.5 kN placed at the free end.

6.9 Two floors separated by a height of 4.5 m are connected by a helicoidal staircase of 35° slope. Design a solid helicoidal staircase slab of width 1.5 m for working load of 4 kN/m^2 using M20 concrete and HYSD-Fe415 bars. The staircase is semi-circular in plan.

6.10 A helicoidal staircase consists of independent steps placed on a rectangular cross-section beam. The subtended angle is 240° and the gradient of the staircase is 1.6. The beam is placed under the middle line of the independent steps. Design the staircase by working stress method using M20 concrete and HYSD-Fe415 bars. The staircase is subjected to a live load of 4 kN/m^2. Design the steps cantiliver on either side of the beam and the helicoidal beam. The width of the beam is to be restricted to 400 mm.

6.11 A girder, circular in plane having a mean radius of 6 m is supported by 10 columns spaced at equal intervals. A load of 20 kN/m is acting on the girder. Design the girder by WSD using M25 concrete and HYSD-Fe415 bars. The width is to be restricted to 400 mm.

6.12 A circular slab of outer radius of 5 m is supported by a beam circular in plan having mean radius of 4 m. The thickness of the slab is 200 mm and it is subjected to a live load of 4 kN/m^2. The circular beam is supported by 6 columns equally spaced in plan. Design the circular beam using M20 concrete and HYSD-Fe415 bars assuming the width of the beam as 300 mm.

Chapter 7

Strength Limit State Design of Beams

7.1 Introduction

A structure should be designed to perform satisfactorily under different forces which are likely to act on it. It should not only serve the purpose for which it is designed but it should also provide adequate safety against failure under most severe load condition. The limit state design considers the different limits, and designs the structure for satisfactory serviceability and adequate safety against failure. Three types of limit states are considered in the design:

1. Strength limit state also referred as limit state of collapse,
2. Serviceability limit state,
3. Durability limit state.

The design based on the strength limit state ensures adequate and reasonable safety of the structure against total or partial collapse. During the service condition, the structure should provide security and serviceability. The limits of serviceability are usually associated with:

a) Deflection tolerances which are controlled either by the appearance of the overall structure or by tolerances of the other elements such as cladding, floor finish and psychological factors.
b) Crack width are to be controlled so as to prevent corrosion of steel, deterioration of concrete, ugly appearance and other detrimental influences.
c) Shrinkage and creep deflections or stresses.
d) Strain limits in working, creep and collapse load conditions.

The durability limit state is associated with the total life and serviceability of the structure when the structure is subjected to environmental or repeated loads. The cover to the reinforcement should be provided to protect the reinforcement from corrosion. Design should be made to provide resistance against wear and tear, and repeated or fatigue type of forces.

There are several uncertainties in the quality of production of the material for construction and the level and occurrence of load on the structure. When the designer specifies a quality of the material, the supervisor must be able to enforce certain desired quality control. There should be an understanding between the designer and the builder, about what is expected and what is feasible. The designer can anticipate or try to enforce a certain probability that the actual strength of the material will not fall below the specified strength. The specified strength of the material based on which the design is made is stated as *characteristic strength*. The characteristic strength is the strength below which not more than five per cent of the test results are expected to fall.

Let f_k = characteristic strength
f_m = population mean value of the actual strengths tested in the field

s = standard deviation of the strengths (population STD)

p_f = accepted or assigned probability that the strength will not fall below the characteristic value.

The probability of failure can then be expressed as:

$$P(f < f_k) = p_f \tag{7.1}$$

Let the frequency distribution of the strength of the materials follow a normal distribution (which is the case in most of the strengths of the materials). Then Eq. (7.1) can be written as:

$$\phi\left(\frac{f_k - f_m}{s}\right) \tag{7.2}$$

In which ϕ is the cumulative distribution function. The above equation is rearranged as

$$\phi^{-1}(p_f) = \frac{f_k - f_m}{s} \tag{7.3}$$

For low probability of failure, the above value will be negative-definite and let it be equal to $-k$. Then

$$\frac{f_k - f_m}{s} = -k \tag{7.4}$$

in which k defines the accepted probability of failure parameter.

The above equation can be rearranged as

$$f_m = f_k + sk \tag{7.5}$$

The mean value of the strength of the materials tested should satisfy the above equation. For 5 per cent accepted probability of failure, the value of k is 1.64. Hence

$$f_m = f_k + 1.64 \, s \tag{7.6}$$

It is, therefore, the responsibility of the builder to ensure that either Eq. (7.6) or (7.5) is satisfied.

By a similar logic, one can establish a relation between the characteristic load and the mean value of the maximum loads acting on the structure, that is

$$F_k = F_m + ks \tag{7.7}$$

in which

F_k = characteristic load specified by the designer

F_m = mean value of the maximum of the observed loads

s = standard deviation in the maximum observed loads

k = a constant which gives the accepted probability that the mean value of the observed maximum loads can (may) exceed the characteristic load

Characteristic load is the value which has a 95 per cent probability of not being exceeded during the life of the structure.

7.2 Design Strengths and Loads

The strength of the material actually adopted in the design computations must take into account other factors such as shape of the structural element, nature of the stress resultants, relation between the strength of the structural element and the control specimen, limit state conditions, etc. The relation between the design and characteristic strengths is connected through partial safety factor as

$$f_d = \frac{f_k}{\gamma_m} \tag{7.8}$$

in which

$\gamma_m = $ partial safety factor applied to the material

$f_d = $ design strength

The partial safety factors applied to concrete and steel in RCC structures are listed in Table 7.1.

Table 7.1 Partial Safety Factors Applied to Materials

Nature of stress resultant	Concrete[b]		Steel[*b]
	k_p	γ_m	γ_s
Flexure	0.67	1.5	1.15
Compression			
Zero eccentricity	0.67	1.5	1.33
Minimum eccentricity	0.67	1.6	1.50

k_p The ratio of the prism strength to the cube strength of the concrete.

[b]Characteristic strength for concrete is associated with that of 150 mm cube tested under saturated moist condition after 28 days of water curing. Characteristic strength of steel is associated with the yield stress or proof stress at 0.2 per cent associated with residual strain under uniaxial tension of the steel with standard gauge length.

In the case of slender members, the strength of the concrete prism is different from that of the concrete control test specimen. The prism strength can be expressed as

$$f_{pb} = k_p f_{ck} \tag{7.9}$$

in which

$f_{pb} = $ concrete prism strength in bending compression

$f_{ck} = $ characteristic strength of the concrete

$k_p = $ prism strength reduction factor ($= 0.67$ for 150 mm cube)

The design strength of the concrete is then

$$f_{cd} = \frac{k_p f_{ck}}{\gamma_m} \tag{7.10}$$

The shape factor does not enter into the picture in case of steel bars (reinforcement) as the test piece is taken from the bar itself. Table 7.2 gives the strength reduction factors for slender elements.

Table 7.2 Strength Reduced Factor (k_p)

$\dfrac{\text{Height}}{\text{Side}}$	0.5	1.0	2.0	3.0	4.00 above
Relative Strength	1.5	1.0	0.8	0.72	0.67

250

Table 7.3 Limits on Strains

	Property	Strain
1.	Bending compression in concrete	0.0035
2.	Axial compression in concrete	0.0020
3.	Combined bending and axial compression in concrete (ϵ_c = least comp. strain on the section)	$0.0035 - 0.75\epsilon_c$
4.	Minimum strain in steel at failure	$0.002 + \dfrac{f_y}{1.15E_s}$
	f_y = characteristic strength	
	E_s = Young's modulus	
5.	Axial tension in concrete	0.00008
6.	Bending tension in concrete	0.00015
7.	Maximum shrinkage strain	0.0003

In addition to the limitations on the design strengths, limits on strains are also placed. These are listed in Table 7.3.

The design loads are obtained multiplying the characteristic loads by the partial safety factors applied to the loads

$$F_d = F_k\gamma_f \qquad (7.11)$$

in which

γ_f = partial safety factor applied to load and given in Table 7.4.

Table 7.4 Partial Safety Factors (γ_f) Applied to the Loads

Load combination	Limit state of					
	Collapse			Serviceability		
	DL	LL	WL	DL	LL	WL
DL + LL	1.5	1.5	0	1.0	1.0	0
DL + WL	1.5[a] or 0.9	0	1.5	1.0	0	1.0
DL + LL + WL	1.2	1.2	1.2	1.0	0.8	0.8

DL = dead load, LL = Live load, WL = wind load or seismic load,

[a]1.5 or 0.9 which is more severe depending on the type of WL and stability requirement.

7.3 Limit Strength of Rectangular Cross-section

The maximum strain at crushing of the concrete is equal to 0.0035. The stress-strain curve at the crushing of the concrete is (idealized) assumed to be parabolic-cum-rectangular as shown in Fig. 7.1a. Let the different notations of Figs. 7.2 and 7.3 be

x_u = depth of the compression block

x_r = depth of the rectangular stress portion

x_c = depth of the C.G. of the compressive stress

From the geometry of the strains of Fig. 7.3b, the depth of the rectangular portion is given by

$$\frac{x_u - x_r}{x_u} = \frac{0.002}{0.0035} \qquad (7.12)$$

or $\quad \dfrac{x_r}{x_u} = \dfrac{0.0035 - 0.002}{0.0035} = \dfrac{15}{35} = \dfrac{3}{7}$

or $\quad x_r = \dfrac{3x_u}{7} \qquad (7.13)$

where 0.002 is the strain at the end of parabolic stress distribution.

The maximum crushing strength of the concrete prism at the extreme fibre $= k_p f_{ck}$. The total area of the stress block which is set equal to the compressive force on the concrete is

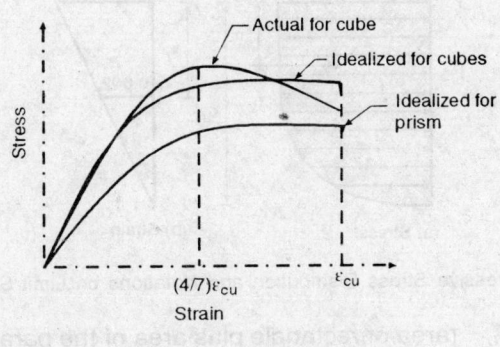

(a) For Concrete

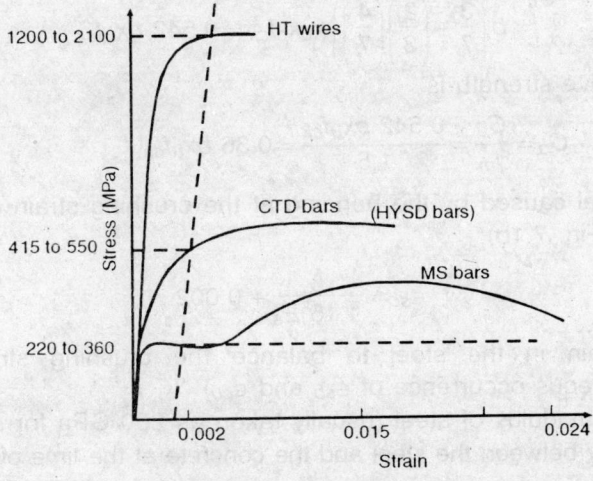

(b) For Steel Bars

Fig. 7.1 Typical Stress-Strain Diagrams for Concrete and Steel.

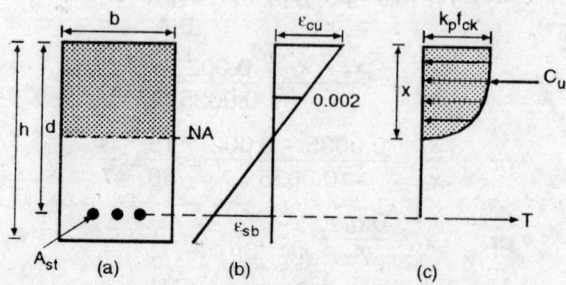

Fig. 7.2 Strain-Stress Distribution on Concrete Beam at Limited State of Strength.

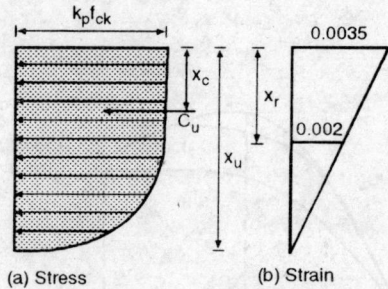

(a) Stress   (b) Strain

Fig. 7.3 Concrete Compressive Stress Distribution and Notations on Limit State Strengths of a Beam.

$$C_u = \text{(area of rectangle plus area of the parabola)}$$

$$= b\,(k_p f_{ck})x_r + \frac{2b}{3}\,(x_u - x_r)\,k_p f_{ck}$$

$$= b\left(\frac{3}{7} + \left(\frac{2}{3}\right)\left(\frac{4}{7}\right)\right)x_u k_p f_{ck} = 0.542\,bx_u f_{ck}$$

The design compressive strength is

$$C_d = \frac{C_u}{\gamma_m} = \frac{0.542\,bx_u f_{ck}}{1.5} = 0.36\,bx_u f_{ck} \tag{7.14}$$

The limit strain in steel caused by the bending at the crushing strain of the concrete is not to be less than (see Fig. 7.1b):

$$\epsilon_{sb} = \frac{f_y}{1.15\,E_s} + 0.002 \tag{7.15}$$

where $\epsilon_{sb}=$ limit strain in the steel to balance the crushing strain in the concrete (simultaneous occurrence of ϵ_{sb} and ϵ_{cu})

$E_s =$ Young's modulus of steel (usually taken as 200 GPa for steel)

The strain compatability between the steel and the concrete at the time of collapse (Fig. 7.2b) is

$$\frac{\epsilon_{cu}}{x_u} = \frac{\epsilon_{sb}}{d - x_u} \tag{7.16}$$

where ϵ_{cu} = crushing strain in concrete.

Let, ϵ_{cu} = 0.0035, then Eqs. (7.15) and (7.16) will result

$$x_u = \frac{\epsilon_{cu}\, d}{\epsilon_{cu} + \epsilon_{sb}} = \frac{0.0035\, d}{0.0035 + 0.002 + f_y/1.15\, E_s}$$

$$= \frac{0.0035\,(1.15)\,(200\,000)\, d}{0.0035\,(1.15)\,(200\,000) + f_y} = \frac{805\, d}{1265 + f_y} \tag{7.17}$$

The distance of the centroid of the compressive force from top (extreme compression) fibre can be obtained by taking moments of the compressive forces about the extreme fibre, that is

$$C_u x_c = b x_r k_p f_{ck}\left(\frac{x_r}{2}\right) + \frac{2}{3}\, b(x_u - x_r)\,(k_p f_{ck})\,(x_u - 5/8)\,(x_u - x_r)$$

On substitution of $k_p = 0.67$, C_u, etc., we have

$$x_c = 0.42 x_u \tag{7.18}$$

The balanced moment capacity of the section can be expressed by taking moment about the centroid of the steel and it is

$$M_{rb} = C_d\,(d - x_c) = 0.36\, b x_u f_{ck}(d - 0.42 x_u) \tag{7.19}$$

A section in which the limiting strains in concrete and steel occur simultaneously is called *Balanced section*. The moment corresponding to the balanced section is called *Balanced design moment*.

The substitution of the value of x_u from Eq. (7.17) in the above equation gives

$$M_{rb} = K b d^2 f_{ck} \tag{7.20}$$

where $K = \dfrac{0.36(805)}{1265 + f_y}\left(1 - \dfrac{0.42(805)}{1265 + f_y}\right)$ (7.21)

The values of x_u and K for different steels are listed in Table 7.5. It can be observed that the moment capacity of a section in which the yield strain in steel and the crushing strain in concrete occur simultaneously is

$$M_{rb} = 0.149\, bd^2\, f_{ck},\ \text{for } f_y = 250\ \text{MPa}$$
$$M_{rb} = 0.138\, bd^2\, f_{ck},\ \text{for } f_y = 415\ \text{MPa}$$
$$M_{rb} = 0.133\, bd^2\, f_{ck},\ \text{for } f_y = 500\ \text{MPa}$$

Equating the design concrete compressive force to the tension force in the steel, we have

$$T = C_d$$

$$\frac{A_{stb}\, f_y}{1.15} = 0.36\, b x_u f_{ck} \tag{7.22}$$

$$\frac{A_{stb}}{bd} = \frac{0.414(805)}{1265 + f_y}\left(\frac{f_{ck}}{f_y}\right) = p_o\left(\frac{f_{ck}}{f_y}\right) \tag{7.23}$$

where A_{stb} = area of the balanced tensile reinforcement ($= A_{sb}$)

The percentage of balanced reinforcement with respect to effective depth can be expressed as

$$p = \frac{100A_{stb}}{bd} = \frac{41.4(805}{1265 + f_y}\left(\frac{f_{ck}}{f_y}\right) = 100p_0\left(\frac{f_{ck}}{f_y}\right) \qquad (7.24)$$

where

$$p_0 = \frac{0.414(805)}{1265 + f_y}$$

The values of p_0 are given in Table 7.5. The approximate percentage of the reinforcement varies from 0.6 to 2.7. The lower limit applies to low strength concretes with high yield steels whereas the upper limit is for higher strength concrete with mild steel reinforcements. In case the beam is under-reinforced, the design moment capacity is

$$M_r = 0.87\ A_{st}f_y(d - 0.42x_u) \qquad (7.25)$$

Table 7.5 Design Parameters for Balanced Section

f_y	$\frac{x_u}{d} = k_u$	j	ϵ_{yp}	K	p_0	K/k_u
For RCC Sections						
250	0.531	0.78	0.0031	0.149	0.220	0.279
360	0.495	0.79	0.0035	0.141	0.205	0.285
415	0.479	0.80	0.0041	0.138	0.198	0.286
500	0.456	0.80	0.0042	0.133	0.189	0.292
For prestressed concrete section						
1500	0.44	0.81	0.0045	0.131	0.186	0.291
1650	0.40	0.83	0.0052	0.123	0.165	0.300

For over-reinforced beams, the moment capacity is restricted to that given in Eq. (7.20). The presence of over-reinforcement can be found if

$$\frac{A_{st}}{bd} > p_0\frac{f_{ck}}{f_y} \qquad (7.26)$$

Area of reinforcement for under-reinforced section is

$$A_{st} = M/jdf_y$$

7.4 Limit Strength of Prestressed Concrete Rectangular Section

The stress-strain relation of the concrete used in prestressed concrete is assumed to be the same as that in RCC as given in Fig. 7.1a. Figure 7.1b also illustrates the stress-strain relation of high tensile steels. The section and steel wires are prestrained in the prestressed concrete. The precompressive strain in the concrete fibre at the level of the pretensioned steel is almost close to zero. Figure 7.4b illustrates the two planes of the section, one at the effective prestrain load condition and the other at the collapse load condition. Line (a) refers to the plane when only the effective prestress acts and on line (b) refers when the section is about to collapse. The total strain in steel at failure is

$$\epsilon_s = \epsilon_{se} + \epsilon_{ce} + \epsilon_{sb} \qquad (7.27)$$

in which $\epsilon_{se} =$ effective prestrain in the steel

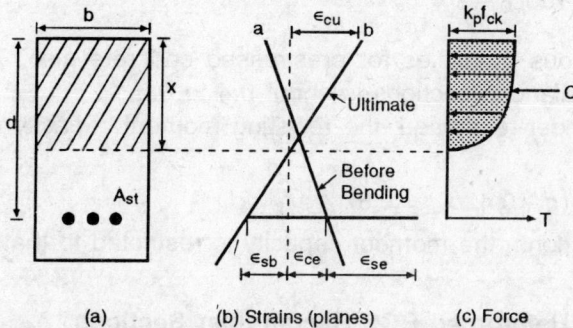

Fig. 7.4 Strain-Stress Distribution Limit Strength of Prestressed Concrete Beam.

$\epsilon_{ce}=$ precompressive strain in the concrete at the steel level. It is negligible when compared with other strains.

$\epsilon_{sb}=$ compatible strain caused by bending corresponding to crushing strain in concrete.

$\epsilon_s=$ total strain in the steel before failure, corresponding to the strain at the proof stress.

Applying a partial safety factor of 1.15 to the proof stress, the value of the strain in steel at failure is

$$\epsilon_s = 0.002 + \frac{f_y}{1.15\, E_s} \tag{7.28}$$

The compatible strain in steel from Eqs. (7.27) and (7.28) is

$$\epsilon_{sb} = 0.002 - (\epsilon_{se} + \epsilon_{ce}) + \frac{f_y}{1.15 E_s} \tag{7.29}$$

For all practical purposes, the effective prestrain in steel can be taken as 0.004. so the above equation reduces to

$$\epsilon_{sb} = -0.002 + \frac{f_y}{1.15 E_s} \tag{7.30}$$

The neutral axis corresponding to the failure of the section is

$$\frac{x_u}{d} = \frac{\epsilon_{cu}}{\epsilon_{cu} + \epsilon_{sb}} = \frac{0.0035}{0.0015 + f_y/1.15 E_s} \tag{7.31}$$

The derivations for the design compressive force, C_d and the moment capacity of the section are similar to those derived for RCC in Section 7.3. These values repeated from the previous expressions:

$$C_d = 0.36\, b\, x_u\, f_{ck} \tag{7.14}$$

$$M_{rb} = K\, bd^2\, f_{ck} \tag{7.20}$$

where $\quad K = (0.36)\left(\dfrac{805}{345 + f_y}\right)\left(\dfrac{1 - 0.42\,(805)}{345 + f_y}\right)$

and $\quad p = \dfrac{100\,A_{st}}{bd} = 100 p_0\,\dfrac{f_{ck}}{f_y}$ (7.24)

Table 7.5 lists the various quantities for prestressed concrete also. The percentage of the reinforcement for the balanced sections is about 0.4 to 0.5.

In case the beam is under-reinforced, the resisting moment capacity of the section is given by

$$M_r = 0.87 A_{st}\,f_y\,(d - 0.42\,x_u) = 0.87\,A_{st}\,f_y \cdot jd$$ (7.25)

For over-reinforced sections, the moment capacity is restricted to that given in Eq. (7.20).

7.5 Design of Doubly Reinforced RCC Rectangular Sections

If the size of the beam size is restricted, some compression reinforcement must be provided in addition to the tensile reinforcement. The design strength of the compressive reinforcement is give by

$$C_{ds} = A_{sc}\,f_{sc}$$ (7.32)

in which $\quad A_{sc}=$ area of the compressive reinforcement

$f_{sc}=$ stress in the reinforcement compatible with the compressive strain in concrete, and it can be expressed as

$$f_{sc} = 0.0035 \left(1 - \dfrac{d_c}{x_u}\right) E_s$$ (7.33)

in which $\quad d_c=$ depth of the compression steel

f_{sc} is subjected to a maximum value of $f_y/1.15$

The value of d_c is usually in the range of $0.1d$ and that of x_u is about $0.45d$ to $0.53d$. Therefore, the compressive stress in the compression reinforcement is

$$f_{sc} = 0.0035(0.8)(200000) = 560\ \text{MPa}$$

It can be seen that the compatible stress in the compression reinforcement works out close to 560 MPa that is based on linear stress-strain relation upto a strain of about 0.003. It indicates that the design stress in the compression steel can be taken close to the design stress in tension for steels having proof stress less than 560 MPa. Therefore,

$$f_{sc} = f_y/1.15$$ (7.34)

will govern over that given in Eq. (7.33).

The area of the reinforcement in tension to balance the force in the compression reinforcement is given by

$$A_{st2} = 1.15\,A_{sc}\,f_{sc}/f_y$$ (7.35)

The total bending resistance of the section is

$$M_r = M_{rb} + A_{sc}\,f_{sc}\,(d - d_c)$$ (7.36)

The total area of the tensile reinforcement is

$$A_{st} = A_{stb} + A_{st2}$$ (7.37)

The design of the beams based on the collapse limit state will be illustrated later.

7.6 Design for Transverse Shear in RCC Beams*

The beams are normally designed for flexure and then tested for shear capacity, depending on which the shear reinforcement is provided. If the nominal shear stress on a section does not exceed the shear strength of the concrete, only a nominal shear reinforcement need to be provided. In case the shear stress in the section exceeds the shear strength of the concrete, shear reinforcement has to be provided subject to the condition that the shear stress is within the maximum allowable shear stress of the concrete with the shear reinforcement. When the actual shear stress on the section is more than the maximum allowable in the concrete along with the shear reinforcement, the section must be increased or redesigned such that the shear stress falls within the maximum limit. The design criterion for shear can, therefore, be stated in three cases:

Case 1: Provide only nominal reinforcement. If

$$\tau_v < k_s \tau_c \tag{7.38}$$

then provide nominal shear reinforcement, in which

τ_v = nominal shear stress on the section
τ_c = shear strength of plain concrete
k_s = coefficient which depends on the shape of the section
The minimum shear reinforcement is governed by

$$s_v = \frac{A_{sv} f_{yv}}{0.4 b_w} \tag{7.39}$$

in which

s_v = maximum spacing of the vertical stirrups
A_{sv} = total cross-section of the stirrup legs effective in shear
b_w = width of the web or the section
f_{yv} = characteristic strength of stirrup steel in N/mm^2

s_v should also be less than 0.75d or 450 mm.

Case 2: Provide shear reinforcement. In case, the nominal shear stress is greater than the shear capacity of the concrete, provided it is less than the maximum allowable, shear reinforcement must be provided to resist the shear force which is in excess of the shear capacity of the section.

The shear capacity of the concrete section is

$$V_c = b d k_s \tau_c \tag{7.40}$$

The net shear force for which the shear reinforcement must be designed is

$$V_s = V - V_c \tag{7.41}$$

After selecting a suitable diameter to the stirrup legs, the spacing of the stirrups can be calculated. The stirrups can be vertical or inclined. Even bent up main bars can provide shear resistance. The shear capacity of bent up bars is

$$V_{cs} = 0.87 \, A_{sv} f_y \sin \phi \tag{7.42}$$

*Chapter 2 discusses some of the derivations on shear stress.

258

where

A_{sv} = area of the cross-section of the bent up bar

ϕ = inclination of the bent up bar with the axis of the member

The shear reinforcement can be designed for the shear force over and above the shear capacity of the concrete section and the inclined bars. The spacing of the vertical stirrups is

$$s_v = \frac{0.87\, A_{sv}\, f_{yv}\, d}{V_s} = \frac{0.87\, A_{sv}\, f_{yv}}{(\tau - \tau_c)\, d} \tag{7.43}$$

The spacing of the inclined stirrups along the axis of the member is

$$s_v = \frac{0.87\, A_{sv}\, f_{yv}\, d}{V_s}\,(\sin \phi + \cos \phi) \tag{7.44}$$

where

s_v = spacing of the stirrups along the length of the member

A_{sv} = total cross-sectional area of the stirrup legs

ϕ = angle between the inclined stirrup with the axis of the member

The maximum allowable shear stress is given in Table 7.8.

Table 7.6 Shear Strength of Concrete (τ_c in MPa)

$\dfrac{100 p_s}{bd}$	Concrete grade					
	M15	M20	M25	M30	M35	M40
0.25	0.35	0.36	0.36	0.37	0.37	0.37
0.50	0.46	0.48	0.49	0.50	0.50	0.51
0.75	0.54	0.56	0.58	0.59	0.60	0.60
1.00	0.60	0.62	0.64	0.66	0.67	0.68
1.25	0.64	0.67	0.70	0.71	0.73	0.74
1.50	0.68	0.72	0.74	0.76	0.78	0.79
2.00	0.71	0.78	0.82	0.84	0.86	0.88
3.00	0.71	0.82	0.92	0.96	0.99	1.01

Case 3: Revision of the section. If the actual shear stress is more than the maximum allowable ($\tau_{c\,max}$), the section be redesigned

The nominal shear stress on a RCC slender section is given by

$$\tau_v = \frac{V}{bd} \tag{7.45}$$

For beams with varying depth, it is

$$\tau_v = \frac{V_s \pm \dfrac{M}{d}\tan \theta}{bd} \tag{7.46}$$

Table 7.7 Shear Strength Coefficient (k_s) (except in flat slabs)

	for solid slabs of overall thickness					
t (mm)upto	150	175	200	250	275	300
k_s	1.3	1.25	1.20	1.10	1.05	1.0

Table 7.8 Maximum Allowable Shear Stress (τ_{max}) with Reinforcement

Concrete grade	M15	M20	M25	M30	M35	M40
τ_{cmax} MPa	2.5	2.8	3.1	3.5	3.7	4.0

where M = bending moment

θ = angle between the top and bottom edges of the beam,

For members subjected to axial force,

$$k_s = 1 + \frac{3P}{A_g f_{ck}} \qquad (7.47)$$

where P = axial force and A_g = gross area of the section

The factor applicable to slabs is discussed in chapter on slabs.

7.7 Design Examples of Rectangular Sections Based on Collapse Limit State

EXAMPLE 7.1 *Design of balanced section.* Design of a simply supported beam of effective span 6 m subjected to the following loads and specifications:

Superimposed load	$w_s = 9.0$ kN/m	
Live load	$w_l = 15.0$ kN/m	
Grade of the concrete	$f_{ck} = 20$ MPa $= 20(10^6)$N/m^2	
Grade of the·main reinforcement	$f_y = 415$ MPa	
Grade of the stirrup steel	$f_{yv} = 250$ MPa	
Span	$L = 6$m	

Let the depth of the beam be assumed in the range of $L/15$ for the purpose of calculation of self-weight. Let $h = 0.55$ m and $b = 0.25$ m

Self-weight $w_g = (0.55(0.25)(25)) \simeq 3.4$ kN/m

$w_d = w_g + w_s = 12.4$ kN/m

The partial safety factors are taken as

$$\gamma_d = \gamma_L = 1.5$$

The maximum bending moment which will force the simply supported beam to collapse will occur at the mid-span and it is equal to

$$M_c = \gamma_d \frac{w_d L^2}{8} + \gamma_L \frac{w_l L^2}{8}$$

$$= 1.5(12.4 + 15)\frac{36}{8} = 184.95 \text{ kNm}$$

Equating the balanced moment capacity to the collapse moment, i.e. $M_{rb} = M_c$, which leads to

$$Kbd^2 f_{ck} = 184.95(1000) \text{ Nm}$$

From Table 7.5 $K = 0.138$, $p_0 = 0.198$.

Let $b = 0.25$ m, then

$$d = \sqrt{\frac{184\,990}{Kbf_{ck}}} = \sqrt{\frac{184\,950}{0.138(0.25)(20)(10^6)}} = 0.52 \text{ m}$$

$$h = 0.52 + 0.04 = 0.56 \text{ m}$$

$$A_{stb} = p_0bd\frac{f_{ck}}{f_y} = 0.198(0.25)(0.52)\left(\frac{20}{415}\right) = 0.001240 \text{ m}^2 = 1240 \text{ mm}^2$$

Provide: $b = 250$ mm, $h = 560$ mm, $d = 520$ mm and $A_c = bh = 0.14$ m^2

The self-weight assumed agrees closely with the actual value.

Provide 4 numbers of ϕ 20 of which two bars can be cranked. Then

$$A_{st} = 1257 \text{ mm}^2$$

Provide 2 numbers of 10 mm bars at top as hanger bars.

Total area of steel $= A_{so} = A_{st} +$ area of hanger bars $= 1257 + 157 = 1414$ mm^2

$$\frac{100\,A_{st}}{bd} = \frac{125\,700}{250(520)} = 0.97\%$$

Design for shear: Maximum design shear force at distance d from the face of support is

$$V = 1.5(12.4 + 15)(3 - 0.52) = 102 \text{ kN}$$

Nominal shear stress is

$$\tau_v = \frac{V}{bd} = \frac{102\,000}{250(520)} = 0.78 \text{ N/mm}^2$$

The shear capacity of M20 concrete for 0.5% reinforcement (from Table 7.6) is $\tau_c = 0.48$ N/mm^2.

Select two-legged 8 mm bars as stirrups, then

The area of stirrup $= 2(50) = 100$ mm$^2 = A_{sv}$

Net design shear force $V_s = V - \tau_c bd = 102\,000 - 62400 = 39600$ N

The spacing of the stirrups

$$s_v = \frac{0.87\,A_{sv}f_{yv}d}{V_s} = \frac{0.87(100)(250)(520)}{39600} = 285 \text{ mm}$$

Spacing based on the minimum shear reinforcement is

$$s_v = \frac{A_{sv}f_{yv}}{0.4b} = \frac{100(250)}{(0.4)(250)} = 250 \text{ mm}$$

or the maximum spacing $= 0.75d$ or 450 mm

Provide 8 mm two-legged stirrups at 250 mm c/c.

The same problem was solved by WSD and the corresponding design details are:

$b = 250$ mm, $h = 800$ mm, $A_{st} = 804$ mm^2, 8ϕ two-legged stirrups at 250 mm spacing.

EXAMPLE 7.2 *Design of under-reinforced section.* Design of a simply supported beam of span 6 m subjected to the following specifications:

$$w_s = 9.84 \text{ kN/m}, \quad w_1 = 15 \text{ kN/m}$$

Size of the beam $b = 0.20$ m, $h = 0.80$ m, $f_{ck} = 20$ MPa, $f_y = 415$ MPa

$$\text{Span} = L = 6 \text{ m}$$

$$\text{Let } d = h - 0.05 = 0.75 \text{ m}$$

$$w_g = (0.2)(0.8)(25) = 4.0 \text{ kNm}$$

$$w_d = w_g + w_s = 13.84 \text{ kN/m}$$

The design load for collapse is

$$w_c = 1.5(w_d + w_1) = 43.26 \text{ kN/m}$$

$$M_c = \frac{w_c L^2}{8} = \frac{43.26(36)}{8} = 194.67 \text{ kNm}$$

Balanced moment capacity of the section is

$$M_{rb} = Kbd^2 f_{ck}$$

$$= 0.138(0.2)(0.75)^2 \frac{(20)(10^6)}{1000} = 310.5 \text{ kNm}$$

The balanced moment capacity of the section is far higher than the collapse design moment; therefore, the section is under-reinforced. The approximate distance of the compression force from the outer fibre of the section is

$$x_c = 0.42 x_u = \frac{0.42 \, A_{st} f_y}{0.36 \, f_{ck} \, b(1.15)} = \frac{A_{st} f_y}{b f_{ck}}$$

Moment capacity of the under-reinforced section is

$$M_r = 0.87 \, f_y A_{st} (d - x_c)$$

$$= 0.87(415)(10^6) \, A_{st} \left(0.75 - \frac{A_{st} f_y}{b f_{ck}} \right)$$

$$= 361.05(10^6) \, A_{st} (0.75 - 106 \, A_{st}) \text{ Nm}$$

$$= 361050 \, A_{st} (0.75 - 106 \, A_{st}) \text{ KNm}$$

Equating the resisting capacity to the collapse moment, we have

$$361050 \, A_{st}(0.75 - 106 \, A_{st}) = 194.67 \text{ kNm}$$

The rearrangement of the above leads to

$$106 \, A_{st}^2 - 0.75 \, A_{st} + 0.539(10^{-3}) = 0$$

or

$$A_{st} = 0.75 - \sqrt{\frac{0.75^2 - 4(106)(0\,000539)}{2(106)}}$$

$$= 0.000810 \text{ m}^2 = 810 \text{ mm}^2$$

Provide 3 numbers of 20 mm dia bars at the bottom straight through and 2 numbers of 8 mm dia bars at the top as hanger bars.

$$\frac{100 \, A_{st}}{bd} = \frac{94200}{200(750)} = 0.628\%$$

Check for shear: Maximum shear force at a distance d from the centre of the support is

$$V = w_c (3 - 0.75) = 97335 \text{ N}$$

Nominal shear stress $= \dfrac{V}{bd} = \dfrac{97335}{200(750)} = 0.65 \text{ N/mm}^2$

Concrete shear capacity $= 0.48 \text{ N/mm}^2$

The shear stress exceeds the allowable value marginally and the minimum shear reinforcement governs the design.

EXAMPLE 7.3 *Design of portico beams.* A portico of size 6 by 3 m is supported by two beams fixed into the wall. The portico is assumed to be supported by two cantilevered beams spaced at 4 m apart.
Data

$$w_l = 1.5 \text{ kN/m}^2, \ f_{ck} = 20 \text{ MPa}, \ f_y = 250 \text{ MPa}$$

Assume the thickness of the slab $= 4/25 = 0.16$ m

Weight of the slab $= 0.16(25) = 4.0 \text{ kN/m}^2$

The design parameters from Table 7.5 for $f_y = 250$ MPa are:

$$x_u = 0.531d, \ K = 0.149 \text{ and } p_0 = 0.22$$

600 mm half brick (115 mm) wall is provided as a parapet along the periphery of the portico.

Design of the slab. The slab is supported by two beams spaced at 4 m apart. Therefore, it can be designed as a one way slab with overhang of 1 m on either side, which means that there is a cantilever of 1 m for the beam, that is (neglect half width of beam)

$$a = 1 \text{ m and } L = 4 \text{ m}$$

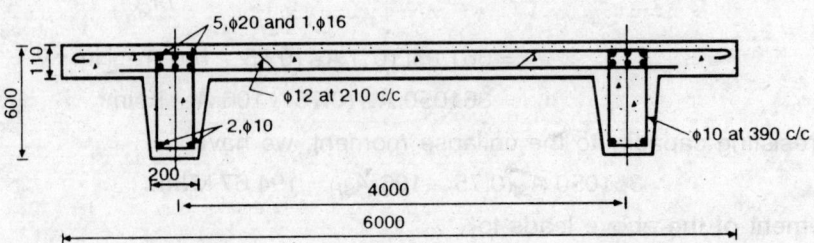

Fig. 7.5 Sectional Details of Portico Slab of Example 7.3 (dimensions in mm).

Figure 7.5 illustrates the cross-section of the portico. The slab is analysed and designed for a width of 1 m.

Self-weight of slab $w_g = 0.16(25) = 4.0 \text{ kN/m}^2$

Let the finished load $= w_s = 1.0 \text{ kN/m}^2$

$$w_d = w_g + w_s = 5 \text{ kN/m}^2$$
$$w_c = 1.5(w_d + w_l) = 1.5(6.5) = 9.75 \text{ kN/m}^2$$

Weight of the parapet $= W = (0.115)(0.6)(20) = 1.38 \text{ kN/m}$

$$W_c = 1.5(1.38) = 2.07 \text{ kN/m}$$

The collapse negative BM on the slab due to self-weight and the concentrated load of the parapet is

$$M_{c1} = \frac{w_c a^2}{2} + W_c a = \frac{9.75}{2} + 2.07 = 6.945 \text{ kNm}$$

The collapse positive BM which occurs at the mid-span is

$$M_{c2} = \frac{w_c L^2}{8} - M_{c1}$$

$$= \frac{9.75(16)}{8} - 6.945 = 12.555 \text{ kNm/m} = 12555 \text{ Nm/m}$$

The section should be designed for the maximum of the above two. Equating the collapse moment to the balanced resisting moment capacity, we have

$$Kbd^2 f_{ck} = M_{c2}$$

$$0.149(1)d^2(20)(10^6) = 12555 \text{ Nm/m}$$

$$d = \sqrt{\frac{12555}{0.149(20)(10^6)}} = 0.065 \text{ m}$$

Let the overall depth of the slab be $h = 0.11$ m which is less than the assumed for weight calculation. Revise the calculation with actual self-weight.

$$w_g = 0.11(25) = 2.75 \text{ kN/m}^2, \ w_g = 1.0, \ w_l = 1.5$$

$$w_c = 1.5(w_g + w_s + w_l) = 7.95 \text{ kN/m}^2$$

$$M_{c2} = \frac{w_c L^2}{8} - \left(\frac{w_c a^2}{2} + W_c a \right)$$

$$= 7.59(2) - (7.95/2 + 2.07) = 9.855 \text{ kNm/m}$$

$$d^2 = \frac{M_{c2}}{Kbf_{ck}} = \frac{9.855(1000)}{0.149(1)(20)(10^6)}$$

or $d = 0.0067$ m

$$A_{st} = p_0 bd \frac{f_{ck}}{f_y} = 0.22(1000)(67) \left(\frac{20}{250} \right) = 480 \text{ mm}^2/\text{m}$$

Provide 12 mm dia bars at 105 mm c/c at the bottom and crank alternate bars at 1 m from the beam.

Provide 0.2% as distribution reinforcement and it is

$$= \frac{0.2}{100} (1000)(110) = 220 \text{ mm}^2/\text{m}$$

Provide 8 mm bars at 200 mm c/c as distribution reinforcement. The negative bending moment is less than half of the positive BM; therefore, the reinforcement which is cranked from the bottom would provide adequate resistance if extended at the top in the overhang portion. The details of the reinforcement are shown in Fig. 7.5.

Design of the beam. Each beam supports half of the slab; therefore, the load of the slab acts as a superimposed load on the beam. The beam has a cantilever span of 3 m supporting 3 m wide slab. The beam section is assumed to be below the slab, therefore it is designed as a rectangular section. The design of T-beam when it is above the slab will be discussed later. Let the web size of the beam for the purpose of self-weight computations be 250×400 mm. The beam would taper down towards the free end. It is also assumed that the portico supports a facia of 600 mm high half brick parapet wall. The weight of the wall should be accounted for in the bending moment computations.

Self-weight of the beam = $w_g = 0.25(0.4)(25) = 2.5$ kN/m

The load from slab including the finishing coat/3 m width and with the partial safety factor is

$$W_a = 7.95(3)(3) = 71.55 \text{ kN}$$

Weight of parapet at free end with 1.5 as partial safety factor is

$$W_{g1} = 3(1.38)(1.5) = 6.21 \text{ kN}$$

Weight of the parapet at the side edge with 1.5 partial safety factor is

$$W_{p2} = 3(1.38)(1.5) = 6.21 \text{ kN}$$

The design collapse moment at the fixed end of the beam is

$$M_c = W_a \left(\frac{L}{2}\right) + W_{p1}(L) + W_{p2}(L/2) + 1.5w_g\left(\frac{L^2}{2}\right)$$
$$= 71.55(1.5) + 6.21(3 + 1.5) + 1.5(2.5)(4.5) = 151.47 \text{ kNm}$$

Equating the moment capacity of the beam to the collapse moment, we have

$$Kbd^2 f_{ck} = M_c = 151\ 470 \text{ Nm then}$$
$$d^2 = \frac{151\ 470}{0.149(0.25)(15)(10^6)}$$

or $d = 0.5206$ m $= 521$ mm

$$A_{st} = \frac{p_0 bd f_{ck}}{f_y} = 0.22(250)(521)\left(\frac{15}{250}\right) = 1719 \text{ mm}^2$$

Provide 5 numbers of 20 mm and 1 numbers of 16 mm diameter bars in two rows.

$$A_{sc} \text{ (provided)} = 1772 \text{ mm}^2$$
$$\frac{100\ A_{st}}{bd} = \frac{177200}{250(521)} = 1.3615$$

Overall depth at the fixed end

$$h = d + 10 + 23 + 20 + 25 = 590 \text{ mm}$$

Depth at free end   = 250 mm

The average depth   = (250 + 590)/2 = 420 mm

The self-weight is approximately equal to that assumed.

Design for shear. Maximum shear force which occurs at the support is

$$V = W_a + W_g + W_{p1} + W_{p2}$$
$$= 71.55 + 1.5(2.5)(3) + 6.21 = 94.770 \text{ kN}$$

Nominal shear stress is

$$\tau_v = \frac{V}{bd} = \frac{94\ 770}{250(520)} = 0.729 \text{ N/mm}^2$$

The shear strength of the M20 concrete from Table 7.6 (for 100 $A_{st}/bd = 1.25$) is

$$\tau_c = 0.67 \text{ N/mm}^2$$

Stirrups have to be designed
The shear capacity V_c is

$$V_c = \tau_c bd = 0.67(250)(520) = 87100 \text{ N}$$

The effective shear force for which shear reinforcement is to be provided is

$$V_s = V - V_c = 94770 - 87100 = 7670 \text{ N}$$

Provide two-legged 10 mm diameter MS bars. Then the spacing of the stirrups is

$$s_v = \frac{0.87\ A_{sv}\ f_{yv}\ d}{V_s}$$

$$= \frac{0.87(157)(250)(520.0)}{7670} = 2315 \text{ mm}$$

Maximum spacing of the stirrups is

$$s_{max} = \frac{A_{sv} f_{yv}}{0.4b} = \frac{157(250)}{2.4(250)} = 392.5 \text{ mm or}$$

$$= 0.75d = 391 \text{ mm}$$

Reinforcement (MS bars) details
    5 numbers of 20 mm dia end
    1 number of 16 mm dia bars at the top from support and
    2 numbers of 20 mm dia bars curtailed at 1.5 m from support
Two-legged 10 mm dia stirrups at 390 mm c/c, (Number of stirrups = 8)
    2 numbers of 10 mm bars at bottom as hanger bars
    Development length of bars

$$L_d = \frac{\phi \sigma_s}{4 f_{bd}} = \frac{\phi(250)}{1.15(4)(1)} = 54\phi$$

Deduct a length of 16ϕ for the hook; then the length is
$$L_{d0} = (54 - 16)\phi = 38\ \phi$$

EXAMPLE 7.4 *Doubly reinforced section:* Design of a simply supported beam of span 6 m subjected to the following loads and specifications:

$$\text{Superimposed load} = w_s = 9.4 \text{ kN/m}$$

$$\text{Live load} = w_l = 15.0 \text{ kN/m}$$
$$f_{ck} = 20 \text{ MPa}, \quad f_y = 415 \text{ MPa}$$
$$b = 0.20 \text{ m}, \, h = 0.4 \text{ m}$$
$$\text{Self-weight of the beam} = w_g = (0.2)(0.4)(25) = 2.0 \text{ kN/m}$$

Partial load factors for dead and live load = $\gamma_f = 1.5$
The collapse moment which occurs at the mid-span is

$$M_c = \gamma_f (w_g + w_s + w_l) \frac{L^2}{8}$$

$$= 1.5(20 + 9.4 + 15) \frac{36}{8} = 178.200 \text{ kN/m}$$

The overall depth of the section is 400 mm and after allowing for a cover of 25 mm, the effective depth can be taken as

$$d = 400 - 25 - 10 = 365 \text{ mm}$$

The design coefficients for a balanced section taken from Table 7.5 are:

$$x_u = 0.497d, K = 0.138, p_0 = 0.198$$

The balanced moment capacity of the section is

$$M_{rb} = Kbd^2 f_{ck} = 0.138(0.2)(0.365)^3 (20)(10^6) = 73540.2 \text{ Nm}$$

The collapse moment is more than the balanced moment capacity of the section; therefore, the section is to be designed as a doubly-reinforced section.

The moment to be resisted by the compression reinforcement is

$$M = M_c - M_{rb} = 178.2 - 73.54 = 104.66 \text{ kNm}$$

Let the centre of compression steel from the extreme compression fibre be

$$d_c = 35 \text{ mm}$$

The strain at the level of the compression steel is

$$\epsilon_{sc} = 0.0035 \left(1 - \frac{35}{x_u}\right) = 0.0035 \left(1 - \frac{35}{0.479d}\right)$$

$$= 0.0035 \left(1 - \frac{35}{0.479(365)}\right) = 0.0028$$

$$f_{sc} = \epsilon_{sc} E_s = 0.0028(200,00) = 560 \text{ N/mm}^2$$

This value is more than f_y which is not allowed. Therefore use,

$$f_{sc} = f_y = 415 \text{ MPa}$$

$$d - d_c = 365 - 35 = 330 \text{ mm} = 0.33 \text{ m}$$

The moment capacity of the compression reinforcement is

$$M_{rc} = 0.87 A_{sc} f_y (d - d_c)$$

Equating the moment capacity of the compression reinforcement to the unbalanced moment, we have

$$0.87 A_{sc} f_y (d - d_c) = M = 104.66 \text{ kNm}$$

$$\text{or } A_{sc} = \frac{104.660}{0.87(415)(0.33)} = 880 \text{ mm}^2$$

The tensile reinforcement needed is

$$A_{stb} = p_0 bd \frac{f_{ck}}{f_y} = 0.198(200)(365)(20)/415 = 697 \text{ mm}^2$$

$$A_{st2} = A_{sc} \frac{f_{sc}}{f_y} = 880 \text{ mm}^2$$

$$A_{st} = A_{stb} + A_{st2} = 1577 \text{ mm}^2$$

Provide: $b = 220$ mm, $h = 400$ mm, $d = 365$ mm, $d_c = 35$ mm

Three numbers of 20 mm bars at top, four numbers of 20 mm bars and two numbers of 16 mm bars at bottom

$$A_{sc} = 942 \text{ mm}^2, \ A_{st} = 1658 \text{ mm}^2$$

$$A_{so} = A_{st} + A_{sc} = 2555 \text{ mm}^2$$

(Design for shear is omitted in this example.)

EXAMPLE 7.5 *Design of prestressed concrete beam:* A simply supported prestressed concrete beam of span 6 m is to be designed for the following specifications :

$$\text{Superimposed load} = w_s = 9.4 \text{ kN/m}$$
$$\text{Live load} = w_l = 15 \text{ kN/m}$$
$$f_{ck} = 35 \text{ MPa}, \ f_y = 1500 \text{ MPa}$$
$$\text{Let the size of the beam} = 0.2 \text{ by } 0.4 \text{ m}$$
$$\text{Self-weight} = w_g = 0.2(0.4)(25) = 2 \text{ kN/m}$$

The maximum moment occurs at the mid-span and the collapse moment with a partial load factor of 1.5 is

$$M_c = \gamma_f (w_g + w_g + w_l) \frac{L^2}{8}$$

$$= 1.5(2.0 + 9.4 + 15) \frac{36}{8} = 178.200 \text{ kNm}$$

The balanced moment capacity of a rectangular section with the partial safety factor is

$$M_{rb} = Kbd^2 f_{ck}$$

The design coefficients can be obtained from Table 7.5 and are:

$$K = 0.131, \ x_u = 0.44d, \ p_o = 0.186$$

Let $b = 0.20$; then

$$d = \sqrt{\frac{M_c}{Kbf_{ck}}} = \sqrt{\frac{0.178\ 200}{0.131(0.2)(35)}} = 0.445 \text{ m}$$

The overall depth $h = 480$ mm
The area of high tensile steel is

$$A_{st} = A_{stb} = p_0 bd \frac{f_{ck}}{f_y} = 0.186(200)(445)35/1500 = 382 \text{ mm}^2$$

Provide $b = 200$ mm, $h = 480$ mm, then
Provide 20 numbers of 5 mm wires, the area of which is

$$A_{st} = 392.7 \text{ mm}^2$$

Provide nominal top bars, 2 numbers of 10 mm as hanger bars

$$A_{sc} = 157 \text{ mm}^2$$

EXAMPLE 7.6 *Comparison of cost calculations:* A simply supported beam of span 6 m is subjected to a superimposed load of 9.4 kN/m and live load of 15 kN/m. Design the following four possible sections and compare the relative costs of the different options. (All are rectangular sections.)
1. Balanced RCC section.
2. Under-reinforced RCC section.
3. Doubly-reinforced RCC section.
4. Prestressed concrete section.

Unit costs of the materials including fixing, or placing or finishing, etc. are:

1 m^3 of M 20 concrete	= Rs 2700/-
1 kN of HYSD reinforcement	= Rs. 2000/-
1 m^3 of M 35 concrete	= Rs 3600/-
1 kN of HT steel	= Rs 6000/-
1 m^2 of formwork RCC	= Rs. 90/-
1 m^2 of formwork for PSC	= Rs. 120/-

Refer Examples 7.1, 7.2, 7.4 and 7.5 in which the beams are designed as balanced, under-reinforced, doubly-reinforced and prestressed beams. Table 7.9 lists the designed parameters alongwith the total quantities of the materials actually provided.

Notations:

A_c = area of concrete section

A_{sc} = total area of steel including nominal hangers bars

S = area of shuttering

C_c = cost of concrete

= $A_c L$ (unit cost of concrete)

C_{st} = cost of steel = $A_{so} (L)$ (unit cost of steel)

Table 7.9 illustrates the cost economics of small span beams. The unit cost of the materials is likely to vary from place and time to time. The values selected are only typical values and the general trend of relative costs should be about the same. The cost of the under-reinforced appears to be the highest in the present set because the actual depth of the section is about 43 per cent more than that of the balance. (With the result the cost has gone higher than that of the balanced section.) The prestressed concrete section cost is about 4 per cent less

Table 7.9 Cost and Relative Costs of different Beams

Section	Areas			Cost in Rs.				
	A_c (m^2)	A_{so} (mm^2)	S (m^2)	C_c	C_{st}	C_s	Total	Relative
RCC balanced	0.14	1414	1.37	2267	1334	740	4342	1
RCC under	0.16	1099	1.8	2592	1037	972	4601	1.06
RCC doubly	0.08	2600	1.0	1296	2453	540	4286	0.99
PSC	0.096	157	1.16	2073	148	836	4169	0.96
		392.7[a]			1111[a]			

[a]Area of HTS wires or the cost

than the balanced RCC section. However, the cost calculations do not include the initial investments towards equipment, etc. The total cost also does not include the cost of shear reinforcement.

7.8 Design of Flanged Sections Based on the Limit State of Strength

Flanged sections such as T-beams are very commonly used in building and bridge construction without any special effort. The beam slab construction in RCC or prestressed concrete often results into a T-beam section. I-beams are not common in RCC construction as only one of the flanges is subjected to compressive force and the other to the tensile force. The flange under tension does not serve any structural purpose except housing the reinforcement. In such a case the flange under tension is more of a burden on the beam rather than of any help. On the other hand, the T-beam sections provide a large concrete cross-sectional area of the flange to resist the compressive force. T-beams are very advantageous in simply supported spans and the inverted T-beam sections have the added advantage in cantilever beams. Whether someone likes it or not, T-section is usually built into the simply supported beam-slab construction.

In case of prestressed concrete beams, the cable force introduces a high compression in the zone where tension is expected in working loads. Therefore, the prestressed concrete beam sections are subjected to reversible type of stresses and the top or the bottom fibre are under compression some time or the other. In such a case, I-sections are better suited. Even in prestressed concrete beams, the magnitude of the compression caused by the working loads is higher than that caused by the prestressing alone. Hence, unequal flange sections are often used in prestressed concrete beams. Bottom flange of simply supported beams is usually smaller than the top flange and the reverse is true in case of cantilever beams.

Figure 7.6a illustrates typical I-sections. It can also be considered as a T-section as the bottom flange does not come into action at the limit state of strength of the beam. Figure 7.6b illustrates a strain variation in the limit state of strength in which the strain at the top fibre is considered as the limiting compressive strain in the concrete. The limiting compressive strain in the concrete is taken as

$$\epsilon_{cu} = 0.0035$$

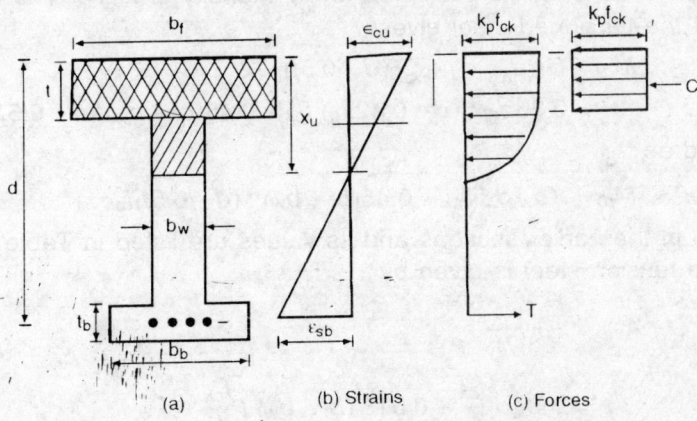

Fig. 7.6 Strain-stress Distribution on Flanged Section.

The neutral axis distance based on the compatibility of strains is

$$\frac{x_u}{d - x_u} = \frac{\epsilon_{cu}}{\epsilon_{sb}} \tag{7.48}$$

in which x_u = neutral axis distance from the top compression fibre

d = effective depth of tensioned steel from the compression fibre

ϵ_{sb}= strain in tensioned steel compatible with compressive strain in concrete

$\quad = \epsilon_s - (\epsilon_{sc} + \epsilon_{ce})$ in prestressed concrete

ϵ_{sc} = effective prestrain in the tensioned steel

ϵ_{ce}= effective precompressive strain in the concrete at the level of steel

Usually the total of the prestrain in steel is taken as :

$$\epsilon_{sc} + \epsilon_{ce} = 0.004 \tag{7.49}$$

Equation (7.48) can be rearranged as

$$x_u = \frac{\epsilon_{cu} d}{\epsilon_{cu} + \epsilon_{sb}} \tag{7.50}$$

It is seen from Table 7.5 that the value of x_u/d is in the range of 0.4 to 0.53. This could mean that the neutral axis is invariably in the web rather than in the compression flange. In certain cases of under-reinforced sections, the neutral axis is likely to fall in the top flange.

The limiting moment capacity of the flanged section can be obtained by taking moment of the compression forces about the centre of the steel. The total compressive foce can be divided into two parts: one coming from the web, and the other coming from the overhang flange. These two design capacities can be expressed as:

$$C_1 = 0.36 \, b_w \, x_u \, f_{ck} \tag{7.51}$$

$$C_2 = \frac{(b_f - b_w) t k_p f_{ck}}{\gamma_m} = 0.45(b_f - b_w) \, t f_{ck} \tag{7.52}$$

in which b_f = width of the flange

b_w = width of the web

t = thickness of the flange and $\gamma_m = 1.5$

It is also assumed that the flange is subjected to an intensity of $k_p f_{ck}$. The bending moment about the centre of the tensioned steel gives

$$M_{rb} = C_1(d - x_c) + C_2(d - 0.5t)$$

$$= 0.36 b_w x_u(d - 0.42 x_u) \, f_{ck} + 0.45(b_f - b_w) t(d - 0.5t) f_{ck} \tag{7.53}$$

It can be expressed as

$$M_{rb} = K b_w \, d^2 \, f_{ck} + 0.45(b_f - b_w) \, t \, (d - 0.5t) f_{ck} \tag{7.54}$$

in which K is given in the earlier sections and its values are listed in Table 7.5.

The area of the tension steel is given by

$$A_{stb} = \frac{C_1 + C_2}{0.87 f_y}$$

$$= p_0 b_w d \frac{f_{ck}}{f_y} + 0.518 \, (b_f - b_w) \, t \frac{f_{ck}}{f_y} \tag{7.55}$$

Very often one is interested in design parameters such as d, A_{st}, etc. rather than the analysis of the section and Eq. (7.54) is not well suited to the design. Therefore, certain approximations are made in the resisting moment capacity of the flanged section. The compressive force contributed by the parabolic curve on the web is small when compared with that coming from the total flange area. Further, its contribution to the moment resistance is very small as the lever arm is small. Hence, the compressive force can be idealized as the force acting on the flange alone and treated as uniformly distributed. The total compresive force is

$$C_d = \frac{b_f t \, k_p f_{ck}}{\gamma_m} = 0.45 b_f t \, f_{ck} \tag{7.56}$$

The moment capacity is

$$M_r = C_d (d - 0.5t) = 0.45 b_f t (d - 0.5t) f_{ck} \tag{7.57}$$

Assuming that the values of b_f and t are known from the slab, the value of the effective depth can be obtained as (equating the moment capacity to the collapse moment, i.e. $M_c = M_r$)

$$d = 0.5t + \frac{M_c}{0.45 b_f t \, f_{ck}} \tag{7.58}$$

The corresponding area of the tension reinforcement is

$$\frac{A_{st} f_y}{1.15} = C_d = 0.45 \, b_f t \, f_{ck}$$

$$\text{or } A_{st} = 0.518 \, b_f \, t \, \frac{f_{ck}}{f_y} \tag{7.59}$$

$$\frac{A_{st}}{b_f t} = 0.518 \frac{f_{ck}}{f_y} \tag{7.60}$$

In case of under-reinforced sections, the area of the steel is given by

$$A_{st} = \frac{1.15 \, M_c}{(d - 0.5t) f_y} \tag{7.61}$$

These expressions will hold good for reinforced or prestressed concrete sections subject to the condition that the neutral axis is in the web, that is

$$x_u = \frac{\epsilon_{cu} d}{\epsilon_{cu} + \epsilon_{sb}} \geq 1.15t \tag{7.62}$$

In case the neutral axis depth is less than the thickness of the flange, the section can be designed as a rectangular section with width equal to b_f.

The depth of the neutral axis in prestressed concrete balanced sections varies from 0.38 to 0.46 times the depth of the steel. Therefore, one can determine whether the section is balanced or not by checking the neutral axis depth. From Eq. (7.62) we can use the design formulae 7.58 and 7.61 provided

$$x_u = 0.42d \geq 1.15t$$

$$\text{or } d > \frac{1.15t}{0.42} = 2.75t \tag{7.63}$$

In case the depth of the beam comes out to be small, which is true in most cases, then it can be increased to about $2.5t$. The area of the tension steel is then obtained from Eq. (7.61). In case of cantilever beams, the thickness of the bottom flange is usually not specified, whereas the top flange is given by other considerations. In such case the thickness of the bottom slab can be assumed as

$$t = 0.4d \tag{7.64}$$

Then Eq. (7.57) reduces to

$$M_r = 0.36\ b_f dt f_{ck} \tag{7.65}$$

$$= 0.144 b_f d^2 f_{ck} \tag{7.66}$$

The depth of the steel or the thickness of the flange can be obtained from Eq. 7.65. The area of steel is given by Eq. (7.61).

7.9 Design of Prestressed Concrete Beam at Transfer Load Condition

At transfer of prestress, the prestressing force acts as a load and causes high compression in the concrete in the zone of the cables. In such cases, the area of the concrete in the zone be designed to resist the compressive force caused by the prestress. The self-weight or the dead load present at that time will relieve some of the compression by introducing tension. The load factor for prestress is usually taken from 1.0 to 1.2 and that for dead load as 0.9. The ultimate prestress is

$$T_{ut} = \gamma_t A_{st} f_y \tag{7.67}$$

where γ_t = load factor (about 1 to 1.2)

The relieving tension caused by the dead load moment is

$$T_{dt} = \frac{0.9 M_d}{d - 0.5t} \tag{7.68}$$

The net compressive force on the concrete is

$$C_{dt} = T_{ut} - T_{dt} \tag{7.69}$$

The area of the concrete needed in the cable zone (bottom flange in case of simply supported beams) is given by

$$C_{dt} = A_{ct}\,(0.45 f_{ck}) \tag{7.70}$$

$$\text{or } A_{ct} = \frac{T_{ut} - T_{dt}}{0.45\ f_{ck}} \tag{7.71}$$

where A_{ct} = area of the concrete to resist the transfer prestress.

f_{ck} = strength of concrete at transfer stage.

7.10 Design Example of Flanged Sections

EXAMPLE 7.7 *Design of singly reinforced concrete simply supported beam.* A floor slab 120 mm thick is supported by simply supported beams spaced 4 m apart. The clear span of the beam is 8 m. The beam is designed for the following:

$$\text{Live load} = w_l = 4 \text{ kN/m}^2$$

$$\text{Concrete grade} = f_{ck} = 20 \text{ MPa}$$

$$\text{Steel grade} = f_y = 415 \text{ MPa}$$

$$t = 120 \text{ mm}$$

Let the **web thickness** = thickness of one-brick wall = b_w = 0.23 m

Let the **effective span** = L = clear span + 5% of the clear span, i.e.

$$L = 8 + 8/20 = 8.4 \text{ m}$$

The flange width is controlled by

centre to centre distance of beams = 4 m

$$= \frac{L}{6} + b_w + 6t = \frac{8.4}{6} + 0.23 + 6(0.12) = 2.316 \text{ m}$$

$$= b_w + 12t = 0.23 + 12(0.12) = 1.67 \text{ m}$$

Select the least of the above, then

$$b_f = 1.67 \text{ m}$$

Let the overall depth of the beam be taken as

$$h = \frac{L}{20} = \frac{8.4}{20} \, 0.42 \text{ m}$$

Web self-weight $= w_g = 0.23(0.42)(25000) = 2420$ N/m
Weight of slab $\ \ = w_s = 0.12(4)(25000) \ \ \ \ = 12000$ N/m
Live load $\quad\quad\quad = w_l = 4000(4) \quad\quad\quad\quad = \underline{16000 \text{ N/m}}$
$\quad\quad\quad\quad\quad\quad\quad\quad\quad\quad$ Total load $= \underline{\underline{30420 \text{ N/m}}}$

Let the total load be $w = 30000$ N/m
The limit load with 1.5 as partial safety factor is

$$w_c = 1.5 \, (30000) = 45000 \text{ N/m}$$

The critical collapse moment on the beam which occurs at the mid-span is

$$M_c = \frac{w_c L^2}{8} = \frac{45000(8.4)^2}{8} = 396\,900 \text{ Nm}$$

The effective depth of the reinforcement can be obtained from Eq. (7.58) as

$$d = 0.5t + \frac{M_c}{0.4 b_f t f_{ck}}$$

$$= 0.06 + \frac{396\,900}{0.45(1.67)(0.12)(20)(10^6)} = 0.06 + 0.32 = 0.28 \text{ m}$$

From Table 7.5, we have

$$x_u = 0.49d = 0.137 \text{ m}$$

The neutral axis falls in the web as x_u is greater than t.
The overall depth $h = d + 0.06 = 0.34$ m

The area of the tensile reinforcement is taken from Eq. (7.59) and it is

$$A_{st} = 0.518 b_f t \frac{f_{ck}}{f_y} = 0.518(1670)(120)(20)/415 = 5000 \text{ mm}^2$$

$$\frac{100 A_{st}}{b_w t} = \frac{5000(100)}{230(280)} = 7.7\%$$

Design for shear. The maximum shear force occurs at a distance d from the support, and it is

$$V = w_c\left(\frac{L}{2} - d\right) = 45000(4.52 - 0.28) = 176\ 400 \ N$$

$$\tau_V = \frac{V}{b_w d} = \frac{0.176\ 400}{(0.23)\ (0.28)} = 2.739 \text{ N/mm}^2$$

The maximum shear strength of M20 concrete from Table 7.7 is

$$\tau_{cmax} = 2.8 \text{ N/mm}^2$$

and the shear capacity $\left(\text{for } \dfrac{100\ A_{st}}{b_w d} = 2.5\right)$ from Table 7.6 is

$$\tau_c = 0.82 \text{ N/mm}^2$$

Shear reinforcement must be provided. The shear capacity of the concrete section is

$$V_c = b_w d \tau_c = 230(280)(0.82) = 52808 \text{ N}$$

The shear force for which the shear reinforcement must be designed is

$$V_s = V - V_c = 176\ 400 - 52808 = 123592 \text{ N}$$

Provide 12 mm MS two-legged stirrups, then the area of the stirrup steel is

$$A_{sv} = 2(113) = 226.0 \text{ mm}^2$$

$$f_{yv} = 250 \text{MPa (for MS bars)}$$

The spacing of the stirrups at the maximum shear force is

$$s_v = \frac{0.87 A_{sv} f_{yv} d}{V_s} = \frac{0.87(226)(250)(280)}{123\ 592} = 111 \text{ mm}$$

Nominal shear reinforcement spacing based on the minimum steel is

$$s_{vmax} = \frac{A_{sv} f_{yv}}{0.4 b_w} = \frac{226(250)}{0.4(230)} = 614 \text{ mm}$$

or $s_{vmax} = 0.75 d = 210 \text{ mm}$

The shear force gradient = 45 000 N/m.
The distance from mid-span at which the shear force is equal to the shear capacity is

$$x_1 = \frac{V_c}{\text{shear force gradient}} = \frac{52\ 808}{45\ 000} = 1.7 \text{ m}$$

Provide the stirrups at different spacing in the different segments as 110 mm c/c in the first one metre from the support and gradually increasing the spacing to 210 mm at 3.1 m from the support.

Provide the two-legged 12 mm bars in following scheme

Spacing	in the segment of		Nos.
	from support		
110 mm	300 mm to 1100 mm		14
140 mm	1100 mm to 2080 mm		7
170 mm	2080 mm to 3100 mm		6
210 mm	3100 mm to 4200 mm		5

Selection of the main reinforcement. Area of steel needed = A_{st} = 5000 mm^2
Provide eight numbers of 25 mm bars and 4 nos. of 20 mm bars
   Area of tension steel provided is

$$A_{st} = 8(490) + 4(314) = 5176 \text{ mm}^2$$

Provide four bars in a row; therefore, the overall depth required is
   h = effective depth + (1/2)(20 mm bar diameter)
      + clear spacing between 20 and 25 bar diameter)
      + 25 mm bar diameter + 30 mm cover
      = 280 + 10 + 30 + 25 + 30 = 375 mm
The depth of the web below the flange = 375 − 120 = 255 mm
   Design details are:
      Web below the flange   = 230 × 255 mm

Four Nos. of 25 mm bars at bottom row are continued straight through, Four Nos. of 20 mm bars in the middle row at (333 − 30 − 10) = 293 mm from top are curtailed at $L/5$ from each support and two out of 4 Nos. of 25 mm bars placed at (293 − 10 − 30 − 13) = 240 mm from top are cranked at $L/4$ from support whereas remaining two are continued to the support.
   Figure 7.7 illustrates the reinforcement details.

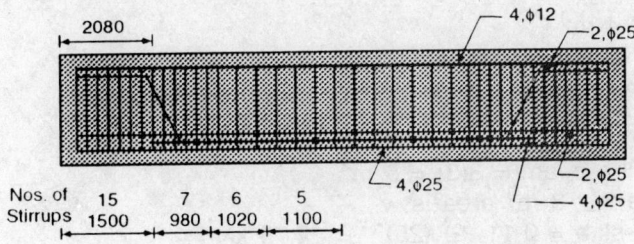

Fig. 7.7 Reinforcement Details of Beam of Example 7.7.

The design of the section is reworked using more accurate Eqs. (7.54) and (7.55). The moment capacity of the section is given by
$$M = (Kb_w d^2 + 0.45(b_f - b_w) t (d - 0.5t))f_{ck}$$
$$= (0.138(0.23)d^2 + 0.45(1.67 - 0.23)(0.12)(d - 0.06)) 20$$
$$= 0.6348d^2 + 1.555(d - 0.06)$$
$$= 0.6348d^2 + 1.555d - 0.093312 \text{ MNm}$$

Equating this value to the collapse moment, we have

$$= 0.6348d^2 + 1.555d - 0.093312 \text{ MNm}$$

or

$$d^2 + 2.45d - 0.77223 = 0$$

$$d = \frac{-2.45 + 3.015}{2} = 0.283 \text{ m}$$

The value of effective depth obtained from the approximate formula was 0.28. The difference between the two results is negligibly small. Such a negligible difference comes when the neutral axis is very close to the bottom of the compression flange. The difference is likely to increase if the neutral axis is far from the bottom of the compression flange, but in any case the difference is of negligible order. The area of the tension reinforcement is obtained by Eq. (7.55), and it is

$$A_{st} = (p_0 b_w d + 0.518 (b_f - b_w) t) \frac{f_{ck}}{f_y}$$

$$= (0.198(230)(283) + 0.518(1670 - 230)(120)) \frac{20}{415}$$

$$= 4935 \text{ mm}^2$$

The value of the area of the reinforcement obtained by the approximate formulae was 5000 mm^2 and that is quite close to the more accurate value of 4935 mm^2.

EXAMPLE 7.8 Design of cantilever beam for portico as given in Example 7.3 (vide statement of the Example 7.3). The portico beam of this example was designed as a rectangular section. However, in this case the web is projecting above the slab surface; therefore, it should be designed as an inverted T-beam.

The slab thickness	= 0.11 m
The slab size	= 6 m × 3 m
Live load	= 1.5 kN/m^2
Roof finish on the slab	= 1.00 kN/m^2

Beams are spaced at 4 m.
Design only the beam as inverted T-section.
Cantilever span of the beam = 3 m

$$f_{ck} = 20 \text{ MPa}, \; f_y = 250 \text{ MPa}$$

Slab area carried by the beam = 3(3) = 9 m^2
Load coming from the 3 × 3 m^3 area is

Weight of the slab = 0.11 (9) (25)		= 24.75 kN
Roof finish	= 1.00 (9)	= 9.00 kN
Live load	= 1.5(9)	= 13.50 kN
	Total load	= 47.25 kN

Let the average size of the beam be 0.25 × 0.4 m^2
Self weight of the web = 0.25(0.4)(25)(3) = 7.5 kN
Weight of the parapet at free end = (0.115)(0.6)(20)(3) = 4.14 kN
UDL dead load = W_d = 24.75 + 9.0 + 7.5 + 4.14 = 45.39 kN
Weight of the parapet at the edge = 0 115(0.6)(20)(3) = 4.14 kN

The design loads for limit state of strength are to be multiplied by the partial safety factor of 1.5 and they are:

Design loads

Load from the slab	$= W_a$	$= 1.5(47.25)$	$= 70.88$ kN
Load from the beam	$= W_g$	$= 1.5(7.5)$	$= 11.25$ kN
Load from end parapet	$= W_{p1}$	$= 1.5(4.14)$	$= 6.21$ kN
Load from edge parapet	$= W_{p2}$	$= 1.5(4.14)$	$= 6.21$ kN

The maximum bending moment corresponding to collapse which occurs at the fixed end is

$$M_c = W_a(L/2) + W_g(L/2) + W_{p1}(L) + W_{p2}(L/2)$$

$$= (L/2)(W_a + W_g + 2W_{p1} + W_{p2})$$

$$= 1.5(70.88 + 11.25 + 3(6.21)) = 151.14 \text{ kN}$$

Assume $b_w = 250$ mm

The effective flange width is

$$b_f = \text{centre to centre distance of the beams} = 3 \text{ m}$$

$$= L/3 + b_w + 6t = 1 + 0.25 + 0.66 = 1.91 \text{ m}$$

$$= b_w + 12t = 0.25 + 1.32 = 1.57 \text{ m}$$

Use $\qquad b_f = 1.57$ m

The design depth of the inverted T-beam is (from Eq. 7.57)

$$d = 0.5t + \frac{M_c}{0.45t \, b_f \, f_{ck}}$$

$$= 0.055 + \frac{157\,140}{0.45(0.11)(1.57)(20)(10^6)} = 0.19 \text{ m}$$

The effective depth when designed as a rectangular section in Example 7.3 was 0.522 m which is far higher than that of the T-beam.

The effective depth obtained is less than $L/15$ and it will cause problem in the shear and deflections; therefore, provide the effective depth as $d = 0.25$ m.

Area of the tension steel is

$$A_{st} = \frac{0.87 \, M_c}{(d - 0.5t) \, f_y} = \frac{0.87(157140)}{(0.25 - 0.055)\,(250)} = 2870 \text{ mm}^2$$

Provide 6 *numbers* of 25 mm dia bars in two rows

$$A_{st} \text{ (provided)} = 490(6) = 2940 \text{ mm}^2$$

Overall depth $\qquad h = d + 12 + 30 + 25 + 30 = 250 + 97 = 347$ or 350 mm

Also provide 2 numbers of 10 mm dia bottom bars.

Total area of steel provided $= A_{st} + A_{sc} = 2940 + 157 = 3097 \text{ mm}^2$

$$\frac{100 \, A_{st}}{b_w d} = \frac{100(2940)}{250(250)} = 4.70\%$$

Design for shear. The total maximum design shear force which occurs at the support is:

$$V = W_a + W_g + W_{p1} + W_{p2}$$

$$= 70.88 + 11.25 + 6.21 + 6.21 = 94.55 \text{ kN}$$

Nominal shear stress in the section is

$$\tau_v = V = \frac{94550}{250\,(250)} = 1.52 \text{ N/mm}^2$$

Shear strength of the M20 concrete with 3% reinforcement taken from Table 7.6 is

$$\tau_c = 0.82 \text{ mm}^2$$

Provide two-legged 10 mm diameter mild steel bars as stirrups; The area of stirrup steel is

$$A_{sv} = 2(78.5) = 157 \text{ mm}^2$$

The spacing of the stirrups is

$$s_v = \frac{0.87\,A_{sv}\,f_{yv}}{b(\tau_v - \tau_c)} = \frac{0.87(157)(250)}{250(0.81)} = 169 \text{ mm}$$

Maximum spacing of the stirrups is

$$s_{max} = \frac{A_{sv}f_{yv}}{0.4b_w} = \frac{157(250)}{0.4(250)} = 392 \text{ mm}$$

$$\text{or } s_{max} = 0.75d = 187 \text{ mm}$$

Provide two-legged 10 mm bars at 165 mm spacing near the support.
The final design parameters are:

Web section at the free end

$$b_w = 250 \text{ mm}$$
$$d_w = 250 - 110 = 140 \text{ mm}$$

Web section at the support end

$$b_w = 250 \text{ mm}$$
$$d_w = 350 - 110 = 240 \text{ mm}$$
$$h = 350 \text{ mm}$$

Provide: Six number of 25 mm bars at the top of which three bars are curtailed at 1500 mm from support.
Two numbers of 10 mm bars at bottom.
Two-legged 10 mm stirrups at 165 mm spacing. (The total of stirrups = 18.)
A_{st} (provided) = 2945 mm^2, A_{sc} = 157 mm^2
Anchor length with hook = 38ϕ

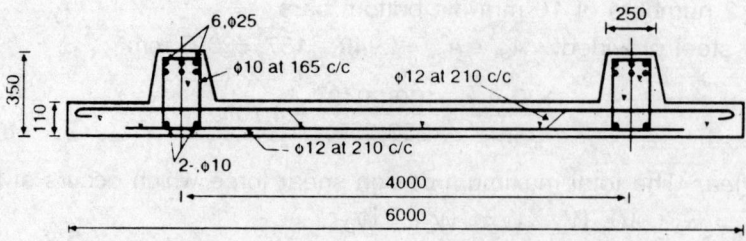

Fig. 7.8 Sectional Details of Portico Slab of Example 7.8 (dimensions in mm).

EXAMPLE 7.9 *Design of T-beam for 15 m span:* The design data of the problem is

$$\text{Clear span} = L_0 = 15 \text{ m}$$

Superimposed load including weight of the slab = 22 kN/m
Live load consists of two concentrated loads of 150 kN each spaced 2 m apart:

$$f_{ck} = 25 \text{ MPa}, \ f_y = 415 \text{ MPa}$$

Width of top flange = b_f = 2m
Thickness of the flange = t = 150 mm

$$\text{Let effective span} = L_0 + \frac{L_0}{15} = L = 16 \text{ m}$$

The maximum bending moment due to the moving loads alone will occur under one of the loads and when the mid-span is at equidistant from both the centroid of the loads and one of the loads. This moment then is

$$M_{l1} = \frac{2W_1}{L}\left(\frac{L}{2} - \frac{a}{4}\right)\left(\frac{L}{2} - \frac{a}{4}\right)$$

$$= \frac{300}{16}(8 - 0.5)^2 = 1054.688 \text{ kNm}$$

This occurs at a distance = $x_1 = \dfrac{L}{2} - \dfrac{a}{4} = 7.5$ m

Let the self-weight of the web be w_g. The bending moments caused by self-weight and the superimposed loads at x_1 are

$$M_{g1} = w_g\frac{L}{2}(x_1) - w_g\frac{x_1^2}{2} = 7.2(8)(7.5) - \frac{7.2(7.5)^2}{2} = 229.5 \text{ kNm}$$

$$M_{s1} = w_s\frac{L}{2}(x_1) - w_s\frac{x_1^2}{2} = 22(7.5)(8 - 3.75) = 701.25 \text{ kNm}$$

Total moment at distance x_1

$$M_{l1} = M_{l1} + M_{g1} + M_{s1} = 1985.438 \text{ kNm}$$

The possible design collapse moment is

$$M_{c1} = 1.5 \ (M_{l1}) = 2978.157 \text{ kNm}$$

There is another possible critical location, as the maximum BM due to DL occurs at mid-span and, therefore, this moment is computed for comparison. When one of the moving loads is at the mid-span, the reaction at the left end support is

$$R_1 = \frac{W}{2} + \frac{W}{L}\left(\frac{L}{2} - a\right)$$

The bending moment at the mid-span is

$$M_{l2} = R_1\left(\frac{L}{2}\right) = \frac{WL}{4} + \frac{W}{2}\left(\frac{L}{2} - a\right)$$

$$= \frac{2WL}{4} - \frac{Wa}{2} = \frac{300(16)}{4} - \frac{150(2)}{2} = 1050 \text{ kNm}$$

The bending moment caused by the superimposed load is

$$M_{s2} = \frac{w_s L^2}{8} = \frac{22(16)(16)}{8} = 704 \text{ kNm}$$

Bending moment due to self-weight

$$M_{g2} = \frac{w_g L^2}{8} = \frac{7.2(256)}{8} = 230.4 \text{ kNm}$$

The collapse of the beam occurs by the failure of the section at mid-span. The collapse moment with partial safety load factor at 1.5, is

$$M_{c2} = 1.5 (M_{g2} + M_{s2} + M_{l2})$$

$$= 1.5(230.4 + 704 + 1050) = 2976 \text{ kNm}$$

From the two critical moments, it can be seen that the BM at mid-span is more critical and it is taken as the design moment.

$$M_c = M_{c2} = 2991 \text{ kNm} \cong 3000 \text{ kNm}$$

and

$$d = 0.5t + \frac{M_c}{0.45 b_{ft} f_{ck}}$$

$$= 0.075 + \frac{3000(10^3)}{0.45(2)(0.15)(25)(10^6)}$$

$$= 0.075 + 0.89 = 0.964$$

The area of the tension reinforcement is obtained from Eq. (7.50) and it is

$$A_{st} = 0.518 \, b_{ft} \frac{f_{ck}}{f_y}$$

$$= 0.518 \, (2)(0.15) \left(\frac{25}{415} \right) = 0.009361 \text{ m}^2$$

Provide   15 numbers of 28 mm bars.

A_{st} (provided) = 15(615) = 9225 mm^2

4 numbers of 10 mm bars at top

$A_{sc} = 2(157) = 314$ mm^2

$A_{so} = A_{st} + A_{sc} = 9225 + 314 = 9539$ mm^2

As the area of steel provided is marginally smaller than that required, increase the effective depth proportionaly, that is

$$d = 0.95 \left(\frac{9361}{9225} \right) = 0.965 \text{ m}$$

The 15 bars are arranged in three rows of five each with clear spacing between the bars as 35 mm; therefore, the overall depth of the beam with 35 mm cover is

$$h = d + 14 + 35 + 28 + 35 = 1077 \text{ m}$$

The depth of the web = $h - t = 967$ mm

Assuming the clear spacing of the bars as 30 mm, the minimum width of the web is computed for accommodating in a row with a cover of 30 mm.

b_w (minimum) = 5(28) + 4(30) + 2(30) = 320 mm

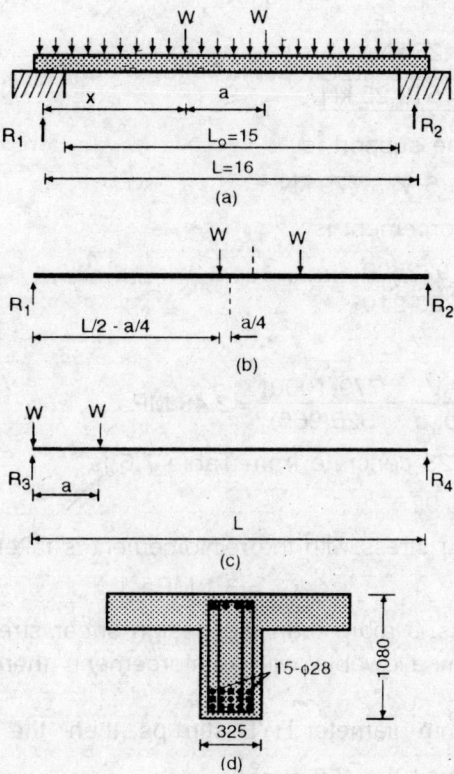

Fig. 7.9 Notation and Sectional Details of Beam of Example 7.10.

Let $b_w = 325$ mm
$$w_g = 0.325(0.947)(25) = \simeq 7.7$$

The assumed weight for the moment calculation is slightly less than the actual; therefore, the moment calculation does not need a revision. The maximum shear force is computed on the actual weight.

Design for shear: The maximum shear force occurs at the support when one of the moving loads is next to the support as shown in Fig. 7.9c. The left end reaction caused by the moving loads is

$$R_3 = W + \frac{W(L-a)}{L} = 2W - \frac{Wa}{L}$$

$$= 300 - \frac{150(2)}{16} = 281.25 \text{ kN}$$

The maximum shear forces due to self-weight, superimposed load and live loads are:

$$V_g = \frac{w_g L}{2} = \frac{7.7(16)}{2} = 62 \text{ kN}$$

$$V_s = \frac{w_s L}{2} = \frac{22(16)}{2} = 176 \text{ kN}$$

$$V_l = 281.25 \text{ kN}$$

Total shear support $= V_t = \overline{519.25 \text{ kN}}$

The design shear force at the support is

$$V = 1.5 \ V_t = 779 \text{ kN}$$

Percentage ratio of the reinforcement is

$$\frac{100 \ A_{st}}{b_w d} = \frac{922500}{300(910)} = 3.4$$

The nominal shear stress is

$$\tau_v = \frac{V}{b_w d} = \frac{779(1000)}{325(965)} = 2.48 \text{ MPa}$$

The shear strength of the M25 concrete from Table 7.6 is

$$\tau_c = 0.92$$

The maximum allowable shear stress with the reinforcement is taken from Table 7.8, and it is

$$\tau_{c max} = 3.1 \text{ MPa}$$

The nominal shear stress is more than the design shear strength without reinforcement and less than the maximum allowable with reinforcement; therefore, shear reinforcement should be designed.

Provide four-legged 12 mm diameter HYD stirrups, then the area of stirrup steel is

$$A_{sv} = 4(113) = 452 \text{ mm}^2$$

The spacing of the reinforcement is

$$s_v = \frac{0.87 \ A_{sv} \ f_{yv}}{b_w (\tau_v - \tau_c)} = \frac{0.87(452)(415)}{325(2.48 - 0.92)} = 323 \text{ mm}$$

The maximum spacing of the stirrups is

$$s_{max} = \frac{A_{sv} f_y}{0.4 b_w} = \frac{452(415)}{0.4(325)} = 1142 \text{ mm, or}$$

$$= 0.75 \ d$$

Provide 6 numbers of four-legged 12ϕ stirrups in the first 2 m from the supports
    4 numbers in the next 2 m and
    6 numbers in the next 4 m upto the mid-span.

The main reinforcement bars can be provided in the following ways:
    Five numbers of bottom row and two numbers of the middle row are continued straight.
    Three numbers of middle row bars curtailed at

$$(L/5 - 20\phi = 16000/5 - 20(28)) = 2640 \text{ mm.}$$

from the support ends.
    Five numbers of the top row bars are cranked up at 4 m from the support.

EXAMPLE 7.10 *Design of prestressed concrete beam*. Design a I-beam for given flange size and bending moment. The data of the problem is $f_{ck} = 35$ MPa and $f_y = 1500$ MPa. The collapse bending moment and shear force including the self-weight are:

$$M_c = 400\ 000\ \text{Nm}$$

$$\text{and}\quad V = 180\ 000\ \text{N}$$

The effective width of the flange and its thickness are:

$$b_f = 1.67\ \text{m and } t = 0.12\ \text{m}$$

The depth of the beam and the area of the steel are computed from the balanced type of beam section using Eqs. (7.58) and (7.59).

$$d = 0.5t + \frac{M_c}{0.45b_f t f_{ck}}$$

$$= 0.06 + \frac{400\ 000}{0.45(1.67)(0.12)(35)(10^4)}$$

$$= 0.06 + 0.127 = 0.187\ \text{m}$$

The thickness of the slab is about 60% of the effective depth; therefore, increase the depth to avoid compression failure. Let d be selected in the range of $t/0.45$, then $d = 0.32$ m From Eq. (7.60), we have the area of the tension reinforcement as

$$A_{st} = \frac{1.15\ M_c}{(d - 0.5t)f_y} = \frac{1.15(400\ 000)}{(0.32 - 0.06)(1500)} = 1179\ \text{mm}^2$$

Provide 32 numbers of 7 mm HTS wires, then

$$A_{st}\ (\text{provide}) = 32\ (38.5) = 1232\ \text{mm}^2$$

Assume four wires are grouped together, thus resulting into 2 rows of groups of wires Let the clear spacing between the groups of wires be 45 mm and the side cover be 30 mm. Let the vertical spacing between the groups be 50 mm, then the minimum web width and depth of the section are:

$$b_w\ (\text{min}) = 2(3) + 4(14) + 3(45) = 251$$

$$h\ (\text{min}) = d + 25 + 14 + 30 = 389$$

$$\text{Use } b_w = 275\ \text{mm and } h = 390\ \text{mm}$$

Design for transfer condition. The ultimate strength of the steel is

$$f_u = 1.2\ f_y\ \text{and}$$

the ultimate tensile force in the wires is

$$T_u = A_{st}\ f_u = 1.2\ A_{st}\ f_y$$

The capacity of the web in compression on either side of the wires is

$$C_d = \left(\frac{k_p f_{ck}}{\gamma_m}\right) b_w(2d_0)$$

Let $\gamma_m = 1$ in which $d_0 = $ effective depth of the web on either side of the wires. This is usually limited to the cover provided to the centre of steel.

$$\text{Therefore, } d_0 = h - d = 390 - 220 = 70\ \text{mm}$$

For structural safety at transfer condition, the capacity of this web region must be more than the ultimate strength of the steel, that is

$$C_u > T_u$$

(Neglecting the self-weight effect in light beams).

$$T_u = 1.2\ A_{st}f_y = 1.2(1232)(1500) = 2217600\ N$$

$$C_u = 0.67\ f_{ck}(2d_0)b_w = 0.45(35)(140)(275) = 902825\ N$$

Since the capacity of the web is less than the ultimate tension in the steel, the web area around the steel should be increased into a bottom flange.

Let b_b = width of the bottom flange, then the capacity of this section is

$$C_d = 0.67\ f_{ck}(t_b)b_b$$

in which t_b = thickness of the bottom flange. Equation the bottom flange capacity to the ultimate tension, we have (let $t_b = 2d_0$)

$$0.67\ f_{ck}(t_b)b_b = T_u = 2217600$$

$$b_b = \frac{2217600}{0.67(35(140)} = 675\ mm$$

Provide bottom flange 675 by 140 mm. Practical section is shown in Fig. 7.10.

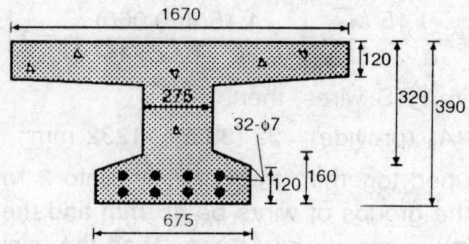

Fig. 7.10 Sectional Details of Prestressed Concrete Beam of Example 7.10.

EXAMPLE 7.11 *Design of 15 m span prestressed concrete simply supported beam.* A simply supported beam is cast integral with a slab of thickness 150 mm. The slab is to be used as top flange and the effective width of the slab which acts as a top flange is 2 m. The weight of the top slab and the other fixed loads transferred to the beam are together 22 kN/m. The effective span of the beam is 16 m. The beam is subjected to two moving loads of 150 kN each which are separated by 2 m. The complete data can be summarized as:

$$L = 16\ m,\ b_f = 2\ m,\ t = 0.15\ m$$
$$w_s = 22\ kN/m$$
$$W_1 = W_2 = W = 150\ kN$$
$$a = 2\ m$$
$$f_{ck} = 35\ MPa\ and\ f_y = 1500\ MPa$$

The maximum bending moment caused by the two moving loads, occurs under one of the two loads when from the mid-span the centroid of the loads is at the same distance as the load from the mid-span. That means that the distance of the load from the left support should be

$$x_1 = \frac{L}{2} - \frac{a}{4} = 8 - 0.5 = 7.5 \text{ m}$$

The corresponding maximum bending moment at $x_1 = 7.5$ m caused by the live loads is

$$M_l = \frac{2W}{L}\left(\frac{L}{2} - \frac{a}{4}\right)\left(\frac{L}{2} - \frac{a}{4}\right)$$

$$= \frac{300}{16}(8 - 0.5)^2 = 1054.688 \text{ kNm}$$

Let the size of the beam web for the purpose of computing self-weight be $= 0.3 \times 1$ m.

$$\text{Self-weight} = w_g = 0.31(1)(25) = 7.5 \text{ kN/m}$$
$$w_s = 22.0 \text{ kN/m}$$
$$\text{Total dead load} = w_d = \overline{29.5 \text{ kN/m}}$$

The maximum bending moment due to the dead load occurs at mid-span. However, the maximum BM due to live load occurs at 7.5 m instead of at the mid span. The moment due to the dead load at distance x_1 is

$$M_d = \frac{w_d L x_1}{2} - \frac{w_d x_1^2}{2}$$

$$= \frac{29.5}{2}(16(7.5) - 7.5^2) = 930.75 \text{ kNm}$$

The collapse design moment with partial safety factor of 1.5 is

$$M_e = 1.5(M_l + M_d)$$

$$= 1.5(1054.688 + 930.75) = 2978.157 \text{ kNm}$$

The maximum BM at the mid-span when one of the loads is at the mid-span is 2991 kNm. Therefore, use

$$M_c = 3000 \text{ kNm}$$

the effective depth of steel computed using Eq. (7.58) is as

$$d = 0.5t + \frac{M_c}{0.45 b_f t f_{ck}}$$

$$= 0.075 + \frac{3000\,000}{0.45(2)(0.15)(35)(10^6)}$$

$$= 0.075 + 0.635 = 0.71 \text{ m}$$

Use
$$d = 0.725 \text{ m}$$

$$A_{stb} = 0.518 b_f t \frac{f_{ck}}{f_y}$$

$$= 0.518(2)(0.15)\frac{(35)}{1500} = 0.003626 \text{ m}^2$$

Select 95 numbers of 7 mm HTS wires, then the area of steel provided is

$$A_{st} = 95(38.5) = 3657.5 \text{ mm}^2$$

Provide   11 cables each containing 8 wires and 1 cable of 7 wires
Total number of cables = 12

Arrange six cables in the central portion and three cables on each side.

The diameter of the duct is = 40 mm
Cover to the cables = 35 mm
The overall depth required would be $h = d + 20 + 50 + 40 + 35 = 870$
use $h = 900$ mm

Design for transfer condition. The prestressing force is to be treated as an external loading at the transfer condition. The load factor for prestressing = 1.2.

The load factor for dead load = 0.9 (as the DL acts opposite to the action of the load).

The ultimate prestressing force at the transfer is

$$T_{ut} = 1.2A_{st}f_y = 1.2(3657)(1500) = 6582\ 600\ N$$

The bending moment due to the dead load at transfer (with load factor of 0.9 is)

$$M_d = 0.9\left(\frac{W_d L^2}{8}\right) = 0.9\frac{(25.5)\ (256)}{8} = 725.76\ kNm$$

Assuming a lever arm $jd = 0.9d$, the net axial tension caused by the dead load at transfer is

$$T_{dt} = \frac{M_{d'}}{jd} = \frac{725\ 760}{0.9(0.725)} = 1112\ 725\ N$$

The net design compressive force on the concrete at the transfer condition is

$$C_{dt} = T_{ut} - T_{dt}$$
$$= 6582\ 600 - 1112\ 275 = 5470\ 325\ N$$

The area of the bottom flange to resist the compressive force at the transfer condition is:

$$A_{cb} = \frac{C_{dt}}{0.67\ f_{ck}} = \frac{5470\ 325}{0.67(35)(10^6)} = 0.2333\ m^2$$

Let the average thickness of the bottom flange be

$$t_b = 0.30\ m$$

Then the average width of the bottom flange should be

$$b_b = \frac{A_{cb}}{t_b} = \frac{0.2333}{0.3} = 0.777\ m$$

The web thickness must be adequate to resist shear force and accommodate the two vertical rows of the cables to be draped up. The shear force in the prestressed concrete sections is not really critical, even though the section is to be designed. A deduction of about 2 to 3% of sectional area is to be accounted for the area taken up by the ducts. Use $b_b = 800$ mm.

Let $b_w = 40 + 40 + 40 + 40 + 50 = 210$ mm

(40 mm cover and 40 mm duct diameter and 50 mm duct spacing)

The maximum width of the flange based on the shear lag is

$$b_f = 12t + b_w = 12(150) + 210 = 2100\ mm$$

whereas the given $b_f = 2000$ mm; therefore provide

$$b_w = 300\ mm$$

The section and the arrangement of the cables is shown in Fig. 7.11.

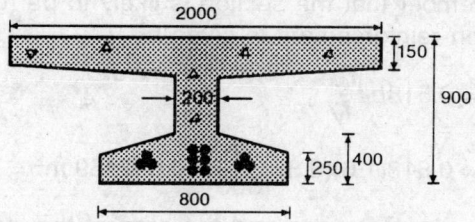

Fig. 7.11 Cross-Sectional Details of Prestressed Concrete Beam of Example 7.11.

EXAMPLE 7.12 *Design of a flanged section prestressed concrete beam.* A simply supported beam of span 50 m is subjected to a live load of 20 kN/m. The width of the top flange is to be less than 2 m and depth to be less than 2.5 m.

$$f_{ck} = 45 \text{ MPa}, \, f_y = 1500 \text{ MPa}$$

b_f can be assumed to be in the range of $L/20$ to $L/40$. Let $b_f = 1.4$ m (= trial width),

$$t = (L/200 \text{ to } L/150) = 0.30 \text{ m and}$$

$$w_g = 21 \text{ kN/m}$$

One can design the width of the top flange for an assumed depth or vice versa. In this example flange width is assumed and the depth of the beam is computed.

The maximum bending moment occurs at the mid-span and the beam will collapse by yielding of the mid-span section.

Let the partial safety factor applied to load = 1.5.

The collapse bending moment which occurs at the mid-span of the simply supported beam is

$$M_c = 1.5(w_g + w_l) \frac{L^2}{8}$$

$$= 1.5(21 + 20) \left(\frac{2500}{2}\right) = 19218.75 \text{ kNm}$$

The depth of the steel is given by

$$d = 0.5t + \frac{M_c}{0.45 \, b_f t f_{ck}}$$

$$= 0.15 + \frac{19218750}{0.45(1.4)(0.3)(45)(10^6)} = 2.40 \text{ m}$$

The overall depth of the beam is likely to exceed the limited value of 2.5 m; therefore, the width of the top flange can be increased to 1.6 m. Let

$$b_f = 1.6 \text{ m; then}$$

$$d = 0.15 + \frac{19218750}{0.45(1.6)(0.3)(45)(10^6)} = 2.1272 \text{ m}$$

Use $d = 2.1$ m then the neutral distance would be

$$x_u = 0.45d = 0.945 \text{ m} > t$$

Therefore, the moment capacity computed from the balanced section is on the safe side. However, one should remember that the section is likely to be (on the) uneconomical (side).

The area of the tension reinforcement is given by

$$A_{st} = 0.518b_t t\frac{f_{ck}}{f_y}$$

$$= 0.518(1.6)(0.3)\frac{(45)}{1500} = 0.007459 \text{ m}^2 = 7459 \text{ mm}^2$$

Provide 194 number of 7 mm HTS wires and the area of the steel provided is

$$A_{st} = 7466 \text{ mm}^2$$

Provide 24 numbers of strands of each 8 wires and 1 number of strand with 4 wires.

Arrange 6 numbers of groups of the strands with 4 four strands in each group and then one strand of 4 wires. Let there be one group in the bottom portion of the web and five groups in the bottom flange at mid-span section.

The overall depth of the beam can be calculated by assuming the diameter of the ducts and spacing of the groups.

Let, the diameter of the duct = 40 mm
the spacing the groups = 70 mm

$$b_w = 2 \times 40 + 2(40) + (70) = 230 \text{ mm}; \text{ use } b_w = 250 \text{ mm}$$

The maximum flange width for shear lag transfer is $12t + b_w$. This value is more than the value assumed. Therefore, the thickness of the web as 250 mm is acceptable.

The approximate overall depth is then computed after selecting the bottom flange size.

Design at transfer condition. The size of the bottom flange is generally governed by the transfer prestress condition. The ultimate tension in the cables is

$$T_u = 1.2A_{st}f_y$$

$$= 1.2(7466)\frac{(1500)}{1000} = 13438.8 \text{ kN}$$

The self-weight of the beam has the tendency to compensate the crushing force on the bottom flange, thus relieving some compression. The load factor applied to the dead load should be 0.9 to compute the design bending moment due to DL.

$$M_d = 0.9\frac{W_dL^2}{8} = 5906.25 \text{ kNm}$$

Let the lever arm $jd = 0.9d$, then the net tension caused by the assumed dead load is

$$T_{dt} = \frac{M_d}{jd} = \frac{5906.250}{0.9(2.1)} = 3125 \text{ kN}$$

The net compression on the bottom is

$$C_{dt} = T_{ut} - T_{dt} = 9332.5 - 3125 = 6207.5 \text{ kN}$$

The minimum compression area needed is

$$A_{cbt} = \frac{C_{dt}}{0.67\ f_{ck}} = \frac{6207\ 500}{0.67(45)(10^6)} = 0.206 \text{ m}^2$$

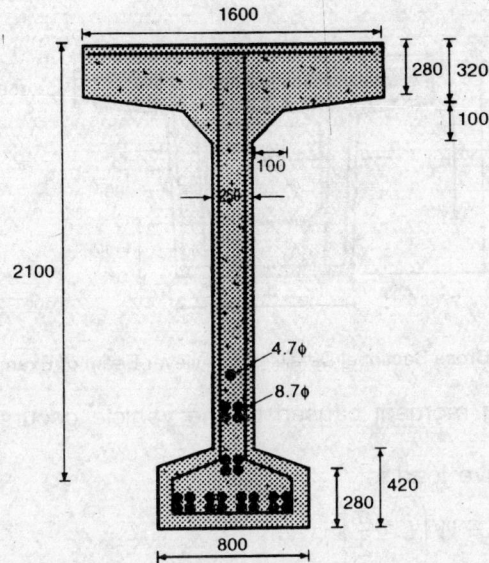

Fig. 7.12 Cross-Sectional Details of Prestressed Concrete Beam of Example 7.12 (Dimension in mm).

One group of the four cables is placed in the web; therefore, some of the web area around the group of the cables will be effective in resisting the compression.

Let this group of the cables be 100 mm above the top edge of the bottom flange, then the effective area of the web in compression is

$$A_{cw1} = b_w (2) (D - d) = 0.18(2)(0.3) = 0.108 \text{ m}^2$$

The area of the bottom flange needed is

$$A_{cb} = A_{sbt} - A_{cw1} = 0.206 - 0.108 = 0.098 \text{ m}^2$$

Let the width of the bottom flange by

$$b_b = (\text{in the range of } b/2) = 0.6 \text{ m}$$

Then the thickness of the bottom flange that is needed is

$$t_b = \frac{A_{cb}}{b_b} = \frac{0.098}{0.06} = 0.16 \text{ m}$$

The design section is shown in Fig. 7.12. The nominal untensioned reinforcement is not discussed here.

EXAMPLE 7.13 *Design of a cantilever beam.* The effective span of a cantilever bridge is 30 m and a 700 kN tracked with a wheel base length of 4 m moves on the bridge girder. Design a single box-type cross-section of the bridge girder with top slab width as 7.5 m and thickness as 150 mm.

$$f_{ck} = 45 \text{ MPa and } f_y = 1500 \text{ MPa}$$

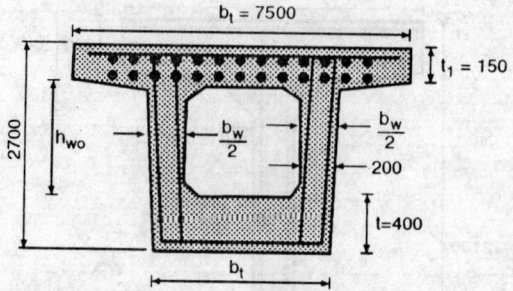

Fig. 7.13 Cross Sectional Details of Cantilever Beam of Example 7.13.

The maximum bending moment caused by the vehicle occurs at the support when the vehicle is at the free end.

Maximum BM due to live load is

$$M_{f1} = W\left(L - \frac{a}{2}\right)$$

where a = base length of the tracked vehicle. The cross-section of the beam is a box type as shown in Fig. 7.13. The depth of the section can be assumed to vary linearly, with the least depth at the free end and maximum at the fixed end. For the purpose of computation of self-weight the following dimensions are assumed:

Width of compression flange at bottom $b_f = 5$ m

$$t = L/60 = 0.4 \text{ m}$$
$$\text{Depth at fixed end } h_{wo1} = L/15 = 2 \text{ m}$$
$$h_{wo2} \text{ (Depth at free end)} = 1 \text{ m}$$

The weight and moment of the slab are:

$$W_g = b_t t_t (L)\,\gamma\left(\frac{L}{2}\right) + b_f t L \gamma\, \frac{L}{2}$$

$$+ \frac{h_{wo1}\,b_w \gamma L}{2}\,\frac{L}{3} + \frac{h_{wo2}\,b_w \gamma L}{2}\,\frac{2}{3}\,L$$

where γ = density of concrete = 25 kN/m³

$$M_g = \frac{\gamma L^2}{2}\left(b_t t_t + b_f t + \frac{b_w}{3}\left(\frac{h_{wo1} + 2h_{wo2}}{2}\right)\right)$$

$$= \frac{25}{2}\,(900)\,(7.5(0.15) + 5(0.4)) + \frac{0.15}{3}\,(4) = 36844 \text{ kNm}$$

Bending moment due to the live load is

$$M_1 = W\left(L - \frac{a}{2}\right)$$

$$= 700(30 - 2) = 19600 \text{ kNm}$$

Total working moment $= 56444$ kNm

The collapse moment $M_c = 1.5(56444) = 84666$ kNm

Let $b_f = 5$ and $t = 0.4$ m (compression flange at the bottom)

Then the effective depth of steel is

$$d = 0.5t + \frac{M_c}{0.45 b_f t f_{ck}}$$

$$= 0.2 + \frac{84.666}{0.45(5)(0.4)(45)} = 2.295 \text{ m}$$

This depth appears to be reasonable

Let $d = 2.35$ m, then

$$A_{st} = \frac{1.15 \, M_c}{(d - 0.5t) f_y} = \frac{1.15(84666000)}{2.15(1500)} = 30190 \text{ mm}^2$$

Provide 66 numbers of 12 wire strands of 7 mm wires

$$A_{st} \text{ (provided)} = 66(12) \, (38.48) = 30476 \text{ mm}^2$$

All cables need not originate from the free end but will have to go through the section at the support.

Self-weight calculation. Weight of top flange (given) = 7.5(0.15) (30)(25) = **844 kN.**
Weight of bottom flange:
($b_f = 5$ m and thickness is 0.4 m at support, and 0.3 m at free end)

$$\frac{5(0.4 + 0.3)}{2} (30)(25) = 1313 \text{ kN}$$

Weight of the web:
(The overall depth at support = $h = d + 0.35 = 2.7$ m, the thickness of each web is taken 0.20 m. The overall depth at free end is taken as 1.5. The clear depth of the web at support = 2.7 − 0.4 − 0.15 = 2.15 m.) The clear depth at the free end is = 1.5 − 0.3 − 0.15 = 1.05 m

$$\text{Weight of web} = 2(0.20) \left(\frac{2.15 + 1.05}{2} \right)(30)(25) = 480 \text{ kN}$$

Total weight = 844 + 1313 + 480 = 2637 kN

The bending moment due to the self-weight of the beam is

$$M_g = 844(15) + 5(0.3)(30)(25)(15)$$

$$+ \frac{5(0.1)(30)(25)}{2} \left(\frac{1}{3} \right)(10)$$

$$+ 2.15 \frac{0.20(30)(25)}{2} \left(\frac{30}{3} \right)$$

$$+ 1.05 \frac{(0.20)(30)(25)}{2} \left(\frac{60}{3} \right)$$

$$= 12656 + 16875 + 625 + 1613 + 1575 = 33344 \text{ kNm}$$

The actual moment caused is slightly less than the assumed value; therefore, no revision is needed in the moment calculations.

Design at transfer condition. The ultimate in the cable with a partial safety factor of 1.2 is

$$T_{ut} = 1.2 \, A_{st} f_y$$

$$= 1.2(36190)\left(\frac{1500}{1000}\right) = 54342 \text{ kN}$$

Let the lever arm be $jd = 0.95 \, d$, then the tension produced by the self-weight is

$$T_{dt} = \frac{0.9 M_g}{jd} = \frac{0.9(33344)}{0.95(2.35)} = 13442 \text{ kN}$$

Net tension to be compensated by the concrete under compression is

$$C_{dt} = T_{ut} - T_{dt} \, 40\,900 \text{ kN}$$

The area of the concrete required at the top flange to resist the transfer force is

$$A_{ct} = \frac{C_{dt}}{0.67 \, f_{ck}} = \frac{40.900}{0.67(45)} = 1.356 \text{ m}^2$$

Area of the top flange = 7.5(0.15) = 1.125 m².
Area of the web required to resist the transfer prestress = 1.333 − 1.125 = 0.207 m²

$$h - d = 2.7 - 2.35 = 0.35 \text{ m}$$

Area of the concrete in web against the force is = $2(h - d)(2b_w)$ = 0.28 m²

This value is more than the required; therefore, the section is safe at transfer of prestress, (see Fig. 7.13). The design for secondary bending moment and nominal reinforcement are not covered here.

7.11 Moment Capacity of a Triangular Section

The depth of the neutral axis is found by compatibility of strain. From Fig. 7.14, we have

$$\frac{x_u}{d} = \frac{\epsilon_c}{\epsilon_c + \epsilon_y}$$

in which $\epsilon_c = 0.0035$ and $\epsilon_y = 0.002 + 0.87 \, f_y/E_s$

Let x_1 = distance where 0.002 strain occurs and it is the common point of parabolic and rectangular stress distribution

$$= (0.002/0.0035)x_u = \frac{4}{7} x_u \qquad (7.72$$

Consider a small element at distance y from the neutral axis, the width of the element is

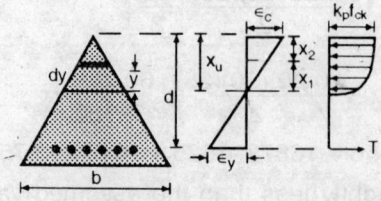

Fig. 7.14 Triangular Section and Stress Block.

$$x = \frac{b(x_u - y)}{d} \qquad (7.73)$$

For $y \le x_1$

The intensity of the stress on element is

$$\sigma_y = \frac{(2x_1 y - y^2)}{x_1^2} k_p f_{ck}$$

$$= \left(\frac{49}{16} \frac{k_p f_{ck}}{x_u^2} \right)(2x_1 y - y^2) \qquad (7.74)$$

The total force contributed from the portion of the parabolic stress variation is

$$C_1 = \int_0^{x_1} x \, \sigma_y \, dy$$

$$= \frac{49(bk_p f_{ck})}{16 x_u^2 d} \int_0^{x_1} (x_u - y)(2x_1 y - y^2) \, dy$$

$$= \frac{49(bk_p f_{ck})}{16 x_u^2 d} \left[2 x_u x_1 \frac{y^2}{2} - (x_u + 2x_1)\frac{y^3}{3} + \frac{y^4}{4} \right]_0^{x_1}$$

$$= \frac{49(bk_p f_{ck})}{16 x_u^2 \, d} \left(\frac{4^3}{7^3} - \frac{5}{7}\frac{4^3}{7^3} + \frac{4^4}{4(7^4)} \right) x_u^4$$

$$= \frac{4}{49} \frac{(bk_p f_{ck})}{d} (7 - 5 + 1) x_u^2$$

$$= \frac{12(bk_p f_{ck} x_u^2)}{49}$$

$$= 0.245 \frac{b x_u^2 f_{ck}}{d} \qquad (7.75)$$

The contribution of the parabolic area towards moment about the centroid of the steel is

$$M_{r1} = \int x \, \sigma_y \, dy(d - x_u - y))$$

$$= \frac{49 \, bk_p f_{ck}}{16 d \, x_u^2} \int (x_u - y) \, (d - (x_u - y))(2x_1 y - y^2) \, dy$$

$$= \frac{49 \, bk_p f_{ck}}{16 d \, x_u^2} \int ((dx_u - x_u^2) 2x_1 y - (dx_u - x_u^2 + 2x_1 d - 4x_u x_1)y^2 - (2x_1 - d + 2x_u)y^3 + y^4)) \, dy$$

The integration is carried out by part as given below

$$M_{r1} = 0.164(1 - 0.67 k_3) \, k_{\frac{2}{3}} \, bd^2 \, f_{ck}$$

The total compressive force from the rectangular area acting on the triangular bit is

$$C_2 = \frac{1}{2}\left(\frac{x_2 b}{d} \right) x_2 k_p f_{ck} = \frac{0.0615}{d} \, b x_u^2 \, f_{ck} \qquad (7.76)$$

The moment contribution of the above force about the CG of steel is

$$M_{r_2} = C_2 \left(d - \frac{2x_2}{3} \right)$$

$$= 0.0615 \left(\frac{bx_u^2 f_{ck}}{d} \right) \left(d - 0.286\, x_u \right) \tag{7.77}$$

The total design compressive force on the section with partial safety factor of 1.5 is

$$C = (C_1 + C_2)/\gamma_m$$

$$= \frac{(0.245 + 0.0165)bx_u^2 f_{ck}}{1.5}\frac{}{d} = 0.204k_{\frac{2}{3}}bdf_{ck} \tag{7.78}$$

The total resisting moment capacity with partial safety factor of 1.5 is

$$M_r = (M_{r1} + M_{r2})/1.5$$

$$= (0.109)\,(1 - 0.67k_3)$$

$$+\ 0.047\,(1 - 0.286k_3)bx_u^2 f_{ck}$$

$$= 0.150(1 - 0.565k_3)\,bx_u^2 f_{ck} \tag{7.79}$$

$$= K_2 bd^2 f_{ck} \tag{7.80}$$

The values of K_2 are given in Table 7.12 where

$$K_2 = 0.15(1 - 0.565k_3)k_{\frac{2}{3}}$$

Equating the tensile force to the compressive force, we have

$$0.87A_{st}f_y = C = 0.204k_{\frac{2}{3}}\,bdf_{ck}$$

or

$$A_{st} = \frac{0.204k_{\frac{2}{3}}}{0.87}\,bd\frac{f_{ck}}{f_y} = 0.234k_{\frac{2}{3}}bd\frac{f_{ck}}{f_y} \tag{7.81}$$

Equation (7.81) can be rearranged as

$$\frac{A_{st}}{bd} = 0.234k_{\frac{2}{3}}\frac{f_{ck}}{f_y} = p_{02}\frac{f_{ck}}{f_y} \tag{7.82}$$

where

$$P_{02} = 0.234k_{\frac{2}{3}} \tag{7.83}$$

The values of P_{02} are given in Table 7.12.

Table 7.12 Design Coefficients for Triangular and Trapezoidal Sections

f_y	$\dfrac{x_u}{d}$	K_2	p_{02}
250	0.531	0.030	0.066
360	0.495	0.026	0.057
415	0.479	0.025	0.025
500	0.450	0.023	0.049

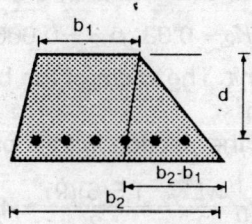

Fig. 7.15 Trapezoidal Section.

7.12 Moment Capacity of Trapezoidal Section (see Fig. 7.15)

Let b_1 = width of the section at top

b_2 = width of the section at the level of steel

The moment capacity of the section can be obtained by the sum of the capacities of a rectangular of width b_1 and the triangular section of width $(b_2 - b_1)$

$$M_r = Kb_1d^2f_{ck} + K_2(b_2 - b_1)d^2f_{ck}$$

$$= ((K - K_2)b_1 + K_2b_2)d^2f_{ck} \qquad (7.84)$$

The values of K and K_2 are given in Tables 7.5 and 7.12 respectively. The amount of reinforcement needed is given by the sum of the quantities required for the rectangular and triangular sections, and it is

$$A_{st} = (p_0b_1d + p_{02}(b_2 - b_1)d)\frac{f_{ck}}{f_y}$$

$$= ((p_0 - p_{02})b_1 + p_{02}b_2)d\frac{f_{ck}}{f_y} \qquad (7.85)$$

The values of p_0 and p_{02} are given in Tables 7.5 and 7.12 or they can also can also be obtained form the expressions given already.

7.13 Design Examples of Non-rectangular Sectioned Beams

EXAMPLE 7.14 *Design of triangular section.* A centilever beam which supports a portico is made of an inverted triangular section. The effective span of the beam is 3 m, and the live and dead loads from the slab over the beam are 1 kN/m, and 3 kN/m respectively. Design the cross-section.

$$\text{Let self-weight} = w_g = 2\text{kN/m}$$
Other data is
$$\text{Weight of the slab} = 3\text{kN/m}$$
$$\text{Live load} = w_l = 1\text{kN/m}$$
$$\text{Total load} = w = 6\text{kN/m}$$
$$\text{Span} = L = 3\text{m}$$
$$\text{Concrete grade} = \text{M15}$$
$$\text{Quality of steel} = f_y = 250 \text{ MPa}$$

The design coefficients for $f_y = 250$ MPa are taken from Table 7.12 and they are:

$$x_u = 0.531d, \quad K_2 = 0.03, \quad p_{02} = 0.066.$$

Design of the Section-collapse limit. The width of the beam at the level of steel be selected in the range of $L/10$, then $b = 0.3$ m.

The design bending moment on the section with a partial safety factor of 1.5 is

$$M_c \, \gamma_f \frac{wL^2}{2} = \frac{1.5(6)(9)}{2} = 40.5 \text{ kNm}$$

Equating the moment capacity to the collapse moment, we have

$$M_r = K_2 b d^2 f_{ck} = M_c = 40\,500\text{Nm}$$

$$\text{or } d = \sqrt{\frac{40\,500}{0.30(0.3)(15)(10^6)}} = 0.548 \text{ m}$$

Use $d = 0.55$ m, and $h = 0.55 + 0.05 = 0.6$ m.

The area of tension reinforcement is

$$A_{st} = p_{02} \, bd \, \frac{f_{ck}}{f_y} = 0.066(0.3)(0.55) \frac{15(10^6)}{250} = 653 \text{ mm}^2$$

The minimum reinforcement required is

$$A_{st} = \frac{0.85A_c}{f_y} = \frac{0.85(300)(550)}{(2)250} = 280 \text{ mm}^2$$

Self-weight is less than that assumed.

Provide 2 numbers of $\phi 20$ and 1 number of $\phi 10$

$$A_s \text{ (provided)} = 706 \text{ mm}^2$$

$$\text{Percentage of reinforcement} = \frac{70\,600}{150(550)} = 0.86$$

Design for shear. The shear force at a distance d from the support with partial safety factor of 1.5 is:

$$V = \gamma_f w(L - d) = 1.5(6000)(3 - 0.55) = 22\,050 \text{ N}$$

The nominal shear stress is

$$\tau_v = \frac{V}{A_c} = \frac{22\,050}{(0.3)(0.55)(0.5)} = 267\,272 \text{ N/m}^2 = 0.267 \text{ MPa}$$

The shear stress is less than the strength; therefore, provide only nominal shear reinforcement.

Select 6 mm two-legged stirrups, then the area of the shear reinforcement is

$$A_{sv} = 2(28) = 56 \text{ mm}^2$$

Maximum spacing of the stirrups is

$$S_{vm} = \frac{A_{sv}f_y}{0.2d} = \frac{56(250)}{0.2(300)} = 233 \text{ mm}$$

Note. The minimum reinforcement in case of traingular section is recommended as

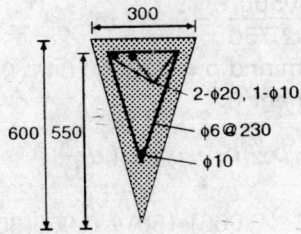

Fig. 7.16 Example 7.14.

$$A_{sv} = \frac{0.2bs_v}{f_v}$$

in which b = width of the triangle at the reinforcement level. The maximum spacing is also restricted by $s_v \leq 0.75\, d$

Provide two-legged 6 mm stirrups at 230 mm spacing. The development length for plain bars. in M15 concrete is 38ϕ when a hook is provided.

EXAMPLE 7.15 *Design of trapezoidal section.* A cantilever beam of 3 m span has a trapezoidal section. The live and dead loads transferred to the beam are 1 kN/m and 3 kN/m respectively. Design the beam.

Let the self-weight of the section be 2 kN/m then the total load on the beam is

$$w = w_g + w_s + w_l = 2 + 3 + 1 = 6 \text{ kN/m}$$

$$f_{ck} = 15 \text{ MPa}, f_y = 250 \text{ MPa}$$

The design coefficients for a trapezoidal section are taken from Tables 7.5 and 7.12 and they are

$$x_u = 0.531d$$

$$K = 0.149,\ p_0 = 0.22$$

$$K_2 = 0.03,\ p_{02} = 0.066$$

Design of section. The collapse moment on the section is

$$M_c = \gamma_f \frac{wL^2}{2} = 1.5(6)\frac{(9)}{2} = 40.5 \text{ kNm}$$

The moment capacity of a trapezoidal section is

$$M_r = ((K - K_2)b_1 + K_2b_2)d^2f_{ck}$$

Let $b_1 = 0.15$ m, $b_2 = 0.30$ m

Then the moment capacity of the section is

$$M_r = ((0.149 - 0.03)(0.15) + 0.03(0.3))(15)a^2 = 0.40275\, d^2 \text{ MNm}$$

Equating the moment capacity to the collapse moment, we have

$$402750\, d^2 = 40500 \text{ Nm}$$

$$d = \sqrt{\frac{40500}{402\ 750}} = 0.317 \text{ m}$$

use $d = 0.32$ m and $h = 0.32 + 0.04 = 0.36$ m

Area of the tensile reinforcement is

$$A_{st} = ((p_0 - p_{02})b_1 + p_{02}b_2)\ d\frac{f_{ck}}{f_y}$$

$$= ((0.22 - 0.066)(150) + 0.066(300))(317)\left(\frac{15}{250}\right)$$

$$= 816 \text{ mm}^2$$

Area of the effective concrete $= \dfrac{(b_1 + b_2)d}{2}$

$$= A_c = \frac{0.45(0.32)}{2} = 0.072 \text{ m}^2$$

Self-weight $= A_c \gamma = (0.072)(1.1)(25) = 1.9 \text{ kN/m}$

(This is less than that was assumed.)

Minimum area of reinforcement is

$$A_{st} = \frac{0.15}{100}\ A_c = \frac{0.15(72000)}{100} = 108 \text{ mm}^2, \text{ or}$$

$$= \frac{0.85}{f_y}\ A_c = \frac{0.85(72000)}{250} = 245 \text{ mm}^2$$

Provide 4 numbers of 16 mm bars

$$A_{st} \text{ (provided)} = 4(201) = 804 \text{ mm}^2$$

$$\% \text{ of reinforcement} = \frac{100\ A_{st}}{A_c} = \frac{80400}{72000} = 1.13$$

Design for shear. The critical shear section is at a distance of d from the support, and the shear force is

$$V = \gamma_f w(L - d)$$

$$= 1.5(6)(3 - 0.32) = 24.12 \text{ kN} = 0.02412 \text{ MN}$$

The nominal shear stress is

$$\tau_v = \frac{V}{A_c} = \frac{0.02412}{0.072} = 0.335 \text{ MPa}$$

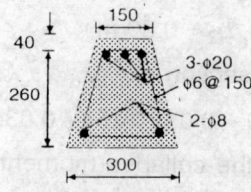

Fig. 7.17 Example 7.15.

The shear strength even for 0.36 per cent is more than 0.37 MPa. Therefore, provide only nominal transverse reinforcement. Select two-legged 6 mm bars, then the area of the shear reinforcement is

$$A_{sv} = 2(28) = 56 \text{ mm}^2$$

The maximum spacing of the stirrups is given by

$$S_v = \frac{A_{sv}f_y}{0.2(b_1 + b_2)} = \frac{56(250)}{0.2(450)} = 155 \text{ mm}$$

Provide two-legged 6 mm bars at 150 mm c/c.

Figure 7.17 illustrates the section

EXAMPLE 7.16 *Inverted trapezoidal section.* A cantilever beam carries a total 4 kN/m and its effective span is 3 m. Design the section.

$$\text{Cantilever span} = L = 3 \text{ m}$$
$$\text{Let the self-weight} = w_g = 2 \text{ kN/m}$$
$$\text{Superimposed load} = w_s = 4 \text{ kN/m}$$
$$\text{Total load} = w = 6 \text{ kN/m}$$
$$f_{ck} = 15 \text{ MPa}, \ f_y = 250 \text{ MPa}$$

The design coefficients taken from Tables 7.5 and 7.12 are

$$K = 0.149, \ p_0 = 0.22$$
$$K_2 = 0.03, \ p_{02} = 0.066$$

Design of the section. The collapse moment on the beam with a partial safety factor of 1.5 for loads is

$$M_c = 1.5 \ \frac{wL^2}{2} = 1.5\frac{(6)(9)}{2} = 40.5 \text{ kNm}$$

Let the width of the section at compression fibre

$$b_1 = 300 \text{ mm}$$

and that at the tension fibre be

$$b_2 = 150 \text{ mm}$$

The moment capacity of the trapezoidal section is

$$M_r = ((K - K_0)b_1 + K_2b_2)d^2 f_{ck}$$

$$= ((0.149 - 0.03)(0.3) + 0.03(0.15))(15)d^2 = 0.603d^2 \text{ MNm}$$

Equating the resisting capacity to the collapse moment, we have

$$0.603d^2 = 0.0405 \text{ MNm}$$

$$d = \sqrt{\frac{0.0405}{0.603}} = 0.259 \text{ m}$$

or

$$\text{use } d = 0.26 \text{ m and } h = d + 0.04 = 0.3 \text{ m}$$

The area of the tension reinforcement is

$$A_{st} = ((p_0 - p_{02})b_1 + p_{02}b_2)d\frac{f_{ck}}{f_y}$$

$$= ((0.22 - 0.066)(300) + 0.066(150))(260)\frac{(15)}{250} = 875 \text{ mm}^2$$

Area of the effective concrete is

$$A_c = \frac{(b_1 + b_2)d}{2} = \frac{0.45(0.26)}{2} = 0.0585 \text{ m}^2$$

The self-weight of the beam is

$$w_g = 0.0585(25) = 1.6 \text{ kN/m}$$

which is less than that was assumed as 2 kN/m.

Provide 3 numbers of 20 mm bars, then

$$A_{st} \text{ (provided)} = 3(314) = 942 \text{ mm}^2$$

The percentage of reinforcement is

$$p = \frac{100A_{st}}{A_c} = \frac{942\,00}{585\,00} = 1.6\%$$

Design for shear. The design shear force at the critical section is

$$V = \gamma_f w(L - d) = 1.5(6)(3 - 0.26) = 24.66 \text{ kN}$$

The nominal shear stress is

$$\tau_v = \frac{V}{A_c} = \frac{24\,660}{58\,500} = 0.42 \text{ MPa}$$

The shear strength in M15 concrete for 1.5% reinforcement is 0.68 MPa; therefore, only nominal transverse reinforcement is needed.

Select two-legged 6 mm wires, then

$$A_{sv} = 2(28) = 56 \text{ mm}^2$$

The maximum spacing of the stirrups

$$s_{vm} = \frac{A_{sv}f_y}{0.2(b_1 + b_2)} = \frac{56(250)}{0.2(450)} = 155 \text{ mm}$$

Provide two-legged 6 mm stirrups at 150 mm c/c.

Bond length. The minimum bond length is

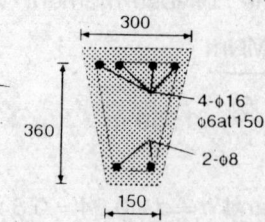

Fig. 7.18 Example 7.17.

$$L_d = \frac{\phi \sigma_s}{4 \tau_{bd}}$$

in which the tensile stress in steel was

$$\sigma_s = \frac{f_y}{1.15} = \frac{250}{1.15}$$

The bond strength of the plain bars for M15 concrete is

$$\tau_{bd} = 1 \text{ MPa}$$

The bond length is

$$L_d = \frac{20(250)}{1.15(4)} = 1087 \text{ mm}$$

The bars must have an anchor length of 1087 mm. A U-hook provides an anchorage length of $16\phi = 320$ mm. The total length beyond the book

$$L_{d0} = L_d - 320 = 1087 - 320 = 767 \text{ mm}$$

Approximately 40ϕ is to be provided with the book.

PROBLEMS
(Use Data as per IS456. LSD = Limit State Design)

7.1 Determine the neutral axes of cross-section for M25 concrete and HYSD-Fe415 bars of $f_y = 415$ M/mm². Use parabolic-cum-rectangular stress on concrete in compression as suggested by IS456, and also the limiting strains in the materials as per IS code.

7.2. Determine the balanced moment capacity of square cross-section with distance of the reinforcement at 0.9 times the size of the cross-section. The moment capacity should be expressed in f_{ck}, f_y and the size of the section.

7.3 Determine the moment capacity of a diamond shaped cross-section with its side a and angle 90°. The depth of reinforcement from the apex to 80% of the overall depth as shown in Fig. 1.

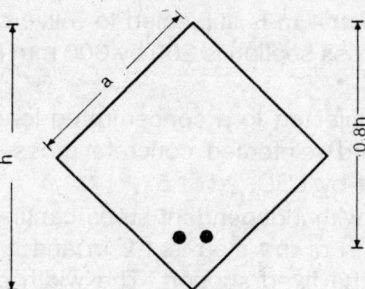

Fig. 1 Problem 7.3.

7.4 A square cross-section has a notch cut out in the compression flange as shown in Fig. 2. The size of the notch is 1/3rd of the size of the section. Derive the balanced moment capacity of the section for the following data: $f_{ck} = 25$ N/mm², $f_y = 415$ N/mm². Use stress-strain relation as per IS456.

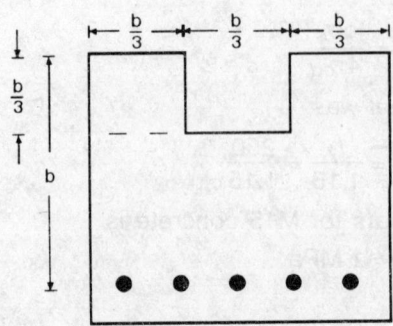

Fig. 2 Problem 7.4.

7.5 Derive the limit state moment capacity of a circular cross-section with reinforcement at a distance of 0.8 time the diameter from the top compression fibre. Calculate the percentage of reinforcement needed for a balanced cross-section. Assume $f_{ck} = 20$ N/mm^2, $f_y = 415$ N/mm^2.

7.6 A simply supported beam of span 6 m is subjected to a live load of 16 kN/m. Design a balanced cross-section reinforced concrete beam by limit state design with width equal to 60% of the effective depth. Sketch the reinforcement details, $f_{ck} = 20$ N/mm^2. $f_y = 415$ N/mm^2, $\gamma_y = 1.5$.

7.7 A simply supported beam of effective span 6 m is subjected to a live load of 26 kN/m. The size of the beam is 400 by 400 mm. Design by limit state design the reinforcement details using M20 concrete and HYSD-Fe415 bars both for tension and shear reinforcement. $\gamma_f = 1.5$.

7.8 A simply supported beam of effective 6 m is subjected to a live load of 26 kN/m, the size of the beam is 250 by 400 mm. Design the reinforcement details by LSD both for bending and shear using M20 concrete and HYSD-Fe415 bars. $\gamma_f = 1.5$.

7.9 A cantilever beam of span 4 m is subjected to a live load of 20 kN/m. Design a balanced reinforced concrete rectangular section using M20 concrete and HYSD-Fe415 bars by LSD. $\gamma_f = 1.5$.

7.10 A cantilever beam of span 4 m is subjected to a live of 20 kN/m. Design reinforcement detail if the size of the cross section is 200 by 300 mm using M20 concrete HYSD-Fe415 bars. $\gamma_f = 1.5$.

7.11 A cantilever beam is subjected to a concentrated load of 5 kN at the free end of 4 m span. Design a balanced reinforced concrete cross-section using M25 concrete and mild steel reinforcement by LSD. $\gamma_f = 1.5$.

7.12 A staircase is provided with independent steps cantilevering from an upright reinforced concrete wall. The length of the step is 1.2 m and it is designed for a load of 1.5 kN placed on 1.1 m from the fixed support. The width of the step is 300 mm. Design a balanced reinforced concrete cross-section using M20 concrete and HYSD-Fe415 bars by LSD. $\gamma_f = 1.5$.

7.13 A staircase slab spans between two walls separated by a distance of 4.8 m. The width of the staircase is 1.2 m and it has two landings, one at each and having a length of 1.2 m. Design the slab for a live load of 4 kN/m^2 using M20 concrete and HYSD-Fe415 bars. The rise and tread of the staircase are 150 and 300 mm respectively.

7.14 A reinforcement concrete cross-section of width 400 mm and effective depth 800 mm is provided with 4 numbers of 20 mm bars as tension reinforcement. The beam is simply supported with an effective span of 8 m and subjected to live UDL of 24 kN/m. Determine whether the section is capable of resisting the limit moment with $\gamma_f = 1.5$. The concrete is of M20 and steel is HYSD-Fe415 bars.

7.15 A reinforcement concrete cross-section of 200 mm width and 400 mm effective depth is provided with 4 numbers of 20 mm bars in tension zone and 3 numbers of 20 mm bars in the compression zone at depth 50 mm from the compression fibre. The cross-section is subjected to a bending moment of 150 kNm. Determine whether the section is capable of resisting limit moment with $\gamma_f = 1.5$. The concrete is of M20 and steel HYSD-Fe415 bars.

7.16 A doubly over-hang beam of total length 8 m is subjected to a live UDL of 16 kN/m. The over-hang span is 1.6 m on each side. Design a reinforced concrete beam with uniform depth throughout using M20 concrete and HYSD-Fe415 bars. The depth of the section should not exceed 400 mm. $\gamma_f = 1.5$.

7.17 An electrical transmission line pole of total 8 m length is embedded into ground to a depth of 1.5 m. A wind load 3 kN acting one the wires is transmitted at 0.5 m from top of the pole. Design a reinforced concrete square cross-section pole using M20 concrete and HYSD-Fe415 bars. Use $\gamma_f = 1.2$.

7.18 A boundary wall of 2 m effective height above the ground level is subjected to wind force of 1.5 kN/m^2. Design a reinforced concrete wall with foundation depth 1 m below ground level. Use M20 concrete and mild steel reinforcement bars. The footing of the wall need not be designed and the wall is assumed to be stable under the wind load condition. Use $\gamma_f = 1.2$. Assume thickness of foundation as 400 mm.

7.19 A boundary wall of 2.5 m high above ground level is subjected to 1.5 kN/m^2 wind load. The wall is designed with pilasters spaced at 2.5 m intervals. Design the boundary wall as well as the pilasters by limit state design in reinforced concrete using M20 concrete the and HYSD-Fe415 bars. Assume that the wall is stable under the wind load condition. Use $\gamma_f = 1.2$. Depth of wall below the ground level is 500 mm.

7.20 A percast reinforced M20 concrete pole of 8 m length and having square cross-section of 300 mm size is provided with 4 numbers of 200 mm HYSD-Fe415 bars one at each corner of the cross-section. The clear cover to the bar is 35 mm. The member is lifted from the ground holding it at its middle. Determine whether the section can withstand the load with a partial safety factor of 1.2 applied to the load.

7.21 Calculate the partial safety factor applied to the load if a rectangular cross-section beam for the following data is designed for a service moment of 100 kNm. $b = 300$ mm, $d = 750$ mm, $A_{st} = 604$ mm^2. The design is based on limit state of strength concrete M20 and HYSD-Fe415 steel.

7.22 A rectangular reinforced concrete section is subjected to an external service moment of 150 kNm. Determine the partial safety applied to loads if the section was designed by limit state of strength with $b = 400$ mm, $d = 800$ mm, $A_{st} = 2400$ mm^2.

7.23 Calculate the limit moment capacity of a rectangular cross-section reinforced beam for the following data : $b = 200$ mm, $d = 600$ mm, $A_{st} = 503$ mm^2, $f_y = 415$ N/mm^2, $f_{ck} = 20$ N/mm^2. State whether the section is over-reinforced or under-reinforced.

7.24 Determine the limit moment capacity of a rectangular cross-section reinforced concrete beam for the following data: $b = 200$, $d = 600$ mm, $A_{st} = 1206$ mm^2, $f_y = 415$ N/mm^2, $f_{ck} = 20$ N/mm^2. State whether the section is over-reinforced or under-reinforced.

7.25 The tension reinforcement in a singly-reinforced balanced rectangular cross-section is increased by 200% of the balanced steel. Determine the enhanced moment capacity of the cross-section as a function of balanced moment capacity assuming all other properties remaining the same.

7.26 The tension reinforcement in a balanced rectangular cross-section beam is reduced to two-thirds of the balanced reinforcement. Determine the revised moment capacity of the section as a function of balanced moment capacity assuming all other data remaining the same.

7.27 Determine the moment capacity of a doubly-reinforced rectangular cross-section beam for the following data: $b = 400$ mm, $d = 700$ mm, $A_{st} = 1206$ mm^2, $f_y = 415$ N/mm^2, $f_{ck} = 20$ N/mm^2, $A_{sc} = 603$ mm^2.

7.28. Determine the moment capacity of a doubly-reinforced rectangular cross-section beam for the following data: $b = 300$ mm, $d = 600$ mm, $A_{st} = 1407$ mm^2, $A_{sc} = 804$ mm^2, $f_y = 415$ N/mm^2, $f_{ck} = 20$ N/mm^2.

7.29 A simply supported reinforced concrete beam is subjected to a working bending moment of 700 kNm and a shear force of 300 kN. Design a T-beam cross-section to resist the above forcs using the follwing data: Maximum width of flange 1000 mm, and thickness of flange 150 mm, use M20 concrete and HYSD-Fe415 bars with $\gamma_f = 1.5$.

7.30 Derive an expression for limit moment capacity of reinforced concrete T-section in terms of width and thickness of the flange, effiective depth and characteristic strength of concrete and steel. Also calculate the percentage of tension steel required with respect to the total web area.

7.31 A reinforced concrete T-beam with the properties given below is subjected to a working moment of 500 kNm. Determine whether the section is capable of withstanding the moment by limit state of strength. $b_f = 900$ mm, $t = 120$ mm, $d = 600$ mm, $A_{st} = 4900$ mm^2, $\gamma_f = 1.5$, $f_{ck} = 20$ N/mm^2, $f_y = 415$ N/mm^2.

7.32 A reinforced concrete T-beam section with the details given below is subjected to an external working bending moment of 320 kNm. The characteristic strengths of concrete and reinforcement are 20 and 415 N/mm^2 respectively. Determine whether the cross-section is structurally safe in limit state of strength for the moment given above, $b_f = 1.6$ m, $t = 0.12$ m, $d = 0.51$ m, $A_{st} = 2918$ mm^2.

7.33 A doubly-reinforced T-beam section with the dimensions given below is subjected to a working bending moment of 300 kNm. Determine whether the section is capable of withstanding in the limit state of strength. $b_f = 1600$ mm, $t = 120$ mm, $b_w = 250$ mm, $d = 330$ mm, $A_{st} = 4600$ mm^2, $A_{sc} = 2938$ mm^2, $d_c = 60$ mm, $f_{ck} = 20$ N/mm^2, $f_y = 415$ N/mm^2.

7.34 Determine the limit moment capacity of a singly-reinforced concrete T-section for the following details; $b_f = 1000$ mm, $t = 120$ mm, $b_w = 250$ mm, $d = 500$ mm, $A_{st} = 1256$ mm^2, $f_{ck} = 20$ N/mm^2, $f_y = 415$ N/mm^2.

7.35 Determine the limit moment capacity of reinforced concrete T-section for the following details: $b_f = 1200$ mm, $t = 150$ mm, $b_w = 400$ mm, $d = 600$ mm, $A_{st} = 4900$ mm^2, $f_{ck} = 20$ N/mm^2, $f_y = 415$ N/mm^2.

7.36 Determine the limit moment capacity of a doubly-reinforced T-section for the following details and specifications: $b_f = 1000$ mm, $t = 150$ mm, $b_w = 400$ mm, $d = 500$ mm, $A_{st} = 4900$ mm^2, $A_{sc} = 2350$ mm^2, $f_{ck} = 20$ N/mm^2, $f_y = 415$ N/mm^2.

7.37 A simply supported beam of an effective span of 10 m is subjected to a uniformly distributed live load of 30 kN/m. The thickness and width of the flange are restricted to 120 and 1200 mm respectively. Design a T-beam section with M20 concrete and HYSD-Fe415 bars by limit state design.

7.38 A simply supported beam of an effective span 8 m is subjected to an uniformly distributed live load of 20 kN/m and of moving load of 100 kN. Design a singly-reinforced T-beam using M20 concrete and HYSD-Fe415 bars. The width and the thickness of the flange are restricted to 1000 and 200 mm respectively.

7.39 A simply supported beam of an effective span 8 m is subjected to uniformly distributed live load of 30 kN/m. The architect has specified the following dimensions for the beam: $b_f = 1000$ mm, $t = 120$ mm, $b_w = 300$ mm, $h = 600$ mm. Design a reinforcement details using M20 concrete and HYSD-Fe415 bars.

7.40 A simply supported beam of an effective span 12 m is subjected to uniformly distributed live load of 20 kN/m. Design a reinforced concrete T-section subjected to limitations: $b_f = 1500$ mm, $t = 160$ mm, $h = 800$ mm, $b_w = 400$ mm, $f_{ck} = 20$ N/mm^2, $f_y = 415$ N/mm^2.

Chapter 8

Serviceability Limit State Design of Beams

8.1 Introduction

Minimum or maximum limits acceptable for satisfactory functioning of a structure are called *limits in serviceability* and a design satisfying such bounds is called *serviceability limit state design*. Normally the factors considered in the serviceability limit states are:

1. Deflection,
2. Cracking which includes crack width and control of crack propagation,
3. Stresses or strains.

In the limit state of durability, one can consider the following:

1. Exposure condition (interior or exterior of a building).
2. Degree of exposure—chemical nature of the environment such as a saline, sulphate or sulphide gas, etc.
3. Wear exposure—surface exposed to movable objects, or traffic, or the nature of traffic.
4. Heat exposure—extreme changes in the temperature including heat generated by fire and accidents.
5. Repeated loads and fatigue.

There are design criteria for serviceability while the durability is usually controlled by specifications.

8.2 Deflection Limit State Design

Deflections in concrete structures are usually small in comparison with those in similar metal structures. This is primarily because the moment of inertia of cross-section of the concrete sections is many times more than that of an equivalent strong metal section. This can also be stated as that the sizes of sections of concrete structures are much larger than those of steel structures designed to resist the same loads. In most cases of concrete structures, the deflections do not govern the design. However, one should ensure that the deflections be within the limits. Setting limits for deflections is due to the following reasons:

1. Large deflections induce large strains even without causing excessive stresses. Such large deflections may be due to loads, creep, shrinkage, etc.
2. Large relative deflections introduce psychological imbalances specially during moving load, and gust wind load conditions.
3. Some deflections may not be too large either for the structural system or for the occupants of the structure. But they may cause damage to the other fixtures

such as ceilings, partition walls or to the service lines such as airconditioning ducts and water mains, etc.

It is true that the deflections cannot be eliminated but one must see that the deflections caused by the external agencies or by the inherent properties of the structure are minimised. The design criterion may be stated as the deflection of the structure or elements must be less than the allowable values:

$$v_t \leq v_a \qquad\qquad (8.1)$$

in which v_t = total deflection

v_a = permissible deflection

The permissible deflections vary with several factors and the nature of the structure. Table 8.1 gives the limiting values of the deflections.

Table 8.1 Maximum Permissible Deflections

Type of structure	$\dfrac{v_a}{\text{span}}$
1. Beams, slabs, including roofs in ordinary buildings measured from the level of casting, including creep and shrinkage	$\dfrac{1}{250}$
2. Beams and slabs of building measured from the level after erecting partition and finishing	$\dfrac{1}{350}$
3. Hand-operated crane girders	$\dfrac{1}{500}$
4. EOT cranes	$\dfrac{1}{1000}$
5. Lateral drift of tall buildings with respect to height	$\dfrac{1}{400}$

The deflections are caused by:
1. External loads such as self-weight, superimposed loads, live loads, wind loads, etc.
2. Shrinkage of concrete.
3. Creep.
4. Temperature variations.
5. Unequal settlements.

In addition they depend on the following factors associated with the structure.

1. Inversely proportional to the modulus of elasticity of the material.
2. Inversely proportional to the effective moment of inertia of the sections.
3. Proportional to the differential shrinkage.
4. Proportional to the creep coefficient.
5. Increases with temperature.
6. Proportional to the square or cube of the effective span.
   The deflection caused by loads in beams can normally be expressed as:

$$v_f = \frac{c_v \, WL^3}{E_c \, l_e} \tag{8.2}$$

where, W = total load on the beam

L = effective span

E_c = short-term modulus of elasticity of the material

l_e = effective moment of the section

v_f = short-term deflection

c_v = coefficient which depends on the boundary conditions of the beam and load distributions. These values for typical cases are given in Table 8.2

Table 8.2 Maximum Deflection Coefficient Beams

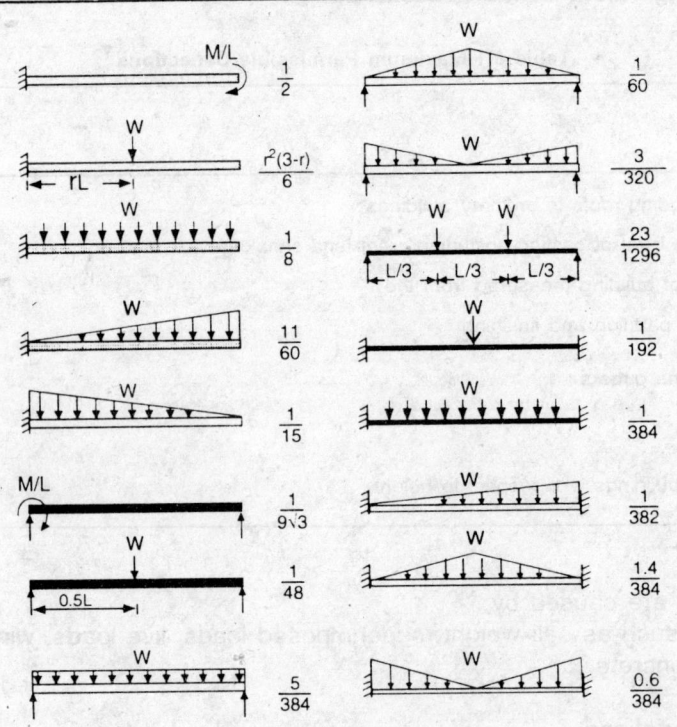

$$E_c = 5700 \sqrt{f_{ck}} \text{ (in N/mm}^2) \tag{8.3}$$

$$l_e = \frac{l_r}{1.2 - \frac{M_{cr}(j)}{M}\left(1 - \frac{x_u}{d}\right)\frac{b_w}{b_f}}$$

and $l_r \le l_e \le l_g$

I_r = moment of inertia of the cracked section
M_{cr} = cracking moment of the section
M = maximum moment under service load
j = ratio of the lever arm to d
d = effective depth of steel
b_w = width of web
b_f = width of compression flange

The approximate moment of inertia of cracked sections about NA and the uncracked RCC sections for rectangular beams are:

$$I_{cr} = \frac{bx_u^2}{3} + m A_{st}(d - x_u)^2 \text{ and} \tag{8.5}$$

$$I_g = \frac{bh^3}{12} \tag{8.6}$$

(The moment of inertia is taken about the NA of cracked section, usually about CG.)

The moment at cracking of the section is given by:

$$M_{cr} = \frac{I_g f_r}{y_t} \tag{8.7}$$

where, f_r = modulus of rupture of concrete = $0.7\sqrt{f_{ck}}$ in N/mm² $\tag{8.8}$

In case of continuous beam the moment of inertia at any section is computed by:

$$X_e = k_1\left(\frac{X_1 + X_2}{2}\right) + (1 + k_1) X_0 \tag{8.9}$$

in which X_e = modified moment of inertia of I_r or I_g or cracking moment M_r
X = appropriate value of I_r, I_g or M_r
X_1, X_2 = values of X at supports
X_0 = value of X at mid-span
k_1 = coefficient given in Table 8.3

The expressions are complicated to incorporate the variable moment of inertia along the length of the beam.

Deflection due to shrinkage can be computed as

$$v_{cs} = \frac{C_{cs} L^2}{R_{cs}} \tag{8.10}$$

Table 8.3 Values of k_1 (continuous beam)

$\dfrac{M_1 + M_2}{M_{f1} + M_{f2}}$ = upto	0.5	0.6	0.7	0.8	0.9	1.0	1.1	1.3
k_1	0	0.3	0.06	0.16	0.3	0.5	0.73	0.97

M_1, M_2 = moments at supports
M_{f1}, M_{f2} = fixed end moments at supports

where v_{cs} = deflection due to shrinkage

c_{cs} = coefficient depending on the support conditions and it is

= 0.5 for cantilever beams

= 0.125 for simply supported beams

= 0.086 for beams continuous at one end

= 0.063 for beams continuous at both ends

R_{cs} = shrinkage radius of curvature of the beam

$$\frac{1}{R_{cs}} = \frac{c_0 \epsilon_{cs}}{h}$$ (8.11)

where $c_0 = \dfrac{0.72(p_t - p_c)}{\sqrt{p_t}} \le 1$, for $0.25 \le p_t - p_c \le 1$

$= \dfrac{0.65(p_t - p_c)}{\sqrt{p_t}} \le 1$, for $p_t - p_c > 1$

$p_t = \dfrac{100\, A_{st}}{bd}$

$p_c = \dfrac{100\, A_{sc}}{bd}$

ϵ_{cs} = ultimate shrinkage strain

= 0.0003 (usually taken as)

Total deflection including creep effect can be calculated as

$$v_{tc} = c_c v_f$$ (8.12)

in which c_c = creep coefficient

= 2.2 for loading after 7 days curing

= 1.6 for loading after 28 days curing

= 1.1 for loading after 1 year

The serviceability limit state design is illustrated through examples.

8.3 Design Examples of Deflection in RCC Beams

EXAMPLE 8.1 A simply supported beam of effective span 6 m is subjected to the following loads and specifications:

$$w_s = 9.4 \text{ kN/m}, \ w_l = 15 \text{ kN/m}, \ f_{ck} = 20 \text{ MPa}, \ f_y = 415 \text{ MPa}.$$

Vide Example 7.1 for limit state design of collapse of the beam. The final design details based on the limit state of collapse are:

$$b = 250 \text{ mm}, \ h = 560 \text{ mm}, \ d = 520 \text{ mm}, \ A_{st} = 1257 \text{ mm}^2.$$

Analysis for deflection

Self-weight = w_g = 0.25(0.56)(25) = 3.5 kN/m

Total service load

$w_t = w_s + w_g + w_l = (9.4 + 3.5 + 15) = 27.90$ kN/m

Since the load is uniformly distributed, the maximum deflection on the simply supported beam occurs at mid-span and is given by

$$v_f = \frac{c_v\,WL^3}{E_c I_c}$$

in which

$$c_v = \frac{5}{384} \text{ for UDL on simply supported beam}$$

$$E_c = 5700\,\sqrt{f_{ck}} = 5700\,\sqrt{20} = 25491 \text{ N/mm}^2$$

Load factor at service load = 1

$$W = w_t L = 27.90(6) = 166.56 \text{ kN}$$

$$M = \frac{w_t L^2}{8} = 27.90\left(\frac{36}{8}\right) = 124.92 \text{ kNm}$$

The design fracture stress is

$$f_{cr} = \frac{0.7\sqrt{f_{ck}}}{1.5} = \frac{3.13}{1.5} = 2.09 \text{ N/mm}^2$$

$$I_e = \frac{I_{cr}}{1.2 - \dfrac{M_{cr}}{M}\,(j)\left(1 - \dfrac{x_u}{d}\right)\dfrac{b_w}{b_f}}$$

$$I_g = \frac{bh^3}{12} = \frac{0.25(0.56)^3}{12} = 3658.7(10^{-6}) \text{ m}^4$$

$$x_u = 0.479d = 0.479(0.52) = 0.2491 \text{ m}$$

The first approximation moment of Inertia is

$$I_{cr} = \frac{bx_u^3}{3} = \frac{0.25(0.2491)^3}{3} = 1288(10^{-6}) \text{m}^4$$

$$M_{cr} = \frac{I_g\,f_{cr}}{y_t} = \frac{3658.7(10^{-6})(2.09)(10^{-6})}{0.26} = 29410.319 \text{ Nm}$$

$$i = 1 - \frac{x_c}{d} = 1 - 0.42\,\frac{x_u}{d} = 0.799$$

$$1 - \frac{x_u}{d} = 1 - 0.479 = 0.521$$

$$\frac{b_w}{b_f} = \frac{0.25}{0.25} = 1$$

$$X = 1.2 - \frac{M_{cr}}{M}\,(j)\left(1 - \frac{x_u}{d}\right)\frac{b_w}{b_f}$$

$$= 1.2 - \frac{29363}{124920}(0.799)(0.521)(1) = 1.1$$

$$I_e = \frac{I_{cr}}{X} = 1.1688 \times 10^{-3} \quad \text{(which is less than } I_{cr})$$

However, I_e must be not less than I_{cr}, so use

$$I_e = I_{cr}$$

The deflection due to the external load is

$$v_f = \frac{c_v\, WL^3}{E_c l_e}$$

$$= \frac{5(166569)(6^3)}{384(25490(10^6)(1288)(10^{-6})} = 0.0142 \text{ m}$$

Allowable deflection $= \dfrac{L}{350} = \dfrac{6}{350} = 0.0171$ m

The actual deflection is less than the allowable value even with small MI. Therefore, the design is acceptable.

EXAMPLE 8.2 *Deflection caused by shrinkage.* Compute the deflection of a beam caused by the shrinkage for the following data of a simply supported beam.

$$\text{Span} = L = 10 \text{ m}$$
$$b = 0.3 \text{ m},\ d = 0.65 \text{ m},\ h = 0.7 \text{ m}$$
$$f_{ck} = 20 \text{ MPa},\ f_y = 415 \text{ MPa}$$
$$\text{Total UDL} = 20 \text{ kN/m}$$
$$\frac{100A_{st}}{bd} = 1.20,\ \frac{100A_{sc}}{bd} = 0.18$$

Shrinkage strain $\qquad\qquad \epsilon_{cs} = 0.00025$

The deflection due to shrinkage is given by

$$v_c = \frac{c_{cs}\, L^2}{R_{cs}}$$

where $c_{cs} = 0.125$ for simply supported beams

$$\frac{1}{R_{cs}} = \text{ curvature due to shrinkage}$$

$$= \frac{c_0\,\epsilon_{cs}}{h}$$

and $\qquad\qquad c_0 = 0.65\, \dfrac{(p_t - p_c)}{\sqrt{p_t}},\ \text{for } p_t - p_c > 1$

$$p_t = \frac{100A_{st}}{bd} = 1.2$$

$$p_c = \frac{100A_{sc}}{bd} = 0.180$$

$$\sqrt{p_t} = 1.095$$

$$c_0 = \frac{0.65(1.2 - 0.18)}{1.095} = 0.605$$

$$\frac{1}{R_{cs}} = \frac{0.605(0.00025)}{0.7} = 2.16(10^{-4})$$

$$v_{cs} = \frac{c_{cs}\, L^2}{R_{cs}} = 0.125(100)(2.16)(10^{-4}) = 0.0027 \text{ m}$$

EXAMPLE 8.3. *Shrinkage deflection.* A cantilever beam of span 5 m has the data given below. Compute the deflection caused by the maximum shrinkage.
Data

$$\text{Span} = L = 5 \text{ m}, \; b = 0.2 \text{ m}, \; d = 0.35 \text{ m}, \; h = 0.4 \text{ m}$$

$$A_{st} = 4 \text{ nos. of 20 mm bars}$$

$$A_{sc} = 2 \text{ nos. of 10 mm bars}$$

$$f_{ck} = 20 \text{ MPa}, \; f_y = 415 \text{ MPa}$$

Maximum shrinkage $= \epsilon_{cs} = 0.0003$

Shrinkage deflection is given by

$$v_{cs} = \frac{c_{cs} L^2}{R_{cs}}$$

where $c_{cs} = 0.5$ for cantilever beam

and $R_{cs} = \dfrac{h}{c_0 \epsilon_{cs}}$

where $c_0 = \dfrac{c_1 (p_t - p_c)}{\sqrt{p_t}}$

$$p_t = \frac{100 \, A_{st}}{bd} = \frac{100(4)(0.000314)}{(0.2)(0.35)} = 1.794$$

$$p_c = \frac{100 \, A_{sc}}{bd} = \frac{100(2)(157)(10^{-6})}{0.2(0.35)} = 0.449$$

$$\frac{p_t - p_c}{\sqrt{p_t}} = \frac{1.794 - 0.009}{\sqrt{1.794}} = 1.005$$

For $p_t - p_c > 1$, the value of c_1 is $c_1 = 0.65$

Therefore $c_0 = \dfrac{c_1 (p_t - p_c)}{\sqrt{p_t}} = 0.65(1.005) = 0.653$

$$R_{cs} = \frac{h}{c_0 \epsilon_{cs}} = \frac{0.4}{0.653(0.0003)} = 2042 \text{ m}$$

$$v_{cs} = \frac{c_{cs} L^2}{R_{cs}} = \frac{0.5(25)}{2042} = 0.00612 \text{ m}$$

EXAMPLE 8.4 *Creep deflection.* A cantilever beam of span 5 m is subjected to loads which cause a deflection of 0.02 m at the free end. Compute the maximum deflection caused by the creep and shrinkage together. The data of the beam is

$$L = 5 \text{ m}, \; b = 0.2 \text{ m}, \; d = 0.35 \text{ m}, \; h = 0.4 \text{ m}$$

$$p_t = 1.794, \; p_c = 0.448$$

Vide Example 8.3 for the data on shrinkage deflection

$$v_{cs} = 0.00612 \text{ m}$$

The creep coefficient (c_c) is taken as 1.6 assuming the beam is loaded after 28 days of casting. The total deflection due to creep alone is

$$v_{cc} = (c_c - 1) \, v_f = (1.6 - 1) \, (0.02) = 0.012 \text{ m}$$

where $v_f = 0.02$ m

The total deflection due to shrinkage and creep is

$$v_{cs} + v_{cc} = 0.00612 + 0.012 = 0.01812 \text{ m}$$

The total deflection including all the effects and load is

$$v_t = v_f + v_{cs} + v_{cc} = 0.02 + 0.01812 = 0.03812 \text{ m}$$

$$\frac{v_t}{L} = \frac{0.03812}{6} = \frac{1}{157}$$

$$v_t = \frac{L}{157} > \frac{L}{250}$$

This is rather a large deflection and not acceptable design.

EXAMPLE 8.5 *Deflection caused by shrinkage in T-beam.* Compute the shrinkage deflection in a T-beam with the following data:

Span = $L = 10$ m
Flange width = $b_f = 1.5$ m
Flange thickness = $t = 0.12$ m
$b_w = 0.3$ m, $d = 0.65$ m, $h = 0.7$ m
$f_{ck} = 20$ MPa, $f_y = 415$ MPa.

Maximum shrinkage = $\epsilon_{cs} = 0.0003$

$$A_{st} = 2513(10^{-6}) \text{ m}^2$$

$$A_{sc} = 618(10^{-6}) \text{ m}^2$$

Beam is simply supported.

Shrinkage deflection is given by

$$v_{cs} = \frac{c_{cs} L^2}{R_{cs}}$$

The value of c_{cs} for simply supported span is

$$c_{cs} = 0.125$$

and $R_{cs} = \dfrac{h}{c_0 \epsilon_{cs}}$

where, $c_0 = \dfrac{c_1(p_t - p_c)}{\sqrt{p_t}}$

$$p_t = \frac{100 A_{st}}{bd} = \frac{2513(10^{-4})}{(0.3)(0.65)} = 1.289$$

$$p_c = \frac{100 A_{sc}}{bd} = \frac{618(10^{-4})}{(0.3)(0.65)} = 0.317$$

$$\frac{p_t - p_c}{\sqrt{p_t}} = \frac{1.289 - 0.317}{\sqrt{1.289}} = 0.856$$

Since $(p_t - p_c) < 1$, $c_1 = 0.72$, then

$$c_0 = 0.72(0.856) = 0.616$$

$$R_{cs} = \frac{h}{c_0 \in_{cs}} = \frac{0.7}{0.616(0.0003)} = 3788 \text{ m}$$

$$v_{cs} = \frac{c_{cs}L^2}{R_{cs}} = \frac{0.125(100)}{3788} = 0.0033 \text{ m}$$

$$= 0.0033\left(\frac{L^2}{L}\right) = \frac{L}{3030}$$

EXAMPLE 8.6 *Total deflection in a beam.* A simply supported beam of effective span 6 m is subjected to the following loads and specifications

$$w_s = 9.4 \text{ kN/m}, \qquad w_l = 15 \text{ kN/m},$$

$$f_{ck} = 20 \text{ MPa}, \qquad f_y = 415 \text{ MPa}.$$

The beam is loaded after 28 days of casting. Compute the total deflection including shrinkage and creep.

Refer Examples 7.1 and 8.1 for the design of the beam based on the collapse limit state and for the computation of deflection caused by the loads. The following values are taken from these two examples:

Design details from Example 7.1 are:

$$b = 0.25 \text{ m}, \qquad d = 0.52 \text{ m},$$

$$h = 0.56 \text{ m}, \qquad A_{st} = 0.001257 \text{ m}^2.$$

$$A_{sc} = 0.000157 \text{ m}^2,$$

Deflection computations from Example 8.1 are:

$$E_c = 25490 \text{ MPa}$$

$$M = 124920 \quad \text{Nm (load factor = 1)}$$

$$f_{cr} = 3.13 \text{ MPa}, \ x_u = 0.2491 \text{ m}$$

$$I_e = 1240 \ (10^{-6}) \text{ m}^4$$

$$v_f = 0.0142 \text{ m}$$

Deflection due to shrinkage. The following data are taken from the code:

Total shrinkage = 0.0003

Shrinkage deflection coefficient for a simply supported beam is

$$c_{cs} = 0.125$$

The deflection due to shrinkage is

$$v_{cs} = \frac{c_{cs}L^2}{R_{cs}}$$

where $R_{cs} = \dfrac{h}{c_0 \in_{cs}}$

$$\text{and } c_0 = \frac{c_t\,(p_t - p_c)}{\sqrt{p_t}} \leq 1$$

$$c_1 = 0.72 \text{ for } 25 \leq (p_t - p_c) \leq 1$$

$$= 0.65 \text{ for } p_t - p_c > 1$$

$$bd = 0.25\,(0.52) = 0.13 \qquad \text{m}^2$$

$$p_t = \frac{100A_{st}}{bd} = \frac{0.001257(100)}{0.13} = 0.967$$

$$p_c = \frac{100A_{sc}}{bd} = \frac{0.000157\,(100)}{0.13} = 0.121$$

$$p_t - p_c = 0.967 - 0.121 = 0.846$$

$$c_0 = \frac{c_1\,(p_t - p_c)}{\sqrt{p_t}} = \frac{0.72(0.846)}{0.9834} = 0.619$$

$$\frac{1}{R_{cs}} = \frac{c_0 \in_{cs}}{h} = \frac{0.619(0.0003)}{0.56} = \frac{1}{3016}$$

The deflection due to shrinkage

$$v_{cs} = \frac{c_{cs}L^2}{R_{cs}} = \frac{0.125\,(36)}{3016} = 0.00149 \qquad \text{m}$$

Deflection including the creep effect

$$v_{tc} = c_c v_f$$

The creep coefficient in case of loading after one year of costing is

$$c_c = 1.1$$

$$v_{tc} = 1.1\,(0.0142) = 0.0156$$

The total deflection including creep and shrinkage is

$$v_t = v_{tc} + v_{cs} = 0.0156 + 0.00149 = 0.01709 \text{ m}$$

Allowable total deflection in beams

$$\frac{L}{250} = \frac{6}{250} = 0.024 \text{ m}$$

The total deflection is less than the allowable deflection.

EXAMPLE 8.7 *Design of doubly reinforced beam.* A simply supported beam of span 6 m is subjected to the following loads and *specifications.*

$$L = 6 \text{ m}, \ w_s = 9.4 \text{ kN/m}, \ w_l = 15 \text{ kN/m}$$

$$f_{ck} = 20 \text{ MPa}, \ f_y = 415 \text{ MPa}$$

$$b = 0.2 \text{ m}, \ h = 0.4 \text{ m}$$

Design the beam by limit state design subject to a deflection limit of $L/250$.

Design for strength limit

$$w_g = 0.2\,(0.4)\,(25) = 2.0 \text{ kN/m}$$

Collapse moment M_c with 1.5 as load factor is

$$M_c = \gamma_f (w_g + w_g + w_l) \frac{L^2}{8}$$

$$= 1.5 (2.0 + 9.4 + 15) \left(\frac{36}{8}\right) = 178,335 \text{ kNm}$$

$$d = h - 0.035 = 0.4 - 0.035 = 0.365 \text{ m}$$

Balanced moment capacity

$$M_{rb} = Kbd^2 f_{ck}$$

$$= 0.138(0.2)(0.365)^2(20)(10^6) = 73007.3 \text{ Nm}$$

The collapse moment is larger than the balanced resistance of the section, so the beam be designed as a doubly-reinforced section. The moment to be resisted by the compression reinforcement is

$$M_{rc} = M_c - M_{rb} = 108\ 328 \text{ Nm}$$

Let $d_c = 35$ mm, then

$$\epsilon_{sc} = \epsilon_{cu}\left(1 - \frac{d_c}{x_u}\right)$$

$$= 0.0035 \left(1 - \frac{0.035}{0.479\ d}\right) = 0.0028$$

$$f_{sc} = \epsilon_{sc}E_s = 0.0028\ (200000) = 560 \text{ N/mm}^2, \text{ limit } f_y = 415$$

$$M_{rc} = 0.87\ A_{sc}\ f_y\ (d - d_c)$$

$$A_{sc} = \frac{108\ 328}{0.87(415)(10^6)(0.33)} = 0.000909 \text{ m}^2$$

$$A_{st} = A_{stb} + A_{st2} = p_0\ bd\ \frac{f_{ck}}{f_y} + A_{sc}\ \frac{f_{sc}}{f_y}$$

$$= 0.198\ (0.2)\ (0.365) \left(\frac{20}{415}\right) + 0.000909 = 0.001606 \text{ m}^2$$

Provide 4, ϕ20 and 2, ϕ16 at bottom; and 3, ϕ20 at top

So $A_{st} = 1658 \text{ mm}^2$, $A_{sc} = 942 \text{ mm}^2$

Design for deflection limit

$$E_c = 5700 \sqrt{f_{ck}} = 25491.1 \text{ MPa}$$

Load factor for service limit state = 1

$$M = \gamma_f (w_g + w_s + w_l) \frac{L^2}{8}$$

$$= 1(2.0 + 9.4 + 15) \frac{36}{8} = 118.44 \text{ kNm}$$

$$f_{cr} = \frac{0.7}{1.5} \sqrt{f_{ck}} = \frac{0.7}{1.5} \sqrt{20} = 2.1 \text{ MPa}$$

$$x_u = 0.479d = 0.479\ (0.365) = 0.1748 \text{ m}$$

Deflection due to loads is

$$v_f = c_v \frac{WL^3}{E_c I_c}$$

in which

$$c_v = \frac{5}{384} \text{ for simply supported beam with UDL}$$

$$W = w_t L = (2.0 + 9.4 + 15)(16) = 158.52 \text{ kN}$$

$$I_e = \frac{I_{cr}}{1.2 - \frac{M_{cr}}{M}(j)\left(1 - \frac{x_u}{d}\right)\frac{b_w}{b_f}}$$

$$I_{cr} = \frac{b(x_u^3)}{3} + mA_{st}(d - x_{cs})^2 = \frac{0.2(0.1748)^3}{3} + 13(0.0016)(0.19)^2$$

$$= 1106.1 \ (10^{-6}) \ m^4$$

$$M_{cr} = \frac{I_g f_{cr}}{y_t} = \frac{(0.2) \ (0.4)^3 (2.10)(10^6)}{6} = 4480 \text{ Nm}$$

$$j = 1 - \frac{x_c}{d} = 1 - 0.42 \frac{x_u}{d} = 0.8 \text{ Nm}$$

$$X = 1.2 - \frac{M_{cr}}{M}(j)\left(1 - \frac{x_u}{d}\right)\frac{b_w}{b_f}$$

$$= 1.2 - \frac{4480}{118890} (0.8) \ (1 - 0.479) \ (1) > 1$$

Use $I_e = I_{cr}$ as I_e must not be less than I_{cr}.

$$v_f = \frac{c_v WL^3}{E_c I_c}$$

$$= \frac{5}{384} \frac{(158520 \ (6^2)}{25490 \ (10^6) \ 1106 \ (10^{-6})} = 0.016 \text{m}$$

Let the beam be loaded after a year, the creep coefficient is $c_s = 1.1$, and the total deflection due to the load and creep is

$$v_{ct} = c_c \ v_f = 1.1 \ (0.016) = 0.018 \text{ m}$$

Deflection due to shrinkage

$$v_{cs} = \frac{c_{cs} L^2}{R_{cs}}$$

in which $\qquad c_{cs} = 0.125$

For simply supported beams:

$$p_t = \frac{100 A_{st}}{bd} = \frac{165800}{200(365)} = 2.27$$

$$p_c = \frac{100 A_{sc}}{bd} = \frac{94200}{200(365)} = 1.29$$

since $p_t - p_c$ is less than 1, so the value of

$$c_1 = 0.72 \quad \text{and}$$

$$c_0 = \frac{c_1(p_t - p_c)}{\sqrt{p_c}} = \frac{0.72(0.98)}{1.507} = 0.47$$

$$R_{cs} = \frac{h}{c_0 \epsilon_{cs}} = \frac{0.4}{0.47(0.0003)} = 2837 \text{ m}$$

The deflection caused by shrinkage is

$$v_{cs} = \frac{c_{cs} L^2}{R_{cs}} = \frac{0.125(36)}{2837} = 1.58 \,(10^{-3}) \text{ m}$$

The total deflection is

$$v_t = v_{cs} + c_c v_f = \frac{(1.58 + 18)\,10^{-3}}{} = 0.0196$$

The allowable deflection is

$$v_a = \frac{L}{250} = \frac{6}{250} = 0.024 \text{ m}$$

The actual deflection is less than the allowable deflection.

Some comments on deflection computations: The magnitude of computations for deflections are far more than that of the design for collapse limit state. The determination of the effective moment of inertia takes undue time and it becomes rather unnecessary, especially when the result comes out to be little consequence to the problems. The deflection caused by the shrinkage was found to be small when compared with that caused by the loads or creep. Unfortunately, the computations of deflection due to shrinkage are rather lengthy. In general, one need not go through the calculations if the deflection is less than 90% of the allowed. Indian code of practice suggests a set of span to depth ratios and if one uses them, the check for deflection need not be carried. These limits are given in Table 8.4.

The ratios given in Table 8.4 are further subject to a modification multiplying factor given in the code depending on the percentage of tension and compression reinforcements. The range of percentage tension reinforcement in most beams is 0.8 to 2 for which the multiplication factors are:

> 1 to 0.8 for deformed bars;
>
> 1.5 to 1.0 far plain bars.

Table 8.4 Span to Depth Ratios (L < 10 m)

Type of structure	L/h
1. Cantilever beams	7
2. Simply supported	20
3. Continuous beams	26

Note: For beams of spans more than 10 m, the values of the ratios can be multiplied by 10/span to find the acceptable ratios.

320

Further, reduction factors are recommended for flanged beams in the range of 0.8 to 1.0 depending on the web width to flange width. Overall, if one looks at the constraints on deflection one is likely to feel that the suggestions made in the introduction of this chapter are not completely true. In the opinion and experience of the author, the design basis of the code for deflections is conservative. The moment of inertia taken for deflection computations is close to the moment of inertia of the cracked sections. This itself is low thus leading to high deflections. The cracks are generated at certain intervals and the moment of inertia of the section at many points on the beam is far higher. Therefore, the beam deflection is likely to be smaller than that computed. In fact, many experimental results indicated that an average moment of inertia is equal to

$$I_e = (I_{cr} + I_g)/2 \tag{8.13}$$

where $I_{cr}^{\prime}$ includes the effect of the reinforcement, has given satisfactory results.

In addition to the deflection limits, the slenderness of the beam should also be restricted if the top flange is not supported laterally. The clear distance between the two lateral supports should not exceed

$$\text{either } L_e < 60b$$

$$\text{or } < 250\,\frac{b^2}{d} \tag{8.14}$$

whichever is less.

In case of cantilever beams, the lateral distance between the free end and the lateral support should not exceed

$$\text{either } L_e \leq 25b$$

$$\text{or } \leq 100\frac{b^2}{d}$$

Effect of the tension reinforcement on the effective moment of inertia. The steel in the tension zone (cracked zone) could be accounted in the effective moment of inertia and in such a case the moment of inertia of the cracked section is

$$I_{cr} = \frac{bx_u^3}{3} + (mA_{st})(d - x_u)^2$$

Let $A_{st} = pbd$, then the above equation reduces to

$$I_{cr} = \frac{bx_u^3}{3} + mp\,bd^3\left(1 - \frac{x_u}{d}\right)^2$$

$$= \frac{bd^3}{3}\left(\left(\frac{x_u}{d}\right)^3 + 3mp\left(1 - \frac{x_u}{d}\right)^2\right) \tag{8.15}$$

By neglecting the effect of the reinforcement, one is likely to overestimate the deflection by about 15 to 150%. Some typical examples are given below.

EXAMPLE 8.8. A simply supported beam of effective span 6 m is subjected to the following loads and specifications.

$$w_s = 9.4 \text{ kN/m}, \ w_1 = 15 \text{ kN/m}$$

$$f_{ck} = 20 \text{ MPa}, \ f_y = 415 \text{ MPa}$$

The beam section was designed for limit state of collapse and the details are:

$$b = 250 \text{ mm}, \quad h = 560 \text{ mm},$$

$$d = 520 \text{ mm}, \quad A_{st} = 1257 \text{ mm}^2.$$

Check for deflection.

(Vide Example 8.1 in which effect of the steel was neglected.)

Self-weight $w_g = 0.25 (0.56)(25) = 3.5$ kN/m

Total load including the self-weight is

$$W = (w_g + w_s + w_l)L = (3.5 + 9.4 + 15)(6) = 166.56 \text{ kN}$$

$$M = \frac{WL}{8} = 166.56 \left(\frac{6}{8}\right) = 124.92 \text{ kNm}$$

$$I_e = \frac{I_{cr}}{1.2 - \dfrac{M_{cr}}{M}(j)\left(1 - \dfrac{x_u}{d}\right)\dfrac{b_w}{b_f}}$$

$$I_{cr} = \frac{bx_u^3}{3} + A_{st} m (d - x_u)^2$$

$$x_u = 0.479, \quad d = 0.479(0.52) = 0.2491 \text{ m}$$

$$m = \frac{280}{f_{ck}} = \frac{280}{20} = 14$$

$$A_{st} = 1257 (10^{-6}) \text{ m}^2$$

$$I_{cr} = \frac{0.25(0.2491)^3}{3} + 1257(10^6)(14)(0.52 - 0.24912)^2$$

$$= 1288 (10^{-6}) + 1291 (10^{-6}) = 2579 (10^{-6}) \text{m}^4$$

The design fracture stress is

$$f_{cr} = \frac{0.7}{1.5} \sqrt{f_{ck}} = 2.09 \text{ N/mm}^2$$

$$I_g = \frac{bh^3}{12} = \frac{0.25(0.56)^3}{12} = 3658.7 (10^{-6}) \text{ m}^4$$

$$M_{cr} = \frac{I_g f_{cr}}{y_t} = \frac{3658.7(10^{-6})(2.09)(10^6)}{0.28} = 29410 \text{ Nm}$$

$$X = 1.2 - \frac{M_{cr}}{M}(j)\left(1 - \frac{x_u}{d}\right)\frac{b_w}{b_f}$$

$$= 1.2 - \frac{29410}{124920}(0.799)(0.521)(1) = 1.1$$

$$I_e = \frac{I_{cr}}{X} \text{ subject to } I_{cr} \le I_e \le I_g$$

Therefore use $\qquad I_e = I_{cr} = 2579(10^{-6}) \text{ m}^4$

It may be seen that the effect of the reinforcement has doubled the moment of inertia of the cracked section. Therefore, the deflection will automatically be reduced by about 50%. We have

$$v_f = \frac{c_r W L^3}{E_c I_e}$$

$$= \left(\frac{5}{384}\right) \frac{(166560)(6^3)}{25490(10^6)(2579)\,10^{-6}} = 0.0075\ m = 7.5\ mm$$

whereas the deflection of the beam without the effect of the steel was 14.2 mm as seen from Example 8.1.

EXAMPLE 8.9 A simply supported beam of span 6 m is having the following design data. Check the deflection limit using the effect of the reinforcement.
Data

$$L = 6\ m,\ b = 0.20\ m,\ h = 0.4\ m$$

$$w_g = 1.92\ kN/m,\ w_s = 9.5\ kN/m,\ w_l = 15\ kN/m$$

$$f_{ck} = 20\ MPa,\ f_y = 415\ N/mm^2$$

$$\text{Let}\quad d = h - 0.035 = 0.365\ m,\ d_c = 0.05\ m$$

$$A_{sc} = 942\ mm^2,\ A_{st} = 1658\ mm^2$$

Deflection computations. The moment of inertia of the cracked section is

$$I_{cr} = \frac{bx_u^3}{3} + A_{st}\,m\,(d - x_u)^2 + A_{sc}\,(m - 1)\,(x_u - d_c)^2$$

$$x_u = 0.479d = 0.479(0.365) = 0.1748\ m$$

$$I_{cr} = \frac{0.20(0.1748)^3}{3}$$

$$+ 1658(10^{-6})(14)(0.365 - 0.1748)^2$$

$$+ 942(10^{-6})(13)(0.1748 - 0.05)^2$$

$$= (356 + 840 + 190)10^{-6} = 1475\ (10^{-6})\ m^4$$

$$I_g = \frac{bh^3}{12} = \frac{0.2(0.4)^3}{12} = 1067(10^{-6})\ m^4$$

The value of I_e is limited by

$$I_{cr} \le I_e \le I_g$$

Therefore, I_e is limited by I_g as I_{cr} worked out larger than I_g. One can question the validity of I_g without including the effect of the reinforcement. This will be a valid question and has considerable influence in the limit state design for deflection limit state of doubly-reinforced sections.

$$E_c = 5700\ \sqrt{f_{ck}} = 25490\ N/mm^2$$

The deflection of the beam is

$$v_f = \frac{5}{384} \frac{W L^3}{E_c I_e}$$

$$W = \gamma_f\,(w_g + w_s + w_l)L$$

$$= (1)(1.92 + 9.5 + 15)(6) = 158.52 \text{ kN}$$

$$v_f = \frac{5(158520)(6^3)}{384(25490)10^6)(1067)(10^{-6})} = 0.0164 \text{ m} = 16.4 \text{ mm}$$

EXAMPLE 8.10 *Deflection computation in T- beam.* A cantilever beam has the following design specifications and details. Compute the deflection including the effect of the steel in the moment of inertia.

Cantilever span $= L = 3$

Thickness of slab $= t = 0.11$ m

Effective flange width $= b_f = 1.57$ m

Total working load (UDL) $= W_a = 59.04$ kN

$b_w = 0.25$ m, $d = 0.25$ m, $h = 0.35$ m

$A_{st} = 2945$ mm², $A_{sc} = 157$ mm²

$f_{ck} = 15$ MPa, $f_y = 250$ MPa

Moment of inertia including the effect of steel (vide Fig. 8.1 for details) of the cracked section about the NA is:

$$I_{cr} = \frac{b_f t^3}{12} + A_f (x_u - 0.5t)^2 + \frac{b_w (x_u - t)^3}{3} + A_{st} \, m \, (d - x_u)^2$$

$$x_u = 0.531d = 0.133 \text{ m}$$

$$I_{cr} = \frac{1.57(0.11)^3}{12} + 1.57(0.11)(0.133 - 0.055)^2$$

$$+ \frac{0.25(0.133 - 0.11)^2}{3}$$

$$+ 2945(10^{-6})(19)(0.25 - 0.133)^2$$

$$= 174(10^{-6}) + 1051 (10^{-6}) + 44.0(10^{-6}) + 766(10^{-6})$$

$$= 2034.9 \, (10^{-6}) \text{ m}^4$$

To compute the gross moment of inertia of the section, one must compute the location of the centroid of the section. The distance of the CG from the extreme compression fibre by taking the moment of areas about middle line of the flange

$$y_c = 0.5t + \frac{250(350 - 110)(120 + 55)}{250(350 - 110) + 1570(110)} = 100 \text{ mm}$$

$$I_g = \frac{1.57(0.11)^3}{12} + 1.57(0.11)(0.1 - 0.055)^2$$

$$+ \frac{0.25(0.24)^3}{12} + 0.25(0.24)(0.12 + 0.1)^2 = 1826(10^{-6}) \text{ m}^4$$

Since the value of I_e is limited by

$$I_{cr} \leq I_e \leq I_g$$

select $\quad I_e = I_g = 1826(10^{-6})$ m^4

The design flexural strength of the concrete is

$$f_{cr} = \frac{0.7 \sqrt{f_{ck}}}{1.5} = 1.8 \text{ MPa}$$

$$E_c = 5700 \sqrt{f_{ck}} = 22076 \text{ N/mm}^2$$

The maximum deflection due to the UDL which occurs at the free end and it is

$$v_f = \frac{WL^3}{8E_e I_e} = \frac{59040(27)}{8(22076)(10^6)(1826)(10^{-6})} = 0.00495 \text{ m} = 4.95 \text{ mm}$$

whereas the value computed by neglecting the effect of the steel in the moment of inertia calculations works out to be 8.21 mm. The allowable deflection is

$$\frac{L}{350} = \frac{3000}{350} = 8.57 \text{ mm}$$

8.4 Deflection and Moment Relation for Limit Values of Span to Depth

In the limit states design the main design is normally governed by the strength limit state while the limit state of serviceability sometimes controls it. One can generate a set of expressions which to an extent can incorporate the two limit states.

The collapse moment can be expressed as

$$M_c = \gamma_f \Sigma (c_{mi} W_i) L \tag{8.16}$$

in which c_{mi} = moment coefficient which is 1/8 for UDL on simply supported span and it is $= \frac{1}{2}$ for UDL on cantilever span, etc.

$\quad$ γ_f = partial safety load factor

$\quad$ W_i = total of the ith type of load on the span

For simplicity all the loads are idealized as UDL. The above equation can then be written as

$$M_c = \gamma_f c_m WL \tag{8.17}$$

The resisting moment capacity of a rectangular section can be expressed as

$$M_{rb} = Kbd^2 f_{ck} \tag{8.18}$$

in which K is defined in Table 7.5 for different types of steels. It varies from 0.11 to 0.131 for HTS, and 0.148 for MS, and 0.133 to 0.138 for HYSD bars including a partial safety factor of 1.5 applied to the concrete.

The deflection of a beam for UDL acting on a beam can be expressed as

$$v = c_d \frac{WL^3}{E_c I_e} \tag{8.19}$$

in which I_e the effective moment of inertia

$\quad$ c_d deflection coefficient which is 5/384 for simply supported span, 1/8 for cantilever span, 1/384 for fixed ended span

The design criteria of the limit states of strength and deflection can be expressed as :

$$M_{rb} \geq M_c \text{ and}$$

$$v \leq v_a$$

$$Kbd^2 f_{ck} \geq \gamma_f c_m \, WL \tag{8.20}$$

$$\text{and} \quad \frac{v_a}{L} \geq \frac{v}{L} = \frac{c_d WL^2}{E_c I_e} \tag{8.21}$$

Equation (8.20) can be expressed as

$$Kbd^2 f_{ck} \geq \gamma_f c_m WL \tag{8.22}$$

The effective moment of inertia of a beam is taken equal to that of a cracked section without the influence of the reinforcement as a first approximation and based on which the slenderness limits are established to start with. A more accurate analysis will be discussed later.

Let $I_e = I_{cr}$

$$= \frac{bx_u^3}{3} = \frac{b}{3}\left(\frac{x_u}{d}\right)^3 d^3 = \frac{bd^3}{3}(k_3)^3 \tag{8.23}$$

in which $\quad \dfrac{x_u}{d} = k_3$

Consider a limiting case in which inequality (8.22) is set as an equality, so that we have

$$\frac{v_a}{L} \geq \frac{c_d WL^2}{E_c I_e} \text{ and} \tag{8.24}$$

$$Kbd^2 f_{ck} = \gamma_f c_m \, WL \tag{8.25}$$

Substitute the value of WL form Eq. (8.25) in Eq. (8.24), then we have

$$\frac{v_a}{L} \geq \frac{c_d L \, Kbd^2 f_{ck}}{E_c I_e \gamma_f c_m}$$

$$\geq \left(\frac{c_d}{\gamma_f c_m}\right) \frac{Kbd^2 L}{I_e}\left(\frac{f_{ck}}{E_c}\right) \tag{8.26}$$

Substitute $I_e = I_{cr} = \dfrac{bd^3 k_3^3}{3}$ in Eq. (8.26), then

$$\frac{v_a}{L} \geq \left(\frac{c_d}{\gamma_f c_m}\right) \frac{3Kbd^2 L}{k_3^3 bd^3}\left(\frac{f_{ck}}{E_c}\right)$$

$$\geq \frac{3c_d K}{\gamma_f c_m k_3^3}\left(\frac{L}{d}\right)\frac{f_{ck}}{E_c} \tag{8.27}$$

We can see that the slenderness of the beam, i.e. L/d controls the deflection limitation for given set of steel and concrete properties. The value of v_a/L can be limited to 1/300 and consequently the inequality (8.27) can be rewritten as

$$\frac{d}{L} \geq 300 \left(\frac{3}{\gamma_f}\right)\left(\frac{f_{ck}}{E_c}\right)\left(\frac{1}{k_3^2}\right)\left(\frac{K}{k_3}\right)\frac{c_d}{c_m} \tag{8.28}$$

For all practical purposes we can take

$$\gamma_f = 1.5$$

$$\frac{K}{k_3} = 0.28 \text{ to } 0.30, \text{ use } 0.29$$

$$E_c = c_{fc}\sqrt{f_{ck}}$$

Inequality Eq. (8.28) can be expressed as

$$\frac{d}{L} \geq 600 \left(\frac{\sqrt{f_{ck}}}{c_{fc}}\right)(0.29)\frac{1}{k_3^2}\left(\frac{c_d}{c_m}\right) \tag{8.29}$$

Most of the concrete structures are loaded after six months of casting, with the result the creep coefficient will be around 1.2 or less. This will enable the designer to limit the deflections to $L/300$ which will lead to a total deflection of $L/250$. Therefore, $L/300$ is selected here.

If the units of stress are N/mm^2, then the value of $c_{fc} = 5700$ and inequality (8.29) is reduced to the one given below:

$$\frac{L}{d} < 28 \frac{k_3^2 c_m}{\sqrt{f_{ck}}\, c_d} \tag{8.30}$$

Table 8.5a Limiting (L/d) Ratios Excluding the Effect of the Reinforcement

Type of beam with UDL	$f_y = 250$ N/mm^3			$f_y = 415$ N/mm^3		
	$f_{ck} = 15$	20	25	15	20	25
1. Simply supported	23.3	20	18	18.7	16.3	14.5
2. Cantilever	9.6	8.3	7.6	7.8	6.8	6.0

Table 8.5b Limiting Ratio (L/d) Including the Effect of the Balanced Reinforcement

Type of beam	$f_y = 250$			$f_y = 415$		
	$f_{ck} = 15$	20	25	15	20	25
1. Simply supported under UDL	55	47	43	44	39	34
2. Cantilever under UDL	23	20	18	18	16	14

Table 8.5 gives maximum values for certain concrete beams. It is rather a serious concern for a designer that the approximate limiting (L/d) ratios of beams satisfying the deflection limit state are rather low. Consequently, the strength limit state is not very effective in determination of the size of the beam. Such limits appear to be uneconomical in many cases. However, if the effect of the reinforcement is included in the computations of the sectional moment of inertia, the limits prescribed by the code will be altered considerably.

8.5 Limitation on Slenderness Ratio Including the Effect of Reinforcement

It was seen in the earlier examples that the moment of inertia of the cracked section is influenced by considering the area of steel. Let us consider the balanced section and the moment of inertia of the cracked section including the effect of reinforcement. We have

$$I_{cr} = \frac{bx_u^3}{3} + mA_{st}(d - x_u)^2$$

$$= \frac{bd^3}{3}\left(\left(\frac{x_u}{d}\right)^3 + 3mp\left(1 - \frac{x_u}{d}\right)^2\right) \qquad (8.31)$$

in which $p = A_{st}/bd$ and m = modular ratio.

On balanced sections, the area of the tension steel can be expressed as

$$p = \frac{A_{st}}{bd} = 0.414\left(\frac{x_u}{d}\right)\frac{f_{ck}}{f_y} \qquad (8.32)$$

Let $\frac{x_u}{d} = k_3$

Then Eqs. (8.31) and (8.32) reduce to

$$I_{cr} = \frac{bd^3}{3}\left(k_3^3 + 1.242m(1 - k_3)^2 \frac{f_{ck}}{f_y}\right) \qquad (8.33)$$

The value of m is usually given by

$$m = \frac{280}{f_{ck}},$$

then Eq. (8.33) reduces to

$$I_{cr} = \frac{bd^3}{d}\left(k_3^3 + \frac{348}{f_y}(1 - k_3)^2\right)$$

$$= \frac{bd^3}{d}k_3^3\left(\left(1 + \frac{348}{f_y}\right) + \frac{348}{f_y}\left(\frac{1}{k_3^2} - \frac{2}{k_3}\right)\right) \qquad (8.34)$$

The value of k_3 varies from 0.42 to 0.53. Therefore, the second part of the expression in Eq. (8.34), namely

$$\frac{1}{k_3^2} - \frac{2}{k_3}$$

is small when compared with other terms. Hence, the moment of inertia of the cracked section can be approximated as

$$I_{cr} = \frac{b(k_3 d)^3}{3}\left(1 + \frac{348}{f_y}\right) \qquad (8.35)$$

The effect of the tension steel enhances the moment of inertia of the cracked section by 2.39 times in case of mild steel and 1.84 times in case of HYSD bars of $f_y = 415$ MPa. Following the derivation similar to that of the previous section, one can arrive at the limiting slenderness ratios of the beams as:

$$\frac{L}{d} < 28 \frac{c_m}{c_d} \frac{k_3^2}{\sqrt{f_{ck}}}\left(1 + \frac{348}{f_y}\right) \qquad (8.36)$$

Before using the modified moment of inertia of the cracked section one should ensure that it is not larger than the value corresponding to the gross section.

$$I_g = \frac{bh^3}{12}$$

$$\frac{I_{cr}}{I_g} = 4k\frac{3}{3}\left(1 + \frac{348}{f_y}\right)\left(\frac{d}{h}\right)^3 \tag{8.37}$$

The largest value that the above ratio can take is less than one. Therefore, the effective moment of inertia will be equal to that of the cracked section.

Table 8.5b gives the approximate upper limit of the L/d ratio with the deflection limit automatically satisfied by including the tension steel into the moment of inertia calculations. This table gives more realistic and economical limits.

8.6 Limit State of Crack Width

There are different theories to predict the spacing of the cracks and limiting the crack width. One of the simpler methods is described in this section. However, one must realize that the computations based on crack width limitation are more involved and idealized or approximated quantities. Concrete develops cracks for several reasons and often at random critical locations. The cracking can be due to the following reasons:

1. The stress caused by the loads exceeds the fracture stress of the concrete.
2. Creep in concrete.
3. Shrinkage of concrete.
4. Rusting of the reinforcement will cause expansion in the diameter of the bars resulting cracking.
5. Aggregate reactions which result in the increase in the volume of the aggregate will result in cracking.
6. Exposure to excessive temperatures during fire.
7. Concrete when exposed to even 100°C will heat the moisture in the concrete to form steam. The steam will burst out of the concrete.
8. Some atmospheric agents can result in cracking.

This section is devoted to the design of RCC beams limited by cracks formed by loads only. Other aspects are discussed under durability considerations. Microcracking occurs at random and at many places under smaller loads. However, only a limited number of these cracks develop and lead to openings. The design is therefore associated with the control of such larger cracks.

Consider a typical beam in a cracked state as shown in Fig. 8.2a. Let the cracks developed be those due to the loads and the spacing of the cracks in the critical section be at $2a$ as shown in Fig. 8.2b.

Let $2a$ = spacing of cracks

$2w$ = width of the crack

Let section CC be the middle section of one typical segment between the cracks and section AA be the cracked section location. The approximate stress variations at the cracked and uncracked sections are shown in Figs. 8.2c and 8.2d. This figure is further expanded along with the stress distributions in Fig. 8.3.

At the uncracked section the moment relation is

$$M_{cr} = f_{ck} Z_t \tag{8.38}$$

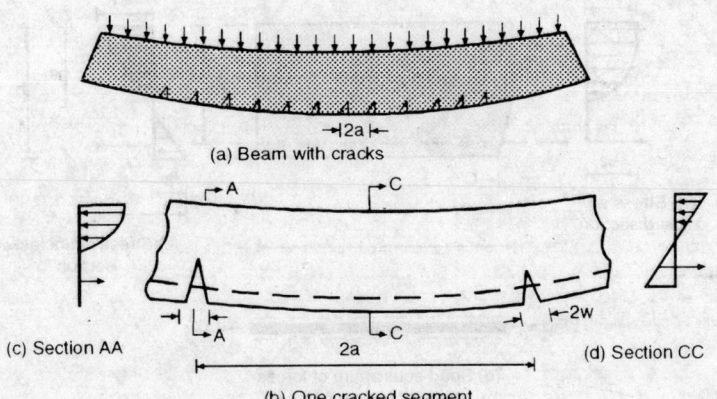

(a) Beam with cracks

(c) Section AA (b) One cracked segment (d) Section CC

Fig. 8.2 Notations of Crack at RCC Section.

in which Z_t = section modulus about the tension fibre

$$Z_t = \frac{I_g}{y_t}$$

where I_g = gross moment of inertia and

y_t = distance of the tension fibre CG

The cracking strain at the extreme fibre is

$$\epsilon_t = \frac{f_{cr}}{E_c} \qquad (8.39)$$

The strain in the tension reinforcement based on the compatibility of the strains is (vide Fig. 8.3c)

$$\epsilon_{s1} = \epsilon_t \frac{(d - kd)}{h - kd} \qquad (8.40)$$

in which kd is the depth of NA of the uncracked section, h is the overall depth and d is the depth of the steel.

The value of d is in the range of $0.9h$. Therefore, Eq. (8.40) can be approximated as:

$$\epsilon_{s1} = 0.9 \, \epsilon_t = \frac{0.9 f_{cr}}{E_c} \qquad (8.41)$$

The stress in steel at the uncracked section can be obtained by multiplying the strain by Young's modulus of the steel and it is

$$\sigma_{s1} = E_s \sigma_{s1} = \frac{0.9 \, E_s \, f_{cr}}{E_c} = 0.9 \, m f_{cr} \qquad (8.42)$$

However, the actual cracking strain in concrete in the presence of the tension reinforcement will be higher than that of the plain concrete. This will automatically increase the actual strain in the steel. Therefore, stress in steel can be modified as at least

$$\sigma_{s1} = m f_{cr}$$

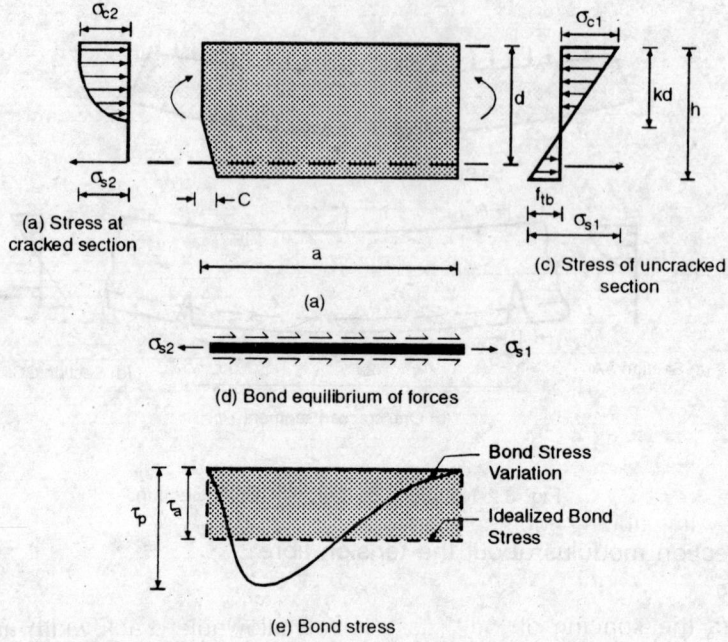

(d) Bond equilibrium of forces

(e) Bond stress

Fig. 8.3 Bond Stress Distribution for a Cracked Section.

This equation overrides Eq. (8.42). The stress in steel at the *cracked* section is influenced by the crack width, the possible slip of the reinforcement in the segment between the cracked and uncracked section. One can also assume no slip in the reinforcement and compute strain in the steel at the cracked section. While total slippage of the reinforcement is not possible, an approximation has to be made either in the bond stress variation or in the slip variation. The bond stress at no crack section is zero and it increases to a peak value near the cracked section. The boundary condition at the cracked section forces zero bond stress at the level of steel. Therefore, the bond stress decreases very rapidly near the cracked section. The bond stress variation with the ribbed or deformed steel bars is closer to uniform variation than that shown in Fig. 8.3c. In the limit one can idealize the bond stress variation as uniform and allow gradual slip of the reinforcement which result into the opening of the crack. The strain in steel at the cracked section can be taken as a mean value of the total strain in the segment between the cracked and uncracked sections. The total elongation of the reinforcement in the segment as mean value of the strains at the two ends of the segment is

$$e_{sm} = \frac{\sigma_{s2} + \sigma_{s1}}{2E_s} \tag{8.44}$$

The stress in the extreme fibre of the concrete at the uncracked section is f_{cr} and the corresponding stress at the cracked section is zero because of the free boundary condition. For the purpose of calculation of total strains in concrete, the average stress is taken. The total tension strain of the concrete in the segment is

$$e_{ct} = \epsilon_t \, a = \frac{f_{cr} \, a}{2E_c}$$ (8.45)

The total slip of the reinforcement which is equal to the opening of the crack is

$$w = e_{sm} - e_{ct}$$ (8.46)

Eqs. (8.44) to (8.46) result in

$$\frac{w}{a} = \frac{\sigma_{s2}}{2E_s} + \frac{\sigma_{s1}}{2E_s} - \frac{f_{cr}}{2E_c}$$

$$= \frac{\sigma_{s2}}{2E_s} + \frac{mf_{cr}}{2E_s} - \frac{f_{cr}}{2E_c}$$

$$\frac{w}{a} = \frac{\sigma_{s2}}{2E_s}$$ (8.47)

Let $\sigma_{s2} = \beta\sigma_{s1}$ (8.48)

Then Eq. (8.47) can be rewritten as

$$a = \frac{2wE_s}{\beta\sigma_{s1}}$$

Since $\sigma_{s1} = mf_{cr}$ the above equation reduces to

$$a = \frac{2wE_s}{\beta mf_{cr}} = \frac{2wE_c}{f_{cr}}$$ (8.49)

Eq. (8.49) gives the spacing of the cracks for an allowable crack width and a known stress in steel at the cracked section. In addition to the compatibility consideration, one should also consider the statical equilibrium consideration of the reinforcement bar between the cracked and uncracked section.

Consider the equilibrium consideration of the reinforcement in the half segment as shown in Fig. 8.3d. The difference in the forces in the reinforcement is resisted by the bond force between the concrete and the steel. For all practical purposes an average bond stress is assumed along the length of the bar which is true when same slip is allowed. The equilibrium of forces on a bar is

$$\frac{\pi}{4} \phi^2 \, (\sigma_{s2} - \sigma_{s1}) = \pi\phi a \, \tau_{bd}$$ (8.50)

in which $\phi =$ diameter of the bar

$\tau_{bd} =$ design bond stress whose values are given in Table 8.7

We already know or assume that

$$\sigma_{s1} = mf_{cr} \text{ and}$$

$$\sigma_{s2} = \beta\sigma_{s1} = \beta mf_{cr}$$

Then Eq. (8.50) reduces to

$$a = \frac{(\beta - 1)\phi mf_{cr}}{4\tau_{bd}}$$ (8.51)

To get an approximate range of the crack spacing one can use the following values

$$m = \frac{280}{3\sigma_{cb}} = \frac{280}{f_{ck}}$$

Table 8.6 Some Design Stresses at Limit State of Serviceability (Stresses in N/mm^2)

Concrete grade	15	20	25	30	40	45
f_{cr}	1.20	2.08	2.33	2.5	2.9	3.1
τ_{bd} (plain bars)	1.0	1.2	1.4	1.5	1.7	1.9
τ_{bd} (deformed bars)	1.6	1.9	2.2	2.4	2.7	3.0

$$\frac{f_{cr}}{\tau_{bd}} = 1.67 \text{ for plain bars and}$$

$$= 1 \text{ for deformed bars}$$

Then *approximate* value of the spacing of the cracks from Eq. (8.51) (*for plain bars*) is:

$$2a_{app} = 2a = 2(\beta - 1)\phi \frac{(1.67)(280)}{4\,f_{ck}}$$

$$= 234 \frac{(\beta - 1)\phi}{f_{ck}} \tag{8.52}$$

For deformed bars

$$2a_{app} = \frac{2(\beta - 1)\phi \cdot (1)(280)}{4\,f_{ck}}$$

$$= \frac{140(\beta - 1)\phi}{f_{ck}} \tag{8.53}$$

The spacing of the cracks are derived in Eqs. (8.49) and (8.51) by compatibility and equilibrium considerations. Equating the above two equations, we have

$$a = \frac{2wE_c}{\beta f_{cr}} = \frac{(\beta - 1)\phi m f_{cr}}{4\tau_{bd}} \tag{8.54}$$

It can be rearranged as

$$\beta(\beta - 1) - \left(\frac{8wE_c\tau_{bd}}{mf_{cr}^2\,\phi}\right) = 0 \tag{8.55}$$

The value of β solved from the above equation is

$$\beta = 0.5 + \sqrt{(1 + 8w_0)} \tag{8.56}$$

in which

$$w_0 = \frac{wE_c\tau_{bd}}{mf_{cr}^2\phi}$$

The approximate range of w_0 can be computed as

$$\frac{E_c\tau_{bd}}{mf_{cr}^2} = \frac{5700(0.28)f_{ck}}{280(0.218)} = 26f_{ck}$$

Allowing a microcracking of crack width equal to 0.02 mm, the value of β is approximately equal to:

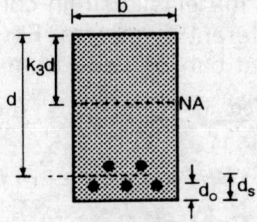

Fig. 8.4 Notations in a Rectangular Cross-section.

For plain bars

$$\beta = 0.5 + \sqrt{1 + \frac{4.16 f_{ck}}{\phi}} \qquad (8.57)$$

For deformed bars

$$\beta = 0.5 + \sqrt{1 + \frac{6.66 \, f_{ck}}{\phi}} \qquad (8.58)$$

where the value of ϕ is in mm.

The value of β indicates that the actual strain in concrete at cracking level is about 2 to 3.5 times the elastic strain. Therefore, the allowable strain in concrete before cracking is

$$\epsilon_{cc} = \beta \epsilon_{ce} \qquad (8.59)$$

$$\text{where } \epsilon_{cc} = \frac{f_{cr}}{E_c}$$

Table 8.7 Some Approximate Values of β for Microcrack Width of 0.02 mm

Concrete f_{ck}	Dia of bars ϕ	Plain bars	Deformed bars
15	10	3.2	3.8
	12	3.0	3.5
	16	2.7	3.2
	20	2.5	3.0
	25	2.4	2.7
	28	2.3	2.6
20	10	3.5	4.3
	12	3.3	4.0
	16	3.0	3.5
	20	2.8	3.3
	25	2.6	3.0
	28	2.5	2.8
25	10	3.9	4.7
	12	3.6	4.3
	16	3.2	3.9
	20	3.0	3.5
	25	2.8	3.3
	28	2.7	3.1

The enhanced cracking strain to the elastic strain corresponding to the fracture stress is indicated in Table 8.7 for bars of different diameters. For a fixed value of β and w, one can find the maximum size of the bar that can be used from Eq. (8.55) and it is

$$\phi = \frac{8w\,E_c\,\tau_{bd}}{mf_{cr}^2\,\beta(\beta-1)} \tag{8.60}$$

8.7 Additional Formulae on Crack Control

There are other approaches which recommend the relation of crack spacing and crack width in relation with the strength of concrete, etc. Typical theoretical and empirical values are listed below:

$$2a = \frac{M_{cr}}{jd\,\tau_{bd}\pi\,\Sigma\phi_i} \tag{8.61}$$

If the bars are of equal diameter and the number of bars equal to N, the above expression can be expressed as:

$$2a \geq \frac{M_{cr}}{jd\tau_{bd}\pi N\phi} \tag{8.62}$$

The crack spacing can, therefore, be controlled by selecting appropriate diameter of the bar. Equation (8.62) can be rewritten as

$$N\phi = \frac{M_{cr}}{2jd\,\tau_{bd}\pi a} \tag{8.63}$$

Normally admissible crack widths under different conditions are given in Table 8.8.

One of the main disadvantage with the two formulae given earlier is that the percentage of the reinforcement or distance of the reinforcement from the cover or the distribution of the reinforcement along the depth is not reflected in. American design practice suggests that in case of steels having yield stress exceeding 40 ksi = (40(6.85) = 275 MPa), the limitation of crack control may be enforced through the following restriction:

$$\sigma_{st}\sqrt[3]{2d_0\,\frac{bd_sA_{st}}{A_{si}}} \leq Z_0 \tag{8.64}$$

Table 8.8 Admissible Crack Widths

	Type of structure	Crack width in mm
1.	For microcracking	0.02
2.	Liquid retaining structures and others for water lightness	0.1
3.	Unprotected ordinary members exposed to aggressive atmosphere	0.2
4.	Protected structural members, indoor and no aggressive atmosphere	0.3

in which

σ_{st} = stress in reinforcement (in ksi)

d_0 = distance of the extreme row of the tension reinforcement from the extreme tension fibre (in)

A_{si} = area of the maximum size bar in tension area (in^2)

A_{st} = total area of the tensile reinforcement (in^2)

d_s = distance of the extreme concrete tension fibre from CG of the steel (inch)

Z_0 = constant which is 175 for interior exposure and 145 for exterior exposure

In case all the tension bars are having the same diameter, then inequality (8.64) can be reduced to

$$\sigma_{st}\sqrt[3]{\frac{2bd_0d_s}{N}} \leq Z_0 \tag{8.65}$$

in which N = total number of tension bars.

In case the total reinforcement consists of only one row of bars, then inequality (8.65) can be reduced to

$$1.26\, d_0\sigma_{st}\sqrt[3]{\frac{b}{Nd_0}} \leq Z_0 \tag{8.66}$$

The above expression can be modified into SI units by appropriate substitution of the units. Inequality (8.66) can be rearranged as:

1 inch = 0.254 m

1 ksi = 6.895 MPa

1 kip/inch = 4480/0.0254 = 176378 N/m = 0.1764 MN/m

175 kip/inch = 30.87 MN/m

145 kip/inch = 25.58 MN/m

$$1.26(0.0254)(d_0)(6.895)(\sigma_{st})\left(\frac{b}{Md_0}\right)^{1/3}$$

$$0.2206\, ad_0\sigma_{st}\left(\frac{b}{Nd_0}\right) \text{ in = MN/m}$$

This can be rearranged as

$$d_0\sigma_{st}\sqrt[3]{\frac{b}{Nd_0}} \leq Z_0\frac{(0.1764)}{0.2206} = Z_{00}$$

The crack limitation in SI units can be stated as:

(a) For single row of the reinforcement

$$d_0\sigma_{st}\sqrt[3]{\frac{b}{Nd_0}} \leq Z_{00} \tag{8.67}$$

(b) For multiple row of the reinforcement

$$\sigma_{st}\sqrt[3]{\frac{bd_0d_s}{N}} \leq Z_{00} \tag{8.68}$$

in which

d_0 = distance of the centre of the reinforcement in metre

σ_{st} = stress in MPa

Z_{00} = in MN/m and it is limited by

Z_{00} = 140 for interior exposed

Z_{00} = 115 for exterior exposed $\tag{8.69}$

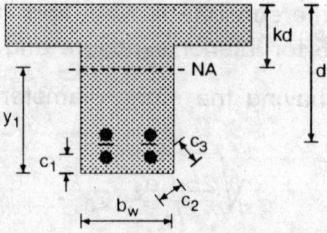

Fig. 8.5 Notations in T-beam for a Crack Width Calculations.

Similarly, inequality (8.65) can be rearranged as

$$\sigma_s \sqrt[3]{\frac{b d_0 d_s}{N}} \leq Z_{00} \tag{8.70}$$

A check of the limitation of crack width is recommended in limit states as well as working stress design.

Based on no slip criterion, one can say that the width of the crack is proportioned to the strain in steel with apparent enhanced strain in the concrete as discussed earlier. The crack width can be expressed as

$$w = \beta c \epsilon_{ac} \tag{8.71}$$

in which

$c=$ distance of the point at which the crack width is to be determined to the surface of the nearest reinforcement bar

$\epsilon_{ac}=$ apparent longitudinal strain in concrete

The apparent strain is given by

$$\epsilon_{ac} = \epsilon_{cc} - \frac{4(10^{-6})}{p} \tag{8.72}$$

$$p = \frac{A_{st}}{b_w d} \tag{8.73}$$

$$\epsilon_{ce} = \frac{y \epsilon_{st}}{d - kd} \tag{8.74}$$

$y=$ distance of the point from NA

$kd=$ NA distance from the compression fibre

$\epsilon_{st}=$ strain in tension steel

(see Fig. 8.5 for notations). Different codes recommend the value of β equal to 2 to 2.5.

8.8 Design of the Number of Bars and Cover to the Reinforcement

The number and size of the reinforcement bars are sometimes governed by the bond capacity. It is also noticed (vide Eq. 8.60) that the crack width is directly proportional to the diameter of the bar. Therefore, one should keep the size of the reinforcement bar as low as possible.

However, small diameter bars occupy more gross area and consequently impose economic limitations and practical constraints. In addition, the crack width is also controlled by the cover to the reinforcement. The farther is the reinforcement from the compression fibre, more efficient is the section in resisting the bending moment. However, at the same time its proximity to the extreme tension fibre increases the crack width. There are thus two conflicting interests in the design. Further, the cover to the reinforcement is also subjected to practical and constructional problems, durability against fire, etc. The limitation of distance of the reinforcement from the tension fibre can be obtained from Eq. (8.68), which yields

$$\frac{bd_0 d_s}{N} \leq \left(\frac{Z_{00}}{\sigma_{st}}\right) = X_0 \qquad (8.75)$$

$$\text{or } d_0 d_s \leq \left(\frac{Z_{00}}{\sigma_{st}}\right)\frac{N}{b} \qquad (8.76)$$

$$\text{where } X_0 = (Z_{00}/\sigma_{st})^3$$

Since the equation is applicable to HYSD bars and others having higher yield stresses, the values of (Z_{00}/σ_{st}) for high yield bars are listed in Table 8.9. The value of σ_{st} is taken equal to $0.6\,f_y$ which is the case for beams of under-reinforced or balanced section. The value of N/b can be approximately connected to the diameter of the bars. In case of beams having a single row of the reinforcement, Eq. (8.76) can be adjusted as

$$d_0^2 < X_0 \frac{N}{b} \qquad (8.77)$$

The approximate value of N/b and d_0 can be expressed by considering the clear spacing between the bars equal to the diameter of the bar. Then,

$$b = (2N + 1)\phi \qquad (8.78)$$

$$d_0 = c + \phi/2 \qquad (8.79)$$

where c = clear cover

The maximum cover limitation is given by

$$\left(c + \frac{\phi}{2}\right)^2 \leq X_0 \frac{N}{(2N+1)\phi}$$

$$\text{or } c \leq \sqrt{\frac{X_0 N}{(2N+1)\phi}} - \frac{\phi}{2} \qquad (8.80)$$

Equation (8.80) can be approximated as

$$c = \sqrt{\frac{X_0}{2\phi}} - \frac{\phi}{2} \qquad (8.81)$$

wherein the values of c and ϕ are in metres.

Table 8.9 Values of X_0-cover Limitation Coefficient

| Proof stress in steel | Exposed condition | |
MPa	Interior	Exterior
415	0.175	0.100
500	0.101	0.506

8.9 Examples on the Design of Limit State Design in Crack

EXAMPLE 8.11 *Fracture moment calculation.* A rectangular cross-sectioned beam of size 300×600 mm is provided with six numbers of 20 mm diameter HYSD bars in two rows. The concrete is of quality M15. Compute the fracture moment.

$$b = 0.3 \text{ m}, \ h = 0.6 \text{ m}$$

$$A_{st} = 6(314) = 1884 \text{ mm}^2$$

$$f_{ck} = 15 \text{ MPa}, \ f_y = 415 \text{ MPa}$$

$$d = h - \text{cover} - \text{dia of bars} - 0.5 \text{ (clear spacing of the bars)}$$

$$= 0.6 - 0.025 - 0.02 - 0.015 = 0.54 \text{ m}$$

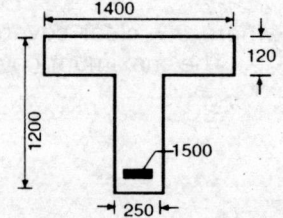

Fracture moment. The design modulus of rupture with partial safety factor of 1.5 is

$$f_{cr} = \frac{0.7}{1.5} \sqrt{f_{cr}} = 1.8 \text{ N/mm}^2$$

Gross moment of inertia is

$$I_g = \frac{bh^3}{12} = \frac{0.3(0.6)^3}{12} = 5.4 \ (10^{-6}) \text{ m}^4$$

Moment corresponding to the modulus of rupture stress is

$$M_{cr} = f_{cr} \frac{I_g}{y_t}$$

$$= \frac{1.8(10^6)(5.4)(10^{-2})}{0.3} = 32400 \text{ Nm}$$

EXAMPLE 8.12 *Fracture moment calculation for T-beam.* Compute the fracture moment of a T-beam with the following data:

$$f_{ck} = 20 \text{ MPa}, \ f_y = 415 \text{ MPa}$$

$$b_f = 1.4 \text{ m}$$

$$t = 0.12 \text{ m}, \ d = 1.2 \text{ m}$$

$$h = 1.2 \text{ m}, \ b_w = 0.25 \text{ m}$$

$$A_{st} = 1500 \text{ mm}^2$$

Moment of inertia calculation: The centroid of the section is computed by taking moments of the area about the middle line of the thickness of the slab. Let y_c be the distance of the CG from the extreme compression flange fibre. Then, we have moment of areas about middle of the top flange:

$$(y_c - 0.5t)(b_f t + b_w(h - t)) = b_w(h - t)\left(\frac{h - t}{2} + \frac{t}{2}\right)$$

$$y_c = 0.5t + \frac{b_w(h - t)(h - t)}{2(b_f t + b_w(h - t))}$$

$$= 0.06 + \frac{0.25(1.08)(1.32)}{2(1.4(0.12) + 0.25(1.08))}$$

$$= 0.06 + \frac{0.3564}{0.876} = 0.467 \text{ m}$$

The moment of inertia about the *CG* is

$$I_g = \frac{b_f t^3}{12} + (y_c - 0.5t)^2 \, b_f t$$

$$+ \frac{b_w(h-t)^3}{12} + b_w(h-t)\left(h - y_c - \frac{h-t}{2}\right)^2$$

$$= \frac{1.4(0.12)^3}{12} + (0.407)^2 (1.4)(0.12)$$

$$+ \frac{0.25(1.08)^2}{12} + 0.25 (1.08)(1.2 - 0.467 - 0.54)^2$$

$$= (201.6 + 27829 + 26244 + 10057) \, 10^{-6} \, m^4 = 64332 \, (10^{-6}) \, m^4$$

The design modulus of rupture with 1.5 as partial safety factor is

$$f_{cr} = \frac{0.7 \sqrt{f_{ck}}}{1.5} = \frac{0.7 \sqrt{20}}{1.5} = 2.08 \, N/mm^2$$

The tensile stress occurs at the extreme fibre of the web and its distance from the CG is

$$y_t = h - y_c = 1.2 - 0.467 = 0.733 \, m$$

Moment at rupture is

$$M_{cr} = \frac{I_g f_{cr}}{y_t} = \frac{64332(2.08)(10^6)(10^{-6})}{0.733} = 182552 \, Nm$$

EXAMPLE 8.13 *Crack width limitation.* Compute the spacing of the microcrack in the beam of Example 8.11 at the fracture moment level. The given data of the problem are :

$$b = 0.3 \, m, \, h = 0.6 \, m$$

$$A_{st} = 4, \, \phi 20$$

$$f_{ck} = 15 \, MPa, \, f_y = 415 \, MPa, \, d = 0.54 \, m$$

The design modulus of rupture

$$f_{cr} = \frac{0.7 \sqrt{f_{ck}}}{1.5} = 1.8 \, N/mm^2$$

$$E_c = 5700 \sqrt{f_{ck}} = 22076 \, N/mm^2$$

$$\tau_{bd} = 1.0(1.6) \, N/mm^2$$

$$m \cong \frac{280}{f_{ck}} \cong 19$$

Microcrack width = $w = 0.02 \, m$ and the diameter bar is $\phi = 20 \, mm$. From Eq. (8.56)

$$\beta = 0.5 + \left(1 + \frac{wE_c\tau_{bd}}{mf_{cr}^2 \phi}\right)^{0.5}$$

$$= 0.5 + \sqrt{1 + \frac{0.02(22076)(1.6)}{19(1.8)^2 (20)}} = 1.75$$

From Eq. (8.49) the spacing of the cracks is

$$2a = \frac{4wE_c}{\beta f_{cr}}$$

$$= \frac{4(0.02)(22076)}{1.75(1.8)} = 560 \text{ mm}$$

EXAMPLE 8.14 *Maximum size of bar.* A RCC beam of rectangular cross-section is designed based on the collapse limit state. The details of the design are:

$$b = 0.3 \text{ m}, \ h = 0.6 \text{ m}$$
$$A_{st} = 2000 \text{ mm}^2$$
$$f_{ck} = 15 \text{ MPa}, \ f_y = 415 \text{ MPa (HYSD bars)}$$

Determine the maximum size of the bar which can be used without causing damageable crack widths in the following exposure conditions:

a) Water tightness.
b) Exposed to outside atmospheric condition,
c) Exposed to indoor.

Specifications and moment at rupture. The design fracture stress (modulus of rupture) for M15 concrete is $f_{cr} = 1.8 \text{ N/mm}^2$.

        Maximum crack width for water tightness= 0.1 mm
        Maximum crack width exposed outdoor   = 0.2 mm
        Maximum crack width exposed indoor   = 0.3 mm

Design bond stress for HYSD bars is = $\tau_{bd} = 1(1.6) \text{ N/mm}^2$
    Moment of inertia of the gross section is

$$I_g = \frac{bh^3}{12} = \frac{0.3(0.6)^3}{12} = 5.4(10^{-3}) \text{ m}^4$$

Let the bars be arranged in two rows, then

$$d = h - 0.06 = 0.54 \text{ m}$$
$$jd = 0.87 (0.54) = 0.4698 \text{ m}$$
$$M_{cr} = I_g \frac{f_{cr}}{y_t} = \frac{5.4(1.8)(10^{-3})(10^6)}{0.3} = 32400 \text{ Nm}$$

The area of the steel provided is = $A_{st} = 2000 \text{ m}^2$
Let N = number of equal size bars, then

$$\frac{N\pi\phi^2}{4} > A_{st} = 2000 \text{ mm}^2$$

$$N\phi^2 \geq \frac{2000(4)}{\pi} = 2546 \text{ mm}^2 \tag{a}$$

a) *Water tightness.* The value of crack width = $w = 0.1 \text{ mm} = 1 (10^{-4}) \text{ m}$

Let the enhanced strain in concrete due to the presence of steel be taken 3.5 times that of the fracture strain. Eq. (8.60) gives the diameter of the bar and it is

$$\phi = \frac{8wE_c\, \tau_{bd}}{mf_{cr}^2\, \beta(\beta - 1)} \qquad \text{(b)}$$

Considering the long-term loadings the Young's modulus of the concrete and the modular ratio are computed and they are:

$$E_c = \frac{5700}{2}\sqrt{f_{ck}} = 2850\sqrt{15} = 11038 \text{ N/mm}^2$$

$$m = \frac{E_s}{E_c} = \frac{200\,000}{11\,038} = 18$$

$$f_{cr} = 1.8 \text{ MPa or N/mm}^2$$

$$\tau_{bd} = 1.6 \text{ MPa or N/mm}^2$$

The substitution of various quantities in the Eq. (b) gives

$$\phi = \frac{8wE_c\tau_{bd}}{mf_{cr}^2\,\beta(\beta - 1)}$$

$$= \frac{8(0.1)(111038)(1.6)}{(18(1.8)^2(3.5)(2.5))} \text{ mm} = 27.6 \text{ mm} \qquad \text{(c)}$$

Let 20 mm dia bars be used in the construction. Then the number of bars can be obtained from Eq. (a):

$$N \geq \frac{2546}{\phi^2} = \frac{2546}{400} = 6.37$$

Provide 7 numbers of $\phi20$ bars for water tightness:

$$A_{st}\,(\text{provided}) = 7(314) = 2198 \text{ m}^2$$

b) *Exposed to outdoor conditions.* The maximum width of the crack permitted in the beams exposed to outdoor conditions is $w = 0.2$ mm

The substitution of this value in Eq. (b) result in

$$\phi = \frac{8wE_c\tau_{bd}}{mf_{cr}^2\,\beta(\beta - 1)}$$

$$= \frac{8(0.2)(11038)(1.6)}{18(1.8)^2\,(3.5)(2.5)} = 55.4 \text{ mm}$$

Any bar upto a diameter of 55.4 mm can be provided. *Provide* 7 numbers of 20 mm bars.

c) *In-door exposed condition.* The allowable strain in concrete exposed to indoor conditions is 0.3 mm. The maximum diameter allowed for a crack width of 0.2 mm is 55.4 mm as seen from the earlier calculations. Therefore, the diameter of the bar can be even larger than 55.4 mm.

Provide 7 numbers of 20 mm bars.

EXAMPLE 8.15 *On crack width.* A simply supported beam of span 6 m is subjected to superimposed load of 9.4 kN/m and live load of 15 kN/m designed with $f_{ck} = 20$ MPa and $f_y = 415$ MPa. The size of the beam and the reinforcement details are shown in Fig. 8.6. Calculate the maximum crack width in the beam section.

$$A_{st} = 1257 \text{ mm}^2, \ b = 250 \text{ mm}, \ d = 520 \text{ mm}$$

The maximum crack width for no slip condition is given by Eq. (8.71), that is

$$w = \beta c \epsilon_{ac}$$

Let $\beta = 2.5$

$$\epsilon_{ac} = \epsilon_{ce} - \frac{4(10^{-6})}{p}$$

$$p = \frac{A_{st}}{bd} = \frac{1257}{250(520)} = 9670(10^{-6})$$

$$\epsilon_{ce} = \frac{y \epsilon_{st}}{d - k_3 d} = \frac{y}{d - k_3 d}\left(\frac{f_{st}}{E_s}\right)$$

$$\frac{f_{st}}{E_s} = \frac{0.6 \, f_y}{E_s} = \frac{0.6(415)}{200\,000} = 1.245 \times 10^{-3}$$

$$kd = 0.479 \,(520) = 249 \text{ mm}$$

$$d - kd = 520 - 249 = 271 \text{ mm}$$

$$\epsilon_{ce} = \frac{y}{d - dk}\left(\frac{f_{st}}{E_s}\right) = \frac{1.245}{271} (10^{-3})y = 4.64 \,(10^{-6}) \, y$$

$$\epsilon_{ac} = 4.6(10^{-6}) \, y - \frac{4}{9670} = (4.6y - 413)10^{-6} \qquad \text{(b)}$$

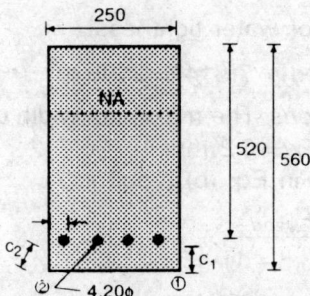

Fig. 8.6 Reinforcement Details of Examples 8.15 and 8.17.

The crack width is computed for different points at which it is likely to be critical. The maximum of the computed values is selected. Consider Fig. 8.6 which shows two possible locations at which the crack is likely to be the highest. The points are marked 1 and 2. The corresponding distances c and y to the points are denoted by c_i and y_i for $i = 1$, and 2.

Consider point 1

$$c_1 = d_s - \frac{\phi}{2} = 40 - 10 = 30 \text{ mm}$$

$$y_1 = h - k_3 d = 560 - 249 = 311 \text{ mm}$$

$$\epsilon_{ac} = (4.6y_1 - 413)10^{-6}$$

$$= (4.6(311) - 413) \, 10^{-6} = 1018 \,(10^{-6})$$

The crack width is

$$w_1 = \beta c_1 \in_{ac} = 2.5(30)(1018)(10^{-6}) = 0.076 \text{ mm}$$

Consider point 2

$$c_2^2 = 30^2 + 30^2$$

$$c_2 = 42.43 \text{ mm}$$

$$y_2 = y_1 = 311 \text{ mm}$$

$$\in_{ac} = 1018(10^{-6})$$

The crack width is

$$w_2 = \beta c_2 \in_{ac} = 2.5(42.43)(1018)(10^{-6}) = 0.10 \text{ mm}$$

On no slip basis the maximum crack width at normal working stress in steel is at point 2 and it is equal to 0.10 mm.

EXAMPLE 8.16 A beam is designed on limit strength basis and the details of the design values are:

$$b = 250 \text{ mm}, \ d = 522 \text{ mm}, \ h = 600 \text{ mm}$$

The reinforcement at bottom is 5 numbers of 20 mm bars and 1 no. of 16 mm bar. The sectional details are given in Fig. 8.7.

Clear cover at bottom = 25 mm

Clear side cover   = 30 mm

$A_{st} = 5(314) + 204$ = 1774 mm^2

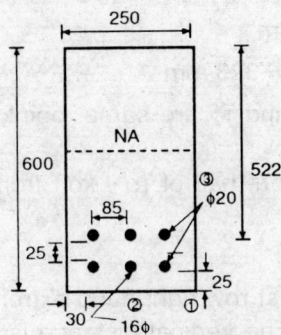

Fig. 8.7 Sectional Details of a Rectangular Cross-section of Examples 8.16 and 8.17.

The crack width is given by

$$w = \beta c \in_{ac}$$

Let $\quad \beta = 2.5$

$$p = \frac{A_{st}}{bd} = \frac{1774}{250(522)} = 0.0136$$

$$kd = 0.53d = 0.53(522) = 277 \text{ mm}$$
$$d - kd = 522 - 277 = 245 \text{ mm}$$
$$\epsilon_{st} = \frac{0.6f_y}{E_s} = \frac{0.6(250)}{200\,000} = 750(10^{-6})$$
$$\epsilon_{ce} = \frac{\epsilon_{st}}{d - kd}y = \frac{750(10^{-6})y}{245} = 3.06y\,(10^{-6})$$
$$\epsilon_{ac} = \epsilon_{ce} - \frac{4(10^{-6})}{p}$$
$$= \left(3.06y - \frac{4}{0.0136}\right)10^{-6} = (3.06y - 294)10^{-6}$$

Three different points on the surface are considered and the crack width at these points is computed. The three points selected are marked 1, 2 and 3 in Fig. 8.7. The distances of the three points from the nearest no slip surfaces are computed below.

Point 1.

$$c_1^2 = 25^2 + 30^2$$
$$\text{or} \quad c_1 = 39 \text{ mm}$$
$$y_1 = h - kd = 600 - 277 = 323 \text{ mm}$$

Point 2. It is the mid-point between 16 mm and 20 mm bars. The clear cover to the bar is 25 + 2 = 27 mm and the horizontal distance of the point is

$$\frac{85}{2} = 42.5 \text{ mm}$$
$$c_2^2 = (42.5 - 8)^2 + 27^2$$
$$\text{or } c_2 = 43.8 \text{ mm}$$
$$y_2 = h - kd = 323 \text{ mm}$$

Since c_2 is larger than c_1, and y_1 and y_2 are same, point 2 is more critical when compared with point 1.

Point 3. Consider a point at about half of $(d - kd)$ from NA.

$$\text{Let } y_3 = \frac{d - kd}{2} = 121.5 \text{ mm}$$

The distance of the point from the first row right-hand extreme reinforcement bar is computed. The horizontal distance = 30 mm. The vertical distance is

$$\frac{d - k_3d}{2} - \frac{25}{2} - \frac{20}{2}$$
$$= 121.5 - 22.5 = 99.0 \text{ mm}$$
$$c_3^2 = 99^2 + 30^2$$
$$\text{or } c_3 = 103 \text{ mm}$$

The crack widths at points 2 and 3 are computed here.

At point 2

$$w_2 = \beta c_2 \epsilon_{ac} = \beta c_2 (3.06 y_2 - 294)10^{-6}$$

$$= 2.5(43.8)(3.06(323) - 294)10^{-6} = 0.076 \text{ mm}$$

At point 3

$$w_3 = \beta c_3 (3.06 y_3 - 294)10^{-6}$$

$$= 2.5(103)(3.06(121.5) - 294)(10^{-6}) = 0.02 \text{ mm}$$

$$w_{max} = w_2 = 0.076 \text{ mm}$$

EXAMPLE 8.17 *Crack control.* A simply supported beam of span 6 m subjected to a superimposed load of 9.4 kN/m and live load of 15 kN/m is designed with $f_{ck} = 20$ MPa and $f_y = 415$ MPa. The size of the beam and the reinforcement details are given in Fig. 8.6. Check whether the section is safe against crack limitation.

$$b = 250 \text{ mm}, \ d = 520 \text{ mm}, \ h = 560 \text{ mm}$$
$$A_{st} = 1257 \text{ mm}^2 \ (4 \text{ numbers of 20 mm bars})$$
$$\phi = 20 \text{ mm}, \ N = \text{number of bars} = 4$$
$$f_{ck} = 20 \text{ MPa}, \ f_y = 415 \text{ MPa}$$

The criterion of the limitation of crack control is taken from Eq. (8.67) and it is

$$d_0 \sigma_{st} \sqrt[3]{\frac{b}{N d_0}} \le Z_{00} \qquad \text{(a)}$$

The values of different quantities in the express for the present problem are:

$$d_0 = 560 - 520 = 40 \text{ mm}$$

σ_{st} = let the allowed stress in tension steel be $0.6 \ f_y = 250$ MPa

Substituting different quantities in Eq. (a) gives

$$0.04(250(10^6))\sqrt[3]{\frac{0.25}{4(0.04)}}$$

$$= 116\ 039\ 72 \text{ N/m} = 11.6 \text{ MN/m}$$

This value is far less than the allowable value ($Z_{00} = 115$) the exterior exposed condition.

PROBLEMS
(Use IS 456 LSD = Limit States Design)

8.1 A cantilever beam of span 5 m is subjected to a working superimposed-dead load of 10 kN/m and live load of 15 kN/m. The beam is made of M20 concrete and HYSD-Fe415 Bars. Design the beam by limit state design with a width equal to 300 mm and compute the deflection due to live load.

8.2 *A simply supported beam of span 8 m is subjected to a working superimposed dead load of 10 kN/m and live load of 25 kN/m. The width, effective depth and overall depth of the beam are 300 mm, 600 mm and 650 mm. The beam is made of M20 concrete with 2000 mm² HYSD-Fe415 steel reinforcement. Determine the deflection due to live load and due to total load. The creep coefficient for the beam is 2.1. Compute the deflections caused by shrinkage and creep.

8.3 A cantilever beam of 5 m span with a rectangular cross-section of width 400 mm, effective depth 500 mm and overall depth 550 mm is made of M20 concrete and 2 nos. of 12 mm HYSD-Fe415 bars at the bottom and 4 numbers of 20 mm bars at top. Estimate deflection caused by shrinkage for a shrinkage strain of 0.0003.

8.4 A T-beam of span 6 m has the following cross-sectional details $b_f = 1.2$ m, $t = 150$ mm, $b_w = 300$ mm, $d = 600$ mm, $h = 650$ mm, $f_{ck} = 20 \, \text{N/mm}^2$, $f_y = 415$ N/mm², $A_{st} = 1570$ mm², $A_{sc} = 226$ mm². If the shrinkage strain is equal to 0.00025, determine the deflection caused by the shrinkage. Assume any reasonable data with reference to cover, etc.

8.5 One way slab of effective span 4 m is made of M20 concrete with overall thickness of 100 mm. 8 mm HYSD-Fe415 bars at 120 mm spacing are placed at an effective depth of 75 mm. Calculate the deflection caused by shrinkage for a shrinkage strain is 0.0003.

8.6 One way simply supported slab of 4 m effective span is made of M20 concrete and has a total thickness of 120 mm, 10 mm HYSD-Fe415 bars at 150 mm spacing are placed at 100 mm below the top face. The slab is subjected to a working uniformly distributed superimposed dead load of 2 kN/m² and the live load of 4 kN/m². Determine the deflections caused by dead and live loads and shrinkage if the shrinkage strain is 0.0003. Also calculate the total deflection caused by loads, shrinkage and creep for a total creep coefficient of 1.6.

8.7 A cantilever slab of span 3 m and 200 m thickness is made of M20 concrete and 16 mm HYSD-Fe415 bars spaced at 200 mm apart at 40 mm from the top fibres. The slab is subjected to dead load of 3 kN/m² and live load of 1.5 kN/m². Shrinkage is assumed as 0.0003. Determine the deflections caused by loads, shrinkage and creep for a creep coefficient of 1.6. Compute the total deflections and check whether the total deflection is less than span/250.

8.8 A doubly-reinforced concrete beam, simply supported with an effective span of 8 m is subjected to a live load of 20 kN/m. The beam is made of M20 concrete and HYSD-Fe415 bars with the following dimensions:

$b = 300$ mm, $d = 500$ mm, $h = 540$ mm, $A_{sc} = 1256$ mm², $A_{st} = 2512$ mm². Determine the deflections due to loads, shrinkage and creep. Assume shrinkage strain as 0.0003 and creep coefficient as 2.1.

8.9 A rectangular cross-sectioned beam of size 400 by 800 mm is provided with 8 nos. of 20 mm HYSD-Fe415 bars in two rows at bottom with a clear cover of 25 mm. The beam is made of M20 concrete. Determine the fracture moment and spacing of the microcracks at that moment. Assume 0.02 mm as microcrack width.

8.10 A rectangular cross-sectioned beam of size 300 by 600 mm is provided with 8 nos. of 20 mm HYSD-Fe415 bars in two rows at bottom fibre and 4 nos. of 20 mm bars in one row at top fibre. The beam is made of M20 concrete. Determine the fracture moment and spacing of cracks at the fracture moment level. Assume a clear cover of 25 mm to the reinforcement on all faces. Assume 0.02 mm microcracks width.

8.11 A T-beam made of M20 concrete and mild steel bars has the following dimensions: $b_f = 1.5$ m, $t = 150$ mm, $d = 1000$ mm, $h = 1050$ mm, $b_w = 300$ mm, $A_{st} = 4$ nos. of 25 mm bars placed in one row. Determine the fracture moment capacity of the section and spacing of microcracks for 0.02 mm crack width.

8.12 A reinforced concrete beam of rectangular cross-section designed based on the collapse limit state of strength has the following details: $b = 400$ mm, $h = 800$ mm, $A_{st} = 2400$ mm^2. Clear cover to reinforcement is 30 mm. The beam is made of M20 concrete and HYSD-Fe415 reinforcement. Determine the maximum size of the bars which can be used if the section is to be water tight. Assume 0.10 mm maximum crack width for water tightness.

8.13 A cantilever beam of 4 m span is made of M20 concrete and HYSD-Fe415 bars has the following details: $b = 300$ mm, $d = 360$ mm, $h = 400$ mm, $A_{st} = 1570$ mm^2 (5 nos. of 20 mm bars in one row at the top fibre). Calculate the maximum crack width if the clear cover to the reinforcement on all faces of beam is 25 mm.

Chapter 9

Limit State Design of RCC Columns

9.1 Introduction

Working stress design of RCC columns was presented in Chapter 5 in which an introduction to the design of the columns was given on the following aspects:

1. Definitions and terminology
2. Types of columns
3. Effective heights and slenderness ratios
4. Minimum percentages of reinforcement and cover requirements
5. Design of transverse reinforcement.

Therefore, the reader is advised to study the first four sections of Chapter 5 which are applicable to the Limit State Design of Columns. An introduction to the limit states design was given in Chapter 7. Some definitions and limitations are reproduced here for quick reference.

$$\text{Slenderness ratio} = k = \frac{L_e}{r} \tag{9.1}$$

$$\text{Slenderness factor} = g = \frac{L_e}{b} \tag{9.2}$$

where L_e = effective length of the column, r = radius of gyration of the section and b = width of the column section.

Sometimes, the slenderness factor is called as slenderness ratio. A vertical compression member is said to be:

$$\text{a pedestal if } L < 2.5b \tag{9.3}$$
$$\text{a stub column if } L < 4b \tag{9.4}$$
$$\text{a short column if } 4b < L_e < 12b \tag{9.5}$$
$$\text{a long column if } 12b < L_e < 60b \tag{9.6}$$

Members with $L_e > 60b$ are not admissible as structural columns.

A reduction factor to load carrying capacity of the long columns is

$$C_r = 1.25 - \frac{L_e}{48b} \tag{9.7}$$

The main structural types of columns are:
a) Tied column.
b) Spiral column.
c) Composite column.
d) Infilled column.

The limits of percentage of the main reinforcement are given by

$$0.8 \leq 100p \leq 6 \tag{9.8}$$

where $100p$ = percentage of the main reinforcement

$$= \frac{100 \, A_s}{A_c} \text{ or } \frac{100 \, A_s}{bd} \tag{9.9}$$

The minimum diameter (ϕ) of the main bars and the minimum number (N) of the bars to be provided in the RCC columns are given by

$$\phi \geq 12 \text{ mm}$$
$$N \geq 4 \text{ for rectangular section} \tag{9.10}$$
$$N \geq 6 \text{ for circular or other shapes}$$

The limits on the transverse reinforcement are:

$$\phi_v \geq \frac{\phi}{4} \tag{9.11}$$
$$\geq 5 \text{ mm}$$
$$s_v \leq b$$
$$\leq 16\phi \tag{9.12}$$
$$\leq 48\phi_v$$

where ϕ_v = diameter of the ties; s_v = spacing of the ties

The pitch of the spiral in a spiral column is governed by

$$\text{pitch} \leq 75 \text{ mm} \tag{9.13}$$
$$\leq \frac{\text{diameter of the core}}{6}$$
$$\geq 30 \text{ mm}$$
$$\geq 3\phi_v \tag{9.14}$$

Normally a minimum cover of 40 mm to the main bars is recommended. Table 5.2 gives detailed requirements while the effective lengths of the columns are also stated in Table 5.3.

As the load on a column increases, it first undergoes an elastic deformation in which the stresses in the reinforcement will be compatible with the strains in concrete. After a nominal load, the rate of stress in the reinforcement increases more rapidly than that corresponding to the compatible strains, so the allowable stress in the compressive steel is taken more than the corresponding compatible stress in the concrete. As the load on the column is further increased, the concrete cover starts spalling at about 0.002 strain. At about the same level of strain or even earlier, the reinforcement bars will start yielding unless the yield stress of the steel is more than 500 MPa. Therefore, the limiting direct compressive strain (ϵ_{cc}) in concrete is recommended to be

$$\epsilon_{cc} = 0.0020 \tag{9.15}$$

whereas the maximum bending compressive strain (ϵ_{cb}) in the concrete is limited to

$$\epsilon_{cb} = 0.0035 \tag{9.16}$$

In case of the columns subjected to combined axial force and bending moment, the limiting compressive strain in the concrete is taken more than 0.002 and less than 0.0035. It is given by

$$\epsilon_{c1} = 0.0035 - 0.75 \, \epsilon_{c2} \tag{9.17}$$

where ϵ_{c1} = maximum compressive strain

ϵ_{c2} = minimum compressive strain

9.2 Ultimate Load Capacity of RCC Tied Columns

Short Columns. The crushing strength of prisms in bending compression is less than the characteristic strength but more than the corresponding value in direct compression. This has already been discussed in Chapter 7 and is

$$f_{cb} = k_p f_{ck} \tag{9.18}$$

where f_{cb} = bending compressive strength of prism

k_p = constant taken equal to 2/3 or 0.67

The direct crushing strength (f_{cc}) of concrete prism is

$$f_{cc} = k_c k_p f_{ck} \tag{9.19}$$

where k_c = constant taken in the range of 0.9.

Finally, one can conclude that the direct crushing strength of the concrete prism as:

$$f_{cc} = k_c k_p f_{ck} = 0.6 f_{ck} \tag{9.20}$$

Similarly, the direct crushing strength of the reinforcement bars can be taken equal to the yield strength or proof strength:

$$f_{sc} = f_y \tag{9.21}$$

In a RCC column, the strains in the concrete and in the steel reach the limiting values at the time of collapse of the column, then the stresses in the concrete and steel are given by Eqs. (9.20) and (9.21). The corresponding capacity of the column is given by the sum of the strengths of the combined materials and it is

$$P_{cu} = A_c f_{cc} + A_s f_{sc} \tag{9.22}$$

where P_{cu} = ultimate compressive capacity of a short column, A_c = area of the concrete section and A_s = total area of the main reinforcement.

The design strength of a concrete column is obtained by dividing the ultimate strength by a partial safety factor applied to materials. Even though the partial safety factor applied to the reinforcement is smaller than that of the concrete in beams, in case of axially loaded columns, because of the uncertainty of the stress in steel reaching yield or the possibility of the bars buckling, a common partial safety factor is applied to both the materials. The problem of spalling of concrete and local buckling of the main bars limit the design strength of reinforcement to a relatively lower value when compared with that in beams.

The *design strength* of a RCC tied short column can be expressed as

$$P_a = \frac{A_c f_{cc} + A_s f_{sc}}{\gamma_m} \tag{9.23}$$

where γ_m = partial safety factor applied to the materials and is taken equal to 1.5.

The substitution of various values in Eq. (9.23) gives

$$P_a = 0.4 f_{ck} A_c + 0.67 f_y A_s \tag{9.24}$$

The allowable load under the live load condition can be obtained from the above equation by dividing P_a by the partial safety applied to the loads, so

$$P_{a2} = \frac{P_u}{\gamma_f} = \frac{0.4}{1.5} f_{ck} A_c + \frac{0.67}{1.5} f_y A_s$$

$$= 0.267 f_{ck} A_c + 0.467 f_y A_s \qquad (9.25)$$

The working stress design method suggests an allowable load as (vide Chapter 5)

$$P_{a1} = \sigma_{cc} A_c + \sigma_{sc} A_s \qquad (9.26)$$

where σ_{cc} = allowable direct compressive stress in concrete and it is approximately equal to

$$f_{ck}/4 = 0.25\, f_{ck}$$

σ_{sc} = allowable compressive stress in steel which is approximately equal to $f_y/2$.

Therefore, the allowable load obtained from the limit state design consideration checks approximately with that from the working stress design. In working stress design, there appears to be a small gain of about 6% in concrete capacity and loss of 6% in the capacity of the steel in the allowable load when compared with the limit state design. Equation (9.24) can be rearranged as

$$P_a = 0.4 f_{ck} A_c \left(1 + 1.67 \frac{A_s f_y}{A_c f_{ck}}\right)$$

$$= 0.4 f_{ck} (A_g - A_s) + 0.67\, f_y A_s$$

$$= 0.4 f_{ck} A_g \left(1 + \left(1.67 \frac{f_y}{f_{ck}} - 1\right) \frac{A_s}{A_g}\right)$$

$$P_a = 0.4 f_{ck} A_g (1 + (1.67\, \alpha - 1) p) \qquad (9.27)$$

where A_g = gross area of the cross-section

$$p = \frac{A_s}{A_g}, \quad \alpha = \frac{f_y}{f_{ck}}$$

The design load from the load analysis can be obtained as

$$P_d = \gamma_{fd} W_d + \gamma_{fl} W_l, \text{ or}$$

$$= \gamma_{fd} W_d + \gamma_{fl} W_w, \text{ or}$$

$$= \gamma_{fc} (W_d + W_l + W_w) \qquad (9.28)$$

where γ_{fd}, γ_{fl} are the partial safety factors applied to the dead and live loads respectively = 1.5

γ_{fc} = combined partial safety factor applied to the combined DL, LL and wind or earthquake load = 1.2

The design criterion is that the load carrying capacity of the column should be more than the design load, that is

$$P_a \geq P_d$$

or

$$0.4 f_{ck} A_g (1 + (1.67\, \alpha - 1) p) \geq P_d \qquad (9.30)$$

or

$$A_g \geq \frac{2.5\, P_d}{f_{ck} (1 + (1.67\, \alpha - 1) p)} \qquad (9.31)$$

Equation (9.31) helps in the direct design for a preassigned percentage of the reinforcement.

Long columns. Design of long columns is very similar to that of the short columns except that a load reduction factor is applied. A column is said to be long if the slenderness factor is greater than twelve. In such a case the load carrying capacity is given by

$$P_{al} = C_r P_a \tag{9.32}$$

where
$$C_r = 1.25 - \frac{L_e}{48b} \tag{9.33}$$

The equivalent short column design load can be obtained by dividing P_d by C_r and then solving for the design details. This method is illustrated through examples.

There are situations where an architect preassigns the size of the column, in such cases the area of the reinforcement can be obtained.

$$A_s = \frac{P_d - 0.4f_{ck}\,A_g}{0.67f_y - 0.4f_{ck}} \tag{9.34}$$

9.3 Design Examples of Tied Columns

EXAMPLE 9.1 *Design of short tied column.* A column resting on an independent footing supports a flat slab. The total load on the column is 590 kN and the height of the slab above the footing is 4.5 m. Design the column with M20 concrete and HYSD bars.

Design of the section. The characteristic strengths and the allowable strengths in the concrete and steel are:

$$f_{ck} = 20 \text{ MPa and } f_y = 415 \text{ MPa}$$

$$f_{cc} = 0.4f_{ck} = 8 \text{ MPa, and } f_{sc} = 0.67f_y = 278 \text{ MPa}$$

Assume about 1% reinforcement in the column and the column as a short one. Then the load carrying capacity of the column is

$$P_a = 0.4A_c f_{ck} + 0.67A_s f_y = 8(A_g - A_s) + 278\,A_s$$

$$= 8A_g + 270A_s = 8A_g + 270\,\frac{(A_g)}{100} = 10.7\,A_g$$

Assuming the self-weight of the column as 10 kN, the design load on the column with 1.5 partial safety factor is

$$P_d = \gamma_f P = 1.5(590 + 10) = 900 \text{ kN}$$

Equating the capacity of column to the design load, we have

$$P_d = P_a = 10.7A_g$$

or
$$A_g = \frac{P_d}{10.7} = \frac{900\,000}{10.70} = 84\,112 \text{ mm}^2$$

Assuming a square section, the column size should be equal to

$$b = D = \sqrt{A_g} = 290 \text{ mm}$$

$$A_s = pA_g = 0.01(84\,112) = 841 \text{ mm}^2$$

Use 4 numbers of 16 mm bars, then the actual area of the steel provided is

$$A_s = 4(201) = 804 \text{ mm}^2$$

Then the requirement of gross area of the section is given by

$$8(A_g - A_s) + 278A_s = P_d$$

or

$$A_g = \frac{P_d - 278A_s}{8} + A_s$$

$$= \frac{900\,000 - 278(804)}{8} + 804 = 86\,100 \text{ mm}^2$$

Use $\quad b = D = 295$ mm then $A_g = 87\,025$ mm^2

The weight of the column is

$$W_g = (0.295)^2(4.5)(25) = 9.8 \text{ kN}$$

So the assumed weight is reasonable.

$$\frac{100\,A_s}{A_g} = \frac{80\,400}{87\,025} = 0.92\%$$

The column can be treated as fixed at one end and hinged at the other end, then the effective length of the column can be taken as

$$L_e = 0.8L = 0.8(4.5) = 3.6 \text{ m}$$

The slenderness factor of the column is

$$L_e = \frac{3.6}{0.295} = 12.2$$

The column can be treated as a short one as assumed.

Design of the ties. The diameter of the tie can be selected about one-fourth of the diameter of the main bars subject to a minimum of 5 mm. Let the diameter of the tie = 6 mm. Then the spacing of the ties should be limited to

$$s_v \leq b = 295 \text{ mm or}$$

$$\leq 16\phi = 16(16) = 256 \text{ mm or}$$

$$\leq 48\phi_v = 48(6) = 288 \text{ mm}$$

Provide $\phi6$ ties at 250 mm spacing

As per the working stress design method, the size of the column is found to be 305 mm and the reinforcement details are exactly the same. There is a saving of 6.5% in the concrete and an overall saving of about 3% by LSD.

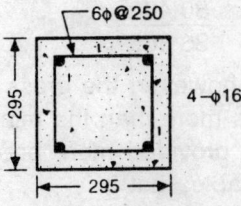

Fig. 9.1 Cross-section of Column in Example 9.1 (dimensions in mm).

EXAMPLE 9.2 *Design of short column*. A column of effective height of 3.6 m is subjected to an axial load of 600 kN including the self-weight, and the size of the column is fixed at 350 mm by 300 mm. Design the column using M20 concrete and HYSD bars.

Design of the main reinforcement. The allowed strengths in the materials are:

$$f_{cc} = 8 \text{ MPa and } f_{sc} = 278 \text{ MPa}$$

The slenderness factor of the column is

$$g = \frac{L_e}{b} = \frac{3.6}{0.3} = 12$$

So the column can be treated as a short one.

$$A_g = bD = 300(350) = 105\ 000 \text{ mm}^2$$

The load carrying capacity of the column is

$$P_a = A_c f_{cc} + A_s f_{sc} = (A_g - A_s)8 + A_s(278)$$
$$= (105\ 000)8 + 270A_s$$

Equating the column capacity to the design load, we have

$$(105\ 000)8 + 270A_s = P_d = 1.5P = 1.5(600\ 000) = 900\ 000 \text{ N}$$

or

$$A_s = \frac{900\ 000 - 840\ 000}{270} = 216 \text{ mm}^2$$

The minimum percentage of steel required is 0.8 and it is given by

$$A_{sm} = 0.008A_g = 0.008(105000) = 840 \text{ mm}^2$$

Try 4 numbers of $\phi16$ bars, then $A_s = 804 \text{ mm}^2$.

The capacity of 4 nos. of $\phi16$ bars is

$$P_s = A_s f_{sc} = 804(278) = 223\ 512 \text{ N}$$

Then the load to be carried by the concrete is

$$P_d - P_s = 900\ 000 - 223\ 512 = 676\ 488 \text{ N}$$

The minimum area of the concrete needed to resist the load is

$$A_{cm} = \frac{P_d - P_s}{f_{cc}} = \frac{676\ 488}{8} = 84561 \text{ mm}^2$$

The corresponding gross area required is

$$A_{gm} = A_{cm} + A_s = 85\ 365 \text{ mm}^2$$

The percentage of the reinforcement on the minimum gross area is

$$p = \frac{100\ A_s}{A_{gm}} = \frac{80\ 400}{85\ 365} = 0.94\%$$

Minimum area of steel is provided; however, the gross area of the section actually provided is $300(350) = 105\ 000 \text{ mm}^2$ which is more than the minimum gross area required. One need not really pay a penalty by having provided more area of the concrete than the minimum required. The design is also acceptable, that is

Provide 4 nos. of $\phi16$ as main bars and ties of $\phi6$ at 250 mm spacing.

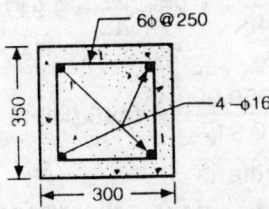

Fig. 9.2 Cross-section of Column in Example 9.2 (dimensions in mm).

EXAMPLE 9.3 *Design of a long tied column.* A column is subjected to an axial load of 700 kN and has an effective height of 5.275 m. Design the column with M20 concrete and HYSD bars.

Design data

$$L_e = 5.275 \text{ m}, \ f_{ck} = 20 \text{ MPa}, \ f_y = 415 \text{ MPa}$$

$$f_{cc} = 8 \text{ MPa}, \ f_{sc} = 278 \text{ MPa}, \ \alpha = f_{sc}/f_{cc} = 34.75$$

Design load = P_d = 1.5(700) = 1050 kN.

Design of the section. Assuming about 1% steel and the column as short one, the load carrying capacity of the section is

$$P_a = A_c f_{cc} + A_s f_{sc}$$

$$= A_g f_{cc} \left(1 + \frac{A_s}{A_g} \left(\frac{f_{sc}}{f_{cc}} - 1 \right) \right) = A_g f_{cc} (1 + p(\alpha - 1))$$

$$= A_g (8)(1 + 0.01(33.75)) = 1.337(8) \ A_g$$

The design condition is $P_a \geq P_d$, then

$$A_g \geq \frac{P_d}{1.337 \ (8)} = \frac{1050 \ 000}{1.337 \ (8)} = 98 \ 168 \text{ mm}^2$$

Assuming a square cross-section, then the size of the column is

$$b = D = \sqrt{A_g} = 313 \text{ mm}. \ \ Try \ b = D = 320 \text{ mm}$$

The slenderness factor for the column is

$$g = \frac{L_e}{b} = \frac{5.275}{0.32} = 16.5$$

The slenderness factor is more than 12, so the column should be treated as a long one and an appropriate reduction factor should be applied. It is better to select an appropriate number of reinforcement bars and then determine the size of the column. Since the size of the column is likely to be around 330 mm, we can use a slenderness factor of 16.

Approximate area of the compression reinforcement is

$$A_s \text{ (approx.)} = 0.01(A_g) = 0.01(98 \ 168) = 982 \text{ mm}^2$$

Provide 4 numbers of 18 mm bars. Then the area of the compression steel provided is

$$A_s = 4(254) = 1016 \text{ mm}^2$$

The load reduction factor is

$$C_r = 1.25 - \frac{g}{48} = 1.25 - \frac{16}{48} = 0.917$$

The equivalent short column load is

$$P_0 = \frac{P_d}{C_r} = \frac{1050\ 000}{0.917} = 1145\ 040\ N$$

The load to be carried by the concrete is

$$P_c = P_o - A_s f_{sc} = 1145\ 040 - 1016(278) = 862\ 590\ N$$

The minimum sectional area of the concrete required is

$$A_c = \frac{P_c}{f_{cc}} = \frac{862\ 590}{8} = 107\ 824\ mm^2$$

$$A_s = 1016\ m^2$$

$$A_g = A_c + A_s = \underline{108\ 840\ mm^2}$$

The size of the square cross-section is

$$b = D = \sqrt{A_g} = \sqrt{108\ 840} = 330\ mm.$$

Use

$$b = D = 330\ mm,\ \ then$$

$$g = \frac{L_e}{b} = \frac{5275}{330} = 16$$

The value of g assumed is 16 against an actual value of 16. Therefore, the design is just right.

Design of transverse reinforcement. Select 6 mm dia ties, then the spacing of the ties is

$$s_v \leq b = 330\ mm\ or$$

$$\leq 16\phi = 16(18) = 288\ mm\ or$$

$$\leq 48\phi_v = 48(6) = 288\ mm$$

Provide 6 mm ties at 280 mm spacing. Figure 9.3 illustrates the section.

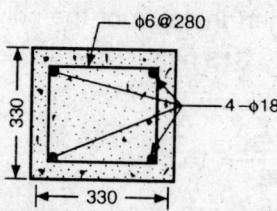

Fig. 9.3 Cross-section of Column in Example 9.3 (dimensions in mm).

EXAMPLE 9.4 *Design of a long-column.* An axial load on a column including the self-weight is 700 kN and its effective length is 5. 275 m. The size of the column is fixed as 300 by 300 mm. Design the column with M20 concrete and HYSD bars.
Design data

$$f_{ck} = 20\ MPa, \qquad f_y = 415\ MPa,$$
$$f_{cc} = 8\ MPa, \qquad f_{sc} = 278\ MPa,$$
$$L_e = 5.275\ m, \qquad b = D = 0.3\ m.$$

357

$$\frac{L_e}{b} = \frac{5.275}{0.3} = 17.6 > 12.$$

The design load $P_d = \gamma_f P = 1.5(700) = 1050$ kN

Design of the section. The slenderness factor of the column is 17.6, so the column is to be treated as long one and the load reduction factor is

$$C_r = 1.25 - \frac{L_e}{48b} = 0.883$$

The equivalent short column load is

$$P_0 = \frac{P_d}{C_r} = \frac{1050}{0.883} = 1189 \text{ kN}$$

Since the size of the cross-section is fixed, one can calculate the load carrying capacity of the concrete by assuming an approximate percentage of steel.

Let the approximate percentage of the reinforcement be two. Then the load carrying capacity of the concrete section is

$$P_c = A_c f_{cc} = 0.98 A_g f_{cc}$$
$$= 0.98(300)(300)(8) = 705\ 600 \text{ N}$$

The load to be carried by the steel is

$$P_s = P_d - P_c = 1189\ 000 - 705\ 600 = 483\ 400 \text{ N}$$

The area of the compression reinforcement required is

$$A_s = \frac{P_s}{f_{sc}} = \frac{483\ 400}{278} = 1739 \text{ mm}^2$$

Provide 6 numbers of φ20 main bars, then the actual area is

$$A_s = 6(314) = 1885 \text{ mm}^2$$

The percentage of the reinforcement is

$$p = \frac{100\ A_s}{A_g} = \frac{188\ 500}{900\ 000} = 2.1\%$$

This value checks well with the assumed percentage.

Select 6 mm ties, then the maximum spacing of the tie is

$s_v \leq b = 300$ mm or
$\leq 16\phi = 320$ mm or
$\leq 48\phi_v = 288$ mm

The *final design* details are:
$b = D = 300$ mm
$A_s = 1885 \text{ mm}^2 = 6 - \phi20$
6 mm ties at 280 mm spacing.

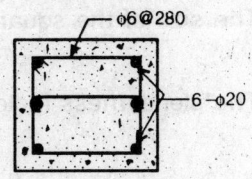

The size of the column is exactly same as that designed by the WSD.

EXAMPLE 9.5 *Design of a column with maximum percentage of steel.* A column is subjected to a load of 700 kN including the self-weight. The effective length of the column is 5.275 m. Design a column of about the smallest possible size with M20 concrete and HYSD bars.

Design data

$$L_e = 5.275 \text{ m, } f_{sc} = 278 \text{ MPa, } f_{cc} = 8 \text{ MPa.}$$

The smallest size column is possible with the largest percentage of the reinforcement or largest slenderness factor. So let $p = 6\%$. The design limit load with 1.5 as the partial safety factor is

$$P_d = 1.5\ P = 1.5(700) = 1050\ kN$$

Design of the section. Since the column will be a long one, we can design the section by trial and correction. First determine the size of the section for a short column with 6% steel. The load capacity on the short column is

$$P = A_c f_{cc} + A_s f_{sc} = A_g f_{cc}\left(1 + \frac{A_s}{A_g}\left(\frac{f_{sc}}{f_{cc}} - 1\right)\right)$$

$$= A_g\ (8)(1 + 0.06\ (32.75)) = 24.2\ A_g$$

The approximate size of the column assuming it to be a square one, is

$$a = \sqrt{A_g} = \sqrt{\left(\frac{P_d}{24.2}\right)} = \sqrt{\left(\frac{1050\ 000}{24.2}\right)} = 208\ mm$$

The slenderness factor of the column is

$$\frac{L_e}{a} = \frac{5275}{208} = 25.2 > 12$$

This leads the design to a long column. Since the load will be enhanced, the size of the column required will also increase and, consequently, the slenderness factor will decrease. Therefore, assume the slenderness factor as 23, then the load reduction factor is

$$C_r = 1.25 - \frac{L_e}{48b} = 1.25 - \frac{23}{48} = 0.77$$

The corresponding equivalent short column load is

$$P_0 = \frac{P_d}{C_r} = \frac{1050\ 000}{0.77} = 1363\ 640\ N$$

The gross area of the column for 6% steel is then given by

$$A_g = \frac{P_0}{24.2} = 56\ 350\ mm^2$$

The size of the square column is

$$a = (b = D) = \sqrt{A_g} = 238\ mm$$

The slenderness factor of the column is

$$g = \frac{L_e}{b} = \frac{5275}{238} = 22$$

The assumed value is 23 against an approximate design value of 22; therefore, the assumption is slightly on the safer side. The actual size can be rounded off after selecting the area of the steel. The minimum required reinforcement is

$$A_s = pA_g = 0.06(238)(238) = 3399\ mm^2$$

The actual area of the reinforcement must be selected less than the above value, so that it does not exceed 6%. Select 8 numbers of 22 mm bars, then

$$A_s = 8(380) = 3040\ mm^2$$

The load carrying capacity of the steel is

$$P_s = A_s f_{sc} = 3040(278) = 845\ 120\ \text{N}$$

The actual slenderness factor of the column can be assumed as 22. Therefore, the reduction factor and the corresponding equivalent short column load are:

$$C_r = 1.25 - \frac{22}{48} = 0.7916$$

$$P_0 = \frac{P_d}{C_r} = \frac{1050\ 000}{0.7916} = 1326\ 430\ \text{N}$$

Load to be carried by the concrete is

$$P_c = P_0 - P_s = 1326\ 430 - 845\ 120 = 481\ 310$$

$$A_c = \frac{P_c}{f_{cc}} = \frac{481\ 310}{8} = 60\ 163\ \text{mm}^2$$

$$A_g = A_c + A_s = 63202\ \text{mm}^2$$

$$b = D = \sqrt{A_g} = 252\ \text{mm}$$

Slenderness assumed is slightly more than the actual, so it is on the safer side.
   The final design details are

$$b = D = 250\ \text{mm}$$

$$A_s = 3040\ \text{mm}^2 = 8 - \phi22$$

6 mm ties at 250 mm in two pairs as shown in Fig. 9.4.
Note: 22 mm bars even though listed but normally not available in market

EXAMPLE 9.6 *Cost comparison of long column with different percentage of steel.* A long column of effective height 5.275 m is to carry a load of 700 kN. Compare its cost of construction with three different percentages of steel.
   Design data

$$L_e = 5.275\ \text{m},\ P = 700\ \text{kN}$$

$$f_{ck} = 20\ \text{MPa},\ f_y = 415\ \text{MPa}$$

Unit cost of different quantities including supplying and placing in position are:

   Cost of M 20 concrete = Rs. 2700/m^3
   Cost of HYSD bars   = Rs. 2000/kN
   Cost of formwork, etc. = Rs. 160/m^2

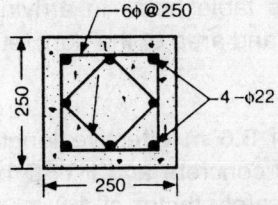

Fig. 9.4 Cross-section of Column in Example 9.5 (dimensions in mm).

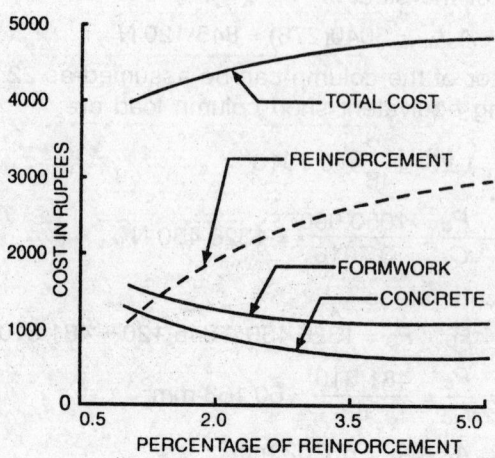

Fig. 9.5 Approximate Cost Comparison Example 9.6.

Cost comparisons. The column has been designed in Examples 9.3 to 9.5 with different percentages of steel. The design details from these examples are listed in Table 9.1. The quantities of the steel for a total length of column of 5.275 and the cost of the different items are listed in the same table. It may be noticed that the amount of the reinforcement varied from 780 N/m^3 of concrete to 2300 N/m^3. The overall cost of the column varied from Rs. 3589 to Rs. 4393 as the percentage of the reinforcement increase from 0.9% to 4.86%. The minimum cost of the column is at 0.9% of the reinforcement. The costs of the concrete, steel and the formwork are comparable to each other and none is negligible. Figure 9.5 illustrates the influence of variation of the percentage of the reinforcement on the cost of different components of the column.

9.4 Design Aid Tables for Axially Loaded Columns

It could be seen from the previous examples that the design is done by trials. Even though the design method is simple, it can be further simplified with the design aids. Table 9.2 gives the recommended main reinforcement and ties for a given area of the concrete. It also gives the load capacities of the reinforcement and the concrete for different characteristic strengths of the steel and the concrete. This table helps in arriving at the desired load capacity by selecting appropriate reinforcement and area of the concrete. The use of the table is illustrated through examples.

EXAMPLE 9.7 Design a column of 3.6 m effective length to carry a total load of 600 kN (including the self-weight) with M20 concrete and HYSD bars.

The design load with a partial safety factor of 1.5 is
$$P_d = 1.5(600) = 900 \text{ kN}$$

Table 9.1 Comparisons of Designs

Details	Percentage of steel		
	0.9	2.0	4.86
Example no.	9.3	9.4	9.5
g	16.0	17.6	21.5
$b = D$ (mm)	330	300	250
A_c (mm^2)	107 830	88 115	59 460
A_s (mm^2)	1 016	1 885	3 040
$\phi 6$ ties numbers	19	19	38
Weight of ties (N)	52	52	100
Weight of main bars (N)	418	776	1251
Cost of concrete (Rs.)	1535	1255	847
Cost of steel (Rs.)	940	1630	2702
Cost of formwork (Rs.)	1114	1013	844
Total cost (by LSD)	3589	3898	4393

One can assume that about 50 to 70% of the load is carried by the concrete. So the approximate load to be carried by the concrete is about 600 kN to 750 kN. Move down in column 10 of Table 9.2, which corresponds to the capacity of M20 concrete, and select a zone of capacity in the range of 600 to 750 kN. In that zone, the sum of the capacities of the concrete and the reinforcement must make upto 900 kN. This happens at

$$A_g = 900 \text{ cm}^2 \quad \text{and} \quad A_s = 4 - \phi 16$$

The corresponding capacities are

$$P_c = 713 \text{ kN}$$
$$P_s = 223 \text{ kN}$$
$$\text{Total } \overline{P_a = 936 \text{ kN}}$$

The total capacity is 936 kN as against 900 kN load. Therefore, modify the concrete area marginally. If a square column is selected, the size of the column should be

$$b = D = \sqrt{A_g} = 300 \text{ mm}$$

$$\text{Slenderness factor} = \frac{3.60}{0.3} = 12$$

Use $b = D = 295$ mm.

Provide 4 numbers of $\phi 16$ as main bars and $\phi 6$ ties at 250 mm spacing as shown in Table 9.2. This is exactly same as the solution given in Example 9.1.

EXAMPLE 9.8 A column of effective height 5.275 m is subjected to a total load of 700 kN. Design a square column with M20 concrete and HYSD bars.
The design load with partial safety factor of 1.5 is :

$$P_d = 1.5(700) = 1050 \text{ kN}$$

If the column is treated as a short column, one can select from Table 9.2 a gross area of 1100 cm^2 against which the capacity of the concrete is 872 kN and the capacity of the reinforcement is 283 kN. The total is 1155 kN which is more than the limit load. The size of the column is

Table 9.2 Capacity Calculation for Column Sections

No.	Gross area of section (cm²)	Longitudinal Reinforcement		Capacity in kN for stress in N/mm²		Ties (in mm)		Concrete capacity (kN)		
								M15	M20	M25
		Nos	Dia(mm)	167.5	278	Dia	Spacing	$f_{cc} = 6$	$f_{cc} = 8$	$f_{cc} = 10$
1	2	3	4	5	6	7	8	9	10	11
1	200	4	12	75	125	5	125	117	156	195
2	250					5	150	147	196	245
3	300					5	150	177	236	295
4	350							207	276	345
5	400							237	316	395
6	450							267	356	445
7	500					5	175	297	396	495
8	550					5	175	327	436	545
9	600	4	14	103	171	5	200	369	492	615
10	650							399	532	665
11	700							429	572	715
12	750							459	612	765
13	800	4	16	134	223	6	250	475	633	792
14	850							505	673	842
15	900							535	713	892
16	1000							595	793	992
17	1100	4	18	170	283	6	275	654	872	1090
18	1200							714	952	1190
19	1300	4	20	210	349	6	300	772	1030	1287
20	1400							832	1110	1387
21	1500							892	1190	1487
22	1600	8	16	269	446	6	250	950	1267	1583
23	1700							1010	1347	1683
24	1800							1070	1427	1783
25	1900							1130	1507	1883
26	2000	8	18	340	566	8	275	1188	1584	1980
27	2200							1308	1744	2180
28	2400							1428	1904	2380
29	2600	8	20	420	698	8	300	1545	2059	2574
30	2800							1665	2219	2774
31	3000							1785	2379	2974
32	3300	4.4	20, 25	538	894	8	300	1960	2614	3268
33	3600							2140	2854	3568
34	3900							2320	3090	3868
35	4200	12	20	630	1047	8	300	2497	3329	4162
36	4600	12	20	630	1047	8	300	2737	3649	4562
37	5000	12	22	763	1267	8	350	2972	3963	4954

$$b = D = \sqrt{A_g} = \sqrt{1100} = 33.17 \text{ cm}$$

The slenderness factor of the column is

$$\frac{L_e}{b} = \frac{5.275}{0.3317} = 15.8$$

Therefore, the column should be designed as a long one. The load reduction factor for a slenderness factor of 16 is

$$C_r = 1.2 - \frac{L_e}{48b} = 1.2 - \frac{16}{48} = 0.917$$

The equivalent short column load is

$$P_0 = \frac{P_d}{C_r} = \frac{1050}{0.917} = 1145 \text{ kN}$$

From the column 6 of Table 9.2, the capacity of 4 numbers of $\phi 18$ bars is:

$$P_s = 283 \text{ kN}$$

The load to be carried by the concrete is

$$P_0 - P_s = 1145 - 283 = 862 \text{ kN}$$

Moving down the column 10 of Table 9.2, the load capacity corresponds to an area of 1100 cm^2 is 872 kN. Therefore, select a gross area of

$$A_g = 1100(862)/872 = 1087 \text{ cm}^2$$

The size of the column is

$$b = D = \sqrt{A_g} = 33.0 \text{ cm}$$

Use 330 by 330 mm column with 4 numbers of $\phi 18$ bars and $\phi 6$ ties at 275 mm spacing.

The solution of Example 9.3 is same as above except the tie spacing here is 275 mm as against 280 mm recommended there.

9.5 Composite Columns

A composite column consists of steel or cast iron core with reinforcement placed around it, and put together in concrete. Here again equilibrium model is selected which need not necessarily satisfy the compatibility of the strains. Such a model can be called a *plastic model* as the stresses in the materials may not have the elastic strain compatibility. The load carrying capacity of the column consists of three parts and each one of them can be expressed as

$$f_{sc} = 167.5$$

$$f_y \geq 240$$

$$P_c = A_c f_{cc} \tag{9.35}$$

$$P_s = A_s f_{sc} \tag{9.36}$$

$$P_{c0} = A_{s0} f_{s0} \tag{9.37}$$

where P_c, P_s and P_{c0} are the load capacities of the concrete, the reinforcement steel and the composite core respectively.

A_{s0} and f_{s0} are the area of the cross-section of the composite core and the design compressive strength in the material of the core.

The design strength in the core material are given in Table 9.3. The total load capacity of a short column is given by

$$P_a = P_c + P_s + P_{s0}$$

Table 9.3 Design Compressive Strengths (in MPa)

f_{ck}	Concrete ($f_{cc} = 0.4 f_{ck}$)				
	15	20	25	30	35
f_{cc}	6	8	10	12	14

Type of steel	Reinforcement ($f_{sc} = 0.67 f_y$)	
	Mild or Medium	HYSD
f_{sc}	167.5	278

Type of metal	Composite–metal core ($f_{sc} = 0.67 f_y$)	
	Cast iron	Mild steel
f_{s0}	95	167.5

$$P_a = A_c f_{cc} + A_s f_{sc} + A_{s0} f_{s0} \tag{9.38}$$

The tendency of spalling of the concrete cover or the buckling of the reinforcement can be eliminated by providing proper bonding and transverse reinforcement. The minimum amount of the transverse reinforcement as prescribed for the RCC columns must be used in the composite columns. In addition, the maximum area of the core must be limited to 20% of the gross area of the cross-section. Sometimes, even the total 20% becomes rather too large for good workability of the concrete while placing. There should be a good working and bonding space between the core and the spiral or the ties. A minimum of 75 mm clearance between the core and the helical reinforcement or 50 mm *clearance* between the core and the ties must be maintained. A reduction factor, as discussed earlier, should be applied to the long columns. Composite columns are normally provided in case of large loads and where the size restriction is severe.

EXAMPLE 9.10 *Design of composite column*. A column of effective length 5.275 m is subjected to a load of 690 kN. Design a composite reinforced column using mild structural steel core and reinforcement bars, and M20 concrete.

   Design data

$$\text{Effective length} = L_e = 5.275 \text{ m}$$

Add extra 10 kN towards self weight then the total load = $P = 690 + 10 = 700$ kN
The design strengths in MPa are:

$$f_{cc} = 8, f_{sc} = 167.5 \text{ and}$$
$$f_{s0} = 167.5, f_{ck} = 20, f_y = 250$$

   The design limit load = $P_d = 1.5(700) = 1050$ kN

Design of the section. Assume that the column is a short one and about 40% of the load is carried by the core. Then the approximate area of the structural steel section is about 400000/167.5 = 2390 mm². Select ISMB 150 as the core composite, then

$$A_{s0} = 1900 \text{ mm}^2.$$

Use only about 0.8% reinforcement, then the load carrying capacity of the section is

$$P_a = A_c f_{sc} + A_s f_{sc} + A_{s0} f_{s0}$$

$$= (A_g - A_s - A_{s0}) f_{cc} + 0.008 A_g f_{sc} + 1900\ (167.5)$$

$$= (0.992\ A_g - 1900)(8) + 0.008\ A_g\ (167.5) + 318\ 250$$

$$= 9.276\ A_g + 303\ 050\ \text{N}$$

The load carrying capacity of the column can be equated to the design load, then we have

$$9.276\ A_g + 303\ 050 = P_d = 1050\ 000\ \text{N}$$

$$\text{or}\ A_g = \frac{1050\ 000 - 303\ 050}{9.276} = 80\ 525\ \text{mm}^2$$

Assuming a square cross-section, the size of the column is

$$b = D = \sqrt{A_g} = 284\ \text{mm}$$

$$\text{Slenderness factor} = \frac{L_e}{b} = \frac{5.275}{0.284} = 18.6$$

The column is a long one so a load reduction factor has to be applied. One should also ensure that there is enough clearance and cover space to the reinforcement. There should be a minimum of 50 mm clearance between the composite core and the ties, and about 40 mm cover to the main reinforcement. This requires the minimum sizes of:

$$b_{min} = 80 + 2(50 + 40) = 260\ \text{mm}$$

$$D_{min} = 150 + 2(50 + 40)\ 330\ \text{mm}$$

whereas the design sizes are 284 mm and 284 mm. Therefore, reduce the core area by selecting a lower structural shape. Select ISMB 125, the area of which is

$$A_{s0} = 1600\ \text{mm}^2$$

and the minimum size of a square column will then be

$$D_{min} = 125 + 2(50 + 40) = 305\ \text{mm}$$

Use $b = D = 320$ mm for calculation of slenderness factor

$$g = \frac{5.275}{0.32} = 16.48$$

$$C_r = 1.25 - \frac{16.48}{48} = 0.906$$

The corresponding equivalent short column load is

$$P_0 = \frac{P_d}{C_r} = \frac{1050\ 000}{0.906} = 1\ 158\ 940\ \text{N}$$

Assuming approximately two per cent reinforcement, the area of the concrete required is

$$A_c = A_g - A_s - A_{s0} = A_g - 0.02\ A_g - A_{s0}$$

$$= 0.98\ A_g - A_{s0} = 0.98(102\ 400) - 1660 = 98\ 692\ \text{mm}^2$$

The load carrying capacity of the section is

$$P_a = A_c f_{cc} + A_s f_{sc} + A_{s0} f_{s0}$$

$$= 98\ 692(8) + A_s(167.5) + 1660(167.5) = 167.5\ A_s + 1\ 067\ 580\ \text{N}$$

Equating the load capacity to the design load, we have

$$A_s = \frac{P_0 - 1067\,580}{167.5} = 545 \text{ mm}^2$$

Select 4 numbers of $\phi 14$ bars. Then the actual area of the reinforcement provided is

$$A_s = 4(154) = 615 \text{ mm}^2$$

and the percentage of the reinforcement is

$$p = \frac{100\,A_s}{A_G} = \frac{615\,00}{102\,400} = 0.6\%.$$

A reduction of 2% was given in the gross area towards the reinforcement; therefore, the design is slightly on the safer side.

Transverse reinforcement. Select ties of 6 mm, then the maximum spacing of the tie is limited to

$$s \leq b = 320 \text{ mm or}$$
$$\leq 16\phi = 16(14) = 224 \text{ or}$$
$$\leq 48\phi_v = 48(6) = 288$$

Final design details are

$$b = D = 320 \text{ mm}$$
$$A_{s0} = 1660 \text{ mm}^2 = (= \text{ISMB } 125)$$
$$A_s = 615 \text{ mm}^2 (= 4 - \phi 14)$$

6 mm ties at 220 mm spacing.
Figure 9.6 illustrates the section.

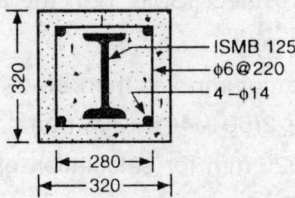

Fig. 9.6 Cross-section of Column in Example 9.10 (dimensions in mm).

Note: The problem of this example is same as those of Examples 9.3 to 9.5. One can see that there is not much gain in providing the composite column for small loads. The total percentage of the steel in this case is 2.2.

9.6 Infilled Columns

Steel pipes filled in with concrete and used as columns are called Infilled columns or concrete filled columns. Concrete when confined in a tube is subjected to triaxial force. The load carrying capacity of the column if filled with concrete can be expressed as:

$$P_a = 1.1\,A_c f_{cc}\left(1 - \frac{0.000025\,L^2}{r_c^2}\right) + A_s f_{sc}$$

where L = unsupported length of the column
$\quad A_s$ = area of section of the shell of the tube
$\quad r_c$ = radius of gyration of the concrete core

The infilled pipes are used for architectural purpose in buildings and in ocean structures. The strength in the confined concrete can easily be taken about 10 to 20% more than the corresponding value of the tied columns.

EXAMPLE 9.10 *Design of infilled column.*
Design data of the problem is:

$$L_e = 5.275 \text{ m}$$

$$f_{cc} = 8 \text{ MPa}, \ P = 700 \text{ kN}; \ f_{sc} = 167.5 \text{ MPa}$$

Let a = radius of the concrete core and t = thickness of the shell
The design limit load is= $P_d = 1.5(700) = 1050$ kN

Design of the section. The radius of gyration of the concrete core is:

$$r_c = \sqrt{\frac{I_c}{A_c}} = \sqrt{\frac{\pi a^4}{4\pi a^2}} = \frac{a}{2}$$

The design has to be carried out by trial. For this purpose, estimate the gross area of the concrete by neglecting the area of the shell to get an approximate idea. Then

$$A_c = \frac{P_d}{1.1 \ f_{cc}} = \frac{1050 \ 000}{8.8} = 119 \ 320 \text{ mm}^2$$

Let a_0 = radius of such a section, then it is given by

$$a_0 = \sqrt{\frac{A_c}{\pi}} = \sqrt{\frac{119 \ 320}{\pi}} = 195 \text{ mm}$$

Let the internal diameter of the pipe be 300 mm, and the thickness of the shell of the pipe be 4 mm. Then the properties of the section are :

$$r_c = \frac{a}{2} = \frac{150}{2} = 75 \text{ mm}$$

$$A_c = \pi 150^2 = 70 \ 685 \text{ mm}^2 \text{ and } A_s = \pi(154^2 - 150^2) \ 3820 \text{ mm}^2$$

The load carrying capacity of the column is

$$P_a = 1.1 \ A_c \left(1 - \frac{0.000025L^2}{r_c^2}\right) f_{cc} + A_s f_{sc}$$

$$= 1.1(70 \ 685)(1 - 0.12)(8) + 3820(167.5) = 1 \ 187 \ 230 \text{ N}$$

The steel pipes are normally rolled in standard inside diameter and three to four different thicknesses of the shell. The grade of the tubes varies from light to heavy which corresponds to the thickness of the shell. The tubes of 175 mm radius will have the thicknesses ranging from 5 mm to 10 mm.

9.7 Spirally Reinforced Columns

In circular cross-sectioned columns the transverse reinforcement can be an independent set of ties or a continuous helical tie known as spiral. A column in which one helical tie is provided to hold the main bars is called a *spirally reinforced* column. The helical tie holds not only the longitudinal bars in position but also the concrete in partial confinement. The capacity of the column is enhanced by the circumferential action of the spiral and, moreover, the ductility of the column is improved. The total compressive strain capacity in the concrete confined by the spiral reinforcement is about four times that corresponding to the tied columns. A typical load deformation behaviour of the tied, spiral and infilled columns have a similar deformation character upto a point close to the maximum load. Beyond this point, there is a substantial difference in their behaviour. The tied columns get spilt and fall off when the load reaches the ultimate load. Whereas in the case of a spirally reinforced column, the concrete crust or the cover over the spiral falls off slowly, thus lowering the resistance upto a point. The helical reinforcement contains the concrete inside the helical area and absorbs the load through hoop action. The load carrying capacity of the spiral columns increases marginally. The composite column behaves similar to the tie column and fails as soon as the spalling of the cover takes place. The infilled concrete column behaviour is similar to that of a spiral column. There is an improved crushing strength of the concrete and ductility in the behaviour.

The load carrying capacity of a spiral column can be expressed as

$$P_a = 1.05 \, (A_c f_{cc} + A_s f_{sc}) \tag{9.39}$$

It can be rearranged in the gross area of the concrete as

$$P_a = 1.05(A_g f_{cc} - A_s f_{cc} + A_s f_{cs})$$
$$= 1.05(A_g f_{cc}(1 + p(\alpha - 1))) \tag{9.40}$$

where $\quad p = \dfrac{A_s}{A_c}$ and $\alpha = \dfrac{f_{sc}}{f_{cc}}$

For a given set of materials and a percentage of the main reinforcement, the gross area of the cross-section of the column can be obtained from the equation.

At least there must be six numbers of the longitudinal bars and the minimum diameter of the bar must be at least 12 mm. The diameter of the spiral reinforcement must be more than one-fourth of the diameter of the bar with a minimum of 5 mm, that is

$$\phi_v \ge \frac{\phi}{4} \text{ or}$$
$$\ge 5 \text{ mm}$$

There are upper and lower bounds to the pitch of the spiral reinforcement so as to provide the confinement to the concrete and at the same time provide workability and interlocking of the cover with the core. This limitations are given by

$$s \le \frac{\text{core diameter of the column}}{6} \text{ or}$$

$$\le 75 \text{ mm or}$$
$$s \ge 30 \text{ mm or}$$
$$\ge 1.5\phi \text{ or}$$
$$\ge \text{maximum size of the aggregate or}$$
$$\ge 3\phi_v$$

The minimum cover and reinforcement and other limitations have already been stated.

EXAMPLE 9.11 *Design of spiral column with minimum reinforcement.* A column of effective length 5.275 m is subjected to a total load of 700 kN. Design a spiral column with M20 concrete and HYSD bars.

Design data

$$L_e = 5.275 \text{ m}, \quad P = 700 \text{ kN},$$
$$f_{ck} = 20 \text{ MPa}, \quad f_{cc} = 8 \text{ MPa},$$
$$f_y = 415 \text{ MPa}, \quad f_{sc} = 278 \text{ MPa}.$$

The design limit load $P_d = 1.5(700) = 1050$ kN.

Design of the section. Let the percentage of the reinforcement be about 0.8. Then the load carrying capacity of the column, assuming it short, is

$$P_a = 1.05(A_c f_{cc} + A_s f_{sc}) = 1.05((A_g - A_s)(8) + A_s(278))$$

$$= 1.05 \, A_g \left(8 + (278 - 8) \frac{A_s}{A_g} \right) = 1.05 \, A_g (8 + 270(0.008)) = 10.67 \, A_g$$

Equating the design load on the column to its capacity, we have

$$A_g = \frac{P_d}{10.67} = \frac{1050\,000}{10.67} = 98\,400 \text{ mm}^2$$

The area of the reinforcement needed is

$$A_s = 0.008 \, A_g = 787 \text{ mm}^2$$

Provide 6 numbers of 14 mm bars. Then the actual area of the reinforcement is

$$A_s = 6(154) = 924 \text{ mm}^2$$

The load carrying capacity of the reinforcement is

$$P_s = 1.05 \, A_s f_{sc} = 1.05(924)(278) = 269\,715 \text{ N}$$

The approximate diameter of the column is

$$D = \sqrt{\frac{4\,A_g}{\pi}} = 356 \text{ mm}$$

Slenderness factor $= \dfrac{L_e}{D} = \dfrac{5275}{356} = 14.8$

The slenderness factor is more than 12 so the column must be designed as a long one, and the corresponding load reduction factor is

$$C_r = 1.25 = \frac{14.8}{48} = 0.94$$

The equivalent short column load is

$$P_0 = \frac{P_d}{C_r} = \frac{1050\,000}{0.94} = 1\,115830 \text{ N}$$

The load to be carried by the concrete is

$$P_c = P_0 - P_s = 1\,115830 - 269\,715 = 846\,115 \text{ N}$$

Area of the concrete $= A_c = \dfrac{P_c}{1.05\,(8)} = 100\,400 \text{ mm}^2$

The gross area of the section is

$$A_g = A_c + A_s = 101\,300 \text{ mm}^2$$

The diameter of the column $D = \sqrt{\dfrac{4\,A_g}{\pi}} = 359 \text{ mm}$

Provide $D = 360$ mm and 6 number of $\phi14$ bars. The value is close to that assumed for slenderness factor and, hence, no revision on the equivalent load is needed.

Transverse reinforcement. The diameter enclosing the main bars is

$$D_c = D - 2(\text{cover}) = 360 - 80 = 280 \text{ mm}$$

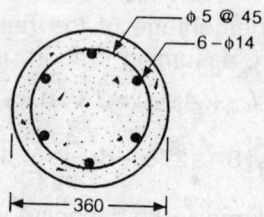

Fig. 9.7 Cross-section of Column in Example 9.11 (dimensions in mm).

Select 5 mm diameter spiral reinforcement. Then the pitch of the spiral is limited by

$$s \le \frac{D_c}{6} = \frac{280}{6} = 47 \text{ mm or}$$

$$\le 75 \text{ mm and}$$

$$\ge 1.5\phi = 1.5(14) = 21 \text{ mm or}$$

$$\ge 3\phi_v = 15 \text{ mm or}$$

$$\ge 30 \text{ mm}$$

Provide 5 mm bar spiral with a pitch of 45 mm

$$p = \frac{100 \, A_s}{A_g} = 0.86$$

9.8 Columns Subjected to Combined Bending and Axial Force: Notations

Columns are invariably subjected to combined axial force and bending moment. The bending moment might be caused either by the eccentricity of the external load, or by the bending moment from a transverse load, or by end moments coming from the interconnected members. Most members in practice act as plane frame members because of symmetry either in load or in the structure. In such cases, the columns are subjected to uniaxial bending moment in addition to the axial force. There are other cases where columns are subjected to biaxial bending moment.

In the limiting case of strength, many columns under combined axial force and bending are treated as cracked sections. Consider a load acting at an eccentricity on a column section as shown in Fig. 9.8. This load causes axial plus bending strains which together form the normal strains on the section. Figure 9.8a illustrates the notations of the sectional details and Figure 9.8b gives the location of load, central line, etc. Figure 9.8c illustrates the strains caused by the eccentrically acting load. For convenience of notations, some tensile strain is shown in the figure, but it is not necessary that all column sections develop tension. The development of the tension in the section is governed by the load, its eccentricity and the sectional properties.

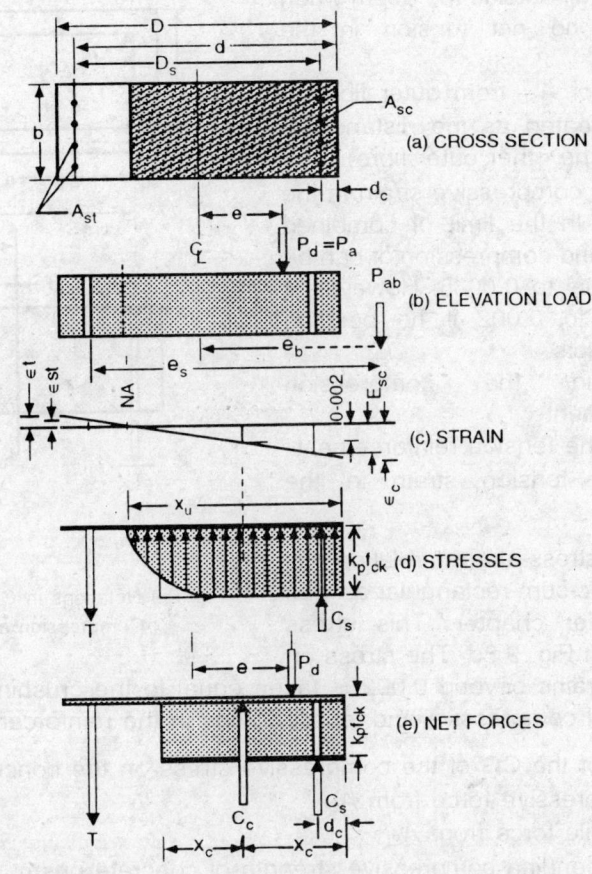

Fig. 9.8 Notations in Column Section under Compression and Bending Moment.

The notations in the column section are shown in Fig. 9.9

D_s = distance between the compression and tension reinforcement

A_{sc} = area of the reinforcement on the compression face. It is also called the area of the compression reinforcement

A_{st} = area of the reinforcement on the tension face. It is also called the tension reinforcement.

In some cases of high compressive load and low bending moment, all the area can be under compression and no tension may exist.

e = eccentricity of the load which is also equal to equivalent eccentricity = M/P. It is measured from the centroid of the gross section although it should really be from the centroid of the transformed section.

x_u = distance of the neutral axis from the extreme compression fibre. The neutral axis is indicated as falling within the depth of the section. Sometimes, it can also fall outside the depth when there is no net tension in the section.

d_c = distance of A_{sc} from outer fibre (it is also treated as the distance of A_{st} from the other outer fibre)

ϵ_c = maximum compressive strain in the concrete. In the limit of combined bending and compression, it can be taken equal to 0.0035. However, it is equal to 0.002 if no bending moment acts.

ϵ_{sc} = strain in the compression reinforcement

ϵ_{st} = strain in the tension reinforcement

ϵ_t = maximum tension strain in the section

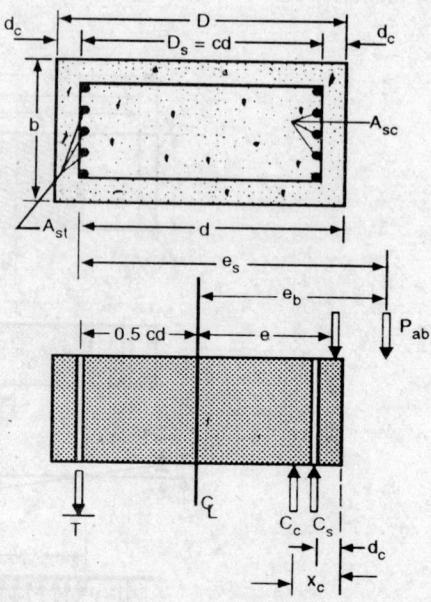

Fig. 9.9 Notations in Column Section under Compression and Bending.

The compressive stress-strain relation is idealized as a parabolic-cum-rectangular as was assumed in an earlier chapter. This stress distribution is shown in Fig. 9.8d. The stress on the concrete for all strains beyond 0.002 is taken equal to the crushing stress. Figure 9.8d also indicates the total compressive and tensile forces in the reinforcements.

Let x_c = distance of the CG of the compressive stress on the concrete

C_s = total compressive force from A_{sc}

T = Total tensile force from A_{st}

$k_p f_{ck}$ = crushing bending compressive strength of concrete prism

C_c = total compression from the concrete

The parabolic-cum-rectangular stress distribution presents problem in the derivation of the expression for C_c if the neutral axis falls outside the depth. There is a more elegant idealization of the compressive stress resulting into statically equivalent rectangular stress. Figure 9.8e shows the idealized rectangular stress block. The centre of the rectangle is placed exactly at x_c which is the CG of the actual stress distribution.

There are several possible modes of failure of a column subjected to combined forces. These are:

1. *Compression failure*
   a) Failure by crushing of the concrete without tension on the other face
      The concrete crushing strain = $(\epsilon_{c1} = 0.0035 - 0.75\,\epsilon_{c2})$

   b) Failure by crushing of the concrete with some tension in the other face
      The concrete crushing strain = $(\epsilon_{c1} = 0.0035)$

2. *Tension failure*
   a) The tension stress in the reinforcement reaches the yield limit before the strain in the compression concrete reaches the crushing strain ($\epsilon_{st} = \epsilon_y$; $\epsilon_c < 0.0035$).

   b) In case of water retaining structures, the tensile strain in concrete when reaches the limiting value, the tension failure in concrete takes place ($\epsilon_t = 0.0002$).

The design criteria of the columns for each of the modes are different; therefore, some limiting values are defined first. Figure 9.10 gives the interaction curve for all modes of failure of the columns. Point *A* indicates the failure load of the column subjected to axial load with nominal eccentricity. All columns must be designed for a nominal eccentricity and the moment caused by this eccentricity is indicated as M_o. As the bending moment on the column increases, the failure is controlled by the crushing strain in the concrete. Line *AB* gives the interaction curve between the moment and the axial force in which the failure of the column is controlled by the compression. Point *B* is called the balanced load point at which the column will fail by the simultaneous occurrence of the limiting strains in concrete and reinforcement. The axial force and bending moment corresponding to this failure are called balanced axial load and bending moment pair. Further increase in the bending moment results into the tension failure. The axial load carrying capacity of the section decreases rapidly in the tension failure zone as the moment increases, which is because of the higher tensile force of reinforcement. Curve *BC* of Figure 9.10 indicates the tension failure.

As the bending moment reaches the pure bending moment capacity, the axial load decreases to zero. Point *C* refers to pure moment zone and point *E* refers to ideally axial load capacity. The two critical load points are denoted by the following coordinates.

   Point C

$$M = M_{rb} = Kbd^2 f_{ck} \text{ and } P = 0 \tag{9.41}$$

   Point E

$$M = 0 \text{ and } P = P_{ao}$$

where P_{ao} = axial load capacity with zero eccentricity and it is

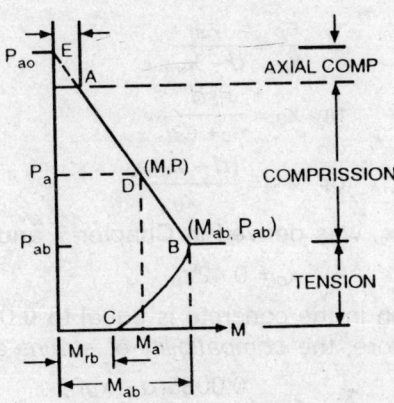

Fig. 9.10 Interaction Curve for Column Design.

$$P_{ao} = f_{cco} A_c + f_{sco} A_s \qquad (9.42)$$

Usually accepted limiting strengths are:

$$f_{cco} = 0.45 f_{ck} \text{ and } f_{sco} = 0.75 f_y$$

Equation (9.42) can be written as

$$P_{ao} = 0.45 f_{ck} (1 - p) A_g + 0.75 f_y A_s \qquad (9.43)$$

It is to be noted that A_s is the total area of the reinforcement in the column. For convenience A_c is taken as $(1 - p)A_g$. However, in most cases the area of the concrete is taken as A_g since area of the steel is small when compared with that of the concrete.

$$A_s = A_{sc} + A_{st} \qquad (9.44)$$

Let M_{ab} and P_{ab} be the bending moment and axial force respectively corresponding to point B. These two constitute a pair for simultaneous occurrence of the limiting strains in the steel and the compression concrete. One can write an interaction formula from the straight line EB. Next section gives the derivation for the balanced failure.

9.9 Limiting Capacity of Cracked Column

Refer to the notations given in Figs. 9.8 and 9.9. The total design reacting forces in the concrete and steel are:

$$C_c = \frac{b(2x_c)k_p f_{ck}(1 - p)}{\gamma_{mc}} \qquad (9.45)$$

$$C_s = \frac{A_{sc} f_y}{\gamma_{ms}} \qquad (9.46)$$

$$T = \frac{A_{st} f_{st}}{\gamma_{ms}} \qquad (9.47)$$

where $\gamma_{mc} = 1.5$ for the concrete and
$\gamma_{ms} = 1.15$ for the steel
$f_{st} =$ tensile stress (strength) in the tension steel

$(1 - p)$ is the factor for deducting the area occupied by the steel.

The stress in steel is computed from the compatibility of the strain. This *compatibility* relation is

$$\frac{\epsilon_c}{x_u} = \frac{\epsilon_{st}}{d - x_u} \qquad (9.48)$$

$$\text{or } x_u = \frac{\epsilon_c d}{\epsilon_c + \epsilon_{st}} \qquad (9.49)$$

$$\text{or } \epsilon_{st} = \frac{(d - x_u)\epsilon_c}{x_u} \qquad (9.50)$$

The relation between x_c and x_u was derived in Chapter 7 and it was shown to be

$$x_c = 0.42 x_u \qquad (9.51)$$

At the time of failure, the strain in the concrete is equal to 0.0035 irrespective of the tension or compression failure. Therefore, the *compatibility* of *strains* at failure is

$$\epsilon_{st} = \frac{0.0035(d - x_u)}{x_u}$$

$$f_{st} = E_s \epsilon_{st} = 0.0035 \, E_s \frac{(d - x_u)}{x_u} \tag{9.52}$$

Subject to the limit that

$$f_{st} \leq f_y \tag{9.53}$$

Substituting Eq. (9.51) in Eq. (9.45) along with the partial safety factor of 1.5 gives

$$C_c = 0.36(1 - p) \, bx_u f_{ck} \tag{9.54}$$

Note: The factor $(1 - p)$ is only an approximate value.
The equilibrium of the design forces on the cross-section is

$$C_c + C_s - T = P_d \tag{9.55}$$

where P_d = design load

$$0.36(1 - p)x_u bf_{ck} + 0.87 \, A_{sc}f_y - 0.87A_{st}f_{st} = P_d \tag{9.56}$$

It may be noted that the compressive reinforcement is assumed to have yielded at the time of failure. This is true in case of steel having a proof stress of the order 600 MPa or less, and when the compression reinforcement is reasonably close to the extreme compression fibre.

The moment equilibrium of the forces taken about the tensile reinforcement gives

$$C_c(d - x_c) + C_s(d - d_c) = P_d(e + 0.5D_s) \tag{9.57}$$

or where $P_d e = M_d$. Rewriting the above equation one can get

$$0.36(1 - p)x_u b(d - 0.42x_u)f_{ck} + 0.87A_{sc}f_y(d - d_c) = P_d e + 0.5P_d D_s. \tag{9.58}$$

Equations (9.56) and (9.58) are the governing equilibrium equations and (9.53) is the compatibility governing condition. One can solve an analysis or design problem from the three governing equations. The known and unknown values are listed in Table 9.5 for different problems.

Balanced force pair. The failure caused by the simultaneous occurrence of the failure strains in the concrete and the steel is called the balanced failure. That means

$$\epsilon_c = 0.0035$$

$$\epsilon_{st} = 0.002 + \frac{f_y}{1.15 \, E_s} \tag{9.59}$$

$$\epsilon_{sc} > 0.002 + \frac{f_y}{1.15 \, E_s} \tag{9.60}$$

The neutral axis distance can be evaluated from Eq. (9.49) and it is

$$x_u = \frac{0.0035 \, d}{0.0055 + 0.87 \, f_y / E_s} \tag{9.61}$$

The value of E_s for most of the steel reinforcement can be taken as

$$E_s = 200 \, 000 \, \text{MPa} \tag{9.62}$$

Then Eq. (9.61) reduces to (where f_y is in MPa)

$$x_u = \frac{805d}{1265 + f_y} \tag{9.63}$$

Table 9.5 Design or Analysis Variables

Problem	Variables		
	Given	To be found	Normally assumed
Design	P, e	$b, d, A_{sc}, A_{st}, d_c, x_u$	$b, A_{sc}/A_{st}, d_c$
Analysis	$b, d, A_{sc}, A_{st}, d_c$	x_u, P, e	

$$x_u = 0.513d \text{ for MS bars of } f_y = 250$$
$$= 0.479d \text{ for HYSD bars of } f_y = 415$$

$$\text{Let} \quad x_u = k_{ub}d \tag{9.64}$$

$$P_a = P_{ab} \tag{9.65}$$

$$e = e_b \tag{9.66}$$

$$\text{or} \quad M_a = M_{ab} \tag{9.67}$$

The equilibrium equations (9.56) and (9.58) can be written as (for simplicity of equations the value of p, which is in the range of 0.01 to 0.03, is neglected in comparison with 1):

$$0.36k_{ub}df_{ck} + 0.87f_y(A_{sc} - A_{st}) = P_{ab} \tag{9.68}$$

$$Kbd^2f_{ck} + 0.87A_{sc}D_sf_y = P_{ab}e_s \tag{9.69}$$

where $K = 0.36K_{ub}(1 - 0.42\,k_u)$ (9.70)

$$e_s = \text{distance of } P_{ab} \text{ from tension steel} = 0.5D_s + e_b$$

$$\text{Let} \quad p_t = \frac{A_{st}}{bd}, \ p_c = \frac{A_{sc}}{bd}; \ q = \frac{A_{sc}}{A_{st}} = \frac{p_c}{p_t} \tag{9.71}$$

$$\alpha = \frac{f_y}{f_{ck}} \tag{9.72}$$

$$c = \frac{D_s}{d} \tag{9.73}$$

Then the governing equations can be expressed as

$$0.36bdf_{ck}(k_{ub} + 2.4\alpha\,(p_c - p_s)) = P_{ab} \tag{9.74}$$

$$bd^2f_{ck}(K + 0.87\,\alpha p_c c) = P_{ab}e_s \tag{9.75}$$

In many problems the reinforcement is *symmetrically distributed*, in which case the above governing equations reduce to

$$P_{ab} = 0.36k_{ub}bdf_{ck} \tag{9.76}$$

$$e_s = \frac{(K + 0.435\,\alpha pc)d}{0.36k_{ub}} = \frac{2.778(K + 0.435\alpha pc)}{k_{ub}} \tag{9.77}$$

The eccentricity of P_{ab} with the centroid of the section is

$$e_b = e_s - 0.5D_s = e_s - 0.5cd \tag{9.78}$$

The value of e_b is the main eccentricity in which we will be interested.

9.10 Design of Columns Failing in Compression (Uncracked)

A column under combined bending and axial force failing in compression of the concrete has its failure point falling on the line EB of Fig. 9.10. Consider any arbitrary point D on the line EB, then from the linear relation, we have

$$\frac{P_{ao} - P_a}{P_{ao} - P_{ab}} = \frac{M_a}{M_{ab}} \tag{9.79}$$

where P_a = capacity of the column when subjected to a bending moment M_a.
The above equation can be rearranged as

$$P_{ao} - P_a = (P_{ao} - P_{ab}) \frac{M_a}{M_{ab}} \text{ or}$$

$$\text{or} \quad P_a = P_{ao} - (P_{ao} - P_{ab}) \frac{M_a}{M_{ab}} \tag{9.80}$$

$$\text{Let} \quad M_a = M_d = P_d e \text{ and } M_{ab} = P_{ab} e_b$$

$$\text{Then} \quad P_a = P_{ao} - (P_{ao} - P_{ab}) \frac{P_d e}{P_{ab} e_b}$$

$$= P_{ao} - \left(\frac{P_{ao}}{P_{ab}} - 1 \right) \frac{e}{e_b} P_d$$

Let the design load $P_d = P_a$, then the above equation reduces to

$$P_a \left(1 + \frac{e}{e_b} \left(\frac{P_{ao}}{P_{ab}} - 1 \right) \right) = P_{ao}$$

$$\text{or} \quad P_a = \frac{P_{ao}}{1 + (P_{ao}/P_{ab} - 1)e/e_b} \tag{9.81}$$

Note: The neutral axis will fall outside the section for uncracked section.

Equation (9.81) can be used for the analysis; however, if one is interested, the design solution has to be obtained by iterations. An attempt is made here to minimise the iterations by some reasonable preliminary assumptions. P_{ao} can be modified as

$$P_{ao} = 0.45 \, bDf_{ck} + 0.75 \, A_s f_y$$

$$= 0.45 bd f_{ck} \left(\frac{D}{d} + 1.67 p\alpha \right) \tag{9.82}$$

(Excluding the area of steel occupied in the concrete)

$$\text{Let} \quad D = d + d_c$$

$$\text{and} \quad \frac{D}{d} = 1 + \frac{d_c}{d} = 1 + u$$

$$\text{where} \quad u = \frac{d_c}{d} \tag{9.83}$$

Equations (9.82) and (9.74) can be expressed as

$$P_{ao} = 0.45 \, bd \, f_{ck} (1 + u + 1.67 p\alpha) \tag{9.84}$$

$$\frac{P_{ao}}{P_{ab}} = \frac{0.45 bd f_{ck} (1 + u + 1.67 p\alpha)}{0.36 bd f_{ck} (k_{ub} + 2.4\alpha(p_c - p_t))}$$

$$= \frac{1.25(1 + u + 1.67p\alpha)}{k_{ub} + 2.4\alpha(p_c - p_t)} \tag{9.85}$$

In case of $p_c = p_t$, we have

$$\frac{P_{ao}}{P_{ab}} = \frac{1.25(1 + u + 1.67p\alpha)}{k_{ub}} \tag{9.86}$$

From Eq. (9.81), we have (for $P_a = P_d$)

$$P_{ao} = P_d(1 + (P_{ao}/P_{ab} - 1)e/e_b) \tag{9.87}$$

The value P_{ao}/P_{ab} is non-dimensionalised.

From Eqs. (9.84) and (9.87), we have

$$bd = \frac{P_d(1 + (P_{ao}/P_{ab} - 1)e/e_b)}{0.45(1 + u + 1.67p\alpha)f_{ck}} \tag{9.88}$$

The quantities on the right-hand side are either known or can be computed for a pre-assigned percentage of steel or could even be estimated. Therefore, the value of bd can be computed. Before using the above equation one has to investigate whether the problem is of tension or compression failure. The compression failure is ensured if

$$P_d > P_{ab} \tag{9.89}$$

$$\text{or} \quad M_d < M_{ab} \tag{9.90}$$

$$\text{or} \quad P_d > 0.36bdf_{ck}(k_u + 2.4\alpha(P_c - p_t)) \tag{9.91}$$

From Eq. (9.90), we have

$$M_d \leq M_{ab}$$

$$\text{or} \quad e \leq \frac{P_{ab}e_b}{P_d} < e_b \tag{9.92}$$

From Eqs. (9.89) and (9.92), the compression failure is assumed if

$$e < e_b \tag{9.93}$$

The design is illustrated through a set of examples.

EXAMPLE 9.12 *Design of short column for combined axial force and bending moment.* A column of effective span 3.6 m is subjected to an axial force of 700 kN and a bending moment of 81.9 kNm in live load condition. Design the column with M20 concrete and HYSD bars.

Design data:

$$L_e = 3.6 \text{ m}, \ P = 700 \text{ kN}, \ M = 81.9 \text{ kNm} = 0.0819 \text{ MNm}$$

$$f_{ck} = 20 \text{ MPa}, \ f_y = 415 \text{ MPa}$$

The design loads are given by

$$P_d = 1.5 \ P = 1.05 \text{ MN}, \ M_d = 1.5 \ M = 0.12285 \text{ MNm}$$

The design coefficients for the grades of the material are

$$k_{ub} = 0.479 \text{ and } K = 0.138$$

$$\alpha = \frac{415}{20} = 20.75$$

Assume the percentage of steel approximately in the lower range.

$$\text{Let} \quad p = 0.01$$

Design of the section. The effective eccentricity of the load is

$$e = \frac{M}{P} = \frac{81.9}{700} = 0.117 \text{ m}$$

As a first trial assume $c = 0.8$ and $u = 0.1$

From Eq. (9.77)

$$e_s = \frac{2.778(K + 0.435pc\alpha)d}{k_{ub}} = 2.778 \frac{(0.138 + 0.435(0.01)(0.8)(20.75))d}{0.479}$$

$$= 1.22d \quad \text{and}$$

$$e_b = {}^=e_s - 0.5cd = (1.22 - 0.4)d = 0.82d$$

The value of e is only 0.117 m whereas e_b is 0.816 d. The compression failure occurs if

$$e < e_b \quad \text{or}$$

$$0.117 < 0.82d \quad \text{or} \quad d > 0.143 \text{ m}$$

The value of d is certainly going to be more than 0.143 m; therefore, the failure will be by compression and the governing equation is Eq. (9.88). It is

$$bd = \frac{P_d(1 + (P_{ao}/P_{ab} - 1)e/e_b)}{0.45(1 + u + 1.67p\alpha)f_{ck}}$$

$$\frac{P_{ao}}{P_{ab}} = \frac{1.25(1 + u + 1.67p\alpha)}{k_{ub} + 2.4\alpha(p_c - p_t)}$$

Let $p_c = p_t = 0.005$ and $u = 0.1$, then

$$\frac{P_{ao}}{P_{ab}} = \frac{1.25(1.1 + 0.346)}{0.479} = 3.77$$

$$bd = \frac{1.05(1 + 2.77(0.117)/0.82d)}{0.45(1.1 + 0.346(20))}$$

$$= 0.0807(1 + 0.4/d)$$

$$\text{or} \quad bd^2 - 0.0807d - 0.032 = 0$$

Let $b = 0.36$ m, then the solution of the above quadratic equation gives

$$d = 0.431 \text{ m}$$

The area of the reinforcement required is

$$A_s = pbd = 0.01(360)(431) = 1552 \text{ mm}^2$$

Use 4 *numbers* of 22 mm bars

$$A_s(\text{provided}) = 4(380) = 1520 \text{ mm}^2$$

$$A_{sc} = A_{st} = 760 \text{ mm}^2 = (2 - \phi22)$$

Select $b = 360$ mm and $d = 435$ mm. Then, we have $d_c = 50$ mm

$$D = d + d_c = 485 \text{ mm}, \quad \text{and} \quad D_s = d - d_c = 385 \text{ mm}$$

$$u = \frac{d_c}{d} = \frac{50}{435} = 0.11 \quad \text{and} \quad c = \frac{D_s}{d} = \frac{385}{435} = 0.89$$

The values of u and c assumed check reasonably well with the actual values. Therefore, the capacity of the column could be verified.

Equation (9.58), gives

$$0.36bd^2 f_{ck}(1 - 0.42\ k_u)k_u + 0.87\ A_{sc}f_y D_s$$

$$= M_d + P_d(d - 0.5D)$$

$$0.36(0.36)(0.189)(20)(k_u - 0.4\ k_u^2)$$

$$+\ 0.87(0.000760(415)(0.385) = 0.12285 + 1.05(0.192)$$

The equation reduces to $0.42\ k_u^2 - k_u + 0.45 = 0$

or $\quad k_u = 1.8$ or 0.6

One value of k_u works out to be more than 1, which means that the neutral axis is outside the section and the entire section is under compression. One value of k_u is less than 1. As e/D is smaller than 0.25 the section can be assumed to be uncracked. The strain at the rear face of the column is

$$\epsilon_{c2} = \frac{0.0035(1.8 - 1)}{1.8} = 0.0016$$

The part of the parabolic distribution of the stress is outside the section which can be seen from Fig. 9.11.

The net compressive force acting on the concrete is equal to the total on the imaginary section of the depth $k_u d$ less the parabolic area force outside the section.

Let C_{co} = total imaginary force

$C_{ce} = 0.36\ bd\ k_u f_{ck}$

C_{xo} = negative force outside the section and is equal to

$$= \frac{2}{3}\ x f_{c1} b$$

$$= \frac{2}{3}\ xb\ \frac{(2a - x)x}{a^2} f_{cc} \tag{9.94}$$

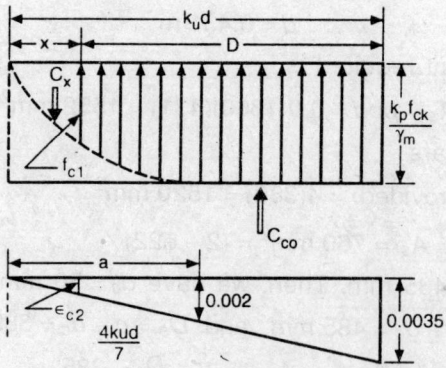

Fig. 9.11 Stress Distribution for Neutral Axis Outside the Section.

$$\text{where } a = \frac{4k_ud}{7} \tag{9.95}$$

$$x = (k_ud - D) \tag{9.96}$$

The net Compressive force from the concrete is

$$C_c = C_{co} - C_{xo} \tag{9.97}$$

Substituting various values in the above expressions gives

$$k_ud = 1.8\,(0.435) = 0.783 \text{ m}$$

$$x = k_ud - D = 0.783 - 0.485 = 0.298 \text{ m}$$

$$a = \frac{4k_ud}{7} = 0.447 \text{ m}$$

$$\frac{x}{a} = \frac{0.298}{0.447} = 0.67$$

$$C_{xo} = \frac{2}{3}\,b(2a - x)\left(\frac{x}{a}\right)^2 f_{ck}\frac{(0.67)}{1.5}$$

$$= \frac{2}{3}\,(0.36)(0.894 - 0.298)(0.67)^2(20)(0.45) = 0.58 \text{ MN}$$

$$C_{co} = 0.36\,bdk_uf_{ck} = 0.36(0.36)(0.435)(1.8)(20) = 2.03\text{MN}$$

$$C_c = C_{co} - C_{xo} = 2.03 - 0.58 = 1.45$$

Let C_{sc} = compression from A_{sc}

C_{st} = compression from $A_{st} = T$

Then the net compression from the steel to be taken is

$$C_s = C_{sc} - T = C_{sc} - C_{st} \tag{9.98}$$

Since $A_{st} = A_{sc}$ and $f_{st} < f_y$

$$C_s > 0$$

The net compression will be

$$P_a = C_c + C_{sc} - C_{st} > 1.05 \text{ MN}$$

The capacity of the column is more than the load acting on it, hence the column is safe.

The same problem was solved by working stress design in Example 5.11. Table 9.6 gives the relative designs.

Table 9.6 Comparison of Design Details

Item	WSD	LSD
b (mm)	365	360
D (mm)	600	485
A_s (mm^2)	(6–ϕ20)	(4–ϕ22)
	1884	1520

The safety of the column can also be checked from the interaction formula. The limiting values of the interaction formulae are :

$$P_{ao} = (0.99)(0.45b\,D\,f_{ck}) + 0.75\,f_y\,A_s = 1.55 + 0.47 = 2.02 \text{ MN}$$

where $A_c = 0.99\,A_g$

$$P_{ab} = 0.36\,bd\,f_{ck}\,(0.99\,k_{ub} + 2.4\alpha(p_c - p_t)) = 0.54 \text{ MN}$$

0.99 refers to about 1% of the area of the concrete occupied by the steel.

$$\frac{P_{ao}}{P_{ab}} = \frac{2.0}{0.54} = 3.7$$

The load carrying capacity of the column is

$$P_a = \frac{P_{ao}}{1 + (P_{ao}/P_{ab} - 1)e/e_b}$$

$$= \frac{2.02}{1 + (3.7 - 1)(0.117)/(0.82)(0.435)} = 1.06 \text{ MN}$$

The design load is 1.05 MN whereas the capacity is 1.06 MN. Therefore, the design is acceptable.

Design of transverse reinforcement
Select 6 mm tie, then the spacing is governed by

$$s_v = b \le 360 \text{ mm or}$$
$$\le 16\phi = 352 \text{ mm or}$$
$$\le 48\phi_v = 288 \text{ mm}$$

Provide 6 mm ties at 280 mm spacing.

EXAMPLE 9.13 *Design of cracked long column.* A column supported at 3.6 m apart with hinge and roller is loaded with 680 kN and 350 kNm of the axial force and bending moment respectively. Design the column with width about 200 mm.

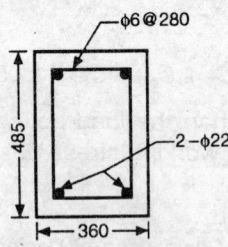

Fig. 9.12 Cross-section of Example 9.16 (dimensions in mm).

Design data and assumptions.

$$L_e = L = 3.6, \text{ m}$$

Let $f_{ck} = 20$ MPa, $f_y = 415$ MPa then

$$\alpha = f_y/f_{ck} = 20.75$$

Superimposed load= 680 kN

Let the self-weight be $= 20$ kN
Total axial load $= P = 700$ kN
Bending moment $= M = 350$ kN
Width of the column $= b = 200$ mm

$$\frac{M}{P} = \frac{350}{700} = 0.5 \text{ m} = 500 \text{ mm}$$

Since the eccentricity is large, the column can be assumed to be a cracked section.
Assume $A_{st} = A_{sc} =$ about 3%
That is, $p_t = p_c = 0.03$

Let $u = \frac{d_c}{d} = 0.15$, $c = \frac{D_s}{D} = 0.75$

Design of the section. Slenderness factor of the column $= \frac{L_e}{b} = \frac{3.6}{0.2} = 18$

The column is to be treated as long one and the load reduction coefficient is

$$C_r = 1.25 - \frac{L_e}{48b} = 0.875$$

The corresponding equivalent short column load is

$$P_o = \frac{P}{C_r} = \frac{700\,000}{0.875} = 800\,000 \text{ N} = 800 \text{ kN}$$

$$e = \frac{M}{P_o} = \frac{350}{800} = 0.4375 \text{ m} = 438 \text{ mm}$$

$$P_d = 1.5\, P_o = 1200 \text{ kN}$$

$$M_d = 1.5\, M = 525 \text{ kNm}$$

The section is assumed to be cracked and the failure mode is by compression of **the concrete.**
The value of e_s from Eq. (9.77) is

$$e_s = \frac{2.778(K + 0.435\, p c\alpha)}{k_u} = 3.1d$$

then $e_b = e_s - 0.5cd = 2.785d$

$$\frac{e}{e_b} = \frac{0.5}{2.785d} = \frac{0.18}{d}$$

$$e_s = \frac{2.778(0.138 + 0.435(0.06)(0.75)(20.75))d}{0.479} = 3.16d$$

$$\frac{P_{ao}}{P_{ab}} = \frac{1.25(1 + u + 1.67 p\alpha)}{k_u + 2.4\alpha(p_c - p_t)} = \frac{1.25(3.229)}{0.479} = 8.43$$

From Eq. (9.87), we have

$$bd = \frac{P_d\,(1 + (P_{ao}/P_{ab} - 1)e/e_b)}{0.45(1 + u + 1.67 p\alpha)f_{ck}}$$

$$= \frac{1.20(1 + 7.43(0.18)/d)}{30.93} = \frac{0.039d + 0.05}{d}$$

$$\text{or } bd^2 - 0.039d - 0.05 = 0$$

Width b has already been selected as 0.2 m, so the solution of the above equation reduces to

$$d = 0.6 \text{ m}$$

$$A_{sc} = 0.03bd = 0.03(200)(600) = 3600 \text{ mm}^2$$

Provide 7 numbers of 25 mm bars in three rows at each face.

$$A_{sc} = A_{st} = 7(490) = 3430 \text{ mm}^2$$

The details are:

$$d = 610 \text{ mm}, d_c = 40 + 25 + 30 + 12 = 107 = 110 \text{ mm}$$

$$D = 720 \text{ mm}, D_s = 500 \text{ mm}$$

$$e_b = 2.785(610) - 250 = 1677 \text{ mm}$$

$$e/e_b = 0.3$$

Since e less than e_b, the compression failure controls the design. The assumed values of u and c check closely with the actual ones,

The capacity of the column

$$P_{ao} = (1 - p)(0.45)bDf_{ck} + 0.75 \, f_y A_s$$

$$= 0.94(0.45)(0.2)(0.72)(20) + 0.75(415)(0.00686) = 3.35 \text{ MN}$$

$$P_{ab} = 0.36bdf_{ck} (1 - p)k_u + 2.4\alpha(p_c - p_t))$$

$$= (0.36)(0.2)(0.61)(20)(0.94)(0.479) = 0.42 \text{ MN}$$

$$\frac{P_{ao}}{P_{ab}} = \frac{3.35}{0.42} = 8$$

The capacity of the column with the moment M_d is

$$P_a = \frac{P_{ao}}{1 + (P_{ao}/P_{ab} - 1)e/e_b} = \frac{3.75}{1 + 7(0.3)} = 1.08 \text{ MN}$$

The capacity is 1.08 MN against a design load of 1.2 MN. Therefore, the section be modified marginally,

The final details are

$$b = 200 \text{ mm}, d = 680 \text{ mm}, D = 800 \text{ mm}$$

$$A_{st} = A_{sc} = 3430 \text{ mm}^2$$

8 mm ties at 200 mm spacing in two rows.

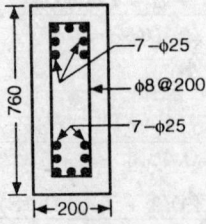

Fig. 9.12 Cross-section of Example 9.13 (dimensions in mm).

Table 9.7 Comparison of the Design Details

Item	WSD	LSD
Example	5.15	9.13
b (mm)	200	200
D (mm)	935	800
A_s (mm^2)	11070	6860
A_g (m^2)	0.1870	0.16

Note: The same problem was solved by the working stress design and the design details are compared in Table 9.7. The LSD gives a smaller section.

9.11 Columns Subjected to Combined Bending and Axial Force: No Crack Design

Some columns are subjected to large bending moments combined with small axial forces. Such columns are dominated by bending moment and so get cracked on the tension face. The columns in water should be designed for no cracking. Therefore, one must impose a no crack constraint. The distance of the neutral axis based on the compatibility of the strains can be written as

$$k_u \geq \frac{\epsilon_c}{\epsilon_c + \epsilon_t} \tag{9.99}$$

where ϵ_t = tension strain in the concrete = 0.0002

ϵ_c = 0.0035

The neutral axis distance which is given by Eq. (9.99) is

$$k_u \geq \frac{35}{37} = 0.95 \tag{9.100}$$

The constraint along with the interaction formula governs the design.

9.12 Design of Cracked Columns with Tension Failure

A concrete column when subjected to large eccentric load will develop tension on one face. If the tensile stress generated in the concrete exceeds the cracking stress, the column is treated as a cracked column. There are two possible modes of failure in such a cracked section:

1. *Compression failure* which is initiated by the concrete reaching the limiting compressive strain. The section is cracked but the stress in the tension steel is less than the proof stress. This failure is also called *primary compression failure.*
2. *Tension failure* is the one which is initiated by the yielding of the tension reinforcement. The strain in tension reinforcement reaches the yield point before the concrete on the compression face reaches the crushing strain. Although the failure is initiated by the yielding of the tension steel, the collapse of the column is finally due to crushing of the concrete. This is because the large deformation of the steel results into larger strain in the compression concrete, finally causing crushing of the concrete. Such a failure is called *secondary compression* failure.

Columns even if cracked but failing in primary compression are designed by the interaction formula discussed in the earlier sections. This section illustrates the design of the concrete columns failing by second mode. The two failure modes and their design criteria are:

1. If $e < e_b$ and $k_u > k_{ub}$, then the failure is by crushing of the concrete. The design criterion is by interaction formula.

2. If $e > e_b$, then the failure is by the yielding of the steel and the design criterion is given as follows:

The compressive capacities of the concrete and steel, and the tensile force capacity of the steel are:

$$C_s = 0.36\, bx_u f_{ck} \tag{9.101}$$

$$C_s = 0.87\, A_{sc}\, f_y \tag{9.102}$$

$$T = 0.87\, A_{st} f_y \tag{9.103}$$

Neglecting the area lost in concrete due to the compression steel.

The equilibrium of forces and moment gives :

$$C_c + C_s - T = P_d \tag{9.55}$$

$$C_c\,(d - x_c) + C_s(d - d_c) = P_d\,(e + d - 0.5D) \tag{9.57}$$

The final governing equations are similar to Eqs. (9.74) and (9.75). These equations are:

$$0.36\, bd\, f_{ck}\,(k_u + 2.4\alpha(p_c - p_t)) = P_d \tag{9.104}$$

$$bd^2 f_{ck}\,(K + 0.87\, p_c\, c\, \alpha) = M_d + P_d\,(d - 0.5D) \tag{9.105}$$

$$\text{or} \quad bd = \frac{P_d}{0.36\, f_{ck}(k_u + 2.4\,\alpha(p_c - p_t))} \tag{9.106}$$

Substitute Eq. (9.106) in (9.105), we then have

$$\frac{P_d\,(K + 0.87\, p_c c \alpha)d}{0.36(k_u + 2.4\alpha(p_c - p_t))} = M_d + P_d\,(d - 0.5\,D) \tag{9.107}$$

One can solve for d from either Eq. (9.107) or from Eqs. (9.105) and (9.106). The columns failing by yielding of the reinforcement can be called as beam-columns.

EXAMPLE 9.14 *Design of beam-column.* A column of effective length 3.6 m is subjected to a bending moment of 100 kNm and axial force of 45 kN. Design the column with M20 concrete and HYSD bars.

Design data and assumptions

Effective length = L_e = 3.6 m
Axial load = 45 kN
Let self-weight = 5 kN
Total load = P = 50 kN
Bending moment = M = 100 kNm
f_{ck} = 20 MPa, f_y = 415 MPa

Let b = 300 mm

Slenderness factor $= \dfrac{3.6}{0.3} = 12$

The column can be treated as a short one.

$$e = \frac{M}{P} = \frac{100}{50} = 2 \text{ m} = 2000 \text{ mm}$$

The eccentricity is very large so one may assume the stress in the tension reinforcement dominates the design.

Design of the section. The design loads are:

$$P_d = 1.5 \ P = 75 \text{ kN}$$

$$M_d = 1.5 \ M = 150 \text{ kN}$$

To solve the two non-linear simultaneous equations, Eqs. (9.105) and (9.106), one must assume certain design parameters.

Let $u = \dfrac{d_c}{d} = 0.15$, $p_c = p_t = 0.01$

$$c = \frac{D_s}{d} = 0.8, \ D = 1.15d$$

The strength ratio $= \alpha = \dfrac{415}{20} = 20.75$

Substitute $D = 1.15d$ in Eq. (9.105), then we have

$$bd^2 f_{ck} (K + 0.87 p_c c \ \alpha) = M_d + P_d (d - 0.5D)$$

$$bd^2 (20)(0.138 + 0.87 \ (0.01)(0.8)(20.75))$$

$$= 0.15 + 0.075(1 - 0.57)d$$

or $(5.65)bd^2 - 0.03225d - 0.15 = 0$

For $b = 300 \text{ mm} = 0.3 \text{ m}$, the above equation gives

$$d = 0.31 \text{ m}$$

$$k_{ud}d = 0.148 \text{ m}$$

$$A_{st} = A_{sc} = 0.01(bd) = 0.01(300)(310) = 930 \text{ mm}^2$$

Use 2 numbers of 25 mm bars on each face, then the actual areas provided are :

$$A_{st} = A_{sc} = 2(490) = 980 \text{ mm}^2$$

Provide $b = 300 \text{ mm},$ $D = 370 \text{ mm},$

$d = 320 \text{ mm},$ $d_c = 50 \text{ mm},$

$$A_s = 4(490) = 1960 \text{ mm}^2.$$

The axial load capacity of the section can be obtained from Eq. (9.104) and it is (allowing for about 1% loss of area of the concrete due to the compression steel).

$$P_a = 0.36 \ bd \ f_{ck} \ ((1 - p)k_{ub} + 2.4 \ \alpha \ (p_c - p_s))$$

$$= (0.36)(0.3)(0.32)(20)(0.99)(0.479) = 0.33 \text{ MN}$$

The capacity is larger than P_d, so the design is safe.

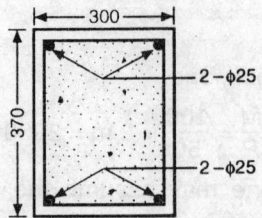

Fig. 9.13 Cross-section of Example 9.14 (dimensions in mm).

It can be concluded that the columns subjected to combined bending and axial forces can be designed as any one of the following problems depending on the constraints:

1. Uncracked column in which compression controls.
2. Uncracked column in which the tensile stress in concrete controls. This is a common case in water retaining structures. The ratio of the eccentricity to the depth of the column is usually in the range of 0.2 to 0.30. It is possible that this ratio can be as large as 0.40. In such cases the width becomes large and the tensile stress in the concrete will control the design.
3. Cracked column sections in which the compressive stress in the concrete controls the design. This is a very common case in frame columns. The ratio of the eccentricity to the depth of the column in such problems is usually in the range of 0.3 to 2.5. There are situations where this value may even exceed 3, but in such cases the width of the columns is usually large. The neutral axis distance from the extreme compression fibre will be more than 0.48*d* in case of HYSD bars and 0.53*d* in case of MS reinforcement.
4. Cracked column sections in which the tensile strength in the reinforcement controls the design. This is not common in columns, but likely to happen in beam-columns. The ratio of the eccentricity to the depth is likely to be more than 3 and the distance of the neutral axis from the extreme compression fibre will be less than 0.53*d* in case of the mild steel reinforcement and 0.48*d* in case of the high yield strength bars.

9.13 Design of Columns Subjected to Wind Load or Seismic Forces

The wind and seismic loads have a tendency to cause overturning of the structure and local bending moment on the columns. The overturning effect will cause additional compressive force in the columns on the leeward side. The live load on the structure under seismic or wind load condition is likely to be about half of that normally accepted in residential and office buildings and it may be close to the full design load in case of warehouses, water tanks and other important structures. The axial forces in the columns under seismic or wind load conditions are in the same range as those under the normal working load conditions. But more important factor is the bending moment in such load conditions. The duration of the severe seismic load can be anywhere from few seconds to two minutes and the duration of gust wind is less than five minutes. Therefore, the partial safety factor applied to the loads in the short duration wind and seismic load conditions is taken 1.2 whereas the corresponding value under live load conditions is 1.5. The structures should be designed to withstand the severe gust and seismic loads with dead load at a partial safety factor of 1.2. The probability of simultaneous occurrence of the gust wind and earthquake is considered to be less than one in ten million. Therefore, one need not design the structures for simultaneous occurrence of the wind and seismic loads. The design methodology is exactly same as was discussed earlier.

EXAMPLE 9.15 *Design of column under seismic load.* A column of effective length 3.6 m is subjected to the following forces:

a) Combined dead and live loads

   Axial force $P_1 = 900$ kN

   Bending moment $M_1 = 50$ kNm

b) Combined dead and live loads and seismic force

   Axial force $P_2 = 1200$ kN

   Bending moment $M_2 = 350$ kNm

c) Combined dead and seismic forces

   $P_3 = 900$ kN

   $M_3 = 300$ kN

d) Combined dead, live and wind loads

   Axial force $P_4 = 1100$ kN

   Bending moment $M_4 = 250$ kNm

Design the column with M20 concrete and HYSD bars.

   Design data and assumptions

$$f_{ck} = 20 \text{ MPa}, \ f_y = 415 \text{ MPa}, \ L_e = 3.6 \text{ m}$$

$$\alpha = \frac{415}{20} = 20.75$$

Let $A_{sc} = A_{st} = 0.005\%$ and $b = 500$ mm.

   Design load analysis. The design forces under the normal load condition are comparable with the values in the seismic load condition, though slightly less. The forces under wind load condition are much less than those under seismic load condition. Therefore, seismic load condition dominates over the wind load. Hence, design the section against seismic load condition.

The design loads are:

For   DL + LL = (Dead Load + Live Load)

   $P_{d1} = 1.5(900) = 1350$ kN, $M_{d1} = 75$ kNm

For   DL + LL + SL(SL = Seismic Load),

   $P_{d2} = 1.2(1200) = 1440$ kN, $M_{d2} = 1.2(350) = 420$ kNm

For   DL + SL, we have

   $P_{d3} = 1.5(900) = 1350$ kN

   $M_{d3} = 1.5(300) = 450$ kN

Design of the section

$$\text{Slenderness factor} = \frac{L_e}{b} = \frac{3.6}{0.5} = 7.2$$

which is less than 12, therefore, the column is treated as a short one. The seismic load can act in either direction; therefore, the reinforcement should be symmetrical.

The design load combinations are:

i) $P_d = P_{d2} = 1440$ kN and

$M_d = M_{d2} = 420$ kNm or

ii) $P_d = P_{d3} = 1350$ kN and

$M_d = M_{d3} = 450$ kNm

The section must be designed for both the conditions. Assuming a compression failure, the column is designed for the first load combination and then checked for the second combination.

Combination 1

$$P_d = 1440 \text{ kN} = 1.44 \text{ MN}$$

$$M_d = 420 \text{ kNm} = \text{MNm}$$

$$e = \frac{M_d}{P_d} = 0.292 \text{ m}$$

Let $p = 0.01$, $c = D_s/d = 0.85$

The properties of the balanced cross-section area:

$$K = 0.138, \ k_u = k_{ub} = 0.479 \text{ and } \alpha = 20.75$$

From Eq. (9.77), we have

$$e_s = \frac{2.778(K + 0.435pc\,\alpha)d}{k_u} = 1.22d$$

$$e_b = e_s - 0.5cd = 1.22d - 0.425d = 0.795d = 0.8d$$

$$\frac{e}{e_b} = \frac{0.292}{0.8d} = \frac{0.356}{d}$$

$$\frac{P_{ao}}{P_{ab}} = \frac{1.25(1 + u + 1.67p\,\alpha)}{k_u + 2.4\alpha(p_c - p_t)} = 3.77$$

From Eq. (9.88), we have

$$bd = \frac{p_d(1 + (P_{ao}/P_{ab} - 1)e/e_b)}{0.45(1 + u + 1.67p\alpha)f_{ck}}$$

$$= \frac{1.44(1 + 2.77(0.365)/d)}{13.02}$$

or $bd^2 - 0.11d - 0.112 = 0$

The solution of the above equation for $b = 0.5$ m is

$$d = 0.60 \text{ m and}$$

$$A_{st} = A_{sc} = 0.005(500)(600) = 1500 \text{ mm}^2$$

Provide four numbers of 22 mm bars on each face, then

$$A_{st} = A_{sc} = 4(380) = 1520 \text{ mm}^3$$

Let $d = 600$ mm $d_c = 50$ mm

$D = 650$ mm, $D_s = 550$ mm

$e_b = 0.8d = 480$ mm, $e/e_b = 0.608$

The neutral axis distance is obtained from the equilibrium of moment, and is

$$0.36bd^2 f_{ck}(1 - 0.42k_u)k_u + 0.87 A_{sc} f_y(d - d_c) = M_d + P_d(0.5D_s)$$

$$0.36(0.5)(0.6)^2(20)(k_u - 42k_u^2)$$

$$+ 0.87(0.001520)(415)(0.55) = 0.42 + 1.44(0.275) = 0.816$$

or $0.42k_u^2 - k_u + 0.4 = 0$

$k_u = 1.87$ or 0.49

The failure is by compression as the value of k_u is greater than k_{ub}. Check for capacity of the column

Check for capacity of the column. The load capacity can be checked by using:

$$P_a = \frac{P_{ao}}{1 + (P_{ao}/P_{ab} - 1)e/e_b}$$

$$P_{ao} = (1 - p)0.45bDf_{ck} + 0.75f_yA_s$$

$$= 0.99(0.45)(0.5)(0.65)(20) + 0.75(415)(2)(0.00152)$$

$$= 3.76 \text{ MN}$$

(Including the loss of area in the concrete due to the steel.)

$$P_{ab} = 0.36bdf_{ck}((1 - p)k_{ub} + 2.4x(p_c - p_t))$$

$$= 0.36(0.5)(0.6)(20)(0.479)(0.99) = 1.02 \text{ MN}$$

$$\frac{P_{ao}}{P_{ab}} = \frac{3.76}{1.02} = 3.66$$

Load combination (seismic load) 1.

$$e = 0.292 \text{ m}, \ e/e_b = 0.608$$

$$P_{a1} = \frac{P_{ao}}{1 + (P_{ao}/P_{ab} - 1)e/e_b} = \frac{3.76}{1 + 2.66(0.608)}$$

$$= 1.44 \text{ MN} = P_{d2} \text{ ; hence safe.}$$

Load combination (seismic load) 2

$$e = \frac{M_{d2}}{P_{d3}} = \frac{450}{1350} = 0.333 \text{ m}$$

$$\frac{e}{e_b} = \frac{0.333}{0.48} = 0.69$$

$$P_{a2} = \frac{P_{ao}}{1 + (P_{ao}/P_{ab} - 1)e/e_b} = \frac{3.76}{1 + 2.66(0.69)}$$

$$= 1.32 \text{ MN} < P_{d3} = 1.35 \text{ MN}$$

Increase the depth marginally as P_{a2} is slightly less than P_{d3}.

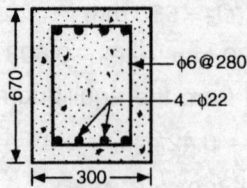

Fig. 9.14 Example 9.15.

Provide 6 mm stirrups at 280 mm spacing.
Final design details are

$$b = 500 \text{ mm}, \ d = 610 \text{ mm}, \ D = 670 \text{ mm}$$
$$A_{sc} = A_{st} = 1520 \text{ mm}$$

9.14 Columns of Circular Cross-section Failing in Compression

The notations used in the formulation of a column of circular cross-section subjected to combined axial and bending moments are:

D = diameter of the section

D_c = diameter of the circle of the location of the reinforcement bars

kD = depth of the neutral axis from the compression fibre

d_c = distance of the extreme compression reinforcement from the outer fibre

d_{ct} = distance of the ith compression reinforcement from the outer compression fibre

y_1 = distance of the extreme tension reinforcement from the outer compression fibre

y_i = distance of the ith tension reinforcement from the outer compression fibre.

The forces in materials are:

C_c = design compression force from the concrete

$$= (1 - p)c_a D^2 \frac{k_p f_{ck}}{\gamma_m} = 0.45(1 - p)c_a D^2 f_{ck} \qquad (9.108)$$

$$C_{st} = A_{sct} f_{sci} \qquad (9.109)$$

$$T_i = A_{sti} f_{sti} \qquad (9.110)$$

where f_{sci} = compression stress in the ith compression bar = $E_s \epsilon_{sci}$

f_{sti} = tensile stress in the ith tensile bar = $E_s \epsilon_{sti}$

a = depth of the compression block = $0.85kD$ $\qquad (9.111)$

c_a = coefficient of area under compression and given by

$$\int_0^\phi \frac{\sin^2\theta \, d\theta}{2} = 0.25 \, (\phi - \sin\phi \cos\phi) \qquad (9.112)$$

$$\cos\phi = \frac{(R - a)}{R} = 1 - \frac{a}{R} \qquad (9.113)$$

R = radius of the column = $0.5D$

$(1 - p)$ = approximate net area coefficient of the concrete

$p = A_s / A_g$

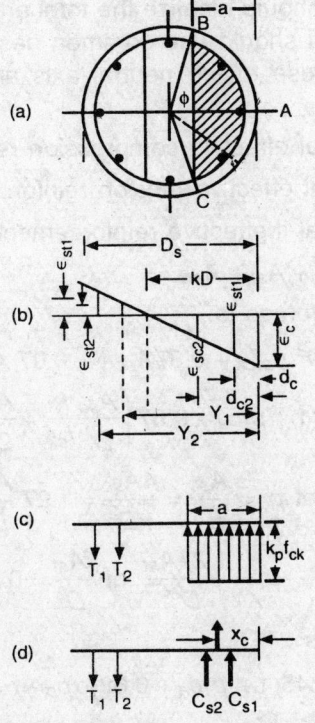

Fig. 9.15 Notations in Circular Cross-section.

The area of the segment ABC of the circle as shown in Fig. 9.15a is given by

$$A_{cc} = \int 2R^2 \sin\theta(\sin\theta)\, d\theta$$

$$= \frac{D^2}{4} \int_0^\phi \sin^2\, d\theta = 0.25\,(\phi - \sin\phi\cos\phi)\, D^2$$

The design compressive force from the concrete area is

$$C_c = A_{cc}\frac{k_p f_{ck}}{\gamma_m}$$

where A_{cc} is expressed above and is set equal to

$$A_{cc} = c_a D^2 \tag{9.114}$$

where the value of c_a is given in Eq. (9.112).

The equilibrium of the forces on the column section is given by

$$C_c + C_s - T = P_d \tag{9.115}$$

where C_s = design compressive force from the reinforcement

$$= \Sigma \, C_{si} = \Sigma A_{sci} \, f_{sci} \tag{9.116}$$

$$\text{Total tension force } T = \Sigma T_i = \Sigma A_{sti} \, f_{sti} \tag{9.117}$$

For convenience of design, one should idealize the total areas of the compressive and tensile reinforcements at two points and should use common design strengths. There will be some reinforcement bars which are close to the neutral axis and which will not be subjected to much stress.

Let A_{sc} = total effective compression reinforcement

A_{st} = total effective tension reinforcement

A_{sd} = total ineffective reinforcement near the NA

$$A_s = A_{sc} + A_{st} + A_{sd} \tag{9.118}$$

The equilibrium of the idealized forces is

$$0.45(1 - p)D^2 \, c_a f_{ck} + 0.87 A_{sc} f_y - 0.87 \, A_{st} f_{st} = P_d \quad \text{or} \tag{9.119}$$

$$D^2 f_{ck} \left(0.45 \, (1 - p) c_a + 0.87 \left(\frac{A_{sc} \, f_y}{D^2 \, f_{ck}} - \frac{A_{st} \, f_{st}}{D^2 \, f_{ck}} \right) \right) = P_d \tag{9.120}$$

$$\text{Let } p_c = \frac{A_{sc}}{A_g} = \frac{4 A_{sc}}{\pi D^2} = 1.27 \, \frac{A_{sc}}{D^2} \tag{9.121}$$

$$p_t = \frac{1.27 A_{st}}{D^2} \quad \text{or} \quad \frac{A_{st}}{D^2} = 0.785 \, p_t \tag{9.122}$$

Eq. (9.120) can be rearranged as

$$D^2 f_{ck} \, (0.45(1 - p) c_a + 0.68(\alpha p_c - \alpha_s p_t)) = P_d \tag{9.123}$$

where $\alpha = \dfrac{f_y}{f_{ck}}, \; \alpha_s = \dfrac{f_{st}}{f_{ck}}$ $\tag{9.124}$

The moment equilibrium taken about the centroid of A_{st} is

$$C_c \, (d - x_c) + C_s D_s = M_d + M_d \, (0.5 D_s) \tag{9.125}$$

where D_s = distance between the idealized centres of the compression and the tension reinforcement

x_c = distance of the centroid of the compression of the concrete from the extreme compression fibre.

The first moment of the area about the extreme fibre is:

$$Q = \int 2R^3 \sin^2\theta(1 - \cos\theta) \, d\theta$$

$$= \frac{D^3}{4} \int_0^\phi (\sin^2\theta - \sin^2\theta \cos\theta) \, d\theta$$

$$= \frac{D^3}{4} \left(\frac{\phi - \sin\phi \cos\phi}{2} - \frac{\sin^2\phi}{3} \right)$$

$$= c_q D^3 \tag{9.126}$$

where $c_q = 0.25 \left(\dfrac{\phi - \sin \phi \cos \phi}{2} - \dfrac{\sin^3 \phi}{3} \right)$ (9.127)

The centroidal distance of the segment from the outer fibre is

$$x_c = \frac{Q}{A_{cc}} = \frac{c_q D^3}{c_a D^2} = c_x D$$ (9.128)

where $\quad c_x = \dfrac{c_q}{c_a} = \dfrac{0.25 \left(\dfrac{\phi - \sin \phi \cos \phi}{2} - \dfrac{\sin^3 \phi}{3} \right)}{0.25(\phi - \sin \phi \cos \phi)}$

$$= \frac{1}{2} - \frac{\sin^3 \phi}{3(\phi - \sin \phi \cos \phi)}$$ (9.129)

For balanced failure Eq. (9.125) can be rewritten as

$$C_c (d - x_{cb}) + C_s D_s = P_{ab}(e_s) = P_{ab}(e_b + 0.5 D_s)$$ (9.130)

The design consideration is that

$$P_a \geq P_d$$

From Eq. (9.81), we have

$$P_{ao} < P_d (1 + (P_{ao}/P_{ab} - 1) e/e_b)$$ (9.131)

9.15 Interaction Formula for Circular Columns

The interaction formula applied to the rectangular sections can be applied directly to the circular sectioned columns. The critical capacities of the columns have to be determined. They are P_{ao}, P_{ab} and, e_b which are given below:

$$P_{ao} = (1 - p)\, 0.45\, f_{ck} A_g + 0.75\, f_y A_s$$ (9.43)

In the case of simultaneous occurrence of the crushing strain in concrete and yielding in the steel, the depth of the compressive stress block is

$$x_u = \frac{0.0035 d}{0.0035 + 0.87\, f_y / E_s}$$

Use $a = a_b = 0.85 x_u = 0.85\, k_{ud} d$

where $k_{ub} = 0.479$ for HYSD bars and

$\qquad\qquad = 0.531$ for MS bars

$$P_{ab} = 0.45 f_{ck} (1 - p) A_{ccb} + 0.87 (A_{sc} - A_{st}) f_y$$

$$= 0.45 f_{ck}(1 - p) c_{ab} D^2 + 0.87(A_{sc} - A_{st}) f_y$$

$$= D^2 f_{ck} (0.45(1 - p) c_{ab} + 0.68(p_c - p_t)\alpha)$$ (9.132)

where $\quad c_{ab} = 0.25(\phi - \sin \phi \cos \phi)_b$ (9.133)

The bending moment equilibrium equation corresponding to the balanced strain failure condition is given in Eq. (9.130) and it can be expressed as

$$D^2 f_{ck}(0.45(1 - p) c_{ab}(1 - u - c_{xb}) + 0.68 p_c \alpha c) = e_s P_{ab}$$ (9.134)

From Eqs. (9.130) and (9.131), we have

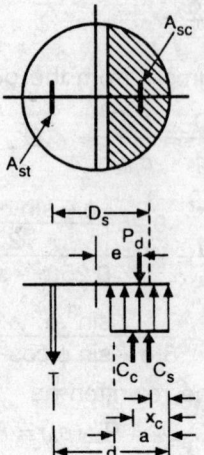

Fig. 9.16 Idealised Column Cross-section and Stress Distribution.

$$e_s = \frac{(0.45(1 - p)(1 - u - c_{xb})c_{ab} + 0.68\, p_c \alpha c)D}{0.45(1 - p)c_{ab} + 0.68(p_c - p_t)\alpha} \qquad (9.135)$$

All the final equations are consolidated here. (For a balanced section)

$$a_b = 0.85\, k_{ub} d$$

where $k_{ub} = 0.479$ for HYSD bars and 0.531 for MS bars

$$\cos \phi_b = 1 - \frac{a_b}{R} = 1 - 0.85\, k_{ub}\, \frac{d}{R}$$

$$= 1 - 0.85\, k_{ub} \left(\frac{D - d_c}{R} \right)$$

$$= 1 - 0.85\, k_{ub} \left(2 - \frac{d_c}{2D} \right)$$

$$= 1 - 0.85 k_{ub} (2 - 0.5u) \qquad (9.136)$$

where $u = \dfrac{d_c}{D}$

Table 9.8 gives some values of ϕ_b and c_{ab}

$$P_{ao} = (0.45(1 - p) + 0.75p\, \alpha) \frac{\pi D^2}{4}\, f_{ck} \qquad (9.137)$$

$$P_{ab} = 0.45(1 - p)c_{ab} f_{ck}\, D^2 + 0.87(A_{sc} - A_{st}) f_y$$

$$= D^2 f_{ck} (0.45(1 - p)c_{ab} + 0.68(p_c - p_t)\alpha) \qquad (9.138)$$

$$c_{ab} = 0.25(\phi_b - \sin \phi_b \cos \phi_b) \qquad (9.139)$$

$$c_{qb} = 0.25 \left(\phi_b - \frac{\sin \phi_b \cos \phi_b}{2} - \frac{\sin^3 \phi_b}{3} \right) \qquad (9.140)$$

Table 9.8 Balanced Cross-sectional Segment (ϕ_b)

u	HYSD			MS		
	ϕ_b	c_{ab}	c_{xb}	ϕ_b	c_{ab}	c_{xb}
0.10	78	0.29	0.23	83	0.33	0.25
0.20	77	0.28	0.22	82	0.32	0.24

$$c_{xb} = \frac{1}{2} - \frac{\sin^3 \phi_b}{3(\phi_b - \cos \phi_b \sin \phi_b)} \tag{9.141}$$

$$e_s = \frac{(0.45(1-p)(1-u-c_{xb})e_{eb} + 0.68\,p_c\alpha c)D}{(0.45(1-p)c_{ab} + 0.68(p_c - p_t)\alpha)} \tag{9.142}$$

$$e_b = e_s - 0.5D_s = e_s - 0.5(D - 2d_c)$$

$$= e_s - 0.5D(1 - 2u) \tag{9.143}$$

From Eqs. (9.131) and (9.137), we have

$$D^2 \geq \frac{P_d(1 + (P_{ao}/P_{ab} - 1)e/e_b)}{0.785(0.45(1-p) + 0.75\,p\,\alpha)} \tag{9.144}$$

9.16 Design Examples of Circular Columns Failing in Compression

EXAMPLE 9.16 *Design of column with small eccentricity*. A column of effective length 3.6 m is subjected to 700 kN axial force and 81.9 kNm bending moment including the effect of the self-weight.

Design data: $L_e = 3.6$ m, $P = 700$ kN, $M = 81.9$ kNm

$$f_{ck} = 20 \text{ MPa}, f_y = 415 \text{ MPa}$$

$$e = \frac{M}{P} = \frac{81.9}{700} = 0.117 \text{ m}$$

$$P_d = 1.5(P) = 1050 \text{ kN} = 1.05 \text{ MN}$$

The design coefficients for the materials and the other assumption be

$$\alpha = \frac{f_y}{f_{ck}} = 20.75$$

Let $p_c = p_t = 0.005$ $p = p_c + p_t = 0.01$

$$u = \frac{d_c}{D} = 0.1 \text{ and } c = \frac{D_s}{D} = 0.8$$

From Table 9.8 for HYSD bars, we have

$$c_{ab} = 0.29, c_{xb} = 0.23$$

Design of the section. The loads are computed from Eqs. (9.137) and (9.138) as follows:

$$P_{ao} = (0.45(1-p) + 0.75\,\alpha p)\frac{\pi D^2}{4}f_{ck} = 9.44\,D^2$$

$$P_{ab} = (0.45(1-p)c_{ab} + 0.68(p_c - p_t)\alpha)D^2 f_{ck}$$

$$= (0.45(0.99)(0.29)(20)D^2) = 2.58\ D^2$$

From Eq. (9.142), we have

$$e_s = \frac{(0.45(1-p)(1-u-c_{xb})\ c_{ab} + 0.68\ p_c\alpha c)D}{(0.45(1-p)c_{ab} + 0.68(p_c - p_l)\alpha)}$$

$$= \frac{(0.45(0.99)(0.9 - 0.23)(0.29) + 0.68(0.005)(20.75)(0.8)D}{0.45(0.99)(0.29)} = 1.23\ D$$

$$e_b = e_s - 0.5D(1 - 2u) = 1.23D - 0.4D = 0.83D$$

$$\frac{e}{e_b} = \frac{0.117}{0.83D} = \frac{0.14}{D}$$

The value of e_b will be greater than e even if D is equal to 0.14 m. Therefore, the column will fail in compression only.

$$\frac{P_{ao}}{P_{ab}} = \frac{9.44}{2.58} = 3.66$$

From Eq. (9.144), we have

$$D^2 = \frac{P_d(1 + (P_{ao}/P_{ab} - 1)e/e_b)}{0.785(0.45(1-p) + 0.75\ p\ \alpha)f_{ck}}$$

$$D^2 = \frac{1.05(1 + 2.66(0.14)/D}{0.785(0.45(0.99) + 0.75(0.01)(20.75))20} = 0.111 + 0.041/D$$

$$\text{or } D^3 - 0.111D - 10.041 = 0$$

The solution of the equation is

$$D = 0.46\ m = 460\ mm$$

The area of the reinforcement is given by

$$A_{st} = A_{sc} = 0.005(A_g) = 0.005(0.785)(460)^2 = 830\ mm^2$$

Provide 6 numbers of 22 mm bars such that two bars can be treated in compression, two in tension and two in the neutral zones. Let

$$D = 475\ mm$$

The value of d_c would be equal to the distance to the centroid of the two bars from the outer fibre.

The diameter of the circle of the location of the reinforcement bars is

$$D_c = D - 80 - 22 = 475 - 102 = 373\ mm$$

$$d_c = \frac{D}{2} - \frac{D_c}{2}\cos 45 = 0.2375 - 0.1319 = 0.106\ m$$

$$u = \frac{d_c}{D} = 0.22$$

$$D_s = D - 2d_c = 0.475 - 0.212 = 0.263\ m$$

$$c = \frac{D_s}{D} = \frac{0.263}{0.475} = 0.55$$

Check for the capacity

$$A_s = 6(380) = 2280 \text{ mm}^2; \ A_g = 0.785 \ D^2 = 0.17711 \text{ m}^2$$
$$A_c = A_g - A_s = 0.1748 \text{ m}^2$$
$$p = 0.012, \ p_c = 0.004$$
$$P_{ao} = 0.45 \ A_c f_{ck} + 0.75 A_s f_y = 2.445 \text{ MN}$$
$$P_{ab} = 0.45(1 - p)c_{ab}D^2 \ f_{ck} = 0.583 \text{ MN}$$
$$\frac{P_{ao}}{P_{ab}} = \frac{2.445}{0.583} = 4.2$$
$$e_s = \frac{(0.45(1 - p)(1 - u - c_{xb})c_{ab} + 0.68p_c\alpha_c)D}{0.45(1 - p)c_{ab} + 0.68(p_c - p_t)\alpha} = 1.02 \ D$$
$$e_b = e_s - 0.5D_s = 0.353 \text{ m}$$
$$\frac{e}{e_b} = \frac{0.117}{0.353} = 0.33$$

The load carrying capacity of the column is

$$P_a = \frac{P_{ao}}{1 + (P_{ao}/P_{ab} - 1)e/e_b} = \frac{2.445}{1 + 3.2(0.33)} = 1.19 \text{ MN}$$

Provide spiral reinforcement of 6 mm bar at a pitch of 50 mm, then the capacity of the column is 1.05 times the above value.

EXAMPLE 9.17 *Design of column with large eccentricity in the load.* A column of effective length 3.6 m is subjected to 680 kN axial force and 350 kNm bending moment. Design the column.

Design data and assumptions. $L_e = 3.6$ m

Let $f_{ck} = 20$ MPa, $f_y = 415$ MPa, $\alpha = 20.75$

$$\text{Axial load} = 680 \text{ kN}$$
$$\text{Self-weight} = 20 \text{ kN}$$
$$\text{Total load } P = 700 \text{ kN} = 0.7 \text{ MN}$$
$$\text{Design load } P_d = 1.5P = 1.05 \text{ MN}$$
$$\text{Design moment} = 1.5M = 1.5(350) = 525 \text{ kNm}$$
$$e = \frac{M}{P} = \frac{0.35}{0.7} = 0.5 \text{ m}$$

Let $p = 0.01$, $p_c = 0.005$, $u = 0.2$, $c = 0.6$

From Table 9.8, the coefficients for HYSD bars are

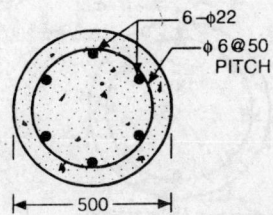

6–ϕ22
ϕ6@50 PITCH
500

Fig. 9.17 Column Cross-section Example 9.16.

$$c_{ab} = 0.28, \quad c_{xb} = 0.22$$

Design of the section. The limit loads are computed from Eqs. (9.137) and (9.138) and are

$$P_{ao} = (0.45(1 - p) + 0.75 \, p\alpha) \frac{\pi D^2}{4} f_{ck} = 9.44 \, D^2$$

$$P_{ab} = (0.45(1 - p)c_{ab} + 0.68(p_c - p_t)\alpha)D^2 f_{ck} = 2.5D^2$$

From Eq. (9.142), we have

$$e_s = \frac{(0.45(1 - p)(1 - u - c_{xb})c_{cb} + 0.68 p_c \alpha c)D}{0.45(1 - p)e_{ab} + 0.68(p_c - p_t)\alpha} = 0.94D$$

$$e_b = e_s - 0.5D(1 - 2u) = 0.64D, \text{ or}$$

$$\frac{P_{ao}}{P_{ab}} = \frac{9.44}{2.5} = 3.78$$

$$\frac{e}{e_b} = \frac{0.5}{0.64D} = \frac{0.78}{D}$$

From Eq. (9.144), we have

$$D^2 = \frac{P_d(1 + (P_{ao}/P_{ab} - 1)e/e_b)}{0.785(0.45(1 - p) + 0.75 p\alpha)f_{ck}}$$

$$= \frac{1.05(1 + 2.78(0.78)/D)}{0.785(0.45(0.99) + 0.75(0.01)(20.75))20}$$

$$= 0.111D - 0.17/D, \text{ or}$$

$$D^3 - 0.111D - 0.24 = 0$$

The solution gives, $D = 0.685$ m select $D = 750$ mm

$$D = 750 \text{ mm}$$

$$D_c = 750 - 110 = 640 \text{ mm}, \quad d_c = 130 \text{ mm}$$

$$u = \frac{d_c}{D} = \frac{130}{750} = 0.17, \quad c = \frac{D_s}{D} = \frac{490}{750} = 0.65$$

$$A_g = 0.4418 \text{ m}^2$$

$$P_{ao} = 0.45(A_g - A_s)f_{ck} + 0.75A_s f_y = 4.89 \text{ MN}$$

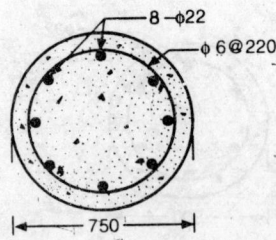

Fig. 9.18 Column Cross-section of Example 9.17.

$$P_{ab} = 0.45(1 - p)c_{ab}D^2 f_{ck} = 1.4 \text{ MN}$$

$$\frac{P_{ao}}{P_{ab}} = \frac{4.89}{1.4} = 3.5$$

$$e_s = \frac{(0.45)0.99(1 - 0.17 - 0.22)(0.28) + 0.68(0.005)(20.75)(0.675))D}{0.45(0.99)(0.28)} = D$$

$$e_b = e_s - 0.5D_s = D - 0.245 = 0.75 - 0.245 = 0.505 \text{ m}$$

$$\frac{e'}{e_b} = \frac{500}{505} = 0.99$$

Since e is less than e_b, there is a compression failure

$$P_a = \frac{P_{ao}}{1 + (P_{ao}/P_{ab} - 1)e/e_b} = \frac{4.89}{1 + 2.5(0.99)} = 1.44 \text{ MN}$$

EXAMPLE 9.18 *Analysis example.* A column is subjected to 1000 kN axial force and a bending moment of 350 kNm. The diameter of the column is 800 mm and is provided with 8 numbers of 25 mm HYSD bars. The concrete is of M20. Check for the structural safety of the column.

Design data.

$$f_{ck} = 20 \text{ MPa}, \ f_y = 415 \text{ MPa}, \ \alpha = 415/20 = 20.75$$

$$D = 800 \text{ mm}, \ A_s = 8(490) = 3920 \text{ mm}^2$$
$$P = 1 \text{ MN}, \ M = 0.35 \text{ MNm}$$
$$P_d = 1.5(1) = 1.5 \text{ MN}$$

$$e = \frac{M}{P} = 0.35 \text{ mm}$$

Analysis

$$A_g = \frac{\pi D^2}{4} = 0.50256 \text{ m}^2, \ A_s = 0.0092 \text{ m}^2$$

$$A_c = A_g - A_s = 0.49864 \text{ m}^2$$

$$p = \frac{A_s}{A_g} = 0.008$$

$$D_c = 800 - 110 = 690 \text{ mm}$$

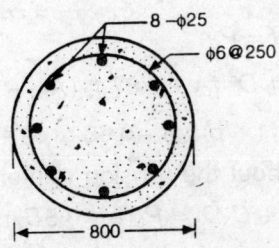

8 –φ25
φ6 @ 250

800

Fig. 9.19 Column Cross-section of Example 9.18.

$$d_c = 0.5(D - D_c \cos 40) = 130 \text{ mm}; \ D_s = D - 2d_c = 540 \text{ mm}$$

$$u = d_c/D = 0.16; \ c = D_s/D = 0.675$$

$$P_{ao} = 0.45 \ A_c f_{ck} + 0.75 A_s f_y$$

$$= 0.45(0.49864)(20) + 0.75(0.00392)(415) = 5.71 \text{ MN}$$

$$P_{ab} = 0.45(1 - p)c_{ab}D^2 f_{ck}$$

The values of c_{ab} and c_{xb} from Table 9.8 are $c_{ab} = 0.28$ and $c_{xb} = 0.22$.

$$P_{ab} = 0.45(1 - 0.008)(0.28)0.64(20) = 1.6 \text{ MN}$$

$$P_{ao}/P_{ab} = 5.71/1.6 = 3.57$$

$$e_s = \frac{(0.45(1 - p)(1 - c_{xb} - u)c_{ab} + 0.68p_c\alpha c)D}{0.45(1 - p)c_{ab}} = D$$

$$e_b = e - 0.5D = D - 0.27 = 0.53 \text{ m}$$

$$\frac{e}{e_b} = \frac{0.35}{0.53} = 0.66$$

Here e is less than e_b; therefore, there is a compression failure. Hence, the capacity of the section is

$$P_a = \frac{P_{ao}}{1 + (P_{ao}/P_{ab} - 1)e/e_b} = \frac{5.71}{1 + 2.57(0.66)} = 2.11 \text{ MN}$$

This value is larger than P_d, hence the column is safe.

9.17 Design of Circular Columns Failing in Tension

A column when subjected to load with large eccentricity greater than e_b, the yielding of the tension reinforcement takes place before the concrete is subjected to the crushing strain. Such a failure is said to be tension failure. More detailed discussion is given in Section 9.12 to which the reader is referred (see Figs. 9.15 and 9.16 for notations).

The compressive and tensile forces on the cross-section are (vide Eqs. (9.108) to (9.110)):

$$C_c = 0.45(c_{ab}D^2 - A_{sc})f_{ck} \tag{9.145}$$

or $\qquad C_c = 0.45 \ (1 - p)c_{ab} \ D^2 \ f_{ck}$

$$C_s = 0.87 \ A_{sc} \ f_y \tag{9.146}$$

$$T = 0.87 \ A_{st} \ f_y \tag{9.147}$$

The equilibrium of forces gives:

$$C_c + C_s - T = P_d$$

or $\qquad 0.45(1/p)c_{ab} \ D^2 \ f_{ck} + 0.87 \ f_y \ (A_{sc} - A_{st}) = P_d \tag{9.148}$

$$D^2 \ f_{ck}(0.45(1 - p)c_{ab} + 0.68 \ \alpha(p_c - p_t)) = P_d \tag{9.149}$$

The equilibrium of the moments about the tension reinforcement gives:

$$C_c \ (d - x_c) + C_s D_s = P_d(e + 0.5D_s)$$

or $\qquad D^3 f_{ck}(0.45(1 - p)(1 - u \ c_{xb})c_{ab} + 0.68p_c \ \alpha \ c) = M_d + 0.5 \ D_s P_d \tag{9.150}$

Solve for D from (Eq. 9.150) after assuming some percentage of the reinforcement, then check for the capacity of the section. The neutral axis of a balanced section is $0.479d$ of HYSD bars, so if the reinforcement is evenly distributed, more number of bars are likely to be in tension than in compression. In case there are eight bars symmetrically placed, then five bars will be in tension and three in compression. It is, therefore, recommended to select the following:

$$p_c = (0.4 \text{ to } 0.5)p$$
$$p_t = (0.6 \text{ to } 0.5)p$$

In the last section we have used fifty-fifty per cent distribution. Here we will use forty and sixty per cent distribution.

EXAMPLE 9.19 *Design of column with load having very large eccentricity.* A column of effective length 3.6 m is subjected to a bending moment of 100 kNm and axial force of 45 kN design the column.

Design data and assumptions

Axial load	= 45 kN
Let self-weight	= 5 kN
Total load	P = 50 kN
	M = 100 kNm

$$e = \frac{M}{P} = \frac{100}{50} = 2 \text{ m}$$

The failure can be expected to be tension in steel as e is large.

$$f_{ck} = 20 \text{ MPa}, \ f_y = 415 \text{ MPa}, \ \alpha = 20.75$$

Design of the section. The design loads are :

$$P_d = 1.5 \ P = 75 \text{ kN}; \ M_d = 1.5 \ M = 150 \text{ kNm}.$$

Let the total percentage of the reinforcement be 2 and

$$p_c = 0.4p = 0.008, \ p_t = 0.6p = 0.012$$

Let

$$u = 0.2 \ \text{and} \ c = 0.65$$

From Table 9.12 we can get the coefficients for balanced cross-section, and they are:

$$c_{ab} = 0.28, \ c_{xb} = 0.22$$
$$D_s = D - 2d_c = D(1 - 2u) = 0.6 \ D$$

Substitute the assumed values in Eq. (9.150), then

$$D^3 f_{ck} \{0.45(1 - p)(1 - u - c_{xb})c_{ab} + 0.68p_c \, \alpha \, c\} = M_d + 0.5 \ D_s P_d$$

$$D^3(20)\{0.45(0.98)(1 - 0.2 - 0.22)(0.28) + 0.68(0.008)(20.75)(0.65)\}$$

$$= 0.15 + 0.5(0.6D)(0.075)$$

$$2.9SD^3 - 0.0225D^2 - 0.15 = 0$$

The cubic equation when solved gives

$$D = 0.38 \text{ m}$$

then $A_s = 0.02(0.785)(380)^2 = 2267$ mm^2

Use 6 numbers of 22 mm bars then, $A_s = 6(380) = 2280$ mm^2

$$A_{sc} = 2(380) = 760 \text{ mm}^2, A_{st} = 4(380) = 1520 \text{ mm}^2$$

Let $\quad D = 400$ mm, then

$$D_c = 400 - 80 - 22 = 298 \text{ mm}, x_c = c_{xb}D = 0.22(400) = 88 \text{ mm}$$

$$d = D_s + 102 = 282 \text{ mm}$$

The forces in the cross-section are:

$$C_s = 0.45 \, (c_{ab}D^2 - A_{sc})f_{ck}$$
$$= 0.45(0.28(0.16) - 0.00076)20 = 0.3964 \text{ MN}$$
$$C_s = 0.87 \, A_{sc}f_y = 0.87(0.00076)(415) = 0.2744 \text{ MN}$$
$$T = 0.87 \, A_{st}f_y = 0.87(0.00152)(415) = 0.5488 \text{ MN}$$

The load carrying capacity is

$$P_a = C_c + C_s - T = 0.122 \text{ MN} < P_d = 0.75$$

The bending moment about the centre of the tension steel is

$$M_a = C_c(d - x_c) + C_s(D_s) - P_d(0.5 \, D_c \cos 60)$$
$$= 0.3964(0.282 - 0.88) + 0.2744(0.18) - 0.075(0.5)(0.298)(0.5) = 0.121 \text{ MNm}$$

This value is less than $M_d = 0.15$ MNm, therefore, the section need to be modified. Keeping the area of the reinforcement same, increase the diameter of the column.

Let $\quad D = 440$ mm

then $x_c = 0.22(440) = 96.8$ mm, $D_c = 440 - 102 = 338$ mm

$$D_s = 0.5D_c(\cos 45 + \cos 60) = 204 \text{ mm}$$

$$d = D_s + 102 = 306 \text{ mm}$$

$$C_c = 0.45(c_{ab}D^2 - A_{sc})f_{ck} = 0.481 \text{ MN}$$

The capacity P_a will be certainly more than P_d, so only the moment capacity is to be verified:

$$M_a = C_c(d - x_c) + C_s D_s - P_d(0.5D_c \cos 60)$$
$$= 0.481(0.306 - 0.0968) + 0.2744(0.204) - 0.075(0.5)(0.338)(0.5)$$
$$= 0.1503 \text{ MNm}$$

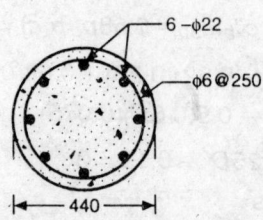

Fig. 9.20 Sectional Details of Example 9.19.

Table 9.9 Comparison of Rectangular and Circular Sectioned Columns (P = 50 KN, M = 100 kNm)

Detail	Rectangular		Circular
	WSD	LSD	LSD
Example No.	5.14	9.14	9.19
b, D (mm)	300.670	300.375	440
A_g (mm^2)	1432	1960	2280
A_g(m^2)	0.201	0.111	0.152

This value is more than the bending moment; hence, the column is safe. Table 9.9 gives the comparison of the design details of a rectangular and circular columns.

PROBLEMS
(Use IS 465, LSD = Limit States Design)

9.1 A stub column fixed at base and free at loaded end is subjected to an axial live load of 1 MN. The cantilever height of the column is 1.2 m. Design a reinforced concrete square column by limit states design using M20 concrete and HYSD-Fe415 bars. Sketch the reinforcement detail indicating main and transverse reinforcement.

9.2 A column fixed at base and hinged at 4 m above the base level is subjected to an axial live load of 5000 kN. Design a reinforced concrete rectangular column with width equal to 400 mm by limit states design using M25 concrete and HYSD Fe415 bars.

9.3 A column fixed at base and free at top having a height of 3 m is subjected to an axial live load of 700 kN. The column size is 200 × 350 mm. Design the reinforcement details of the column by limit state method using M20 concrete and HYSD-Fe415 bars.

9.4 A column hinge connected at both ends having a supported height of 5 m is subjected to an axial live load of 800 kN. Design a reinforced concrete square cross-sectioned column with minimum possible size by limit state method using M20 concrete and HYSD Fe415 bars.

9.5 A reinforcement concrete column fixed at a base and roller supported at the top of height of 4 m is subjected to an axial live load of 800 kN. Design a reinforced concrete column of square cross-section with minimum possible percentage of reinforcement by limit state method using M20 concrete and HYSD-Fe415 bars.

9.6 Design a reinforcement rectangular tied column of an effective height 8 m with breadth not exceeding 400 mm and having 3% of reinforcement by limit state method using M20 concrete and HYSD-Fe415 bars. The column is subjected to an axial live load of 1 MN.

9.7 A free standing wall of height 4 m is subjected to an axial live load of 400 MN/m. Design a reinforced concrete wall by limit state design with 1% HYSD-Fe415 steel and M20 concrete.

9.8 A column of an effective height 6.5 m is subjected to an axial live load of 2.5 MN. Design a rectangular composite column with ISMB 400 and HYSD-Fe415 bars. The area of ISMB 400 is 7846 mm^2. Yield stress of structural steel is 250 N/mm^2. Design the column by limit states design using M20 concrete and HYSD-Fe415 bars and sketch the reinforcement details.

9.9 A column is subjected to an axial live load of 4 MN. The effective height of the column is 6 m and the size of the column is to be restricted by 500 × 900 mm. Design a composite reinforced concrete column by limit states design using rolled steel beam section, M20 concrete and HYSD-Fe415 bars. Use 2 nos. of ISMB 400 where the area of each beam section is 7846 mm^2.

9.10 A column is subjected to an axial live load 700 kN. Its effective height is 3 m. Design an infilled concrete columns by limit state design using M20 concrete and mild steel structural tube. Assume the thickness of the tubes as 5 mm for all outer diameter of the tubes varying from 250 mm to 400 mm.

9.11 A column fixed at base and free at top having a height of 3 m is subjected to a working axial live load of 700 kN. Design a circular spiral reinforced concrete column by limit states design using M20 concrete and HYSD-Fe415 bars. Design two columns, one with minimum and another with maximum permissible reinforcements. Sketch the reinforcement details.

9.12 A column of effective height 5.5 m is subjected to a working axial live load of 700 kN. Design a reinforced concrete circular spiral column by limit states design using M20 concrete and HYSD-Fe415 bars subject to a 450 mm as the maximum diameter of the column.

9.13 A column of effective height of 4 m is subjected to a working axial force of 800 kN and bending moment of 100 kNm. Design a rectangular sectioned column by limit state design using M20 concrete and HYSD-Fe415 bars such that the width of the column should not exceed 400 mm. Use close to minimum percentage of reinforcement.

9.14 A column of 6 m effective length is subjected to a working axial live load of 800 kN and bending moment of 70 kNm under wind load condition. Design the column with M20 concrete and HYSD-Fe415 bars by limit state design, assuming width of the column not to exceed 400 mm. Also the percentage of reinforcement to be kept within 1%.

9.15 A column of effective length of 5 m is subjected to a working axial live load of 900 kN and bending moment of 80 kNm. The size of the column and its reinforcement details are given below. Determine whether the column when made of M20 concrete and HYSD-Fe415 bars can withstand the above load $b = 350$ mm, $D = 500$ mm, $A_s = 1708$ mm^2 $D_s = 420$ mm. The reinforcement is equally divided on two opposite faces.

9.16 A reinforced concrete column is subjected to a working axial live load of 800 kN. The effective height of the column is 5 m and its size and reinforcement details are: $b = 425$ mm, $D = 500$ mm, $A_s = 1708$ mm^2 equally divided on the two faces. The column is made of M20 concrete and HYSD bars. Investigate what is the maximum bending moment that the column can resist under wind load condition with the same axial live load. Check the result by limit states design.

9.17 A reinforced concrete column of effective height 6 m having $b = 500$ mm and $D = 700$ mm is provided with 4 Nos. of 25 mm bars at each of the opposite faces of the depth of the column. The column is made of M20 concrete and HYSD-Fe415 bars. Investigate by limit states design whether the column is capable of resisting 1.0 MN axial live load alongwith the bending moment of 200 kNm under wind load condition. Assume the clear cover to the reinforcement as 40 mm.

9.18 A column of effective height equal to 3.2 m is subjected to a working axial force of

600 kN and bending moment of 400 kNm. Design the column with M20 concrete and HYSD-Fe415 bars by LSD. Assume the width of the column as 500 mm. Investigate whether the column is cracked or uncracked section. Provide unequal reinforcement at the tension and compression faces.

9.19 A long reinforced concrete column of 7 m effective height is subjected to a working axial live load of 600 kN and bending moment of 300 kNm. Design a column by LSD with width not exceeding 300 mm using M20 concrete and HYSD-Fe415 bars. Provide unequal reinforcement at the tension and compression faces of the column.

9.20 A column of effective height 8 m is subjected to a working axial force of 40 kN and bending moment of 100 kNm. Design a reinforced rectangular sectioned column with width as 300 mm by LSD using M20 concrete and HYSD-Fe415 bars.

9.21 A reinforced concrete column section is subjected to a working axial force of 1000 kN and bending moment of 100 kNm under live load condition. Design an uncracked concrete column section by LSD using M20 concrete and mild steel reinforcement. Provide minimum reinforcement.

Chapter 10

Yield Line Theory and Design of Slabs

10.1 Introduction

As the load on a slab increases, the stresses in the reinforcement and in the concrete increase more or less proportionately upto a load level corresponding to the yield stress in the reinforcement. Further increase in the load will cause excessive deformations and rapid increase in the strains. The deflections of the slab will be elasto-plastic upto a level called *limit load*. At the limit load, the slab will continue to deform without any additional load leading to total collapse. In this process, cracks will progress at all the critical moment zones and the pattern of cracks will form a set of continuous crack lines called the *yield lines*. As the bending moment along the yield lines approaches the moment capacity of the section, a set of yield lines results into a mechanism leading to total collapse of the slab. A typical load deflection curve of a slab is shown in Fig. 10.1. The load at which the slab collapses is called the

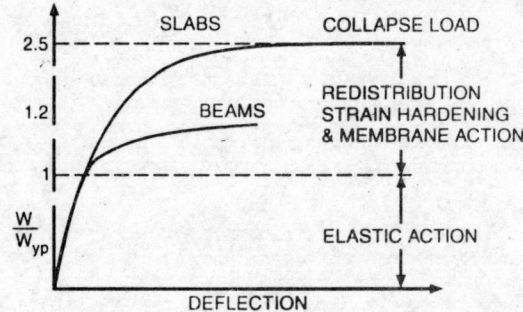

Fig. 10.1 Forces Deformation Relation.

collapse load and is same as the limit load. Three types of phenomena take place after the elastic load. They are: redistribution of bending moment, strain hardening effect of steel and membrane action of the plate. The membrane action is usually absent in beams.

The progress of the cracks is governed by the following factors. The yield line is initiated at a point where the external bending moment reaches the yield moment capacity of the section. Usually it is at the maximum elastic moment section. Then the crack progresses in a direction of least resistance of the overall slab such that the yield line pattern develops in mechanism of collapse. One is interested in the final pattern of the yield lines in the limit design for strength. The following assumptions are used in selecting the collapse yield line pattern.

1. The elastic deformations are negligible when compared with the plastic deformations.
2. Plastic hinge lines are formed along the yield lines and all the plastic deformations are lumped along the yield lines. The segments of slabs between the yield lines are treated as plane rigid elements.

3. Yield lines must terminate at the boundary of the slab or at the intersection of other yield lines.
4. Yield lines are almost straight lines (which is actually based on the second assumption). As the slab elements within the yield lines are treated as plane elements, the intersecting lines of such planes must be straight lines.
5. Each of the segments of the slab will tend to rotate in a rigid body motion. The axes of rotation lie along the lines of supports or over the column supports.
6. For a mechanism to develop, the yield line must pass through the intersection of the axes of rotation of adjacent segments.
7. Yield lines must end at the axes of rotation or at the support lines.

Since this chapter is meant to be an introduction, only simple problems are presented.

10.2 Typical Yield Line Patterns and Failure Mechainsms

The collapse load on a slab can be obtained by two basic methods, namely

1. Equilibrium method (statical method)
2. Mechanism method (virtual work).

In the equilibrium method, the equilibrium of the forces and the moments are to be satisfied subject to the condition that the moment at any point is less than or equal to the capacity of the section. Therefore, one selects any statically admissible field and maximizes the moments till a failure pattern is obtained. Such a statical method gives a *lower bound* collapse load. A lower bound collapse predicts a lower failure load when compared with the actual value. In a design problem, the loads are fixed and a reverse process is applied to find the moment capacities required to resist the load. Therefore, the lower bound in the analysis problem gives a safe design. Normally, any statical method gives a safe design. An elastic analysis could be taken as a particular solution of an equilibrium method.

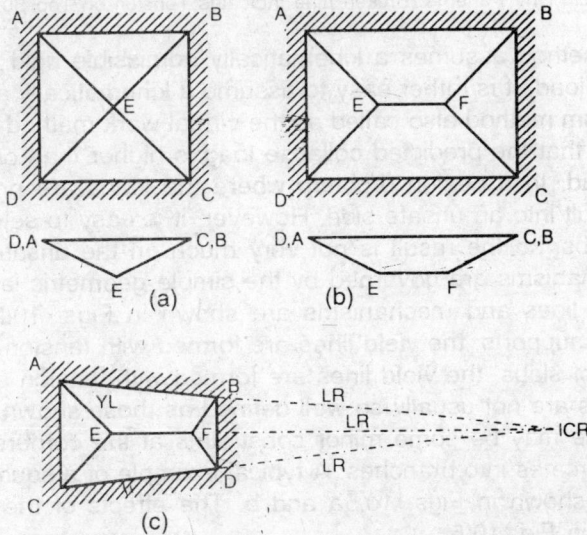

Fig. 10.2 Yield Line Pattern and Mechanisms.

410

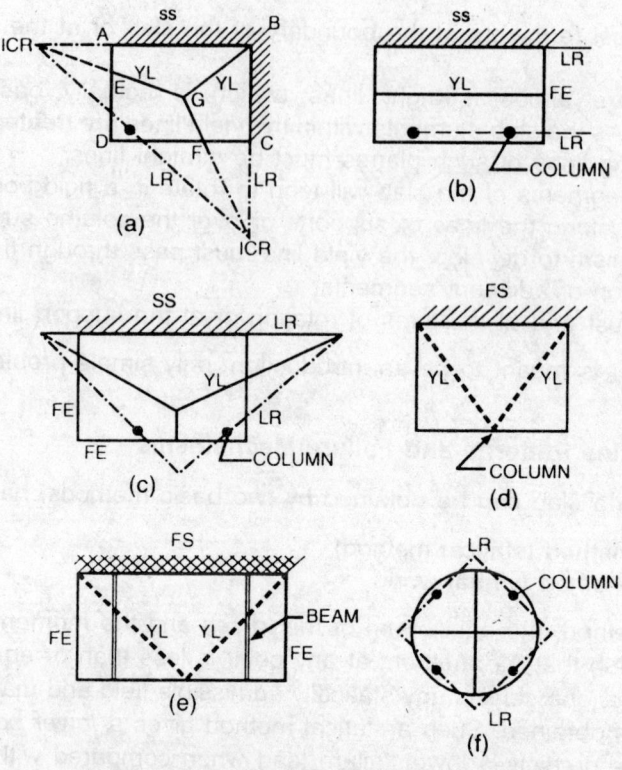

Notation : FE = Free Edge, FS = Fixed Support
ICR = Instantaneous Centre of Rotation
LR = Line of Rotation, SS = Simply supported
YL = Yield Line (Firm line indication compression at top face
and broken line for tension at top face.)

Fig. 10.3 Typical Yield Line Patterns (Broken Line Indicates Tension on Top Surface of the Slabs.)

The mechanism method assumes a kinematically admissible rigid body mechanism and estimates the collapse load. It is rather easy to assume a kinematically admissible mechanism in slabs. The mechanism method also called as the virtual work method gives an *upper bound* solution, which means that the predicted collapse load is higher than or at the most equal to the actual collapse load. In a design problem where the loads are preassigned, the upper bound solution will result into an unsafe side. However, it is easy to select the most probable true mechanism in slabs, so the result is not very much on the unsafe side. The formation of yield lines and mechanisms are governed by the simple geometric laws as already stated. Typical idealized yield lines and mechanisms are shown in Figs. 10.2 to 10.5. In case of cantilever or continuity supports, the yield lines are formed with tension at the top fibres, and in the case of interior of slabs, the yield lines are formed with tension at the bottom surface. In reality, the yield lines are not usually so well defined as those shown in Figs. 10.2 to 10.5. In the actual slab, there may be some minor constraints at the corners of the slab with the result the yield line bifurcates into branches. A typical example of a square slab with idealized and real yield lines is shown in Figs. 10.5a and b. The effects of the corners on the yield line patterns is shown in Fig. 10.5.

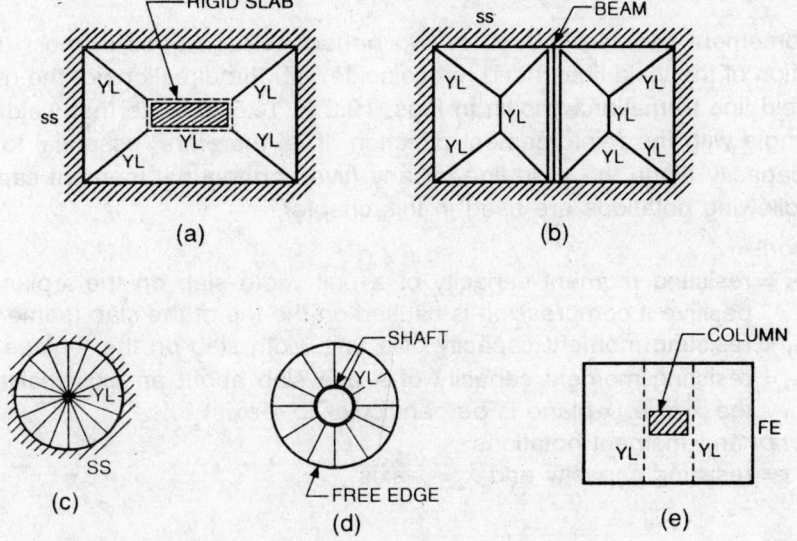

Fig. 10.4 Typical Yield Line Patterns.

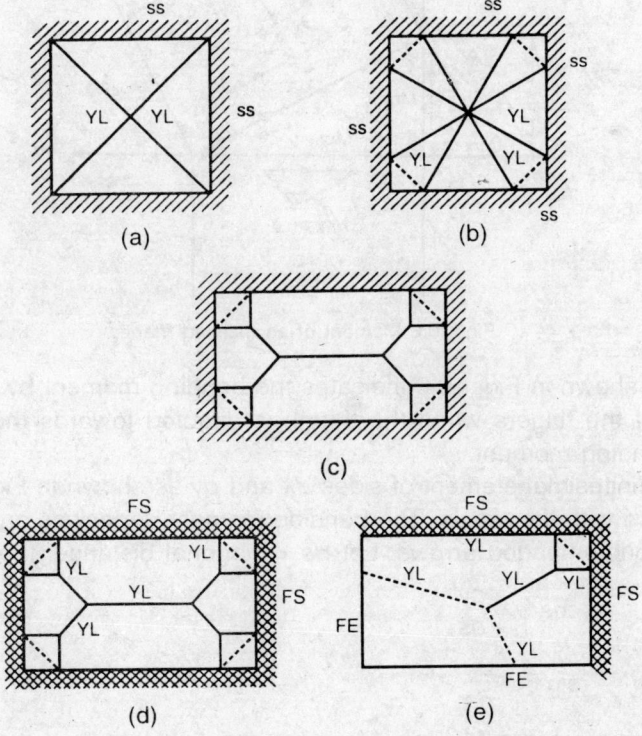

Fig. 10.5 Typical Yield Line Patterns with Corner Effects.

412

10.3 Moment Capacity Along an Yield Line

The reinforcement bars are placed in two orthogonal directions in most of the RCC slabs. The direction of the yield lines need not coincide with the directions of the reinforcement bars. Typical yield line formations shown in Figs. 10.2 to 10.5 indicate that yield line can be at an inclined angle with the reinforcement direction. It is, therefore, essential to find the resisting moment capacity along the yield line for any given orthogonal moment capacities.

The following notations are used in this chapter.

Notations

M_{rx} = resisting moment capacity of a unit width slab on the x-plane. It is measured positive if compression is caused on the top of the slab (same as the load face).

M_{ry} = resisting moment capacity of a unit width slab on the y-plane

$M_{r\phi}$ = resisting moment capacity of a unit slab about an axis making ϕ-degress with the x-axis (x-plane is perpendicular to x-axis)

Subscript and moment notations

r = resisting capacity and x = x-axis

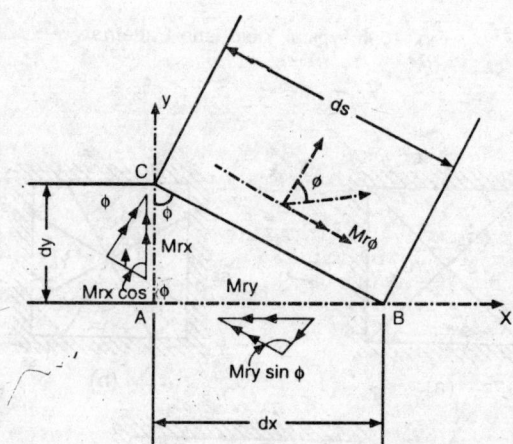

Fig. 10.6 Moment of an Inclined Plane.

A double arrow as shown in Fig. 10.6 indicates the bending moment by the right-hand thumb rule. The curling of the fingers when the thumb is directed towards the arrow indicates the direction of the bending moment.

Consider an infinitesimal element of sides dx and dy as shown in Fig. 10.6. Let the plane *CB* be at an angle ϕ with the x-axis. The bending moment capacities on the three planes are indicated by the double-headed arrows. Let ds = diagonal distance, then

$$\cos \phi = \frac{dy}{ds} \tag{10.1}$$

$$\sin \phi = \frac{dx}{ds} \tag{10.2}$$

Resolve the bending moments M_{rx} and M_{ry} along the ϕ-direction.

The sum of the components of the two capacities on x- and y-planes is equated to that on the ϕ plane whicn gives

$$(M_{rx} \cos \phi)dy + (M_{ry} \sin \phi) \, dx = M_{r\phi} \, ds \tag{10.3}$$

Dividing the above equation by ds, we have

$$(M_{rx} \cos \phi) \frac{dy}{ds} + (M_{ry} \sin \phi) \frac{dx}{ds} = M_{r\phi} \tag{10.4}$$

The substitution of Eqs. (10.1) and (10.2) in Eq. (10.4) gives

$$M_{r\phi} = M_{rs} \cos^2 \phi + M_{ry} \sin^2 \phi \tag{10.5}$$

In case $\phi = 45°$, the above equation reduces to

$$M_{r45} = \frac{(M_{rx} + M_{ry})}{2} \tag{10.6}$$

Let $cM_{rx} = M_{ry}$, then Eqs. (10.5) and (10.6) reduce to $\tag{10.7}$

$$M_{r\phi} = M_{rx} (c \sin^2 \phi + \cos^2 \phi) \text{ and} \tag{10.8}$$

$$M_{r45} = M_{rx} (1 + c)/2 \tag{10.9}$$

10.4 Method of Virtual Work

The virtual work concept is a powerful tool in solving the collapse loads of slabs. A most probable yield line pattern is assumed and a virtual displacement compatible with the boundary conditions alongwith the yield line formation is assumed. The segments of the slab within the yield lines go through rigid body displacements with the collapse load acting on the structure. The virtual work done by the load is, therefore, equal to the load multiplied by the virtual displacement.

Let $v^* =$ virtual displacement

$w =$ load acting during the displacement

then the virtual work done by the load is

$$V^* = \int v^* w \, dx \, dy \tag{10.10}$$

where $V^* =$ virtual work done

The asterisk indicates the virtual work.

The main difference between the virtual work and the actual work by a load is the magnitude of the load acting while the displacement takes place. The total load acts through the virtual displacement whereas in a real situation where the load causes the deformation, as the load increases the displacement increases. Therefore, the total work done is equal to mean load multiplied by the displacement. The actual work done by a load in a linear force deformation relation is

$$V = \frac{1}{2} \int vw \, dx \, dy \tag{10.11}$$

Figure 10.7 illustrates the difference in the virtual work and real work. While the virtual displacement is taking place in the structure, a relative rotation of the surfaces takes place about the yield lines. For simplicity let the resisting moment capacity along a yield line be assumed to be constant, then the internal resisting work is

414

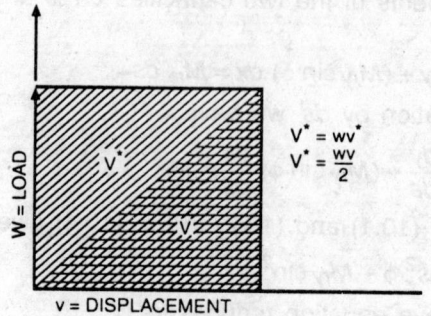

Fig. 10.7 Virtual Work.

$$U^* = \Sigma \, \theta_i^* \, M_{ri} \, s_i \qquad (10.12)$$

where $\qquad \theta_i^* =$ relative virtual rotation of the surface at the ith yield line

$M_{ri} =$ moment capacity per unit width along the ith yield line

$s_i =$ length of the ith yield line

From the principle of conservation of energy, we have

$$\sum_{t=1}^{N} \theta_i^* \, M_{ri} \, s_i = \int^{A} v^* \, w \, dA \qquad (10.13)$$

In a conservative and linear system, the rate of change of the total work done results into Eq. (10.13) which gives the collapse load of a system.

The design load w in the limit state of collapse is given by

$$w = \gamma_{fd} \, W_d + \gamma_{fl} \, W_l \qquad (10.14)$$

$$\text{or} \quad w = \gamma_{fd_1} \, W_d + \gamma_{fl_1} \, W_1 + \gamma_{fw} \, w_w \qquad (10.15)$$

where $\gamma_{fd}, \gamma_{fl} \dots$ are the partial safety factors applied to loads.

10.5 Resisting Moment Capacity of a Section and Method of Design

The balanced moment capacity of a section based on limit state of strength using the rectangular-cum-parabolic stress distribution in bending compression in concrete was derived in Chapter 7. It is given by

$$M_{rb} = Kbd^2 \, f_{ck} \qquad (10.16)$$

The corresponding area of the tension reinforcement is given by

$$A_{st} = p_0 bd \frac{f_{ck}}{f_y} \qquad (10.17)$$

where $p_0 =$ constant given in Table 10.1.

In case of under-reinforced section, the moment capacity of the section is given by

$$M_r = 0.87 \, A_{st} \, f_y \, (d - 0.42 \, x_u) \qquad (10.18)$$

Table 10.1 Design Parameters in LSD

f_y	x_u	K	p_0
250	0.531	0.149	0.220
415	0.479	0.138	0.198
500	0.456	0.133	0.139

The balance moment capacities of different depths of slab are given in Table 10.2 using the effective depth of the slab as:

$$d = t - 20 \text{ mm for } t \leq 150 \text{ mm}$$
$$= t - 25 \text{ mm for } t > 150 \text{ mm}$$

The corresponding area of the tension reinforcement per one metre width is also given in the same table.

Table 10.2 Balanced Moment Capacities and Area of Tension Steel for
$b = 1000$ mm, M_{rb} in Nm/m, A_{st} in mm^2/m

	M15		M20		M25	
	M_{rb}	A_{st}	M_{rb}	A_{st}	M_{rb}	A_{st}
			$f_y = 250$ MPa			
80	8046	792	10728	1056	13410	1320
90	10952	924	14602	1232	18253	1540
100	14304	1056	19072	1408	23840	1760
105	16148	1122	21531	1496	26913	1870
110	18104	1188	24138	1584	30173	1980
115	20171	1254	26895	1672	33618	2090
120	22350	1320	29800	1760	37250	2200
125	24641	1386	32855	1848	41068	2310
130	27044	1452	36058	1936	45073	2420
135	29558	1518	39411	2024	49263	2530
140	32184	1584	42912	2112	53640	2640
145	34922	1650	46563	2200	58203	2750
150	37772	1716	50362	2288	62953	2860
155	37772	1716	50362	2288	62953	2860
160	40733	1782	54311	2376	67888	2970
165	43806	1848	58408	2464	73010	3080
170	46991	1914	62655	2552	78318	3190
175	50288	1980	67050	2640	83813	3300
180	53696	2046	71595	2728	89493	3410
185	57216	2112	76288	2816	95360	3520
190	60848	2178	81131	2904	101413	3630
195	64592	2244	86122	2992	107653	3740
200	68447	2310	91263	3080	114078	3850
205	72414	2376	96552	3168	120690	3960
210	76493	2442	101991	3256	127488	4070
215	80684	2508	107578	3344	134473	4180
220	84986	2574	113315	3432	141643	4290
225	89400	2640	119200	3520	149000	4400

(Contd.)

Table 10.2 (Contd.)

	M15		M20		M25	
	M_{rb}	A_{st}	M_{rb}	A_{st}	M_{rb}	A_{st}
			f_y = 415 MPa			
80	7452	429	9936	573	12420	716
90	10143	501	13524	668	16905	835
100	13248	573	17664	763	22080	954
105	14956	608	19941	811	24926	1014
110	16767	644	22356	859	27945	1073
115	18682	680	24909	907	31136	1133
120	20700	716	27600	954	34500	1193
125	22822	751	30429	1002	38036	1252
130	25047	787	33396	1050	41745	1312
135	27376	823	36501	1097	45626	1372
140	29808	859	39744	1145	49680	1431
145	32344	895	43125	1193	53906	1491
150	34983	930	46644	1240	58305	1551
155	34983	930	46644	1240	58305	1551
160	37726	966	50301	1288	62876	1610
165	40572	1002	54096	1336	67620	1670
170	43522	1038	58029	1384	72536	1730
175	46575	1073	62100	1431	77625	1789
180	49732	1109	66309	1479	82886	1849
185	52992	1145	70656	1527	88320	1908
190	56356	1181	75141	1574	93926	1968
195	59823	1217	79764	1622	99705	2028
200	63394	1252	84525	1670	105656	2087
205	67068	1288	89424	1718	111780	2147
210	70846	1324	94461	1765	118076	2207
215	74727	1360	92636	1813	124545	2266
220	78712	1396	104949	1861	131186	2326
225	82800	1431	110400	1908	138000	2386

Method of design: The steps in the method of design of a slab based on limit state of strength are suggested below.

1. Estimate the thickness of the slab in the range of $L/40$ to $L/20$ depending on the live load and aspect ratio of the slab for the purpose of estimating the self-weight.
2. Calculate the design load using the appropriate partial safety factors.
3. Assume a relative resisting moment capacity factor μ.

Use
$$\mu = \left(\frac{L_y}{L_x}\right)^2 \tag{10.19}$$

where L_x = effective span along x-axis (also equal to short span)
L_y = effective span along the y-axis (also equal to long span)

4. Assume a most probable yield line formation compatible with the boundary conditions and the possible collapse mode. If possible, investigate the most probable collapse mode.
5. Using the virtual work governing criterion, determine the required moment capacities along the yield lines.

6. Design the section for the required moment capacities and check whether the originally assumed depth of the section compares well with the designed value. In case it does not check well, redesign the section.
7. Check for shear capacity of the section and redesign the slab in case the nominal shear stress exceeds the allowable value.
8. Check for the minimum percentage of reinforcement.
9. Check for the development lengths.
10. Detailing of the reinforcement.

10.6 Design for Limit State of Serviceability

The limitation on the deflection and crack width are the two main limit states of serviceability in the design of RCC slabs. In case, the depth of the slab is more than $L/20$ in simply supported slabs, $L/7$ in cantilever slabs and $L/30$ in continuous slabs, one need not go into the details of deflection calculations. In most cases, the deflections of slabs are within the allowable limits; therefore, one may not have to be much concerned about the deflection computations. The crack width is not a serious concern in the design of slabs provided reinforcement detailing is made carefully. Select smaller diameter bars. (Let the diameter of the bar be in the range of $t/15$ to $t/8$ and the spacing of the bars be in the range of t to $3t$.)

10.7 Collapse Load of Simply Supported Rectangular Slab

Rectangular slab is one of the most commonly constructed floor slabs; therefore, the collapse load of such a slab is derived by virtual work method.

Let $$L = L_x \text{ and } L_y = rL$$

and $$M_{ry} = cM_{rx} = cM_r \qquad (10.20)$$

A most possible yield line configuration of a rectangular slab under UDL is shown in Fig. 10.8

Let $v^* =$ virtual deflection of the yield line EF. From symmetry of the slab and the load, the yield line EF will be at the mid-span. The rotation of the segments will also be symmetrical. The magnitudes of the rotations of segments S_1 and S_3 about the y-axis are same.

$$\theta_{y1} = \theta_{y3} = \frac{v^*}{0.5L} = \frac{2v^*}{L}$$

Similarly, the other rotation of the segments are:
$$\theta_{x1} = \theta_{x3} = 0 \text{ and } \theta_{y2} = \theta_{y4} = 0$$

$$\theta_{x2} = \theta_{x4} = \frac{v^*}{\alpha rL}$$

where $\alpha rL = \alpha L_y =$ distance of point E from the edge.

As the segments rotate, the work done by the resisting moment is equal to

$$U^* = \Sigma \, \theta_{xi} \, M_{ry} \, s_{xi} + \Sigma \, \theta_{yi} \, M_{rx} \, s_{yi}$$
$$= (\Sigma \, \theta_{xi} \, M_{ry} \, x_i + \Sigma \, \theta_{yi} \, M_{rx} \, y_i) \qquad (10.21)$$

where $s_{xi} = x_i =$ projected length of the yield line along the x-axis.
Similarly $s_{yi} = y_i$ along y-axis

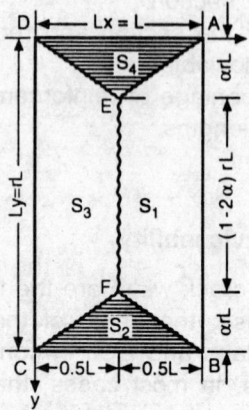

Fig. 10.8 Yield Line Pattern in Rectangular Slab.

The projected length of yield lines *AE*, *EF* and *FB* on the *y*-axis is equal to $L_y = rL$. Similarly, the projected length on the *x*-axis is L_x. Then

$$U^* = (\theta_{x1} L + \theta_{x3}L + \theta_{x2}(L) + \theta_{x4}(L)) M_{ry} + (\theta_{y1}(L_y) + \theta_{y3}(L_y) + \theta_{y2}(2\alpha L_y) + \theta_{y4}(2\alpha L_y)) M_{rx}$$

$$= \frac{v^*}{\alpha rL}(0 + 0 + L + L) M_{ry} + \frac{2v^*}{L}(rL + rL + 0 + 0) M_{rx}$$

$$= \frac{v^*}{\alpha r}(2M_{ry}) + 4v^* rM_{rx}$$

$$= v^*\left(\frac{2cM_r}{\alpha r} + 4rM_r\right)$$

$$= 2\frac{(c + 2r^2 \alpha)}{\alpha r} v^* M_r \tag{10.22}$$

The virtual work done by the external load acting on any segment can be calculated by multiplying the intensity of the load by the virtual displacement at the centroid of the load on the segment. The segment S_1 can be divided into two triangular segments *AEB* and *EFB*. The centroids of these triangles will be at a distance one-third of the altitude from the base. The virtual work done by the load acting on the segment S_1 is

$$V_i^* = (\text{Load on segment } AEB)\left(\frac{v^*}{3}\right)$$

$$+ (\text{Load on segment } EFB)\frac{2v^*}{3}$$

$$= \frac{w_c}{2}(L_y)\left(\frac{L}{2}\right)\frac{v^*}{3} + \frac{w_c}{2}(1 - 2\alpha) L_y \frac{L}{2}\frac{2v^*}{3}$$

$$= \frac{w_c}{12}(rL^2 + 2r L^2)(1 - 2\alpha)) v^*$$

$$= \frac{w_c}{12}(3 - 4\alpha) rL^2 v^*$$

where w_c = collapse load.

   Similarly, the total virtual work done including the other three segments by the load can be obtained as

$$V^* = 2w_c\left(\frac{rL^2}{12}(3 - 4\alpha)\,v^* + \frac{1}{2}(L)(\alpha rL)\frac{v^*}{3}\right)$$

$$= w_c\left(\frac{(3 - 4\alpha)\,rL^2}{6} + \frac{r\alpha L^2}{3}\right)v^*$$

$$= \frac{(3 - 2\alpha)}{6}\,rL^2\,w_c\,v^* \tag{10.23}$$

Equating the internal and external virtual works of the system, we have from Eqs. (10.22) and (10.23):

$$\frac{(3 - 2\alpha)}{6}\,rL^2\,w_cv^* = \frac{2(c + 2r^2\alpha)}{\alpha r}\,v^*\,M_r$$

or
$$w_c = \frac{12\,(c + 2r^2\,\alpha)}{\alpha r^2\,(3 - 2\alpha)L^2}\,M_r \tag{10.24}$$

Or, if one is interested in the moment capacity required for a given collapse load, the above equation can be rearranged as

$$M_r = \frac{\alpha(3 - 2\alpha)\,r^2\,L^2\,w_c}{12\,(c + 2r^2\,\alpha)} \tag{10.25}$$

   The value of α is still unknown as it is arbitrarily selected. One can obtain the extreme value of the collapse load or the capacity of the section by setting

$$\frac{dM_r}{d\alpha} = 0$$

which gives

$$(3 - 4\alpha)\,(c + 2r^2\,\alpha) - \alpha\,(3 - 2\alpha)\,(2r^2) = 0$$

or
$$3c + 6r^2\alpha - 4\alpha c - 8r^2\alpha^2 - 6r^2\alpha + 4r^2\alpha^2 = 0$$

or
$$\alpha^2\,(4r^2 - 8r^2) + \alpha\,(6r^2 - 6r^2 - 4c) + 3c = 0$$

or
$$4r^2\alpha^2 + 4\alpha c - 3c = 0$$

$$\alpha = \frac{-4c + \sqrt{(16c^2 + 48r^2c)}}{8r^2} \tag{10.27}$$

or
$$2r^2\alpha = (\sqrt{(c^2 + 3r^2c)} - c) \tag{10.28}$$

The substitution of above in Eq. (10.25) gives

$$M_r = \left(\frac{\sqrt{c^3 + 3r^3c} - c}{12(2r^2)}\right)\left(3 - \frac{\sqrt{c^2 + 3r^2c} - c}{r^2}\right)\left(\frac{r^2\,L^2\,w_c}{c + \sqrt{(c^2 + 3r^2c)} - c}\right)$$

$$= \frac{(\sqrt{(c^2 + 3r^2c)} - c)\,(c + 3r^2 - \sqrt{(c^2 + 3r^2c)})\,(r^2\,L^2\,w_c)}{(24r^4)\,\sqrt{(c^2 + 3r^2c)}}$$

$$= \frac{(\sqrt{(c^2 + 3r^2c)} - c)^2}{24r^2c}\,L^2w_c$$

$$M_r = \frac{1}{24}\left(\frac{\sqrt{(c + 3r^2)} - \sqrt{c}}{r}\right)^2 L^2 w_c \tag{10.29}$$

$$w_c = \frac{24\,M_r}{L^2}\left[\frac{r}{\sqrt{(c + 3r^2)} - \sqrt{c}}\right]^2 \tag{10.30}$$

Equation (10.30) can also be obtained by minimizing $V^* - U^*$ with respect to the variable α. $V^* - U^*$ is the net virtual work on the system.

Consider the limiting cases of the rectangular slab:

Case 1—One-way slab: In case of one-way slab one can assume that the moment capacity of the section normal to the long span (y-axis) could be treated as negligibly small, which means

$$M_{ry} = 0$$

or $$M_{ry} = cM_r = 0 \text{ or } c = 0$$

Equation (10.30) can be reduced to

$$M_r = \frac{w_c L^2}{8} \tag{10.31}$$

Case 2—Square slab: When $L_x = L_y = L$ and $c = 1$, Eq. (10.31) reduces to

$$M_r = \frac{w_c L^2}{24} \tag{10.32}$$

Case 3—Rectangular slab with

$$\frac{M_{rx}}{M_{ry}} = \left(\frac{L_y}{L_x}\right)^2 = r^2$$
$$r^2 = 1/c$$

or

Equation (10.30) reduces to

$$M_r = c\,(\sqrt{(c + 3/c)} - \sqrt{c})^2 \frac{L^2\,w_c}{24}$$

$$= (\sqrt{(c^2 + 3)} - c)^2 \frac{L^2\,w_c}{24}$$

10.8 Collapse Load on a Rectangular Slab with Fixed Boundary

Consider the case of rectangular slab shown in Fig. 10.8 in which all the boundaries are fixed.

Let M_m = negative moment capacity $x = 0$, and $x = L$

M_{rp} = positive moment capacity along x-direction

Let the negative and positive moment capacities on the x-plane be equal to cM_m and cM_{rp}. The negative yield lines (tension on the top surface) are formed all along the boundary in addition to the yield line shown in Fig. 10.8. Assuming a virtual displacement of v^* for the line EF, the virtual work done by the external load is same as that of simply supported slab, and it is

$$V^* = \left(\frac{3 - 2\alpha}{6}\right) rL^2\,w_c\,v^* \tag{10.23}$$

The negative moment along the boundary will contribute towards the internal virtual work in addition to the work done along the positive yield lines. The total internal virtual work done is

$$U^* = \frac{2(c + 2r^2\alpha)}{\alpha r} M_{rp} v^*$$

$$+ 2rLM_{rn}\left(\frac{2v^*}{L}\right) + 2LcM_{rn}\frac{v^*}{\alpha rL}$$

$$= 2\left(\frac{(c + 2r^2\alpha)}{\alpha r} M_{rp} + \left(\frac{2r^2\alpha + c}{r\alpha}\right) M_{rn}\right) v^*$$

$$= 2(c + 2r^2\alpha)(M_{rp} + M_{rn})\frac{v^*}{r\alpha} \qquad (10.33)$$

Equating the virtual work done by the internal and external forces, we have

$$\frac{(3 - 2\alpha) rL^2}{6} w_c = \frac{2(c + 2r^2\alpha)}{r\alpha}(M_{rp} + M_{rn})$$

or $\quad (M_{rp} + M_{rn}) = \dfrac{(3 - 2\alpha) r^2 \alpha L^2 w_c}{12(c + 2r^2\alpha)} \qquad (10.34)$

Case 1—Long slab in which $c = 0$, $r^2 = $ large, and $\alpha \to 0.0$

$$M_{rn} = M_{rp} = M_r, \text{ we have}$$

$$M_r = \frac{w_c L^2}{16} \qquad (10.35)$$

Case 2—Square slab where $c = 1$, and $r = 1$ and let

$$M_{rn} = M_{rp} = M_r, \text{ then}$$

$$M_r = \frac{w_c L^2}{48} \qquad (10.36)$$

Equation (10.34) has to be minimized with respect to α so as to obtain the value of α as done in Section 10.7.

10.9 Equilibrium to Mechanism Method

The collapse load can be calculated from the equilibrium of forces of the segments of a mechanism instead of virtual work. Both the approaches should result into the same upperbound collapse load. Consider the yield line diagram shown in Fig. 10.8 and take the equilibrium of any typical segments. The yield line is symmetrical so select segments *AB*, *FE* and *CBF*. The forces acting on the segments are shown in Fig. 10.9. The segment *ABFE* can be treated as consisting of two triangles *ABE* and *BFE*, and the loads acting on the triangular portions are:

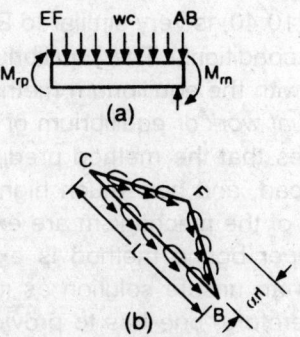

Fig. 10.9 Free Body of a Diagram of a Slab Segment.

$$W_1 = \frac{(rL)}{2}\frac{(L)}{2}\, w_c = \frac{rL^2}{4}\, w_c$$

$$W_2 = \frac{rL(1-2\alpha)}{2}\frac{L}{2}\, w_c = \frac{rL^2(1-2\alpha)}{4}\, w_c$$

The equilibrium of moment taken about line AB of the segment ABFE gives (Fig. 10.9a)

$$rL\,(M_{rp} + M_{rn}) = W_1\frac{L}{6} + W_2\frac{2L}{6} = \frac{rL^3}{24}(1 + 2(1-2\alpha))w_c$$

or $\quad M_{rp} + M_{rn} = \dfrac{L^2}{24}(3 - 4\alpha)$ $\hfill$ (10.37)

Similarly, by considering the segment BCF and its moment equilibrium about line CB gives

$$L(cM_{rp} + cM_{rn}) = \frac{\alpha rL}{2}\,(L)\left(\frac{\alpha rL}{3}\right) = \frac{\alpha^2\, r^2\, L^3}{6}$$

or $\quad (M_{rp} + M_{rn}) = \dfrac{\alpha^2 r^2 L^2}{6c}$ $\hfill$ (10.38)

From Eqs. (10.37) and (10.38), we have

$$\frac{L^2}{24}(3 + 4\alpha) = \frac{\alpha^2 r^2 L^2}{6c}$$

or $\quad 4r^2\alpha^2 + 4\alpha c - 3c = 0$ $\hfill$ (10.39)

This equation is exactly same as Eq. (10.28) which was obtained from the virtual work concept. The solution of the above equation is

$$\alpha = \left(\frac{\sqrt{(c^2 + 3r^2 c)} - c}{2r^2}\right)$$ $\hfill$ (10.28)

Substitution of Eq. (10.28) in Eq. (10.38) gives

$$(M_{rp} + M_{rn}) = \frac{(\sqrt{(c^2 + 3r^2 c)} - c)^2}{24cr^2}\, L^2$$

or $\quad M_{rp} + M_{rn} = \dfrac{(\sqrt{(c + 3r^2)} - \sqrt{c})^2}{24r^2}\, L^2$ $\hfill$ (10 40)

Equation (10.40) is very similar to Eq. (10.29), and reduces to Eq. (10.29) for simply supported boundary conditions. The equilibrium of the forces in the mechanism method should not be confused with the equilibrium method of analysis of the plates. The *mechanism* method using either *virtual work* or equilibrium of forces gives the *upper bound* solution of the collapse load. This implies that the method predicts a higher collapse load when compared with the actual collapse load, and how much higher, depends on the assumed mechanism. If the assumed yield lines of the mechanism are exactly same as the actual ones, the collapse load predicted by the upper bound method is exact, otherwise, it is on the higher side. An upper bound solution is an unsafe solution as it estimates the load carrying capacity higher than the real value. Therefore, one has to provide extra margin of safety in design if the yield line pattern is not well defined in the analysis.

10.10 Effect of Restrained Corners

Case 1—Fixed all round the boundary: In most of the slabs, the corners are restrained from lifting up due to either a wall constructed over the supports or other loads and continuity. In such cases forking (Y-fork) yield line pattern is generated as shown in Fig. 10.5d. The collapse load corresponding to such a boundary condition will be different from that already discussed. Consider a simple case of a square plate of span L and having constrained corners. A typical yield line pattern is shown in Fig. 10.10a.

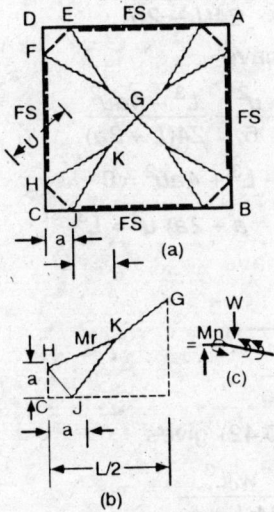

Fig. 10.10 Equilibrium of Segment with Forking in the Yield Lines.

Let M_{rp} = resisting positive moment capacity of the slab in the x- and y-directions (isotropic slab)

 a = side of the corner of the plate which is held firm

Let M_{rn} = negative moment capacity of the plate at the corners and at the boundary.

The collapse load can be calculated by the equilibrium of a symmetrical element as shown in Fig. 10.10b. The shear force on the edges of the yield line is zero from the symmetry of the problem.

The total load acting on the segment is

$$W_1 = \frac{1}{2} \sqrt{2} a(u) \, w_c = \frac{au \, w_c}{\sqrt{2}} \tag{10.41}$$

Taking equilibrium of the moments about yield line HJ, we have

$$M_{rp}(HJ) + M_{rn}(HJ) = W_1 \frac{u}{3}$$

or

$$(M_{rp} + M_{rn}) \sqrt{2} a = \frac{au \, w_c}{\sqrt{2}} \frac{u}{3}$$

or

$$(M_{rp} + M_{rn}) = \frac{u^2 \, w_c}{6} \tag{10.42}$$

The equilibrium of the forces on the segment $HKGF$ and HC gives (assuming M_m on the boundary)

$$(M_{rp} + M_{rn})(L - 2a) = \frac{w_c}{2} L \left(\frac{L}{2}\right)\left(\frac{L}{6}\right) - W_1 \frac{u}{3\sqrt{2}} \qquad (10.43)$$

$$\text{(approximate)}$$

or

$$(M_{rp} + M_{rn})(L - 2a) = \frac{1}{6}\left(\frac{L^3}{4} - au^2\right) w_c$$

or

$$(M_{rp} + M_{ru}) = \frac{(L^3 - 4au^2) w_c}{24(L - 2a)} \qquad (10.44)$$

From Eqs. (10.42) and (10.44), we have

$$\frac{u^2}{6} = \frac{L^3 - 4au^2}{24(L - 2a)}$$

or

$$4u^2(L - 2a) - L^3 + 4au^2 = 0$$

$$4(L - a + 2a) u^2 = L^3$$

or

$$u^2 = \frac{L^3}{4(L + a)} \qquad (10.45)$$

or

$$u = \frac{1}{2}\sqrt{\frac{L^3}{L + a}} \qquad (10.46)$$

Substitution of Eq. (10.46) in Eq. (10.42) gives

$$M_{rp} + M_{rn} = \frac{w_c L^3}{24(L + a)} \qquad (10.47)$$

The collapse load is least for $a = 0$ and is given by

$$w_c = \frac{24(M_{rp} + M_{rn})}{L^2} \qquad (10.48)$$

Case 2—Top reinforcement provided at the corners only: In case of a slab with top reinforcement only at the corners, the equilibrium Eq. (10.43) need to be modified by substituting $M_m = 0$ along the boundary. This gives

$$M_{rp} = \frac{(L^3 - 4au^2)}{24(L - 2a)} w_c \qquad (10.49)$$

Let $M_{rn} = cM_{rp}$ at the corners then from Eqs. (10.42) and (10.49), we have

$$M_{rp} = \frac{u^2 w_c}{6(1 + c)} = \frac{(L^3 - 4au^2) w_c}{24(L - 2a)}$$

or

$$4u^2(L - 2a) = (1 + c)(L^3 - 4au^2)$$

$$u^2(4L - 8a + (1 + c) 4a) = (1 + c)L^3$$

or

$$u^2 = \frac{(1 + c)L^3}{4(L + a(c - 2))} \qquad (10.50)$$

Substituting Eq. (10.50) in Eq. (10.42), we get

$$(M_{rp} + M_m) = \frac{(1 + c)\, L^3 w_c}{24(L + a(c - 2))} \tag{10.51}$$

$$M_r = M_{rp} = \frac{L^3\, w_c}{24(L + a(c - 2))} \tag{10.52}$$

Usually a small portion of the corner, say, $a = 0.2L$ to $0.25L$ is provided with the reinforcement at the top face. Equation (10.52) which governs the design can be solved by assigning different values to a.

Case 1: $c = 1$

Let $M_m = M_{rp} = M_r$; that is, $c = 1$, then Eq. (10.52) reduces to

$$M_r = \frac{w_c\, L^3}{24(L - a)} \tag{10.53}$$

or the collapse load is

$$w_c = \frac{24(L - a)\, M_r}{L^3} \tag{10.54}$$

Case 2: $c = 0$, we have

$$M_r = \frac{w_c L^3}{24(L - 2a)} \tag{10.55}$$

or the collapse load is

$$w_c = \frac{24(L - 2a) M_r}{L^3} \tag{10.56}$$

Assuming $a = 0.2L$, we have

$$M_r = \frac{w_c L^2}{24(1 - 0.4)} - \frac{w_c L^2}{(0.6)(24)} \tag{10.57}$$

The derivation is only an approximate one and more accurate collapse load requires more elaborate procedure. The above expressions give an approximate idea of the effect of the yield line at the corners. The ratio of the collapse loads with and without corner cracking is given by

$$\frac{w_{cc}}{w_{c0}} = \frac{L - a}{L} \quad \text{for } c = 1 \tag{10.58}$$

$$\frac{w_{cc}}{w_{c0}} = \frac{L - 2a}{L} \quad \text{for } c = 3 \tag{10.59}$$

where w_{cc} = collapse load with corner cracking

w_{c0} = collapse load with diagonal yield line

It may, therefore, be observed that the yield pattern will tend corner cracking pattern. The collapse load without top reinforcement ($c = 0$) with respect to that with top reinforcement at the corners is $(L-2a)/(L-a)$. The collapse load without top corner reinforcement is less than that corresponding to a value with top reinforcement and the ratio is estimated in the range of 0.95. In general, the slabs are designed with corner top reinforcement.

10.11 Design Examples

EXAMPLE 10.1 *Design of one-way slab*: A simply supported slab of clear spans 4 m by 10 m is subjected to a live load of 2.5 kN/m^2.
   Design data

$$\text{Long span} \quad L_y = 10 \text{ m}$$
$$\text{Short span} \quad L_x = 4\text{m}$$
$$\text{Live load} \quad w_l = 2.5\text{kN/m}^2$$
$$f_{ck} = 20 \text{ MPa; } f_y = 415 \text{ MPa}$$

The design coefficients for a balanced section are:

$$x_u = 0.479d, \quad j = 0.80$$
$$K = 0.138, \quad p_0 = 0.145$$

Design for bending. Let the effective spans be 4.1 and 10.1 m, then the aspect ratio of the slab is

$$\frac{L_y}{L_x} = \frac{10.1}{4.1} = 2.46$$

The slab has to be designed as a one-way slab as the aspect ratio is more than 2. For the purpose of computing the bending moment, one has to have the self-weight of the slab. Let the thickness of the slab be selected in the range of $L_{xo}/24$, say, $t = 175$ mm.
   The self-weight of the slab is

$$w_g = \gamma_c t = 25(0.175) = 4.375 \text{ kN/m}^2$$

Total working load on the slab is

$$w_t = w_g + w_l = 6.875 \text{ kN/m}^2$$

In case of one-way slabs, the yield line is generated at the middle of the short span and parallel to 10 m span. The corresponding collapse moment is same as $w_c L^2/8$.
   The maximum collapse bending moment which occurs at the middle of the short span is

$$M_c = \frac{\gamma_f w_t L^2}{8} = (1.5) \frac{6875(4.1)^2}{8} = 21669 \text{ Nm/m} = 0.021669 \text{ MNm/m}$$

Equating the moment capacity to the external bending moment, we have

$$Kbd^2 f_{ck} = M_c$$
$$d = \sqrt{\frac{M_c}{Kbf_{ck}}} = \sqrt{\frac{0.021669}{0.138(20)}} = 0.09\text{m}$$

Let
$$d = 0.11\text{ m}$$

The overall thickness of the slab is
$$t = d + 0.02 = 0.13 \text{ m}$$
$$w_g = 0.13(25000) = 3250 \text{ N/m}^2$$
$$w_t = 5750 \text{ N/m}^2$$

This value is less than the assumed one so the bending moment is on the safer side. The collapse BM is

$$M_c = 5750(1.5)\frac{(4.1)^2}{8} = 18123 \text{ Nm/m}$$

Area of the tension reinforcement is

$$A_{st} = \frac{1.15M_c}{jdf_y} = \frac{1.15(18123)}{0.80(0.11)(415)} = 560 \text{ mm}^2/\text{m}$$

Use 12 mm bar at 200 mm spacing, then the tension reinforcement provided is

$$A_{st} \text{ (provided)} = \frac{1000(113)}{200} = 566 \text{ mm}^2/\text{m}$$

$$\frac{100A_{st}}{bd} = \frac{56\,500}{1000\,(110)} = 0.50\%$$

Design for shear. The critical shear plane is at a distance *d* from the support, and the shear force on this plane is

$$V_c = \gamma_f w_t\left(\frac{L}{2} - d\right) = 1.5(5750)(2.05 - 0.11) = 16733 \text{ N/m}$$

The nomial shear stress in the slab is

$$\tau_v = \frac{V_c}{bd} = \frac{0.016733}{0.11} = 0.15 \text{ MPa}$$

The allowable shear stress for 0.5% tension reinforcement is more than 0.15 MPa. Provide the minimum 0.12% distribution reinforcement in the longitudinal direction.

$$\text{Minimum steel} = \frac{0.12\,(130)(1000)}{100} = 156 \text{ mm}^2/\text{m}$$

Provide 8 mm HYSD-Fe415 bars at 320 mm spacing as distribution steel.

The problem was solved by the working stress method and the corresponding quantities in both the methods are given in Table 10.3.

EXAMPLE 10.2 *Design of simply supported two-way slab with unrestrained corners.* A simply supported slab with clear spans of 3.5 and 4.5 m is subjected to a live load of 3 kN/m^2.

Design data. Clear spans, live load and the material strengths are:

$$L_{x0} = 3.5 \text{ m}, L_{y0} = 4.5 \text{ m}; w_l = 3 \text{ kN/m}^2$$

$$f_{ck} = 20 \text{ MPa}, f_y = 415 \text{ MPa}$$

Table 10.3 Comparison of Details of One-Way Slab by WSD and LSD

Item	WSD	LSD
Thickness	170 mm	130 mm
A_s (x-direction)	470 mm^2/m	565 mm^2/m^2
A_s (y-direction)	210 mm^2/m	154 mm^2/m
Concrete (4.3 × 10.3 m)	7.529 m^2	5.7577 m^3
Steel	2.349 kN	2.467 kN
Cost of concrete (at Rs. 2700 m^3)	Rs. 20326	Rs. 15546
Cost of steel (at Rs. 2000/kN)	Rs. 4698	Rs. 4934
Total Cost	Rs. 25024	Rs. 20480
Relative cost	1.22	1.0

The design coefficents for a balanced section are:

$$x_u = 0.479 \, d, \, j = 0.80, \, k = 0.138$$

Corners of the slab are assumed to be free to lift up.

Design for bending. Let the thickness of the slab be 150 mm for the purpose of computing the self-weight.

Self-weight $\qquad w_g = 25 \, (0.15) = 3.75 \text{ kN/m}^2$

Total load on the slab $\qquad w_t = w_g + w_l = 6.750 \text{ kN/m}^2$

Let the effective spans be $L_y = 4.65$ m, $L_x = 3.65$ m.

The aspect ratio of the slab is

$$r = \frac{L_y}{L_x} = \frac{4.65}{3.65} = 1.27$$

Since the aspect ratio is less than 2 the slab must be designed as a two-way slab.

Assume the moment capacities ratio as

$$\frac{M_{rx}}{M_{ry}} = r$$

or $\qquad M_{ry} = \dfrac{M_{rx}}{r} = 0.787 \, M_r$, that is, $c = 0.787$

The collapse load is

$$w_c = \gamma_f \, w_t = 1.5 \, (6.75) = 10.125 \text{ kN/m}^2$$

The yield line pattern is same as that shown in Fig. 10.8 and the resisting moment capacity required is given by Eq. (10.29). It is

$$M_r = \frac{(\sqrt{(c + 3r^2)} - \sqrt{c})^2}{24r^2} \, w_c \, L^2 = 7682 \text{ Nm/m}$$

$$M_{ry} = cM_r = 6045 \text{ Nm/m}$$

Equating the moment capacity to the larger of the two bending moments, we have

$$Kbd^2 f_{ck} = M_r$$

$$d = \sqrt{\frac{M_r}{Kbf_{ck}}} = \sqrt{\left(\frac{0.007687}{0.138(1)(20)}\right)} = 0.053 \text{ m}$$

Use $\qquad d = 0.080$ m and $t = 0.105$ m.

The self-weight $\qquad = 25000 \, (0.105) = 2625 \text{ N/m}^2$

Total load $\qquad w_t = 2625 + 3000 = 5625 \text{ N/m}^2$

Use $\qquad w_t = 5700 \text{ N/m}^2$, $L_x = 3.6$ m

The bending moments are recomputed from Eq. (10.29) and are

$$M_x = 7095 \text{ Nm/m and } M_y = 5584 \text{ Nm/m}$$

Area of the tension reinforcement in the short span direction is

$$A_{sx} = \frac{1.15M_r}{jdf_y} = \frac{7095}{0.80(0.08)(415)} = 267 \text{ mm}^2/\text{m}$$

Provide 10 mm bars at 250 mm spacing in the x-direction. The actual reinforcement provided is

$$A_{st} \text{ (provided)} = \frac{1000(78.5)}{250} = 314 \text{ mm}^2/\text{m}$$

The total depth of the slab with 15 mm clear cover to the reinforcement is

$$h = d + 5 + 15 \text{ mm} = 80 + 5 + 15 = 100 \text{ mm}$$

The precentage of reinforcement in the short span direction is

$$p = \frac{100 A_{sx}}{bd} = \frac{31400}{1000(80)} = 0.39\%$$

The effective depth available for steel in the y-direction is

$$d_y = 80 - 10 = 70 \text{ mm}$$

The area of the tension reinforcement in the y-direction is

$$A_{sy} = \frac{1.15 \, M_{ry}}{jdf_y} = \frac{1.15(5584)}{0.80(0.07)(415)} = 276 \text{ mm}^2/\text{m}$$

Minimum reinforcement required in the slab is taken as 0.12% and is

$$A_{sm} = 0.0012 \, bh = 120 \text{ mm}^2/\text{m}$$

Provide 8 mm bars at 180 mm spacing in the y-direction, then the actual area of steel provided is

$$A_{sy} \text{ (provided)} = \frac{50(1000)}{180} = 278 \text{ mm}^2/\text{m}$$

Check for shear stress. The load distribution in the short span direction at the middle line can be taken as $w_x = w_t$.

The maximum shear force which occur at a distance d from the support is

$$V_c = w_x (0.5 \, L_x - d) \, \gamma_f$$
$$= (5700) (1.8 - 0.08) (1.5) = 14706 \text{ N/m}$$

The nominal shear stress on the section is

$$\tau_v = \frac{V_c}{bd} = \frac{0.014706}{1 \, (0.08)} = 0.18 \text{ MPa}$$

The nominal shear stress is far lower than the shear capacity of the concrete.

Reinforcement details. The slab in each direction is divided into one middle strip of 0.75L and two-edge strips of each 0.125L. The reinforcement computed is that for middle strip. The reinforcement in the edge strip is usually governed by that required for torsion on each face. The spacing of the reinforcement at the bottom face can be at 1.33 times that of the short space reinforcement. In small span slabs, edge span width is so small that it is inconvenient to change the spacing of the reinforcement. Therefore, provide the same spacing of the reinforcement in the entire length of the slab. Figure 10.11 shows the reinforcement details. Also note that the assumed effective spans are slightly larger than the actual ones. The effective span is $L_{x0} + d = 3.58$ m against 3.6 m assumed. The bending moments computed are slightly more than the actual ones because of the assumed larger spans.

430

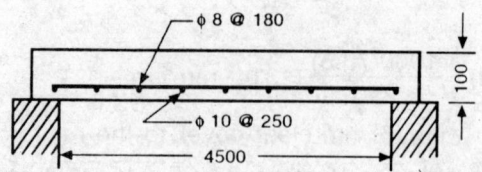

ϕ 8 @ 180

ϕ 10 @ 250

100

4500

Fig. 10.11 Reinforcement Details of Slab, Example 10.2.

EXAMPLE 10.3 *Analysis of two-way slab with two adjacent sides continuous.* Before actually designing the section, one must calculate the collapse load on the structure. Although the load is symmetric, the boundary conditions are not symmetric; therefore, the yield line pattern will not be symmetric. Figure 10.12 illustrates a typical yield line formation. There are three variables, namely α_1, α_2 and β as seen from the figure. The collapse load can be determined by considering the equilibrium of the four segments at a time. For convenience and simplicity, the following assumptions are made

$$M_{rp} = cM_{rn}$$

$$M_{rx} = M_{ry} = M_{rp} = M_r$$

Consider the equilibrium of the segments *ABFE*, *BCF*, *CDEF* and *DAE* respectively. Divide *ABFE* into two triangular parts *ABE* and *EFB*. The equilibrium of the moment about line *AB* gives.

$$M_r\,(rL) = \left(\frac{rL\,(\beta L)}{2}\,\frac{\beta L}{3} + rL\,\frac{(1 - \alpha_1 - \alpha_2)\,(\beta L)}{2}\,\frac{2\beta L}{3} \right) w_c$$

or

$$M_r = \frac{w_c}{6}\,(1 + 2\,(1 - \alpha_1 - \alpha_2))\,\beta^2\,L^2$$

$$= \frac{w_c\,(3 - 2(\alpha_1 + \alpha_2))\,\beta^2 L^2}{6} \qquad (10.60)$$

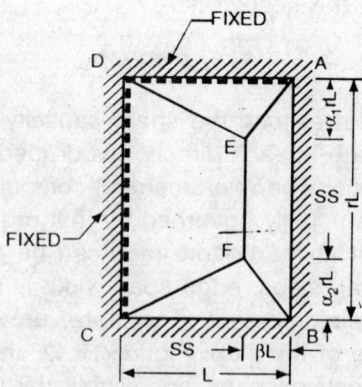

Fig. 10.12 Yield Line Pattern of Slab, Example 10.3.

The equilibrium of the triangular segment BCF is

$$M_r(L) = \frac{L}{2}\,(\alpha_2\, rL)\left(\frac{\alpha_2\, rL}{3}\right)w_c$$

or
$$M_r = \frac{w_c\,(\alpha_2^2\, L^2)}{6} \tag{10.61}$$

Similarly, the equilibrium of the other segments gives

$$M_r = \frac{w_c\,(3-2)\,(\alpha_1 + \alpha_2)\,(1-\beta)^2\, L^2}{6\,(1+c)} \tag{10.62}$$

$$M_r = \frac{w_c\,(\alpha_1^2\, L^2)}{6\,(1+c)} \tag{10.63}$$

The values of α_1, α_2 and β can be solved with the help of Eqs. (10.60) to (10.63). From Eqs. (10.60) and (10.62), we have

$$(3-2\,(\alpha_1 + \alpha_2))\,\beta^2 = \frac{(3-2\,(\alpha_1 + \alpha_2))\,(1-\beta)^2}{1+c}$$

or
$$\beta^2\,(1+c) = (1-\beta)^2 \tag{10.64}$$
or
$$\beta\,\sqrt{(1+c)} = 1-\beta$$

$$\beta = \frac{1}{1 + \sqrt{(1+c)}} \tag{10.65}$$

Similarly, from Eqs. (10.61) and (10.63), we have

$$\alpha_2^2\,(1+c) = \alpha_1^2 \tag{10.66}$$

or
$$\alpha_2 = \frac{\alpha_1}{\sqrt{(1+c)}} \tag{10.67}$$

From Eqs. (10.60) and (10.61), we have

$$(3-2\,(\alpha_1 + \alpha_2))\,\beta^2 = \alpha_2^2 \tag{10.68}$$

or
$$\left(3 - 2\alpha_1\left(1 + \frac{1}{\sqrt{(1+c)}}\right)\right)\left(\frac{1}{1 + \sqrt{(1+c)}}\right)^2 = \frac{\alpha_1^2}{1+c} \tag{10.69}$$

One can see that the equations become more tedious and so one typical case is illustrated here.

Consider an isotropic plate with $M_{rp} = M_m$, that is $c = 1$, then the above equations can be reduced to:

$$\beta = \frac{1}{1 + \sqrt{2}} = 0.414 \tag{10.70}$$

$$\alpha_2 = \frac{\alpha_1}{\sqrt{2}} = 0.707\,\alpha_1 \tag{10.71}$$

$$\left(3 - 2\alpha_1\left(1 + \frac{1}{\sqrt{2}}\right)\right)\left(\frac{1}{1 + \sqrt{2}}\right) = \frac{\alpha_1^2}{2}$$

or

$$\alpha_1^2 + 4\,\alpha_1 \left(\frac{1+\sqrt{2}}{\sqrt{2}}\right)\left(\frac{1}{1+\sqrt{2}}\right) - \frac{6}{1+\sqrt{2}} = 0$$

$$\alpha_1^2 + 2\sqrt{2}\,\alpha_1 - 6/(1+\sqrt{2}) = 0$$

or $\qquad\qquad \alpha_1 = 0.586$

Therefore, $\alpha_2 = 0.414$ and $\qquad\qquad\qquad\qquad\qquad\qquad\qquad\qquad\qquad\qquad$ (10.72)

$\qquad\qquad \beta = 0.414$

The substitution of above in any one of the moment equilibrium equations will give the collapse load relation. From Eq. (10.63), we have

$$M_r = \frac{w_c\,(\alpha^2)\,L^2}{6(1+1)} = 0.029\,w_c\,L^2 \qquad\qquad\qquad (10.73)$$

The collapse load of a simply supported square slab would be given by the equations as:

$$M_r = w_c L^2/24 = 0.0417\ w_c L^2$$

which is close to the real value.

EXAMPLE 10.4 A floor slab of a room clear size 4 by 5 m is continuous along the two adjacent supports and discontinuous at the other two supports. The slab is subjected to a live load of 3 kN/m^2.

Design data

$$L_{xo} = 4\ \text{m},\ L_{yo} = 6\ \text{m},\ w_l = 3000\ \text{N/m}^2$$
$$f_{ck} = 20\ \text{MPa},\ f_y = 415\ \text{MPa}$$
$$K = 0.138,\ x_u = 0.479d,\ j = 0.80$$

Design for bending. Let the thickness of the slab be taken as 0.15 m for the purpose of computation of the self-weight.

$$\text{Self-weight} = 25000(0.15) = 3750\ \text{N/m}^2$$

Total load on the slab is

$$w_t = w_g + w_l = 6750\ \text{N/m}^2$$

The effective collapse design load with partial safety factor of 1.5 is

$$w_c = 1.5(6750) = 10125\ \text{N/m}^2$$

The effective spans can be taken as

$$L_x = 4 + 0.15 = 4.15\ \text{m} = L$$
$$L_y = 5 + 0.15 = 5.15\ \text{m} = rL$$

The aspect ratio of the slab is

$$r = \frac{L_y}{L_x} = \frac{5.15}{4.15} = 1.24$$

The subscript n refers to negative BM and similarly the subscript p refers to the positive BM.

The thickness of the slab is governed by the absolute maximum bending moment in the entire slab.

The absolute maximum bending moment is

$$M = \alpha_{xn} \, w_c L_x^2 = 0.029 \, w_c L^2 = 0.029(10125)(4.15)^2 = 5057 \text{ Nm/m}$$

The effective depth fo the slab is obtained by equating the moment capacity to the maximum BM, and is given by

$$d = \sqrt{\frac{M_r}{f_{ck}Kb}} = \sqrt{\frac{0.005057}{20(0.138)(1)}} = 0.043 \text{ m}$$

As the assumed depth of the slab is more than the actual one, the self-weight of the slab and the effective spans are modified by using $h = 0.125$ m.

$$w_g = 25000 \, (0.125) = 3125 \text{ N/m}^2$$

$$w_t = w_g + w_l = 6125 \text{ N/m}^2$$

Use $w_t = 6200 \text{ N/m}^2$ and $w_c = 1.5(6200) = 9300 \text{ N/m}^2$

$$L_x = 4.105 \text{ m} = L$$

The negative and positive BMs are:

$$M_{rp} = M_{rn} = M_r = 0.029 \, w_c L^2$$

The maximum effective depth needed is

$$d = \sqrt{\frac{0.004545}{20(0.138)}} = 0.048$$

Use $d = 0.08$ m and $h = 0.10$ m

The assumed depth was 0.125 m; therefore, the computed bending moments are close to the actual ones and are on the safe side. The tension reinforcement along the axis is

$$A_{sx} = \frac{1.15 \, M_r}{jd \, f_y} = \frac{(1.15)4545}{0.80(0.08)(415)} = 196 \text{ mm}^2/\text{m}$$

The effective depth of the steel in the long span direction is

$$d_y = 0.08 - 0.01 = 0.07 \text{ m}$$

The area of the reinforcement is

$$A_{sy} = \frac{(1.15)(4545)}{0.80(0.07)(415)} = 215 \text{ mm}^2/\text{m}$$

Use 8 mm diameter bars in both the directions, then the required spacings in the x and y directions of the bars at the bottom face of the slab are:

$$s_x = 180 \text{ mm and } s_y = 155 \text{ mm}$$

Crank alternative bars in both the directions and then provide extra bars of $\phi 8$ bars at 300 mm c/c in the x- and y-directions at top over the support.

The minimum reinforcement is

$$A_{sm} = \frac{0.12}{100} \, (100)(1000) = 120 \text{ mm}^2/\text{m}$$

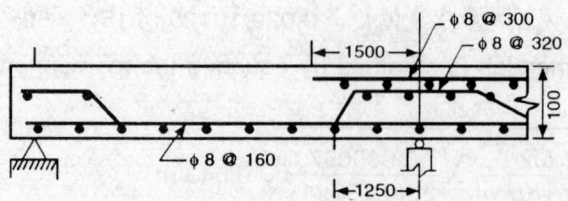

Fig. 10.13 Reinforcement Details of Slab, Example 10.4.

This value is less than that provided. There is no need to check for shear stress in this type of slabs as the shear stress will be only nominal and would not require revision or the shear reinforcement. Figure 10.13 illustrates the reinforcement details.

The assumption of equal moment capacity in both the directions results into higher percentage of the reinforcement in the long span direction mainly because the lever arm available for this reinforcement is less than the one for short span reinforcement. This looks little odd; however, the solution is for one of the possible mechanisms. Therefore, it is recommended that the reinforcement spacing in the short span be kept same as that of the long span one.

EXAMPLE 10.5 *Cantilever slab.* A portico slab has a free cantilever of 5 m and width 6 m. Design the slab with HYSD-Fe415 steel and M20 concrete.

Design data

$$\text{Width of the portico} \quad B = 6 \text{ m}$$
$$\text{Cantilever span} \quad L = 5$$
$$\text{Live load} \quad w_l = 1.5 \text{ kN/m}^2$$
$$\text{Facia and finish load} \quad w_2 = 1 \text{ kN/m}^2$$
$$f_{ck} = 20 \text{ MPa and} \quad f_y = 415 \text{ MPa}$$

Design of the section. Let the average thickness of the slab be assumed as $L/15 = 0.32$ m for the purpose of computing the self weight. (The thickness of the slab at free edge be 0.15 m only.)

Self-weight $w_g = 25(0.32) = 8\text{kN/m}^2$

$$w_l = w_s = 2.5 \text{ kN/m}^2$$

Total load $w_t = w_g + w_l + w_2 = 10.5 \text{ kN/m}^2$

$$w_c = 1.5 \, w_t = 15.75 \text{ kN/m}^2$$

The maximum bending moment at the support is

$$M_c = \frac{w_c L^2}{2} = \frac{15.75(25)}{2} = 196.875 \text{ kNm/m}$$

The moment capacity of the section is equated to the external BM, then we have

$$d = \sqrt{\frac{M_c}{bKf_{ck}}} = \sqrt{\frac{0.196875}{1(0.138)(20)}} = 0.268 \text{ m}$$

Use d = 0.32 m

$$A_{st} = \frac{1.15M}{jdf_{ck}} = \frac{(1.15)196875}{0.80(0.32)(415)} = 2131 \text{ mm}^2$$

Provide ϕ20 bars at 145 mm c/c at top, then the actual reinforcement provided is

$$A_{st} \text{ (provided)} = \frac{1000(314)}{145} = 2165 \text{ mm}^2$$

$$\frac{100A_{st}}{bd} = \frac{216500}{1000(320)} = 0.67$$

The shear capacity of the concrete for the above per cent of the reinforcement is $\tau_c = 0.5$ MPa.

Let the overall depth of the slab at the support be

$$h = d + 0.5\phi + 0.025 = 0.355 \text{ m}$$

Overall depth at the free end = 0.15 m

Average depth = (0.355 + 0.15)/2 = 0.253 m

The assumed average thickness is more than the actual value; therefore, the design is slightly on the safe side. The distribution reinforcement be provided at 0.12%, that

$$A_{sx} = \frac{0.12}{100}(253)(1000) = 304 \text{ mm}^2/\text{m}$$

Provide ϕ12 bars at 350 mm c/c along the width of the slab.

$$w_g = (0.253) = 6.4 \text{ kN/m}^2$$

$$w_t = 6.4 + 2.5 = 8.9 \text{ kN/m}^2$$

Check for shear. The critical section for shear force is at a distance *d* from the support and the shear force at this section is

$$V_c = \gamma_f w_t(L - d) = 1.5(8.9)(5 - 0.32) = 62.478 \text{ kN/m}$$

The nominal shear stress is

$$\tau_v = \frac{V_c}{bd} = \frac{62478}{1000(320)} = 0.20 \text{ MPa}$$

The shear strength is higher than the nominal shear stress, so the section is safe in shear.

Check for deflection. Moment of inertia of the cracked section considering the effect of the reinforcement is

$$I_{cr} = mA_{st}(d - x_u)^2 + \frac{b(x_u)^3}{3} = 0.00061 + 0.0012 = 0.00181 \text{ m}^4$$

The gross moment of inertia of the section is

$$I = \frac{bh^3}{12} = \frac{1(0.355)^3}{12} = 0.00373 \text{ m}^4$$

The effective sectional moment of inertia can be taken as

$$I_e = 0.5 (I + I_{cr}) = 0.0028 \text{ m}^4$$

The Young's modulus of concrete is

$$E_c = 5000 \sqrt{20} = 22360 \text{ MPa}$$

The deflection caused by the permanent load is

$$v_d = \frac{w_d L^4}{8 E_c I_e} = \frac{0.0074(625)}{8(22360)(0.0028)} = 0.009 \text{ m}$$

The deflection caused by the live load is

$$v_l = \frac{w_l L^4}{8 E_c I_e} = \frac{0.0015 (625}{8 (22360)(0.0028)} = 0.03 \text{ m}$$

Let the creep coefficient be $c_c = 2.1$, then the total deflection is

$$v_t = c_c v_d + v_l = 0.018 \text{ m}$$

Allowable deflection including the creep effect is

$$v_a = \frac{L}{250} = \frac{5}{250} = 0.02 \text{ m}$$

The deflection is within the limit.

  Development length. Design of the development length is important in all structures, and more so in the cantilever beams. The development length is given by

$$L_d = \frac{\phi \, \sigma_s}{4 \tau_{bd}} = \frac{20 \, (415)}{4 \, (1.2) \, (1.6)} = 1080 \text{ mm}$$

All the tension bars should have a development length of 1080 mm, and care must be taken to provide this embedded length at the support. Curtail alternate bars at 3 m from the support.

  This design does not include the design of the support either for bending or for stability. In case of cantilever slabs or beams, careful attention be paid for the design of the support and development length.

10.12 Flat Slabs

Beam and slab construction has the advantage of providing intermediate supports to the slabs, thus reducing the effective span of the slabs. But the beams require larger depths thus leading to more heights of the buildings. In some situations, specially in warehouses, it is desirable to have larger clear ceiling heights. Flat slabs are the slabs which rest directly on the columns without beams and thereby provided a larger clear ceiling height for the same given total height of a building. In addition, the form-work requirement is also reduced when compared with the beam and slab construction. Flat slabs are invariably two-way slabs and rest on several columns. Sometimes the top of the columns are widened so as to provide wider base to support the slab. Such widened portions are called *column heads*. There is a limit to which one can treat the widened portion as a part of the column. The width must be limited to the portion within 90° of the segment as shown in Figs. 10.14b and c. Any projection beyond the column head should really be treated as a part of the slab rather than that of the columns.

In other words, it should be treated as thickening of the slab at the column head. Such thickened portions of the slab are called *drops*. The drops are sometimes known as capital of the columns. The drop when provided should be rectangular or circular and need to be about one-third of the panel length.

In each panel, the slab is divided into column and middle strips in each direction. The width of the *column strip* is equal to half of the column spacing and it is placed half on either side of the column line. In case of unequal spans, it can be taken equal to half of the average. In addition, it should also be restricted to 0.5 times the column spacing in any direction. *The middle* strip is the one bounded by the column strips and its width is equal to the spacing of the columns minus the width of the column strip. The width is usually equal to or greater than half of the spacing of the columns.

Let b_{cl} = width of column strip in the lth column row

   b_{ml} = width of the middle strip

   L_{xl} = spacing of the columns in the x-direction in the lth panel

   L_{yl} = spacing of the columns in the y-direction in the lth panel

Then width of the column and middle strips spacing in the x-direction is given by

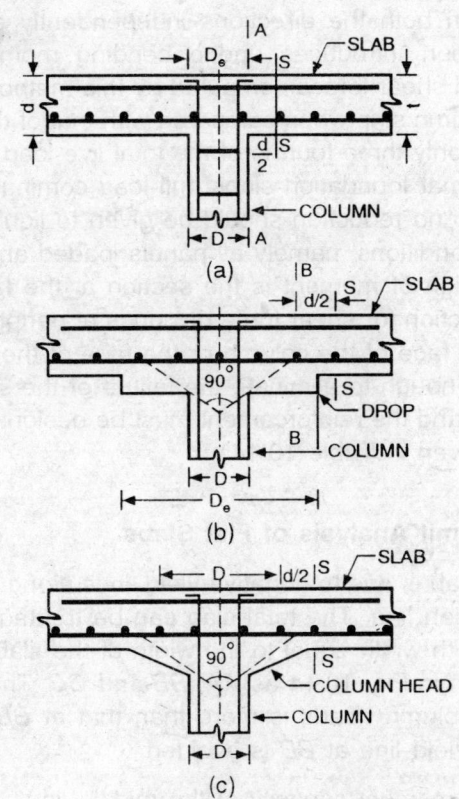

Fig. 10.14 Typical Flat Slabs.

Table 10.4 Distribution of BM Across the Panel

Strip and its boundary conditions	Per cent of the total BM
1. *Column strip*	
a) Negative BM at exterior support	100
b) Negative BM at the interior support	75
c) Positive BM	60
2. *Middle strip*	
The difference between the panel moment and the column strip moment.	

$$b_{cl} = 0.25 (L_{yl-1} + L_{yl}) < 0.25 (L_{xi} + L_{xl+1})$$
$$b_{ml} = L_{yi} - 0.25 (L_{yi} + L_{yi} + 1)$$

Similarly, for the widths of the strips in the *y*-direction. In most situations, the spacing of the columns in one direction is same in all the panels. Therefore, the calculation of the widths of the strips is normally a trivial exercise.

The slabs can be analysed as equivalent frames having idealized continuous wall supports along the transverse column lines. The stiffness of the column is divided by the panel width, and is considered as the stiffness of the vertical element per unit width of the frame. The analysis has to be done in both the directions independently as two sets of independent frames. Such an idealization introduces undue bending moments into the middle strip; therefore, the moments and shear forces computed by this method must be proportioned with higher weightage to the column strip when compared with that of the middle strip. The analysis is to be carried by loading only three-fourths of the total live load in each panel and full dead load. However, in case of mat foundation slabs, full load coming from the column should be taken as the load, and also no reduction should be given to liquid loads. The frames should be analysed for two load conditions, namely all panels loaded and alternative panels loaded. The critical section for design of moment is the section at the face of the column or at the face of drop. The critical section for shear force design is at peripheral line around the column at a distance $0.5d$ from the face of the column or the face of the drop. Usually the thickness of the drop is taken large enough to eliminate the failure of the section around the periphery of the column. The section and the reinforcement must be designed to withstand the weighted proportioned moment as given in Table 10.14.

10.13 Direct Method of Limit Analysis of Flat Slabs

The collapse mode in flat slab is due to negative yield lines along the column line and positive yield lines along the mid-span line. The total slab can be treated as resting as a continuous support in each direction with width equal to the width of the slab. The three critical sections in the outer span are shown in Fig. 10.14 as *AA, BB* and *CC*. The resisting moment capacity at section *AA* across the column heads is more than that at *BB*. The failure mechanism of the drop *D*, such that the yield line at *BB* is avoided.

Let M_{rp} = total positive moment capacity at the middle line

      $M_{r\,nc}$ = total negative moment capacity at the face of the column or drop

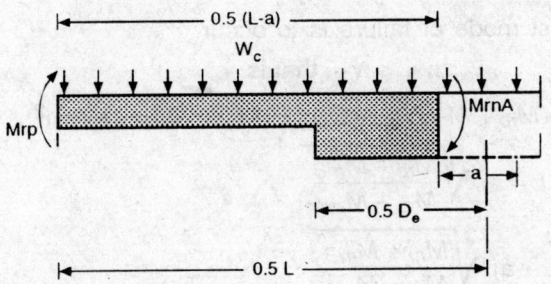

Fig. 10.15 Critical Moments in Segment of a Flat Slab.

$B =$ width of the panel
$L =$ span of the panel
$a =$ column size
$D_c =$ drop size

Figure 10.15 illustrates the average resisting moment capacities along with the other dimensions of a segment between the yield lines. The equilibrium of forces about the face of the column line gives

$$(M_{rp} + M_{r\,nA}) = 0.5\,(L - a)\,B\,(0.5)\,\frac{(L - a)}{2}\,w_c$$

or
$$M_{rp} + M_{r\,nA} = \frac{B(L - a)^2}{8}\,w_c \qquad (10.74)$$

The above equation gives the sum of the magnitudes of the positive and negative bending moment capacities which is approximately equal to the maximum BM on a simply supported beam of span of $(L-a)$. One can select a relation between the positive and the negative moment capacities and then establish the desired sections. Let

$$M_{r\,nA} = c_A M_{rp} = c_A M_r \qquad (10.75)$$

then Eq. (10.74) gives

$$M_r = \frac{B(L - a)^2\,w_c}{(1 + c_A)} \qquad (10.76)$$

Consider another possible case of yield line generating at section BB instead of at AA. Let

$$M_{r\,nB} = \text{moment capacity at } BB$$

The equilibrium equation of the segment CB gives

$$M_{rp} + M_{r\,ns} = \frac{B(L - D_c)^2}{8}\,w_{c2} \qquad (10.77)$$

The collapse load based on the first and second mechanisms can be expressed as

$$w_{c1} = \frac{8(M_{rp} + M_{r\,nA})}{B(L - a)^2} \qquad (10.78)$$

$$w_{c2} = \frac{8(M_{rp} + M_{mB})}{B(L - D_c)^2}$$ (10.79)

The condition for the first mode of failure is to occur,

$$w_{c1} < w_{c2}, \text{ that is}$$

$$(M_{rp} + M_{mA})(L - D_c)^2 < (M_{rp} + M_{mB})(L - a)^2$$ (10.80)

or $\quad L - D_c < (L - a) \sqrt{\dfrac{M_{rp} + M_{mB}}{M_{rp} + M_{mA}}}$

or $\quad D_c > L - (L - a) \sqrt{\dfrac{M_{rp} + M_{mB}}{M_{rp} + M_{mA}}}$ (10.81)

The fraction under the square root will be less than one and it will be in the range of 0.7. The first mode of failure is likely to occur if.

$$D_c \geq (0.16L + 0.85a)$$ (10.82)

Assuming a is in the range of $L/15$, the width of the drop to be maintained for the first mode of failure is sure to occur if.

$$D_c \geq 0.25L$$ (10.83)

There are situations when the first mode of failure is likely to occur even if D_c is in the range of 0.20L. The design of the flat slab is best illustrated through examples.

Interior span: In the interior spans, three yield lines will be formed; one at each face of the column lines and the other one at the mid-span. The corresponding collapse load relation can be obtained as

$$M_{rl} = \frac{B(L - a)^2 \, w_c}{8(1 + c_B)}$$ (10.84)

where $M_{mB} = c_B M_r$

The design procedure consists of computing the average moment capacity required from the collapse load Eq. (10.76). Then the BMs on the column and middle strips can be obtained by assigning appropriate weightages as given in Table 10.4. The overall depth of the slab is controlled by the bending moment in the exterior span of the column strip. The depth of the slab is first designed and then the reinforcements required at various places are computed.

Some of the design considerations are:

1. The end span should not be larger than the interior spans.
2. The ratio of the successive span lengths should be within 0.75 to 1.33.
3. A cantilever projection of about one-third of the exterior span can be permitted with appropriate modification in the bending moments.
4. The design live load should not be more than three times the dead load. This is a major constraint, and when ignored it may effect the magnitude of the positive bending moment.

The sum of the magnitudes of the positive and average of the negative bending moments is equal to

$$M_0 = \frac{W L_0}{8}$$ (10.85)

where M_0 = sum of the magnitudes of the positive and average of the negative bending
moments, and it is $M_0 = M_3 + 0.5 (M_1 + M_2)$ (10.86)

M_3 = positive bending moment

M_1 and M_2 = magnitudes of negative bending moments at the face of the two columns
on either side of the span (vide Fig. 10.16)

L_0 = clear span extending from face to face of the columns or capitals. It should
be equal to larger of the two unequal adjacent spans and > 0.65 spacing
of the columns.

W = design load on the clear span = $w_c B (L - a)$

The relative magnitudes of the negative and positive bending moments can be calculated
using the ratios given in Table 10.5

The magnitude of the bending moment can be obtained as

$$M_t = c_i M_0 \qquad (10.87)$$

where the moment coefficients, c_i's are listed in Table 10.5.

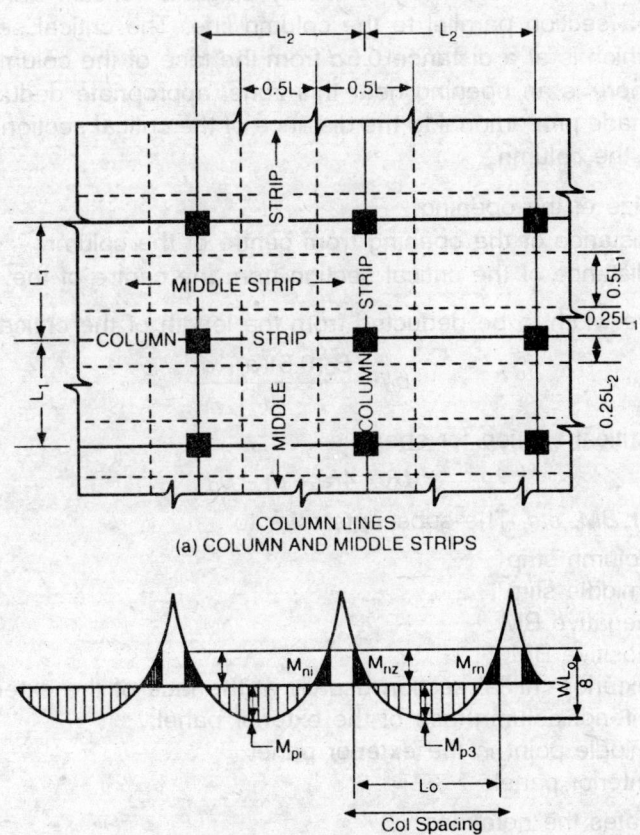

(a) COLUMN AND MIDDLE STRIPS

(b) TYPICAL BENDING MOMENT

Fig. 10.16 Notation in Flat Slab.

Table 10.5 Relative Values of Positive and Negative Bending Moments BM

Bending moment and location	c_i
1. *Interior spans*	
a) Negative BM = M_{ni}	$0.65\ M_0$
b) Positive BM = M_{pi}	$0.35\ M_0$
2. *End span*	
a) Exterior negative BM = M_{n1}	0 to $0.1\ M_0$
b) Positive BM = M_{p3}	$0.60\ M_0$
c) Interior negative BM = M_{n2}	$0.70\ M_0$

The distribution of the moments at a section between the column and middle strips should be as per Table 10.4. The value of c_B to be used in the equation is $0.65/0.35 = 1.85$.

For convenience let $c_B = \infty$

       or $c_A = \infty$ depending the failure mode.

Design for shear: The slabs are likely to fail by diagonal tension around the column faces rather than along a section parallel to the column line. The critical section for shear is the peripheral plane which is at a distance $0.5d$ from the face of the column or the column head or drop. In case there is an opening near this zone, appropriate deduction in the peripheral length should be made proportional to the distance of the critical section to that of the opening from the centre of the column.

Let B_0 = size of the opening

     d_1 = distance of the opening from centre of the column

$0.5(a + d)$ = distance of the critical section from the centre of the column

Then the ineffective width to be deducted from the length of the critical section is given by

$$b_d = \frac{B_0(0.5)(a + d)}{d_1} \qquad (10.88)$$

the length of the critical section for shear is

$$b_0 = 4(a + d) - b_d \qquad (10.89)$$

Typical notation for BM, etc. The subscripts refer to:

    c = column strip

    m = middle stirp

    n = negative BM

    p = positive BM

    1 = exterior critical section usually at the face of the exterior columns or drops

    2 = interior column face of the exterior panel

    3 = middle point in the exterior panel

    i = interior panels

Figure 10.16 illustrates the notations.

    M_{n1} = negative BM at section 1 on the panel

    M_{nc1} = negative BM at section 1 on the column strip

M_{pi} = positive BM at the interior panel
M_{pci} = positive BM at the interior panel on the column strip

The other notations are defined similarly.

10.14 Design Examples of Flat Slabs

EXAMPLE 10.6 *Flat roof slab without column heads*: A roof slab is supported on columns spaced at 5 m apart in two perpendicular directions. The size of the square column is 440 mm and the live load on the roof is 1500 N/m². The load of the waterproof treatment course on the slab is 2000 N/m². Design a flat slab without drops or column heads. Height of the column above the mat foundation is 6 m.

Design data

Spacing of the columns $L_1 = L_2 = L = B = 5$ m
Size of the column $a = 0.44$ m
Live load $w_l = 1.5$ kN/m²
Superimposed load $w_s = 2.0$ kN/m²
$f_{ck} = 20$ MPa and $f_y = 415$ MPa

The limit state strength coefficients are $K = 0.138$, $j = 0.80$.

Design of the section for moment: The design is done by using direct design method. For the purpose of estimating the self-weight of the slab, let thickness of the slab be assumed in the range of $L/20$ for the sales without column heads.

Let the thickness of the slab $t = 0.24$ m
Self-weight = 0.24(25) $= w_g = 6.0$ kN/m²
Total dead load $w_d = w_g + w_s = 8.0$ kN/m²

Clear spacing between the columns is
$$L_0 = L - a = 5 - 0.44 = 4.56 \text{ m}$$

The total design load in a panel is
$$W_c = w_c L L_0 = \gamma_f (w_d + w_l) L L_0 = (1.5)9.5(5)(4.56) = 324.9 \text{ kN}$$

Sum of the magnitudes of the bending moments in the panel between the faces of the columns
$$M_0 = \frac{w_c L_0}{8} = \frac{324.9(4.56)}{8} = 185.193 \text{ kNm} = 0.185193 \text{ MNm}$$

Magnitude of the negative BM at the face of the columns in the interior panels is
$$M_{ni} = 0.65 M_0 = 0.12038 \text{ MNm}$$

The column is resting on a mat foundation; therefore, the base of it can be treated as fixed. The end walls of the hall restrain the free lateral movement of the roof slab; therefore, the top of the column can be assumed as fixed in position and rotation. Hence the effective height of the column is
$$L_c = 0.65 H = 0.65(6) = 3.9 \text{ m}$$

The exterior negative and positive bending moment coefficients and the interior negative bending moment coefficients are taken from Table 10.5 and are:

$$c_1 = 0.1, \quad c_3 = 0.60, \quad c_2 = 0.70$$

The corresponding bending moments are:

$$M_{n1} = c_1 M_0 = 0.01852 \quad \text{MNm}$$

$$M_{p3} = c_3 M_0 = 0.11112 \quad \text{MNm}$$

$$M_{n2} = c_2 M_0 = 0.12964 \quad \text{MNm}$$

The bending moments on the column strip are computed from the total moment after applying the distribution factors given in Table 10.4. They are:

$$M_{nc1} = M_{n1} = 0.01852 \text{ MNm}$$

$$M_{pc3} = 0.6 \, M_{p3} = 0.06667 \text{ MNm}$$

$$M_{nc2} = 0.75 \, M_{n2} = 0.09723 \text{ MNm}$$

$$M_{ncl} = 0.75 \, M_{ni} = 0.090285 \text{ MNm}$$

The thickness of the slab is constant and without any column heads or drops. The maximum moment occurs at the interior column face of the exterior panel. The effective depth of the slab is given by

$$d = \sqrt{\frac{M_{nc2}}{bKf_{ck}}} = \sqrt{\frac{0.09723}{2.5 \, (0.138) \, (20)}} = 0.12 \text{ m}$$

Use $d = 0.15$ m and $t = 0.18$ m.

The slab was assumed to be 0.24 m for the purpose of computing self-weight. Since the thickness actually provided is 0.18 m, the moments are revised.

$$w_g = 0.18 \, (25) = 4.5 \text{ kN/m}^2$$

$$w_t = 4.5 + 2 + 1.5 = 8.0 \text{ kN/m}^2$$

The actual value of w_t is $(8/9.5) = 0.842$ times the load that was assumed. Therefore, the final moments are computed by multiplying the previous moments by 0.842. The design moments are (all in MNm):

$$M_0 = 0.842 \, (0.185193) = 0.15593$$
$$M_{n1} = 0.1 \, M_0 = 0.015593; \; M_{p3} = 0.09356$$
$$M_{n2} = 0.10916, \; M_{nc1} = M_{n1} = 0.015593$$
$$M_{pc_3} = 0.05614, \; M_{nc2} = 0.08187$$
$$M_{nci} = 0.07602, \; M_{pci} = (0.35)(0.75)(M_0) = 0.04093$$

Normally no shear reinforcement is provided in the slabs. It is, therefore, desirable that the adequacy of the depth of the section against shear should be checked at the earliest.

The shear strength for diagonal tension failure is

$$\tau_c = 0.25 \sqrt{f_{ck}} = 1.12 \, \text{MPa}$$

The critical shear plane is the peripheral plane which is at a distance $0.5d$ from the face of the column. The length of the critical section is

$$b_0 = 4 \, (a + d) = 4 \, (0.44 + 0.15) = 2.36 \text{ m}$$

The shear force on the plane is

$$V_c = w_t (L_1 L_2 - (a + d)^2) \gamma_f$$

$$= 8.0(25 - 0.64^2) (1.5) = 295 \text{ kN}$$

The nominal shear stress is

$$\tau_v = \frac{V_c}{b_0 d} = \frac{0.295}{(2.36)(0.15)} = 0.83 \text{ MPa}$$

The nominal shear stress is less than the shear capacity of the concrete; therefore, there is no need for transverse reinforcement or thickening of the slab.

Design of reinforcement
Column strip: The positive BM at mid-span of exterior panel is

$$M = M_{pc3}$$

The area of the tension steel at the bottom at the mid-span is

$$A_{st3} = \frac{1.15 M_{pc3}}{jd f_y} = \frac{(1.15)\,56140}{(0.80)(0.15)(415)} = 1296 \text{ mm}^2$$

Provide 12 numbers of $\phi 12$ bars in the column strip at the mid-span.

$$A_{st} \text{ (provied)} = 1356 \text{ mm}^2$$

Half of the bars are to be cranked.
The negative BM at exterior support $M = M_{nc1}$
The area of the reinforcement at top near the column line is

$$A_{st1} = \frac{1.15 M_{nc1}}{jd f_y} = \frac{(1.15)\,15593}{(0.80)(0.15)(415)} = 360 \text{ mm}^2$$

The minimum area of the reinforcement required is

$$A_{stm} = \frac{0.12\,bt}{100} = 540 \text{ mm}$$

Needed: 5 *numbers of 12* mm bars in the 2.5 m width of the column strip at top near the column line (reinforcement as required). The cranked bars from positive reinforcement are adequate. The positive BM in the column strip of the *interior panel* is

$$M_{pci} = (0.35)\,(0.75)\,M_0 = 0.04093 \text{ MNm}$$

The area of the tension reinforcement required is

$$A_{sti} = \frac{(1.15)40930}{(0.80)(0.15)(415)} = 945 \text{ mm}^2$$

Needed: 9 *numbers of* $\phi 12$ bars at the bottom ($A_{st} = 1017 \text{ mm}^2$) of which four bars are cranked. The negative bending moment at the interior column of the end panel is

$$M_{nc2} = 0.08187 \text{ MNm}$$

The area of the tension reinforcement required is

$$A_{st2} = \frac{(1.15)\,81870}{(0.80)\,(0.15)\,(415)} = 1891 \text{ mm}^2$$

Extra: 6 *numbers* 12 mm bars at top over the column line ($A_{st} = 1921 \text{ mm}^2$).

446

Design of middle strip: The bending moment in the middle strip are obtained by subtracting the BM on the column strip from the total BM on the panel. One can calculate the number of bars required by simple proportion of the bending moments subjected to the minimum reinforcement.

1. The negative BM at the exterior support is zero; however, a nominal reinforcement is provided at top.

Needed: 5 *numbers of* φ12 bars at top in the middle strip at the edge.

2. The positive BM in the exterior panel

$$(0.35)(0.25)(M_0) = 0.01364 \text{ MNm}$$

The area of the reinforcement needed for this moment is 314 mm², and it is less than the minimum required. So *provide* 5 *numbers of* φ12 at the bottom.

It can be seen that the bending moments in the middle strip are only nominal and the minimum tension reinforcement governs the design. This minimum reinforcement is 5 nos. of 12 mm bars in 2.5 m width. The reinforcement details are given in Fig. 10.17. The curtailment and cranking of the bars is recommended as per the normal practice.

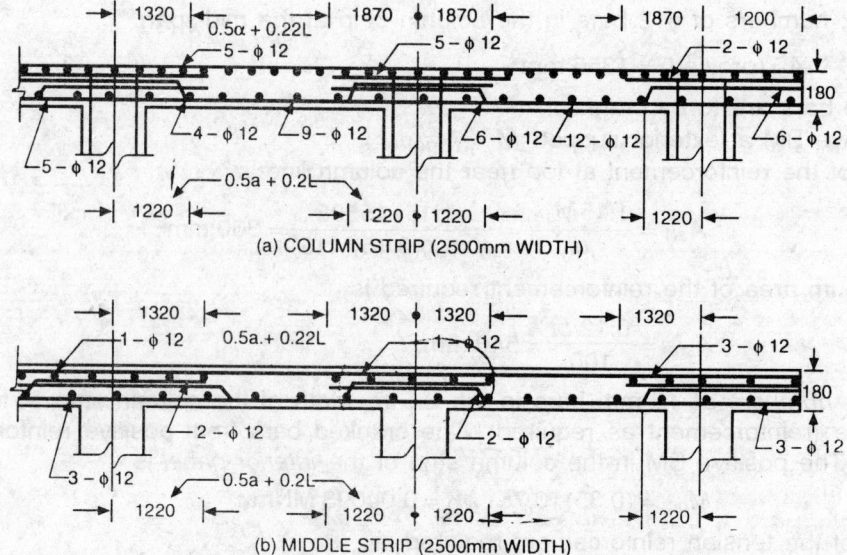

Fig. 10.17 Reinforcement Details in Flat Slabs of Example 10.6.

Development length is given by

$$L_d = \frac{\phi f_y}{4\, \tau_{bd}} = \frac{12\,(415)}{4\,(1.2)\,(1.6)} = 648 \text{ mm}$$

EXAMPLE 10.7. *Flat slab with column heads.* A slab is supported on columns spaced at 5 m apart in both directions. The sizes of the column and column head are 440 by 440 mm and 760 by 760 mm respectively. The superimposed live and dead loads are 1.5 and 2 kN/m² respectively. The height of the column above the mat foundation is 6 m, including 160 mm high column head.

Table 10.6 Reinforcement Details (all bars of 12 mm dia) (Example 10.6)

	No. of bars and length		Location	Strip
1.	2,3740	extra	Over the exterior column line and at top face	Column
2.	5,3740	extra	Over the interior columns line of the end panel and at top face	"
3.	6,5000	straight	At the mid-span at bottom face in the	"
	6,7070	cranked	the end way	
4.	5,5000	straight	At the mid-span at bottom face in	"
	4,7070	cranked	the interior bay	
5.	.3,2640	extra	Over the exterior column line and at top face	Middle
6.	1,2640	extra	Over the interior column line and at top face	"
7.	3,5000	straight	At the mid-span and at bottom face	"
	2,6520	cranked		

Design data

Spacing of the columns $= L_1 = L_2 = L = B = 5$ m

Size of the column $= a = 440$ mm

Size of the column head $= a_1 = 760$ mm

$w_s + w_l = 3500$ N/m^2

$f_{ck} = 20$ MPa, $f_y = 415$ MPa

The design coefficients are: $K = 0.138$, $j = 0.80$.

Design of the section. For the purpose of estimating the self-weight of the slab, let the overall thickness of the slab be assumed in the range of $L/20$.

Net thickness of the slab $= t = 0.24$ m

Self-weight $= w_g = 0.24(25) = 6$ kN/m^2

The total dead load $= w_d = 6 + 2 = 8$ kN/m^2

The total design load $= w_t = 8 + 1.5 = 9.5$ kN/m^2

Clear spacing between the column heads is

$$L_0 = L - a_1 = 5 - 0.76 = 4.24 \ m \nleq 0.65L$$

The total design load in a panel is

$$W = \gamma_f w_t L L_0 = (1.5)(9.5)(5)(4.24) = 302.1 \text{ kN}$$

The sum of the magnitudes of the positive and negative bending moments in a panel is

$$M_0 = \frac{WL_0}{8} = \frac{302.1\,(4.24)}{8} = 160.113 \text{ kNm}$$

The magnitude of the negative BM at the face of the column head in the interior panel is

$$M_{ni} = 0.65 \, M_0 = 0.65\,(160.113) = 10.4.07 \text{ kNm}$$

The coefficients of exterior negative and positive BMs, and the interior bending moment are taken from Table 10.5; they are: $c_1 = 0.10$, $c_3 = 0.60$, $c_2 = 0.70$

The corresponding bending moments are:

$$M_{n1} = c_1 M_0 = 0.016 \text{ MNm}$$

$$M_{p3} = c_3 M_0 = 0.09606 \text{ MNm}$$

$$M_{n2} = c_2 M_0 = 0.1121 \text{ MNm}$$

The bending moments on the *column strip* of 2.5 m width at different locations are computed using Table 10.4. These moments are:

Negative BM at the exterior face of the exterior panel:

$$M_{nc1} = M_{n1} = 0.016 \text{ MNm}$$

$$M_{pc3} = (0.6) M_{p3} = 0.05763 \text{ MNm}$$

The negative BM in the interior column face of the exterior panel is :

$$M_{nc2} = 0.75 \, M_{n2} = 0.08408 \text{ MNm}$$

The negative BM in the interior panels is

$$M_{nci} = 0.75 \, M_{ni} = 0.07865 \text{ MNm}$$

The values listed are only the magnitudes of the moments and the thickness of the slab is governed by the largest magnitude of the BMs. The effective depth of the slab needed is given by

$$d = \sqrt{\frac{M}{K b f_{ck}}} \sqrt{\frac{M_{nc2}}{K b f_{ck}}} \sqrt{\frac{0.08408}{0.138(2.5)(20)}} = 0.120 \text{ m}$$

Use $d = 0.14$ m, then the overall thickness of the slab is

$$t = 0.14 + 0.03 = 0.17 \, \text{m}$$

The thickness of the slab was assumed as 0.24 m for the purpose of self-weight; therefore, the earlier moment computed is on the safer side.

Self-weight $w_g = 0.17 (25) = 4.5 \text{ kN/m}^2$

$$w_t = 4.5 + 2.0 + 1.5 = 8.0 \text{ kN/m}^2$$

Design for shear. The shear strength for diagonal tension is

$$\tau_c = 0.25 \sqrt{f_{ck}} = 1.12 \text{ MPa}$$

The critical shear plane is the peripheral plane which is at a distance $0.5d$ from the face of the column head. The total length of the critical shear plane section is

$$b_0 = 4 (a_1 + d) = 4 (0.76 + 0.14) = 3.6 \text{ m}$$

The total shear force on this plane is

$$V_c = \gamma_f \, w_t \, (L_1 \, L_2 - (a_1 + d)^2)$$

$$= (1.5)(9)(25 - 0.90^2) = 326.56 \text{ kN}$$

The nominal shear stress is

$$\tau_v = \frac{V_c}{b_0 d} = \frac{326560}{2600 \, (140)} = 0.65 \text{ MPa}$$

The nominal shear stress is within the allowable value.

Design of reinforcement

Column strip: The positive BM at mid-span = M_{pc3} = 0.05763 MNm and the area of the reinforcement needed is

$$A_{st3} = \frac{1.15 M_{pc3}}{jdf_y} = \frac{(1.15)57630}{(0.80)(0.14)(415)} = 1426 \text{ mm}^2$$

Provides 13 numbers of 12 mm bars and crank 6 bars from each side. Then the actual steel provided is A_{st} = 1469 mm². The negative BM at the exterior support is M_{nc1} and the area of the reinforcement which is to be laid at the top near the column is

$$A_{st1} = \frac{1.15 M_{nc1}}{jdf_y} = \frac{(1.15)\,16000}{(0.80)(0.14)(415)} = 396 \text{ mm}^2$$

6 bars are already cranked, so add two extra bars.

The negative bending moment at the interior of the end panel is M_{nc2} = 0.08408 MNm, and the area of the reinforcement is

$$A_{st2} = \frac{(1.15)84080}{0.80(0.14)(415)} = 2080 \text{ mm}^2$$

Provide 13 *numbers of* 12 mm *bars at top* (extra).

A_{st} provided is 2147 mm² (inclusive of 6 cranked bars from each direction)

Design of middle strip. The bending moments in the middle strip are obtained by subtracting the bending moments of the column strip from the total bending moments in the panel at the corresponding section. The minimum reinforcement at any section can be taken as 0.12% of the area of the concrete, and is:

$$A_{smin} = \frac{0.12 \, bt}{100} = \frac{0.12(2500)(170)}{100} = 510 \text{ mm}^2$$

Therefore, the minimum number of 12 mm bars at any given section is 5. The bending moment on the middle strip is two-thirds of that of the column strip in the positive BM zone and it is one-third of the negative BM zone. It can be observed that the amount of the reinforcement needed in the middle strip is governed by the minimum requirement rather than by the bending moments. Figure 10.18 illustrates the reinforcement details. The development length of the bars is

$$L_d = \frac{\phi f_y}{4 \, \tau_{bd}} = \frac{12(415)}{4(1.2)(1.6)} = 691 \text{ mm}$$

Comment about the desirability of column heads or drops: The provision of column heads or drops decrease the effective span and consequently the bending moment. Any decrease in the concrete due to smaller thickness of the slab is usually compensated by the addition of the concrete at the column heads and extra formwork. The length of the cranked bars or top bars is increased in the case of slabs with column heads, so the gain in the decrease of the area of the reinforcement is compensated. The drops or column heads become almost unavoidable if the nominal shear stress exceeds the allowable value. In general, such a situation arises when (i) the spacing of the columns is large, say, more than 6 m, (ii) the size of the column is very small compared with the panel size, say, the size of the column is less

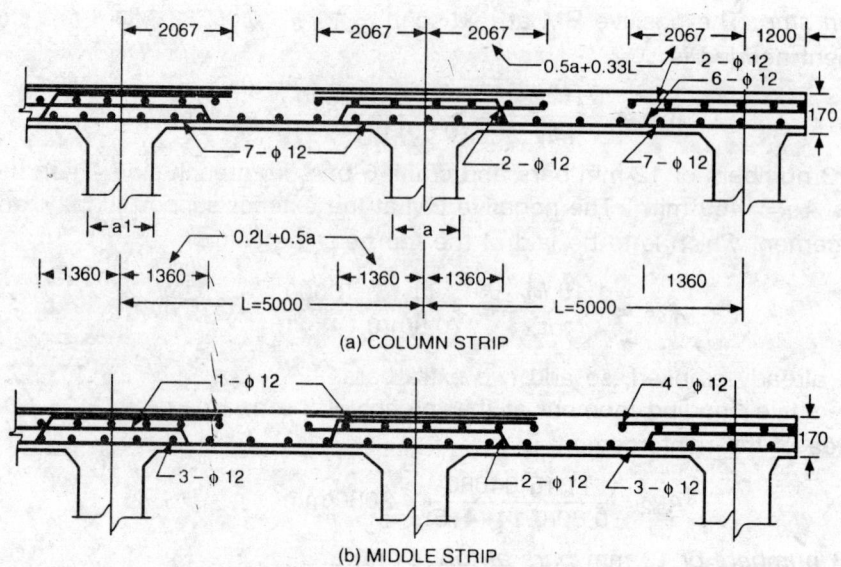

Fig. 10.18. Reinforcement Details of Flat Slab of Example 10.7.

than $L/15$; and (iii) the total and intensity on the slab is high, say, it is more than 30 kN/m². Otherwise, it is desirable to design the flat slab without column heads which is not only economical but helps in aesthetic and functional aspects also.

10.15 Grid Floors (Waffle Slab)

An assembly of intersecting beams placed close to each other and interconnected to a slab is called grid floor. The two sets of beams can intersect at right angle to each other, then such a grid is called *orthogrid*. In case these beams are placed parallel to the diagonals, then it is called a diagrid. Grid slabs or floors are found to be economical and praticable for clear spans of 8 to 25 m. Large unobstructed covered space can be obtained with flat floor surface without taking away much of the height of the floor by the beams. The spacing of the beams in each direction can be different, and it depends on the architectural considerations. A minimum grid spacing of the metre and a maximum spacing of two metres appears to be reasonable. In case the beams are unequally spaced in both the directions, then one of the spacings can be less than even one metre. A span can be divided into 5 to 12 grids so as to present a pleasant appearance. About six to eight grids in each direction will result into a minimum bending moment per unit width, so one can provide about six to eight grids in each direction. The design bending moments are computed using the same procedure as used in the design of two-way slabs. The design procedure is illustrated through an example.

EXAMPLE 10.8 *Design of grid floor.* A hall of dimensions 12.2 m by 18.3 m is to be covered by a flat roof. Design a grid floor.

Design data

L_x = 12.2 m,	L_y = 18.3 m
f_{ck} = 20 MPa,	f_y = 415MPa
Roof finish load	w_s = 2kN/m²
Live load	w_l = 1.5 kN/m²

Selection of grid and grid properties: Let the floor be divided into eight grids in the short span direction and 12 in the long span direction. This makes the spacing of the grids in both the directions same.

$$\text{Aspect ratio} = r = \frac{18.3}{12.2} = 1.5$$

$$\text{Grid spacing} = a = \frac{12.2}{8} = \frac{18.3}{12} = 1.525 \text{ m}$$

The thickness of the slab is usually selected nominal and in the range of a/15, subjected to a minimum of about 60 mm. Similarly, the depth of the web of the grid beam can be selected in the range of L_x/25 to L_x/15 depending on the load intensity on the floor.

Let the following dimensions be selected.

Thickness of the slab t = 100 mm

Thickenss of the web b_w = 150 mm

Depth of the web h_w = 600 mm

The grid slab can be considered as a T-beam section in each direction with flange thickness equal to the thickness of the slab and the width of the flange equal to the grid spacing. The flange width should be limited to $15t + b_w$ in this case. The properties of one grid beam, the beam being a T-section, are computed first:

$$b_f = a = 1525 \text{ mm}$$

Design of the section:

Roof finish load	$= w_s = 2$ kN/m²	
Slab weight	$= 0.1(25) = w_{gs} = 2.5$ kN/m²	
Weight of the web	$= w_g = \dfrac{(1.525 + 1.375)(0.15)(0.6)(25)}{(1.525)(1.525)} = 2.7$ kN/m²	
Live load	$= w_l = 1.5$ kN/m²	
Total load	$= 8.9$ kN/m²	

Use $\qquad w_t = 9$ kN/m² and $w_c = 1.5(9) = 13.5$ kN/m²

The collapse mechanism is same as that of a flat plate with aspect ratio equal to 1.5. The desired resisting moment capacity per grid in the short span direction is given by Eq. (10.20).

$$M_r = \frac{a \left(\sqrt{(c + 3r^2)} - \sqrt{c} \right)^2 w_c L^2}{24 r^2}$$

where a = grid spacing and the value of c can be selected as equal to $1/r$. Thus, $c = 0.67$. Then, from the above expression, we have

$$M_r = 206.02 \text{ kNm}$$

The resisting moment capacity in the long span direction would be equal to

$$M_{ry} = cM_r = 138.06 \text{ kNm}$$

The bending moment per unit of one grid spacing would be equal to the product of the grid spacing and the bending moment per unit width. The effective depth of the reinforcement from the top fibre can be taken as

$$d = h - 0.05 = 0.7 - 0.05 = 0.65 \text{ m}$$

The area of the tension reinforcement needed per each grid spanning in the short span (same as the x-axis direction) is

$$A_{stx} = \frac{1.15\,M_r}{jdf_y} = \frac{1.15(206020)}{0.8(0.65)(415)} = 1098 \text{ mm}^2$$

Provide 2 numbers of 20 and 18 mm bars in each beam at the bottom. The area of the reinforcement provided is

$$A_{stx} = 2(314 + 254) = 1136 \text{ mm}^2$$

The percentage of the reinforcement in the T-section is

$$p_x = \frac{100 A_{stx}}{A_c} = \frac{113600}{242500} = 0.47\%$$

The bending moment in the grid along the long span direction is about two-thirds of that in the short span direction.

Provide 3 numbers of 18 mm bars in the grid in the long span direction.

$$A_{sty} = 762 \text{ mm}^2$$

Check for shear stress: The load in the middle strip level is almost transferred to the short span. Therefore, the total load at the middle grid beam spanning in the short span is

$$W = w_t a L_x$$

The design force on a grid beam at a distance d from the end support is

$$V_c = \left(\frac{W}{2} - w_t a d\right)\gamma_f = w_t a \left(\frac{L_x}{2} - d\right)\gamma_f$$
$$= 9(1.525)(6.1 - 0.65)(1.5) = 112.2 \text{ kN}$$

The nominal shear stress in the beam is

$$\tau_v = \frac{V}{b_w d} = \frac{112200}{150(650)} = 1.15 \text{ MPa}$$

The percentage of reinforcement in the web is

$$p_t = \frac{100\,A_{stx}}{b_w d} = \frac{113600}{150\,(650)} = 1.2\%$$

The shear strength in M20 concrete for 1.2% reinforcement is 0.67 MPa. The allowable shear force on the section is

$$V_a = b_w d\tau_c = 150\,(650)(0.67) = 65325 \text{ N}$$

The design shear force on the beam is

$$V_s = V - V_a = 112200 - 65325 = 46875 \text{ N}$$

Select two-legged 8 mm HYSD stirrups, then the area of the stirrup is

$$A_{sv} = 100 \text{ mm}^2$$

The spacing of the stirrups is

$$s_v = \frac{A_{sv} f_y d}{1.15 \ V_s} = \frac{100(415)(650)}{1.15(50775)} = 462 \text{ mm}$$

The maximum spacing of the stirrups is

$$s_{max} = \frac{A_{sv} f_y}{0.4 \ b_w} = \frac{100(415)}{0.4(150)} = 691 \text{ mm}$$

Provide 2-legged 8 mm HYSD-Fe415 stirrups at 450 mm spacing in the short and long spans beams.

The deflections in the slab and particularly in grid floor are negligibly small; therefore, there is no need to check for the deflection limits.

Design of slab. The slab over the grids is considered continuous in both directions, square in plane and supported by the grid beams at the spacing a. The load on the slab including its weight is

$$w_t = w_{gs} + w_s + w_l = 5.9 \text{ kN/m}^2$$

Use

$$w = 1.5 w_t = 1.5(5.900) = 8.85 \text{ kN/m}^2$$

The maximum negative bending moment in the slab is

$$m = \frac{wa^2}{24} = \frac{(8.85)(1.525)^2}{24} = 0.858 \text{ kNm/m}$$

The flexural stress in the plain concrete slab is

$$\sigma_{cr} = \frac{m}{Z} = \frac{858(6)}{(100)^2} = 0.5 \text{ MPa}$$

The flexural strength of the concrete is

$$f_{cr} = 0.7 \sqrt{f_{ck}} = 3.1 \text{ MPa}$$

The flexural tension is far less than the strength, therefore, provide only nominal reinforcement in the slab. About 0.12% reinforcement in each direction must be adequate in the slab.

$$A_s = \frac{0.12(100)(1000)}{100} = 120 \text{ mm}^2/\text{m}$$

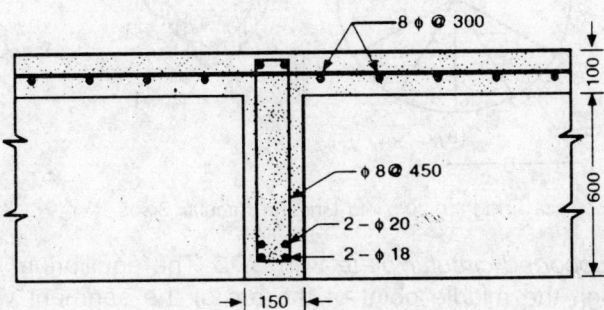

Fig. 10.19 Reinforcement Details of Waffle Slab, Example 10.18.

Table 10.7 Comparison of the Design Details of Grid Floor 15 × 18 m Clear with 1.525 m Grid Spacing Thickness of the Slab 100 mm

	Item	WSD	LSD
1.	Reinforcement in the:		
	a) Short span beams	4 − φ20	2 − φ20, 2 − φ18
	b) Long span beams	3 − φ20	3 − φ18
2.	Stirrups	φ8 @ 350 mm	8φ @ 450 mm

Provide 8 mm HYSD-Fe415 bars at 300 mm spacing in the middle level of the slab in each direction. Figure 10.19 illustrates the typical reinforcement details of the grid slab. Table 10.7 gives a comparison of the design details of the grid floors designed by WSD and LSD. The size of the grid beam is kept same for convenience of the comparison.

10.16 Design of Circular Slabs

Circular slabs are not so common in civil engineering construction when compared with the rectangular ones. These slabs are used in wells, circular water tanks, footings and sometimes in buildings. A typical yield line pattern is shown in Fig. 10.20a for a simply supported plate.

Let M_R = BM on the plane normal to the radius per unit width

M_θ = BM on the θ-plane which is parallel to the radius per unit width

Whether the reinforcement is provided in the radial and circumferential directions or in the *x*- and *y*-directions, one can assume an isotropic character in many cases. Therefore,

$$M_R = M_\theta \qquad (10.90)$$

The circular slab deforms into a conical shape, thus developing radial yield lines generating from the centre of the plate. The circumferential moment plays an important role of the failure load. Consider a small segment bounded by two radial yield lines as shown in Fig. 10.20b.

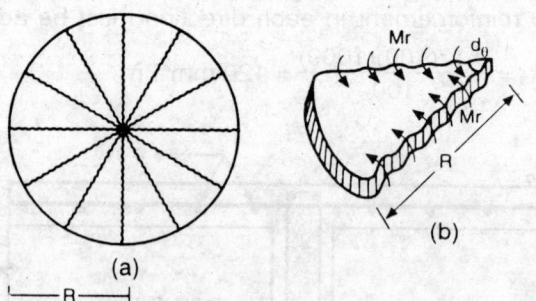

Fig. 10.20 Yield Lines in Circular Slabs.

Case 1: *Simply supported circular plate with UDL.* The equilibrium of the forces about a tangent passing through the middle point of the arc of the segment with included angle of 'dθ' (of a simply supported slab) is

$$M_R \, (Rd\,\theta) = w_c \, (Rd\,\theta) \left(\frac{R}{2}\right) \frac{R}{3}$$

or
$$M_R = \frac{w_c R^2}{6} \qquad\qquad (10.91)$$

The collapse load is given by

$$w_c = \frac{6M_R}{R^2} \qquad\qquad (10.92)$$

Case 2: Fixed ended circular plate with UDL. Similarly, the collapse load of a circular slab *fixed* along the support is given by

$$w_c = \frac{6M_R \, (1 + c)}{R^2} \qquad\qquad (10.93)$$

where cM_R = negative BM capacity of the support.

Case 3: Simply supported circular plate with central load. In case of a circular slab with a *concentrated load* at mid-point, the collapse load can be shown as

$$w_c = 2\pi M_R \qquad\qquad (10.94)$$

Case 4: Circular footing with wall at the periphery. The collapse load of a foundation slab of a well with uniform contact soil reaction can be shown to be equal to

$$w_c = \frac{6M_R \, (1 + c)}{R^2} \qquad\qquad (10.95)$$

This is exactly same as that of a plate with fixed boundary conditions. The derivation is simple and is same as shown in case 1.

Case 5: Circular footing with circular wall on the periphery and resting on cohesionless soil. The soil pressure in case of a footing resting on cohesionless soil is not uniform. It is zero at the edges and reaches maximum at the centre. Assuming the pressure variation is also circular, the collapse load can be shown to be

$$w_c = \frac{30 \, M_R \, (1 + c)}{7R^2} \qquad\qquad (10.96)$$

There are many forms of supports and loads on circular or semicircular slabs, therefore, only few simple cases are given in this book. The design of the slab is illustrated through examples.

EXAMPLE 10.9 *Design of circular slab with UDL:* A circular slab of effective radius 3 m is subjected to a uniformly distributed live load of 5 kN/m^2. The slab is simply supported all along the boundary. Design the plate with M20 concrete and HYSD-Fe415 bars.

Design data and analysis

Radius of the plate $\qquad\qquad R = 3\,\text{m},\ w_l = 5\,\text{kN/m}^2$
$$f_{ck} = 20\,\text{MPa}, f_y = 415\,\text{MPa}$$

The design coefficients are: $\quad K = 0.138$ and $j = 0.8$.

The slab will develop radial yield lines symmetrically. Assume isotropic reinforcement in the slab, then the collapse load on the circular plate is given by Eq. (10.92). It is

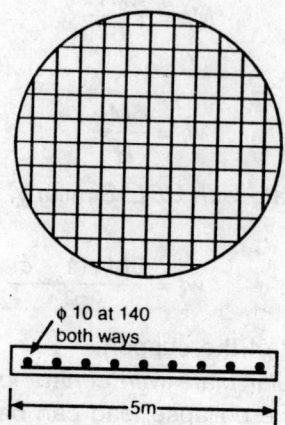

φ 10 at 140
both ways

|←————— 5m —————→|

Fig. 10.21 Reinforcement Details in a Circular Slab of Example 10.9.

$$w_c = \frac{6M_r}{R^2}$$

or

$$M_R = \frac{w_c R^2}{6}$$

Design of the section: For the purpose of estimating the self-weight of the slab, assume the thickness of the slab as $R/20$, say, $t = 150$ mm.

Self-weight	$w_g = 0.15(25) = 3.75$ kN/m^2
Live load	$w_l = 5$ kN/m^2
Total load	$w_t = 8.75$ kN/m^2
Collapse load	$w_c = 1.5\, w_t = 13.125$ kN/m^2

Moment capacity required is

$$M_R = \frac{w_c R^2}{6} = \frac{13.125(9)}{6} = 19.688 \text{ kNm/m}$$

The effective depth required is

$$d = \sqrt{\frac{M_R}{bKf_{ck}}} = \sqrt{\frac{0.019688}{1(0.138)(20)}} = 0.084 \text{ m}$$

Use $d = 0.12$ m and $t = 0.12 + 0.02 = 0.14$ m

The revised collapse load and moments are:

$$w_c = 1.5(0.14(25) + 5) = 12.75 \text{ kN/m}^2$$

$$M_R = 0.019125 \text{ MNm/m}$$

$$A_{st} = \frac{1.15M}{jdf_y} = 552 \text{ mm}^2/\text{m}$$

Provide 10 mm bars at 140 mm spacing in each direction at the bottom face. The same example when solved by WSD resulted in $t = 170$ mm and 12 mm bars at 200 mm spacing.

EXAMPLE 10.10 *Design of a well cap:* A RCC circular well of internal and external radii of 7.20 and 7.80 m is provided with a solid well cap built integral with the RCC well steining. The load on the well cap is 65 kN/m². Design the well cap with M20 concrete and HYSD-Fe415 bars.

Design data and analysis

Outer radius of the well = 7.8 m
Inner radius of the well = 7.2 m
Mean radius of the well = 7.5 m

$$f_y = 415 \text{ MPa}, \ f_{ck} = 20 \text{ MPa}, \ w_l = 65 \text{ kN/m}^2$$

Design coefficients are: $K = 0.138$, $j = 0.8$.

Assume isotropic reinforcement at bottom and also at top near the supports. As the slab is cast integral with the wall, the support can be assumed to be fixed. Let the moment capacity at the support be about 50% more than that at the mid-span. That is, the negative BM capacity of the slab is

$$M_{Rn} = 1.5 \ M_R, \ \text{for} \ c = 1.5$$

The collapse load of the slab is given by Eq. (10.93) and is

$$w_c = 6M_R \frac{(1 + c)}{R^2}$$

or

$$M_R = \frac{w_c R^2}{6(1 + c)}$$

Design of the section: For the purpose of self-weight calculations, let the slab thickness be assumed in the range of $R/15$.

Let

$$t = 0.60 \text{ m}$$
$$w_g = 0.6(25) = 15 \text{ kN/m}^2$$
$$w_l = 65 \text{ kN/m}^2$$
$$\underline{w = 80 \text{ kN/m}^2}$$
$$w_c = 1.5 \ w_t = 120 \text{ kN/m}^2$$

The resisting moment capacity of the section should be at least

$$M_R = \frac{w_c R^2}{6(1 + c)} = \frac{80(7.5)^2}{6(1 + 1.5)} = 300 \text{ kNm/m}$$

The effective depth required is

$$d = \sqrt{\frac{M_R}{bKf_{ck}}} = \sqrt{\frac{0.300}{1 \ (0.138) \ (20)}} = 0.33 \text{ m}$$

Let

$$t = 0.48 \text{ m and } d = 0.40 \text{ m}$$

Then the collapse load is

$$w_c = 1.5 \ (0.48(25) + 65) = 115.5 \text{ kN/m}^2$$

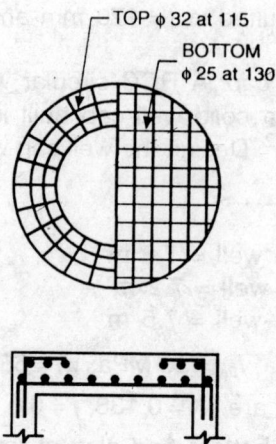

Fig. 10.22 Reinforcement Details in a Well Cap of Example 10.10.

and the collapse BM is

$$M_R = \frac{w_c\,R^2}{6(1+c)} = 4.33.125 \text{ kNm/m}$$

The area of tension reinforcement required is

$$A_{st} = \frac{1.15 M_R}{jdf_y} = \frac{1.15\,(433.125) \times 100}{0.8\,(0.40)\,(415)} = 3750 \text{ mm}^2/\text{m}$$

Provide 25 mm bars at 130 mm spacing in both the directions at bottom face with a clear cover of 40 mm.

$$d = 480 - 40 - 25 - 25/2 = 402 \text{ mm}$$

Therefore, the effective depth provided is adequate.

$$A_{st} \text{ (Provided)} = \frac{1000}{130}\,(490) = 3769 \text{ mm}^2/\text{m}$$

The negative bending moment capacity of the section near the support is assumed to be 1.5 M_R. The slab is safe if full redistribution takes place. Since the full redistribution is not possible in RCC slabs, assume about 80% redistribution. The design moment capacity at the supports should be

$$M_{rn} = \frac{cM_R}{0.8} = \frac{1.5\,(433.125)}{0.8} = 812.1 \text{ kNm/m}$$

The area of the top reinforcement required is

$$A_{stn} = \frac{1.15\,M_{rn}}{jdf_y} = \frac{1.15(812100)}{0.8\,(0.4)(415)} = 7032 \text{ mm}^2/\text{m}$$

Provide 32 mm bars at 115 mm spacing at top.

The development length of the bars is

$$L_d = \frac{\phi f_y}{4\tau_{bd}} = \frac{32\,(415)}{4(1.2)\,(1.6)} = 1729 \text{ mm}$$

Provide the top reinforcement up to $R/3 = 2.5$ from the support.

PROBLEMS
(Use IS 456, LSD = limit states design)

10.1 A rectangular slab is simply supported on four edges with effective spans 3 m by 8 m. It is subjected to a live load of 4 kN/m². Design a reinforced concrete slab using M20 concrete and HYSD-Fe415 bars by LSD method. $\gamma_f = 1.5$.

10.2 A long slab of effective spans 3 m by 8 m is simply supported on 3 m long edges and is continuous at 8 m long supports. The slab is subjected to a live load 4 kN/m². Design a RCC slab using M25 concrete and mild steel bars by LSD method. Sketch the reinforcement and constructional details. $\gamma_f = 1.5$.

10.3 A verandah slab of 10 m length, cantilevers from a fixed wall to 2 m cantilever span. The live load on the slab is 4 kN/m². Design a RCC slab with M25 concrete and HYSD-Fe415 bars by LSD method. Sketch the reinforcement details and indicate the minimum embedment length. $\gamma_j = 1.5$.

10.4. A slab is simply supported on effective spans 4.5 m by 5.5 m and subjected to a live load of 3 kN/m². Design a reinforcement concrete slab, assuming that the corners of the slab are allowed to warp and using M20 concrete and HYSD-Fe415 bars. Design the slab by LSD method giving reinforcement details both in plan and in cross-section. $\gamma_f = 1.5$.

10.5 A simply supported slab of effective spans 4.5 m by 5.5 m is subjected to a live load of 3 kN/m². Design a RCC slab assuming that the corners of the slab are fixed against the warping. Design the slab by using M20 concrete and HYSD-Fe415 bars by LSD method. Sketch the reinforcement details.

10.6 A square slab simply supported at two adjacent edges and having continuity connection at the other two adjacent supports is subjected to a live load of 3 kN/m². Design a reinforced concrete slab by LSD method using M20 concrete and HYSD-Fe415 bars. Sketch the reinforcement details. $\gamma_f = 1.5$.

10.7 A rectangular slab of 4 m by 6 m is simply supported on the two opposite walls in short span and continues on the two long walls. The slab is subjected to a live load of 3 kN/m². Design a reinforced concrete slab by LSD method using M20 concrete and HYSD bars. $\gamma_f = 1.5$.

10.8. A flat slab is supported by a set of columns spaced at 4.5 m intervals in two perpendicular directions. The size of the column is 400 mm square. The slab is subjected to a live load of 1.5 kN/m². Design a flat slab by LSD method using M20 concrete and HYSD-Fe415 bars. Assume no drops or column heads in the construction. The effective height of the column could be taken as 3.2 m.

10.9 A flat slab is supported by a series of columns spaced at 6 m in two perpendicular directions. The size of the column is 500 mm square and provided with a column head of size of 600 mm square. The slab is subjected to a live load of 5 kN/m². Design a flat slab by LSD method M25 concrete and HYSD-Fe415 bars.

10.10 A flat slab with drop panels having square drop panels of 1 m square is suppored by columns spaced at 6 m apart in both directions. The size of the column is 500 mm square. The slab is subjected to a live load of 60 kN/m². Design a reinforced concrete flat slab by LSD using M20 concrete and HYSD-Fe415 bars. Assume an effective height of the column as 4 m. Sketch the reinforcement details.

10.11 A floor slab circular in plan with 4 m radius is subjected to a live load of 3 kN/m². Design a reinforced concrete slab by LSD using M20 concrete and mild steel reinforcement. Sketch the reinforcement details.

10.12 A column carrying a load of 1000 kN is supported by a circular footing resting on a soil having a net safe bearing capacity of 100 kN/m². Assume the column diameter as 500 mm square. Design a RCC circular footing by LSD using M20 concrete and mild steel reinforcement.

10.13 A water tank with circular cylindrical shaft of mean radius 6 m is resting on circular raft foundation. The total load carried by the shaft is 22 MN and the thickness of the shaft wall is 120 mm. The safe bearing capacity of the soil is 70 kN/m². Design a RCC raft by LSD using M20 concrete and HYSD-Fe415 bars. Sketch the reinforcement details (vide a bending moment coefficient as given in the chapter.)

Chapter 11

Design of Reservoirs and Water Tanks

11.1 Introduction

Storage reservoirs and overhead tanks are used to store water, liquid petroleum, petroleum products and similar liquids. The force analysis of the reservoirs or tanks is about the same irrespective of the chemical nature of the product. All tanks are designed as crack-free structures to eliminate any leakage. Water or raw petroleum retaining slab and walls can be of reinforced concrete with adequate cover to the reinforcement. Water and petroleum do not react with concrete and, therefore, no special treatment to the surface is required. Industrial wastes can also be collected and processed in concrete tanks with few exceptions. The petroleum products such as petrol, diesel oil, etc. are likely to leak through the concrete walls, therefore such tanks need special membranes to prevent leakage. The present chapter deals with the design of water tanks. Reservoir is a common term applied to liquid storage structure and it can be below or above the ground level. Reservoirs below ground level are normally built to store large quantities of water whereas those of overhead type are built for direct distribution by gravity flow and are usually of smaller capacity.

The reservoirs can be made with a minimum M20 concrete so as to provide not only strength but also higher density to prevent leakage. A well-graded aggregate with water cement ratio less than 0.5 is desired for making impervious concrete. As the head of water increases, there is a tendency of seepage; however, water heads upto 15 m should not present any problem of seepage provided crack-free design methods are applied. Mild or high yield strength steel reinforcement bars can be used in the storage tanks. The permissible stress in the reinforcement is controlled by the strain and crack widths rather than by the strengths. Plain concrete may fracture at about 0.0002 tension strain but when it is strengthened by reinforcement, the level of the cracking strain is decreased by one-third to one-fifth, depending on the nature of the bond between the concrete and the reinforcement. Deformed bars or ribbed steels improve the level of the cracking capacity in the concrete. A crack width of 0.1 mm has been accepted as a permissible value in water retaining structures. This crack width is not a real crack width but is an imaginary value which really does not cause apparent cracking. In view of the complexities and uncertainties associated with the crack widths, a simplified approach through allowable stresses is presented. The fracture strength of the concrete is given by

$$f_{cr} = 0.7 \sqrt{f_{ck}} \tag{11.1}$$

where f_{cr} = flexural fracture strength in N/mm^2 (= MPa)

f_{ck} = characteristic strength of 150 mm cube in N/mm^2

The allowable direct and bending tensile stresses in the concrete can be expressed as

$$\sigma_{cat} = 0.27 \sqrt{f_{ck}} \tag{11.2}$$

$$\sigma_{cbt} = 0.37 \sqrt{f_{ck}} \tag{11.3}$$

where σ_{cat} = allowable axial tensile stress in concrete in MPa

σ_{cbt} = allowable bending tensile stress in concrete in MPa

The above values are for reinforced cement concrete but not for plain concrete. Table 11.1 gives recommended allowable tensile stresses in reinforced concrete to ensure no leakage.

Table 11.1 Allowable Tensile Stresses in Reinforced Concrete (MPa) on Water Face

Stress	Grade of concrete				
	M15	M20	M25	M30	M35
Direct tension	1.1	1.2	1.3	1.5	1.6
Bending tension	1.5	1.7	1.8	2.0	2.2

The face of the concrete away from the water face can be designed as a cracked section, provided a minimum of 115 mm uncracked section is available on the water face for a head of 5 m. The allowable stresses in the reinforcement bars are listed in Table 11.2.

Table 11.2 Allowable Stresses in Reinforcement (MPa)

Nature of stress	Plain mild steel	High yield deformed*
1. Tension in steel placed within 225 mm from water face	100	150
2. Tension in steel placed beyond 225 mm from water face	120	190
3. Compression	120	190

*Permissible stresses in deformed mild steel can be increased by 20% over and above the values listed against plain bars.

It should be stated that all the tensile force either due to hoop tension or bending tension must be resisted by the reinforcement with the allowable stresses listed in Table 11.2. The allowable tensile stress in the reinforced concrete corresponds to the uncracked composite section. The *minimum* recommended covers to the reinforcement are:

Cover = 20 mm in direct tension or the diameter of the bar

= 25 mm in bending tension

= 35 mm in alternating wetting and drying condition

Table 11.3 Minimum Reinforcements (HYSD-Fe415)

Nature	Percentage
1. Dummy concrete with no tension at all	0.12
2. Concrete upto thickness of 100 mm	0.24
3. Concrete with thickness more than 425 mm	0.16
4. Concrete below the ground level	0.12
5. Concrete subjected to freezing and thawing	0.24

Note: For concrete with thickness between 100 mm and 425 mm a linear interpolation from 0.24 to 0.16% be used. The percentage be increased by 25% in case of mild steel reinforcement.

A minimum reinforcement in the concrete is must since the allowed stresses exceed the normal limits of plain concrete, to minimize cracking due to shrinkage and temperature. Concrete submerged under water or under soil is less sensitive to shrinkage and temperature effects when compared with that exposed to alternative drying and wetting.

11.2 Balanced Moment and Tension Capacities – Design Criteria

The balanced moment capacities of the rectangular sections with full permissible stresses were derived earlier in Chapters 2 and 7. In case of concrete cross-sections where the tension occurs on the fibres away from the water face, the capacities derived earlier will hold good; however, in case the tension face is on the water side, then the moment capacities are to be modified as per the allowable stresses. The neutral axis distance from the strain compatibility is given by (see Chapter 2).

$$\frac{\sigma_{cb}}{nd} = \frac{\sigma_{st}}{(d-nd)m} \tag{11.4}$$

or

$$n = \frac{1}{1 + \sigma_{st}/m\sigma_{cb}} \tag{11.5}$$

The value of modular ratio is $m = 280/3\sigma_{cb}$
Therefore Eq. (11.5) works out to be

$$n = \frac{1}{1 + 3\sigma_{st}/280} \tag{11.6}$$

The balanced moment capacity of the section is

$$M_{rb} = \frac{n}{2}\left(1 - \frac{n}{3}\right)bd^2\sigma_{cb} = Kbd^2\sigma_{cb} \tag{11.7}$$

where

$$K = \frac{n}{2}\left(1 - \frac{n}{3}\right) \tag{11.8}$$

The lever arm is given by

$$j = 1 - \frac{n}{3} \tag{11.9}$$

and the amount of reinforcement required is

$$A_{st} = \frac{M}{jd\sigma_{st}} \tag{11.10}$$

In case of balanced sections, the reinforcement is

$$A_{st} = \frac{Kbd^2\sigma_{cb}}{jd\sigma_{st}} = \frac{nbd\sigma_{cb}}{2\sigma_{st}}$$

or

$$\frac{A_{st}}{bd} = \frac{n\sigma_{cb}}{2\sigma_{st}} \tag{11.11}$$

The allowable compressive stress in concrete on the water face is same as that recommended in ordinary cases. Table 11.4 gives the design constants of balanced sections in which tension is on the water face.
The design criterion for bending is given by

$$\tag{11.12}$$

Table 11.4 Design Coefficients of Balanced Sections

Quantity	Tensile stress on			
	Water face		Away from water face	
	HYSD-Fe415	MS	HYSD-Fe415	MS
σ_{st} (MPa)	150	100	230	140
n	0.383	0.483	0.289	0.4
j	0.872	0.839	0.904	0.867
K	0.167	0.202	0.130	0.173

$$Kbd^2\sigma_{cb} \geq M$$

and
$$A_{st} \geq \frac{M}{j d \sigma_{st}} \tag{11.13}$$

The design of the concrete section is controlled by

$$\frac{M}{Z} \leq \sigma_{cbt}$$

where Z = modulus of the section

The design criterion for direct tension can be expressed as

$$A_s \geq \frac{T}{\sigma_{st}} \tag{11.14}$$

and
$$A_c + (m-1)A_{st} \geq \frac{T}{\sigma_{cat}}$$

or
$$A_c \geq \frac{T}{\sigma_{cat}} - (m-1)A_s \tag{11.15}$$

where T = axial tension and
A_s = area of the steel equally distributed in the cross-section

In case the axial tension and the bending moment are acting on the same plane, then the section be designed for combined action as discussed in Chapter 6.

The design criterion for limited tension in the concrete is given by

$$\frac{M}{Z_t} + \frac{T}{A_t} \leq \sigma_{cbt} \tag{11.16}$$

where Z_t and A_t are the section modulus and area of the transformed cross-section.

11.3 Rectangular Water Tanks

The walls of rectangular water tanks are subjected to hoop tension and bending moments. Usually the bending moment plays a dominant role in the design of the thickness of the section. The corner joints between the two adjacent walls can be treated as fixed as the pressure on the walls provide balancing type of reaction. The top edge of the walls which supports a relatively light roof slab can be treated as hinged or can be constructed as a hinged joint. The bottom edge of the walls built as an integral part of the base slab is treated as fixed. There are situations where mastic pads, etc. are provided between the wall and the bottom slab, then the joint is treated as hinged. Figure 11.1b shows the base fixed boundary condition of a wall Fig. 11.1a shows the variation of the bending moment along the height of

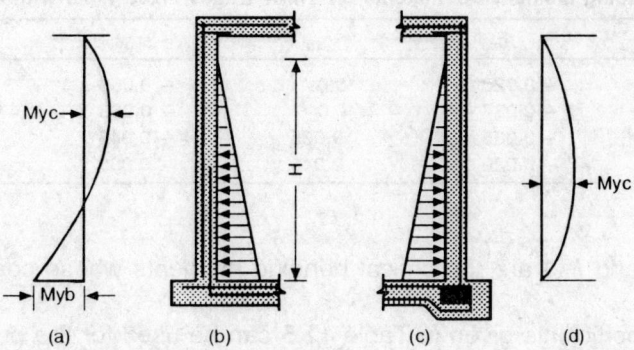

Fig. 11.1 Bending Moment and Hydrostatic Pressure on Vertical Walls.

the wall for fixed base condition. Similarly, Figs. 11.1c and d give the hinged base condition with the variation of the bending moment. Figure 11.2a shows a wall spanning between the two side walls subjected to lateral pressure. Figure 11.2b shows the bending moment variation along the length at about the mid-height of the tank. A tank wall is idealized as fixed on three edges and hinged at top as indicated in Fig. 11.2c. The bending moments are critical at points A, B C as shown in the figure. The critical bending moments can be expressed as:

$$M_{xa} = c_{xa}\,\gamma H^3 \qquad (11.17)$$
$$M_{xc} = c_{xc}\,\gamma H^3 \qquad (11.18)$$
$$M_{yb} = c_{yb}\,\gamma H^3 \qquad (11.19)$$
$$M_{yc} = c_{yc}\,\gamma H^3 \qquad (11.20)$$

where H = depth of the liquid = depth of the tank

$\dot{\gamma}$ = density of the liquid

c_{ij} = bending moment coefficient

and the subscripts $x = x$-direction, $y = y$-direction and a, b and c refer to points, A, B, and C respectively.

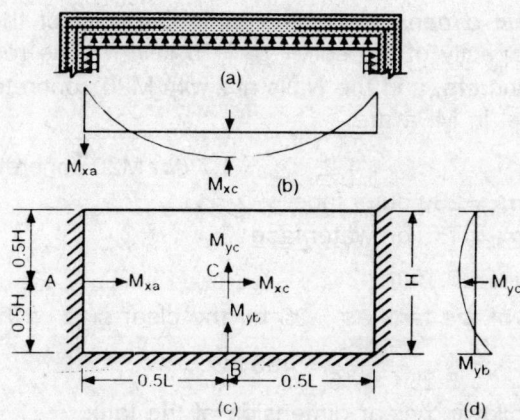

Fig. 11.2 Location of Critical Moments and BM Profile on Walls.

Table 11.5 Bending Moment Coefficients on Three Edges Fixed Plate with Hydrostatic Load

H/L	c_{xa}	c_{xc}	c_{yb}	c_{yc}
0 to 0.3	− 0.020	0.0	− 0.063	0.03
0.5	− 0.037	0.0	− 0.063	0.03
0.75	− 0.035	0.005	− 0.048	0.02
1.0	− 0.029	0.010	− 0.035	0.00

M_{xa}, M_{xx}, M_{yb} and M_{yc} are the critical bending moments whose coefficients are given in Table 11.5.

The moment coefficients given in Table 11.5 can be used for the purpose of design. The bending moment which causes compression on the load face is treated as positive; otherwise it is negative. The negative bending moment normally dominates and controls the wall thickness requirement. Unfortunately, the negative bending moment creates tension on the load face and consequently the permissible stress in the steel is on the lower side. In addition to the bending moment, the walls are also subjected to direct tension which is coming from the hydrostatic thrust on the walls which are normal to the wall under design. The design for direct tension is not very critical as the tensile force is usually small. The design for bending and tension acting on the same plane can be done separately and the reinforcement required can be superimposed.

The bending moment increases proportionately with the cube of the depth of the tank; therefore, it is desirable to keep the depth of the tank as small as possible. The depth of the tanks may be varied from 3 m to 8 m in which the upper limit is for large tanks. Smaller depths demand larger ground area and consequently the size of the roof slab increases. In most cases the height to side ratio of the walls is less than 0.5 and so the walls act as propped cantilever beams. In case of the walls having more than 6 m height, extra inclined props or abutments be provided to the wall to reduce the bending moment on the walls.

EXAMPLE 11.1 *Rectangular water tank above ground level.* A rectangular water tank is to be designed to store 2500 kl water. The tank is to be made just above the ground level and the safe bearing capacity of the soil is 75 kN/m^2. Design the water tank.

Design data and main dimensions of the tank. Capacity of the tank = Q = 2500 kl = 2500 m^3; safe bearing capacity of the soil = p_a = 75 kN/m^2. The roof and bottom slabs, and columns are with M20 concrete, and the walls are with M20 concrete. The allowable stresses in M20 concrete and steel in MPa are:

$$\sigma_{cb} = 7, \ \sigma_{cat} = 1.2, \ \sigma_{cbt} = 1.7 \ \text{(for M20 concrete)}$$
$$\sigma_{st} = 230 \ \text{outer face}$$
$$\sigma_{st} = 150 \ \text{on water face}$$

Free board = 0.15 m

Assuming a clear height of the tank as 5.25 m, the clear size of the tank works out to be

$$\text{Clear base} = \frac{2500}{5.25 - 0.15} = 490 \ \text{m}^2$$

Therefore, assume the following clear dimension of the tank.

Height = 5.4 m, length = 24.6 m, width = 19.6 m

Let the roof slab be supported by columns spaced 5 m apart in both directions. Let the centre-to-centre distance between the walls and the head of water be:

$$L_x = 24.6 + 0.4 = 25 \text{ m}$$
$$L_y = 19.6 + 0.4 = 20 \text{ m}$$
$$H = 5.4 - 0.15 = 5.25 \text{ m}$$

assuming the thickness of the wall at base as 0.4 m.

Let the size of the column be assumed as 0.44 m square, the capacity of the tank is:

Inside volume (gross volume)

$$= (2.46(19.6)(5.25) = 2531.3 \text{ m}^3$$

Less the volume of the 12 columns

$$= 12(5.25)(0.44)(0.44) = 13 \text{ m}^3$$
$$\text{Net volume} \qquad = 2518.3 \text{ m}^3$$

Design of roof slab. (M20 concrete and cracked section). The roof slab is designed as a flat slab with columns spaced at 5 m apart. Let the thickness of the slab be 0.24 m for self-weight purpose and the loads on the slab are:

Live load	$= 1.5 \text{ kN/m}^2$
Surface finish load	$= 2.0 \text{ kN/m}^2$
Self-weight = 0.24(25)	$= 6.0 \text{ kN/m}^2$
Total load w	$= 9.5 \text{ kN/m}^2$

Centre to centre of the panel $L = 5$ m

Clear panel $L_0 = 5 - 0.44 = 4.56$ m

Total design load on the panel is

$$W = wLL_0 = 9.5(5)(4.56) = 216.6 \text{ kN}$$

The sum of the magnitudes of the positive and negative bending moments in a panel is

$$M_0 = \frac{WL_0}{8} = \frac{216.6(4.56)}{8} = 123.462 \text{ kNm}$$

The effective height of the wall be $L_w = 5.4$ m

The relative stiffness of the wall is

$$K_w = \frac{I_w}{L_w} = \frac{Lt^3}{12(5.4)} = \frac{5(0.22)^3}{12(5.4)} = 8(10^{-4})$$

where the average thickness of the wall is taken as 0.22 m. The relative stiffness of the panel slab by assuming the thickness 0.22 is

$$K_s = \frac{I_s}{L} = \frac{Lt^3}{12(L)} = \frac{0.22^3}{12} = 8.9(10^{-4})$$

The ratio of the relative stiffness of the wall and the roof slab panel is

$$\alpha_c = \frac{K_w}{K_s} = \frac{8}{8.9} = 0.9$$

The ratio of the live load to the dead load is

$$\frac{w_l}{w_d} = \frac{1.5}{8.0} = 0.1875$$

The design bending moments in the flat slab are computed using the moment coefficients given in Tables 4.6 and 4.7. The reader is advised to see Chapter 4 for notations and design of the flat slab. The same notations are used in this section.

$$\alpha = 1 + \frac{1}{\alpha_c} = 2.111$$

The negative and positive bending moment coefficients in the exterior panel and the interior panels are computed using Table 4.7. These are:

$$c_1 = \frac{0.65}{\alpha} = \frac{0.65}{2.111} = 0.308$$

$$c_3 = 0.63 - \frac{0.28}{\alpha} = 0.497$$

$$c_2 = 0.75 - \frac{0.1}{\alpha} = 0.702$$

$$c_i = 0.65$$

The corresponding magnitudes of the bending moments (in MNm) in a panel are:

$$M_{n1} = c_1 M_0 = 0.038$$

$$M_{p3} = c_3 M_0 = 0.0613$$
$$M_{n2} = c_2 M_0 = 0.08667$$
$$M_{ni} = c_i M_0 = 0.0825$$

where the subscripts n and p refer to the negative and positive bending moments respectively. Subscripts 1, 2, 3 and i refer to the locations as follows:

$\quad$ 1 : the inside face of the outer wall
$\quad$ 2 : the face at the inner column of the outer panel
$\quad$ 3 : the middle point of the outer panel
$\quad i$: the interior panel

The bending moments on the panel are distributed between the column and the middle strips as per Table 4.6. The bending moments in the column strip which are indicated by a subscript c are:

$$M_{nc1} = M_{n1} = 0.038 \text{ MNm}$$

$$M_{pc3} = 0.6\, M_{p3} = 0.03673 \text{ MNm}$$

$$M_{nc2} = 0.75\, M_{n2} = 0.065 \text{ MNm}$$

$$M_{nci} = 0.75\, M_{ni} = 0.06188 \text{ MNm}$$

(100% of the negative BM in the end support, 75% of the negative BM in the interior supports and 60% of the positive BM are assigned to the column strip.) In case of an end wall support 75% can be assigned to the column strip. The effective depth of the section is calculated from the maximum negative bending moment, and is:

$$d = \sqrt{\frac{M}{K b\, \sigma_{cb}}} = \sqrt{\frac{M_{nc2}}{0.13(2.5)(7)}} = 0.17 \text{ m}$$

(*Note:* The roof slab is designed on cracked section basis as it is not in direct contact with the water, however uncracked section is most desirable.)

Check for shear stress: The allowable shear stress based on the diagonal tension is

$$\tau_c = 0.16 \sqrt{f_{ck}} = 0.67 \text{ MPa}$$

The critical shear plane is the peripheral plane which is at a distance $0.5d$ from the face of the column. The length of the plane is

$$b_0 = 4(a + b) = 4(0.44 + 0.2) = 2.56 \text{ m}$$

The shear force on the plane is

$$V = w(L^2 - (a + d)^2) = 9.5(25 - 0.64^2) = 233.6 \text{ kN}$$

The nominal shear stress on the plane is

$$\tau_v = \frac{V}{b_0 d} = \frac{0.2336}{2.56(0.2)} = 0.46 \text{ MPa} < \tau_c$$

This shear stress is within the allowable value, therefore, the depth is adequate against shear stress.

The overall depth of the slab can be

$$t = d + 0.03 = 0.23 \text{ m}$$

This value checks well with the assumed value for self-weight.

Design of reinforcement in the column strip: The area of the reinforcement is obtained from

$$A_{st} = \frac{M}{jd\sigma_{st}} \quad \text{(for 2.5 m strip width)}$$

The required areas of the reinforcement at different sections are:

$$A_{st1} = \frac{M_{nc1}}{jd\,\sigma_{st}} = \frac{38\,000}{0.904(0.2)(230)} = 914 \text{ mm}^2$$

$$A_{st2} = \frac{M_{nc2}}{jd\,\sigma_{st}} = 1563 \text{ mm}^2$$

$$A_{st3} = \frac{M_{pc3}}{jd\,\sigma_{st}} = 883 \text{ mm}^2$$

$$A_{sti} = \frac{M_{nci}}{jd\,\sigma_{st}} = 779 \text{ mm}^2$$

The minimum area of the reinforcement needed is (0.12%):

$$A_{stm} = \frac{0.12bt}{100} = \frac{0.12(2500)(230)}{1000} = 690 \text{ mm}^2$$

There is direct tension in the slab as it supports the vertical wall. This tensile force is

$$\frac{1}{4}\left(\frac{\gamma H^2}{2}\right) = \frac{10(5.25)^2}{8} = 34.5 \text{ kN/m}$$

Direct tension per 2.5 m panel is

$$T = 34.5(2.5) = 86.25 \text{ kN}$$

The area of the tensile reinforcement for direct tension is

$$A_s = \frac{T}{\sigma_{st}} = \frac{86\ 250}{230} = 375\ \text{mm}^2$$

At any given section the total area of the reinforcement is equal to the sum of the areas needed for bending and direct tension. The direct tensile stress caused in the concrete is

$$\sigma_t = \frac{T}{A_c} = \frac{86\ 250}{2500\ (230)} = 0.15\ \text{MPa}$$

which is very small.

Design of the reinforcement in the middle strip: The positive bending moment on the middle strip in the exterior panel is

$$M_{mp3} = 0.35(0.25)M_0 = 0.0108\ \text{MNm}$$

The area of the reinforcement is

$$A_{st} = \frac{10800}{0.904(0.2)(230)} = 259\ \text{mm}^2$$

This value is less than the minimum required. Similarly, the area of the reinforcement in the middle strip is controlled by the minimum value. The reinforcement details are shown in Fig. 11.3.

Design of column (M20 concrete): The column are spaced at 5 m and are subjected to axial force only. The load from the roof slab on each column is

$$P_s = wL^2 = 9.5(25) = 237.5\ \text{kN}$$

$$\text{Let the self-weight} = 32.5\ \text{kN}$$

$$\text{Total load}\ \overline{P = 270\ \text{kN}}$$

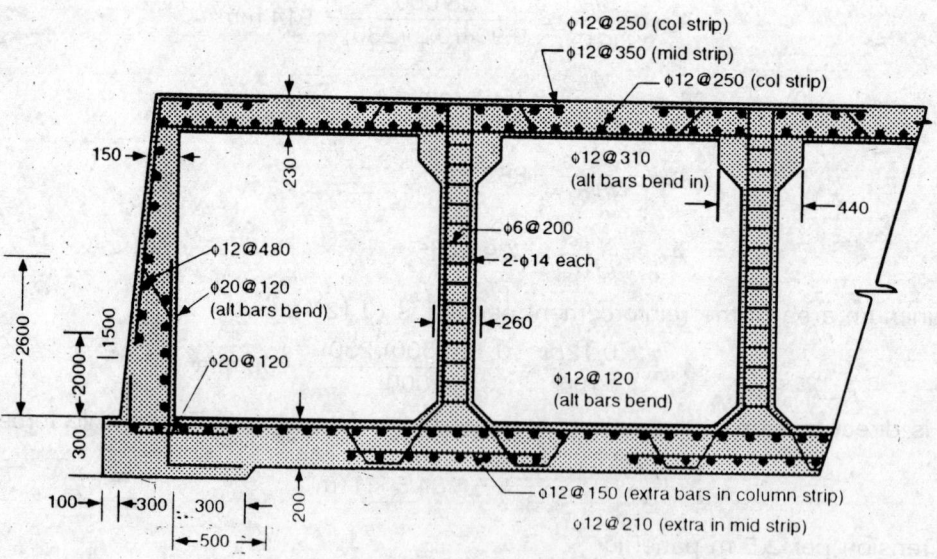

Fig 11.3 Reinforcement Details of Columns and Slabs.

Assume only 0.8% of reinforcement in the column, the capacity of the column is then given by:

$$P_a = A\sigma_{cc} + A_s\sigma_{sc}$$
$$= (A_g - A_s)\sigma_{cc} + A_s\sigma_{sc}$$
$$= A_g((1 - 0.008)\sigma_{cc} + 0.008\sigma_{cc})$$
$$= A_g(0.992(5) + 0.008(190)) = 6.48\,A_g$$

The design criterion is

$$P_a \geq P$$

or

$$6.48\,A_g \geq 0.27\ \text{MN}$$

or

$$A_g \geq \frac{0.27}{6.48} = 0.042\ \text{m}^2$$

The size of the column required is

$$a = \sqrt{A_g} = 0.21\ \text{m}$$

Provide 260 mm square columns with four numbers of 14 mm bars in each column. Also provide 6 mm ties at 200 mm spacing. A column head of size 440 mm square is to be provided so as to match with the assumptions made in the design of the roof slab. The column is a long one and its capacity is more than P.

Design of the vertical wall: The lengths of the walls (20 m and 25 m) are large when compared with the height. The walls will act as cantilevered from the base with some end constraints. The bending moment coefficients are taken from Table 11.5 for H/L less than 0.3 and are

$$c_{xa} = -0.02, \quad c_{yb} = -0.063 \quad \text{and} \quad c_{yc} = 0.03$$

The corresponding bending moments are:

$$M_{xa} = c_{xa}\gamma H^3 = -0.02(10)(5.25)^3 = -28.94\ \text{kNm/m}$$
$$M_{yb} = c_{yb}\gamma H^3 = -0.063\,(10)(5.25)^3 = -91.16\ \text{kNm/m}$$
$$M_{yc} = c_{yc}\gamma H^3 = 0.03\,(10)(5.25)^3 = 43.41\ \text{kNm/m}$$

The axial load coming on the wall from the roof slab for the 2.5 m width is

$$P_s = \frac{wL}{2} = 9.5(2.5) = 23.75\ \text{kN/m}$$

Let the self-weight of the wall for 300 mm average thickness be

$$P_g = 5.4(1)(0.3)(25) = 40.5\ \text{kN/m}$$

Total axial force $P = P_s + P_g = 64.25\ \text{kN/m}$

Use $P = 65\ \text{kN/m} = 0.065\ \text{MN/m}$

At the bottom of the wall the bending causes tension in the concrete on the water face, so the design of the section is controlled by

$$\frac{M}{Z} - \frac{P}{A} \leq \sigma_{cbt}$$

or

$$\frac{M}{Z} - \frac{P}{A} = \frac{6M}{t^2} - \frac{P}{t} \leq \sigma_{cbt}$$

The bending moment and the axial force at the base are

$$M = (- M_{yb}) = 0.09116 \text{ MNm/m}$$
$$P = 0.065 \text{ MN/m}$$

Therefore, substituting the above quantities in the governing equation, we get

$$\frac{6(0.09116)}{t^2} - \frac{0.065}{t} \le 1.7$$

or
$$1.7t^2 + 0.065t - 6(0.09116) \ge 0$$
or
$$t \ge 0.265 \text{ m}$$

The average thickness assumed was 300 mm as against a desired value of 265 mm. A reduction in axial load will increase the thickness of the wall. Therefore, assume the thickness of the wall at the base as

$$t = 300 \text{ mm}$$

The effective thickness of the wall is also obtained for comparison by the cracked theory. The balanced moment coefficients of a cracked section with allowable tensile stress in steel as 150 MPa are taken from Table 11.4.

$$j = 0.872, \ K = 0.167.$$

Neglecting the effect of the axial force, we have

$$d = \sqrt{\frac{M}{Kb\sigma_{cb}}} = \sqrt{\frac{0.09116}{0.167(1)(7)}} = 0.279 \text{ m}$$

Providing a clear cover of about 25 mm, the overall thickness of the wall needed is about 310 mm. However, the smaller value is selected.

Let
$$t = 300 \text{ mm} \ \text{ and } \ d = 270 \text{ mm}$$

then the area of the tension reinforcement is

$$A_{st} = \frac{M}{jd\sigma_{st}} = \frac{91160}{0.872(0.27)(150)} = 2581 \text{ mm}^2/\text{m}$$

Provide 20 mm *bars at* 120 mm *spacing* on the inner face at the bottom of the wall. As the bending moment at the mid height of the tank is positive, all the bars can be curtailed or cranked at mid-height. The development length of the bars is

$$L_d = \frac{\phi\sigma_{st}}{4\tau_{bd}} = \frac{150\,\phi}{4(1.2)(1.6)} = 20\phi = 400 \text{ mm}$$

The development length is small so the curtailment of the bars at mid-height is satisfactory.

Reinforcement at mid-height: The bending moment on the vertical fibres at mid-height of the tank is

$$M_{yc} = 43\,410 \text{ Nm/m}$$

The thickness of the wall at the top is assumed to be 150 mm and it is increased to 300 mm at the base. So the thickness at the mid-height is 0.5(150 + 300) = 225 mm. the corresponding effective depth to the reinforcement is 225 − 35 = 190 mm. Therefore,

$$t_c = 225 \text{ mm}, \ d = 190 \text{ mm}$$

The area of the tension reinforcement at the mid-height (on the outer fibres) is

$$A_{st} = \frac{M_{yc}}{0.904\,(0.19)(230)} = 1099 \text{ mm}^2/\text{m}$$

The alternate negative reinforcement bars are cranked, so the net reinforcement needed is only nominal.

Design of the horizontal reinforcement: The positive bending moment on the horizontal fibres of the wall is zero; therefore, only nominal reinforcement be provided. However, there is negative bending moment at the corners of the wall. Its value at about the mid-height of the wall is

$$M_{xa} = -28940 \text{ Nm/m}$$

There is an axial tension coming from the hydrostatic force acting on the walls normal to this wall. This force, at mid-height is of $H/2$ width of the wall. Therefore, the tension force can be taken approximately as

$$T = \frac{\gamma H^2}{2} = \frac{10(5.25)^2}{2} = 13.8 \text{ kN/m}$$

The tensile stress due to the combined action on the vertical plane of the wall is

$$\sigma_t = \frac{M}{Z} + \frac{T}{A} = \frac{6M}{t^2} + \frac{T}{t}$$

The thickness of the wall at mid-height is 225 mm so

$$\sigma_t = \frac{6(28\,940)}{225(225)} + \frac{13800}{225(1000)} = 3.43 + 0.06 = 3.49 \text{ MPa}$$

The axial tension contribution is negligible but the tension caused by bending moment exceeds the allowable value of 1.7 MPa. Therefore, a fillet be provided at the corners so as to reduce the tension.

Provide 150 mm thick haunch at the corners of the wall. The overall thickness at the joint including the fillet is 225 + 150 = 375 mm and the bending stress is

$$\sigma_t = \frac{6M}{t^2} = \frac{6(28\,940)}{375(375)} = 0.5 \text{ MPa} < \sigma_{cbt}$$

The effective depth of the section at this point can be taken as 375 − 100 = 275 mm and the area of the tension reinforcement is

$$A_{st} = \frac{M}{jd\,\sigma_{st}} = \frac{28940(1000)}{0.872(275)(150)} = 525 \text{ mm}^2/\text{m}$$

The minimum reinforcement for 225 mm thickness is about

$$A_{sm} = \frac{0.2}{100}\,(225)\,(100) = 450 \text{ mm}^2/\text{m}$$

Provide 12 mm bars at 250 mm spacing on the inner face of the wall and then add extra bars as shown in Fig. 11.4.

Design of the base slab: The bottom slab is resisting on the soil and it supports the water and the columns. The weight of the water is directly transferred to the soil; therefore, the bearing capacity has to be checked. The load from the column is transferred through the

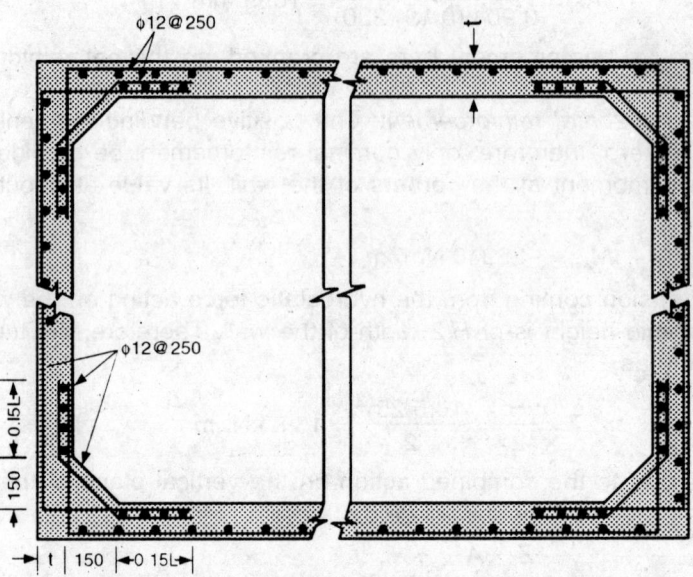

Fig. 11.4 Reinforcement Details in Walls in Plan.

bottom slab. The slab need to be designed for tank empty condition as a two-way slab subjected to net pressure from the soil. First the bearing pressure on the soil is computed.

Load from the roof	$= 9.5$ kN/m^2
Load of the columns, etc.	$= 1.5$ kN/m^2
Load of the water	$= 52.5$ kN/m^2
Load of the bottom slab	$= 6.0$ kN/m^2
(It is assumed 240 mm thick)	
Total load	$= 69.5$ kN/m^2

The bearing pressure is 69.5 kN/m^2 as against a safe bearing capacity of 75 kN/m^2. Therefore, the load is within the limits.

Structural design of the bottom slab: The net bearing pressure on the soil in the tank empty condition is that due to the roof load and the column load and it is

$$w = 9.5 + 1.5 = 11 \text{ kN/m}^2$$

The bottom slab is designed as a flat slab subjected to the soil pressure and supported by the columns spaced at 5 m apart in tank empty condition. Assume a widening of the column as shown in Fig. 11.3. The widened size of the column base be same as the column head and it is 440 mm square.

Effective span of the slab $L = 5$ m
Clear span $L_0 = L - a = 4.56$ m

The design of the bottom slab is similar to that of the roof slab except that this slab has to be designed for a load of 11 kN/m^2 less the weight of slab. The depth of the slab is proportioned to square root of the moments.

The reinforcement of the bottom slab is also proportioned with respect to an allowable stress of 150 MPa and the depth. They are shown in Fig. 11.3. The bearing pressure on the soil with 280 mm thickness of the bottom slab is within the limits.

The wall is resting on the slab with a cantilever of 500 mm; therefore, the bending moment from the wall is to be distributed between the cantilever and the inside of the slab. For all practical purposes, the base slab can be treated as fixed because of the load of the water on it with equal soil reaction. Therefore, additional reinforcement must be provided at the edges to resist the bending moment and the tension from the hydrostatic force.

The hydrostatic force at the base is

$$T = \frac{3}{4}\left(\frac{\gamma H^2}{2}\right) = \frac{3}{8}(10)(5.25)^2 = 104 \text{ kN/m}$$

Area of the tension steel in the bottom slab is

$$A_s = \frac{T}{\sigma_{st}} = \frac{104000}{150} = 693 \text{ mm}^2/\text{m}$$

The reinforcement details are shown in Fig. 11.3.

EXAMPLE 11.2 *Design of a large water tank with counterforts.* A reservoir of 31 Ml is to be designed as a rectangular underground water tank. The soil is clayey type whose density can be taken as 16 kN/m^3 and angle of repose is 27°. The ground area available is 55 m by 85 m and the top of the tank is to be 500 mm below the soil. The tank is to be designed in two equal compartments. Use $f_{ck} = 20$ MPa, and $f_y = 415$ MPa.

Design data and main dimensions

Capacity of the water tank	= 31 Ml = 31000 m^3
Ground area available	= 55 m by 85 m
Let the inside clear dimensions	= 50 m by 80 m
Inside gross area	= 50 × 80 = 4000 m^2

Let the tank be rectangular and the roof is supported by column spaced at 5 m apart in both the directions and a partition wall in the middle. Let the size of the column be 400 mm square for volume computations. The number of columns are = 9 × 15 = 135

Area occupied by the columns	= 135(0.16) = 21.6 m^2
Area occupied by the middle wall	= 0.4(50) = 20 m^2
Net area available for water	= 4000 − 20 − 21.6 = 3958.6 m^2
Head of water	$= \dfrac{31.000}{3958.6} = 7.831$ m

Use head of water 7.85 m

Overall height inside the tank   = 8.1 m

Where the free board made available is = 0.25 m

The height of the tank is rather large. Therefore, it is desirable to design the vertical walls with counterfort.

476

The design of the water tank consists of:

1. Design of roof slab as a flat slab with columns spaced at 5 m bothways.
2. Design of the columns.
3. Design of the wall with counterforts at 5 m.
4. Design of middle wall.
5. Design of the base slab as a flat slab.

The water tank is constructed below the ground level by replacing the soil. The unit weight of the soil is 16 kN/m^3 whereas the weight of the water is 10 kN/m^3. The load at the base of the soil is going to be less than the weight of the soil which was replaced. Therefore, the bearing pressure on the soil is going to be less than the safe allowable value.

The allowable stresses in MPa are:

$$\sigma_{cb} = 7, \quad \sigma_{cat} = 1.2, \quad \sigma_{cbt} = 1.7$$
$$\sigma_{st} = 230 \text{ away from water face}$$
$$\sigma_{st} = 150 \text{ on the water face}$$

Design of the roof slab (flat slab). The reader should see Chapter 4 for design of flat slab for notations.

The effective span in both the directions $L = 5$ m
Column head is assumed to be $\quad a = 0.44$ m
Clear span of the panel $\quad L_0 = L - a = 4.56$ m
Live load on the roof $\quad = 1.5$ kN/m^2
Water proof treatment etc. $\quad = 2.0$ kN/m^2
Weight of the soil $= (16)\,(0.5) \quad = 8.0$ kN/m^2
Self-weight $= 0.26\,(25) \quad = 6.5$ kN/m^2
(let $t = 0.26$ m)

$$\text{Total load } w = \overline{18.0 \text{ kN/m}^2}$$

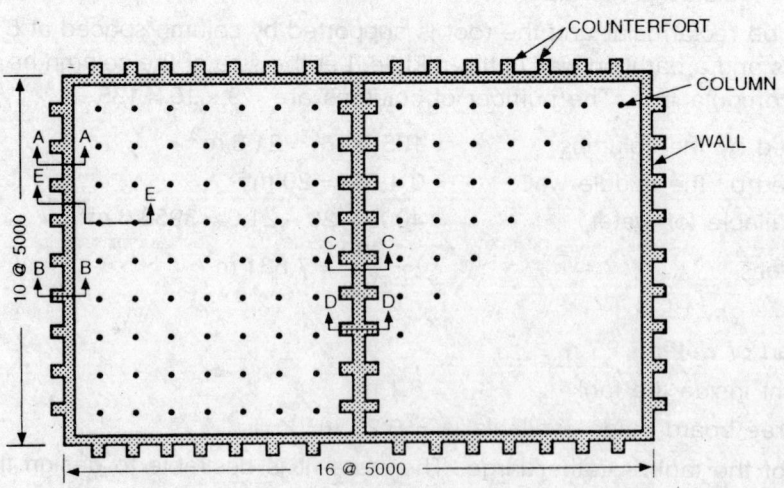

Fig. 11.5 Arrangement of Columns, Walls and Abutments. Example 11.2.

Design load on a panel $W = wLL_0 = 18(5)(4.56) = 410.5$ kN. The sum of the magnitudes of the positive and negative bending moments in a panel are:

$$M_0 = \frac{WL_0}{8} = \frac{410.5(4.56)}{8} = 234 \text{ kNm}$$

Clear height of the column $= 8.1$ m
Effective height of the wall $= 0.7(8.1) = 5.61$ m

Relative stiffness of the wall is

$$K_w = \frac{I_w}{L_w} = \frac{5(0.2)^3}{12(5.61)} = 5.95(10^{-4})$$

where 0.2 m is assumed as the average thickness of the wall. The relative stiffness of the slab is

$$K_s = \frac{I_s}{L_s} = \frac{Lt^3}{12L} = \frac{0.26^3}{12} = 14.6(10^{-4})$$

$$\alpha_c = \frac{K_w}{K_s} = \frac{5.5}{14.6} = 0.38$$

$$\alpha = 1 + 1/\alpha_c = 3.63$$

The negative and positive bending moment coefficients in the exterior and in the interior panels are computed using Table 4.7. They are

$$c_1 = \frac{0.65}{\alpha} = \frac{0.65}{3.63} = 0.018$$

$$c_3 = 0.63 - \frac{0.28}{\alpha} = 0.55$$

$$c_2 = 0.75 - \frac{0.1}{\alpha} = 0.722$$

$$c_i = 0.65$$

The corresponding magnitudes of the bending moments (in MNm) in a panel are:

$$M_{n1} = c_1 M_0 = 0.0042$$
$$M_{p3} = c_3 M_0 = 0.1287$$
$$M_{n2} = c_2 M_0 = 0.169$$
$$M_{ni} = c_i M_0 = 0.1521$$

The bending moments on the columns strip are computed using the coefficients given Table 4.6 and are (in MNm)

$$M_{nc1} = M_{n1} = 0.0042$$
$$M_{pc3} = 0.6 \, M_{p3} = 0.0772$$
$$M_{nc2} = 0.75 \, M_{n2} = 0.1268$$
$$M_{nci} = 0.75 \, M_{ni} = 0.1141$$

The effective depth of the steel is computed from the maximum of the magnitudes of the bending moments, and it is

$$d = \sqrt{\frac{M}{Kb\sigma_{cb}}} = \sqrt{\frac{M_{nc2}}{0.13(2.5)(7)}} = 0.236 \text{ m}$$

The total depth of the slab can be taken as

$$t = d + 0.029 = 0.265$$

The self-weight of the slab is 6.63 kN/m² as against the assumed 6.5 kN/m². The total design load is 18.13 instead of 18 kN/m². The increase in the bending moments is about 0.7% and so the reinforcement is adjusted to the higher side.

Let the design load $w = 18.25$ kN/m²

Design for shear force. The allowable shear stress based on the diagonal tension is

$$\tau_c = 0.16 \sqrt{f_{ck}} = 0.71 \text{ MPa}$$

The critical shear plane is the peripheral plane which is at a distance $0.5d$ from the face of the column. The length of the plane is

$$b_0 = 4(a + b) = 4(0.44 + 0.236) = 2.704 \text{ m}$$

The shear force on the plane is

$$V = w L^2 - (a + b)^2) = 18.25(25 - 0.676^2) = 448 \text{ kN}$$

The nominal shear stress is

$$\tau_v = \frac{V}{b_0 d} = \frac{0.448}{2.704(0.236)} = 0.7 \text{ MPa} < \tau_c$$

The depth of the slab is just adequate to resist the shear force.

Design of the reinforcement in the column strip. The required areas of the reinforcements at different sections for 2.5 m width slab are:

$$A_{st1} = \frac{M_{nc1}}{jd\sigma_{st}} = \frac{4200}{0.904(0.236)(230)} = 86 \text{ mm}^2$$

$$A_{st2} = \frac{126\,800}{0.904(0.236)(230)} = 2584 \text{ mm}^2$$

$$A_{st3} = \frac{77\,200}{0.904(0.236)(230)} = 1573 \text{ mm}^2$$

$$A_{sti} = \frac{114\,100}{0.904(0.230)(230)} = 2325 \text{ mm}^2$$

The minimum area required is

$$A_{stm} = \frac{0.12}{100}(bt) = 0.0012(2500)(265) = 795 \text{ mm}^2$$

Direct tension coming from the counterfort which is treated as hinged at top and distributed over 2.5 m width is

$$T = \frac{L}{4}\frac{(\gamma H^2)}{2} = \frac{5(10)(7.85)^2}{8} = 385 \text{ kN}$$

Extra tension reinforcement in the column strip is

$$A_{st} = \frac{T}{\sigma_{st}} = \frac{385\,000}{230} = 1674 \text{ mm}^2$$

This amount of extra reinforcement is placed equally at top and bottom in the column strip.

Design of columns. Load from the roof $= wL^2 = 18.25(25) = 456.25$ kN

Let the self-weight $\qquad\qquad\qquad\qquad\qquad\qquad\qquad = 39.75$ kN

$$\text{Total load } P = 495 \text{ kN}$$

Assume only 0.8% reinforcement in the column, then its capacity is

$$P_a = A_c \sigma_{cc} + A_s \sigma_{sc} = A_g((1-p)\sigma_{cc} + p\sigma_{sc})$$
$$= A_g(0.992(5) + 0.008(190)) = 6.48\, A_g > P$$

Therefore, $\qquad\qquad A_g > \dfrac{P}{6.48} = \dfrac{0.495}{6.48} = 0.0764 \text{ m}^2$

Use size of the column as 300 mm by 300 mm. Then the slenderness factor of the column is

$$g = \frac{L_c}{a_0} = \frac{0.65(8.1)}{0.3} = 18$$

The slenderness factor is more than 12, therefore, the column is to be treated as a long one. The load reduction factor to the column is

$$C_r = 1.25 - \frac{18}{48} = 0.875$$

The equivalent short column load is

$$P_o = \frac{P}{C_r} = \frac{0.495}{0.875} = 0.566 \text{ MN}$$

From Table 5.4, the capacity of a column with 900 cm^2 area and 4 numbers of 16 mm bars in the main reinforcement is

$$P_a = 446 + 152 = 598 \text{ kN} > P_0$$

Therefore, *provide 300 mm* square column with 4 numbers of 16 mm bars and 6 mm ties at 250 mm spacing. Let there be a square column head of 440 mm.

Design of the wall. The wall is to be designed for two critical load conditions: (1) Full hydrostatic pressure with no earth pressure and (2) tank empty with full earth pressure load. The vertical wall is supported by counterforts spaced at 5 m apart and connecting top and bottom slabs. It is designed as a continuous slab over four edges and subjected to hydrostatic load. The bending moments at critical points in the plate can be expressed as

$$M_{yb} = c_{yb} w_0 H^2 \qquad\qquad\qquad\qquad (11.21)$$
$$M_{yc} = c_{yc} w_0 H^2 \qquad\qquad\qquad\qquad (11.22)$$
$$M_{xa} = c_{xa} w_0 H^2 \qquad\qquad\qquad\qquad (11.23)$$
$$M_{xc} = c_{xc} w_0 H^2 \qquad\qquad\qquad\qquad (11.24)$$

where the points A, B and C are indicated in Fig. 11.2, the moment coefficients are given in Table 11.6 and $w_0 = \gamma H = $ maximum pressure at base. Let the width of the counterfort $b_w = 0.4$ m.

In the present case

$$\frac{H_0}{L_0} = \frac{8.1}{4.6} = 1.76$$

where $\qquad\qquad\qquad\qquad L_0 = $ clear span in which the thickness of the counterfort is assumed

Table 11.6 Bending Moment Coefficients on a Plate with all Edges Fixed and Subjected to Hydrostatic Load

H/L	c_{yb}	c_{yc}	c_{xa}	c_{xc}
0 – 0.5	– 0.05	0.006	– 0.033	0.021
0.67	– 0.046	0.010	– 0.030	0.018
1.0	– 0.033	0.012	– 0.018	0.012
1.5	– 0.019	0.008	– 0.007	0.005
2.0	– 0.012	0.005	– 0.003	0.002

as 0.4 m.

Design for full water head and no earth pressure out side:

The intensity of load $\quad = w_0 = \gamma H = 10(7.85) = 78.5 \text{ kN/m}^2$

$$w_0 H^2 = 0.0785(7.85)^2 = 4.837 \text{ MNm/m}$$

The corresponding bending moments are computed using the moment coefficients of Table 11.6 (all moments in MN m/m)

$$M_{yb} = -0.016\, w_0 H^2 = -0.016(4.837) = -0.0774$$
$$M_{yc} = 0.007\, w_0 H^2 = 0.007(4.837) = 0.0339$$
$$M_{xa} = -0.005\, w_0 H^2 = -0.005(4.837) = -0.0242$$
$$M_{xc} = 0.0035\, w_0 H^3 = 0.0035(4.837) = 0.0169$$

The maximum magnitude of the bending moment which occurs at the base is

$$M = 0.0774 \text{ MNm/m}$$

This being negative with small axial force present, the tensile stress occurs on the water face, therefore, the design criterion is

$$\sigma = \frac{M}{Z} \le \sigma_{cbt} = 1.7 \text{ MPa}$$

or $\qquad\qquad Z \ge \dfrac{M}{1.7} = \dfrac{0.0774}{1.7}$

or the thickness of the wall at base is

$$t_b \ge \sqrt{\frac{6(0.0774)}{1.7}} = 0.523 \text{ m}$$

The thickness of the wall at mid-height is controlled by M_{xa} and the thickness is

$$t_a \ge \sqrt{\frac{6(0.0242)}{1.7}} = 0.292 \text{ m}$$

The thickness of the slab at the base is taken as 525 mm and reduced to 125 mm at top. By this way, the thickness at mid-height of the wall is

$$t_a = 0.5(525 + 125) = 325 \text{ mm}$$

which is more than the desired 0.292 m.

As the thickness of the wall is gradually reduced, the spacing of the horizontal

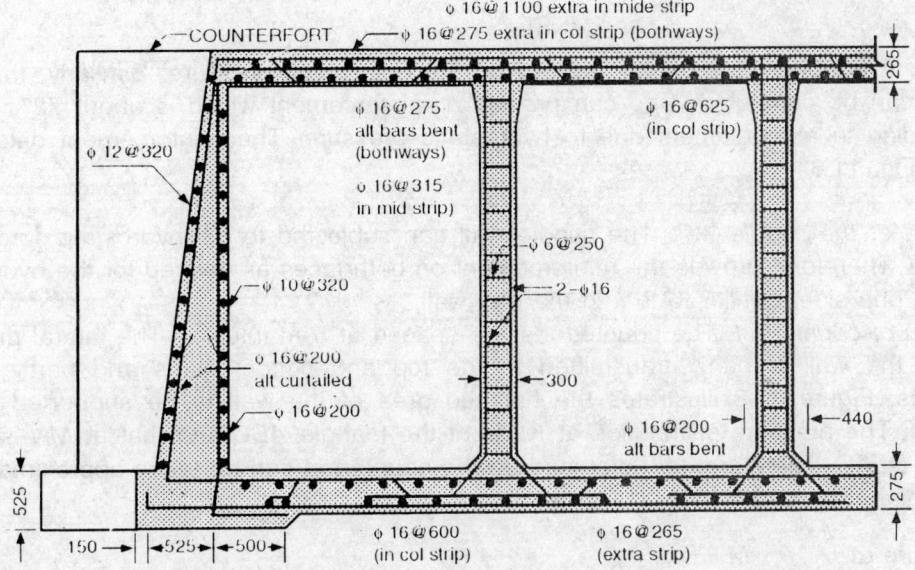

φ 16 @ 1100 extra in mide strip

φ 16 @ 275 extra in col strip (bothways)

COUNTERFORT

φ 16 @ 275
alt bars bent
(bothways)

φ 16 @ 315
in mid strip)

φ 16 @ 625
(in col strip)

φ 12 @ 320

φ 6 @ 250

φ 10 @ 320

2 – φ16

φ 16 @ 200
alt curtailed

300

φ 16 @ 200

φ 16 @ 200
alt bars bent

440

525

150

525

500

275

φ 16 @ 600
(in col strip)

φ 16 @ 265
(extra strip)

Fig. 11.6 Reinforcement Details in Column and Walls etc. Example 11.2.

reinforcement can be maintained constant. The area of the reinforcement needed at the base of the wall and at the mid-height are:

$$A_{sty} = \frac{M_{yb}}{jd\,\sigma_{st}} = \frac{77400}{0.872(0.49)(150)} = 1207 \text{ mm}^2/\text{m}$$

$$M_{stx} = \frac{M_{xa}}{jd\sigma_{st}} = \frac{24200}{0.872(0.30)(150)} = 597 \text{ mm}^2/\text{m}$$

Bending moment due to earth pressure. The wall must be designed for the active earth pressure acting from outside when there is no water inside. The active earth pressure intensity at the base is computed as

$$w_s = \frac{(1 - \sin \phi)}{1 + \sin \phi}\,\gamma_s H_s \qquad\qquad (11.25)$$

where

$\gamma_s = 16 \text{ kN/m}^3 = \text{density of soil}$

$H_s = \text{head of the soil} = 8.1 + 0.265 + 0.5 = 8.865 \text{ m}$

$\phi = \text{angle of repose} = 27°$

The earth pressure distribution is very similar to that of the hydrostatic pressure distribution. The active earth pressure at the base of the wall is

$$w_s = \frac{(1 - \sin 27)}{1 + \sin 27}\,16(8.867) = 53.26 \text{ kN/m}^2$$

The corresponding hydrostatic pressure was $w_0 = 78.5 \text{ kN/m}^2$. The bending moments caused by the earth pressure are almost similar to those by the hydraulic pressure. The bending moment can be computed as

$$M_{yb} = -0.016\,w_sH_0^2 = -0.016(53.26)(8.1)^2 = -55.9\text{ kNm/m}$$

$$= -0.0559\text{ MNmm}$$

The moment is about 72% of that caused by the hydrostatic pressure. Similarly, the other moments can be computed. One can provide a reinforcement which is about 72% of that corresponding to reinforcement due to hydrostatic pressure. The reinforcement details are shown in Fig. 11.6.

Design of the middle wall. The middle wall can subjected to full hydrostatic force from either face. Therefore, provide the reinforcement on both faces as desired for the hydrostatic force from one side. Similar to that in the end wall.

Design of counterfort. The counterforts are spaced at 5 m interval. The lateral pressure acting on the wall is partly transmitted to the top and bottom slabs and partly to the counterforts. Figure 11.8 illustrates the hatched area of the wall to be supported by the counterfort. The pressure distribution at KL is of the triangle ABC and that at MN is of the trapezium DEFG. Therefore, the total load on the counterfort can be taken approximately as a mean value. The details are

Pressure at $\qquad M = DG = \dfrac{\gamma L}{2}$

Pressure at $\qquad N = EF = \left(H - \dfrac{L}{2}\right)\gamma$

The total pressure on section MN is

$$P_1 = 0.5(DG + EF)MN$$

$$= 0.5\left(\frac{L}{2} + H - \frac{L}{2}\right)\gamma(H - L) = \frac{\gamma H}{2}(H - L)$$

The total pressure on section KL is

$$P_2 = \frac{\gamma H^2}{2}$$

The total pressure on the counterfort can be approximated as:

$$P = \frac{L(P_1 + P_2)}{2} = \frac{\gamma H}{4}(H - L + H)L = \frac{\gamma H}{4}(2H - L)L \qquad (11.26)$$

The counterfort can be treated as partially constrained at both ends and subjected to a lateral force of P. The maximum bending moment on the counterfort can be estimated as

$$M = \frac{PH_e}{9} \qquad (11.27)$$

where H_e = effective height of the counterfort

Let $\quad H_e = H_o + 0.3 = 8.4\text{ m}$

The substitution of different values in the equations gives

$$P = \frac{\gamma H}{4}(2H - L)L = \frac{10(7.85)}{4}(15.7 - 5)(5) = 1050\text{ kN}$$

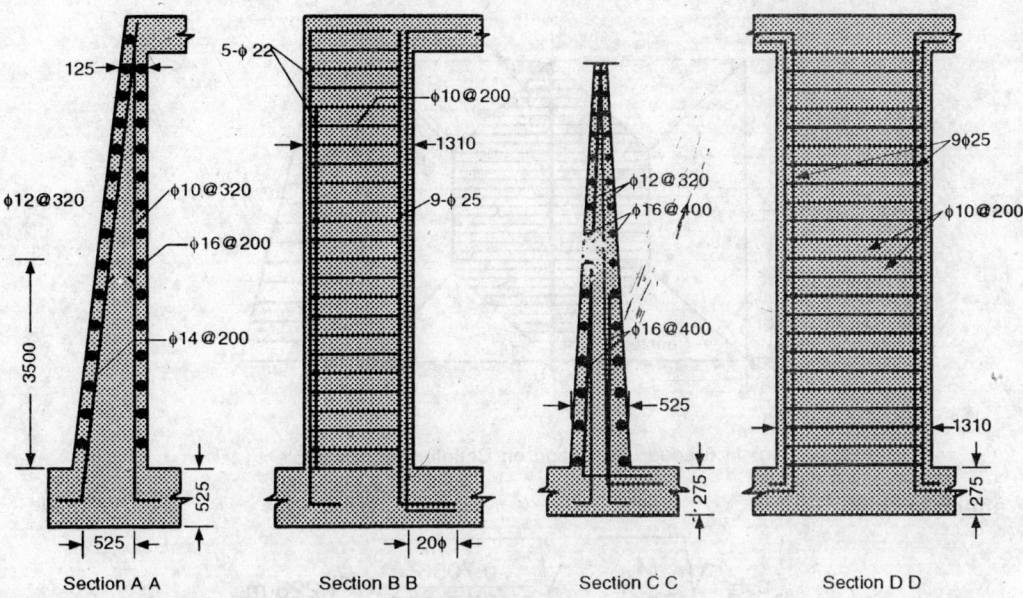

Fig. 11.7 Reinforcement Details of Walls and Abutments, Example 11.2.

$$M = \frac{PH_e}{9} = (1050) \frac{(8.4)}{9} = 980 \text{ kNm} = 0.98 \text{ MNm}$$

The bending moment is maximum at about $0.6H$ from the top of the tank. The counterfort acts as a T-beam. Therefore, the thickness and the width of the flange are:

$$t = 0.35 \text{ m (wall thickness at mid height)}$$
$$b_f = 12t + b_w = 4.6 \text{ m}$$

The effective depth of the beam is (from Eq. 3.13)

$$d_1 = \frac{t}{2n} + \frac{M}{b_f t j \sigma_{cb}}$$

$$= \frac{0.35}{2(0.289)} + \frac{0.98}{4.6(0.35)(0.904)(7)} = 0.702 \text{ m}$$

The earth pressure acting on the outer face of the wall causes a force or bending moment which is equal to 72% of that due to hydrostatic. The bending moment due to earth pressure causes tension on the water face and the section will be rectangular one. Therefore, the counterfort is to be designed as a rectangular section with allowable tension in the steel as 150 MPa. The corresponding balanced section design coefficients are taken from Table 11.4. These are:

$$j = 0.872 \quad \text{and} \quad K = 0.167$$

The bending moment caused by the earth pressure is prorated and it is

$$M_s = 0.72 M = 0.72(0.98) = 0.7056 \text{ MNm}$$

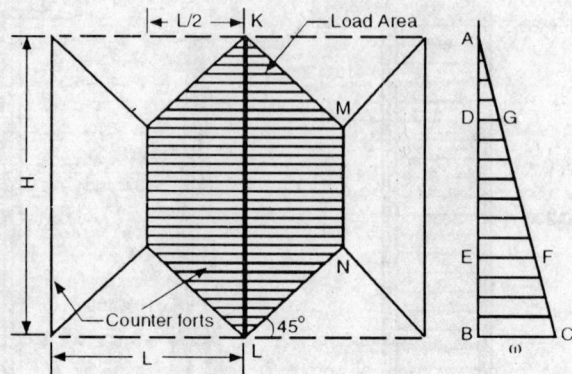

Fig. 11.8 Load Distribution on Counterforts, Example 11.2.

The effective depth is:

$$d_2 = \sqrt{\frac{M_s}{Kb\sigma_{cb}}} = \sqrt{\frac{0.7056}{0.167(0.4)(7)}} = 1.228 \text{ m}$$

To keep the reinforcement to a minimum level and to satisfy both the load conditions, select

$$d = 1.25 \text{ m}$$

The total depth and the depth of the web are:

$$h = d + 0.06 = 1.31 \text{ m}$$
$$h_w = h - t = 1.31 - 0.525 = 0.785 \text{ m}$$

The area of the reinforcements on the outer and inner faces are:

$$A_{sto} = \frac{M}{jd\sigma_{st}} = \frac{980\,000}{0.904(1.25)(230)} = 3770 \text{ mm}^2$$

$$A_{sti} = \frac{M_s}{jd\sigma_{st}} = \frac{705\,600}{0.872(1.25)(150)} = 4315 \text{ mm}^2$$

Provide 10 numbers of 22 mm bars on the outer face and 9 numbers of 25 mm bars on the inside face of the counterfort.

Design of counterfort of the middle wall. Since the middle wall is subjected to hydrostatic pressure from either face or on both faces, the total depth of the counterfort is taken as 1.3 m and the area of the reinforcement on each face is that corresponding to that on the water face. That is 9 numbers of 25 mm bars on each face.

Design for shear. The shear force at the bottom of the counterfort is

$$V = 0.6 \; P = 0.6(1.05) = 0.63 \text{ MN}$$

The nominal shear stress is

$$\tau_v = \frac{V}{bd} = \frac{0.63}{0.4(1.25)} = 1.25 \text{ MPa}$$

The percentage of tension steel is

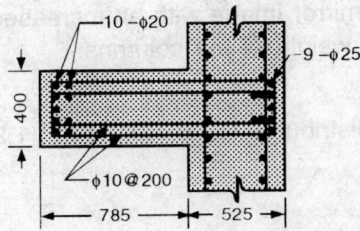

Fig. 11.9 Outer Wall Counterfort, Example 11.2.

$$\frac{100 A_{st}}{bd} = \frac{100(10)(380)}{400(1250)} = 0.76$$

The allowable shear stress is

$$\tau_c = 0.35 \text{ MPa}$$

Use four-legged 10 mm stirrups, then

$$A_{sv} = 4(78.5) = 314 \text{ mm}^2$$

The spacing of the stirrups is

$$s_v = \frac{A_{sv}\sigma_{sv}}{b(\tau_v - \tau_c)} = \frac{314(230)}{400(1.25 - 0.35)} = 200 \text{ mm}$$

Provide 4-legged 10 mm stirrups at 200 mm spacing.

Design of the base slab. The base slab or the bottom slab supports the wall and the columns, and rests on soil. The weight of the water is directly transferred to the soil by bearing. The axial force and the bending moment from the columns or the wall are partly transferred to the bottom slab partly and through a couple from the differential bearing pressure. The counterfort is assumed partially constrained at the bottom and top. Therefore, it transmits some bending moment to the slab. The horizontal tensile force between the abutment and the base slab which is developed by the lateral thrust of water need to be resisted by tension reinforcement near the end walls. This tension in the slab is transmitted to the group through frictional force. Therefore, the bottom slab be designed for the following:

1. Soil reaction on the panels when the tank is empty.
2. The bending moment from the walls.
3. Tension from the walls.

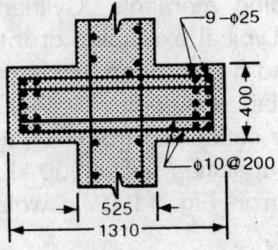

Fig. 11.10 Middle Wall Counterfort, Example 11.2.

Thickness of the wall is governed by the bending moment in a panel. The bottom slab acts very similar to the roof slab as a mirror image with an increased load intensity. The increase in the load intensity is due to the weight of the columns.

Load from the roof $= 18.25$ kN/m^2

Load of the columns evenly distributed is $8.1(0.09)(25)(1.2)/25$ $= 0.88$ kN/m^2

Intensity of soil reaction $= 19.13$ kN/m^2

Use the load as 19.25 kN/m^2

The roof slab is designed for a load of 18.25 kN/m^2 whereas the bottom slab is to be designed for 19.25 kN/m^2. The depth of the slab can be proportioned to square root of the ratio of the loads and similarly the area of the reinforcement.

$$t_b = t_r \sqrt{\frac{19.25}{18.25}} = 265(1.027) = 272 \text{ mm}$$

where $t_b =$ thickness of the bottom slab, $t_r =$ thickness of roof slab

Use $t_b = 275$ mm.

The reinforcement details of this slab are obtained by dividing the spacing of the reinforcement in the roof slab by 1.027. These details are shown in Fig. 11.6.

The reinforcement in the vertical wall is continued into the slab to resist the bending moment from the wall. Additional reinforcement must be provided at the counterforts to resist the tension force.

Tension from the counterfort $= T = 0.6P = 0.63$ MN

Area of the reinforcement is

$$A_s = \frac{630\,000}{150} = 4200 \text{ mm}^2$$

9 *numbers* of 25 mm bars at the inner face counterforts be continued into the bottom slab to a length of $20\phi = 500$ mm.

11.4 Design of Cylindrical Water Tanks

A circular cylinder when subjected to a radial force will develop hoop force in the shell. If the force is applied from inside, then there is hoop tension and if applied from outside, there is hoop thrust in the shell. The bending moments in the shell are generated either due to boundary effects or due to unsymmetrical loads. A proper selection of the boundary connections can minimize the bending moments. Cylindrical water tanks are commonly employed in overhead water tanks. Typical examples of the tanks are shown in Fig. 11.11. The shapes shown in Figs. 11.11a and b are commonly used for small capacity tanks of upto 100 kl. The roof and the bottom slab are of flat slab or of beam and slab construction. Figures 11.11c to e are employed for capacities from 100 kl to 700 kl and shapes shown in Figs. 11.11f and g can be used for capacities upto 3000 kl. For capacities higher than 2500 kl one can resort to the tanks shown in Fig. 11.12 in which large number of columns are provided.

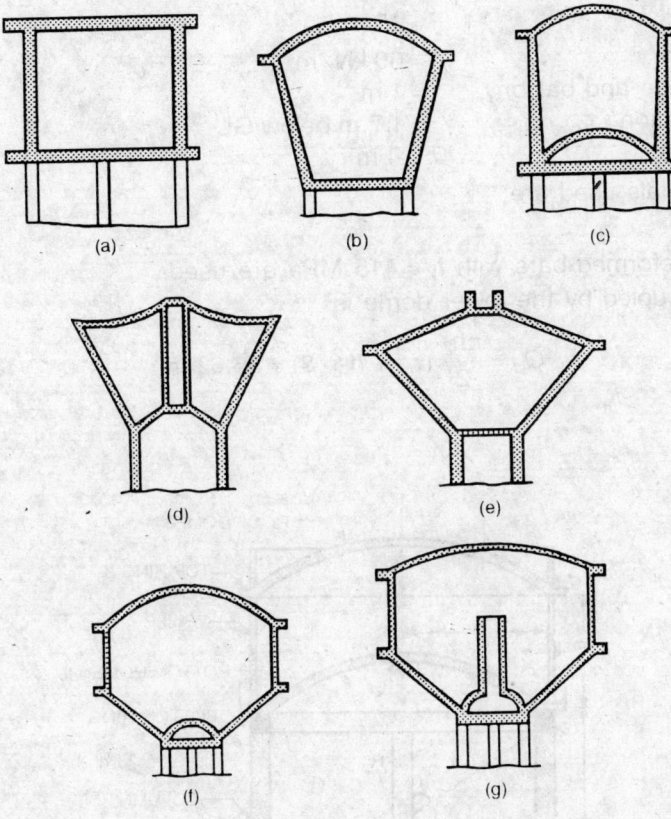

Fig. 11.11 Typical Shapes of Overhead Water Tanks.

EXAMPLE 11.3 *Design of* 125 kl *water tank:* A water tank of 125 kl capacity with 15 m staging is to be designed.

Design data and capacity calculations: Figure 11.13 gives the general arrangement of the various structural elements. The main elements of design are: Roof dome, top ring beam, wall, bottom ring beam, columns, and foundation.

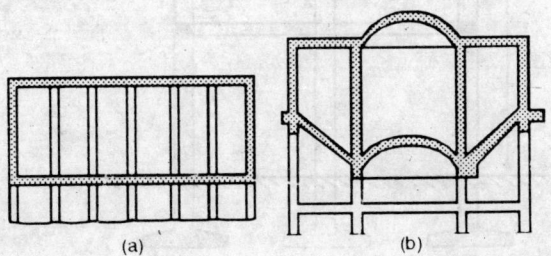

Fig. 11.12 Multiple Columns in Water Tank.

Tank capacity   = 125 kl
Staging   = 15 m
Bearing capacity   = 60 kN/m^2
Width of staircase and balcony   = 1 m
Depth of foundation   = 1.7 m below GL
Let inside radius $r = 4$ m

Concrete and materials used are:

M20 concrete

High strength deformed bars with $f_y = 415$ MPa are used.

The volume occupied by the lower dome is

$$Q_d = \frac{\pi h_2}{2}(r^2 + h_2^2/3) = 36.62 \text{ m}^3$$

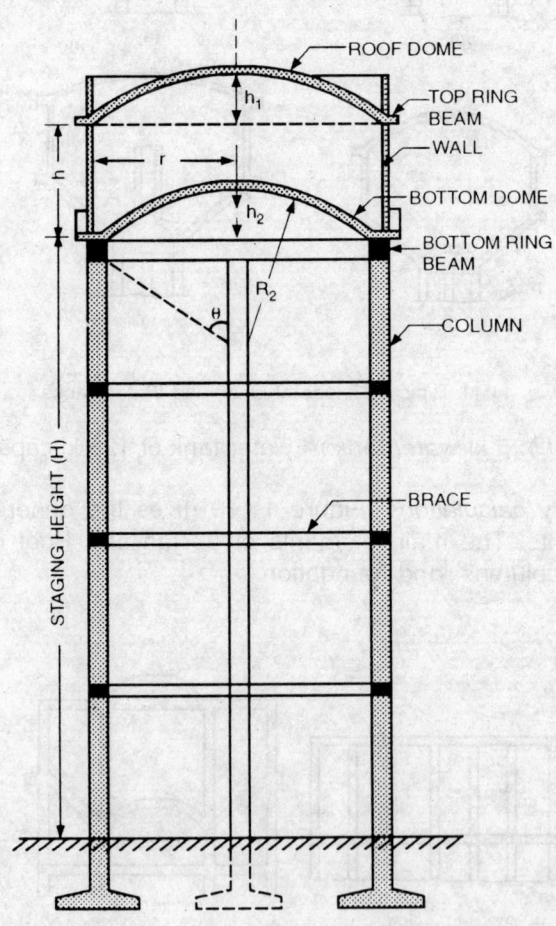

Fig. 11.13 Notation in OHWT, Example 11.3.

where $\qquad h_2$ = rise of the bottom dome = 1.4 m

Surface area of the dome

$$S = \pi(r^2 + h_2^2) = 56.42 \text{ m}^2$$

Assume free board $\qquad = 0.15$ m

Volume of the water $\qquad = \pi r^2(h - 0.15) - Q_d = 125 \text{ m}^3$

where h = height of the vertical wall

Therefore, overall height of the cylindrical wall works out to be

$$h = (125 + 36.62)/\pi r^2 + 0.15 = 3.36 \text{ m}$$

Design of roof dome: Radius of the chord and height of the dome are:

$r = 4m$, $h_1 = 1.4$ m, then the radius of the dome is given by

$$r^2 = (2R_1 - h_1)h_1$$

or $\qquad\qquad 4^2 = (2R_1 - 1.4)\, 1.4$ then

$$R_1 = 6\,414 \text{ m}$$

and semicentral angle $\qquad \theta = 38.6°$ (from $\sin \theta = r/R_1$)

Let the thickness of the dome be 80 mm.

The intensity of self-weight $\qquad = 0.08(25) = 2 \text{ kN/m}^2$

Live load $\qquad\qquad = 0.75 \text{ kN/m}^2$

Total design load $\qquad q = 2.75 \text{ kN/m}^2$

Total dome load $\qquad W_1 = 2.75\ S = 155$ kN

$$\cos\theta = 0.782, \ \sin \theta = 0.624$$

Meridional thrust $= N_\phi = \dfrac{qR_1}{(1 + \cos \theta)} = \dfrac{2.75(6.41)}{1.782} = 9.9 \text{ kN/m}$

$$\text{Stress} = \dfrac{9900}{80(1000)} = 0.12 \text{ MPa}$$

Provide only nominal reinforcement as the stress is very small.

Provide 8 mm bars at 200 mm spacing both ways

Design of the cylindrical wall

$$r = 4 \text{ m} \text{ and } h = 3.36 \text{ m}$$

T = hoop tension at bottom of the wall = $\gamma hr = 134.4$ kN/m

$$A_s = \dfrac{T}{\sigma_{st}} = \dfrac{134\,400}{150} = 896 \text{ mm}^2/\text{m}$$

Provide at the rate of 8 no. of 12 mm bars per metre.

$$A_s \text{ (provided)} = 2(452) = 904 \text{ mm}^2/\text{m}$$

Let t be the thickness of the wall, then the tension stress criterion gives

$$\text{Tension stress} = \dfrac{T}{A_c + (m - 1)A_s} < \sigma_{act}$$

$$= \dfrac{134400}{1000t + 12(904)} < 1.2$$

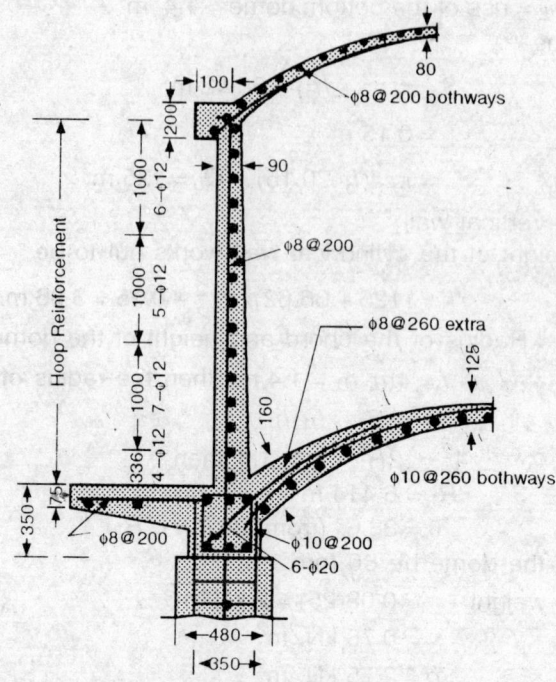

Fig. 11.14 Structural Details of Water Tank.

or $\qquad$ $t = 101$ mm

Use 110 mm wall thickness at base and reduce it to 90 mm at top. The minimum reinforcement required is 0.24%

$\qquad$ Minimum steel $\qquad$ = 240 mm²/m

Use nominal vertical reinforcement of 8 mm bars at 200 mm spacing.

Design of bottom dome: The geometrical properties of the bottom dome are selected similar to those of the top dome to maintain the formwork identical.

$$R_2 = 6.414 \text{ m}, \ h_2 = 1.4 \text{ m}, \ r = 4 \text{ m}$$

Let the self-weight $\qquad q_g = 4 \text{ kN/m}^2$

Uniform radial pressure $\quad q_1 = (h - h_2)\gamma = (3.36 - 1.4)(10) = 19.6 \text{ kN/m}^2$

Hydrostatic pressure at the base

$$q_2 = 1.4(10) = 14 \text{ kN/m}^2$$

Meridional thrust

$$N_0 = \frac{q_g R_2}{1 + \cos \theta} + \frac{q_1 R_2}{2} + \frac{\gamma R_2^2 (1 + \cos \theta - 2 \cos^2 \theta)}{6(1 + \cos \theta)}$$

$$= \frac{4(6.414)}{1.782} + \frac{19.6(6.414)}{2} + \frac{10(6.414)^2(1.782 - 1.223)}{6(1.782)}$$

$$= 13.5 + 62.9 + 21.4 = 97.8 \text{ kN/m}$$

Provide 125 mm thick dome with 10 mm bars at 260 mm spacing in both directions. (This is only nominal reinforcement as the stress is far below the allowable value.)

Design of the top ring beam: The horizontal component of the meridional thrust from the top dome is

$$F = N_\phi \cos \theta = 9.9(0.782) = 7.74 \text{ kN/m}$$

Hoop tension $\qquad T = Fr = 7.74(4) = 30.96 \text{ kN}$

Area of steel required

$$A_s = \frac{T}{\sigma_{st}} = \frac{30960}{150} = 260 \text{ mm}^2$$

Use 2 numbers of 12 mm bars, then the actual area is

$$A_s = 226 \text{ mm}^2$$

$$A_c = \frac{T}{\sigma_{cat}} - (m-1)\, A_s = \frac{30900}{1.2} - 12(226) = 23040 \text{ mm}^2$$

Use 120×200 mm beam section

The weight of the ring beam is

$$W_2 = 2\pi(r+t)\, A_c\, \gamma_c$$

$$= 2(3.141)(4 + 0.5(0.09 + 0.012))(0.024)(25) = 16 \text{ kN}$$

The weight of the vertical wall is

$$W_3 = 2\pi(r+t)(h)(t)(\gamma_c)$$

$$= 2(3.141)(4.1)(3.36)(0.1)(25) = 216 \text{ kN}$$

Let the bottom ring beam weight be

$$W_4 = 3.141(4.65)(2)(0.235)(25) = 172 \text{ kN}$$

Bottom dome weight is

$$W_5 = 56.42\,(0.125)(25) = 176 \text{ kN}$$

Weight of concrete + LL =

$$W_6 = W_1 + W_2 + W_3 + W_4 + W_5 = 735 \text{ kN}$$

Weight of the water = $W_7 = 1250 \text{ kN}$

Total load at the bottom ring beam is

$$W_8 = W_6 + W_7 = 1985 \text{ kN}$$

Design of bottom ring beam for bending: Provide four columns to support the ring beam.

The bending moment coefficients for circular beam with four supports are taken from Chapter 6 and they are:

For negative moment $c_n = 0.137$
For positive moment $c_p = 0.07$
For twisting moment $c_t = 0.021$
Radius of the ring beam = $R_0 = r + 0.5t = 4.05$ m

Vertical load on the beam per unit length $w = \dfrac{W}{2\pi R_0}$

or $$w = \frac{1985}{2\pi (4.05)} = 78 \text{ kN/m}$$

Maximum negative BM $$= M_n = c_n w R_0^2$$

$$= 0.215(78)(4.05)^2 = 275.3 \text{ kNm}$$

The equilibrium of moments and forces give

$$\frac{jnbd^2}{2}\sigma_{cb} + \frac{(n-0.1)(1.5\,m)}{n}\sigma_{cb}qA_s d - 0.45Td = M$$

$$\frac{nbd}{2}\sigma_{cb} + \frac{(n-0.1)(1.5\,m)}{n}\sigma_{cb}qA_s - A_s\sigma_s + T = 0$$

in which $\quad d = $ depth of steel, $b = $ width, $A_{sc} = q\,A_s = $ compression steel

$\quad 0.1d = $ cover to the centre of steel, $n = $ neutral axis/d

$\quad m = $ modular ratio, $q = A_{sc}/A_s$

$\quad T = $ hoop tension in the ring beam caused by meridional thrust from the lower dome, and it is

$\quad T = RN_\phi \cos\theta = (4)(97.8)(0.782) = 306 \text{ kN}$

$$e = \frac{M}{T} = \frac{0.2333}{0.306} = 0.9 \text{ m}$$

The ring beam is just below the cylindrical wall and the balcony. Therefore, the actual section that is subjected to bending is not only the beam but also the wall and the balcony. The stiffness of the wall is high.

As a preliminary exercise, select the depth of the section exclusively for BM, then (let b = thickness of wall = 0.12)

$$d = \sqrt{\frac{M_n}{Kb\sigma_{cb}}} = \sqrt{\frac{0.2753}{0.13(0.12)(7)}} = 1.57 \text{ m}$$

The depth of the wall is more than adequate so all the tension from the hoop must be resisted by the reinforcement in the ring beam. Select a nominal size of 350×350 mm $\sigma_{st} = (150 + 190)/2 = 170$.

$$A_s = \frac{T}{\sigma_{st}} = \frac{306000}{170} = 1800 \text{ mm}^2$$

Provide 6 numbers of 16 mm bars at top and bottom.

If 100 cm is the width of the balcony attached to the beam, its projection beyond the face of the beam = cantilever span = 100 − (35 − 10.5)/2 = 87.75 cm.

Assume 10 cm thick cantilever, bending moment on the cantilever

$$= (0.1)(0.8775)(25000)(87.75)/2 + 1500(87.75) = 230\,000 \text{ Ncm}$$

Use 100 mm thick slab with 8 mm bars at 200 mm spacing.

Since the vertical wall also resists the shear force, the total shear is to be distributed across the depth of the wall.

Maximum shear force $= V = \dfrac{W_8}{8} = 250 \text{ kN}$

$$\tau_v = \frac{V}{bd} = \frac{0.25}{0.1(3.7)} = 0.67 \text{ MPa}$$

Provide 10 mm stirrups at 200 mm spacing.

Wind and earthquake forces

Let the wind pressure = 1.5 kN/m^2

Wind load on the tank

$$2(4.1)(3.36 + 0.35 + 1.4/3)(1.5)(0.7) = 36 \text{ kN}$$

Wind diagonal to the braces would result in maximum compression in the column. Wind load diagonal on the columns = $3(0.5)(1.41)(1.5) = 47$ kN, where the size of the column is assumed as 0.5 m and height as 15 m.

Let the depth of the brace be 0.5 m, then

Wind force on braces = $3(8)(0.5)(1.5) = 18$ kN

Total wind force $F_w = 36 + 47 + 18 = 101$ kN

It has been found from the tower analysis, using the importance coefficient for earthquake the seismic coefficient works out to be 0.05 (normally allowed coefficient for the region is 0.04).

The force due to earthquake on the tower

$$\alpha_h W_8 = 0.05(1985) = 100 \text{ kN}$$

whereas the wind force is 101 kN, therefore, the wind force dominates over the earthquake.

Let the depth of the foundation be 1.7 m.

Overturning moment caused by wind force

$$M = 36(15 + 1.7) + (47 + 18)(7.5) = 1089 \text{ kNm}$$

For the purpose of column design, the wind force along the diagonal is critical, therefore, the axial force on the column due to wind load is

$$P_1 = \frac{M}{a} = \frac{1089}{8.2} = 133 \text{ kN}$$

where $a = 8.2$ m is the diagonal distance between the columns.

Design of columns

Total load of the overhead tank	= 1985 kN
Load on each column = 1985/4	= 496 kN
Let the self-weight of each column	= 90 kN
Let the weight of the braces on each column	= 76 kN
Total load on the column	= 662 kN
Use maximum $DL + LL$ on each column	= 665 kN
Force due to wind load	$P_1 = 133$ kN
Load on column under wind load condition	$P = 798$ kN

Maximum shear force on each columns at

the bottom panel = $36/4 + 0.5(15 - 1.875)(1.5) + 18/4 = 24$ kN

c/c of the panel height = $15/4 = 3.75$ m

Bending moment on the column $M = (3.75 - 0.25)(24)/2 = 42$ kNm

M/P is small, therefore design the column as uncracked section.

Use 48 × 48 cm with 4 nos of 20 mm and 4 nos of 12 mm bars, then

$$A_s = 1708 \text{ mm}^2$$

Transformed sectional properties are

$$A = 0.25 \text{ m}^2$$

$$I = 0.00524 \text{ m}^4$$

Direct stress $= \dfrac{P}{A} = \dfrac{0.798}{0.25} = 3.2 \text{ MPa}$

Bending stress $= \dfrac{My}{I} = \dfrac{0.0042(0.24)}{0.00524} = 1.92 \text{ MPa}$

Safety condition under wind load is

$$\frac{3.2}{8} + \frac{1.92}{7} = 0.91 \text{ less than 1.33 therefore safe.}$$

Use 8 mm stirrups at 300 mm spacing

Design of braces: Maximum bending moment occurs in the lower most brace, and it is

$$M = (24 + 20)(3.75)/2 = 78.75 \text{ kNm}$$

Use beam section of 0.20 by 0.54 m with 2 numbers of 20 mm main bars at top and bottom faces.

Provide stirrups of 8 mm bars at 400 mm spacing

Design of Foundation

DL + LL on each column	= 496 kN
Column weight = (0.23)(15)(25)	= 86 kN
Weight of brace = 3(5.3)(0.26)(25)	= 104 kN
Load at base	= 686 kN
Difference of foundation concrete and soil weights	= 58 kN
Total net load	= 744 kN
Direct load from each column	= 744 kN
Axial force due to wind	= 299 kN
Design load	= 1043 kN

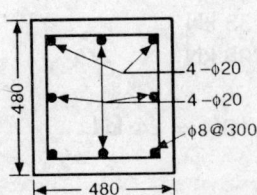

Fig. 11.15 Column Section of Ex. 11.3.

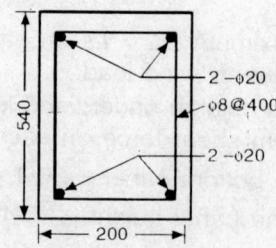

Fig. 11.16 Brace Section of Ex. 11.3.

Provide an independent footing for each column

Foundation area required = (1043)/(1.25)(60)　　= 13.8 m^2

(Allowable bearing pressure under wind load condition is 1.25 (60) kN/m^2)

Provide 3.75 × 3.75m footing A = 3.75^2 = 14.04 mm^2

Net bearing pressure = 1043/14.04 = 74.1 kN/m^2

Effective pressure = $74.1 - \dfrac{58}{14.04}$ = 70.5 kN/m^2 cantilever span = 1.625 m

BM on the footing slab

$$M = 70.5(1.625)(3.750)/2 = 214.8 \text{ kNm}$$

Effective depth needed = $d = \sqrt{\dfrac{M}{Kb\,\sigma_{cb}}} = \sqrt{\dfrac{0.2148}{0.179(3.75)(7)}} = 0.21$ m

Use 270 mm thick slab at the face of the column with 10 numbers of 20 mm bars.

Shear stress = 0.0705(3.75^2 – 1^2)/4(0.21) = 1.1 MPa

Since this is high and not admissible, modify the section. Provide a splay in the concrete column starting from 65 cm above the base of the footing. Widening of 50 cm extra on each side of the column.

Width of punching　　= (480 + 500 + 500 + 200) = 1680 mm

Shear stress　　　　= 0.0705(3.75^2 – 1.68^2)/4(1.68)(0.21) = 0.60 MPa

Allowable shear stress τ_c = 0.16 $\sqrt{f_{ck}}$ = 0.67 MPa

Design of stair case. The staircase is placed from ground to brace with landings projecting beyond the beams. This will provide a horizontal distance of (572 – 50) = 522 cm for a rise of 375 cm. This is acceptable as it is a service staircase. The landing is 100 cm width, of which 50 cm is on the brace and 50 cm cantilevered.

Width of staircase　　　　　　　= 1 m

Let DL including railing, etc.　　= 7 kN/m^2

　　　　　　　　　LL　　　= 1.5 kN/m^2

Maximum BM = 8.5 (5.3)(5.3)/10　　= 23.8 kNm

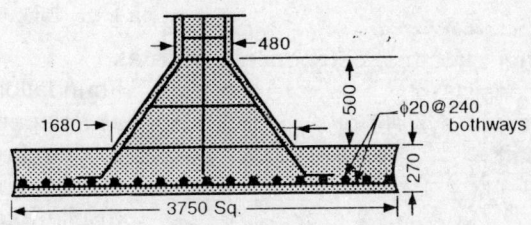

Fig. 11.17 Column-footing Details of Ex. 11.3.

496

Provide 220 mm thick slab with 6 numbers of 12 mm longitudinal bars.

Design of the cantilever landing

Live load = 3 kN

Cantilever beyond the brace = 500 mm = 0.5 m

$$M = 3(0.5) + 0.5(0.25)(25) = 4.57 \text{ kNm}$$

Use thickness = 100 mm and bend three main bars of the staircase slab to the top for the cantilevers plus 2 nos. of 12 mm projecting from the column.

Bending analysis of the tank. The size of the tank is small and the members are slender. Therefore, the bending stresses generated are negligible. However, a nominal increase in the thickness of the lower dome at the interface along with additional steel is provided.

11.5 Design of Intze Tank

Out of the several shapes used in the overhead water tanks, Intze tank has gained some popularity because of its dominant membrane action. Structurally and architecturally it has some advantages making the best use of circular shapes. The intersections between the elements when made symmetrical will result in only minor perturbations. Figure 11.18 illustrates a typical intze tank whose main structural components are:

1. Roof dome
2. Top ring beam
3. Vertical wall
4. Middle ring beam
5. Conical dome
6. Lower dome
7. Bottom ring beam

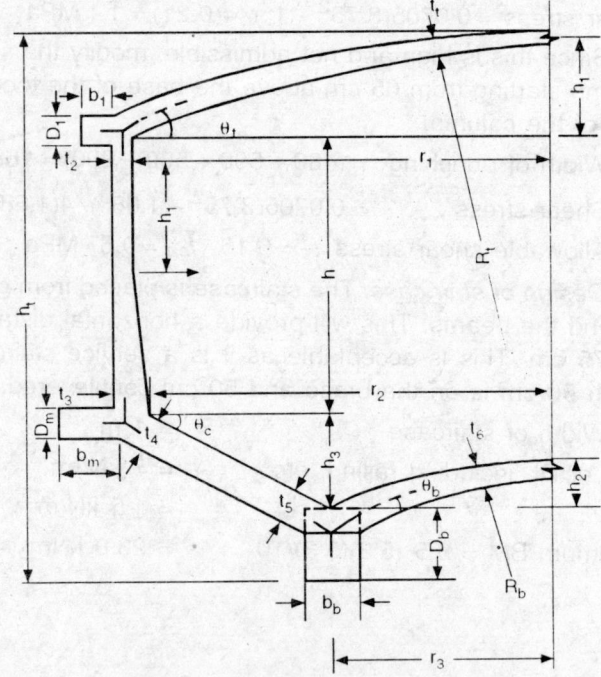

Fig. 11.18 Intze Tank Notations

In addition there are other structural components such as:

1. Staging which consists of
   a) column and brace system or
   b) cylindrical shaft

2. Foundations which can be of
   a) independent footings or
   b) ring strip of beam and slab
   c) mat foundation or
   d) pile foundation

3. Balcony, staircase, etc. minor items

The designer has to select the structural system of the foundation depending on the soil conditions. The Intze tanks are commonly employed for capacities ranging from 100 kl to 3000 kl.

Membrane analysis of each of the elements is normally applied in the design of the Intze tanks. This analysis is discussed in detail in this chapter and it is applicable in the following cases:

1. The thicknesses of the shells are small when compared with the radii of the shells. R/t ratio greater than 30 is the safe acceptable limit. However, the thin shell theory with minor edge perturbations can be accepted in some of the elements even if R/t is in the range 20.

2. The boundary membrane forces of different components must be properly balanced by ring beams so as to minimize the transverse shear forces on the shell elements. The ring beams must be placed properly at the correct locations with no eccentricities. The eccentricity of the radial or meridional forces with respect to the centroid of the ring beam will cause end moments on the shells.

3. The reinforcement details play an important role. As far as possible small diameter bars be provided at closer intervals. Proper anchor lengths, minimization of congestion of reinforcement and overlaps be maintained. Proper bending of the bars be applied so as to have smooth transmission of forces.

The membrane theory is an equilibrium method; therefore, improper geometrical alignments and poor detailing, though structurally strong, can lead to cracking and leakage. The notations used in the design are given in Fig. 11.18.

Notations

t = thickness of shell or wall, h = height
r = radius in plan, R = radius of curvature of the shell
b = width of the beam, D = depth of the beam
γ = density of water, γ_c = density of concrete

The subscripts refer to

1 or t = top dome or top ring beam, w = wall
   m = middle ring beam,
   b = bottom dome or bottom ring beam

Geometric properties and capacity calculations: Let

r_1 = radius of the inside face of the cylindrical wall at top
h_1 = rise of the top or roof dome
θ_t = semicentral angle of the top dome
R_t = radius of curvature of the top dome
t_1 = thickness of the top dome

Usually the radius r and the rise h_1 are selected a priori. The rise of the top dome can be selected in the range of $r_1/6$ to $r_1/3$ depending on the size of the tank. The relation between the radius of the dome and that of the cylindrical shell is:

$$(2R_t - h_1)h_1 = r_1^2 \tag{11.28}$$

Therefore,
$$R_t = \frac{r_1^2 + h_1^2}{2h_1}$$
(11.29)

The semi-central angle is given by

$$\sin \theta_t = \frac{r_1}{R_t}$$
(11.30)

$$\cos \theta_t = 1 - \frac{h_1}{R_t}$$
(11.31)

Volume of the liquid in the cylindrical portion of the tank (excluding the free board)

$$Q_1 = (\pi/2)(r_1^2 + r_2^2)(h - h_6)$$
(11.32)

where h_6 = free board, r_2 = radius at the base of the wall. Volume of a frustrum of a cone is

$$Q_2 = (\pi/3)(r_2^3 + r_2 r^3 + r_3^2)h_3$$
(11.33)

where r_3 = inside radius of the cone which is same as the radius of the circle of the bottom dome

h_3 = rise of the cone

The volume of a segment of a spherical dome is

$$Q_3 = (\pi/6)(3r_3^2 + h_2^2)h_2$$
(11.34)

The net capacity of the tank is

$$Q = Q_1 + Q_2 - Q_3$$

$$= (\pi/6)\Big[3(r_1^2 + r_2^2)(h - h_6) + 2(r_2^2 + r_2 r_3 + r_3^2)h_3$$

$$- (3r_3^2 + h_2^2)h_2 \Big]$$
(11.35)

For a given capacity, one can assume r_3, h_2, r_2, h_3, h_6 and r_1, and then determine the height of the tank. Or select the height h and determine r_1 by assuming r_3 and r_2 as a fraction of r_1. The free board height h_6 is normally taken from 150 mm to 250 mm depending on the capacity of the tank. The forces in the members are proportional to square of h and linearly connected with r_1. Whereas the volume of the tank is linearly connected with h and proportional to square of the radius. Therefore, it is advisable to select height of the tank proportional to square root of r_1.

If the slope of the conical wall is about 45° then

$$h_3 = r_2 - r_3$$

$$\frac{h_3}{r_1} = \frac{r_2}{r_1} - \frac{r_3}{r_1}$$
(11.36)

EXAMPLE 11.4 Design of Intze tank of 2500 kl capacity. Let the height of the wall is selected as square root of the radius of the cylindrical wall. Free board = 0.15 m.
Capacity calculations

$$Q = 2500 \text{ m}^3$$

Let $\quad r_2 = r_1, r_3 = 0.7r_1, h_2 = h_3 = 0.3r_1$

$h = 1.3\sqrt{r_1}$ then the volume equation becomes:

$$Q = \frac{\pi}{6}(6r_1^2(1.3\sqrt{r_1} - 0.15) + 2(1 + 0.49 + 0.7)r_1^3(0.3) - (3(0.49) + 0.09)(0.3)r_1^3$$

$$2500 = \frac{\pi r_1^2}{6}(7.8\sqrt{r_1} - 0.9 + 1.314r_1 - 0.468r_1)$$

or $\quad r_1^2(7.8\sqrt{r_1} + 0.846r_1 - 0.9) = 4774.65$

Try $r_1 = 11$ m, then LHS of the above equation is

$\qquad$ LHS $= 4147.35 < 4774.65$

Therefore, *try* $r_1 = 11.8$ m, then

$\qquad$ LHS $= 4995 > 4774.65$

The solution of the above equation is for

$\qquad r_1 = 11.6$ m,

and $\qquad h = 1.3\sqrt{r_1} = 4.426$ m

$\qquad h_2 = h_3 = 3.48$ m, $r_3 = 8.12$ m

One can observe that the value of r_1 is not too sensitive to relative ratios even if the shape of the tank is considerably different.

Design of a roof dome: The roof dome is subjected to live load and self-weight, and is supported by the cylindrical wall and the top ring beam. The live load on the roof can be taken 750 N/m^2. The centre line of the shell boundary must be at the intersection of the ring beam and the wall such that the horizontal component of the thrust is resisted by the ring beam and the vertical component by the wall.

The radius of the dome is given by

$$R_t = \frac{r_1^2 + h_1^2}{2h_1} \tag{11.29}$$

The surface area of the dome is

$$S_t = 2\pi R_t h_1 \tag{11.37}$$

The membrane stress resultants in the dome are:

$$N_\phi = \frac{w_t R_t}{1 + \cos\theta} \tag{11.38}$$

$$N_\theta = w_t R_t\left(\cos\theta_r - \frac{1}{1 + \cos\theta}\right) \tag{11.39}$$

where N_ϕ = meridional thrust

$\qquad N_\theta$ = circumferential thrust

$\qquad \theta$ = half angle subtended by the segment at the centre of sphere of the dome.

Both the stress resultants are considered positive if they are compressive in nature.

N_ϕ is maximum at $\theta = \theta_t$ and N_θ is maximum at $\theta = 0$. These values are:

$$N_\phi = \frac{w_t R_t}{1 + \cos \theta_t} \tag{11.40}$$

$$N_\theta = w_t R_1 (\cos \theta_t - 0.5) \tag{11.41}$$

The circumferential stress changes from compressive to tension if the semicentral angle exceeds 51.8°. It is desirable to keep the semicentral angle less than 45° so to avoid double form shuttering to the shell. Since the forces on the shell surface are compressive and cause only nominal stress, the thickness of the dome is primarily controlled by the practical considerations. The thickness is normally selected as

$$80 \le t_1 \le 140 \text{ mm} \tag{11.42}$$

Provide nominal reinforcement of 0.3% in each direction, and select the bar size such that the spacing of the reinforcement is less than three times the thickness.

The total live and dead loads from the shell can be computed as

$$W_1 = w_l S_t = 2\pi \, R_t h_1 w_l \tag{11.43}$$

$$W_2 = 2\pi \, R_t h_1 w_{tg} \tag{11.44}$$

Design of the top ring beam: The top ring beam is provided so as to resist the horizontal component of the meridional thrust of the top dome. The middle plane of the ring beam is aligned with the central line of the dome. The hoop tension produced in the ring is

$$T_t = (N_\phi \cos \theta_1)(r_1 + 0.5t_2) \tag{11.45}$$

where t_2 = thickness of the wall at top, usually limited by constructional aspect.

T_t = hoop tension in the top ring beam

It is desirable to design the top ring beam as an uncracked section so as to avoid possible corrosion of the reinforcement and also to provide good stiffness to support the roof dome.

The area of the hoop reinforcement is given by

$$A_{st} = \frac{T_t}{\sigma_{st1}} \tag{11.46}$$

where σ_{st1} = allowable stress in steel on water face.

The size of the cross-section of the beam is governed by the concrete tensile strength criterion, and is

$$\sigma = \frac{T_t}{A_c + (m-1)A_{st}} \le \sigma_{ct} \tag{11.47}$$

or

$$A_c = \frac{T_t}{\sigma_{ct}} - (m-1) \, A_{st} \tag{11.48}$$

The value of A_s used in the above expression is the actual area of the steel provided but not that given by Eq. (11.46). The gross size of the ring beam is

$$b_1 D_1 = A_c + A_{st}$$

The width (b_1) of the beam can be made larger than the depth (D_1).
The weight of the ring beam is

$$W_3 = 2\pi(r_1 + 0.5b_1)\, b_1 D_1 \gamma_c \tag{11.49}$$

Design of the cylindrical wall: The cylindrical wall is designed to resist the hoop tension caused by the hydrostatic pressure. The maximum hoop tension occurs at the bottom of the wall and it is

$$T_w = \gamma h r_2 \tag{11.50}$$

The area of the tensile steel at the base is given by

$$A_{sw} = \frac{T_w}{\sigma_{st1}} \tag{11.51}$$

where A_{sw} = area of the hoop steel in the wall per unit height

The thickness of the wall can be kept constant upto a height from top and then it can be increased linearly towards the base. The thickness at the base is controlled by the tensile stress criterion in concrete, and it is

$$t_3 = \frac{T_w}{\sigma_{ct}} - (m-1)\, A_{sw} \tag{11.52}$$

One has to be consistent with the units in these equations. The area of the hoop reinforcement can be decreased gradually to the minimum towards the top of the wall. There is possibility of some bending moment at the base of the wall. The maximum negative bending moment possible on the wall is

$$M_{n1} = \frac{\gamma r_1 h t_2}{\sqrt{12}}\left(1 - h\sqrt{\frac{r_1 t_3}{1.532}}\right) \tag{11.53}$$

The vertical reinforcement at the base of the wall could be controlled by the bending moment listed above. In most cases, the actual bending moment at the base is less than that mentioned above.

The area of the cross-section of the wall is

$$A_w = t_2 h_0 + 0.5\,(t_2 + t_3)(h - h_0)$$

where h_0 = height of constant thickness (t_2) of the wall
The total weight of the wall is

$$W_4 = 2\pi(r_1 + 0.5 t_2)\gamma_c\, A_w \tag{11.54}$$

Design of the middle ring beam: The purpose of the middle ring beam is to provide a horizontal support to the conical dome. The total load at the top of the conical dome is transmitted along the inclined line. Therefore, a horizontal component of the inclined force is resisted by the hoop tension in the ring beam. Assume a cross-section of the ring beam and compute the total load upto the top of the cone. In case there is a balcony, the load of the balcony must also be added. The balcony can also act as a ring beam. The weight of the water in the trapezoidal area resting on the wall should also be accounted.

Let $\quad\quad\quad\quad\quad\quad W_5$ = weight of the balcony slab
$\quad\quad\quad\quad\quad\quad\quad\; W_6$ = live load on the balcony
$\quad\quad\quad\quad\quad\quad\quad\; W_7$ = extra load due to railing etc. on the balcony

The balcony can be designed for a live load of 1.5 kN/m².
The weight of the middle ring beam is:

$$W_8 = 2\pi(r_1 + t_2 + 0.5b_2)\, b_2 D_2 \gamma_c \tag{11.55}$$

The weight of the water in the trapezoidal portion above the wall is

$$W_9 = \frac{\pi}{2}(r_1^2 - r_1^2)(h + h_6)\,\gamma \tag{11.56}$$

Total load at the top of the conical wall is

$$W_{10} = W_1 + \ldots + W_9 \tag{11.57}$$

The load per unit length on the cone is

$$w_{10} = \frac{W_{10}}{2\pi(R + 0.5t_2)} \tag{11.58}$$

The horizontal component of the thrust on the cone is

$$H_m = W_{10}\frac{(r_2 - r_3)}{h_3} \tag{11.59}$$

The hoop tension in the middle ring is

$$T_m = H_m(r_1 + 0.5t_2) \tag{11.60}$$

$$= \frac{W_{10}(r_1 + 0.5t_2)(r_2 - r_3)}{2\pi(r_1 + 0.5t_2)h_3}$$

$$= \frac{W_{10}(r_2 - r_3)}{2\pi h_3} \tag{11.61}$$

The area of the tension reinforcement in the middle ring is

$$A_{sm} = \frac{T_m}{\sigma_{st2}} = \frac{W_{10}(r_2 - r_3)}{2\pi h_3 \sigma_{st2}} \tag{11.62}$$

where σ_{st2} = allowable stress in steel (usually that corresponds to the value away from the water face)

The size of the beam can be obtained from the criterion of the allowable tensile stress in the concrete.

$$A_{cm} = \frac{T_m}{\sigma_{ct}} - (m-1)\,A_{sm}$$

$$= \frac{W_{10}(r_2 - r_2)}{2\pi h \sigma_{ct}} - (m-1)\,A_{sm} \tag{11.63}$$

Design of the conical dome: The approximate inclined length of the cone is

$$L_c = \sqrt{((r_1 - r_3)^2 + h_3^2)} \tag{11.64}$$

Self-weight of the frustum of the cone is

$$W_{11} = \pi(r_2 + r_3)\,t_5 L_c \gamma_c \tag{11.65}$$

where t_5 = average thickness of the conical shell (dome)

The weight of the water over the conical shell is

$$W_{12} = \pi(r_2^2 - r_3^2)(h + 0.5h_3)\gamma \tag{11.66}$$

The total load on the conical wall is

$$W_{13} = W_{10} + W_{11} + W_{12} \tag{11.67}$$

There are two stress resultants in the conical dome. The meridional thrust is controlled by the total load on the shell and the hoop tension is by the radial pressure. The total load acting on the wall at its base is resisted by an inclined thrust which is equal to

$$N_\phi = \frac{W_{13} \cosec \theta_c}{2\pi r_3} \tag{11.68}$$

where θ_c = slope of the conical shell

The hoop tension in the shell is caused by the radial pressure. It varies nonlinearly along the length of the cone; therefore, it is computed at a height y from the base of cone (see Fig. 11.19).

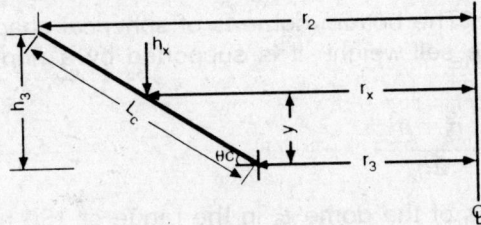

Fig. 11.19 Notations of Conical Wall.

Head of water at that point $= h_x = h + (h_3 - y)$ $\qquad$ (11.69)
Radius of the cone at the point $= r_x = r_3 + y \cot \theta_c$ $\qquad$ (11.70)

The force acting normal to the inclined wall of the cone is

w_x = water pressure + component of weight of the shell

$$= \gamma h_x + \gamma_c t_5 \cos \theta_c \tag{11.71}$$

Hoop tension in the shell is

$$T_c = w_x r_x \cosec \theta_c$$
$$= (\gamma(h + h_3 - y) + \gamma_c t_5 \cos \theta_c)(r_3 + y \cot \theta_c) \cosec \theta_c \tag{11.72}$$

The maximum hoop force is likely to occur at the base of the cone, and it is for $y = 0$

$$T_{co} = (\gamma(h + h_3) + \gamma_c t_5 \cos \theta_c) r_3 \cosec \theta_c \tag{11.73}$$

The shell thickness and the reinforcement are designed to resist N_ϕ and T_c. As N_ϕ is compressive, it does not govern the design. The design is controlled by T_c. Calculate T_c at top, mid-height and at bottom and then take the mean value of the three to compute the area of steel. Let T_1, T_2 and T_3 are the hoop tensions for $y = h_3$, $0.5h_3$ and 0. Then the total hoop tension on the entire length is

$$T_{cm} = \frac{(T_1 + 2T_2 + T_3)L_c}{4} \tag{11.74}$$

The area of the hoop steel needed is

$$A_s = \frac{T_{cm}}{\sigma_{st1}}$$

The thickness of the shell controlled by the tensile stress in the concrete is given by

$$\sigma = \frac{T_{cm}}{L_c t_5 + (m-1)A_s} < \sigma_{ct}$$

or

$$t_5 = \frac{T_{cm}}{L_c \sigma_{ct}} - \frac{(m-1)A_s}{L_c} \qquad (11.76)$$

The thickness of the shell can be varied linearly from top to bottom having the smaller size at top such that the mean thickness is not less than the t_c.

The assumed thickness of the conical shell must be compared with actual thickness and then compute the actual weight of the cone W_{11}. The appropriate minimum area of the reinforcement along the length of the shell be placed. The reinforcement can be placed at both the faces of the shell.

Design of bottom dome: The bottom dome is of spherical shape and is subjected to the weight of the water and the self-weight. It is supported by a ring beam. The radius of the curvature of the dome is

$$R_b = \frac{r_3^2 + h_2^3}{2h_2}$$

Assume a nominal thickness of the dome t_6 in the range of 150 mm to 250 mm depending on the size of the dome, then the self-weight of the dome is given by

$$W_{14} = 2\pi R_b h_2 t_6 \gamma_c \qquad (11.77)$$

The weight of the water over the dome is

$$W_{15} = \pi\gamma(r_3^2(h + h_3) - (3r_3^2 + h_3^2)h_2/6) \qquad (11.78)$$

The total load on the dome including self-weight is

$$W_{16} = W_{14} + W_{15}$$

The semicentral angle of the dome is given by

$$\sin\theta_b = \frac{r_3}{R_b} \qquad (11.79)$$

The total load is transmitted to the ring beam by meridional thrust. The circumferential stress is maximum at top and it is compressive throughout for ϕ_b less than 52°. Even the meridional thrust causes small compressive stress on the shell thickness. The meridional thrust at the support is

$$N_0 = \frac{W_{16}R_b}{2\pi r_3^2} \qquad (11.80)$$

The horizontal and vertical components of the meridional thrust on the ring beam are:

$$F = N_0 \cos\theta_b = \frac{W_{16}(R_b - h_2)}{2\pi r_3^2} \qquad (11.81)$$

$$F_v = N_\phi \sin \theta_b = \frac{W_{16}}{2\pi r_3} \tag{11.82}$$

The maximum compressive stress in the shell is given by

$$\sigma = \frac{N_\phi}{t_6} \tag{11.83}$$

Provide nominal reinforcement in both the directions of the shell and thicken the end portion by about 30% so as to resist the secondary bending moment. The reinforcement is to be provided at top and bottom faces of the shell near the supports.

Design of bottom ring beam: The bottom ring beam is subjected to the total vertical load of the tank and the water including self-weight in the vertical direction. In addition, it is also subjected to the radial force which is the difference of the horizontal components of the meridional thrusts from the conical and bottom domes. The net radial thrust on the ring beam is

$$F_{hb} = T_{c0} \cos \theta_c - \frac{W_{16}(R_b - h_2)}{2\pi r_3^2}$$

$$= (\gamma(h + h_3) + \gamma_c t_5 \cos \theta_c) r_3 \cot \theta_c - \frac{W_{15}(R_b - h_2)}{2\pi r_3^2} \tag{11.84}$$

In case of shallow spherical domes, the net radial force F_{hb} on the ring might become negative indicating a hoop tension. The hoop force on the ring beam is given by

$$P_b = F_{bh} r_3$$

$$= (\gamma(h + h_3) + \gamma_c t_5 \cos \theta_c - \frac{W_{16}(R_b - h_2)}{2\pi r_3} \tag{11.85}$$

In case of the ring beam supported on a shaft, the beam need to be designed only for the hoop force and if it is supported on a set of columns, it is to be designed as a ring beam subjected to bending and axial force. The design of the staging depends on the type of staging, therefore, no general discussion is given here. The design is illustrated by typical examples.

EXAMPLE 11.5 *Design of 770 kl overhead water tank on cylindrical shaft.* An overhead water tank for a capacity of 770 kl with 18 m staging is to be provided over a soil where safe net bearing capacity is 100 kN/m^2.

Data and basic dimensions
    Capacity = 770 kl
    Staging = 18 m
    Soil profile: Top 1.5 m depth is medium plastic clay underlain by plastic clayey silt upto
          5 m depth.
    Safe bearing capacity at 2 m depth   = 100 kN/m^2
    Depth of foundation $h_f = 2$ m
    Type of the tank: Intze tank
    Type of staging: Cylindrical shaft
    Concrete: M20

Reinforcement: High yield deformed bars with proof stress of 415 N/mm^2

Inside surface dimensions of the tank

Some dimensions are assumed as below:

Radius of the vertical cylindrical wall at top	$r_1 = 6.40$ m
Radius of the vertical wall at base	$r_2 = 6.25$ m
Rise of the top dome	$h_1 = 1.5$ m
Rise of the bottom dome	$h_2 = 1.7$ m
Centre to centre radius of the staging	$r_3 = 4.5$ m
Rise of conical shell	$h_3 = 1.75$ m
Free board	$= 0.15$ m

Capacity calculations

Total volume of the water is

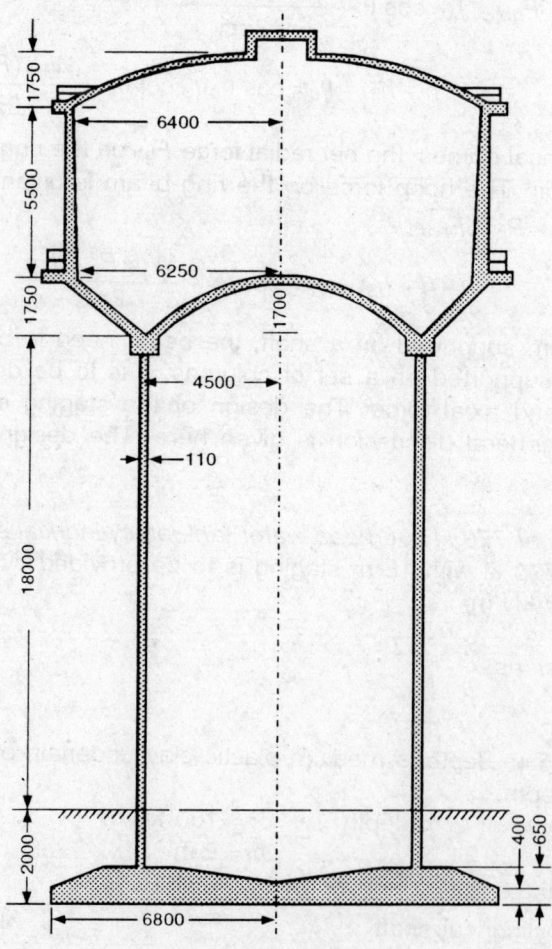

Fig. 11.20 Notations of OHWT Example 11.5.

$$V = \pi(0.5(r_1^2 + r_2^2)(h - 0.15) + (r_2^2 + r_2 r_3 + r_3^2) h_2/3$$
$$- (3r_3^2 + h_2^2)h_2/6 = 770 \text{ m}^2$$

Substituting of different quantities in the above equation gives

$$\pi(0.5(6.4^2 + 6.25^2)(h - 0.15) + (6.25^2 + 6.25(4.5) + 4.5^2)1.75/3$$
$$- (3(4.5)^2 + 1.7^2)(1.7)/6))$$
$$= 125.7(h - 0.15) + (51.005 - 18.031)\pi$$
$$= 125.7h + 14.119 = 770 \text{ m}^2$$

The solution of the above equation gives

$$h = 5.456 \text{ m} \text{ then provide } h = 5.5 \text{ m}$$

This results in a free board = 0.20 m

Allowable Stress and other Code specification
Allowable stresses in RCC water retaining concrete M20 are:

$$\sigma_{ct} = 1.2, \ \sigma_{cbt} = 1.7$$

The reinforcement is HYSD-Fe415 bars only
Allowable stresses in HYSD-Fe415 bars

$$\sigma_{st1} = 230 \text{ (in ordinary RCC)}$$
$$\sigma_{st2} = 150 \text{ (up to 225 mm from water face)}$$
$$\sigma_{st3} = 190 \text{ (beyond 225 mm from water face)}$$

Permissible stresses under wind or seismic load condition are increased by 33.3% of the allowable stresses.

Seismic coefficients are:

Soil structure interaction for raft in soft soils $= \beta = 1.0$
Importance factor $= I = 1.5$
Seismic zone factor (Zone III) $= F_0 = 0.2$
Damping $= 5\%$
Basic wind pressure $= 1.5 \text{ kN/m}^2$

Design by Membrane Analysis
Design of Roof Dome

Let the thickness $= t_1 = t = 0.075 \text{ m}$
Rise of the dome $= h_1 = 1.5 \text{ m}$
Chord radius $= r_1 = 6.4 \text{ m}$
Surface radius

$$R = 0.5(r_1^2/h_1 + h_1)$$
$$= 0.5(6.4^2/1.5 + 1.5) = 14.403 \text{ m}$$
$$\cos \phi = 1 - 5/14.403 = 0.896 \text{ and}$$
$$\phi = 26.36$$

Let the density of the M20 concrete be 25 kN/m^2

Live load = w_l = 0.75 kN/m^2
Dead load = w_d = 0.075(25) = 1.8 kN/m^2

Total live load w = 2.55 kN/m^2

Total live load on the shell is:

$$W_1 = 2\pi R h_1 w = 2\pi (14.403)(1.5)(0.75) = 102 \text{ kN}$$

Self-weight of the dome

$$W_2 = 2\pi R h_1 w_d = 2\pi (14.403)(1.5)(1.8) = 244 \text{ kN}$$

Meridional thrust

$$N_\phi = \frac{wR}{1 + \cos\phi} = \frac{2.55(14.403)}{1 + 0.896} = 19.37 \text{ kN/m}$$

Compressive stress $= \dfrac{N_\phi}{t} = \dfrac{19.37\,(10^{-3})}{0.075} = 0.258 \text{ MPa}$

$$N_\theta = wR\left(\cos\phi - \frac{1}{1 + \cos\phi}\right)$$

$$= 2.55\,(14.403)(0.896 - 0.527) = 13.53 \text{ kN/m}$$

The hoop stress is compressive over the entire domain and it is less than the meridional stress. The actual compressive stress is far less than the allowable compressive stress; therefore, provide a nominal reinforcement.

Minimum reinforcement in either direction is

$$\frac{0.15\,(bt)}{100} = \frac{0.15\,(1000)(75)}{100} = 112.5 \text{ mm}^2/\text{m}$$

Provide 8 mm *bars at* 225 mm c/c *bothways*
(reinforcement spacing not more than three times the thickness)
Actual area of steel provided = 222 mm^2/m

Percentage of steel $= \dfrac{22200}{75000} = 0.3\%$

Design of Top Ring Beam (M20 concrete). Hoop thrust on the ring beam is same as the horizontal component of the meridional thrust from the top dome. The hoop tension in the ring beam is, therefore, equal to

$$T = (N_0 \cos\phi)r_1 = 19.37\,(0.896)(6.4) = 111.08 \text{ kN}$$

Area of the tension steel required is

$$A_s = \frac{T}{\sigma_{st2}} = \frac{111080}{150} = 740.53 \text{ mm}^2$$

Provide 2 *numbers of* 16 mm *and* 3 *numbers of* 12 mm *bars.*

$$A_s \text{ (provided)} = 741 \text{ mm}^2.$$

Let the width of the beam $b_1 = 240$ mm
Let the depth required be D_1

The tensile stress in the concrete is

$$\frac{T}{b_1 D_1 + (m-1) A_{st}} = \sigma_{ct}$$

or $\quad D_1 = \left(\frac{T}{\sigma_{ct}} - (m-1) A_{st}\right)\bigg/ b_1 = \left(\frac{111080}{1.2} - 18(741)\right)\bigg/ 240 = 330.11 \text{ mm}$

Provide. 2 numbers of 16 mm and 3 numbers of 12 mm bars with 6 mm ties at 300 mm c/c.

Weight of the ring beam is

$$W_3 = 2\pi \left(r_1 + \frac{b_1}{2}\right) b_1 D_1 \gamma_c$$

$$= 2\pi(6.52)(0.24)(0.33)25 = 81 \text{ kN}$$

Design of the Vertical Wall of the Tank

Wall height = h	= 5.5 m
Maximum hoop tension is at h	= 5.5 m
$T_1 = \gamma r_2 h = 10(6.25)(5.5)$	= 343.75 kN

Hoop tension at 4.5 m from top

$$T_1 = 10(6.25)(4.5) = 281.25 \text{ kN}$$

Total hoop tension at the bottom 1 m depth

$$T = (T_1 + T_2)/2 = 312.5 \text{ kN}$$

Area of tension steel required in the bottom 1 m depth of the wall

$$A_s = \frac{T}{\sigma_{st2}} = \frac{312500}{150} = 2083 \text{ mm}^2/\text{m}$$

Provide 10 numbers of 12 mm bars on each face.

$$A_s \text{ (provided)} = 2260 \text{ mm}^2$$

The tensile stress in the concrete is

$$\frac{T}{A_c + (m-1) A_s} = \sigma_{ct}$$

or $\quad A_c = \frac{T}{\sigma_{ct}} - (m-1) A_s$

$$= \frac{312500}{1.2} - 12(2260) = 233297 \text{ mm}^2$$

Thickness required $\quad = \dfrac{233297}{1000} = 234 \text{ mm}$

Provide

Thickness at bottom = t_3 = 250 mm
Thickness at top $\quad$ = t_2 = 100 mm
The average thickness between 4.5 and 5.5 m depth is

$$100 + \frac{(250-100)(5)}{5.5} = 236 \text{ mm}$$

Table 11.7 HYSD-Fe415 Reinforcement 12 mm Dia in there Vertical Wall

Distance from top (m)	T (kN)	Reinforcement	
		Required mm^2	Numbers of $\phi 12$
0 to 1.5	72	480	6 (middle)
1.5 to 2.5	128	853	4 (each face)
2.5 to 3.5	190	1270	6 "
3.5 to 4.5	252	1680	8 "
4.5 to 5.5	313	2083	10 "

The reinforcement details in the vertical wall are given in Table 11.7.
Weight of the wall

$$W_4 = 2\pi \frac{(r_1 + r_2)}{2} \frac{(t_2 + t_3)(h - D_1)\gamma_c}{2}$$
$$= 2\pi(12.65) \frac{(0.1 + 0.25)}{2} (5.5 - 0.33)(25) = 898 \text{ kN}$$

Average thickness of the wall = 175 mm·

Minimum percentage of mild steel is 0.3% for 100 mm thick wall and 0.2% for 450 mm walls. The linear interpolation of the percentage for 175 mm thick is

$$0.3 - \frac{(175 - 100)(0.1)}{350} = 0.28$$

Minimum area of the reinforcement needed is
   (80% of the mild steel for HYSD-Fe415 bars)

$$0.8\frac{(0.28)(175)(1000)}{100} = 392 \text{ mm}^2/\text{m}$$

Provide 8 mm bars at 250 mm c/c on both faces

$$A_s ((\text{Provided}) = 402 \text{ mm}^2/\text{m}$$

Design of Middle Gallery

Width of the walking gallery = 1 m

The gallery is used by workmen; therefore, the working load is taken as that of a roof balcony. The live loads on roofs is taken as 1500 N/m^2 or 1000 N placed near the tip of the balcony (say about 75 mm from the free end of the balcony). The load that governs the design is usually the later case.

Let width of the ring beam be equal to 500 mm so as to serve partly as a balcony at this height. So the cantilever span of the slab = 1 − 0.5 = 0.5 m.

Let the thickness of the slab be	= 75 mm
Self-weight of the slab = 0.075(1)(25000)(0.5)	= 940 N/m
Let the railing load (at 75 mm from tip)	= 750 N/m

Bending moment due to self-weight $= 940\frac{(0.5)}{2}$ = 235 Nm/m

Bending moment due to the railing

$$750(0.5 - 0.075) = 320 \quad \text{Nm/m}$$

Bending moment due to UDL (live load)

$$1500 \frac{(0.5)^2}{2} = 188 \text{ Nm/m}$$

or the bending moment due to end live load

$$1000(0.5 - 0.075) = 425 \text{ Nm/m}$$

The later case of live load governs the design.
The total design bending moment is

$$M = 225 + 320 + 425 = 980 \text{ Nm/m}$$

The working moment capacity of the section is

$$M_r = 0.896 bd^2 \geq M = 980\,000 \text{ Nmm/m}$$

$$d = \sqrt{\frac{980000}{0.896(1000)}} = 33 \text{ mm}$$

The slab thickness provided = 75 mm
The effective depth = 75 − 30 = 45 mm

Area of steel needed is $= A_s = \dfrac{980000}{0.906(45)(230)} = 105 \text{ mm}^2$

Provide 8 mm radial bars at 200 mm c/c and anchored into the ring beam.

$$A_s \text{ (provided)} = 250 \text{ mm}^2/\text{m}$$

Distribution reinforcement is $= 0.15(75) \dfrac{(500)}{100} = 56.25 \text{ mm}^2$

Provide 2 numbers of 8 mm bars in 500 mm projection.

$$A_s \text{ (provided)} = 100.5 \text{ mm}^2$$

Total weight of the slab

$$W_5 = 2\pi(6.25 + 0.25 + 0.5)(0.5)(0.075)25 = 43 \text{ kN}$$

Total live load on the gallery

$$W_6 = 2\pi(6.25 + 0.25 + 0.5)(1)(1.5) = 66 \text{ kN}$$

Total railing load

$$W_7 = 2\pi(7.5)(0.750) = 35 \text{ kN}$$

Design of Middle Ring Beam: Let the size of the beam for the purpose of computing self weight be 500 × 500 mm.

Self-weight = 0.5(0.5)(25) = 6 kN/m

(All loads are listed in kN)

Total self-weight

$$W_8 = 2\pi(6.25 + 0.25 + 0.25)6 = 255$$

LL from the top dome $W_1 = 102$

Weight of the top dome $W_2 = 244$

Weight of the top ring beam $W_3 = 81$

Weight of the vertical wall $\quad W_4 = 898$
Weight of the gallery slab $\quad W_5 = 43$
LL on the gallery $\quad\quad\quad\quad W_6 = 66$
Weight of the railing $\quad\quad\quad W_7 = 35$

Weight of the water on the slating edge of the vertical wall

$$W_9 = \pi(6.4^2 - 6.25^2)(5.5)\frac{(10)}{2} = 164$$

Total load transferred to the conical wall is equal to the weight of the components of the tank up to the top of the cone plus the live loads etc., and it is

$$W_{10} = W_1 + \ldots + W_9 = 1883 \text{ kN}$$

Hoop tension on the ring beam is

$$T = \frac{W_{10}(\cot\theta)}{2\pi} = \frac{1883}{2\pi} \cot 45 = 300 \text{ kN}$$

Area of tension steel in the ring beam is

$$A_s = \frac{T}{\sigma_{st2}} = \frac{300(1000)}{150} = 2000 \text{ mm}^2$$

Provide 10 numbers of 16 mm bars

$$A_s \text{ (provided)} = 10(201) = 2010 \text{ mm}^2$$

The actual tensile strength in the ring beam must be limited to the allowable tension in the concrete. Therefore,

$$\frac{T}{A_c + (m-1)A_s} \leq 1.2 \text{ N/mm}^2$$

$$A_c \geq \frac{T}{1.2} - (m-1) A_s = \frac{300\,000}{1.2} - 12(2010) = 226330 \text{ mm}^2$$

Provide the ring beam of size = 500×450 mm and 8 mm *ties* at 450 mm c/c.
The assumed weight of the ring beam checks well with actual size of the beam. The actual weight of the beam is

$$W_8 = 2(6.25 + 0.25 + 0.25) \times (0.5)(0.45)(25) = 239 \text{ kN}$$

Therefore, revised total load transferred to the conical wall is

$$W_{10} = W_1 + \ldots + W_9 = 1867 \text{ kN}$$

Design of Conical Shell: There are two membrane stress resultants in the conical shell. The meridional thrust is maximum at the base of the cone whereas hoop tension varies along the depth.

Let the average thickness of the shell be $\quad$ = 450 mm
The slope of the wall = θ $\quad\quad\quad\quad\quad\quad$ = 45°
Height of the cone = h_3 $\quad\quad\quad\quad\quad\quad$ = 1.75 m
Self-weight of the slab = 0.450(25) $\quad\quad$ = 11.25 kN/m²
Length of the slab = 1.414(1.75) $\quad\quad\quad$ = 2.475 m
Weight of the conical wall

$$W_{11} = 2\pi\gamma_c \frac{(r_2 + r_3)(t)(L)}{2}$$

$$= \pi(25)(6.25 + 4.5)(0.45)(2.475) = 937 \text{ kN}$$

Weight of the water over the conical wall is

$$W_{12} = \pi\gamma(r_2^2 - r_3^2)(h + 0.5h_3)$$

$$= \pi(10)(6.25^2 - 4.5^2)(6.375) = 3768 \text{ kN}$$

Total load on the conical slab at its base

$$W_{13} = W_{10} + W_{11} + W_{12}$$

$$= 1876 + 937 + 3768 = 6572 \text{ kN}$$

Meridional thrust in the slab of the cone

$$N = \frac{W_{13} \cosec \theta}{2\pi r_3} = \frac{(6575)(1.414)}{2\pi(4.5)} = 328.6 \text{ kN/m}$$

Horizontal component of the thrust $H_1 = N \cos\theta = 233$ kN/m
Assume the thickness of the conical slab at base as $t_5 = 400$ mm
The meridional thrust

$$\frac{N}{bt} = \frac{328\,600}{1000(400)} = 0.82 \text{ MPa}$$

This stress is much smaller than the allowable compressive stress; however, the **actual** thickness of the slab is governed by the hoop tension rather than the meridional compression.

The hoop tension is computed at a height (y) from the base of the cone. The horizontal radius of the cone at height y from base $= r_x = (4.5 + y)$ m.
Height of the water at this level

$$h_x = h + (h_3 - y) = (7.25 - y) = (7.25 - y) \text{ m}$$

Normal load on the slanting slab

$$p_x = \text{ water pressure} + \text{component of weight of the slab}$$

$$= (7.25 - y) 10 + 11.25 \cos\theta$$

$$= 7.25 + (11.25/1.414) - 10y = (80.46 - 10y)$$

Hoop tension in the slab

$$T = (p_x \cosec \theta)(r_x)$$

$$= (80.46 - 10y)(1.4142)(4.5 + y)$$

Hoop tensions in the slab at depths $y = 0$, $y = 0.85$ and $y = 1.75$ m are computed and are given below:

$$T_1 = 510 \text{ kN/m}$$

$$T_2 = 542 \text{ kN/m}$$

$$T_3 = 554 \text{ kN/m}$$

The maximum hoop tension in the slab occurs at the intersection of the middle ring beam. The total hoop tension in the slab

$$T = \frac{(T_1 + 2T_2 + T_3)}{2}\left(\frac{L}{2}\right)$$

$$= \frac{(510 + 2(542) + 544)(2.475)}{4} = 1329 \text{ kN}$$

Total area of hoop tension steel required is

$$A_s = \frac{T}{\sigma_{st2}} = \frac{1329\,(1000)}{150} = 8860 \text{ mm}^2$$

Provide 16 mm *dia bars*, 21 *at the inner face and* 24 *at the outer face of the slab*
Total area of the steel provided is

$$A_s = 45(201) = 9045 \text{ mm}^2$$

The area of the concrete required is governed by the allowable tensile stress in the concrete. The hoop tension in the concrete is

$$\frac{T}{A_c + (m-1)(A_s)} \leq \sigma_{ct} = 1.2 \text{ N/mm}^2$$

or $\qquad A_c \geq \dfrac{T}{1.2} - (m-1)\,A_s = \dfrac{1329000}{1.2} - 12(9045) = 998960 \text{ mm}^2$

Average thickness of the slab

$$t = \frac{A_c}{L} = \frac{998960}{2475} = 404 \text{ mm}$$

Provide the thickness of the slab at top as $t_4 = 450$ mm
and at base as $\qquad t_5 = 375$ mm
The average thickness of the slab is 412.5 mm. Revised weight of the conical wall

$$W_{11} = \pi(25)(6.25 + 4.5)(0.4125)(2.475) = 862 \text{ kN}$$

Therefore, total load on the cone at its base

$$W_{13} = W_{10} + W_{11} + W_{12} = 1867 + 862 + 3768 = 6497 \text{ kN}$$

Minimum reinforcement in the radial direction is

0.24% for 100 mm thick and 0.16 for 450 mm thick.

Percentage of reinforcement required

$$= 0.24 - \frac{(0.24 - 0.16)}{(450 - 100)}(412.5 - 100) = 0.17$$

$$A_s = 0.17\,(412.5)\,\frac{(1000)}{100} = 702 \text{ mm}^2/\text{m}$$

Provide 12 mm *bars at* 300 mm *c/c in the radial direction at each surface.*

$$A_s = \text{(provided)} = 753 \text{ mm}^2/\text{m}$$

Design of Bottom Dome
Half chord length $\qquad\qquad r_3 = 4.5$ m
Rise of the dome $\qquad\qquad\quad h_2 = 1.7$ m
Let the thickness of the shell $\quad t_6 = t = 0.15$ m
Radius of the dome

$$= R = 0.5(r_3^2 + h_2^2)/h_2 = 0.5((4.5^2 + 1.7^2)/1.7 = 6.806 \text{ m}$$

Self-weight of the dome is

$$W_{14} = (2\pi R h_2 \gamma_c)$$

$$= 2\,(3.146)(6.806)(1.7)(0.15)(25) = 273 \text{ kN}$$

Weight of the water over the dome

$$W_{15} = \pi\gamma\,(r_3^2\,(h + h_3) - (3r_3^2 + h_2^2)\,h_2/6)$$

$$= 3.146\,(10)(4.5^2)\,(5.5 + 1.75)$$

$$- (3(4.5)^2 + 1.7^2)(1.7/6)) = 4046 \text{ kN}$$

Total weight on the dome

$$W_{16} = W_{14} + W_{15} = 273 + 4046 = 4319 \text{ kN}$$

Semicentral angle $\qquad \sin\phi = \dfrac{r_3}{R} = \dfrac{4.5}{6.806} = 0.66$

Meridional thrust

$$N_\phi = \frac{W_{16}R}{2\pi r_3^2} = \frac{4319(6.806)}{2(3.1416)(4.5)^2} = 231 \text{ kN/m}$$

Compressive stress $= \dfrac{N_\phi}{bt} = \dfrac{231000}{1000(150)} = 1.54 \text{ N/mm}^2$

The compressive stress is only nominal, therefore, provide minimum reinforcement. The hoop compression is also in the same range. The minimum reinforcement is

$$A_s = 0.23\,\frac{(150)1000}{100} = 345 \text{ mm}^2/\text{m}$$

Provide 10 mm dia bars at 225 mm c/c.

Horizontal thrust $\qquad H_2 = N_\phi \cos\theta = 231\,(0.7513) = 173.6 \text{ kN/m}$

Design of the Bottom Ring Beam: The horizontal component of the thrust from the shell at the ring beam level is

$$H_1 = 233 \text{ kN/m}$$

The horizontal component of the meridional thrust from the bottom dome at the beam level is

$$H_2 = 173.6 \text{ kN/m}$$

The net force on the ring beam is

$$H_3 = H_1 - H_2 = 233 - 173.5 = 59.5 \text{ kN/m}$$

The net force is inwards and causes compression on the ring beam. The hoop compression is

$$T = H_3 r_3 = 59.5\,(4.5) = 267.8 \text{ kN}$$

Provide a ring beam of cross-section of 400 × 400 mm.

The compressive stress $\qquad = \dfrac{267\,800}{400 \times 400} = 1.6 \text{ N/mm}^2$

The minimum reinforcement be 0.24%

$$A_s = 0.24 \frac{(400 \times 400)}{100} = 384 \text{ mm}^2$$

Provide 4 numbers of 16 mm bars and 8 mm ties at 300 mm c/c

Weight of the beam $\qquad W_{17} = 2\pi r_3 (0.4)(25)$

$$= 2(3.1416)(4.5)(0.4)(0.4)(25) = 114 \text{ kN}$$

Design of the Cylindrical Shaft

Height of the shaft above GL = staging of the tank = $h = 18$ m

Radius of the centre line of the shaft $\qquad = r_3 = 4.5$ m

Depth of foundation $\qquad = 2$ m

Let the thickness of the shaft wall be $\qquad = t = 0.11$ m

Note: Minimum thickness of shaft shell be 150 mm

The shaft is having a constant thickness from 1 m below the ground and then it is increased to about 300 mm thick at the top of the foundation slab. The loads acting at 1 m below GL on the shaft are listed here in kN.

Weight of the shaft upto 1 m below GL

$$W_{18} = (2\pi r_3 t) h_5 \gamma_c$$

$$= 2\pi(4.5)(0.11)(18 + 1 - 0.4)(25) = 1446$$

Weight of the bottom ring beam $\qquad W_{17} = 114$

Weight of the bottom dome $\qquad W_{14} = 273$

Weight of the conical shell $\qquad W_{11} = 862$

Weight of the middle ring beam $\qquad W_8 = 239$

Weight of the gallery slab $\qquad W_5 = 43$

Weight of the vertical wall $\qquad W_4 = 898$

Weight of the top ring beam $\qquad W_3 = 81$

Weight of the top dome $\qquad W_2 = 244$

Total weight of the concrete upto 1 m below GL = W_{20} $\qquad = 4200$ kN

Live load $W_1 + W_6 = 1.2 + 66$ $\qquad = 168$ kN

Railing weight W_7 $\qquad = 35$

Weight of water including free board $\qquad = 7700$

Total water, live and railing load = W_{21} $\qquad = 7903$

Weight of staircase (approximate) $\qquad = 1(0.3)(0.2)\dfrac{(19)}{0.3} 25 = 95$ kN

Staircase railing and partial live load, etc. $\qquad = 168.0$ kN

Total load from staircase $\qquad = 263$ kN

Total load at shaft base = $W_{22} = 4200 + 7903 + 263 = 12366$ kN

Use the total load as $\qquad W_{22} = 12400$ kN

Area of cross-section of the shaft

$$A = 2\pi r_3 t = 2(3.1416)(4.5)(0.11) = 3.11 \text{ m}^2$$

Moment of inertia of the section

$$I = \pi r_3^3 t = \pi (4.5)^3 (0.11) = 31.49 \text{ m}^4$$

Axial stress on the shaft wall

$$\frac{W_{22}}{A} = \frac{124\,00000}{3.11(10^6)} = 4.0 \text{ N/mm}^2$$

The shaft is treated as a cylindrical shell subjected to axial force and bending. The allowable buckling compressive stress in the shell is given by

$$\sigma_{cr} = \frac{0.25\, f_{ck}}{1 + f_{ck}/F_{cr}}$$

in which

$$F_{cr} = \frac{0.2Et}{R} = \frac{0.2(5700\sqrt{20})(0.11)}{4.5} = 124.6 \text{ N/mm}^2$$

Allowable buckling stress is

$$\sigma_{cr} = \frac{0.25(20)(1.2)}{1 + (20/1245)1.2} = 5.03 \text{ N/mm}^2$$

(including the age effect of 1.2 to the concrete)

Actual stress under normal load condition is already computed as 4 MPa and it is less than the allowable so only nominal reinforcement is required. The minimum percentage of reinforcement of HYSD-Fe415 bars in shell slab is

$$A_s = 0.22 \frac{(1000)(110)}{100} = 242 \text{ mm}^2/\text{m}$$

Provide 8 mm bars at 200 mm spacing in longitudinal as well as circumferential direction (bars staggered on the faces).

$$A_s \text{ (provided)} = 251 \text{ mm}^2/\text{m}$$

Wind force

Basic wind pressure upto 30 m = $p = 1.5 \text{ kN/m}^2$

Projected effected area of the OHWT on vertical plane is:

$$A_1 = (2/3)((6.4 + 0.1)(2)(1.5 + 0.075))$$

$$+ (6.4 + 0.1)(2)(5.5) + \frac{2}{3}(6.4 + 01)(2)(1.7 + 0.4)$$

$$= 13(0.67(1.575) + 5.5 + 0.67(2.1)) = 103.5 \text{ m}^2$$

Vertically projected area of the shaft

$$A_2 = 2(4.5 + 0.055)(18 - 0.4) = 9.11(17.6) = 160.4 \text{ m}^2$$

The shape of the tank and the shaft in plan is circular; therefore, the shape factor of the structure against wind is 0.7. The net force caused by the wind is obtained by multiplying the basic wind pressure by the shape factor. The maximum bending moment caused by the wind load about 1 m below the GL is

$$= 0.7p \ A_1(19 + 1.75 + 2.75) + A_2 \frac{(18.6)}{2}$$

$$= 0.7(1.5((103.5)\left((23.5) + 160.4\left(\frac{18.6}{2}\right)\right) = 4120 \text{ kNm}$$

Add extra for gallery, railing effects, then the maximum bending moment is

$$M_W = 4120 + 60 = 4280 \text{ kNm}$$

Maximum shear caused by the wind load is at the GL and it is :

$$= 0.7 \, p(A_1 + A_2) = 1.05(103.5 + 160.4) = 277 \text{ kN}$$

Add extra gallery and railing effects, then

$$V_W = 277 + 13 = 290 \text{ kN}$$

The moment and shear caused by the wind are compared with that of the seismic effect to determine the critical force.

Seismic Force. The design load for seismic effect is taken equal to the dead load, plus weight of the water plus 25% of the live load.

Load from the OHWT including water load and 25% of the live load is

$$W_{23} = (W_{20} - W_{18}) + W_7 + 7700 + 0.25(168)$$

$$= (4200 - 1446) + 35 + 7700 + 42 = 10\,531 \text{ kN}$$

Weight of the shaft including the staircase

$$W_{24} = W_{18} + 263 = 1446 + 263 = 1709 \text{ kN}$$

Effective equivalent load $= W_{25} = W_{24} + \dfrac{W_{24}}{3} = 10480 \text{ kN}$

The lateral deflection at the top of the water tank due to equivalent load is

$$v = \frac{W_{25}L^3}{3EI}\left(1 + \frac{3(0.5h)}{2L}\right)$$

in which

$$L = 18 + 1.75 + \frac{5.5}{2} = 22.5 \text{ m}$$

$$E = 5700\sqrt{20} = 25500 \text{ N/mm}^2 = 25.5 \text{ GPa}$$

$$\frac{1.5h}{2L} = \frac{1.5(5.5)}{2(22.5)} = 0.183$$

$$I = 31.49 \text{ m}^4 \text{ Therefore}$$

$$v = \frac{11100(10^3)(22.5)^3(1.183)}{3(25.5)(10^9)(31.49)} = 0.062 \text{ m}$$

Natural period of the tank

$$T = 2\pi \sqrt{\left(\frac{v}{g}\right)}$$

$$= 2\pi \sqrt{\left(\frac{0.062}{9.81}\right)} = 0.5 \text{ sec}$$

Average acceleration coefficient for $T = 0.5$ sec and for 5% damping from IS : 1893 is

$$\frac{S_a}{g} = 0.17$$

Seismic coefficient

$$\alpha_h = \beta I F_0 \frac{S_a}{g} = 1(1.5)(0.2)(0.17) = 0.051$$

Moment caused by the seismic forces about 1 m below the ground level is

$$M_q = \alpha_h \left(W_{23} \left(19 + 1.75 + \frac{5.5}{2} \right) + W_{24} \left(\frac{18.6}{2} \right) \right)$$
$$= 0.051(10531(23.5) + 1709(9.3)) = 13491 \text{ kNm}$$

Add extra for small earth pressure for 1 m depth.

Use
$$M_q = 13491 + 509 = 14000 \text{ kNm}$$
$$V_q = 0.051(10531 + 1709) = 624 \text{ kN}$$

Use
$$V_q = 630 \text{ kN}$$

Design check for wind load or seismic load conditions. It can be seen from the two previous sections that the seismic load condition dominates over the wind load condition. The critical bending moment and shear forces at 1 m below the ground level (neglecting the passive resistance of the earth) are

$$M_q = 1400 \text{ kNm}$$
$$V_q = 630 \text{ kN}$$

The thickness of the shaft wall is increased at 1 m below GL up to the foundation slab. The stresses are, therefore, critical at 1 m below GL. The stresses in the concrete are computed for seismic load condition.

The axial force in this load condition is

$$P = W_{23} + W_{24} = 10531 + 1709 = 12240 \text{ kN}$$

The maximum compressive stress on the concrete is

$$\sigma_c = \frac{P}{A} + \frac{My}{I}$$
$$= \frac{12240(10^3)}{3.11(10^6)} + \frac{14000(10^3)(4.555)}{31.49(10^6)} = 3.94 + 2.03$$
$$= 5.97 \text{ N/mm}^2$$

No tension is developed in the section.

Allowable stress under seismic load condition is

$$\sigma_a = 5.03(1.33) = 6.69 \text{ N/mm}^2$$

The actual stress is less than the allowable value without considering the area of the reinforcement.

$$\text{Average of shear stress} = \frac{V_q}{A} = \frac{630\,000}{3.11(13^6)} = 0.2 \text{ N/mm}^2$$

520

This value is far less than the allowable stress. The nominal circumferential reinforcement specified earlier is adequate.

Design of Foundation

Depth of foundation below GL = 2.0 m

Bearing capacity at 2 m below GL $p_a = 100$ kN/m^2

Total load up to 1 m below GL $W_{22} = 12400$ kN

Weight of the 40 cm splay at the bottom
of the shaft (average thickness = 0.2 m)

$$W_{27} = 2\pi(4.5)(0.2)(0.4)(25) = 56 \text{ kN}$$

Let the average thickness of the foundation be equal to 0.3 m and the average width of the strip be 5.5 m. Then the weight of the foundation is

$$W_{28} = 2\pi(4.5)(5.5)(0.3)(25) = 1166 \text{ kN}$$

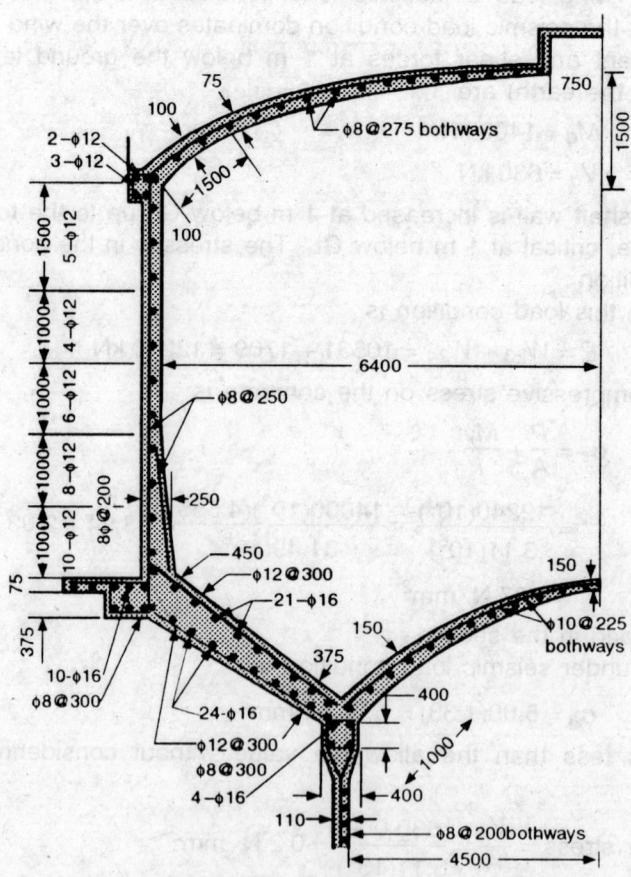

Fig. 11.21 Reinforcement Details of Intze Tank, Example 11.5.

Total working load on the foundation is

$$W_{29} = W_{22} + W_{27} + W_{28} = 12400 + 46 + 1166 = 13632 \text{ kN}$$

Use $\qquad W_{29} = 13700 \text{ kN}$

Minimum bearing area needed is

$$A_f = \frac{W_{29}}{p_a} = \frac{13700}{100} = 137 \text{ m}^2$$

Let the radius of a mat foundation be equal to 6.8 m. Then the area and moment of inertia of the area of the foundation are:

$$A_f = \pi R^2 = \pi(6.8)^2 = 145.27 \text{ m}^2$$

$$I_f = \frac{\pi R^4}{4} = \frac{\pi(6.8)^4}{4} = 1679.3 \text{ m}^4$$

The bearing pressure on the soil under normal live load condition is

$$\sigma = \frac{W_{29}}{A_f} = \frac{13700}{145.27} = 94.3 \text{ kN/m}^2$$

Axial load under seismic load condition is

$$W_{30} = P + W_{27} + W_{28} = 13462 \text{ kN}$$

The bearing pressure under the seismic load condition is

$$\sigma = \frac{W_{30}}{A_f} + \frac{M_q R}{I_f} = \frac{13462}{145.27} + \frac{14000\,(6.8)}{1679.3}$$

$$= 927 + 56.7 = 149.4 \text{ kN/m}^2$$

The safe bearing pressure under live load condition is 100 kN/m^2 and the under seismic load condition in case of mat foundation is 1.5 (100) = 150 kN/m^2. These values are slightly more than the actual pressure computed; therefore, the foundation size is just right.

Design of the foundation slab. A raft foundation with outer radius equal to 6.8 m and subjected to circular load from the shaft is designed.

Let $\quad c$ = radius of the shaft

$\qquad R$ = outer radius of the foundation slab

Radial and circumferential bending moments are introduced in the raft slab. These moments are given by

Due to W

for $r < c$ = radius of the shaft

$$M_r = \frac{W}{16\pi} [3((r/R)^2 - 1) + 2(\ln(R/c) + 1 - (c/R)^2)]$$

$$M_\theta = \frac{W}{16\pi} [((r/R)^2 - 3) + 2(\ln(R/c) + 1 - (c/R)^2)]$$

for $r > c$

$$M_r = \frac{W}{16\pi} [3((r/R)^2 - 1) + 2(2\ln(R/r) - (c/R)^2 + (c/r)^2)]$$

$$M_\theta = \frac{W}{16\pi}\left[((r/R)^2 - 3) + 2(2ln(R/r) - (c/R)^2 - (c/r)^2 + 2)\right]$$

Due to M

for $r < c$

$$M_r = \frac{Mr}{4\pi R^2 c^2}(3R^2 - 2c^2 - c^4/R^2) - \frac{5Mr}{12\pi R^4}(R^2 - r^2)$$

$$M_\theta = \frac{Mr}{24\pi c^2 R^2}(3R^2 - 2c^2 - c^4/R^2) - \frac{5Mr}{36\pi R^4}(5Rr - 3r^2)$$

for $r > c$

$$M_r = \frac{M}{8\pi R^2 r}(2(R^2 - r^2) + c^2(R^4 - r^4)/r^2 R^2) - \frac{5Mr}{12\pi R^4}(R^2 - r^2)$$

$$M_\theta = \frac{M}{24\pi R^2 r}(2(3R^2 - r^2) - c^2(3R^4 + r^4)/R^2 r^2) - \frac{Mr}{36\pi R}(5Rr - 3r^2)$$

where $R = 6.8$ m

r = distance of the point from centre

W = net load on the slab and it is

$W_{31} = W_{25} + W_{27} = 11100 + 56 = 11156$ kN

The combined effect of the axial force and bending moment are computed and given in Table 11.8 for $W = 11156$ kN, $M_q = M = 14000$ kNm.

The actual live load is assumed to come after about 12 months of curing of the foundation concrete. The effective thickness of the slab required for the maximum bending moment is

$$d = \sqrt{\frac{M_r}{Kb\sigma_{cb}}} = \sqrt{\frac{0.325}{(0.13)(1)(7)(1.33)}} = 0.52 \text{ m}$$

Use $d = 0.6$ m

The critical section for shear force is the circular ring at distance $d/2$ from the face of the shaft wall. The radial distance of the circle is

$$r = c + d/2 = 4.5 + 0.3 = 4.8 \text{ m}$$

The shear force on the outer ring strip is

$$V = W_{31}\left(1 - \frac{r^2}{R^2}\right) = 11156\left(1 - \frac{4.8^2}{6.8^2}\right) = 5597 \text{ kN}$$

Peripheral distance $= 2\pi r = 30.16$ m

Nominal shear stress $= \dfrac{5.597}{30.16(0.6)} = 0.3$ MPa

This value is admissible for 0.5% of the reinforcement. Let the thickness of the slab at free end be 0.4 m and its effective depth be 0.32 m. The same thickness is provided at the centre of the foundation slab. This will provide adequate resistance against the bending moments.

Design of reinforcement At $r = 0$

$$A_s = \frac{54000}{0.904(0.320\ (230)(1.33)} = 610 \text{ mm}^2/\text{m}$$

Table 11.8 Effect of Axial Force and Bending Moment (Moments in kNm/m)

	$r = 0$	2.25	4.65	6.8 m
M_r	– 54	165	325	0
M_0	– 54	9	116	85

Provide ϕ12 bars at 120 mm c/c at top both radial and circumferential upto 2.5 m from the centre.

At $r = 4.65$ m, radial reinforcement

$$A_s = \frac{325\ 000}{0.904(0.6)\ (230)\ (1.33)} = 1958\ \text{mm}^2/\text{m}$$

Provide ϕ20 *bars* at 160 mm c/c at bottom (spacing measured at $r = 4.68$ m). Curtail alternate bars at 2.0 m from the centre and the remaining bars at 1 m from the centre.

Circumferential reinforcement

$$A_s = \frac{116\ 000}{0.904\ (0.32)\ (230)\ (1.33)} = 1311\ \text{mm}^2/\text{m}$$

Provide The moment at $r = 2.25$ m is 9 kNm/m; therefore, the reinforcement required at $r = 2.25$ m is nominal.
At $r = 6.8$ m

$$A_s = \frac{85\ 000}{0.904(0.32)\ (230)\ (1.33)} = 960\ \text{mm}^2/\text{m}$$

Design of staircase (Design by limit state of strength). RCC cantilever steps are provided inside the shaft up to 16 m above GL and then it is taken out through an opening in the shaft.

Rise of the step	= 200 mm
Tread of the step	= 300 m
Width of the step	= 900 mm

These steps are individually cantilevered from the shaft wall.

Landings are provided for every 15 steps.	
Landing length	= 1000 mm
Width of the landing	= 900 mm

A landing is provide at 16 m above GL which projects both inside and outside of the shaft. The outside landing supports a steel ladder which is placed up to the middle ring beam gallery.

a) *Design of RCC step*

Let the thickness = 90 mm
Self-weight of each step
$W_g = 0.3(0.9)(0.09)25$ = 0.60 kN
Weight of railing/step = 0.12 kN
LL placed near free end = 1.0 kN
(uniformly distributed live load placed on the step does not govern the design)

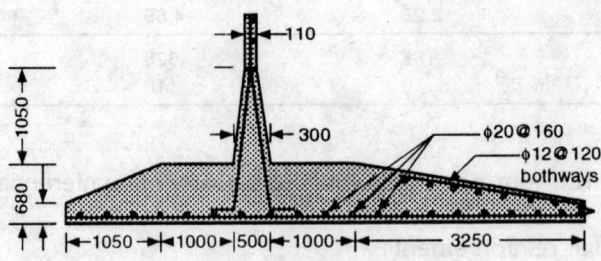

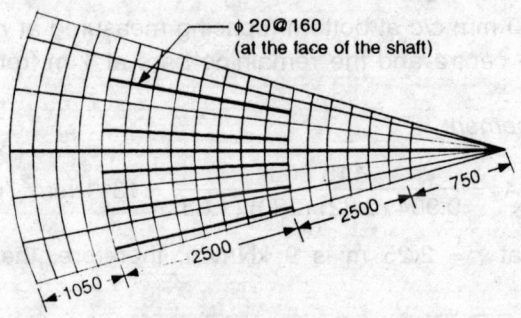

Fig. 11.22 Reinforcement Details of Mat Foundation, Example 11.5.

The collapse moment of each step is

$$M_c = 1.5 \left[(0.60) \left(\frac{0.90}{2} \right) + 1.12(0.9 - 0.05) \right] = 1.66 \text{ kNm}$$

Resisting moment capacity of RCC section with HYSD-Fe415 bars is

$$M_r = 0.130bd^2 f_{ck} \geq M = 1.66$$

Therefore, $\qquad d = 78 \text{ mm}$

The overall thickness of 100 mm is adequate and gives a cover of 20 mm.
Area of steel required for each step is

$$A_s = \frac{1660}{0.804(0.078)(230)} = 113 \text{ mm}^2$$

Provide 3 numbers of 8 mm bars at top and anchor them into the shaft wall.

$$A_s \text{ (provided)} = 150 \text{ mm}^2$$

Also provide three numbers of 8 mm distribution bars.

Design of Landing

Length of landing	= 1000 mm
Width of landing	= 900 mm
Let the thickness of landing	= 100 mm
Self-weight = 1(0.9)(0.1)(25)	= 2.25 kN

Railing weight   = 0.3 kN

Live load 1.5 kN near free end or 1.5 kN/m^2. The concentrated load near the free end governs the design.

Design free end load   = 1.5 kN

Weight of the steel ladder   = 2 kN

The collapse moment on the slab

$$M = 1.5\left((2.25)\frac{(0.9)}{2} + (0.3 + 1.5 + 2)(0.9 - 0.05)\right) = 6.303 \text{ kNm}$$

The effective depth of steel from bottom fibre is

$$d^2 = \frac{6.303\ (10^6)}{0.138\ (1000)(15)}$$

or $d = 78$ mm

Provide 100 mm thick landing slab. The area of steel required is

$$A_s = \frac{6303}{0.904(0.078)(230)} = 397 \text{ mm}^2$$

Provide 8 numbers of 8 mm bars on top and continued into the inside landing. Provide three numbers of 8 mm distribution bars over the 900 mm length.

In case of only the inside landing, the top bars are to be anchored into the shaft wall.

Check on the moment capacity of the shaft wall for cantilever moment coming from the landing.

The bending moment from the landing on the shaft wall is more critical than that caused by each step.

The bending moment from the landing = 6.303 kNm/m

The maximum moment on the shaft wall is distributed in the ratio of the heights of the wall above and below the landing slab. The critical case is at the first landing which is at 1.5 m above GL.

The maximum BM on the shell wall from the landing is

$$M = 6.303\ \frac{(18 - 1.5)}{18} = 5.78 \text{ kNm/m}$$

(Neglecting the curvature effect to be on the safer side.)

Flexural stress caused by the cranked in moment is (bending stress)

$$\sigma_b = \frac{M}{Z} = \frac{6(5.78)(10^6)}{1000(110)^2} = 2.87 \text{ N/mm}^2$$

Maximum axial stress on the shaft was computed as

$$\sigma_a = 5.79 \text{ N/mm}^2 \text{ (1 m below GL)}$$

The allowable axial stress under seismic load condition is:

$$\sigma_{ac} = 6.69 \text{ N/mm}^2$$

Allowable bending compression in M20 concrete is

$$\sigma_{bc} = 7(1.33) = 9.31 \text{ N/mm}^2$$

The design criterion is based on the interaction formula and is

$$\frac{\sigma_a}{\sigma_{ac}} + \frac{\sigma_b}{\sigma_{bc}} \leq 1$$

in earthquake load condition.

The substitution of the different quantities gives

$$\frac{5.97}{6.69} + \frac{2.87}{9.31} = 0.89 + 0.3 = 1.13$$

Since the design criterion is not satisfied, reinforcement must be provided to resist the cranked in BM. Two rows of bars, one on each face of the shaft wall are placed. The lever arm between the reinforcement is taken as $110 - 50 = 60$ mm. The axial force on the reinforcement bars is

$$F = \frac{M}{60} = \frac{5.78(10^6)}{60} \quad 96330 \text{ N/m}$$

Allowable compressive stress in the compression reinforcement is

$$\sigma_{sc} = 130 \text{ N/mm}^2$$

Area of steel required on each face

$$A_{sc} = \frac{96330}{130} = 741 \text{ mm}^2/\text{m}$$

Nominal steel provided = 126 mm²/m

Net area of steel to be provided = 615 mm²/m

Provide 8 mm HYSD-Fe415 bars at 80 mm c/c. On each face of the shaft near the landing and also at the steps. The length of each bar is 1000 mm extending 750 mm upward and downward. This reinforcement is in addition to the nominal reinforcement provided.

Maximum local tension produced near the landing is

$$\sigma_t = -3.72 + 1.88 + 2.87 = 1.03 \text{ N/mm}^2$$

Allowable flexural stress in M20 concrete is

$$\sigma_{at} = \frac{0.7}{2} \sqrt{20} = 1.56 \text{ N/mm}^2$$

The allowable flexural tension in normal load condition is higher than that caused under seismic load condition. Therefore, no extra steel is required for local tensile stress.

Openings in the shaft. Whenever an opening is made in the shaft for door or for ventilation, the area of reinforcement and the part of the area of the concrete lost due to the opening must be substituted in equivalent area of steel placed on the boundaries of the opening. In the bottom zone of the shaft, where the stresses are maximum, the area of the concrete taken out because of the opening, must be replaced either by thickening with concrete and steel at the boundaries of the opening or by equivalent area of the steel. The size of the door opening at the base can be as much as 1200 mm by 2200 mm. The reinforcement details around the opening are given in Figs. 11.23 and 11.24.

Continuity Correction. The shell is designed by membrane theory which is a lower bound solution. The solution is on the safer side as no crack theory is used in the membrane analysis. The continuity correction is applicable to local effects only. The length of the slanting slab of the conical done is rather small and application of continuity correction with such short

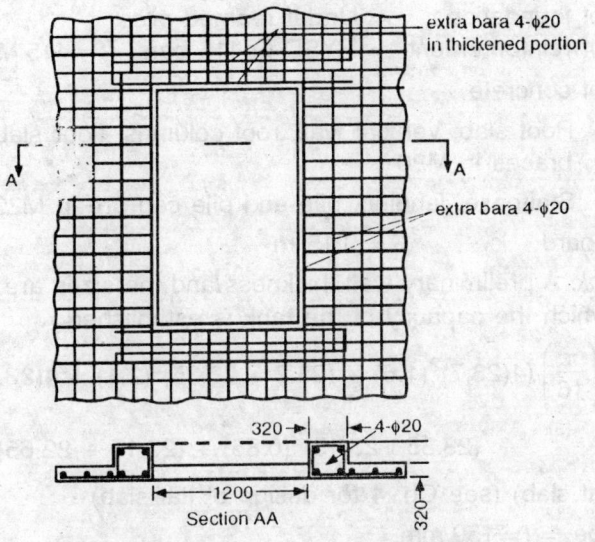

extra bara 4-φ20
in thickened portion

extra bara 4-φ20

320 ─┤   ├─ 4-φ20

1200

320

Section AA

Fig. 11.23 Reinforcement Details at Opening Shaft, Example 11.5.

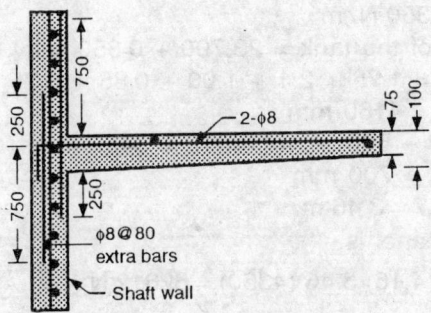

750

250

75
100

2-φ8

750

250

φ8 @ 80
extra bars

Shaft wall

Fig. 11.24 Reinforcement Details of Stair Case Step in Shaft, Example 11.5.

elements is not advisable. One has to go into more complicated bending analysis of shells is such thick and short length shells. The saving if any that might result in such short shells in going to be marginal and, therefore, not applied here. As the method applied is only a statical one, the solution is a lower bound and, therefore, structurally sound design. The thickness of the shells near the edges are increased as per IS code on design of RCC shell structures.

EXAMPLE 11.6 *Design of 2500 kl water tank*

Design data

Capacity of the tank   = 2500 kl
Staging height   = 18 m
Type of tank   = Circular in plan with flat roof and flat floor

Type of staging   = Columns and braces
Type of foundation   = Under reamed piles
Type of reinforcement   = HYSD-Fe415 bars, $f_y = 415$ MPa
Type of concrete

1. Roof slab, vertical wall, roof columns, floor slab, columns and braces in M20

2. Staircase, landing, pile and pile cap are in M20

Free board   = 0.15 m

Capacity of the tank. A preliminary wall thickness and the sizes are shown in Figs. 11.25 and 11.26 based on which the capacity of the tank is established.

Volume of the tank $= \left(\dfrac{\pi}{16}\right) (4(23.7)^2(1.6) + (23.7 + 23.35)^2(2.1) + 4(23.35)^2(1.0) +$

$$(23.35 + 23.15)^2(0.85) + (23.15 + 22.65)^2(0.25)) = 2510 \text{ m}^3$$

Design of roof slab (flat slab) (see Ch. 4 for design of flat slab)

Let the thickness be $= t = 150$ mm
Live load $= w_l = 750$ N/m^2
Self-weight $= 3600$ N/m^2
Total load intensity $= 4350$ N/m^2
Outer to outer diameter of the tank = 23.700 + 0.350 = 24.05 m
Clear height in the tank = 1.75 + 2.1 + 1.00 + 0.85 + 0.25 = 5.95 m
The main panel size be = 4160 mm
End panel size = (24050 − 500) − 4(4160)/2 = 3455 mm
Size of the column head = 700 mm
Clear span $L_0 = 4.16 - 0.7 = 3.46$ m
Total load on the main panel is

$$W = (4.16)(3.46)(4350) = 62612 \text{ N}$$

Use $W = 63000$ N

Total bending moment

$$M_0 = \frac{WL_0}{8} = \frac{63000(3.46)}{8} = 27750 \text{ Nm}$$

Supporting column size = 230 by 230 mm
Effective height of the column

$$= 0.7(5.95) = 4.165 \text{ m}$$

The relative stiffness of the column

$$K_c = \frac{(0.23)^4}{4.165} = 0.00068$$

$$= \text{ relativeness of the wall (assumed)}$$

The relative stiffness of the slab

$$K_s = \frac{4.16(0.15)^3}{4.16} = 0.00338$$

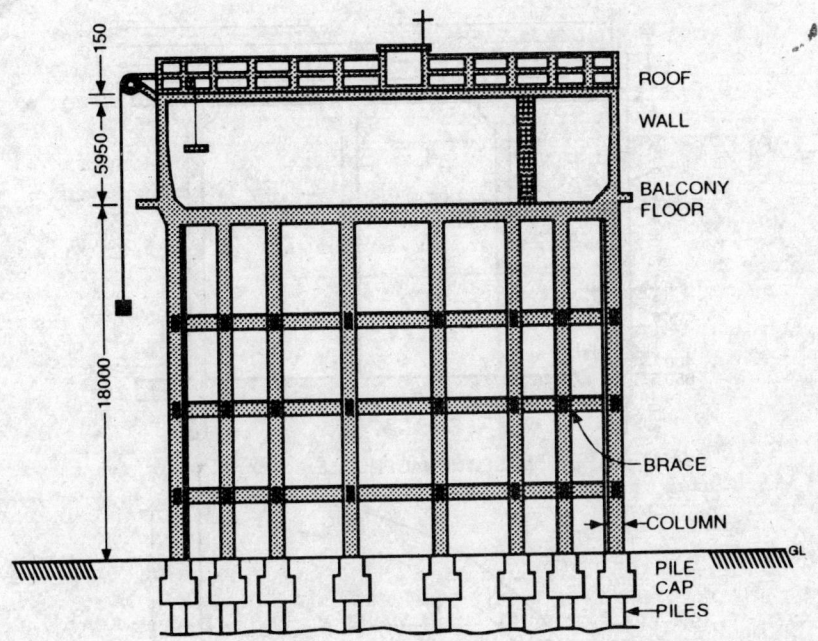

Fig. 11.25 Column, Brace and Foundation Arrangement of Water Tank, Example 11.6

$$\alpha_c = \frac{0.00068}{0.00338} = \left(\frac{K_c}{K_s}\right) = 0.2$$

$$\alpha = 1 + 1/\alpha_c = 6$$

The bending moment coefficients in the end span are
Interior negative BM coefficient

$$c_{n3} = 0.75 - 0.1/6 = 0.733$$

Positive design BM coefficient

$$c_{p1} = 0.63 - 0.28/6 = 0.583$$

Exterior negative BM coefficient

$$c_{n1} = 0.65/6 = 0.11$$

Design moment coefficient in interior span

$$c_{ni} = 0.65 \text{ and } c_{pt} = 0.35$$

The bending moment coefficients on the column and middle strips are:

$$c_{n3c} = 0.75 c_{n3} = 0.55$$

$$c_{n1c} = c_{n1} = 0.11$$

$$c_{p1c} = 0.6 c_{p1} = 0.350$$

$$c_{nic} = 0.75 c_{ni} = 0.488$$

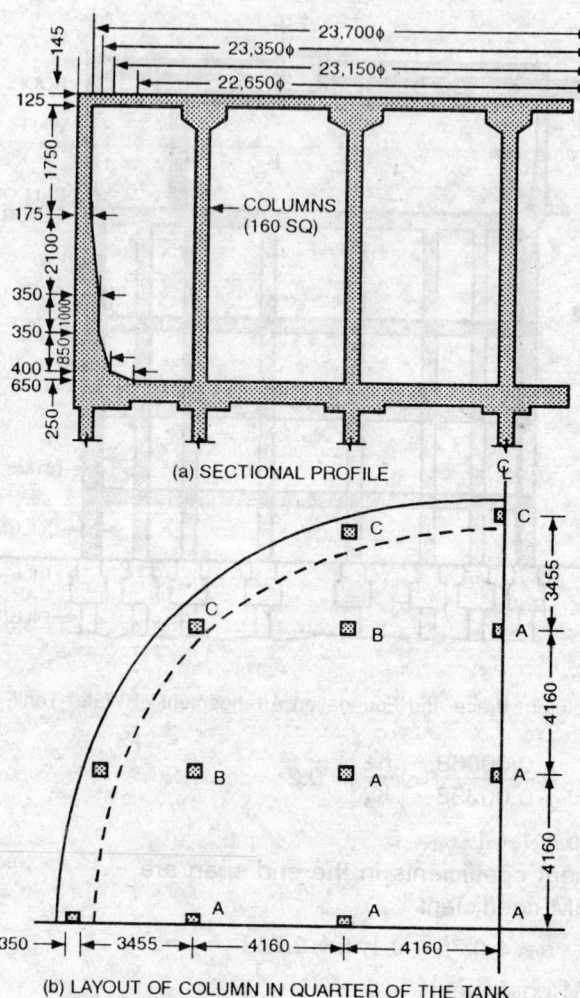

(a) SECTIONAL PROFILE

(b) LAYOUT OF COLUMN IN QUARTER OF THE TANK

Fig. 11.26 OHWT Size and Layout of Columns, Example 11.6

$$c_{pic} = 0.6c_{pt} = 0.21$$
$$c_{n3m} = 0.146$$
$$c_{n1m} = 0$$
$$c_{p1m} = 0.233$$
$$c_{nim} = 0.162$$
$$c_{pim} = 0.14$$

Maximum bending moment on the column strip is
$$M = 0.55\ M_0 = 15260\ Nm$$

Effective depth of the section required

$$d = \sqrt{\frac{M}{bR}} = \sqrt{\frac{15260}{2.08 \,(1.2)}} = 78 \text{ mm}$$

Provide $\qquad d = 125$ mm and $D = 145$ mm.

Maximum area of the reinforcement required at top in the column strip is

$$A_{st_1} = \frac{15\,260\,000}{0.87\,(125)\,(150)} = 935 \text{ mm}^2$$

Maximum area of the reinforcement required at the bottom in the column strips is

$$A_{st_2} = \frac{0.35\,(27750)\,(1000)}{0.87\,(125)\,(150)} = 596 \text{ mm}^2/2.08 \text{ m}$$

Minimum steel $= 0.23\,(2080)\,(145)/100 = 693 \text{ mm}^2/2.08 \text{ m}$

Provide 8 mm bars at 150 mm spacing at bottom and the alternate bars bent up. Add 8 mm extra bars at top at 175 mm spacing in the column strip.

Also provide 8 mm bars at 150 mm spacing at bottom in the middle strip with alternate bars bent.

Check for shear stress

Shear force $\qquad = 4350(4.16^2 - 0.825^2) = 72\,319 \text{ N}$

Shear stress $\qquad = \dfrac{72\,319}{4\,(825)\,(125)} = 0.17 \text{ MPa}$

This is admissible with nominal reinforcements.

Design of Columns Supporting the Roof Slab

Load on the column $\qquad\qquad\qquad\qquad\qquad\qquad\qquad W = 75\,280 \text{ N}$

Let the column size be 200 × 200 mm

Self-weight of the column $= 0.04\,(5.95)\,(25000) \qquad\qquad = 5\,950 \text{ N}$

Weight of the column head $= 0.5\,(0.49 - 0.04)\,(0.225)\,(25000) = 1266 \text{ N}$

$\qquad\qquad\qquad\qquad\qquad\qquad$ Total design load $\qquad = 82\,496 \text{ N}$

Effective length of the column $\quad = 0.7\,(5.95) = 4.165 \text{ m}$

Slenderness factor $\qquad\qquad = \dfrac{4165}{200} = 20.8$

The load reduction factor

$$C_r = 1.25 - \frac{20.8}{48} = 0.8177$$

Equivalent short column load

$$P_0 = \frac{82\,496}{0.8177} = 100\,900 \text{ N}$$

The bending moment on the column is

$$M = \frac{0.08(0.5w_1)L_0L^2}{\alpha}$$

$$= \frac{0.04(750)\,(4.16)\,(3.46)^2}{6} = 249 \text{ Nm}$$

$$\frac{M}{P} = 0.002 \text{ m}$$

The bending moment on the column is small and the column is designed as an uncracked section.

Provide the minimum reinforcement of 4 numbers of $\phi12$ bars.

$$A_{sc} = 4(113) = 452 \text{ mm}^2$$

$$A_t = A_g + 1.5\,(m-1)\,A_{sc} = 40000 + 1.5\,(12)\,(452) = 48\,136 \text{ mm}^2$$

$$Z = 200^3/6 = 1333\,333 \text{ mm}^3$$

The actual stresses in the section are

$$\sigma_{cc} = \frac{P_0}{A_t} = \frac{100\,900}{48\,136} = 2.10 \text{ MPa}$$

$$\sigma_{cb} = \frac{M}{Z} = 0.19 \text{ MPa}$$

Since the stresses are only nominal, let the column *section be* 160 mm *by* 160 mm.

$$A_t = 25600 + 8136 = 33\,736 \text{ mm}^2$$

$$Z = 682\,667 \text{ mm}^3$$

$$\text{Slenderness factor } = \frac{4165}{16} = 26$$

$$\text{Load reduction factor } = 1.25 - 26/48 = 0.71$$

$$\text{The equivalent short column load } = \frac{82496}{0.71} = 116\,190\text{N}$$

The actual stresses are

$$\sigma_{cc} = 116\,190/33\,736 = 3.44 \text{ MPa}$$

$$\sigma_{cb} = 249\,000/682\,667 = 0.36 \text{ MPa}$$

$$\frac{\sigma_{cc}}{\sigma_{acc}} + \frac{\sigma_{cb}}{\sigma_{acb}} = \frac{3.44}{5} + \frac{0.36}{7} = 0.74 < 1$$

Provide 6 mm ties at 160 mm spacing.

Design of cylindrical wall. The design load coefficients are taken from the IS 3370-1967 part IV. The base of the wall is resting on a thick slab which is subjected to dead and water load. So the base of the wall is treated as fixed. The top of the wall is treated as free. The force on the wall thus obtained will be on the safer side. The wall is variable in thickness so an equivalent thickness is taken as average value of the total area.

$$t = (0.150(1.75) + 0.5\,(0.175 + 0.35)\,(2.1) + 0.35\,(1.00)$$

$$+ 0.5(0.35 + 0.5)(0.85) + 0.5(0.5 + 0.75)(0.25))/5.95$$

$$= 0.29 \text{ m}$$

$$K = \frac{48H^4}{D^2t^2} = \frac{48(5.97)^4}{(24.05 - 0.58)^2(0.29)^2} = 1284$$

$$\log 1284 = 3.1$$

$$\frac{H^2}{Dt} = \frac{5.97^2}{(24.05 - 0.58)(0.29)} = 5.2$$

The coefficients are selected from code and are given in Table 11.8 along with the forces.

$$\text{Hoop tension} = \text{coefficient } (wRH)$$
$$\text{Moment} = \text{coefficient } (wH^3)$$

The forces, the area of the hoop tension and the actual stresses in the concrete are calculated in the tabular form using the following formulae:

Check for hoop tension

Area of the tension reinforcement is

$$A_{st} = \frac{T}{\sigma_{st}} = \frac{T}{150} \text{ mm}^2$$

Tension in the concrete is

$$\sigma_t = \frac{T}{A_c + (m-1)A_{st}}$$

$$= \frac{T}{1000t + 12 A_{st}} < 1.2 \text{ MPa}$$

Check for bending stresses is made by neglecting the effect of the reinforcement to be on the safer side.

$$\sigma_{cr} = \frac{M}{Z} = \frac{6M}{1000t^2} < 1.7 \text{ MPa}$$

The allowable stresses may be increased by about 10% due to the age of the concrete in the particular element. However, no increase in the allowable stresses of the wall, roof, and roof columns is permitted in this design. The allowable stresses in the floor and columns of the concrete may be given an age factor of 1.1 and that in foundation a factor of 1.15.

In can be seen that the hoop stress normally controls the thickness of the wall and the vertical reinforcement is controlled by that minimum required. The details of the reinforcement in the wall is shown in Fig. 11.27.

Table 11.9 Hoop Tension and Reinforcement (per metre depth)

Depth	Coefficient	Force (N)	A_{st} (mm^2)	t (mm)	σ_t (MPa)
0.2H	0.242	169 400	1129	155	1.0
0.4H	0.431	301 900	2013	288	1.19
0.6H	0.478	334 900	2233	328	0.94
0.8H	0.267	187 100	1247	350	0.51
0.9H	0.094	65 900	440	380	0.17
1.0H	0	0	0	650	0

Table 11.10 Bending Moment (Nm/m and Stress (MPa)

Depth	Coefficient	Moment	t(mm)	σ_{cr}
0.2H	0.0007	1489	155	0.37
0.4H	0.0027	5745	228	0.66
0.6H	0.0056	11915	328	0.66
0.8H	0.0028	5958	350	0.29
0.9H	− 0.0055	− 11702	380	0.48
1.0H	− 0.0216	− 45960	650	0.65

Design of floor slab. The floor slab is designed as a flat slab with drops. The columns are spaced at 4.16 m apart except at the peripheral segments. Another set of columns are placed below the circular wall as shown in Fig. 11.26. For the purpose of computing self-weight, let the thickness of the slab be assumed as 350 mm and that of the drop as 550 mm. The size of the drop be assumed as 1600 by 1600 mm for the purpose of weight calculations.

Weight of the slab $= 0.35(25000)$ $= 8750$ N/m^2

Weight of the drop $= \dfrac{2.56(9.2)(25000)}{(4.16)(4.16)}$ $= 740$ N/m^2

Weight of the water $= 59500$ N/m^2

Total Load $w = 68990$ N/m^2

use $w = 69000$ N/m^2

Clear span between the column heads

$$= 4.16 - 1.6 = 2.56 \text{ m}$$

Minimum design clear span $= 0.65L = 2.704$ m

Use $\qquad L_0 = 2.704$ m

Total design load on the panel

$$W = wLL_0 = 69000(4.16)(2.704)$$

$$M_0 = \frac{WL_0}{8} = 262340 \text{ Nm}$$

Assume the following for the purpose of computing the relative stiffnesses of the column and the slab.

Size of the floor column $= 400 \times 400$ mm
Height of the column $= 4.1$ m
Effective height $= 2.9$ m
Relative stiffness of the floor column $= 0.0088$ m^4
Relative stiffness of the roof column $= 0.00016$
Relative stiffness of the slab $= 0.06125$

$$\alpha_c = \frac{0.0088 + 0.00016}{0.06125} = 0.146$$

$$\alpha = 1 + \frac{1}{\alpha_c} = 7.85$$

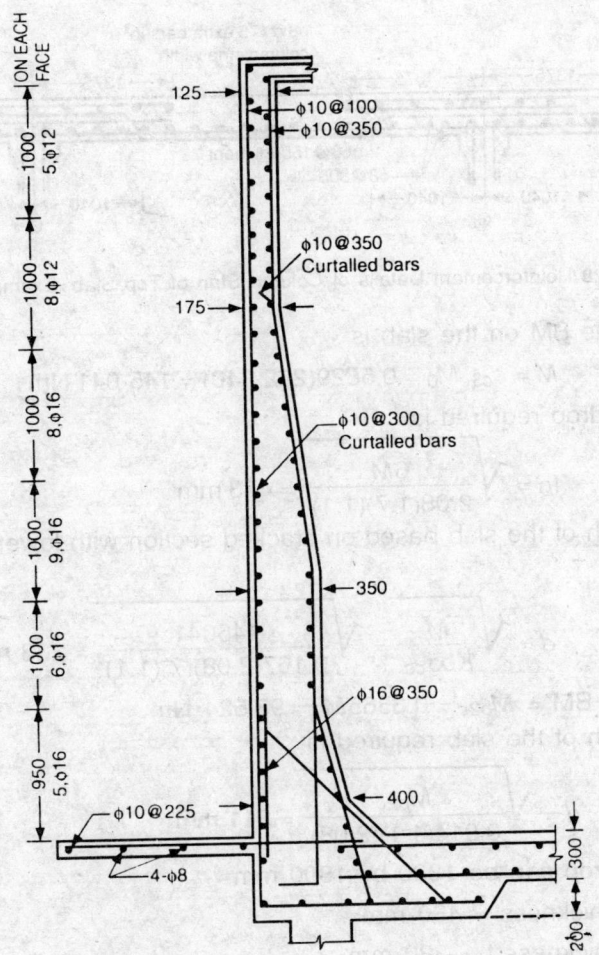

The following labels appear in the figure:

ON EACH FACE

1000 5,φ12

1000 8,φ12

1000 8,φ16

1000 9,φ16

1000 6,φ16

950 5,φ16

125

φ10@100
φ10@350

φ10@350
Curtailed bars

175

φ10@300
Curtailed bars

350

φ16@350

400

φ10@225

4-φ8

300

200

Fig. 11.27 Reinforcement Detail of Cylindrical Wall, Example 11.6.

The bending moment coefficients in the design of the flat slab are:

End span

$$c_{n1} = 0.65/7.85$$
$$c_{p2} = 0.63 - 0.28/7.85$$
$$c_{n2} = 0.75 - 0.1/7.85$$

Interior span $c_{ni} = 0.65, \ c_{pi} = 0.35$

The BM coefficients on the column strip are:

$$c_{n1c} = c_{n1} = 0.0828$$
$$c_{p2c} = 0.6c_{p1} = 0.3565$$
$$c_{n3c} = 0.75c_{n2} = 0.5529$$
$$c_{nic} = 0.75c_{ni} = 0.4875$$
$$c_{pic} = 0.60c_{pi} = 0.21$$

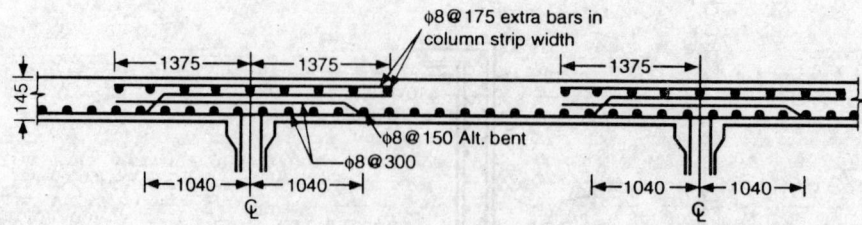

Fig. 11.28 Reinforcement Details of Column Strip of Top Slab, Example 11.6.

The maximum possible BM on the slab is

$$M = c_{c3c}M_0 = 0.5529(262\ 340) = 145\ 041\ Nm$$

Thickness of the drop required is

$$t_D = \sqrt{\frac{6M}{2.08(1.7)(1.1)}} = 473\ mm$$

The effective depth of the slab based on cracked section with lower permissible stress in steel is

$$d = \sqrt{\frac{M}{Kb\sigma_{cb}}} = \sqrt{\frac{145041}{0.167(2.08)(7)(1.1)}} = 233\ mm$$

Maximum positive BM = $M_{p2c} = 0.3565M_0 = 93\ 524$ Nm

The effective depth of the slab required is

$$d = \sqrt{\frac{M_{p2c}}{0.914(1.1)(2.08)}} = 211\ mm$$

Use The size of the drop can be 1800 by 1800 mm

Drop thickness = 480 mm

Slab thickness t = 330 mm

$$d \doteq 290\ mm$$

Design of reinforcement in the column strip per 2.08 m

Top reinforcement

$$A_{st} = \frac{M_{n3c}}{150jd} = 2526\ mm^2$$

Bottom reinforcement

$$A_{stb} = \frac{M_{p2c}}{230jd} = 1550\ mm^2$$

Provide 14 numbers of 12 mm bars at the bottom face of the slab and the alternate bars can be bent. 9 numbers of 16 mm bars at top face over the column support-extra.

Reinforcement in the middle strip per 2.08 m

Maximum negative BM is

$$M = 0.25_{n3}M_0 = 48\ 354 \text{ Nm}$$

The top face reinforcement required is

$$A_{st} = \frac{M}{150jd} = 842 \text{ mm}^2$$

Maximum positive BM is

$$M = 0.4c_{p1}\ M_0 = 62\ 349 \text{ Nm}$$

The bottom face reinforcement required is

$$A_{stb} = \frac{M}{230jd} = 1034 \text{ mm}^2$$

The minimum required top and bottom face reinforcement areas are: 1597 mm^2 and 1240 mm^2 respectively.

Provide 12 numbers of 12 mm bars at bottom face and let alternate bars be bent up. Five numbers of ϕ16-bars extra be added at top face.

It can be observed that the reinforcement in the middle strip is controlled by the minimum percentage even in the most critical zone. Even the bottom reinforcement in the column strip is also controlled by the minimum percentage. The bending moment coefficients in the other segments are smaller than those used in these calculations. Hence, the same reinforcement is extended through all the panels.

Check for shear stress. The shear stress is computed for a drop size of 1800 mm square. The shear force in the periphery of the drop at distance 0.5d from the face of the drop is

$$V = 69000\ (4.16^2 - 2.18^2) = 866\ 180 \text{ N}$$

Nominal shear stress is

$$\frac{866180}{4(2180)(330)} = 0.30 \text{ MPa}$$

This value is less than the allowable stress.

Size of the column head = 1000 mm

Shear force over the periphery of the column

$$V = 69000(4.16^2 - 1.48^2) = 1042\ 950 \text{N}$$

Nominal shear stress

$$= \frac{1042\ 950}{4(1480)(480)} = 0.37 \text{ MPa}$$

This value is allowable.

Allowable diagonal shear $= 0.16\ \sqrt{f_{ck}} = 0.71 \text{MPa}$

Design of Columns. Age factor of 1.15 is given to the concrete in the columns.

Column supporting 4.16 square panel

    Type-A (4.16 m square panel)

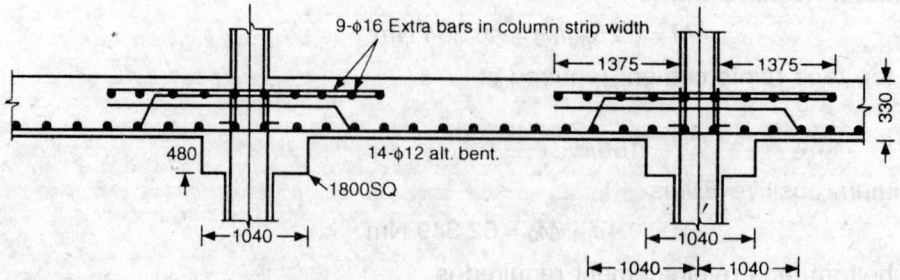

Fig. 11.29 Reinforcement Details of Bottom Slab, Example 11.6.

Area of the panel $A_A = 17.31$ m^2

Load from the roof column   = 82 496 N

Load of the floor slab   = 142 771 N

Load of the drop   = 4500 N

Load of the water   = 1 021 030 N

Weight of the column (450 mm)   = 91 125 N

Load of the braces   = 63 000 N

(3×2 braces of $0.3 \times 0.4 \times .3.5$)

                   Total load   = 1 404 922 N

Use total load $W_A = 1 410 000$ N

Provide 8 numbers of 16 mm bars, then the capacity of the bars is

$$P_s = 8(201)(190) = 305\ 520\ \text{N}$$

Load to be resisted by the concrete section

$$= 1\ 410\ 000 - 305\ 520 = 1\ 104\ 480\ \text{N}$$

Area of the concrete required is

$$A_c = \frac{W_A - P_s}{5(1.15)} = 192\ 082\ \text{mm}^2$$

$$A_g = A_c + A_{sc} = 193\ 691\ \text{mm}^2$$

Size of the square column   = 440 mm

Percentage of steel   = $p = 160800/400^2 = 0.83\%$

Provide 440 by 440 mm size column with 8 numbers of 16 mm main bars and 6 mm stirrups at 250 mm spacing.

    This column design is to be checked for seismic load condition.

Type B column: Column close to the wall

Area of the floor supported by the column is

$$A_B = \frac{(4.16 + 3.455)^2}{4} = 14.5\ \text{m}^2$$

The load on the column from the floor slab is prorated based on the area of the support. Therefore, it is

$$w_A \frac{A_B}{A_A} = 1\ 410\ 000 \frac{(14.5)}{17.31} = 1\ 181\ 000 \text{ N}$$

Add extra for the column weight, etc. = 69000 N

Use W_B = 1250,000 N

Let the size of the column = 420 × 420 mm.

Provide 4 numbers of 16 mm bars and 4 numbers of 12 mm bars with 6 mm ties at 250 mm spacing.

Type C column (under the wall)

Approximate maximum spacing along the periphery	= 4.9 m
Segment width of slab = 1.75 + 0.25	= 2 m
Load from the roof = 4.9(2)(4350)	= 42 630 N

Load of the wall

$$(25000)\ (4.9)\ (0.175(1.75)) + (0.263)\ (2.1) + (0.35)\ (1.0)$$
$$+ (0.375)\ (0.85) + (0.525)\ (0.25) = 203\ 172 \text{ N}$$

Load from the floor slab (25000) (4.9) (2) (0.40)   = 98 000 N

Load of the water

$$= 10000\ (4.9)\ (1.825)\ (1.75) + 1.737\ (2.1) + 1.65\ (1)$$
$$+ 1.625\ (0.85) + 1.475\ (0.25) = 501\ 831$$

Weight of the balcony including the LL = 4000 (4.9) 0.9)	= 17 640 N
Weight of the column = 25000 (0.16)(17.52)	= 70 080 N
Weight of the braces and staircase = 25000 (0.16) (3.4)	= 13 600 N
Total	= 946 953 N

The total load on the column computation is made on the safer side by taking a larger projected area.

Therefore, use the total load on the column as

$$W_c = 940\ 000 \text{ N}$$

Provide 4 numbers of 16 mm and 4 numbers of 12 mm bars, then

$$A_{sc} = 1256 \text{ mm}^2$$

$$P_s = 1256\ (190) = 238\ 640 \text{ N}$$

$$W_c - P_s = 701\ 360 \text{ N}$$

then $A_c = 121\ 976 \text{ mm}^2$

$$A_g = 123\ 232 \text{ mm}^2$$

Size of the column be   = 350 × 350 mm

Provide 4 numbers of 16 and 12 mm bars as main reinforcement and 6 mm ties at 250 mm spacing.

Total Load Calculation. All loads are in kN units.

Gross area = 3.141(12.025)²   = 454.28 m²

Weight of:

Roof slab = 65.8705(25)   = 1646.8

Roof columns(21 numbers)	= 126
Column heads and drops-roof	= 39.3
Tank wall	= 3015.6
Floor slab	= 3747.8
Drops of floor slab	= 600.3
Balcony	= 173.2
Columns above GL	
13 nos. 440 mm	= 1102.3
8 nos. 420 mm	= 618.1
16 nos. 350 mm	= 858.5
Braces (estimated at 0.13 m^2)	= 2300.0
Staircase	= 100.0

Total weight above staging $\quad W_1 = 9223$ kN

Total weight of the staging $\quad W_2 = 4979$ kN

$$\text{Total weight } W_3 = 14202 \text{ kN}$$

Weight of water $\qquad\qquad W_4 = 25100$ kN

Live load $\qquad\qquad\qquad W_5 = 350$ kN

Total superimposed live load $\quad W_6 = 25450$ kN

Approximate amount of concrete

use in the structure = 592 m^2

concrete = 0.247 m^3/kl of water.

Seismic and Wind Forces: Effective seismic load is

$$W_7 = W_1 + W_4 + 0.5W_5 + \frac{W_6}{3} = 36\ 158 \text{ kN}$$

Inertia of the columns about the main diagonal is

$$I_f = (0.35)^2(8.32 + 3.0)^2 + 10(0.42)^2(8.32)^2\frac{(8.32)}{11.775}$$

$$+ 14(0.44)^2(4.16)^2\frac{(4.16)}{11.775} = 196.75 \text{ m}^4$$

Use $\qquad\qquad E_c = 25000$ MPa

Length of the column when provided with three intermediate braces is

$$L_e = (17.52 + 0.5)/4 = 4.5 \text{ m}$$

Translational stiffness of all the columns is

$$\frac{12EI}{L_e^3} = \frac{E_c}{L_a^3}(16(0.35)^4 + 8(0.42)^4 + 13(0.44)^4) = 268 \text{ MN/m}$$

Lateral deflection caused by W_7 is

$$v = \frac{4(36.158)}{268} = 0.73 \text{ m}$$

Free period of oscillations of the structure is

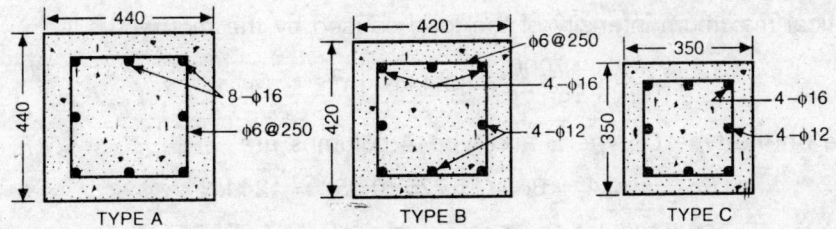

Fig. 11.30 Sectional Details of Columns, Example 11.6.

$$T = 2\pi \sqrt{\frac{v}{g}} = 1.7 \text{ seconds}$$

The corresponding average acceleration coefficient for 5% damping from code is

$$\frac{S_a}{g} = 0.06$$

The coefficient of dependence upon the soil foundation system is

$$\beta = 1.2$$

Importance factor $\qquad I_\alpha = 1.5$

Seismic zone response spectrum factor $F_0 = 0.2$

The seismic coefficient to the structure is

$$\alpha_h = \beta I_\alpha F_0 \frac{S_a}{g} - = 1.2(1.5)(0.2)(0.06) = 0.0216$$

This implies that the structure is flexible and responses favourably to the seismic loads.

Use $\qquad \alpha_h = 0.022$

The horizontal seismic loads are:
Corresponding to the overhead tank

$$F_{1s} = \alpha_h(w_1 + w_4 + w_3) = 762.8 \text{ kN}$$

Corresponding to the staging is

$$F_{2s} = \alpha_h(w_2) = 109.5 \text{ kN}$$

Assume a basic wind pressure of 1.5 kN/m^2, then the wind force on the tank and the staging are:

$$F_{1w} = 0.7pDH = 0.7(1.5)(24.05)(6.575) = 166 \text{ kN}$$

$$F_{2w} = (9(0.42)(17.52) + 4(24.05 - 9(0.42)(0.5))\, 1.5p = 240 \text{ kN}$$

The seismic forces dominate over the wind load; therefore, the net shear and bending moment caused by the seismic load at the foundation level are

$$V = F_{1s} + F_{2s} = 762.8 + 109.5 = 872.3 \text{ kN}$$

$$M_f = F_{1s}(18 + 5.95(0.5)) + F_{2s}(9) = 16985 \text{ kNm}$$

Use $\qquad V = 900 \text{ kN}$

$$M_f = 17000 \text{ kNm}$$

The hypothetical maximum intensity of the load caused by the moment is

$$w = \frac{M_f}{I_f} = \frac{17000}{196.75} = 86.4 \text{ kN/m}^3$$

Seismic force on the type C, type B and type A columns are

$$P_{Cs} = wd_cA_c = 86.4(11.775)(0.35)^2 = 124 \text{ kN}$$

$$P_{Bs} = 86.4(8.32)(0.42)^2 = 126 \text{ kN}$$

$$P_{As} = 86.4(4.16)(0.44)^2 = 70 \text{ kN}$$

The total area of cross-section of the columns is

$$A_t = 16(0.35)^2 + 8(0.42)^2 + 13(0.44)^2 = 5.888 \text{ m}^2$$

Clear height of each column panel $L_0 = 4$ m

The bending moments on the columns of types C, B and A in the lowest panel are:

$$M_C = \frac{V}{AE} A_c \frac{L_0}{2} = \frac{900}{5.888}(0.35)^2(0.5)(4) = 37.5 \text{ kNm}$$

Similarly

$$M_B = 54 \text{ kNm}$$

$$M_A = 59 \text{ kNm}$$

Maximum forces acting on the different types of columns in the normal and seismic load condition are listed in Table 11.11 with the equivalent eccentricities, etc.
The seismic load varies from 5 to 12% of the normal axial load on the columns, and the eccentricity ratio to the side of the column is about 10%. But the allowable stress under seismic load condition is 33% more than the normal load condition. Therefore, the seismic load does not govern the design of the columns.

Design of Column Braces. The maximum bending moment on the column braces occurs at the outer most end of the lowest brace, and it is

$$M_1 = 1.56 \ M_C = 0.05876 \text{ MNm}$$

The corresponding bending moment at the other end of the brace is

$$M_2 = M_B = 0.04793 \text{ MNm}$$

The maximum shear force on the brace is

$$V = \frac{M_1 + M_2}{a} = \frac{0.10669}{4.16} = 0.02546 \text{ MN}$$

Assume an age factor to the concrete as 1.15, and width of the brace as 200 mm, then the effective depth of the brace based on the maximum BM and balanced section is

$$d = \sqrt{\frac{M_1}{1.15 \ bK\sigma_{cb}}} = \sqrt{\frac{0.05876}{1.15(0.20)(0.914)}} = 0.48 \text{ m}$$

Since the brace will be provided with top and bottom reinforcement and the critical BM is at the face of the column, the depth computed above gives a conservative estimate. Area of the

Table 11.11 Forces on Columns (all in MN and m units)

Type	Axial force (MN)			BM(MNm)	$\frac{BM}{P} = e$	$\frac{e}{D}$	$\frac{P_s}{W_1}$
	LL	Seismic	Total		M/P (m)		
	W_1	P_s	P	M			
A	1.41	0.070	1.48	0.059	0.04	0.09	0.05
B	1.25	0.126	1.376	0.054	0.04	0.10	0.1
C	0.94	0.124	1.064	0.0375	0.035	0.1	0.13

tension reinforcement is (use $d = 0.48$ m)

$$A_{st} = \frac{M_1}{jd\sigma_{st}} = \frac{58760}{0.904(0.48)(230)(1.33)} = 445 \text{ mm}^2$$

Shear force on the brace at a distance d from the face of the column is

$$V_s = 25600 \left(\frac{2.08 - 0.48 - 0.18}{2.08} \right) = 17600 \text{ N}$$

The nominal shear stress on the brace is

$$\tau_v = \frac{V_s}{bd} = 0.19 \text{ MPa}$$

The shear stress is less than the allowable value; therefore, provide only nominal transverse reinforcement. Select two-legged 6 mm stirrups, then the maximum spacing is

$$s_v = \frac{A_{sv}f_y}{0.4 \, b} = \frac{2(28)(415)}{0.4(200)} = 290 \text{ mm}$$

Provide 250 by 500 mm size brace with 4 numbers of 12 mm bars each at top and bottom 6 mm stirrups at 280 mm spacing.

Design of Foundations: The seismic or wind load conditions are not critical in the design of columns, so they are also not critical in the design of the foundations. An independent set of piles is designed to support each of the columns.

Foundation to type A columns

Load from the column = W_A	= 1410 kN
Weight of pile cap = (6) (0.5) (25)	= 75 kN
Weight of pedestal = (0.9) (0.9) (0.4) (25)	= 8kN
Add extra for over burden, etc.	= 27 kN
Total load on piles	= 1520 kN

Provide 5 numbers of double under-reamed piles of 300 mm diameter

$$\text{Load on each pile} = \frac{1520}{5} = 304 \text{ kN}$$

Double under-reamed 5 m long, 300 mm pile has a capacity of 310 kN
Assuming the size of the capping slab = 2.3 × 2.3 m
Distance of the pile from end = 250 mm

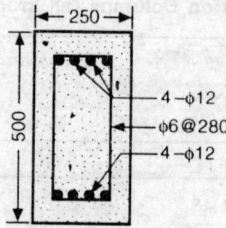

Fig. 11.31 Sectional Details of Brace, Example 11.6.

The clear spacing available between the piles

$$= \sqrt{2}(1150 - 550 + 150) - 150 = 850 \text{ mm}$$

Size of the pedestal = 900 × 900 mm

Moment arm $\qquad = \dfrac{2.3 - 0.9}{2} - 0.25 = 0.45 \text{ m}$

Maximum bending moment is

$$M = 2(304)(0.45) - \frac{110}{3} \times \frac{(1.15)}{3} = 260 \text{ kNm}$$

Quality of concrete $\quad$ = M20

The effective depth of the section is

$$d = \sqrt{\frac{M}{bK\sigma_{cb}}} = \sqrt{\frac{0.260}{2.3(0.130)(7)}} = 0.36 \text{ m}$$

Provide overall and effective depth of the pile cap as

D = 500 mm, d = 415 mm.

The area of the tension reinforcement is

$$A_{st} = \frac{M}{jd\,\sigma_{st}} = \frac{260000}{0.904(0.415)(230)} = 3013 \text{ mm}^2$$

Provide 15 numbers of 16 mm bars in each direction over the width of the pile cap.

The critical section for shear is at a distance d from the face of the column, and this distance is 0.415 + 0.15 + 0.3 = 0.865 m from free end.

This line passes through the pedestal column section, therefore, there is no critical shear stress problem.

The development length of the bars is

$$L_d = \frac{\phi\sigma_s}{4\tau_{bd}} = \frac{16(230)}{4(1.6)} = 575 \text{ mm}$$

The distance of the critical moment zone from the free and is 600 mm; therefore, the length of the bar with an end cover of 75 mm is not adequate. Provide a bend at the end of the bars with 75 mm leg.

Foundation to type B columns

Load from the column	= 1250 kN
Weight of pile cap, etc.	= 70 mN
Total load	= 1320 kN

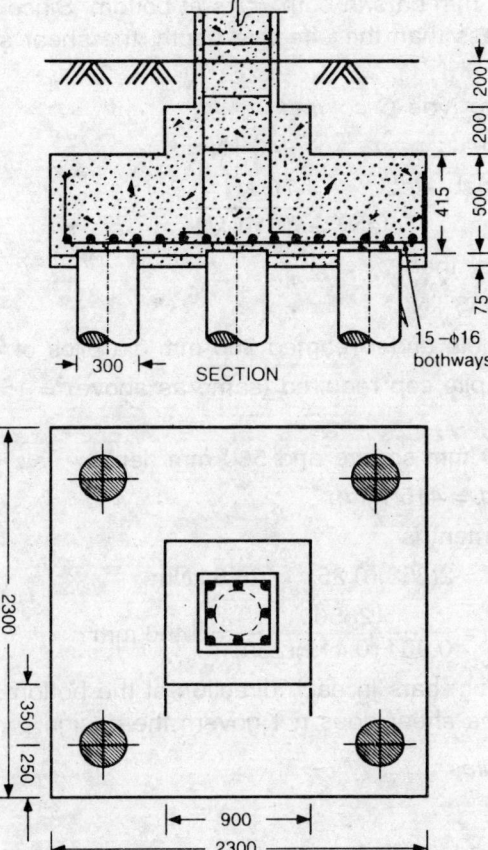

The section shows dimensions: 200, 200, 200, 415, 500, 75, 300, and label "15-φ16 bothways". The plan view shows dimensions 2300, 350, 250, 900, 2300.

Fig. 11.32 Details of Pile Cap under a Typical Column, Example 11.6.

Using four piles, the load on each pile is 330 kN.
Provide 4 numbers of double under-reamed piles of 300 mm diameter and 6 m long.

Distance of pile from free edge = 250 mm
Minimum width of pile cap is

$$B_m = 2(250) + 300 + 750 = 1550 \text{ mm}$$

Provide 1600 by 1600 mm pile cap with 600 by 600 by 200 mm pedestal. **The bending moment on the pile cap is**

$$M = 2(330)(0.8 - 0.3 - 0.25) = 165 \text{ kNm}$$

Provide overall and effective depths of the pile cap as

$$D = 500 \text{ mm}, d = 415 \text{ mm}$$

The area of the tension steel is

$$A_{st} = \frac{165\,000}{0.904(0.415)(230)} = 1922 \text{ mm}^2$$

Provide 10 numbers of 16 mm bars in both ways at bottom. Since the clear spacing between the pedestal and pile is less than the effective depth, the shear stress does not govern the design.

Design of foundation to type C column

Load from the column	= 940 kN
Add extra for caping slab etc.	= 70 kN
Total	= 1010 kN

Provide 4 numbers of piles, then

Load on each pile = 253 kN

Use 4 numbers of double under-reamed 300 mm dia piles of 4000 mm depth.

Minimum width of the pile cap required (same as above) = 1550 mm

Provide a pile cap of 1600 mm square and 500 mm depth

$$d = 415 \text{ mm.}$$

Maximum bending moment is

$$M = 2(253)(0.25) = 126.5 \text{ kNm}$$

$$A_{st} = \frac{126500}{0.904 \, (0.415)(230)} = 1466 \text{ mm}^2$$

Provide 8 numbers of 16 mm bars in each direction at the bottom. As seen from the design of the previous pile cap, the shear does not govern the design.

Design of under-reamed piles

Double under-reamed

Diameter of the pile = 300 mm

Type of soil silty-clay with about 20% fine sand

Capacity of 3.5 m double under-reamed pile	= 240 kN
Capacity of 1.5 m extra length at 140 kN/3 m	= 70 kN
Total capacity	= 310 kN

Design of Balcony

Width of the balcony	= 0.9 m
Weight of the balcony (0.1) (2500)	= 2500 N/m^2
Extra load for railing, etc.	= 500 N/m^2
Live load	= 1500 N/m^2
Total load	= 4500 N/m^2

Bending moment $= \dfrac{4500 \, (0.9)^2}{2} = 1830 \text{ Nm/m}$

$$d = \sqrt{\frac{1830}{0.91}} = 45 \text{ mm}$$

Use $\qquad D = 100$ mm, $d = 75$ mm

$$A_{st} = \frac{1830\,000}{0.904\,(75)\,(230)} = 117\ \text{mm}^2/\text{m}$$

Provide 10 mm bars at 225 mm spacing at top of the slab cantilevered into the tank wall. Also provide 4 numbers of 8 ϕ as distribution reinforcement over the width.

Design of Staircase. The staircase is provided along the periphery of the staging. Landings are provided on the outer columns and the staircase slab spans between the landings as a continuous one-way slab.

Maximum span of the staircase	= 4.8 m	
Average width of landing	= 600 mm	
Width of staircase	= 900 mm	
Rise of the steps	= 230 mm	
Tread of the steps	= 253 mm	
Let the thickness of the slab	= 150 mm	
Weight of the slab	= 0.15 (1.36) (0.9) (25000)	= 4590 N/m
Weight of the steps		= 2875 N/m
Railing load, etc.		= 135 N/m
	Total dead load	= 7600 N/m
	Live load	= 2250 N/m
	Total load = w	= 9850 N/m

Maximum negative and positive BMs are

$$M_1 = \frac{wL^2}{10} = 22695\ \text{Nm}$$

$$M_2 = \frac{wL^2}{12} = 18912\ \text{Nm}$$

The corresponding collapse moment are (with partial safety factor of 1.5)

$$M_{c1} = 1.5\ (22\,695)\ \text{Nm}$$

$$M_{c2} = 1.5\ (18\,912)\ \text{Nm}$$

Effective depth required is

$$d = \sqrt{\frac{M_{c1}}{(0.9)(0.130)(20)}} = 126\ \text{mm}$$

Use $\qquad\qquad D = 150$ mm.

The areas of the tension reinforcement for negative and positive BMs are

$$A_{st1} = \frac{1.15 M_{c1}}{jdf_y} = 913\ \text{mm}^2$$

$$A_{st2} = 761\ \text{mm}^2$$

Provide 7 numbers of 12 mm bars at bottom of which 3 bars bent up at 1300 mm from the support. Add 6 numbers of 12 mm bars at top in the supports. Also provide 8 mm bars at 150 mm spacing as distribution reinforcement.

Design of landing

The superimposed load on the landing

$$\frac{(9850)\ (4.8)}{} = 47\ 280\ N$$

Self-weight = 0.25 (0.9) (0.65) (25000) = 36 56 N

Total load W = 50 936 N

Cantilever span = L = 0.9 m

Maximum collapse moment

$$M_c = \frac{1.5 WL^2}{2} = 30944\ Nm$$

Effective depth required

$$d = \sqrt{\frac{M_c}{bKf_{ck}}} = \frac{30944000}{400\ (0.138)\ (20)} = 168\ mm$$

Use

$$D = 250\ mm\ and\ d = 220\ mm$$
$$A_{st} = 478\ mm^2$$

Maximum shear stress in the landing occurs at a distance *d* from the support and at this point the staircase slab comes into effect; therefore, the shear stress is negligible.

11.6 Recommended Dimension for Intze Tank and Examples

Since the Intze water tank is one of the most commonly used overhead water tanks approximate dimensions and reinforcement details are worked out for different capacities of the tanks. The recommended dimensions are for the following conditions:

Concrete M20 and steel is of HYSD-Fe415 bars
The thickness of the tank wall at top is 100 mm.

For notations see Fig. 11.18.

Table 11.12 gives the recommended dimensions of the tank, Table 11.13 gives the details of top dome and ring beam, and Tables 11.14 and 11.15 give the details of tank wall, middle ring beam and conical wall. The design details for the bottom ring beam are not listed as the number of supports and the type of support depends on the soil and other considerations.

EXAMPLE 11.7 *Design of 1000 kl Overhead water tank with 20 m staging.*

Design data and capacity calculations

Capacity of the tank = 1000 kl, free board = 0.15 m
Staging of the tank = 20 m

The tank is to be constructed on a foundation which is already located. There are two sets of columns placed on circles of radii 6.15 m and 9.65 m.

The reinforcement is high yield strength deformed bars of f_y = 415 MPa.
Concrete in foundation is f_{ck} = 20 MPa
Concrete in tank is f_{ck} = 20 MPa
Concrete in columns and braces is f_{ck} = 25 MPa

Table 11.12 Typical Dimensions of Intze Water Tank (all dimensions in mm)

No.	Capacity (kl)	r_1	h	h_1	h_2	r_3	h_3
1	300	4500	4450	1205	1400	3375	1125
2	400	5000	4650	1340	1550	3750	1250
3	500	5500	4700	1475	1710	4125	1375
4	600	6000	4750	1600	1860	4500	1500
5	700	6500	4800	1740	2010	4875	1625
6	800	7000	4850	1875	2170	5250	1750
7	900	7250	4900	1940	2250	5440	1810
8	1000	7500	5100	2010	2330	5624	1875
9	1100	7750	5250	2080	2400	5810	1940
10	1200	7800	5350	2140	2480	6000	2000
11	1300	8250	5400	2210	2560	6190	2060
12	1400	8500	5500	2280	2640	6375	2125
13	1500	8750	5550	2340	2720	6560	2180
14	1600	9000	5550	2410	2800	6750	2250
15	1700	9250	5550	2480	2870	6940	2310
16	1800	9500	5550	2550	2870	7125	2375
17	1900	9750	5250	2610	3030	7310	2440
18	2000	10000	5550	2680	3100	7500	2500
19	2100	10250	5500	2750	3180	7690	2560
20	2200	10450	5550	2800	3250	7840	2610
21	2300	10700	5500	2870	3320	8025	2675
22	2400	11000	5400	2950	3420	8250	2750
23	2500	11200	5450	3000	3480	8400	2800

Safe net bearing capacity of the soil = 70 kN/m^2
(see Fig. 11.33 for notations)
Let the inside radius of the cylindrical wall of the tank at top be $r_1 = 9.6$ m and that at base $r_2 = 9.4$ m.

Chord radius of the dome $r_3 = 6.15$ m

The clear height of the wall $= h$

Rise of the top and lower domes $= h_1 = h_2 = 2$ m

Gross volume of the tank $= \pi(r_1^2 + r_2^2)h/2 = 283.56h$

Volume of the free board $= \pi r_1^2 (0.15) = 43.4$ m^3

Volume of the lower dome $= \pi(3r_3^2 + h_2^2)h_2/6 = 123$ m^3

Volume of the inside eight columns$= \pi(0.15)^2(h)8 = 0.56\,h$

The net volume of the tank is where the diameter of the column is assumed as 0.3 m and there are eight such columns in the tank.

$$283.56h - 43.4 - 123 - 0.56h > 1000 \text{ m}^2$$

or $h > (1000 + 43.4 123)/(283.56 - 0.56) = 4.12$ m

Table 11.13 Structural Details of Top Dome and Ring Beam (All Dimensions are in mm) Intze Tank

No.	Capacity (kl)	Top dome			Top ring beam			Reinforcement		
		t	o	s	b	D	o	N	ϕ_v	s
1	300	80	8	225	170	350	12	4	8	200
2	400	80			190	380	12			
3	500	80			210	420	14			
4	600	80			230	460	14			
5	700	80			250	480	16			
6	800	90			265	535	16			
7	900	90	8	225	265	535	18	4	8	200
8	1000	90	8	200	285	570	18	4	10	200
9	1100	90			305	610	18			
10	1200	90			305	610	18			
11	1300	90			315	630	20			
12	1400	100			325	650	20			
13	1500	100			325	650	20	4	10	200
14	1600	100	8	200	335	670	18	6	10	180
15	1700	100			340	690	20			
16	1800	100			340	690	22			
17	1900	100			350	710	22			160
18	2000	100			360	725	25			
19	2100	100			360	725	25			
20	2200	100			375	740	25			
21	2300	100			380	760	25			
22	2400	100			380	760	25			
23	2500	100	8	200	380	760	25	6	10	160

o = Diameter of main bar, N = Number of main bars, $\rightarrow$ Implies applicable range, ϕ_v = Diameter of ties, s = Spacing of reinforcement, t = Thickness, b = Width, D = Depth.

Use height of the tank wall = h = 4.15 m.

Design of top dome :

Radius of the chord = r_3 = 6.15 m

Rise of the shell = h_1 = 2 m

Radius of the shell surface = $R = (r_3^2 + h_1^2)/2h_1 = 10.45$ m.

Semi-central angle is given by

$\sin \theta = 6.15/10.45 = 0.59$, that is, $\theta = 36°$

Let the thickness of the dome be assumed as 0.08 m, then the self weight of the dome is

$$w_g = 0.08(25) = 2 \text{ kN/m}^2$$

Live load $w_1 = 0.75$ kN/m^2

Total load $w = 2.75$ kN/m^2

Table 11.14 Structural Details of Wall, Middle Ring Beam (All Dimensions in mm) Intze Tank

No.	Capacity (kl)	Details of wall at bottom						Middle ring beam			
		t	Hoop		Vertical		b	D	Main		Stirrups
			φ	s	φ	s			φ	s	φ$_v$	s
1	300	155	16	110	10	130	550	150	20	4	8	150
2	400	185	16	100	10	130	700	150	20			
3	500	210	16	100	10	130	825	150	20			
4	600	225	16	100	10	130	995	150	22			
5	700	240	18	100	10	125	1000	160	22			
6	800	250	18	10	10	125	1000	170	22			
7	900	275	18	100	10	125	1000	195	22	6	8	150
8	1000	295	20	100	10	122	1000	215	22	6	10	200
9	1100	315	20	100	10	125		240	22			
10	1200	330	20	100	12	160		265	25			
11	1300	345	20	100	12	160		285	25			
12	1400	360	22	110	12	150		305	25			
13	1500	375	22	110	12	150		325	28	6	10	200
14	1600	385	22	110	12	130	1000	345	28	6	12	300
15	1700	400	22	110	12	125		365				
16	1800	410	22	100				385				
17	1900	420	22	100				385				
18	2000	430	22	100			1000	405		8		300
19	2100	435	25	110	12	125	1100	405	28	8	12	300
20	2200	450	25	110	12	110	1100	425		10		
21	2300	455	25	110			1100	445		10		
22	2400	460	25	100			1150	440		10		
23	2500	470	25	100			1150	460	28	10		300

Note: The spacing of the reinforcement listed is for a single layer.

$$\text{Weight of the dome} = W_1 = 2\pi R h_1 w_g = 263 \text{ kN}$$

$$\text{Live load on the dome} = W_2 = 2\pi R h_1 w_1 = 98 \text{ kN}$$

$$\text{Total load} = 361 \text{ kN}$$

$$\text{Meridional thrust} = N_\phi = \frac{wR}{1 + \cos\theta} = 15.87 \text{ kN/m}$$

$$\text{Stress} = \frac{0.01587}{0.08} = 0.2 \text{ MPa}$$

As the stress is only nominal, provide the minimum reinforcement

$$A_{sm} = 0.24(80)(1000)/100 = 192 \text{ mm}^2/\text{m}$$

Provide 8 mm *bars* at 260 mm in both ways.

Design of the roof slab. The annular roof slab is supported by the wall on the outer face and by the ring beam at inner face. The span of the slab is

$$L = 9.7 - 6.15 = 3.55 \text{ m}$$

Table 11.15 Structural Details of Bottom Dome and Conical Wall (All Dimensions in mm) Intze Tank

No.	Capacity (kl)	Bottom dome (both ways)			Conical dome				
		t	φ	s	t	Hoop		Inclined	
						φ	s	φ	s
1	300	150	8	150	215	22	120	12	160
2	400	150	8	150	255				
3	500	160	8	125	290	↓	↓		
4	600	160	8	125	315				
5	700	160	8	125	335	25	140		125
6	800	160	10	200	350	↓	140	↓	
7	900	160	10	200	380	↓	110	12	↓
8	1000	180	10	200	410		110	12	125
9	1100	180	10	200	435	↓	110		
10	1200	200	10	200	460	28	120		
11	1300	200	10	200	480		120		
12	1400	200	10	200	500	↓	120	↓	↓
13	1500	225	12	225	520	↓	110	12	125
14	1600	225	12	225	535	28	110	12	125
15	1700				550		110		
16	1800				565		110		
17	1900	↓	↓	↓	580	↓	100	↓	↓
18	2000				590		100		
19	2100	250	12	225	605	28	100	12	125
20	2200				620		100		
21	2300				630		90		
22	2400	↓	↓	↓	635	↓	90	↓	↓
23	2500	250	12	225	650	28	90	12	125

Note: The spacing of the reinforcement listed above is for single layer.

Let the thickness of the slab be assumed as 80 mm, then the self-weight is

$$w_g = 2 \text{ kN/m}^2, \text{ live load } w_1 = 0.75 \text{ kN/m}^2$$

Total load $w = 2.75 \text{ kN/m}^2$

Maximum bending moment

$$M = \frac{wL^2}{10} = 3.47 \text{ kNm/m}$$

$$d = \sqrt{\frac{M}{Kb\,\sigma_{cb}}} = \sqrt{\frac{0.00347}{0.13(1)(7)}} = 0.062 \text{ m}$$

$$A_{st} = \frac{M}{jd\sigma_{st}} = \frac{3470}{0.904(0.062)(230)} = 269 \text{ mm}^2/\text{m}$$

Provide 8 mm bars at 160 mm spacing with alternate bars bent.

Total load from the roof slab

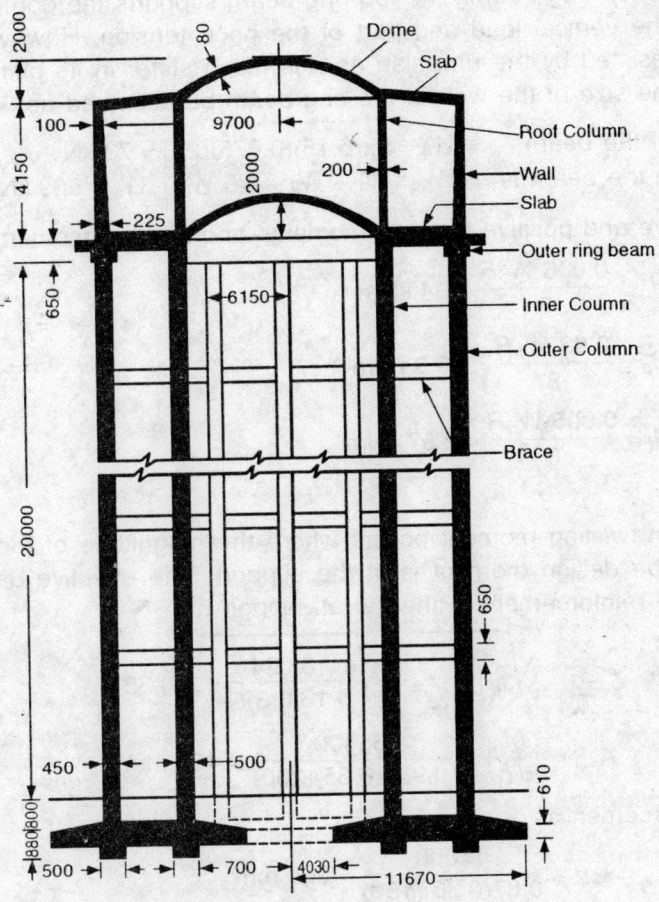

(See Fig. 11.34 for column layout)

Fig. 11.33 Overall Dimension Example 11.7.

$$W_3 = \pi(r_1^2 - r_3^2)\, w = 470 \text{ kN}$$

The hoop tension caused by the roof dome is resisted by the roof slab along with the beam.

Hoop tension = $T = N_\phi\, (\cos \phi)(r_3) = 79$ kN

Hoop tension in the conerete of the slab is

$$\frac{T}{tL} = \frac{79}{80(3.55)} = 0.28 \text{ MPa}$$

This value is computed neglecting the depth of the ring beam and it is far below the allowable value.

$$\text{Area of hoop tension steel} = \frac{T}{\sigma_{st}} = \frac{79000}{230} = 344 \text{ mm}^2$$

Provide 8 mm bars at 300 mm spacing in the circumferential direction in the slab.

Design of the top ring beam. The top ring beam supports the dome and the roof slab. It is subjected to the vertical load and part of the hoop tension. However, most of the hoop tension will be resisted by the ring slab as it is much stiffer in its plane compared with the ring beam. Let the size of the web of the ring beam be assumed as 300 by 250 mm.

Self-weight of the beam $= W_4 = 2\pi(6.15)(0.075)(25) = 73$ kN

Total load on the beam is $= W_5 = W_1 + W_2 + 0.5\,W_3 + W_4 = 669$ kN

Maximum negative and positive bending moments and twisting moment in the beam are:

$$M_n = \frac{0.006\;W_5 R}{8} = 34 \text{ kNm}$$

$$M_p = \frac{0.03\;W_5 R}{8} = 15.3 \text{ kNm}$$

$$M_t = \frac{0.005\;W_5 R}{8} = 2.57 \text{ kNm}$$

where $R = 6.15$ m.

The maximum twisting moment occurs where the magnitude of the bending moment is least; therefore, the design moment is at the support. The effective depth of the beam for $b = 0.3$ m and the reinforcement at the top at support are:

$$d = \sqrt{\frac{M_n}{K b\,\sigma_{cb}}} = \sqrt{\frac{0.034}{0.13(0.3)(7)}} = 0.35 \text{ m}$$

$$A_{st1} = \frac{M_n}{j d\,\sigma_{st}} = \frac{34400}{0.904(0.35)(230)} = 460 \text{ mm}^2$$

The bottom reinforcement is

$$A_{st2} = \frac{17000}{0.87(0.35)(150)} = 366 \text{ mm}^2$$

The maximum shear force on the beam at the support is

$$V = \frac{W_5}{16} = \frac{669}{16} = 42 \text{ kN}$$

The critical section of shear stress is at a distance d from the column face. At this section there is some twisting moment. Therefore, the net design shear at the critical section will be about the same as the shear force at the centre of the support. The nominal shear stress in the section is

$$\tau_v = \frac{V}{bd} = \frac{0.042}{0.3(0.35)} = 0.4 \text{ MPa}$$

Allowable shear stress $\tau_c = 0.3$ MPa

Select two-legged 8 mm stirrups, then the spacing of the stirrups is

$$s_v = \frac{100(230)}{300(0.4 - 0.3)} = 766 \text{ mm}$$

The maximum allowed spacing is

$$s_{v max} = \frac{100(230)}{0.4(300)} = 191 \text{ mm}$$

Provide two-legged 8 mm *stirrups at* 190 mm spacing and

$$b = 300 \text{ mm}, \quad D = 350 + 25 = 375$$

2 numbers of 18 mm bars at top and bottom each.
Weight of the web of the beam is

$$W_4 = 2\pi r_3 (0.3)(0.295)(25) = 86 \text{ kN and}$$

$$W_5 = W_1 + W_2 + 0.5 W_3 + W_4 = 682 \text{ kN}$$

Design of the roof column

Clear height of the column $= 4.05 + 0.1 = 4.15$ m

Effective height $= 0.7(4.15) = 2.905$ m

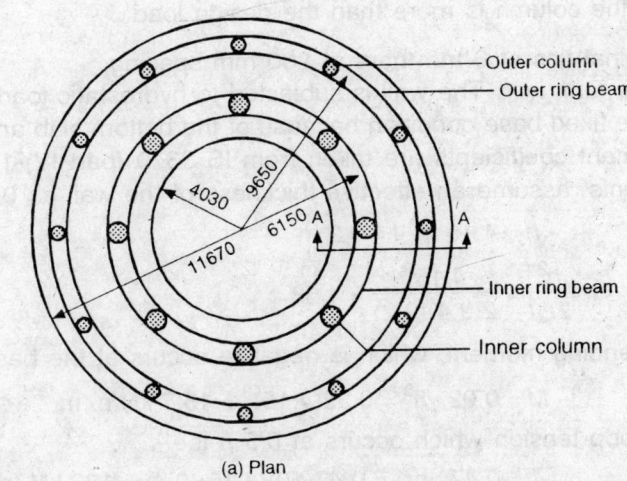

(a) Plan

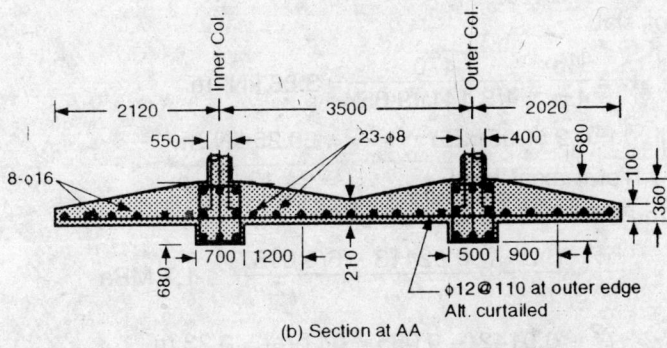

(b) Section at AA

Fig. 11.34 Foundation Details, Example 11.7.

Provide 200 mm diameter column, then the self-weight

$$W_g = \pi(0.04)(4.15)(25)/4 = 3 \text{ kN}$$

Load from the roof	$= 682/8 = 85$ kN
Total load	$P = 88$ kN

Slenderness factor is $\quad C_r = 1.25 - \dfrac{2.905}{48(0.2)} = 0.95$

Design load $\quad = P_0 = \dfrac{88}{0.95} = 93 \text{ kN} = 0.093 \text{ MN}$

Provide the minimum longitudinal reinforcement in the column, that is, 6 numbers of 12 mm bars.

$$A_{sc} = 6(113) = 678 \text{ mm}^2$$

Allowable capacity of the column is

$$P_a = A_c\sigma_{cc} + A_{sc}\sigma_{sc}$$
$$= (0.0307)(5) + (0.000678)(190) = 0.282 \text{ MN}$$

The capacity of the column is more than the design load.

Provide nominal ties of 6 mm bars at 200 mm spacing.

Design of the tank wall. The wall is subjected to hydrostatic load and supported at base. The wall will have fixed base condition because of the bottom slab and the support. The hoop tension and moment coefficients are taken from IS 3370 (part IV)-1967. For the purpose of moment coefficients, assume an effective thickness of the wall as 0.2 m.

$$h = 4.15 \text{ m}, \ t = 0.2 \text{ m}$$

$$\frac{h^2}{2r_2t} = \frac{4.15^2}{2(9.4)(0.2)} = 4.58$$

The maximum bending moment, which is negative occurs at the base, is

$$M = 0.02 \ \gamma h^3 = 0.22(4.15)^3 = 15.7 \text{ kNm/m}$$

The maximum hoop tension which occurs at 0.5 h is

$$T = 0.47 \ \gamma h r_2 = 0.47(10)(4.15)(9.4) = 183 \text{ kN/m}$$

Let the thickness of the wall at the top be 0.1 m.

Load from the roof slab

$$w = \frac{W_3}{4\pi r} = \frac{470}{4(3.141)(9.65)} = 3.88 \text{ kN/m}$$

Weight of the wall	$= 0.2 (4.05)(25)$	$= 20.25$ kN/m
	Total axial load w	$= 24.13$ kN/m

The tensile stress in the section is governed by

$$-\frac{w}{t} + \frac{6M}{t^2} = -\frac{0.02413}{t} + \frac{6(0.0157)}{t^2} \leq 1.7 \text{ MPa}$$

or $\quad\quad\quad\quad t^2 + 0.0142t - 0.055 = 0, \ \text{or} \ t = 0.23 \text{ m}$

Use the thickness of the wall at base as $t = 250$ mm.

Area of the steel to be placed on the inside face is

$$A_{st} = \frac{M}{j d \sigma_{st2}} = \frac{15700}{0.9(0.22)(150)} = 529 \text{ mm}^2/\text{m}$$

Area of the tension steel required at the mid-height is

$$A_s = \frac{T}{\sigma_{st2}} = \frac{183\,000}{150} = 1210 \text{ mm}^2/\text{m}$$

Provide 12 mm bars at 180 mm on each face as circumferential reinforcement. Then

$$A_s \text{ (provided)} = 1255 \text{ mm}^2$$

Thickness of the wall at the mid-height is

$$t = (100 + 250)/2 = 175 \text{ mm}$$

Tensile stress in the concrete at this section is

$$\sigma = \frac{T}{bt + (m-1)A_s} = \frac{0.183}{0.175 + (12)(0.001225)} = 0.96 \text{ MPa}$$

This value is less than allowable. The reinforcement details are shown in the Fig. 11.35.

$$\text{weight of the wall} = W_6$$
$$= 2\pi\,(9.60)(4.15)(0.175)(25) = 1100 \text{ kN}$$

Design of the bottom dome

Radius of the chord $= 6.15$ m

Rise of the shell $h_2 = 2.0$ m

The properties of this dome are same as those of top dome and are $R = 10.45$ m, $\theta = 36°$

Let the thickness of the dome be

$$t = 0.15 \text{ m}$$

Self-weight $w_g = 0.15(25) = 3.75 \text{ kN/m}^2$

Total self-weight $W_7 = 2\pi R h_2 w_g = 492$ kN

Weight of the water over the dome

$$W_8 = \pi\gamma(r_3^2 h - (3r_3^2 + h_2^2)h_2/6) = 3720 \text{ kN}$$

Total load on the dome

$$W_9 = W_8 + W_7 = 4212 \text{ kN}$$

Meridional thrust from the dome is

$$N\phi = \frac{W_9 R}{2\pi r_3^2} = 185 \text{ kN/m}$$

$$\text{stress} = \frac{0.185}{0.15} = 1.23 \text{ MPa}$$

Provide 10 mm *bars at* 225 mm *both ways*

Horizontal thrust from the dome is

$$H_2 = N\phi \cos\theta = 185(0.81) = 150 \text{ kN/m}$$

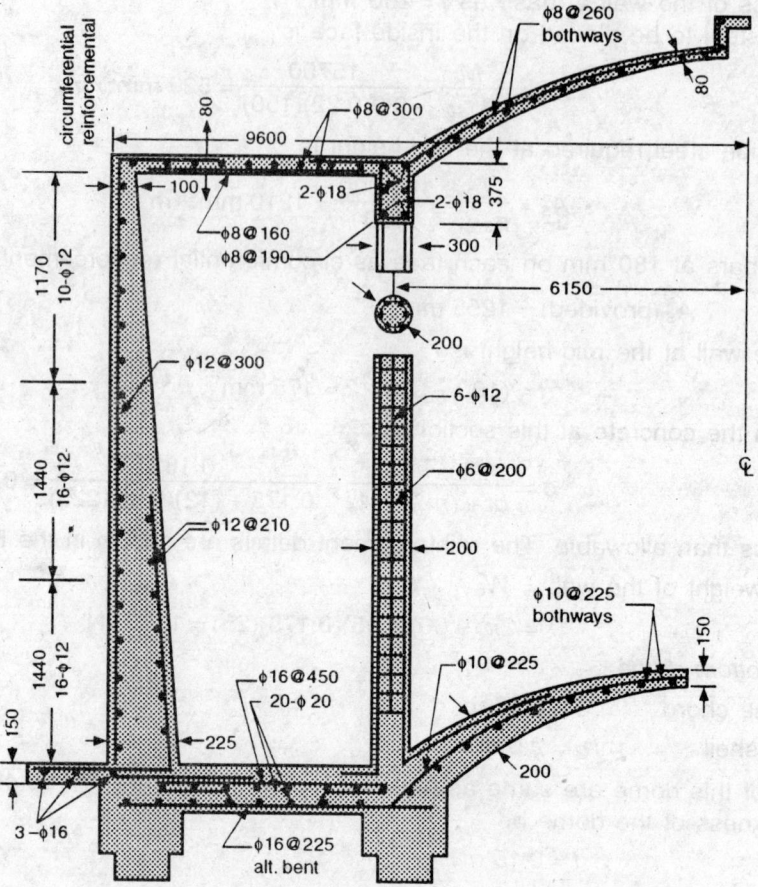

Fig. 11.35 Structural Details of Tank, Example 11.7.

Design of the bottom slab. The bottom slab is supported by two ring beams which are at 6.15 m and 9.65 m radii. The slab is interconnected with the wall and the dome. Therefore, it can be treated as a one-way slab with continuous edges.

The effective span $= 9.65 - 6.15 = 3.5$ m

Assume 400 mm thickness slab for the purpose of computing self-weight.

Self-weight	$= 0.4(25) = 10$ kN/m²
Water load	$= 4.15(10) = 41.5$ kN/m²
Total load	$w = 51.5$ kN/m²
Clear span	$L = 3$ m

The maximum bending moment is

$$M = \frac{wL^2}{12} = \frac{51.5(9)}{12} = 38.62 \text{ kNm/m}$$

The total depth of the section based on the uncracked section is

$$t = \sqrt{\frac{6M}{\sigma_{cbt}}} = \sqrt{\frac{6(0.03862)}{1.7}} = 0.37 \text{ m}$$

The area of the tension reinforcement is

$$A_{st} = \frac{M}{jd\,\sigma_{st2}} = \frac{38620}{0.9(0.34)(150)} = 842 \text{ mm}^2/\text{m}$$

Provide 16 mm *bars at* 225 mm spacing with alternate bars cranked up and extra top bars of 16 mm at 450 mm spacing.

Since the slab is stiff, most of the hoop tension will be resisted by the slab and partly by the ring beams.

The hoop tension $T = H_2 R = 0.923$ MN

Area of the tension steel $A_s = \dfrac{T}{\sigma_{st2}} = 6150 \text{ mm}^2$

Provide 20 numbers of 20 mm bars in the circumferential direction.

Neglecting the web of the beams, the direct tensile stress in the concrete is

$$\sigma = \frac{T}{0.37(3.5) + 12(0.00628)} = 0.68 \text{ MPa} < \sigma_{act}$$

Weight of the slab $W_{10} = \pi(9.4^2 - 6.15^2)(0.37)(25) = 1470$ kN

Design of the balcony

Width of balcony	= 900 mm
Let the ring beam width	= 600 mm
Clear cantilever span	= 900 − (300 − 100) = 700 mm
Live load	= 3 kN/m²
Self-weight, etc	= 2 kN/m²
w	= 5 kN/m²

Maximum bending moment

$$M = \frac{wL^2}{2} = \frac{5(0.49)}{2} = 1.225 \text{ kN/m}$$

$$d = \sqrt{\frac{M}{Kb\sigma_{cb}}} = \sqrt{\frac{1225}{0.13(7)}} = 37 \text{ mm}$$

Use $\qquad d = 80 - 25 = 55 \text{ mm}$

$$A_{st} = \frac{M}{jd\sigma_{st}} = \frac{1225}{0.904(0.055)(230)} = 170 \text{ mm}^2/\text{m}$$

Let the vertical reinforcement at the inner face of the wall be bent to the top surface of the balcony.

W_{11} = weight of the balcony = $2\pi(9.65 + 0.45)(0.08)(0.7)(25) = 90$ kN

Live load on the balcony $\qquad W_{12} = 2\pi(9.65 + 0.45)(0.9)(3) = 171$ kN

Design of the outer ring beam. The outer ring beam rests on 12 columns. The diameter of the column is assumed as 400 mm.

Radius of the circle of the columns $= R_0 = r_1 + 0.05 = 9.65$ m

Load from the wall and balcony is:

$$W_{13} = 0.5 \, W_3 + W_6 + W_{11} + W_{12} = 235 + 1100 + 90 + 171 = 1596 \text{ kN}$$

Load of the water over the bottom slab is

$$W_{14} = \pi(9.5^2 - 6.15^2)(4.15)(10) = 6836 \text{ kN}$$

Load transferred from the bottom slab is

$$W_{15} = 0.5(W_{10} + W_{14}) = 0.5(1470 + 6836) = 4153 \text{ kN}$$

Self-weight of the ring beam be

$$W_{16} = 2\pi(9.65)(0.6)(0.6)(25) = 546 \text{ kN}$$

Design load on the outer ring beam is

$$W_{17} = W_{13} + W_{15} + W_{16} = 1596 + 4153 + 546 = 6295 \text{ kN}$$

Load on each outer column is

$$W_{18} = W_{17}/12 = 525 \text{ kN}$$

The maximum negative, positive and twisting moments on the beams are:

$$M_n = 0.045 \, W_{18} \, R_0 = 228 \text{ kNm}$$

$$M_p = 0.017 \, W_{18} \, R_0 = 86 \text{ kNm}$$

$$M_t = 0.002 \, W_{18} \, R_0 = 10.1 \text{ kNm}$$

The depth of the section required is

$$d = \sqrt{\frac{M_n}{K b \sigma_{cb}}} = \sqrt{\frac{0.228}{0.13(0.6)(7)}} = 0.65 \text{ m}$$

The area of the tension steel at top face near the support is

$$A_{sn} = \frac{M_n}{jd \, \sigma_{st}} = \frac{228\,000}{0.904(0.65)(230)} = 1687 \text{ mm}^2$$

The area of the tension steel required at the mid-span

$$A_{sp} = \frac{M_p}{jd\sigma_{st}} = \frac{86\,000}{0.904(0.65)(230)} = 636 \text{ mm}^2$$

Provide four numbers of 20 mm bars each at bottom and top faces of the beam and add extra two numbers of 20 mm bars at top near the support. The maximum torsional moment is at a distance of 1.5 m from the support. The shear force along with the torsion is critical.

Intensity of the load on the beam is

$$w = \frac{W_{17}}{2\pi R_0} = \frac{6295}{2\pi(9.65)} = 103.8 \text{ kN/m}$$

The critical section for transverse shear force is at a distance d from the face of the support. Its distance from the centre of the support is $0.5a + d = 0.2 + 0.65 = 0.85$ m.

The shear force at that section is

$$V = \frac{W_{18}}{2} - w(0.85) = 174.3 \text{ kN}$$

Torsional force at this section is

$$M_{t0} = \frac{M_t}{1.5}(0.85) = 5.7 \text{ kNm}$$

The equivalent design shear force is

$$V_e = V + 1.6\,\frac{M_{t0}}{b} = 0.1743 + \frac{1.6}{0.6}(0.0057) = 0.189 \text{ kN}$$

The nominal shear stress on the section is

$$\tau_v = \frac{V_e}{bd} = \frac{0.189}{0.6(0.65)} = 0.48 \text{ MPa}$$

The tension reinforcement available at this section is

$$A_{st} = 6(314) = 1884 \text{ mm}^2$$

$$\frac{100 A_{st}}{bd} = \frac{188\,400}{600(650)} = 0.48\%$$

The allowable shear stress for 0.5% tension reinforcement is

$$\tau_c = 0.3 \text{ MPa}$$

Since the nominal shear stress is more than the shear capacity of the concrete, stirrups are to be provided. Select two-legged 10 mm stirrups, then the spacing of them is governed by

$$A_{sv} \geq \left(\frac{M_{t0}}{b_1 d_1 \sigma_{sv}} + \frac{V}{2.5 d_1 \sigma_{sv}} \right) s_v$$

where

$$A_{sv} = 157 \text{ mm}^2$$
$$b_1 = 600 - 60 = 540 \text{ mm}$$
$$d_1 = 650 - 40 = 610 \text{ mm}$$

Therefore

$$157 \geq \left(\frac{5700\,00}{540(610)(230)} + \frac{174\,300}{2.5(610)(230)} \right) s_v$$

or

$$s_v < 275 \text{ mm}$$

The spacing is also controlled by

$$s_v < \frac{A_{sy}\,\sigma_{sv}}{(\tau_v - \tau_c)b} = \frac{157(230)}{0.14(600)} = 430 \text{ mm}$$

Provide two-legged 10 mm *stirrups at 270 mm spacing.*

Overall depth of the beam = 685 mm

Weight of the ring beam $= W_{16} = 2\pi(9.65)(0.6)(0.685)(25) = 623 \text{ kN}$

Design of inner ring beam. The inner ring beam is supported by eight columns on circle of radius 6.15 m. It supports the bottom dome and the slab.

The load transferred from the dome $W_9 = 4212$ kN

The load transferred from the slab $W_{15} = 4153$ kN

Self-weight $W_{19} = 2\pi r_3(0.7)(0.8)(25)$ $\qquad = 540$ kN

Total load $\qquad\qquad\qquad\qquad W_{20} = \overline{8905\ kN}$

(assumed dimension of the beam is 0.7 by 0.8 m)

Load transferred to the column $\qquad W_{21} = \dfrac{W_{20}}{8} = 1113$ kN

Use $W_{21} = 1115$ kN

The maximum negative, positive and twisting moments on the ring beam are:

$$M_n = 0.066\ W_{21}\ R = 447\ kNm$$

$$M_p = 0.03\ W_{21}\ R = 203\ kNm$$

$$M_t = 0.005\ W_{21}\ R = 34\ kNm \quad \text{where}\ R = r_2 = 6.15\ m$$

The effective depth of the beam required is

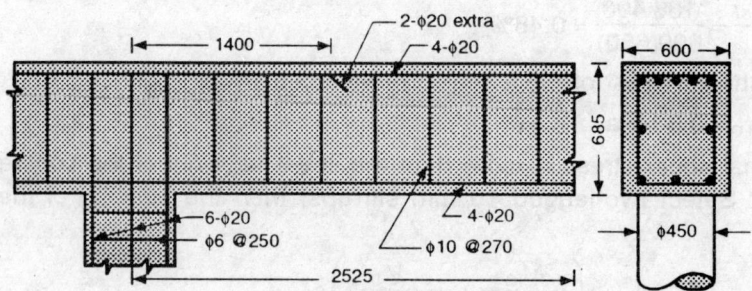

Fig. 11.36 Sectional Details of Outer Ring Beam and Outer Column, Example 11.7.

$$d = \sqrt{\dfrac{M_n}{Kb\sigma_{cb}}} = \sqrt{\dfrac{0.447}{0.13(0.7)(7)}} = 0.838\ m$$

Use $d = 0.84$ m and $D = 0.88$ m.

The area of the tension steel required at the top face near the support is

$$A_{sn} = \dfrac{M_n}{jd\sigma_{st2}} = \dfrac{447\ 000}{0.904(0.84)(150)} = 3924\ mm^2$$

The area of the tension reinforcement required at the bottom face at mid-span is

$$A_{sp} = \dfrac{M_p}{jd\sigma_{st}} = \dfrac{203\ 000}{0.904(0.84)(230)} = 1163\ mm^2$$

Provide 4 numbers of 22 mm bars continuous both at top and bottom face and, 5 numbers of 25 mm bars *extra* at top near the supports, each 2.8 m length.

(Segment length of the beam $L_1 = \pi R/4 = 4.83$ m.)

The shear force is critical at a distance d from the face of the column, that is, 0.275 + 0.88 = 1.155 m from the centre of the support.

Intensity of the load on the beam is

$$w = \frac{W_{21}}{L_1} = \frac{1115}{4.83} = 231 \text{ kN/m}$$

The shear force at a distance 1.155 from the centre of the column is

$$V = \frac{W_{21}}{2} - w(1.155) = \frac{1115}{2} - 231(1.155) = 291 \text{ kN}$$

Distance where the torsion is maximum is 9.5° from support and is $\pi R(9.5)/180 = 1.02$ m. Distance from the mid-span is $0.5L_1 - 1.02 = 1.374$ m. The torsional moment at the critical shear zone is

$$M_{t0} = \frac{M_t}{1.374}(2.414 - 1.155) = \frac{34(1.259)}{1.374} = 31 \text{ kNm}$$

The equivalent design shear force is

$$V_e = V + \frac{1.6 \, M_{t0}}{b} = 0.291 + \frac{1.6(0.031)}{0.7} = 0.362 \text{ MN}$$

The nominal shear stress at the critical section is

$$\tau_v = \frac{V_e}{bd} = \frac{0.362}{0.7(0.84)} = 0.62 \text{ MPa}$$

The allowable shear stress is about $\tau_c = 0.3$ MPa
Select four-legged 10 mm stirrups, then the design criterion is given by

$$A_{sv} \geq \left(\frac{M_{t0}}{b_1 d_1} + \frac{V}{2.5 d_1} \right) \frac{s_v}{\sigma_{sv}}$$

where
$$b_1 = 0.7 - 0.06 = 0.64 \text{ m}$$
$$d_1 = d - 0.04 = 0.8 \text{ m}$$
$$A_{sv} = 4(78.5) = 314 \text{ mm}^2$$

Therefore,
$$0.000\,314 \geq \left(\frac{0.031}{0.64(0.8)} + \frac{0.291}{2.5(0.8)} \right) \frac{s_v}{230}$$

or
$$s_v < 0.34 \text{ m} = 350 \text{ mm}$$

Provide four-legged 10 mm stirrups at 350 mm spacing.
Design of staging

Number of inner columns	$N_i = 8$
Number of outer columns	$N_0 = 12$
Height of staging	= 20 m
Depth of column below GL	= 0.8 m
Depth of inner ring beam	= 0.88 m
Depth of outer ring beam	= 0.685 m

Inner column
- Total length of inner column   = 20 + 0.8 − 0.88 = 19.92 m
  Total length of the outer column   = 20 + 0.8 − 0.685 = 20.115 m
  Assume the depth of the brace as 0.6 m.

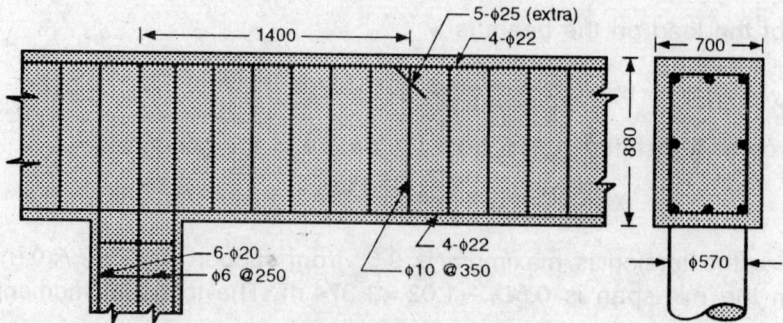

Fig. 11.37 Sectional Details of Inner Ring Beam and Inner Column, Example 11.7.

Provide the following clear panel heights

Upper most	= 2.72 m
Next four panels each at 3.7 m	= 4 × 3.7 = 14.8 m
Total depth of the braces	= 4 × 0.6 = 2.4 m
Total height of the inner column	= 19.92 m

Angle subtended at centre by two successive inner columns $\theta_1 = 45°$

Centre to centre distance between the two successive columns on the inner ring is

$$L_3 = 2R \sin(\theta_1/2) = 12.3 \sin 22.5° = 4.71 \text{ m}$$

Clear distance $\quad L_4 = 4.71 - 0.55 = 4.16 \text{ m}$

Weight of four braces = $W_{22} = 4(4.16)(0.12)(25) = 50$ kN

(size of the brace = $0.2 × 0.6$ m^2)

Weight of the inner column = $W_{23} = \dfrac{\pi(0.55)^2}{4}(19.92)(25) = 118$ kN

Total load on each inner column at its base is

$$W_{21} = W_{21} + W_{22} + W_{23} = 1283 \text{ kN}$$

Provide minimum percentage of the main reinforcement in the column which is about 0.8%. Provide 8 numbers of 18 mm bars, then

$$A_{sc} = 8(254) = 2032 \text{ mm}^2$$

Net area of the concrete section

$$A_c = \frac{\pi(550)^2}{4} - 2032 = 235550 \text{ mm}^2$$

Axial load capacity of the column is

$$P_a = A \sigma_{acc} + A_{sc} \sigma sc$$

$$= 0.2355(6.0) + 0.002032(190) = 1.8 \text{ MN} = W_{24}$$

The section is adequate for direct load. Provide 6 mm ties at 250 mm spacing.

Outer columns

Number of column = 12

The diameter of the column = 400 mm

Chord distance between the columns = 19.3 sin 15 = 4.995 m
Clear length of the brace = 4.995 − 0.4 = 4.595 m

(size of the brace = 0.2 × 0.6 m^2)

Weight of the braces on one column

$$W_{25} = 4(4.595)(0.12)(25) = 56 \text{ kN}$$

Weight of the column

$$W_{26} = \pi(0.04)(20.115)(25) = 63 \text{ kN}$$

Total load at the base of the column

$$W_{27} = W_{18} + W_{25} + W_{26} = 644 \text{ kN}$$

The minimum number of bars to be provided in a circular cross-sectioned column is six, and minimum percentage of the reinforcement is 0.8%.

Provide 6 *numbers* of 16 mm bars in the longitudinal direction *and* 6 mm *ties at* 250 mm *spacing.*

$$A_{sc} = 6(201) = 1206 \text{ mm}^2 \text{ and } A_c = 124\,454 \text{ mm}^2$$

Axial load capacity of the column is

$$P_a = A_c\,\sigma_{acc} + A_{sc}\,\sigma_{st} = 0.975 \text{ MN}$$

The axial load capacity of the column is more than the axial force on it. The columns will be analysed for seismic or wind load condition later.

Design for seismic or wind load condition

Transformed area of the inner columns = A_i = 0.256 m^2

Transformed area of the outer columns = A_0 = 0.140 m^2

The moment of inertia of the columns about the diagonal of the base is computed by the equivalent ring method.

The moment of inertia of the columns about the main diagonal is

$$I_c = \Sigma A_i y_i^2$$
$$= 4(0.256)(6.15)^2 + 6(0.14)(9.65)^2 = 11.36 \text{ m}^4$$

Total load from the tank excluding the live load is

$$W_{28} = 12W_{18} + 8W_{21} - W_2 - \frac{0.75}{2.75}\,W_4 = 14994 \text{ kN}.$$

Use the total load from the tank as = W_{28} = 15 000 kN.

Weight of the staging

$$W_{29} = 8(W_{22} + W_{24}) + 12(W_{25} + W_{26}) = 2772 \text{ kN}$$

Effective load

$$W_{30} = W_{28} + \frac{W_{29}}{3} = 15924 \text{ kN}$$

Let E_c = 28500 MPa

Effective length of the column in each panel for translation is taken equal to the clear height of the panel, $L_e = 3.7$ m.

The translational stiffness of all the columns is

$$K_c = \frac{12\,E_c I_c}{L_e^3} = \frac{12 E_c \pi}{3.7^3}\,(12(0.4)^4 + 8(0.55)^4)/64$$

$$= \frac{12 E_c}{3.7^3}\,(0.051) = 344.4 \text{ MN/m}$$

There are five panels of which four each equal to 3.7 m and the top panel height is smaller. The total lateral deflection of the tank due to W_{30} acting on the tank is

$$v = \frac{5\,W_{30}}{K_c} = \frac{5(15.924)}{344.4} = 0.23 \text{ m}$$

Free period of oscillations of the structure is

$$T = 2\pi \sqrt{\frac{v}{g}} = 0.96 \text{ sec.}$$

The corresponding average acceleration coefficient for 5% damping is taken from IS 1893-84 and is

$$\frac{S_a}{g} = 0.11$$

The coefficient of dependence upon the soil foundation system and importance factor are

$$\beta = 1.2 \text{ and } I_u = 1.5.$$

The seismic zone response spectrum factor is $F_0 = 0.2$.
The seismic coefficient to the structure is

$$\alpha_h = \beta I_u F_0 \frac{S_a}{g} = 1.2(1.5)(0.2)(0.11) = 0.0396$$

Use $\alpha_a = 0.04$.

The seismic loads corresponding to the overhead tank and staging are:

$$F_{1s} = \alpha_h W_{28} = 0.04(15000) = 600 \text{ kN}$$

$$F_{2s} = \alpha_h W_{29} = 0.04(2772) = 110 \text{ kN}$$

Assuming a basic wind pressure of 1.5 kN/m², the forces due to the wind on the tank and on the staging are:

$$F_{1w} = 0.7pDh_t = 0.70(1.5)(19.4)(2.0 + 4.15 + 0.88) = 143 \text{ kN}$$

$$F_{2w} = 0.7p((7(0.4) + 5(0.55))18 + 4(0.6)(12.3)(1.5)) = 149.2 \text{ kN}$$

The forces due to the seismic load exceed those of the wind load. Therefore, seismic forces govern the design. Total shear force at the foundation level is

$$V = F_{1s} + F_{2s} = 710 \text{ kN}$$

The total area of the columns is

$$A_{c0} = 12A_0 + 8A_i = 12(0.14) + 8(0.256) = 3.728 \text{ m}^2$$

The bending moment caused by the seismic load at base is

$$M_q = F_{1s}(H + 0.5h) + F_{2s}(0.5H)$$
$$= 600(20 + 2.075) + 110(10) = 14345 \text{ kNm}$$

Use
$$V = 710 \text{ kN} = 0.71 \text{ MN}$$
$$M_q = 14345 \text{ kNm or } 14.4 \text{ MNm}$$

The maximum axial force caused on the outer and inner columns due to the moment caused by the seismic load is

$$P_{10} = \frac{M_q}{I_{c_0}} y(A_0) = \frac{14.4(9.65)}{0.117} (\pi(0.2)^2) = 342 \text{ kN}$$

$$P_{1i} = \frac{14.4(6.15)}{0.117} \frac{(\pi(0.55)^2)}{4} = 412 \text{ kN}$$

The combined axial load on the outer and inner columns due to weight and seismic loads are:

$$P_0 = W_{27} + P_{10} - \text{live load effect} = 870 \text{ kN}$$
$$P_i = W_{24} + P_{1i} - \text{live load effect} = 1530 \text{ kN}$$

The axial force on the column during seismic load condition are more than 1.33 times that under the live load condition; therefore, the seismic load condition governs the design. The transverse force on the outer and inner columns at the bottom panel are:

$$V_0 = \frac{VA_0}{A_{c0}} = \frac{0.71(0.14)}{3.728} = 0.0266 \text{ MN}$$

$$V_i = \frac{V A_i}{A_{c0}} = \frac{0.71(0.256)}{3.728} = 0.0488 \text{ MN}$$

The seismic load on the lower most panel members is

$$F_4 = 0.04(25) \left[\pi \left(12(0.04) + \frac{8(0.55)^2}{4} \right) 3.7 + 2\pi(9.65 + 6.15)(0.02) \right]$$
$$= 24.5 \text{ kN}$$

The horizontal force resisted by the outer and inner columns are:

$$F_{40} = \frac{F_4 A_0}{A_{c0}} = \frac{0.0245(0.14)}{3.728} = 0.0009 \text{ MN}$$

$$F_{4i} = \frac{F_4 A_i}{A_{c0}} = \frac{0.0245(0.256)}{3.728} = 0.0017 \text{ MN}$$

Design of the outer column. The critical axial force and bending moments are on the lower most panel. The axial force and the bending moment on the outer column are:

$$P_0 = 0.87 \text{ MN}$$

$$M_0 = V_0 \frac{(L_e)}{2} = 0.0266 \frac{(3.7)}{2} = 0.0492 \text{ MNm}$$

$$e = \frac{M_0}{P_0} = 0.057 \text{ m}$$

$$\frac{e}{D} = \frac{0.054}{0.4} = 0.145$$

568

The eccentricity ratio on the column is only 0.145; therefore, the column can be designed as an uncracked section. The modulus of the section of the column is

$$Z_0 = \frac{\pi D^3}{32} = \frac{\pi (0.4)^3}{32} = 0.00628 \text{ m}^3$$

The stresses caused by the axial and bending moment in the section of the column are:

$$\sigma_{cc} = \frac{P_0}{A_0} = \frac{0.87}{0.14} = 6.19 \text{ MPa}$$

$$\sigma_{cb} = \frac{M_0}{Z_0} = \frac{0.0492}{0.00628} = 7.88 \text{ MPa}$$

The allowable stresses in M25 concrete at the age of 6 months under the seismic load condition are:

$$\sigma_{acc} = (1.1)6(1.33) = 8.78 \text{ MPa}$$

$$\sigma_{acb} = (1.1)8.5(1.33) = 12.44 \text{ MPa}$$

The acceptable design criterion is

$$\frac{\sigma_{cc}}{\sigma_{acc}} + \frac{\sigma_{cb}}{\sigma_{acb}} \leq 1$$

substitution, we have

$$\frac{6.19}{8.78} + \frac{7.88}{12.44} = 0.705 + 0.634 = 1.339$$

which is more than one; therefore, the size of the column need t be increased. Provide 450 m mm diameter column with 6 numbers of 20 mm bars, then the properties of the section are:

$$A_{sc} = 6(314) = 1884 \text{ mm}^2$$

$$A_0 = A_g + (m-1)A_{sc} = \frac{\pi (0.45)^2}{A_g} + 9(0.001884) = 0.176 \text{ m}^2$$

$$I_0 = \frac{\pi D^4}{64} + \frac{(m-1)A_{sc}}{8}(D_c^2) = \frac{\pi (0.45)^4}{64} + 9(0.001884)(0.35)^2$$

$$= 0.002273 \text{ m}^4$$

$$Z_0 = \frac{2I_0}{D} = 0.0101 \text{ m}^2$$

The axial and bending stresses in the concrete of the column are:

$$\sigma_{cc} = \frac{P_0}{A_0} = \frac{0.870}{0.176} = 4.93 \text{ MPa}$$

$$\sigma_{cb} = \frac{M_0}{Z_0} = \frac{0.0492}{0.101} = 4.87 \text{ MPa}$$

$$\frac{\sigma_{cc}}{\sigma_{acc}} + \frac{\sigma_{cb}}{\sigma_{acb}} = \frac{4.93}{8.78} + \frac{4.87}{12.44} = 0.95$$

This value is slightly less than one; therefore, the section is suitable.

Provide 450 mm diameter column with six numbers of 20 mm bars and 6 mm ties at 250 mm spacing.

Design of the inner columns. The forces acting on the column are:

$$P_i = 1.530 \text{ MN}, \ M_i = \frac{V_i L_e}{2} = \frac{0.0488(3.7)}{2} = 0.0903 \text{ MNm}$$

$$e = \frac{M_i}{P_i} = \frac{0.0903}{1.530} = 0.59 \text{ m}$$

The eccentricity of the load is small; therefore, the column is designed as uncracked column. The moment of inertia and the modulus of the column section are:

$$I_i = \frac{\pi D^4}{64} + \frac{(m-1)A_{sc}}{8} D_c^2 = \frac{\pi(0.55)^2}{64} + \frac{9(0.002032)(0.45)^2}{8}$$

$$Z_i = \frac{2I_i}{D} = 0.018 \text{ m}^3$$

where $D = 0.55$ m and $D_c = 0.45$ m

$$\sigma_{cc} = \frac{P_i}{A_i} = \frac{1.530}{0.256} = 5.96 \text{ MPa}$$

$$\sigma_{cb} = \frac{M_i}{Z_i} = \frac{0.0903}{0.018} = 5.02 \text{ MPa}$$

$$\frac{\sigma_{cc}}{\sigma_{acc}} + \frac{\sigma_{cb}}{\sigma_{acb}} = \frac{5.96}{8.78} + \frac{5.02}{12.44} = 1.08$$

The value is slightly more than one which is not admissible; therefore; use 8 numbers of 20 mm bars in place of 18 mm bars and 570 mm diameter column. The properties of the section then are:

$$A_i = \frac{\pi D^2}{4} + (m-1)A_{sc} = \frac{\pi(0.57)^2}{4} + 9(0.002512) = 0.278 \text{ m}^2$$

$$Z_i = 0.0204 \text{ m}^3$$

$$\sigma_{cc} = \frac{1.530}{0.278} = 5.5 \text{ MPa}$$

$$\sigma_{cb} = \frac{0.0903}{0.0204} = 4.43 \text{ MPa}$$

$$\frac{\sigma_{cc}}{\sigma_{acc}} + \frac{a_{cb}}{\sigma_{acb}} = 0.63 + 0.35 = 0.98$$

This value is about two per cent less than admissible value.

Provide 570 mm diameter column with eight numbers of 20 mm bars and 6 mm ties at 250 mm spacing.

Total weight of inner and outer columns are:

$$W_{23} = \frac{\pi(0.57)^2}{4}(19.92)(25) = 127 \text{ kN}$$

$$W_{26} = \frac{\pi(0.45)^2}{4}(20.115)(25) = 80 \text{ kN}$$

Design of brace beams on the outer circle (M25 concrete).

The bending moment on each brace in the seismic load condition is

$$M = \frac{1}{2}(V_0 + V_0 - F_{40})\frac{L_e}{2}\cosec 15°$$

$$= \frac{(0.0521)(3.7)3.86}{4} = 0.186 \text{ MNm}$$

where 15° is the half of the subtended angle by the brace segment. The torsional capacity of the section is neglected.

Let $b = 0.3$ m, then the effective depth of the section is

$$d = \sqrt{\frac{M}{Kb(1.1)(1.33)\sigma_{cb}}} = \sqrt{\frac{0.186}{0.13(0.3)(1.1)(1.33)(8.5)}} = 0.62 \text{ m}$$

Use $d = 0.62$ m, $D = 0.65$ mm

$$A_{si} = \frac{M}{jd\sigma_{st}(1.33)} + \frac{186\ 000}{0.904(0.62)(230)(1.33)} = 1085 \text{ mm}^2$$

Provide three numbers of 16 mm bars each at the top and bottom and extra one number of 25 mm bars at the top and bottom near the support only. *The size of the beam is* 300 mm *by* 650 mm. This is larger than the assumed value; therefore, the weights to be recomputed.

Clear length of the brace = $L_0 = 4.995 - 0.45 = 4.545$ m

The weight of the braces on each column is

$$W_{25} = 4(4.545)(0.30)(0.65)(25) = 87 \text{ kN}$$

Since the columns have been designed with margin of 5% on the total load, increase in the weight of the brace will not effect the column design.

The transverse shear force on the brace is

$$V = \frac{2M}{L_0} = \frac{2(0.186)}{4.545} = 0.082 \text{ MN}.$$

$$\tau_v = \frac{V}{bd} = \frac{0.082}{0.3(0.620)} = 0.44 \text{ MPa}$$

The allowable shear stress for 0.5% reinforcement is 1.33 (0.29) MPa. Therefore, transverse reinforcement is to be provided. Use two-legged 8 mm stirrups, then

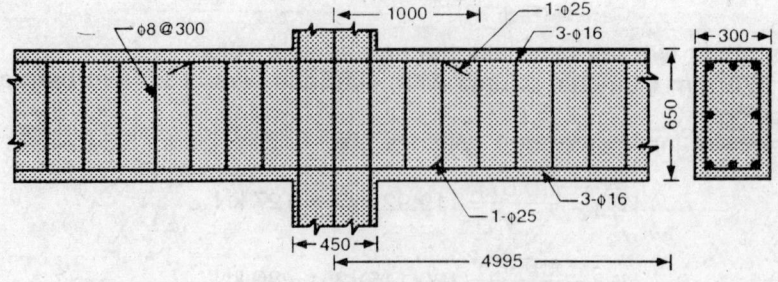

Fig. 11.38 Sectional Details of Brace Outer Column, Example 11.7.

$$A_{sv} = 100$$

The spacing of the stirrups is controlled by

$$s_v = \frac{A_{sv}\sigma_{sv}}{b(\tau_v - \tau_c)} = \frac{100\,(230)\,(1.33)}{300\,(0.15)} = 710 \text{ mm}$$

or

$$s_{max} = \frac{A_{sv}f_y}{0.4b} = \frac{100\,(415)}{0.4\,(300)} = 345 \text{ mm}$$

Provide two-legged 8 mm *stirrups at* 300 mm spacing.

Design of the brace on the inner columns. Half of the subtended angle of the segment of the brace is

$$\theta = \frac{360}{2(8)} = 22.5°$$

The bending moment on the brace is

$$M = \frac{1}{2}\,(V_i + V_i - F_{4i})\,\frac{L_e}{2}\,\text{cosec } 22.5$$

$$= \frac{((0.0488)2 - 0.0017)(3.7)(2.61)}{4} = 0.231 \text{ MNm}$$

Let $b = 0.38$ m, then the effective depth of the section is

$$d = \sqrt{\frac{0.231}{0.13(0.38)(1.1)(1.33)(8.5)}} = 0.62 \text{ m}$$

Use $b = 0.38$ m, $d = 0.62$ m, $D = 0.65$ m

$$A_{st} = \frac{M}{jd\sigma_{st}(1.33)} = \frac{231\,000}{0.904\,(0.62)(1.33)(230)} = 1347 \text{ mm}^2$$

Provide three numbers of 20 mm bars each at the top and bottom and extra one *bar of* 25 mm at the support.

The self-weight of the four braces on one column is

$$W_{22} = 4(4.71 - 0.57)(0.38)(0.67)(25) = 105 \text{ kN}$$

The transverse shear force on the brace is

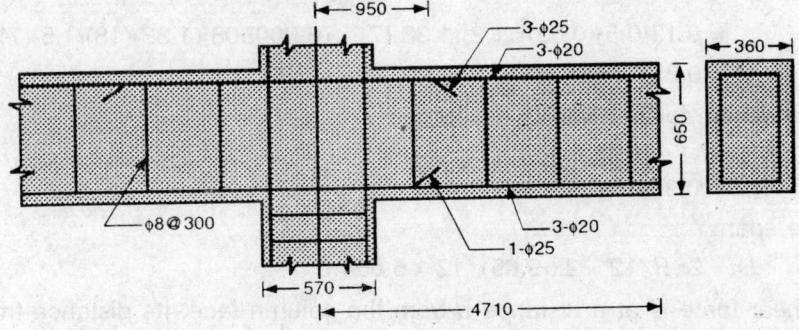

Fig. 11.39 Sectional Details of Brace Inner Column, Example 11.7

572

$$V = \frac{2M}{L_{oi}} = \frac{0.462}{4.14} = 0.1116 \text{ MN}$$

$$\tau_v = \frac{0.1116}{0.38\,(0.62)} = 0.47 \text{ MPa}$$

Provide two-legged 10 mm stirrups, then the spacing of the stirrups is controlled by the maximum allowable spacing.

Provide 8 mm stirrups at 300 mm spacing.

Design of the foundation ring beam for the outer columns.
   (M20 concrete)

The load at the top of the column	$W_{18} = 525$ kN
Weight of the column	$W_{26} = 80$ kN
Weight of the braces	$W_{25} = 87$ kN
Seismic effect	$P_{10} = 342$ kN
Max. Load from each column	$W_{31} = 954$ kN

The maximum negative, positive and twisting moments on the ring beam computed on higher side are:

$$M_n = 0.045 \ W_{31} \ R = 0.264 \text{ MNm}$$
$$M_p = 0.017 \ W_{31} \ R = 0.1 \text{ MNm}$$
$$M_t = 0.002 \ W_{31} \ R = 0.0117 \text{ MNm}$$

   where $R = 9.65$ m.

The actual sectional properties provided are:

$$b = 0.5 \text{ m}, \ D = 0.68 \text{ m}, \ d = 0.62 \text{ m}$$
$$A_{st} = 8(254) = 2032 \text{ mm}^2, \ A_{sc} = 508 \text{ mm}^2$$

(a cover of 50 mm is allowed).

   The moment capacity of the section with an age factor of 1.2 and seismic load condition is

$$M_r = Kbd^2(1.2)(1.33)\sigma_{cb} + A_{sc}(1.5 \text{ m})(1.33)\,\sigma_{cb}\,(d - d_c)$$
$$= 0.13(0.5)(0.62)^2(1.2)(1.33)(7) + (0.000508)(1.33)(18)(1.5)(7)(0.58)$$
$$= 0.351 \text{ MNm} > M_n$$

Therefore, the design for bending is acceptable.

$$A_{st} \text{ (required)} = \frac{264\,000}{0.904(0.62)(1.33)(230)} = 1540 \text{ mm}^2$$

Length of one span

$$L_5 = 2\pi R/12 = 2\pi(9.65)/12 = 5.05 \text{ m}$$

The critical shear force is at a distance d from the column face. Its distance from the centre of the column is $0.225 + 0.62 = 0.845$ m.

The shear force is

$$V = \frac{0.954}{2} - \frac{0.954}{5.05}(0.845) = 0.319 \text{ MN}$$

$$\tau_v = \frac{V}{bd} = \frac{0.319}{0.5\,(0.62)} = 1.03 \text{ MPa}$$

Allowable shear stress is

$$\tau_c = 1.2(1.33)(0.29) = 0.46 \text{ MPa}$$

Provide 2-legged 16 mm stirrups.

$$A_{sv} = 402 \text{ mm}^2$$

$$s_v = \frac{A_{sv}\,\sigma_{sv}\,(1.33)}{b\,(\tau_v - \tau_c)} = \frac{402\,(230)(1.33)}{500\,(0.57)} = 431 \text{ mm}$$

Provide 2-legged 16 mm stirrups at 400 mm spacing near the support and increase the spacing to 450 mm at the middle. The reinforcement at the middle of the span can be about 50% of that at the support.

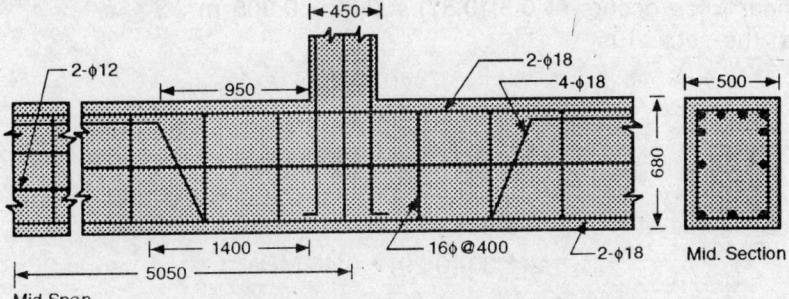

Fig. 11.40 Details of Foundation Outer Ring Beam.

Design of the foundation ring beam under inner columns

Load from the tank $W_{21} = 1115$ kN

Load of the column $W_{23} = 127$ kN

Load of the braces $W_{22} = 100$ kN

Load due to seismic effect $P_{1i} = 412$ kN

Total load $W_{32} = \underline{1754 \text{ kN}}$

The maximum negative bending moment is

$$M_n = 0.066\ W_{32}R = 0.066(1.754)(6.15) = 0.712 \text{ MNm}$$

where $R = 615$ m

The section that is provided is

$$b = 0.7 \text{ m}, D = 0.68 \text{ m}, d = 0.62 \text{ m}$$

$$A_{st} = 8(490) = 3920 \text{ mm}^2, A_{sc} = 3\,(490) = 1470 \text{ mm}^2$$

The capacity of the *T*-section with an age factor of 1.2 under seismic load condition is

$$M_n = KB_f d^2 (1.2)(1.33)\sigma_{cb} + A_{sc}(1.5\ m)(1.33)\ \sigma_{cb}\ (d - d_c)$$

$$= 0.13\ (1.61)\ (0.62)^2\ (1.2)\ (1.33)\ (7)$$

$$+ 0.00147\ (1.5)\ (18)\ (1.33)\ (7)\ (0.58)$$

$$= 1.1\ \text{MNm}$$

where
$$b_f = \frac{L}{3} = \frac{4.83}{3} = 1.61\ m$$

and
$$t = 0.36 m > nd = 0.283\ (0.62)$$

$$A_{st} = \frac{712000}{0.9(0.62)(1.33)(230)} = 4171\ \text{mm}^2$$

A_{st} provided is 3920 mm^2 with extra reinforcement of 2 numbers of $\phi 12$ at mid-depth. The check on the requirement is made assuming a uniform load on the ring beam, whereas the loading is of mixed type. Hence the reinforcement provided is adequate.

Length of one span $= L_6 = \dfrac{2\pi R}{8} = \dfrac{2\pi(6.15)}{1} = 4.83$ m

The critical shear force occurs at 0.5 (0.57) + 0.62 - 0.905 m
The shear force at the section is

$$V = \frac{1.754}{2} - \frac{1.754(0.905)}{4.83} = 0.548\ \text{MN}$$

$$\tau_v = \frac{V}{bd} = \frac{0.548}{0.7(0.62)} = 1.26\ \text{MPa}$$

The allowable shear stress at 0.5% steel is

$$\tau_c = 1.2(1.33)(0.29) = 0.46\ \text{MPa}$$

Therefore, provide $\phi 16$ mm stirrups,

$$A_{sv} = 402\ \text{mm}^2$$

$$s_v = \frac{A_{sv}\,\sigma_{sv}\,(1.33)}{b\,(\tau_v - \tau_y)} = \frac{402\,(230)(1.33)}{700(1.26 - 0.46)} = 219\ \text{mm}$$

Design of foundation. An annular circular raft is provided under the two rows of the columns projecting on either side of the two ring beams. The existing layout of the foundation is shown in Fig. 11.42.

Load from the outer column (normal load condition)

$$W_{33} = W_{18} + W_{26} + W_{25} = 692\ \text{kN}$$

Load from the inner column (normal load condition)

$$W_{34} = W_{21} + W_{22} + W_{23} = 1342\ \text{kN}$$

Weights of the outer and inner ring beams are:

$$W_{35} = 2\pi R_{n0}\ (0.5)\ (0.68)\ (25) = 516\ \text{kN}$$

$$W_{36} = 2\pi R_i\ (0.7)\ (0.68)\ (25) = 460\ \text{kN}$$

Total loads on the outer and inner ring beams are:

$$W_{37} = 12\ W_{33} + W_{35} = 8820\ \text{kN}$$

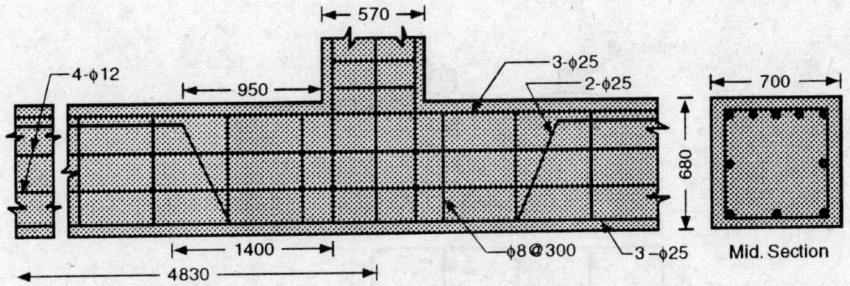

Fig. 11.41 Details of Foundation Inner Ring Beam, Example 11.7.

$$W_{38} = 8\,W_{34} + W_{36} = 11196 \text{ kN}$$

Cross-sectional area of the raft is

$$A = (1.1 + 0.36)\,(1.77)/2 + 2\,(0.21 + 0.36)\,(1.15)/2$$

$$+ (0.1 + 0.36)\,(1.77)/2 = 1.47 \text{ m}^2$$

Weight of the raft $\quad W_{39} = 2\pi\,(7.95)\,A(25) = 1836 \text{ kN}$

Net safe bearing capacity of the soil is given as 70 kN/m²; therefore, the net load on the soil is computed. The difference in the density of the concrete and the soil is taken as 25–16 = 9 kN/m³.

Under normal load condition

Load from the outer column $\quad W_{31} = 692 \text{ kN}$

Load from the inner column $\quad W_{34} = 1342 \text{ kN}$

Load on the outer and inner ring beams are:

$$W_{37} = 12 \times 692 + 516 = 8820 \text{ kN}$$

$$W_{38} = 8\,W_{34} + W_{36} = 8\,(1342) + 460 = 11196 \text{ kN}$$

Weight of raft $\quad W_{39} = 1836 \text{ kN}$

The load per metre run of the outer and inner ring beams with and without self-weights are:

$$w_{37} = \frac{8820}{2\,(3.141)\,(9.65)} = 145.47 \text{ kN/m}$$

$$w_{38} = \frac{11196}{2\,(3.141)\,(6.15)} = 289.74 \text{ kN/m}$$

$$w_{33} = \frac{12\,(692)}{2\pi\,(9.65)} = 136.96 \text{ kN/m}$$

$$w_{34} = \frac{8\,(1342)}{2\pi(6.15)} = 257.34 \text{ kN/m}$$

Consider a segment of the foundation slab in which the length at the outer circle is 1 m. Then the subtended angle at the centre and the corresponding lengths of the arcs at the inner face, at the inner ring beam and at the outer ring beam are:

$$\theta = 1/11.67 = 0.0857 \text{ rad or } 4.9°$$

$$s_1 = 4.03\theta = 0.345 \text{ m}$$

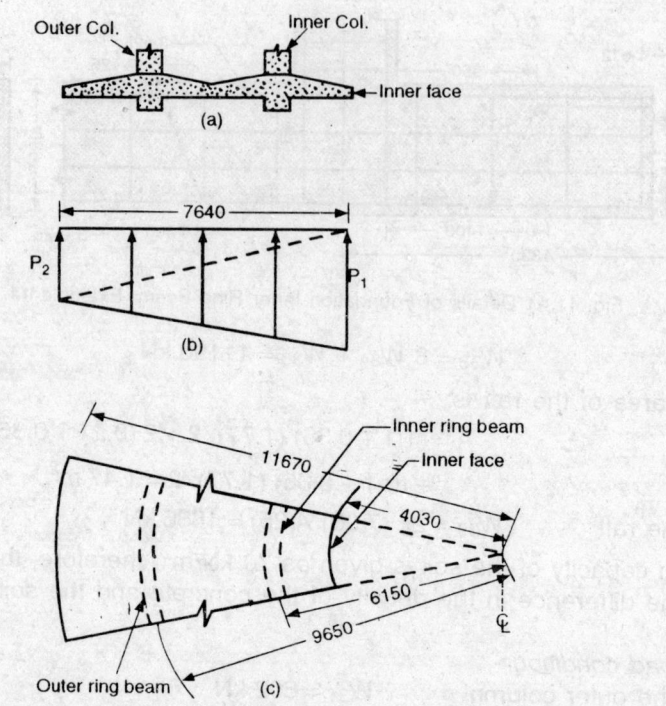

Fig. 11.42 Foundation Slab-Segment and Pressure Distribution, Example 11.7.

$$s_t = 6.15\theta = 0.527 \text{ m}$$

$$s_0 = 9.65\theta = 0.827 \text{ m}$$

The corresponding gross loads at the outer and inner ring beams of the segment are:

$$P_{0g} = s_0 \, w_{33} = 120.3 \text{ kN}$$

$$P_{ig} = s_i w_{38} = 152.69 \text{ kN}$$

Plan area of the segment is

$$A_{fw} = 0.5 \, (s_1 + s_2) \, (R_2 - R_1)$$

$$= 0.5 \, (0.345 + 1) \, (7.64) = 5.138 \text{ m}^2$$

Self-weight of the slab is

$$w_g = \frac{1.47 \, (0.5) \, (0.345 + 1) \, (25)}{7.64} = 3.23 \text{ kN/m}^2$$

The equilibrium of forces on the segment of the foundation are:

$$0.5 \, (p_1 + p_2) \, A_{fs} + P_{0g} = P_{ig} + w_g A_{fs}$$

or

$$p_1 + p_2 = 106.26 + 6.46 = 112.72 \text{ kN/m}^2$$

and the moment equilibrium is

$$A_{fs} \frac{(7.64)}{6} (p_1 + 2p_2) = P_{ig}(6.15 - 4.03) + P_{0g}(9.65 - 4.03) + w_g(7.95 - 4.03)$$

or

$$p_1 + 2p_2 = 154.75$$

Therefore,

$$p_2 = 42.03 \text{ kN/m}^2$$

$$p_1 = 70.69 \text{ kN/m}^2$$

$$(p_1 + p_2)/2 = 56.36 \text{ kN/m}^2$$

The maximum pressure on the soil is

$$p_1 = 70.69 \text{ kN/m}^2$$

(add the surcharge of soils over the foundation)

The net reaction of the soil at the inner face of the foundation slab is

$$p_i = p_1 - w_g = 70.69 - 3.23 = 67.46 \text{ kN/m}^2$$

Net pressure at the inner face of the inner ring beam is

$$p_4 = p_1 - \frac{(p_1 - p_2)(1.78)}{7.64} - w_g = 60.78 \text{ kN/m}^2$$

Clear span of the cantilever slab at the inner ring beam is

$$L = 2.13 - 0.35 = 1.78 \text{ m}$$

Length of the arc at the face of the inner ring beam is

$$s_4 = \theta(6.15 - 0.35) = 0.497 \text{ m}$$

The area of the slab of the clear span cantilever is

$$A_4 = (0.497 + 0.345)(0.5)(1.78) = 0.749 \text{ m}^2$$

Maximum bending moment on the slab is

$$M = \frac{A_4}{6}(2p_3 + p_4) L = 43.48 \text{ kNm and}$$

$$d = \sqrt{\frac{M}{Kb\sigma_{cb}}} = \sqrt{\frac{0.04348}{0.13(0.497)(7)}} = 0.32 \text{ m}$$

Providing a minimum clear cover of 50 mm, let the overall depth at the face of the beam be 0.43 m and d = 0.37 m

$$A_{st} = \frac{M}{jd\sigma_{st}} = \frac{43\,480}{0.904(0.37)(230)} = 565 \text{ mm}^2 = 1137 \text{ mm}^2/\text{m}$$

Provide 20 mm bars at 275 mm spacing at the bottom face radially.

Provide a minimum thickness of 150 mm at the free ends and a circumferential reinforcement of 12 mm bars at 250 mm spacing.

The critical shear force is at a distance d from the face of the beam. The area on which the force is acting is

$$A_{vs} = 0.5(0.345)(1 + 5.43/4.03)(1.78 - 0.37) = 0.57 \text{ m}^2$$

The shear force is

$$V = 0.57(67.46 + 61.78)/2 = 36.8 \text{ kN}$$
$$b = 0.345(5.43)/4.03 = 0.465 \text{ m}$$

Nominal shear stress $= V/bd = 0.21$ MPa

The bending moment at the middle point between the ring beams is likely to be critical. The area of the inside half segment is

$$A_2 = 0.5(0.345)(1 + 7.85/4.03)(7.85 - 4.03) = 1.94 \text{ m}^2$$

The pressure at the mid-point is

$$p_5 = 51.26 \text{ kN/m}^2$$

The bending moment at the mid-span is

$$M = \frac{1.94}{6}(2(67.46) + 51.26)(3.82) - 257.34(7.85 - 6.15)(0.527)$$

$$= -0.6 \text{ kNm}$$

Provide only nominal reinforcement at top.

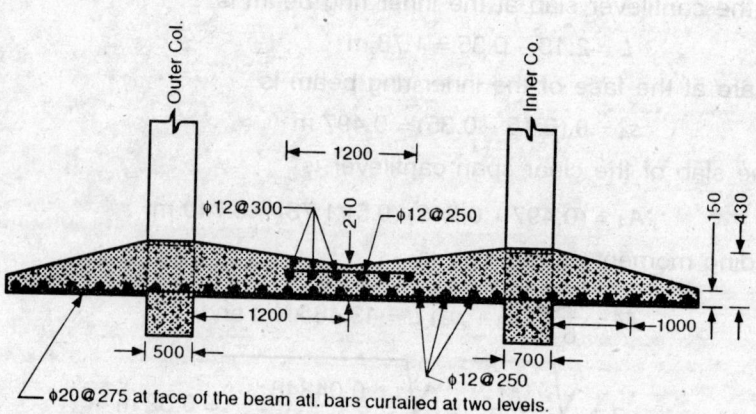

Fig. 11.43 Reinforcement Details in the Foundation Raft, Example 11.7.

Reinforcement details are given in Fig. 11.43.

PROBLEMS

11.1 *Square water tank above ground level:* A square water tank is to be designed to store 2000 kl water with a total height not exceeding 5 m. The tank is resting above ground level where the safe net bearing capacity of the soil is 90 kN/m². Design the water tank for the following specifications:

Concrete = M20, Reinforcement = HYSD-Fe415 bars, free board = 150 mm. The tank is to be covered with a reinforced concrete slab on columns spaced at about 4 m. The live load on the roof slab is 1.5 kN/m². Sketch the reinforcement details.

11.2 *Square water tank below ground level:* A square water tank is to be designed to store 2000 kl water with a total height not exceeding 5 m. The tank is below ground level resting on soil with safe bearing capacity of 120 kN/m^2. The roof of the tank is 1 m below ground level. Design the water tank for the following specifications: Concrete = M20, reinforcement = HYSD-Fe415 bars, free board = 150 mm. The tank is to be covered with a reinforced concrete slab having columns spaced at about 4 m. Sketch the reinforcement details.

11.3 *Counterfort walled water tank above the ground level:* A large reservoir to store 10 million litres of water is placed above the ground level on a soil of safe net bearing capacity of 200 kN/m^2. The inside clear height of the water tank is 6 m and the spacing of the counterfort or the column inside the water tank should not exceed 5 m. The live load on the roof of the tank is 1.5 kN/m^2. Design the rectangular reservoir with M20 concrete and HYSD-Fe415 bars.

11.4 *Cylindrical water tank:* A cylindrical water tank with 3 m inside radius is to be designed to store 75 kl water. The tank is resting on ground with spherical roof dome. Design the water tank with M20 concrete and HYSD-Fe415 bars. Assume a free board of 150 mm and live load on roof as 0.75 kN/m^2.

11.5 *Submerged cylindrical water tank:* A cylindrical water tank submerged in a lake, placed resting on the bottom of the lake is to be designed for a capacity of 120 kl. The maximum head of water on the tank is 10 m. Design a cylindrical water tank with 8 m inside diameter and having a hemispherical dome at the top. Use M20 concrete and HYSD-Fe415 reinforcement bars.

11.6 *Overhead cylindrical water tank:* Design a overhead cylindrical water tank for the following data and specifications: Capacity = 200 kl, staging = 15 m, inside diameter of the tank = 6 m. The tank has flat bottom slab and spherical roof dome with a rise of 1.2 m. Provide 6 nos. of columns with 3 intermediate bracings. Assume a seismic coefficients of 0.04. Provide annular raft slab foundation resting on a soil of safe bearing capacity of 90 kN/m^2. Use M20 concrete and HYSD-Fe415 reinforcement bars throughout the construction.

11.7 *Overhead cylindrical water tank with flat bottom slab:* A concrete reservoir is to be designed for a capacity of 2000 kl with 20 m staging. The tank consists of spherical roof dome of rise 2 m, cylindrical vertical wall having an inside diameter of 16 m. Flat bottom slab supported by a number of columns not only under the periphery but also under the slab. There are 12 columns under the cylindrical wall spaced at equal intervals and another set of 8 columns placed on a circular line of 8 m diameter. Provide a circular ring beam over the inside columns to support the slab. There are four intermediate bracings spaced at equal intervals. The foundation to the tank is made of a raft slab resting on a soil having a safe allowable bearing pressure of 100 kN/m^2. Use M20 concrete and HYSD-Fe415 reinforcement bars throughout the construction.

11.8 *Design of an Intze tank with column staging:* An overhead water tank of capacity of 900 kl with 6 m staging is to be designed as an Intze tank. The diameter of the cylindrical wall is 12 m and the diameter of the bottom ring beam is 8 m. The tank is resting on 8 nos. of columns spaced at equal intervals provided with 3 intermediate bracings. The foundation is made of solid circular raft slab resting on a soil having a safe bearing pressure of 70 kN/m^2. Design the entire water tank using M20 concrete and HYSD-Fe415 bars.

11.9 *Design of an Intze tank with shaft staging:* An overhead water tank with 900 kl capacity and 16 m staging is to be designed as an Intze tank resting on cylindrical shaft. The diameter of the tank 12 m and the diameter of the shaft is 8 m. The rises of the roof and bottom domes are 1.5 m each. The circular cylindrical shaft is resting on annular raft foundation over a soil of safe bearing capacity of 80 kN/m^2. Design the tank using M20 concrete and the staging and foundation with M20 concrete. The HYSD-Fe415 reinforcement bars are to be used throughout the construction.

11.10 *Design of an Intze water tank on pile foundation:* Redesign the problem 11.9 with pile foundation instead of raft foundation. Assume the capacity of the pile as 300 kN.

Problems on Design of Elements in Water Tanks

The following problems are associated with the design of one or two elements in water tanks. The time required for the solution is about 20 to 30 minutes.

11.11 *Design of ring beam:* A ring beam resisting on a cylindrical wall of 18 m diameter supports a spherical dome of 3 m rise. Design the ring beam when the dome is subjected to a live load of 750 k/N/m^2 and the thickness of the dome is 100 mm. The beam is made of M20 concrete and HYSD-Fe415 reinforcement bars.

11.12 *Design of a ring beam:* A circular ring beam of 5 m radius is supporting an Intze water tank and it is subjected to outward radial force of 100 kN/m and vertical load of 800 kN/m. The beam is supported by 10 nos. of columns spaced at equal intervals. Design a rectangular cross section ring beam using M20 concrete and HYSD-Fe415 reinforcement bars.

11.13 *Design of ring beam supported on shaft:* A circular ring beam of 5 m radius supports an Intze water tank and resting on cylindrical shaft. The beam is subjected to outward radial pressure of 150 kN/m and vertical load of 900 kN/m. Design the ring beam using M20 concrete ad HYSD-Fe415 reinforcement bars.

11.14 *Design of ring beam:* A ring beam is provided in an Intze water tank at the intersection of the bottom end of conical shell and bottom dome. The diameter of the ring beam is 10 m. The conical shell transfers 400 kN/m at an angle of 40 with the horizontal and the dome exerts meridional thrust of 300 kN/m at an an inclination of 30° with the horizontal. Design the ring beam assuming that it is supported on 8 columns spaced at equal intervals. Use M20 concrete and HYSD-Fe415 bars. The width of the ring beam should not exceed 500 mm.

11.15 *Design of a cylindrical wall of water tank:* The diameter of cylindrical water tank is 18 m and the head of water in the tank is 6 m. Assume that the top of the wall is hinged and bottom of the wall is fixed, design the cylindrical wall with M20 concrete and HYSD-Fe415 bars. The bending moment coefficients can be taken either from the book or from codes.

11.16 *Design of columns of water tank:* An overhead water tank is supported by 10 columns spaced equally on a circle of 6 m radius. The height of the staging is 15 and it is subjected to a total axial load of 20,000 kN, bending moment of 8,000 kNm and a shear force of 1000 kN. The staging is provided with 3 intermediate braces equally spaced. Design the columns of square cross section using M20 concrete and HYSD-Fe415 reinforcement bars. The forces listed above are the forces in seismic load condition.

11.17 *Design of brace:* Vide Problem 11.16, design brace of the staging using M20 concrete and HYSD-Fe415 reinforcement bars. Assume a rectangular cross-section beam with width equal to 250 mm.

11.18 *Design of a circular shaft:* A 5.5 m circular cylindrical shaft supports an overhead water tank. The axial load, shear force and bending moment about the base of the shaft are 18,000 kN, 900 kN and 700 kNm respectively under seismic load condition. Design the shaft using M 25 concrete and HYSD-Fe415 reinforcement bars if the height of the shaft is 15 m.

Chapter 12

Ultimate Strength Design of Beams

12.1 Introduction

The method of design of reinforced concrete structure by the ultimate strength considerations has taken an important place in many countries from 1950s till about 1970. At one time, the method was the most preferred one due to the economy and partly the simplicity in its approach. It also had the simplest rational which could easily be understood and practised by engineers. The method had a disadvantage of not being very clear under the service load conditions. The method which emphasizes on the strength consideration has been replaced by the limit states design in many countries during the seventies. It is, therefore, desirable to be familiar with the ultimate strength design, and with this objective a brief discussion is presented in this chapter.

The ultimate strength design method defines a load called the *ultimate load* which is a hypothetical load obtained by raising the working or service loads after multiplying them by some safety factors. These safety factors are called *Load factors*. The ultimate load on a structure is given by

$$W_u = N_{d1}\, W_d + N_l W_l, \text{ or} \tag{12.1}$$
$$= N_{d1}\, W_d + N_l W_{l2} + N_e W_e, \text{ or} \tag{12.2}$$
$$= N_{d1}\, W_d + N_l W_{l2} + N_l\, W_e, \text{ or} \tag{12.3}$$
$$= N_{d2} W_d + N_l W_{l2} + N_e W_e, \text{ or} \tag{12.4}$$
$$= N_c\, (W_d + W_{l2} + W_c) \tag{12.5}$$

where W_u = ultimate load, W_d = dead load

W_l = full live load

W_{l2} = most probable live load which is likely to occur with gust wind or seismic load.

W_e = equivalent dynamic force caused by wind load or seismic load (environmental load).

N_{d1} = load factor to be applied to dead loads if the dead load is additive to the environmental loads. Its value is usually taken in the range of 1.1 to 1.5.

N_{d2} = load factor applied to the dead load if the environmental load has the opposite effect of the dead load, such as the uplift, overturning stability, etc. Its value is usually taken as 0.9.

N_l = load factor applied to the full live load and its value is in the range of 1.4 to 2.5. The higher value is used in case of moving type of load, and the lower value can be used in the fixed type of load such as those in warehouses and less important structures.

N_e = load factor applied to wind or seismic loads. It is usually in the range of 1.4 to 2.5.

N_c = combined load factor applied to all the loads put together and its value is in the range of 1.25 to 1.8.

Table 12.1 Load Factors

Load combination		Loads		
		DL	LL	WL
DL and LL		1.5	2.2	0
DL and WL		1.5[a]	0	2.2
	or	0.9		
DL, LL and WL		1.5	2.2	0.5

[a] 1.5 or 0.9 which is more sever depending on the type of WL and stability requirement.
DL = dead load, LL live load, WL = wind load or seismic load

The load factors vary from country to country. One should take account of the specifications, allowed tolerances and type of supervision available in the construction.

The structure is subjected to the most severe ultimate load with the appropriate load locations, and analysed by using the elastic-linear structural analysis. One can question the elastic behaviour of the structure under the ultimate loads. The ultimate loads are only hypothetical loads and the elastic solution is an equilibrium one. Therefore, the application of the method is on the safer side, though behaviour of the real structure under the action of the ultimate loads will not be elastic and linear. The order of safety factor has some association with the factors of safety but not a true indication. A structure is not likely to collapse even if it is subjected to the ultimate load. In that sense the load is not collapse load but it is a load which when acts on the structure, it is close to the collapse.

At a load corresponding to the ultimate load, the critical sections of the members of the structure are expected to reach their limits of strengths. The sections must be designed so as to resist the stress resultants caused by the ultimate load acting on the structure. The design criterion can be expressed as

$$M_r \geq M_u \tag{12.6}$$

where M_r = resisting moment capacity of the section
M_u = ultimate bending moment caused by the ultimate load acting on the structure

Similar design criterion can be extended to the other properties such as shear, compression, torsion, etc. Hence the design criterion can be stated in more general terms as

$$F_r \geq F_u \tag{12.7}$$

where. F_r = the resisting capacity of the section
F_u = ultimate active force on the section

The symbol F can refer to BM, compression, tension or torsion, etc.

When a section is subjected to more than one stress resultant, such as compression and bending, the design criterion is established through an interaction formula. There are linear and nonlinear interaction formulae which are beyond the scope of this chapter. Table 12.1 gives the load factors used in this book when applied to buildings.

12.2 Ultimate Strength of Rectangular Cross-section

The maximum strain at crushing of the concrete in bending is equal to 0.0035. The stress-strain curve at crushing of the concrete is (idealised) assumed to be rectangular as shown in Fig.

584

12.1. Figures 12.1 and 12.2 give different notations and stress-strain curves of concrete and steel.

Let x_u = depth of the neutral axis

x_c = depth of the CG of the compressive stress

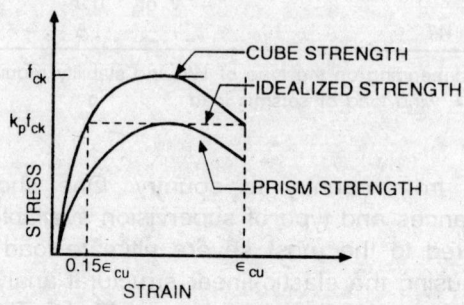

Fig. 12.1 Typical Stress-strain Curve of Concrete

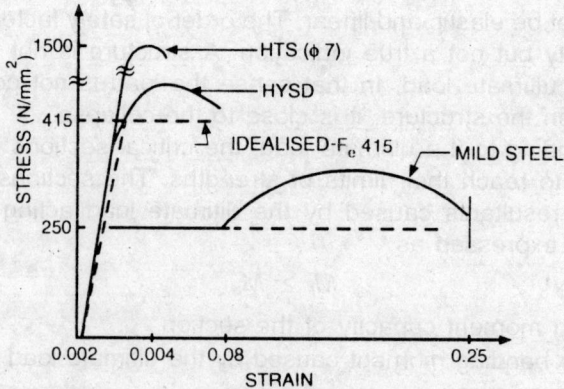

Fig. 12.2 Typical Stress-strain Relation in Steel.

Whitney has suggested an idealised rectangular stress block at the ultimate load level such that the statics of both the diagrams match.

Figure 12.3a illustrates the rectangular cross-section which has reached its capacity in bending. Figure 12.3b indicates the strains across the depth of the section assuming the plane section before bending remains plane even at the time of failure.

Let ϵ_{cu} = crushing strain in concrete in bending

ϵ_{sb} = strain in the steel caused by the bending of the section

Figure 12.3c illustrates the actual stress distribution on the compression face of the concrete whereas Fig. 12.3d gives the idealised form of the rectangular stress block. The rectangular stress block is so selected such that the total force acting on the section is same as that due to the actual stress distribution and the centre of the compression from the two distributions is about the same distance from the neutral axis. This idealization leads to

$$a = 0.85 \ x_u \tag{12.8}$$

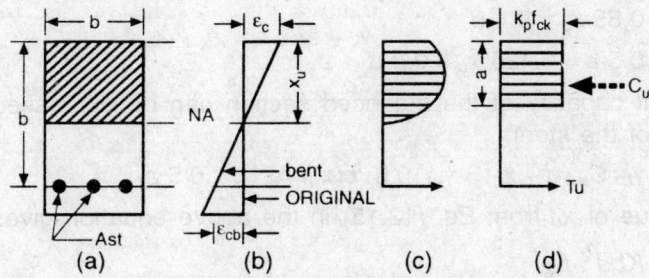

Fig. 12.3 Stress Distribution on Rectangular Section.

$$x_c = 0.5a \tag{12.9}$$

The maximum crushing strength of the concrete prism at the extreme fibre is given by $k_p f_{ck}$. The total area of the stress block which is equal to the compressive force on the concrete is

$$C_u = abk_p f_{ck} = 0.85b\, x_u\, k_p\, f_{ck} \tag{12.10}$$

$$C_u = 0.6b\, x_u\, f_{ck} \tag{12.11}$$

where $\quad k_p = 0.67$

The strain in steel caused by bending at crushing strain of concrete is not to be less than

$$\epsilon_{sb} = \frac{f_y}{1.15\, E_s} + 0.002 \tag{12.12}$$

where $\quad \epsilon_{sb}$ = strain in the steel reinforcement corresponding to the crushing strain in concrete

$\quad E_s$ = Young's modulus of steel (usually taken as 200 GPa for steel)

The strain compatibility between steel and concrete at the time of collapse (see Fig. 12.3) is

$$\frac{\epsilon_{cu}}{x_u} = \frac{\epsilon_{sb}}{d - x_u} \tag{12.13}$$

Let $\quad \epsilon_{cu} = 0.0035$, then Eqs. (12.12) and (12.13) will result

$$x_u = \frac{\epsilon_{cu} d}{\epsilon_{cu} + \epsilon_{sb}}$$

$$= \frac{0.0035\, d}{0.0035 + 0.002 + f_y/1.15\, E_s}$$

$$= \frac{0.0035\,(1.15)\,(200\,000)\, d}{0.0035\,(1.15)(200\,000) + f_y}$$

$$= \frac{805\, d}{1265 + f_y} \tag{12.14}$$

Sometimes for convenience, the strain in steel is equated to ϵ_y to 0.004. In that case the neutral axis distance is given by

$$x_u = \frac{0.0035\, d}{0.0075} = 0.47\, d \tag{12.15}$$

One can use Eq. (12.15) in most cases, then

$$a = 0.85 \, x_u = 0.4 \, d \tag{12.16}$$

$$x_c = 0.5 \, a = 0.425 \, x_u = 0.2 \, d$$

The design moment capacity of the balanced section can be expressed by taking moment about the centroid of the steel.

$$M_{rb} = \gamma_m \, C_u \, (d - x_c) = \gamma_m \, (0.6) \, b(x_u) \, f_{ck} \, (d - 0.5 \, a) \tag{12.17}$$

Substituting the value of x_u from Eq. (12.15) in the above equation gives

$$M_{rb} = K b d^2 \, f_{ck} \tag{12.18}$$

where
$$K = 0.6(0.47)(0.8) \, \gamma_m = 0.225 \, \gamma_m = 0.196 \tag{12.19}$$

$$\gamma_m = \text{material reduction coefficient } (= 0.87)$$

The material reduction coefficient is usually taken in the range of 0.8 to 0.95. It is taken as 0.87 in this book.

Equating the total compressive force to the tension force in steel, we have

$$T = \gamma_m \, C_u$$

$$A_{stb} \, f_y = 0.6 \, \gamma_m \, b x_u \, f_{ck} \tag{12.20}$$

$$\frac{A_{stb}}{bd} = 0.6 \, \gamma_m \, \frac{x_u}{d} \, \frac{f_{ck}}{f_y} = 0.24 \, \frac{f_{ck}}{f_y} \tag{12.21}$$

where A_{stb} = area of the balanced tensile reinforcement (= A_{st})

The percentage of the balanced reinforcement with respect to effective depth can be expressed as

$$p = \frac{100 \, A_{stb}}{bd} = 100 \, p_0 \left(\frac{f_{ck}}{f_y} \right) \tag{12.22}$$

where $p_0 = 0.24$

The approximate percentage of the reinforcement varies from 0.6 to 2.4. The lower limit applies to low strength concretes with high yield steels, whereas the upper limit is for higher strength concretes with mild steel reinforcement. In case the beam is under-reinforced, the design moment capacity is

$$M_r = A_{st} \, f_y \, (d - 0.425 \, x_u) = 0.8 \, A_{st} \, x_y \, d \tag{12.23}$$

For over-reinforced beams, the moment capacity is restricted to that given in Eq. (12.18). The presence of over-reinforcement can be found if

$$\frac{A_{st}}{bd} > 0.24 \, \frac{f_{ck}}{f_y} \tag{12.24}$$

12.3 Limit Strength of Prestressed Concrete Rectangular Section

The stress-strain relation of concrete used in prestressed concrete is assumed to be the same as that in RCC as given in Figure. 12.1a. Figure 12.1b also illustrates the stress-strain relation of high tensile steels. The pre-compressive strain in fibre of the concrete at the level of pretensioned steel is almost close to zero. Figure 12.4b illustrates the two planes of the section, one at the effective prestrain load condition and the other at the collapse load condition. Line *a* refers to the plane when only the effective prestress acts and line *b* refers the section which is about to collapse. The total strain in the steel at failure is

$$\epsilon_s = \epsilon_{se} + \epsilon_{cc} + \epsilon_{sb} \tag{12.25}$$

in which

ϵ_{se} = effective prestrain in the steel (usually about 0.004)

ϵ_{ce} = precompressive strain in the concrete at the steel level (it is usually negligible)

ϵ_{sb} = compatible strain caused by bending, corresponding to crushing strain in the concrete

ϵ_s = total strain in the steel before failure, corresponding to the strain at proof stress in the balanced section.

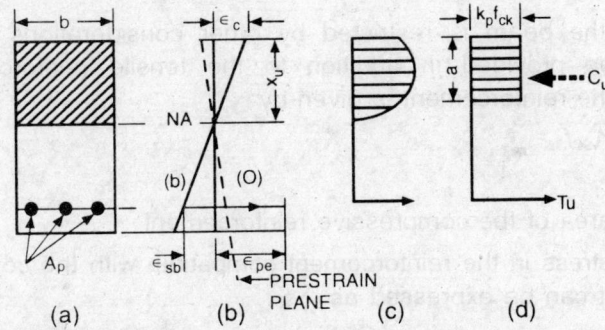

Fig. 12.4 Ultimate Capacities in Bending and Stress Block of Prestressed Concrete.

Applying a partial safety factor of 1.15 to the proof stress, the value of the strain in steel at failure is

$$\epsilon_s = 0.02 + \frac{f_y}{1.15\,E_s} \tag{12.26}$$

The compatible strain in steel from Eqs. (12.25) and (12.26) is

$$\epsilon_{sb} = 0.002 - (\epsilon_{sr} + \epsilon_{ce}) + \frac{f_y}{1.15\,E_s} \tag{12.27}$$

For all practical purposes, the effective prestrain in steel can be taken as 0.004. Thus the above equation reduces to

$$\epsilon_{sb} = -0.002 + \frac{f_y}{1.15\,E_s} = 0.004 \text{ (approximated)} \tag{12.28}$$

The neutral axis corresponding to the failure section is

$$\frac{x_u}{d} = \frac{\epsilon_{cu}}{\epsilon_{cu} + \epsilon_{sb}} = \frac{35}{75} = 0.47 \tag{12.29}$$

The derivations for ultimate compressive force (C_u) and moment resisting capacity of the section are exactly same as those derived for RCC section of the previous article. These values repeated from the previous expressions are:

$$C_u = 0.6bx_u f_{ck} \tag{12.11}$$

$$M_{rb} = Kbd^2 f_{ck} \qquad (12.19)$$

where
$$K = 0.225\, \gamma_m$$

and
$$p = \frac{100 A_{st}}{bd} = 100 p_0 \frac{f_{ck}}{f_y} \qquad (12.21)$$

In case the beam is under-reinforced, the resisting moment capacity of the section is given by

$$M_r = \gamma_m A_{st} f_y (d - 0.425 x_u) = 0.8 \gamma_m A_{st} f_y d \qquad (12.23)$$

For over-reinforced sections, the moment capacity is restricted to that given in Eq. (12.18).

12.4 Design of Doubly Reinforced RCC Rectangular Sections

In case the size of the beam is restricted by other considerations, some compression reinforcement must be provided in addition to the tensile reinforcement. The design compressive force in the reinforcement is given by

$$C_{us} = A_{sc} f_{sc} \qquad (12.30)$$

in which

A_{sc} = area of the compressive reinforcement

f_{sc} = stress in the reinforcement compatible with the compressive strain and it can be expressed as

$$f_{sc} = 0.0035 \left(1 - \frac{d_c}{x_u}\right) E_s \qquad (12.31)$$

in which

d_c = depth of the compression steel

and
f_{sc} = is subjected to a maximum of f_y

The value of d_c is usually in the range of $0.1d$ and that of x_u is about $0.47d$. Therefore, the compressive stress in the compression reinforcement is

$$f_{sc} = 0.0035(0.8)(200\,000) = 560 \text{ MPa}$$

It can be seen that the compatible stress in the compression steel works out close to 560 MPa which is based on linear stress-strain relation upto a strain of about 0.0035. It indicates that the design stress in the compression steel can be taken close to the design stress in tension, as f_{sc} cannot be more than f_y. Therefore,

$$f_{sc} = f_y \qquad (12.32)$$

The reinforcement in tension to balance the compression reinforcement is given by

$$A_{st2} = A_{sc} f_{sc} / f_y \qquad (12.33)$$

The total bending resistance of the section is

$$M_r = M_{rb} + A_{sc} f_{sc} (d - d_c) \qquad (12.34)$$

The corresponding total tensile reinforcement is

$$A_{st} = A_{stb} + A_{st2} \qquad (12.35)$$

12.5 Design for Transverse Shear in RCC Beams

The beams are normally designed for flexure and then tested for shear capacity, depending on which, the shear reinforcement is provided. If the nominal shear stress on a section does not exceed the shear strength of the concrete, then only nominal shear reinforcement need to be provided. In case the shear stress on the section exceeds the shear strength of the concrete, then shear reinforcement has to be provided subject to the condition that the shear stress is within the maximum shear strength of the concrete with shear reinforcement. When the actual shear stress on the section is more than the maximum strength in the concrete along with the shear reinforcement, the section must be redesigned such that the shear stress falls within the maximum limits. The design criterion for shear can therefore be stated in four cases as follows:

Case 1: No transverse reinforcement is needed in case
$$\tau_v < 0.5\, \tau_c$$
Case 2: Only nominal reinforcement. Provide nominal shear reinforcement if
$$0.5\tau_c < \tau_y < k_s\, \tau_c \tag{12.36}$$

in which τ_v = nominal shear stress on the section

τ_c = shear strength of the concrete (given in Table 12.2)

k_s = coefficient which depends on the shape of the section (given in Table 12.3)

Table 12.2 Shear Strength of Concrete (τ_c in MPa)

$\dfrac{100\, A_s}{bd}$	Concrete grade					
	M15	M20	M25	M30	M35	M40
0.25	0.52	0.54	0.54	0.56	0.56	0.56
0.50	0.69	0.71	0.72	0.72	0.73	0.73
0.75	0.81	0.83	0.85	0.87	0.88	0.88
1.00	0.90	0.92	0.96	1.0	1.0	1.0
1.50	1.05	1.06	1.10	1.15	1.18	1.20
2.00	1.05	1.20	1.24	1.26	1.30	1.32
3.00	1.05	1.20	1.38	1.45	1.50	1.60

The maximum spacing of the shear reinforcement is governed by
$$s_{vm} = \frac{A_{sv} f_y}{0.4 b_w} \tag{12.37}$$

in which s_{vm} = maximum spacing of the vertical stirrups

A_{sv} = total area of the cross-section of the stirrup legs effective in shear

b_w = width of the web

f_y = characteristic strength of steel in N/mm^2

s_{vm} *should also be less than 0.75d or 450 mm.*

Case 3: Shear reinforcement. In case the nominal shear stress is greater than the shear capacity of the concrete, provided it is less than the maximum limit, shear reinforcement must

Table 12.3 Shear Strength Coefficient (k_s) (except in flat slabs)

	For solid slabs of overall thickness					
t (mm) upto	150	175	200	250	275	300
k_s	1.3	1.25	1.20	1.10	1.05	1.0

be provided to resist the shear force which is in excess of the shear capacity of the section.

The shear capacity of a section is

$$V_c = bdk_s \tau_c \tag{12.38}$$

The net shear force for which the shear reinforcement must be designed is

$$V_s = V - V_c \tag{12.39}$$

After selecting a suitable diameter of the stirrup legs, the spacing of the stirrups can be calculated. The stirrups can be vertical or inclined. Even bent up main bars can provide shear resistance. The shear capacity of the bent up bars is

$$V_{cs} = A_{sv} f_y \sin \phi \tag{12.40}$$

where A_{sv} = area of the cross-section of the bend up bars

ϕ = inclination of the bent up bar with the axis of the member.

The shear reinforcement can be designed for the shear force over and above the shear capacity of the concrete section and the inclined bars.

The spacing of the vertical stirrups is

$$s_v = \frac{A_{sv} f_y d}{V_s} = \frac{A_{sv} f_y}{b_w (\tau_v \quad \tau_c)} \tag{12.41}$$

The spacing of the inclined stirrups along the axis of the member is

$$s_v = \frac{A_{sv} f_y d}{V_s} (\sin \phi + \cos \phi) \tag{12.42}$$

(The reader is referred to Chapter 2 for the derivation of Eq. (12.40))

The maximum shear strength with transverse steel is given in Table 12.4.

The nominal shear stress on a RCC slender section is given by

$$\tau_v = \frac{V}{bd} \tag{12.43}$$

For beams with varying depth, it is

$$\tau_v = \frac{V_s + M \tan \theta / d}{bd} \tag{12.44}$$

where M = BM

θ = angle between the top and bottom edges of the beam.

Table 12.4 Maximum Shear Strength with Reinforcement

Concrete grade	M15	M20	M25	M30	M35	M40
τ_{cmax} (MPa)	3.8	4.2	4.7	5.2	5.6	6.0

For members subjected to axial force

$$k_s = 1 + \frac{3P}{A_g f_{ck}}$$

(12.45)

where $\qquad$ P = axial force

$\qquad$ A_g = gross area of the section

12.6 Design Examples of Rectangular Sections

EXAMPLE 12.1 *Design of a balanced section:* Design a simply supported beam of effective span 6 m subjected to the following loads and specifications:

Superimposed load	$w_s = 9.0$ kN/m
Live load	$w_l = 15.0$ kN/m
Grade of concrete	$f_{ck} = 20$ MPa $= 20(10^6)$ N/m^2
Grade of the main reinforcement	$f_y = 415$ MPa
Grade of the stirrup steel	$f_y = 250$ MPa
Span	$L = 6$ m

Let depth of the beam be assumed in the range of $L/15$ for the purpose of calculation of the self-weight.

Let $\qquad$ $h = 0.50$ m and $b = 0.25$ m

Self-weight $\qquad$ $w_g = (0.5)(0.25) = 3.4$ kN/m

$\qquad\qquad\qquad$ $w_d = w_g + w_s = 12.4$ kN/m

The load factors are taken as

$$N_{d1} = 1.5 \text{ and } N_l = 2.2$$

The ultimate bending moment which will occur at the mid-span is equal to

$$M_u = 1.5 \frac{w_d L^2}{8} + 2.2 \frac{w_l L^2}{8}$$

$$= (18.6 + 33)\left(\frac{36}{8}\right) = 232.2 \text{ kNm}$$

The balanced moment capacity equal to the ultimate moment leads to

$$Kbd^2 f_{ck} = 232.2(1000) \text{ Nm}$$

where $\qquad$ $K = 0.225 \, \gamma_m = 0.225(0.87) = 0.196$, $\quad p_0 = 0.24$, then

$$d^2 = \frac{0.232\,200}{0.196(0.25)(20)} = 0.237 \text{ m}^2$$

or $\qquad$ $d = 0.487$ m

and $\qquad$ $h = 0.487 + 0.04 = 0.527$ m

$$A_{st} = p_0 bd \frac{f_{ck}}{f_y} = 0.24(250)(487)\frac{(20)}{415} = 1407 \text{ mm}^2$$

Provide $b = 250$ mm, $h = 530$ mm, $d = 490$ mm
Four numbers of $\phi 20$ and 1 number of $\phi 16$ of which half the bars can be cranked.

$$A_{st} \text{ (provided)} = 1457 \text{ mm}^2$$

Provide 2 numbers of 10 mm bars at top as hanger bars.

Total area of steel $A_{so} = A_{st} +$ area of hanger bars

$$= 1457 + 157 = 1614 \text{ mm}^2$$

$$\frac{100 \, A_{st}}{bd} = \frac{161400}{250(520)} = 1.25$$

Design for shear. Maximum design shear force at a distance d from the centre of support is

$$V_u = ((1.5)(12.4 + 2.2(15))(3/-0.49) = 128 \text{ kN}$$

Nominal shear stress $\quad \tau_v = \dfrac{V_u}{bd} = \dfrac{128\,000}{250(490)} = 1.05 \text{ N/mm}^2$

The shear capacity of M20 concrete for 1.25% reinforcement (from Table 12.2) is

$$= \tau_c = 1.01 \text{ N/mm}^2$$

(Two-legged 8 mm bars as stirrups)

Area of the stirrup $\quad = 2(50) = 100 \text{ mm}^2 = A_{sv}$

$$= V_s = V_u - \tau_c bd = 128\,000 - 1.01(490)(250)$$

$$= 4275 \text{ N}$$

Spacing of the stirrups $\quad s_v = \dfrac{A_{sv} f_y d}{V_s} = \dfrac{(100)(250)(490)}{4275} = 2870 \text{ mm}$

Spacing based on the minimum shear reinforcement is

$$s_{vm} = \frac{A_{sv} f_y}{0.4 b} = \frac{100(250)}{(0.4)(250)} = 250 \text{ mm}$$

Maximum spacing $\quad = 0.75d$ or 450 mm

Provide 8 mm *two-legged* stirrups at 250 mm c/c.

EXAMPLE 12.2 *Design of under-reinforced section.* Design of a simply supported beam of span 6 m subjected to the following specifications:

$$w_s = 9.6 \text{ kN/m} \quad \text{and} \quad w_l = 15 \text{ kN/m}$$

Size of the beam $\quad b = 0.20 \text{ m}$ and $h = 0.80 \text{ m}$

$$f_{ck} = 20 \text{ MPa} \quad \text{and} \quad f_y = 415 \text{ MPa}$$

Let

$$d = h - 0.05 = 0.75 \text{ m}$$

$$w_g = (0.2)(0.8)(25) = 4.0 \text{ kN/m}$$

$$w_d = w_g + w_s = 14.0 \text{ kN/m}$$

$$w_u = 1.5 w_d + 2.2 w_l = 53.76$$

$$M_u = \frac{w_u L^2}{8} = \frac{53.46(36)}{8} = 241.92 \text{ kNm} = 241\,920 \text{ Nm}$$

Balanced moment capacity of the section is

$$M_{rb} = K b d^2 f_{ck}$$

$$= 0.196(0.2)(0.75)^2 \frac{(20)(10^6)}{1000} = 441.0 \text{ kNm}$$

The balanced moment capacity of the concrete section is higher than the ultimate design moment; therefore, the section can be designed as under-reinforced.

Moment capacity of the section is

$$M_r = f_y A_{st} (d - x_c)$$

$$= (415)(10^6) A_{st} \left(0.75 - \frac{A_{st} f_y}{b f_{ck}} \right)$$

$$= 415(10^6) A_{st}(0.75 - 103 A_{st}) \text{ Nm}$$

$$= 415\,000 A_{st} (0.75 - 103 A_{st}) \text{ kNm}$$

Equating the resisting capacity to the ultimate moment, we have

$$415\,000 A_{st} (0.75 - 103 A_{st}) = 241.920 \text{ kNm}$$

The rearrangement of the above equation leads to

$$103 A_{st}^2 - 0.75 A_{st} + 0.5391 (10^{-3}) = 0$$

or

$$A_{st} = \frac{0.75 - \sqrt{0.75^2 - 4(103)(0.00058)}}{2(103)} = 0.000880 \text{ m}^2$$

Provide 3 numbers of 20 mm dia bars at the bottom and 2 numbers of 10 mm dia bars at the top as hanger bars

Area of concrete $\qquad A_c = 0.2(0.8) = 0.16 \text{ m}^2$

Total area of steel $\qquad = A_{st} + A_{sc} = 942 + 157 = 1099 \text{ mm}^2$

$$\frac{100 A_{st}}{bd} = \frac{94200}{200(750)} = 0.628\%$$

Check for shear. Maximum shear force at a distance d from the centre of the support is

$$V_u = w_u (3 - 0.75) = 120.96 \text{ kN} = 120\,960 \text{ N}$$

Normal shear stress $\qquad = \dfrac{V_u}{bd} = \dfrac{120\,960}{200(750)} = 0.81 \text{ N/mm}^2$

Concrete shear capacity $\qquad = 0.80 \text{ N/mm}^2$

The minimum shear reinforcement governs the design as seen from Example 12.1.

EXAMPLE 12.3 *Doubly-reinforced section.* Design a simply supported beam of span 6 m subjected to the following loads and specifications:

Superimposed load $w_s = 9.5 \text{ kN/m}$

Live load $\qquad w_l = 15.0 \text{ kN/m}$

$f_{ck} = 20 \text{ MPa}, \quad f_y = 415 \text{ MPa}$

$b = 0.20 \text{ m}, \qquad h = 0.4 \text{ m}$

Self-weight of the beam $w_g = (0.2)(0.4)(25) = 2 \text{ kN/m}$

Ultimate moment which occurs at the mid-span

$$M_u = (1.5 w_d + 2.2 w_l) \frac{L^2}{8} = (1.5(2.0 + 9.5) + 2.2(15)) \frac{36}{8}$$

$$= 225.585 \text{ kNm} = 225\,585 \text{ Nm}$$

The overall depth of the section is 400 mm, and after allowing for a cover of 25 mm the effective depth can be taken as

$$d = 400 - 25 - 10 = 365 \text{ mm}$$

The balanced moment capacity of the concrete section is

$$M_{rb} = K\,bd^2\,f_{ck}$$

$$= 0.196(0.2)(0.356)^2(20)(10^6) = 104\,448 \text{ Nm}$$

The ultimate moment is more than the balanced moment capacity; therefore, the section is to be designed as a doubly-reinforced section.

The moment to be resisted by the compression reinforcement is

$$M = M_u - M_{rb} = 225\,585 - 104\,448 = 121\,137 \text{ Nm}$$

Let the centre of the compression steel from the extreme compression fibre be

$$d_c = 35 \text{ mm}$$

The strain at the level of the compression steel is

$$\epsilon_{sc} = 0.0035\left(1 - \frac{35}{x_u}\right) = 0.0035\left(1 - \frac{35}{0.425d}\right)$$

$$= 0.0035\left(1 - \frac{35}{0.425(365)}\right) = 0.0028$$

and
$$f_{sc} = \epsilon_{sc}E_s = 0.0028(200\,000) = 560 \text{ N/mm}^2$$

This value is more than f_y, which is not allowed; therefore,

$$f_{sc} = f_y = 415 \text{ MPa}$$

$$d - d_c = 365 - 35 = 330 \text{ mm} = 0.33 \text{ m}$$

Moment capacity of the compression reinforcement is

$$M_{rc} = A_{sc}\,f_y\,(d - d_c)$$

Equating the moment capacity of the compression reinforcement to that of the unbalanced, we have

$$A_{sc}\,f_y\,(d - d_c) = M = 121\,137 \text{ Nm}$$

or
$$A_{sc} = \frac{121\,137}{(415)(0.33)} = 884 \text{ mm}^2$$

The tensile reinforcements corresponding to the balanced and compression steels are:

$$A_{stb} = p_0\,bd\,\frac{f_{ck}}{f_y} = 0.24(200\,(365)\left(\frac{20}{415}\right) = 844 \text{ mm}^2$$

and
$$A_{st2} = A_{sc}\,\frac{f_{sc}}{f_y} = 884 \text{ mm}^2$$

$$A_{st} = A_{stb} + A_{st2} = 1741 \text{ mm}^2$$

Provide three numbers of 20 mm bars at top, five numbers of 20 mm bars and one number of 16 mm bars at bottom

$$A_{sc} = 942 \text{ mm}^2 \text{ and } A_{st} = 1771 \text{ mm}^2$$

$$A_{so} = A_{st} + A_{sc} = 2713 \text{ mm}^2$$

Design for shear is not included in the example.

EXAMPLE 12.4 *Design of prestressed concrete beam.* A simply supported prestressed concrete beam of span 6 m is to be designed for the following specifications:

Superimposed load $\qquad w_s = 9.4 \text{ kN/m}$

Live load $\qquad w_l = 15 \text{ kN/m}$

$$f_{ck} = 35 \text{ MPa}, \quad f_y = 1500 \text{ MPa}$$

Let the size of the beam for the purpose of self-weight be

$$= 0.2 \times 0.4 \text{ m}^2$$

Self-weight $\qquad w_g = 0.2(0.4)(25) = 2 \text{ kN/m}$

The maximum moment occurs at the mid-span and the ultimate moment with load factors of 1.5 and 2.2 is

$$M_u = (1.5(w_g + w_s) + 2.2 w_l)\frac{L^2}{8}$$

$$= (1.5(11.40) + 2.2(15))\left(\frac{36}{8}\right) = 224.91 \text{ kN}$$

The balanced moment capacity of a rectangular section

$$M_{rb} = K b d^2 f_{cks} \text{ where } K = 0.196, \text{ then}$$

$$d = \sqrt{\frac{M_u}{K b f_{ck}}} = \sqrt{\frac{0.22491}{0.196(0.2)(35)}} = 0.405 \text{ mm}$$

The overall depth $\qquad h = 405 + 35 = 440 \text{ m}$

The area of high tensile steel is

$$A_{st} = A_{stb} = p_0 b d \frac{f_{ck}}{f_y} = 0.24 (200) (450\frac{(35)}{1500} = 454 \text{ mm}^2$$

Provide $\qquad b = 200 \text{ mm}, h = 440 \text{ mm}, A_c = 0.088 \text{ mm}^2$

24 numbers of 5 mm wires, the area of which is

$$A_{st} = 470 \text{ mm}^2$$

Provide nominal top bars of 2 numbers of 10 mm as hanger bars

$$A_{sc} = 157 \text{ mm}^2$$

12.7 Design of Flanged Sections Based on the Ultimate Strength Design

Flanged sections such as T-beams are very commonly used in building and bridge construction without any special effort. Figure 12.5a illustrates typical I-sections. It can also be considered as a T-section as the bottom flange does not come into action at the limit state of strength of the beam. Figure 12.5b illustrates a strain variation in the limit state of strength, in which the strain at the top fibre is considered as the limiting compressive strain in the concrete.

The neutral axis distance based on the compatibility of strains is

$$\frac{x_u}{d - x_u} = \frac{\epsilon_{cu}}{\epsilon_{sb}} \qquad (12.46)$$

Usually the total of the prestrain in steel is taken as:

$$\epsilon_{se} + \epsilon_{ce} = 0.004 \qquad (12.47)$$

Equation (12.46) can be rearranged as

$$x_u = \frac{\epsilon_{cu} \, d}{\epsilon_{cu} + \epsilon_{sb}} \qquad (12.48)$$

It was seen that the value of x_u/d is 0.47 for balanced section. This could mean that the neutral axis is likely to be in the web rather than in the compression flange. In certain cases of under-reinforced sections, the neutral axis is likely to fall in the flange.

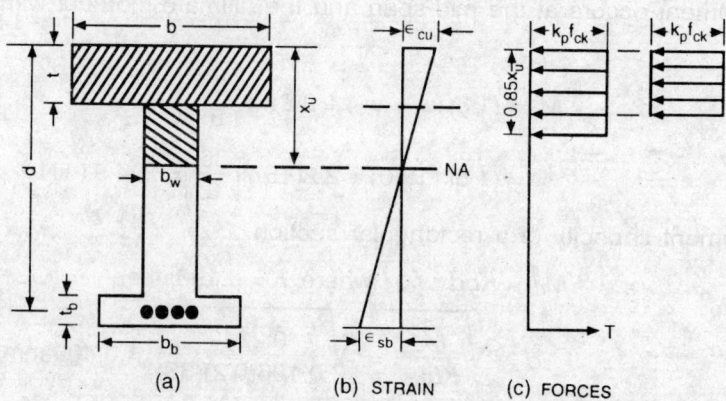

(a) (b) STRAIN (c) FORCES

Fig. 12.5 Stress Distribution of Flanged Section.

The limiting moment capacity of the flanged section can be obtained by taking moment of the compression forces about the centre of steel. The total compressive force can be divided into two parts: one coming from the web and the other coming from the overhang flange. These two design capacities can be expressed as

$$C_1 = 0.6 b_w x_u f_{ck} \qquad (12.49)$$

$$C_2 = (b_f - b_w) t k_p f_{ck} = 0.67(b_f - b_w) f_{ck} \qquad (12.50)$$

It is also assumed that the flange thickness is subjected to the intensity of $k_p f_{ck}$. The bending moment about the centre of the tension steel gives

$$M_{rb} = \gamma_m (C_1(d - x_c) + C_2(d - 0.5t))$$

$$= 0.87 \, (0.6 b_w x_u (d - 0.425 x_u) f_{ck}$$

$$+ \, 0.45(b_f - b_w) t (d - 0.5t) f_{ck}) \qquad (12.51)$$

It can be expressed as $\quad M_{rb} = K b_w d^2 f_{ck} + 0.58(b_f - b_w) t (d - 0.5t) f_{ck} \qquad (12.52)$

in which K is given in the earlier sections and its value is $K = 0.196$.

The area of the tension steel is given by

$$A_{stb} = \frac{C_1 + C_2}{f_y}$$

$$= p_0 bd \frac{f_{ck}}{f_y} + 0.58(b_f - b_w)\, t\, \frac{f_{ck}}{f_y} \tag{12.53}$$

Very often one is interested in designing the parameters such as d, A_{st}, etc. rather than the analysis of the section. In such cases Equation (12.52) is not well suited to the design. Therefore, certain approximations are made in the resisting moment capacity of a flanged section. The compressive force contributed by the web thickness below the flange is small when compared with that coming from the total flange area: Further, its contribution to moment resistance is much less as the lever arm is small. Hence, the compressive force can be idealised as acting on the flange alone and uniformly distributed into a rectangular block. The total compressive force is

$$C_u = b_f t k_p f_{ck} = 0.67 b_f t f_{ck} \tag{12.54}$$

The moment capacity is $M_r = \gamma_m C_u\,(d - 0.5t) = 0.58 b_f t\,(d - 0.5t) f_{ck}$ \hfill (12.55)

Assuming that the values of b_f and t are known from the other considerations, the value of the effective depth can be obtained as (by equating the moment capacity to the ultimate moment, i.e. $M_u = M_r$):

$$d = 0.5t + \frac{M_u}{0.58 b_f t f_{ck}} \tag{12.56}$$

The corresponding area of the tension reinforcement is

$$A_{st} f_y = \gamma_m\, C_u = 0.58 b_f\, t f_{ck}$$

or $$A_{st} = 0.580 b_f t \frac{f_{ck}}{f_y} \tag{12.57}$$

$$\frac{A_{st}}{b_f t} = 0.580 \frac{f_{ck}}{f_y} \tag{12.58}$$

In case of under-reinforced beams, the area of steel is

$$A_{st} = \frac{M_u}{(d - 0.5t) f_y} \tag{12.59}$$

The expressions will hold good for reinforced or prestressed concrete sections, subject to the condition that the neutral axis is in the web, that is

$$x_u = \frac{\epsilon_{cu} d}{\epsilon_{cu} + \epsilon_{sb}} \geq 1.15t \tag{12.60}$$

$$a = 0.85 x_u = 0.425 d$$

In case the neutral axis depth is less than the thickness of the flange, the section can be designed as a rectangular section with width equal to b_f.

The depth of the neutral axis in prestressed concrete balanced sections varies from 0.38 to 0.46 times the depth of the steel. Therefore, one can find whether the section is balanced or not by checking the neutral axis depth. From Eq. (12.60), we can use the design formulae provided

$$a = 0.425 d \geq 1.15t$$

or
$$d > \frac{1.15\,t}{0.425} = 2.75\,t \tag{12.61}$$

In case the depth of the beam comes out to be small, which is true in most cases, then the depth can be increased to about 2.5t.

The area of tension steel is then obtained as

$$A_{st} = \frac{M_u}{(d - 0.5t)f_y} \tag{12.62}$$

In case of cantilever beams, the thickness of the bottom flange is usually not specified whereas the top flange is given by other considerations. In some cases the thickness of the bottom slab can be assumed as

$$t = 0.4d \tag{12.63}$$

Then Eq. (12.55) reduces to

$$M_r = 0.185 b_f d^2 f_{ck} \tag{12.64}$$

12.8 Design of Prestressed Concrete Beam at Transfer Load Condition

At transfer of prestress, the prestressing force acts as a load and causes high compression in the zone of the cables. In such a case, the area of the concrete in the zone can be designed to resist the compressive force caused by the prestress. The self-weight of the dead load present at that time will relieve some of the compression by introducing tension. The load factor for prestress is usually taken from 1 to 1.2 and that for dead load as 0.9. The ultimate prestress is

$$T_{ut} = N_f A_{st} f_y \tag{12.65}$$

where $\qquad N_f$ = load factor (about 1.0 to 1.2).

The relative tension caused by the dead load moment is

$$T_{dt} = \frac{0.9 M_d}{d - 0.5t} \tag{12.66}$$

The net compression on the concrete is

$$C_{dt} = T_{ut} - T_{dt} \tag{12.67}$$

The area of the concrete needed in the cable zone (bottom flange in case of simply supported beams) is

$$C_{dt} = A_{ct}(0.45 f_{ck}) \tag{12.68}$$

or
$$A_{ct} = \frac{T_{ut} - T_{dt}}{0.45\,f_{ck}} \tag{12.69}$$

12.9 Design Example of Flanged Sections

EXAMPLE 12.5 *Design of RCC singly reinforced simply supported beam.* A floor slab 120 mm thick is supported by simply supported beams spaced 4 m apart. The clear span of the beam is 8 m. The beam is designed for the following:

Live load $= 4 \text{ kN/m}^2$
Concrete grade $f_{ck} = 20 \text{ MPa}$
Steel grade $f_y = 415 \text{ MPa}, t = 120 \text{ mm}$
Let the web thickness $b_w = 0.23 \text{ mm}$
Effective span $L = $ clear span + 5% of the clear span
$= 8 + 8/20 = 8.4 \text{ m}$

The flanged width b_f is controlled by

$b_f = $ centre to centre distance of beams = 4 m or

$$b_f = \frac{L_0}{6} + b_w + 6t = \frac{8.4}{6} + 0.23 + 6(0.12)$$

$$b_f = 2.35 \text{ m, or}$$

$$b_f = b_w + 12t = 0.23 + 12(0.12) = 1.67 \text{ m}$$

Select the flange width as the least of the above i.e. $b_f = 1.67 \text{ m}$
Let the overall depth of the beam be taken as

$$h = \frac{L}{20} = \frac{8.4}{20} = 0.42 \text{ m}$$

Self-weight of web $w_g = 0.23(0.42)(25000) = 2400 \text{ N/m}$
Weight of slab $w_s = 0.12(4)(25000) = 12000 \text{ N/m}$
Live load $w_l = 4000(4) = 16000 \text{ N/m}$

The ultimate load with 1.5 and 2.2 as load factor is

$$w_u = 1.5(14400) + 2.2(16000) = 56800 \text{ N/m}$$

The ultimate moment on the beam which occurs at the mid-span is

$$M_u = \frac{w_u L^2}{8}$$

The effective depth of steel can be obtained from Eq. (12.56) as

$$d = 0.5t + \frac{M_u}{0.58 \ b_f t \ f_{ck}}$$

$$d = 0.60 + \frac{501\,000}{0.58(1.67)(0.12)(20)(10^6)} = 0.06 + 0.21 = 0.27 \text{ m}$$

Use $d = 0.28 \text{ m then}$
$a = 0.425 \ d = 0.12 \text{ m}$

The neutral axis falls just at the bottom of the flange.

Overall depth $= h = d + 0.06 = 0.34 \text{ m}$

Area of the tensile reinforcement is taken from Eq. (12.57) and it is

$$A_{st} = 0.58 b_f t \frac{f_{ck}}{f_y} = 0.58(1670)(120)\left(\frac{20}{415}\right) = 5600 \text{ mm}^2$$

$$\frac{100 A_{st}}{b_f t} = 58\left(\frac{20}{415}\right) = 2.8\%$$

Design for shear. The maximum shear force occurs at a distance d from the support, and it is

$$V_u = w_u\left(\frac{L}{2} - d\right) = 56800(4.2 - 0.28) = 222\ 000\ \text{N}$$

$$\tau_v = \frac{V_u}{bd} = \frac{0.222}{(0.23)(0.28)} = 3.4\ \text{MPa}$$

The maximum shear strength from Table 12.4 is

$$\tau_{cmax} = 4.2\ \text{N/mm}^2$$

and the shear capacity of the concrete (for $100\ A_{st}/bd = 3.0$) from Table 12.2 is

$$\tau_c = 1.2\ \text{N/mm}^2$$

Provide two-legged 12 mm MS stirrups, then the area of the stirrup steel is

$$A_{sv} = 2(113) = 226.0\ \text{mm}^2$$

$$f_y = 250\ \text{MPa (for MS bars)}$$

The spacing of the stirrups at the maximum shear force is

$$\frac{A_{sv}f_y}{b_w(\tau_v - \tau_c)} = 110\ \text{mm}$$

Provide the stirrups at 110 mm c/c near the support.

Shear force gradient = 56800 N/m

Distance from the mid-span at which the shear force is equal to the shear capacity of the concrete is

$$x_1 = \frac{V_s}{\text{shear force gradient}} = \frac{77\ 280}{56\ 800} = 1.39\ \text{m}$$

Nominal shear reinforcement spacing based on the minimum steel is:

$$s_{vmax} = \frac{A_{sv}f_y}{0.4b_w} = \frac{226(250)}{0.4(230)} = 614\ \text{mm or}$$

$$= 0.75\ d = 210\ \text{mm}$$

Provide the stirrups at different spacings in the different segments at 110 mm c/c in the first one metre from support and gradually increasing the spacing to 210 mm at 3.1 from the support.

Provide two-legged 12 mm bars in the following scheme:

Spacing	Segment from support	No. of stirrups
110 mm	– 400 mm to 1100 mm	15
140 mm	1100 mm to 2080 mm	17
170 mm	2080 mm to 3100 mm	6
210 mm	3100 mm to 4200 mm	5

Selection of main reinforcement

Area of steel needed   = 5600 mm^2

Provide 11 numbers of 25 mm bars and 1 number of 20 mm bars.

Area of tension steel provided

$$A_{st} = 11(490) + (314) = 5704\ \text{mm}^2$$

Provide four bars in a row; therefore, the overall depth required is

$$h = \text{effective depth}$$
$$+ 1/2 \ (20 \text{ mm bar diameter})$$
$$+ \text{clear spacing between 20 and 25 mm bars}$$
$$+ (25 \text{ mm bar diameter})$$
$$+ 30 \text{ mm cover}$$
$$= 280 + 10 + 30 + 25 + 30 = 375 \text{ mm}$$

· Web depth below flange = 375 − 120 = 255 mm

The design details are:

web below the flange = 230 × 255 mm
2 nos of 25 mm bars at 333 mm from top are continued,
4 nos of 25 mm bars at middle but curtailed at $L/5$ from support and
3 nos of 25 mm bars and one 20 mm bar are cranked.

Figure 12.6 illustrates the reinforcement details.

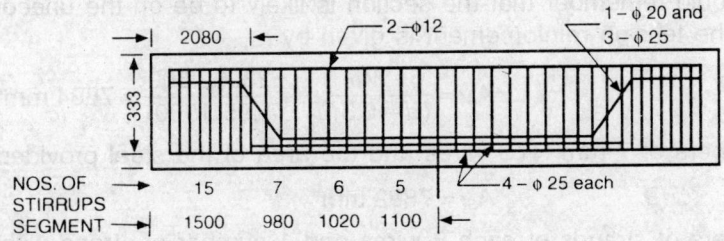

Fig 12.6 Reinforcement Details in Example 12.5.

Check for bond stress. The development length of the bar is given by

$$L_d = \frac{\phi f_y}{4 \tau_{bd}} = \frac{250 \phi}{4(1.2)} = 52 \ \phi$$

Deduct 16 φ for the hook, then the net length for development for full bond.

$$= (52 - 16)\phi = 36 \ \phi = 36(25) = 900 \text{ mm}$$

EXAMPLE 12.6 *Design of a prestressed beam.* A simply supported beam of span 50 m presented is subjected to a live load of 20 kN/m. The width of the top flange is to be less than 2 m and the depth of the beam to be less than 2.5 m.

Assume $f_{ck} = 45$ MPa, $f_y = 1500$ MPa

b_f can be assumed to be in the range of $L/40$ to $L/20$

Let $b_f = 1.6$ m and $t = (L/200 \text{ to } L/150) = 300$ mm

Let $w_g = \dfrac{WL}{24000}$ (approximately)

in which L is span in metres and W is the total live load on the span

$$w_g = \frac{20 \ (25000)}{24000} = 21 \text{ kN/m}$$

One can design the width of the top flange for an assumed depth or vice versa. In this example flange width is assumed and the depth of the beam is computed.

Let the combined load factor = 1.8. The ultimate bending moment which occurs at mid-span of the simply supported beam is

$$M_u = 1.8(w_g + w_i) \frac{L^2}{8} = 1.8(21 + 20) \frac{(2500)}{8} = 23062.5 \text{ kNm} = 23.0625 \text{ MNm}$$

The depth of the steel is given by

$$d = 0.5t + \frac{M_u}{0.58 b_f t f_{ck}}$$

$$= 0.15 + \frac{23.0625}{0.58 \, (1.6)(0.3)(45)} = 1.99$$

Use $d = 2.1$ m, then the neutral distance would be

$$a = 0.425d = 0.89 \text{ m} > t$$

Therefore, the moment capacity computed from the balanced section is on the safer side. However, one should remember that the section is likely to be on the uneconomical side.

The area of the tension reinforcement is given by

$$A_{st} = \frac{M_u}{(d - 0.5t)f_y} = \frac{23062\,500}{1.95(1500)} = 7884 \text{ mm}^2$$

Provide 205 numbers of 7 mm HTS wires and the area of the steel provided is

$$A_{st} = 7892 \text{ mm}^2$$

Provide 25 numbers of strands of each 8 wires and 1 number of strand with 5 wires.

Let there be one group in the bottom portion of the web and five groups in the bottom flange.

The overall depth of the beam can be calculated by assuming the diameter of the ducts and spacing of the groups.

Let the diameter of the duct   = 40 mm
Let the spacing of groups   = 50 mm
Let $b_w = 50 + 2\,(40) + 50$ = 180 mm

The maximum flange width for shear lag transfer is $12t + b_w$. This value is more than the value assumed. Therefore, the thickness of the web as 180 mm satisfactory. The approximate overall depth is then computed after selecting the bottom flange size.

Design at transfer condition. The size of the bottom flange is generally governed by the transfer prestress condition. The ultimate tension in the cables is

$$T_{ui} = A_{st}f_y = \frac{(7892)(1500)}{10^6} = 11.83 \text{ MN}$$

The self-weight of the beam has the tendency to compensate the compression on the bottom flange thus relieving some compression. The load factor applied should be 0.9 to compute the design bending moment due to DL.

$$M_d = 0.9 \, \frac{w_d L^2}{8} = 5.9 \text{ MNm}$$

Let the lever arm $jd = 0.9d$, then the net tension caused by the dead load (the assumed dead load) is

$$T_{dt} = \frac{M_d}{jd} = \frac{5.9}{0.9(2.1)} = 3.12 \text{ MN}$$

The net compression on the bottom is

$$C_{dt} = T_{ut} - T_{dt} = 11.83 - 3.12 = 8.71 \text{ MN}$$

The minimum compression area needed is

$$A_{cbt} = \frac{C_{dt}}{0.58 \, f_{ck}} = \frac{8.71}{0.58(45)} = 0.333 \text{ m}^2$$

Two groups of the cables are placed in the web; therefore, some of the web area around the group of the cables will be effective in resisting the compression. Let this group of the cables be 100 mm above the top of the bottom flange. Then the effective area of the web in compression is

$$A_{cw1} = b_w(2)(0.1) = 0.18 \,(0.2) = 0.036 \text{ m}^2$$

The area of the bottom flange is

$$A_{cb} = A_{cbt} - A_{cw1} = 0.333 - 0.036 = 0.297 \text{ m}^2$$

Let the width of the bottom flange be $b_b = 800$ mm.

Then the thickness of the bottom flange is

$$t_b = \frac{A_{cb}}{b_b} = \frac{0.297}{0.8} = 0.371 \text{ m}$$

Allowing some area of the section to be occupied by the ducts, the bottom flange is selected as

$$b_b = 800 \text{ mm} \quad \text{and} \quad t_b = 380 \text{ mm}$$

The areas of different elements in the I-section are:

Top flange $\qquad\qquad A_{cf} = 1.6(0.3) = 0.48 \text{ m}^2$

Bottom flange $\qquad\; A_{cb} = 0.8(0.38) = 0.304 \text{ m}^2$

Web $\qquad\qquad\qquad A_{cw} = b_w(d - t - 0.5t_b) = 0.290 \text{ m}^2$

Total area of the section $\quad A_c = 1.074 \text{ m}^2$

The self-weight of the section is

$$w_g = A_c \,(25) = 1.074(25) = 26.85 \text{ kN/m}$$

The assumed self-weight is 21 kN/m, whereas the designed weight is 26.85 kN/m. Hence, the section is modified while selecting the actual section. The design section is shown in Fig. 12.7. The effective depth is increased to 2.2 m and the overall depth is 2.5 m.

12.10 Moment Capacity of a Triangular Section

The depth of the neutral axis based on the compatibility of strains using Fig. 12.12 is

$$\frac{x_u}{d} = \frac{\epsilon_{cu}}{\epsilon_{cu} + \epsilon_{sb}} \qquad\qquad (12.13)$$

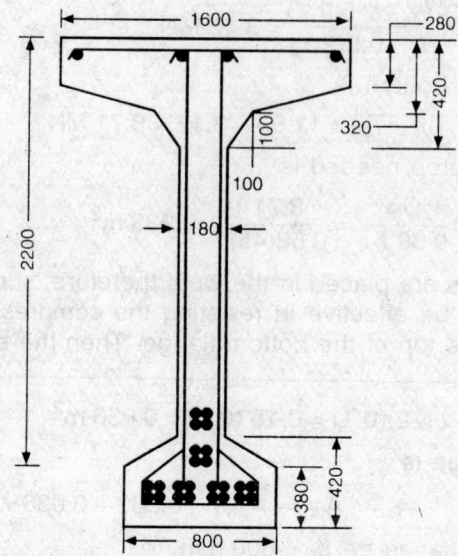

Fig. 12.7 Sectional Details of PSC Beam in Example 12.6.

The total compressive force from the rectangular stress distribution acting on the triangular bit is

$$C_u = \frac{1}{2}\left(\frac{x_2 b}{8}\right) x_2 k_p f_{ck}$$

$$= \frac{1}{2d}(bk_p f_{ck})x_u^2$$

$$= \frac{0.214}{d} bx_u^2 f_{ck} = 0.0473\, bdf_{ck} \qquad (12.70)$$

where $x_2 = 0.8x_u$ and $x_u = 0.47d$

The moment contribution of the above force about the CG of steel is

$$M_r = C_u\left(d - \frac{2x_u}{3}\right)$$

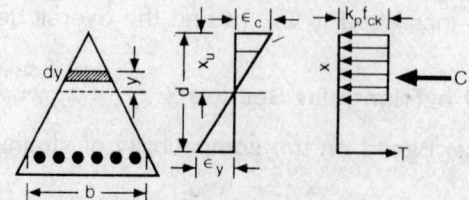

Fig. 12.8 Triangular Section and Stress Distribution.

$$= 0.214 \left(\frac{bx_u^2 f_{ck}}{d} \right)(d - 0.54x_u) = K_2bd^2 f_{ck} = 0.035bd^2 f_{ck} \tag{12.71}$$

Equating the tensile force to the compressive force, we have

$$0.87A_{st} f_y = C_u = 0.0473bdf_{ck}$$

or
$$A_{st} = 0.054bd \frac{f_{ck}}{f_y} \tag{12.72}$$

or
$$A_{st} = p_{02}bd \frac{f_{ck}}{f_y} \tag{12.73}$$

12.11 Moment Capacity of Trapezoidal Section (Fig. 12.9)

Let b_1 = width of the section at top
 b_2 = width of the section at the level of steel

The moment capacity of the section can be obtained by the sum of the capacities of a rectangular section of width b_1 and the triangular section of base width $(b_2 - b_1)$.

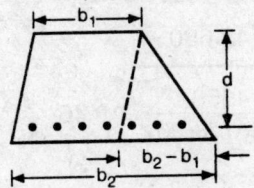

Fig. 12.9 Trapezoidal Section.

$$M_r = Kb_1d^2 f_{ck} + K_2(b_2 - b_1)d^2 f_{ck}$$
$$= ((K - K_2)b_1 + K_2b_2)d^2 f_{ck} \tag{12.74}$$

or
$$M_r = ((0.19 - 0.035)b_1 + 0.035b_2)d^2f_{ck}$$
$$= (0.155b_1 + 0.035)b_2)d^2 f_{ck} \tag{12.75}$$

The area of the reinforcement needed is given by the sum of the quantities required for the rectangular and triangular sections, and it is

$$A_{st} = (p_0b_1d + p_{02}(b_2 - b_1)d) \frac{f_{ck}}{f_y} \text{ or}$$

$$A_{st} = ((p_0 - p_{02}) b_1 + p_{02}b_2)d \frac{f_{ck}}{f_y} \tag{12.76}$$

$$= (0.186b_1 + 0.054b_2)d \frac{f_{ck}}{f_y} \tag{12.77}$$

12.12 Design Examples of Non-rectangular Sectioned Beams

EXAMPLE 12.7 *Design of triangular section.* A cantilever beam which supports a portico is made of an inverted triangular section. The effective span of the beam is 3 m and the live and dead loads from the slab over the beam are 1 kN/m, and 3 kN/m respectively. Design the cross-section.

Let
$$\text{Self-weight} \qquad w_g = 2 \text{ kN/m}$$
$$\text{Weight of the slab} \qquad = 3 \text{ kN/m}$$
$$\text{Live load} \qquad w_l = 1 \text{ kN/m}$$
$$\text{Total load} \qquad w = 6 \text{ kN/m}$$
$$\text{Span} \qquad L = 3 \text{ m}$$
$$\text{Concrete grade} \qquad = \text{M20}$$
$$\text{Quality of steel} \qquad f_y = 415 \text{ MPa}$$
$$\text{Load factors:} \qquad N_{d1} = 1.5, \; N_1 = 2.2$$

Design of the section. Let the width of the beam at the level of steel be selected in the range of $L/10$ then $b = 0.3$ m.

The ultimate bending moment on the section with a load factors of 1.5 and 2.2 is

$$M_u = (1.5(5) + 2.2(1)) \frac{9}{2} = 43.65 \text{ kNm}$$

Equating the moment capacity to the design moment, we have

$$M_r = K_2 b d^2 f_{ck} \ge M_u = 0.04365 \text{ MNm}$$

or
$$0.035 d^2 f_{ck} = 43\,650$$

or
$$d = \sqrt{\frac{0.4356}{0.035(0.3)(20)}} = 0.46$$

The area of tension reinforcement is

$$A_{st} = p_{02} \, bd \, \frac{f_{ck}}{f_y}$$

$$= 0.054(0.3)(0.46) \frac{15(10^6)}{415} = 360 \text{ mm}^2$$

The minimum reinforcement required is

$$A_{st} = \frac{0.85 \, A_c}{f_y} = 0.85 \frac{(300)(460)}{(2)(415)} = 141 \text{ mm}^2$$

The self-weight is less than that assumed; therefore, no revision is needed in the moment computation.

Provide 2 numbers of $\phi 20$ then

$$A_s \text{ (provided)} = 628 \text{ mm}^2$$

$$\text{Percentage of reinforcement} = \frac{62800}{150(530)} = 0.8\%$$

Design for shear. The shear force at a distance d from the support is

$$V = ((1.5)(5) + 2.2(1)) \, (L - d) = 24 \text{ kN}$$

The nominal shear stress is

$$\tau_v = \frac{V}{A_c} = \frac{0.024}{(0.3)(0.53)(0.5)} = 0.3 \text{ MPa}$$

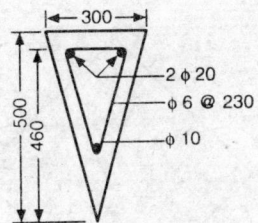

Fig. 12.10 Sectional Detail of Example 12.7.

The shear stress is less than the shear strength of the concrete; therefore, provide only nominal shear reinforcement.

Provide two-legged $\phi6$ mm stirrups, then the area of the shear reinforcement is

$$A_{sv} = 2(28) = 56 \text{ mm}^2$$

Maximum spacing of the stirrups is

$$s_{vm} = \frac{A_{sv}f_y}{0.2b} = \frac{56(250)}{0.2(300)} = 233 \text{ mm}$$

Note: The minimum shear reinforcement in case of triangular section is recommended as

$$A_{sv} = \frac{0.2bs_v}{f_y}$$

The maximum spacing is also restricted by

$$s_v < 0.75d$$

Provide two-legged 6 mm stirrups at 230 mm spacing.

The development length for plain bars in M20 concrete is 38ϕ when a hook is provided.

$$h = d + 0.05 = 0.58 \text{ m}$$

The same problem was solved by the limit state design, in which $b = 300$ mm, $d = 500$ mm, $A_{st} = 706$ mm^2.

PROBLEMS

(USD= Ultimate Strength Design)

(Use $\epsilon_c = 0.0035$ and $\epsilon_y = 0.003$ in all problems)

(γ_f = combined load factor)

12.1 Determine the neutral axis of balanced rectangular cross-section for M25 concrete and HYSD-Fe415 bars of $f_y = 415$ N/mm^2. Assume rectangular stress block on concrete in compression.

12.2 Determine the balanced moment capacity of square cross-section with distance of the reinforcement at 0.9 times depth of the cross-section. The moment capacity should be expressed in f_{ck}, f_y and size of the section.

12.3 Determine the moment capacity of a diamond-shaped cross-section with its side a and

608

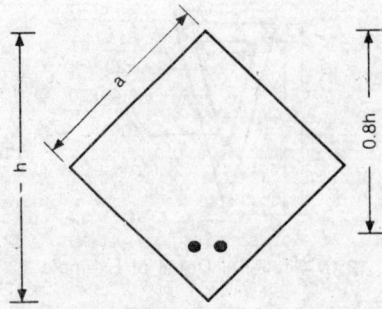

Fig. 1 Problem 12.3.

angle 90°. The depth of reinforcement from the apex to 80% of overall depth as shown in Fig. 1.

12.4 A square cross-section has a notch cut out in the compression flange as shown in Fig. 2. The size of the notch is 1/3rd of the size of the section. Derive the balanced moment capacity of the section for the following data: $f_{ck} = 20$ N/mm^2, $f_y = 415$ N/mm^2.

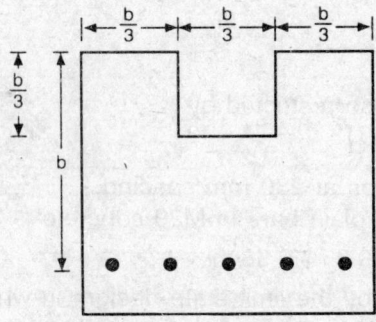

Fig. 2 Problem 12.4.

12.5 Derive balanced moment capacity of a circular cross-section with reinforcement at a distance of 0.8 time the diameter from the top compression fibre. Calculate the percentage of reinforcement needed for a balanced cross-section. Assume $f_{ck} = 20$ N/mm^2, $f_y = 415$ N/mm^2.

12.6 A simply supported beam of span 6 m is subjected to a live load of 16 kN/m. Design a balanced cross-section reinforced concrete beam by USD with width equally to 60% of the effective depth. Sketch the reinforcement details, $f_{ck} = 20$ N/mm^2, $f_y = 415$ N/mm^2, $\gamma_f = 2.0$.

12.7 A simply supported beam of effective span 6 m is subjected to a live load of 16 kN/m. The size of the beam is 400 by 400 mm. Design by USD the requirement details using M20 concrete and HYSD-Fe415 bars both for tension and shear reinforcement. $\gamma_f = 2.0$.

12.8 A simply supported beam of effective span 6 m is subjected to a live load of 16 kN/m.

The size of the beam is 250 by 400 mm. Design the reinforcement details by USD both for bending and shear using M20 concrete and HYSD-Fe415 bar. $\gamma_f = 1.8$.

12.9 A cantilever beam of span 4 m is subjected to a live load of 20 kN/m. Design by USD a balanced reinforced concrete rectangular section using M20 concrete and HYSD-Fe415 bars. $\gamma_f = 1.8$, $b = 300$ mm

12.10 A cantilever beam of span 4 m is subjected to a live load of 20 kN/m. Design reinforcement detail by USD if the size of the cross-section is 200 by 300 mm, using M20 concrete HYSD-Fe415 bars. $\gamma_f = 1.8$.

12.11 A cantilever beam is subjected to a concentrated load of 5 kN at the free end of 4 m span. Design a balanced reinforced concrete cross-section using M25 concrete and mild steel reinforcement by USD. $\gamma_f = 1.8$, $b = 300$ mm.

12.12 A staircase is provided with independent steps cantilevering from an upright reinforced concrete wall. The length of the step is 1.2 m and it is designed for a load of 1.5 kN placed on 1.1 m from the fixed support. The width of the step is 300 mm. Design a balanced reinforced concrete cross-section using M20 concrete and HYSD-Fe415 bars by USD. $\gamma_f = 1.8$.

12.13 A staircase slab spans between two walls separated by a distance of 4.8 m. The width of the staircase is 1.2 m and it has two landings one at each end and having a length of 1.2 m. Design by USD the slab for a live load of 4 kN/m^2 using M20 concrete and HYSD-Fe415 bars. The rise and thread of the staircase are 150 and 300 mm respectively.

12.14 A reinforced concrete cross-section of width 400 mm and effective depth 800 mm is provided with 4 nos. of 20 mm bars as tension reinforcement. The beam is simply supported with an effective span of 8 m and subjected to UDL to 24 kN/m. Determine whether the section is capable of resisting the ultimate moment with $\gamma_f = 1.8$. The concrete is of M20 and steel is HYSD-Fe415 bars.

12.15 A reinforced cross-section of 200 mm width and 400 mm effective depth is provided with 4 nos. of 20 mm bars in tension zone and 2 nos. of 20 mm bars in the compression zone at depth 50 mm from the compression fibre. The cross-section is subjected to a bending moment of 150 kNm. Determine whether the section is capable of resisting ultimate moment with $\gamma_f = 1.8$. The concrete is of M20 and steel HYSD-Fe415 bars.

12.16 A doubly-overhang beam of total length 8 m is subjected to a UDL of 16 kN/m. The overhang span is 1.6 m on each side. Design a reinforced concrete beam with uniform depth throughout using M20 concrete and HYSD-Fe415 bars. The depth of the section should not exceed 400 mm. $\gamma_f = 1.8$, $b = 300$ mm.

12.17 An electrical transmission line pole of total 8 m length is embedded into ground to a depth of 1.5 m. A wind load of 3 kN acting on the wires is transmitted at 0.5 m from top of the pole. Design a reinforced concrete square cross-section pole using M20 concrete and HYSD-Fe415 bars by USD. Use $\gamma_f = 1.5$.

12.18 A boundary wall of 2 m effective height above the ground level is subjected to wind force of 1.5 kN/m^2. Design a reinforced concrete wall with foundation depth 1 m below ground level. Use M20 concrete and mild steel bars. The footing of the wall need not be designed and the wall is assumed to be stable under the wind load condition. Use $\gamma_f = 1.5$.

12.19 A boundary wall of 2.5 m high above ground level is subjected to 1.5 kN/m^2 wind load.

The wall is designed with pilasters spaced at 2.5 m intervals. Design the boundary wall as well as the pilasters by USD in reinforced concrete using M20 concrete and HYSD-Fe415 bars. Assume the wall is stable under the wind load condition. Use $\gamma_f = 1.5$.

12.20 A precast reinforced concrete pole of 8 m length and having square cross-section of 300 mm size of provided with 4 nos. of 20 mm HYSD-Fe415 bars one on each corner of the cross-section. The clear cover to the bar is 25 mm. The member is lifted from the ground holding it at its middle. Determine by USD whether the section can withstand the load with a load factor of 1.5 applied to the load.

12.21 Calculate for USD the load factor applied to a rectangular cross-section beam for the following data which was designed for working moment of 100 kNm. $b = 300$ mm. $d = 750$ mm, $A_{st} = 604$ mm^2.

12.22 A rectangular reinforced concrete section is subjected to an external service moment of 150 kNm. Determine the combined load factor applied to loads if the section was designed by USD. $b = 400$ mm, $d = 480$ mm, $A_{st} = 804$ mm^2.

12.23 Calculate the ultimate moment capacity of a rectangular cross section reinforced beam for the following data: $b = 400$ mm, $d = 600$ mm, $A_{st} = 503$ mm^2, $f_y = 415$ N/mm^2, $f_{ck} = 20$ N/mm^2. State whether the section is over-reinforced or under-reinforced?

12.24 Determine the ultimate moment capacity of a rectangular cross-section reinforced concrete beam for the following data: $b = 200$ mm, $d = 600$ mm, $A_{st} = 1206$ mm^2, $f_y = 415$ N/mm^2, $f_{ck} = 20$ N/mm^2. State whether the section is over-reinforced or under-reinforced?

12.25 The tension reinforcement in a singly reinforced balanced rectangular cross-section is increased by 200% of the balanced steel. Determine the enhanced ultimate moment capacity of the cross-section as a function of balanced moment capacity assuming all other properties remaining the same.

12.26 The tension reinforcement in a balanced rectangular cross-section beam is reduced to 2/3rd of the balanced reinforcement. Determine the revised moment capacity of the section as a function of balanced moment capacity assuming all other data remaining the same.

12.27 Determine the ultimate moment capacity of a doubly-reinforced rectangular cross-section beam for the following data:
$b = 400$ mm, $d = 700$ mm, $A_{st} = 1206$ mm^2, $f_y = 415$ $f_{ck} = 20$ N/mm^2, $A_{sc} = 603$ mm^2.

12.28 Determine the ultimate moment capacity of a doubly-reinforced rectangular cross section beam for the following data:
$b = 300$ mm, $d = 600$ mm, $A_{st} = 1407$ mm^2, $A_{sc} = 804$ mm^2, $f_y = 415$ N/mm^2, $f_{ck} = 20$ N/mm^2.

12.29 A simply supported reinforced concrete beam is subjected to a working bending moment of 700 kNm and a shear force of 300 kN. Design a T-beam cross section by USD to resist the above forces using the following data: Maximum width of flange 1000 mm, thickness of flange 150 mm. Use M20 concrete and HYSD-Fe415 bars with $\gamma_f = 1.8$.

12.30 Derive an expression for ultimate moment capacity of reinforced concrete T-section in terms of width and thickness of the flange, effective depth and characteristic strengths of concrete and steel. Also calculate the percentage of tension steel required with respect of the total web area.

12.31 A reinforced concrete T-beam with the properties given below is subjected to a working moment of 500 kNm. Determine whether the section is capable of withstanding the moment by USD. $b_f = 900$ mm, $t = 120$ mm, $d = 600$ mm, $A_{st} = 4900$ mm^2, $\gamma_f = 1.8, f_{ck} = 20$ N/mm^2, $f_y = 415$ N/mm^2.

12.32 A reinforced concrete T-beam cross-section with the details given below is subjected to an external working bending moment of 320 kNm. The characteristic strengths of concrete and reinforcement are 20 and 415 N/mm^2 respectively. Determine whether the cross-section is structurally safe by USD for the moment given above.

$b_f = 1.6$ m, $t = 0.12$ m, $d = 0.51$ m, $A_{st} = 2918$ mm^2, $\gamma_f = 1.8$.

12.33 A doubly-reinforced T-beam section with the dimensions given below is subjected to a maximum working bending moment of 300 kNm. Determine whether the section is capable of withstanding in the ultimate strength. $b_f = 1600$ mm, $t = 120$ mm, $b_w = 250$ mm, $d = 330$ mm, $A_{st} = 4600$ mm^2 $A_{sc} = 9938$ mm^2, $d_c = 60$ mm $f_{ck} = 20$ N/mm^2 $f_y = 415$ N/mm^2, $\gamma_f = 2.0$

12.34 Determine the ultimate moment capacity of a singly-reinforced concrete T-section for the following details.

$b_f = 1000$ mm, $t = 120$ mm, $b_w = 250$ mm, $d = 500$ mm, $A_{st} = 1256$ mm^2, $f_{ck} = 20$ N/mm^2, $f_y = 415$ N/mm^2.

12.35 Determine the ultimate moment capacity of reinforced concrete T- section for the following details; $b_f = 1200$ mm, $t = 150$ mm, $b_w = 400$ mm, $d = 600$ mm, $A_{st} = 4900$ mm^2, $f_{ck} = 20$ N/mm^2, $f_y = 415$ N/mm^2 $A_{st} = 4900$ mm^2.

12.36 Determine the ultimate moment capacity of a doubly-reinforced T-section for the following details and specifications: $b_f = 1000$ mm $t = 150$ mm, $b_w = 400$ mm, $d = 500$ mm, $A_{sc} = 4900$ mm^2, $A_{st} = 7350$ mm^2, $f_{ck} = 20$ N/m^2, $f_y = 415$ N/mm^2.

12.37 A simply supported beam of an effective span 10 m is subjected to a uniformly distributed live load of 30 kN/m. The thickness and width of the flange are restricted to 120 and 1200 mm respectively. Design a T-beam section with M20 concrete and HYSD-Fe415 bars by USD. $\gamma_f = 1.8$.

12.38 A simply supported beam of an effective span 8 m subjected to an uniformly distributed live load of 20 kNm and a moving load of 150 kN. Design a singly-reinforced T-beam by USD using M20 concrete and HYSD-Fe415 bars. The width and thickness of the flange are restricted to 1000 and 200 mm respectively. $\gamma_f = 1.8$.

12.39 A simply supported beam of an effective span 8 m is subjected to a uniformly distributed live load of 30 kN/m. The architect has specified following dimensions for the beam: $b_f = 1000$ mm, $t = 120$ mm, $b_w = 300$ mm, $h = 600$ mm. Design reinforcement details by USD using M20 concrete and HYSD-Fe415 bars. $\gamma_f = 1.8$.

12.40 A simply supported beam of an effective span 12 m subjected to a uniformly distributed live load of 20 kN/m. Design a reinforced concrete T-section subjected to limitations: $b_f = 1500$ mm, $t = 160$ mm, $h = 800$ mm, $b_w = 400$ mm, $f_{ck} = 20$ N/mm^2, $f_y = 415$ N/mm^2. $\gamma_f = 1.8$.

PART II

DESIGN OF FOUNDATIONS

Chapter 13

Working Stress Design of RCC Shallow Foundations

13.1 Introduction

This chapter is devoted to the design of RCC elements in the foundation, therefore, offers no explanation on the soil mechanics. The important characteristics of soils that one is expected to know for the design of the RCC foundations are:

1. Type of soil.
2. Bearing capacity.
3. Settlement at different pressures.
4. Water table depth.
5. Friction angle of soil.

In special foundations, one may have to know more detailed soil characteristics, but in case of normal buildings such as residential, office and industrial buildings, the characteristics listed above should be adequate. The vibration characteristics become important when the foundations are subjected to dynamic forces. This chapter is also devoted to the analysis of soil pressures under simple load conditions and the consequent pressures on the foundations (Fig. 13.1). The importance of the foundation needs no special emphasis as performance of

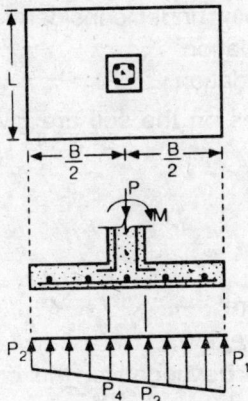

Fig. 13.1 Pressure Distribution and Notations under Isolated Symmetrical Footing.

the structure is very much influenced by the strength and settlement of the foundation. The primary design criteria in shallow foundations are:

1. Maximum pressure on the soil should not exceed the allowable pressure.
2. Unequal settlements either due to loads or environmental conditions must be limited to the allowable values.

3. The minimum depth of foundation should be provided so as to protect it from local failure of soil either by strength or by the environmental factors such as erosion, etc.
4. No tension is allowed on soil.

In addition, the structural components of the elements must satisfy the requirements laid down by the appropriate materials. Some of the minimum requirements in the RCC foundations are:

1. Minimum cover to the reinforcement should be 50 mm in case, the concrete is resting on nominal lean concrete, or 75 mm when either no lean concrete base is provided or in aggressive environment even with the lean concrete.
2. The temperature and shrinkage variations in depths of 1 m with respect to ground level are less when compared with those in the superstructure. Therefore, a nominal reinforcement of 0.15% should be adequate under normal environmental conditions.

13.2 Size of Foundation

The size of foundation is controlled by the allowable bearing capacity and to some extent by the settlement properties. The bearing pressure on soil can be expressed as

$$p = \frac{P}{A} \pm \frac{My}{I} \tag{13.1}$$

in which

p = bearing pressure on the soil
P = axial load on the foundation
A = area of the foundation = BL
M = bending moment
I = sectional moment of inertia of the base area = $LB^3/12$
y = distance of the point under consideration form the C.G. of the foundation
B = width of the foundation
L = length of the foundation

The maximum and minimum pressures on the soil are given by

$$p_1 = \frac{P}{A} + \frac{MB}{2I} \tag{13.2}$$

$$p_2 = \frac{P}{A} - \frac{MB}{2I} \tag{13.3}$$

in which $\qquad p_1$ = maximum pressure
$\qquad\qquad\quad p_2$ = minimum pressure ≥ 0

In case the value of p_2 works out to be negative, the equations mentioned above are not applicable as the soil is not capable of resisting the tension. Therefore, the equations need modification. The equilibrium of forces on the foundation soil with partial contact of the foundation slab with the soil gives (vide Fig. 13.2).

$$\frac{LHp_1}{2} = P \tag{13.4}$$

in which H = length of the foundation slab in contact with the soil and is less than B. The moment equilibrium about a horizontal axis passing through the controid gives

$$\frac{LHp_1}{2}\left(\frac{B}{2}-\frac{H}{3}\right)=M \tag{13.5}$$

The values of H and L can be computed from the above two equations for a given allowable bearing loads. If p_a be the allowable bearing pressure on the soil then one must ensure that the maximum pressure on the soil is less than the bearing capacity of the soil. The allowable bearing capacity specified is invariably the *net bearing capacity* of the soil and consequently the weight of the soil above the base of the foundation can be added to the bearing capacity to compute the gross bearing capacity of the soil. The size of the foundation can therefore be determined by the allowable pressure criterion and is given by

$$p_1 \le p_a \tag{13.6}$$

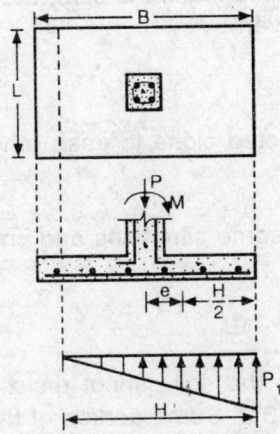

Fig. 13.2 Pressure Distribution under Eccentrically Loaded Footing.

Case 1: *Full length in contact of the soil.* If the total base of the foundation is in contact with the soil, that is, p_2 is more than zero, Eqs. (13.2) and (13.6) give

$$p_1 = \frac{P}{A} + \frac{MB}{2I} \le p_a \tag{13.7}$$

or, by substituting

$$A = BL, \text{ and } I = \frac{LB^3}{12}$$

in Eq. (13.7), we have

$$\frac{P}{BL} + \frac{6M}{B^2 L} \le p_a \tag{13.8}$$

or

$$BL \ge \frac{P}{p_a} + \frac{6M}{B} \tag{13.9}$$

One may now ask, how does one know that the value of p_2 is greater than zero? Such a question is not only relevant but rather essential in this design as B and L are yet the unknown design variables. To start with, one presupposes that p_2 is greater than zero, but later on it is to be checked whether such an assumption is valid. One can obtain a design by forcing

p_2 equal to zero if there are no other constrains. This need not necessarily result in an economical design. From Eq. (13.3), we have

$$p_2 = \frac{P}{LB} - \frac{6M}{LB^2} = 0$$

(13.10)

or

$$B = \frac{6M}{P}$$

(13.11)

The value of L can be determined from Eq. (13.9) as

$$L \geq \frac{P}{Bp_a} + \frac{6M}{B^2}$$

(13.12)

In case the bending moment on the foundation is zero, the problem is considerably simplified and the size of the foundation is given by

$$BL \geq \frac{P}{p_a}$$

(13.13)

The values of B and L can be selected close to each other if no constraints on any one of the design variables are fixed.

Case 2: Partial uplift from soil. For some situations and combinations of P and M, it is likely that

$$p_2 = \frac{P}{A} - \frac{MB}{2I} < 0$$

which implies tension on soil. Since the soil cannot resist tension, the pressure on the soil gets redistributed as shown in Fig. 13.3. Some portion of the slab will not be contact with the soil. From the equilibrium Eq. (13.4) and the criterion of design, we have

$$p_1 = \frac{2P}{LH} \leq p_a$$

(13.14)

or

$$LH \geq \frac{2P}{p_a}$$

(13.15)

From Eqs. (13.4) and (13.5), we have

$$\frac{LHp1}{2}\left(\frac{B}{2} - \frac{H}{3}\right) = P\left(\frac{B}{2} - \frac{H}{3}\right) = M$$

or

$$\frac{B}{2} - \frac{H}{3} = \frac{M}{P}$$

(13.16)

Equations (13.15) and (13.16) will give a set of values for L and H for a given value of B. In case $M = 0$, the problem reduces to that discussed in Case 1.

Case 3: A symmetrical foundation. The axial load on the foundation of the two previous cases was considered to be acting at the centroid of the foundation. In case of a foundation where the moment can act in either direction, it is designed symmetrical. There may be some situations where the bending moment is fixed and acts only in one direction. In such cases the foundation slab need not be centred with the axial force but could be laid eccentric. The width of the foudation can be made equal to H. Figure 13.3 illustrates the slab location. The equilibrium of forces gives

$$\frac{LHp_1}{2} = P \tag{13.4}$$

and the moment equilibrium gives

$$\frac{LHp_1}{2}\left(e + \frac{H}{2} - \frac{H}{3}\right) = M$$

or

$$\frac{LHp_1}{2}\left(e + \frac{H}{6}\right) = M \tag{13.17}$$

or

$$e + \frac{H}{6} = \frac{M}{P} \tag{13.18}$$

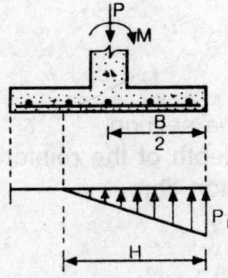

Fig. 13.3 Pressure Distribution under Footing Partially in Contact with Soil.

13.3 The Depth of Foundation

The depth of foundation is measured from the ground level to the bottom surface of the lean concrete. A lean cement concrete of 1:4:8 or lime concrete of 1:3:6 can be used as a surface finish on the soil. The thickness of the lean concrete can be from 75 mm to 100 mm in which the lower value is used in small residential type of buildings and the upper value can be adopted in heavy column foundations. The depth of the foundation should be taken so as to avoid any damage to the foundation concrete and to protect the soil below the foundation. The depth of the foundation depends on the nature of the soil. The following guidelines are recommended for depth of foundation D:

1. $D \geq 400$ mm, in rocky soil
   ≥ 1000 mm, in clay or sandy soil
2. In case there is a soil filling, part of the loose soil be replaced either by compacted sand or by gravel on which the footing can rest. Alternatively, part of the filling can also be replaced by lean concrete in machine or heavy foundations. The compacted sand or the lean concrete must develop the same bearing capacity as that of the base soil. Settlements must be within the limits.
3. In case of sloping soils, the depth of the foundation should be selected such that the minimum depth of the foundation nromal to the sloping ground should be equal to 400 mm in rocky soils and 1000 mm in clay or sandy soils.

4. In case of depths of adjacent foundations having different reduced levels, the difference between levels should not exceed half the clear spacing of the foundation slabs in sandy or clayey soils and it should not exceed the clear spacing in rocky soils.

5. The foundation slab or footing should be protected by an apron placed at the ground level over the compacted filling. The width of the apron should be at least 500 mm and it can be of PCC or brick in pointing. Such an apron must be capable of resisting normal wear and tear.

13.4 Structural Design of RCC Sections

The structural design of the slabs of the footing is exactly the same as that of any RCC slab or beam. The design is illustrated with examples. The balanced working moment capacity of a RCC section is given by

$$M_r = Kbd^2\sigma_{cb} \tag{13.19}$$

in which

b = width of the section

d = effective depth of the reinforcement from the extreme compression fibre

$K = nj/2$

jd = lever arm

σ_{cb} = allowable bending compressive stress in concrete

nd = neutral axis distance from compression fibre

$$= d/\left(1 + \frac{\sigma_{st}}{m\sigma_{cb}}\right)$$

σ_{st} = allowable tension in the steel

m = modular ratio = $E_s/E_c = 280/3\sigma_{cb}$

The area of reinforcement required in a balanced cross section is given by

$$A_{st} = \frac{M}{jd\sigma_{st}} \tag{13.20}$$

The design criterion of the section against bending is given by

$$M_r = Kbd^2\sigma_{cb} \geq M \tag{13.21}$$

or

$$bd^2 \geq \frac{M}{K\sigma_{cb}} \tag{13.22}$$

Nominal shear stress in a section is:

$$\tau = \frac{V}{bd} \tag{13.23}$$

in which V = shear force.

The design criterion for shear without web reinforcement is

$$\tau = \frac{V}{bd} \leq \tau_c \tag{13.24}$$

where τ_c = allowable average shear stress.

The reinforcement bars should have adequate bond length so as to prevent bond failure. The bond length is given by

$$L_d = \frac{\phi\sigma_s}{4\tau_{bd}}$$

(13.25)

where

σ_s = actual stress in steel

ϕ = diameter of the bar

τ_{bd} = design bond stress

Some of the values such as K, etc. are given in Table 13.1 for ready reference.

Table 13.1 Some Working Design Constants

σ_{st} (N/mm^2)	Parameter	Grade of concrete (σ_{cb} in N/mm^2)	
		M15,(5)	M20,(7)
140	n	0.400	0.400
	j	0.867	0.867
	K	0.173	0.173
	$\frac{100\,A_{st}}{bd}$	0.714	1.000
230	n	0.289	0.289
	j	0.904	0.904
	K	0.13	0.13
	$\frac{100\,A_{st}}{bd}$	0.314	0.440

13.5 Design of Foundations to Wall

The following are some of the typical RCC foundations for wall and isolated columns.
1. A strip foundation:
   a) Axially loaded wall.
   b) Wall subjected to eccentric load or subjected to combined axial force and bending moment.
2. Isolated footing with constant thickness:
   a) Axially loaded column.
   b) Column subjected to axial force and moment or eccentric load with symmetrical foundation.
   c) Same as in case (b) with unsymmetrical load.
3. Isolated footings with variable depth:
   In all of the above cases, there are two critical sections in the foundation slab, a section where the bending moment is maximum and another section where the shear force is maximum. The foundation slab cantilevers from the face of the column. The bending moment is maximum at the face of the column (AA) as shown in Fig. 13.4. However, the critical section

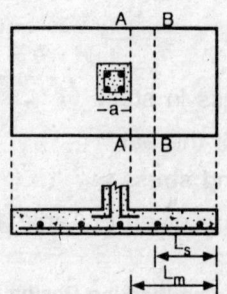

Fig. 13.4 Critical Sections and Notations in Footing Slabs.

where the maximum diagonal tension occurs is at $0.5d$ from the force of the column. The section which is subjected to critical shear, then corresponding to a section at a distance d from the face of the column, where d is the effective depth of the reinforcement (Fig. 13.4).

Let L_m = cantilever span for critical bending moment

L_s = cantilever span for critical shear force

Let the net intensity of pressure at the face of the column be equal to q_3, which can be computed as

$$q_3 = q_1 - \frac{(q_1 - q_2) L_m}{B} \tag{13.26}$$

The bending moment developed at the face of the column is

$$M = \frac{BL_m}{6} (q_3 + 2q_1) \tag{13.27}$$

The weight of the footing and the soil over the footing or the surcharge over the footing are balanced by the bearing pressure corresponding to the loads. Therefore, this part of the load does not enter the moment or the shear force calculation. The values of q are computed exactly the same way as of p except that the value of P is equal to the superimposed load at the ground level istead of gross P. Similarly, the critical maximum shear force is given by

$$V = \frac{BL_s}{2} (q_5 + q_1) \tag{13.28}$$

where

$$q_5 = q_1 - \frac{(q_1 - q_2) L_s}{B} \tag{13.29}$$

The value of q_1 and q_2 are computed as

$$q_1 = \frac{P_s}{A} + \frac{MB}{2I} \tag{13.30}$$

$$q_2 = \frac{P_s}{A} - \frac{MB}{2I} \tag{13.31}$$

where P_s = superimposed load on the footing.

EXAMPLE 13.1. *Strip foundation to wall under axial force.* A brick wall carrying a live load of 60 kN/m and dead load of 80 kN/m (including the weight of the wall) is to be provided with a reinforced concrete slab foundation. The thicnkess of the wall next to the foundation is 340 mm. The allowable bearing capacity of the soil is 120 kN/m^2.

Design data:

Dead load $\qquad = P_d = 80$ kN/m, Live load $= P_l = 60$ kN/m

Bearing capacity $\qquad = p_a = 120$ kN/m^2

Thickness of wall $\qquad = a = 0.34$ m

Select the following materials for construction

Concrete $\qquad$ = M20 grade and Steel = HYSD-Fe415 bars

The allowable stresses are taken from the code and they are:

$$\sigma_{cb} = 7 \text{ MPa}, \ \sigma_{st} = 230 \text{ MPa}$$

For a balanced sectional design, the design coefficient taken from Table 13.1 are:

$$n = 0.289, \ j = 0.904 \ K = 0.13.$$

The allowable average shear stress for 0.5% of tension reinfrocement is: $\tau_c = 0.30$ MPa.

Foundation design. Let the depth of foundation be 1 m. The net load on the soil is equal to the superimposed load on the footing plus the difference in the weight of the footing and the soil replaced.

Difference in densities of concrete and soil = 8 kN/m^3

Estimate the volume of the concrete in the footing and in the pedestal or stem and add corresponding differences in the weight. This weight can be estimated in the range of 5% to 15% of the superimposed load to start with and then checked. The higher value of 15% is selected for heavy raft foundations. Superimposed load on the footing.

$$P_s = P_d + P_l = 80 + 60 = 140 \text{ kN/m}$$

Extra load due to the difference in weight of the concrete and soil

$$= \text{about } 0.05 \ (P_d + P_l) = 5 \text{ kN/m}$$

Net load on the foundation = P = 145 kN/m

As the foundation is a strip foundation for the wall, it is designed for 1 m length of the wall. So, the length of foundation L = 1 m. The area of the foundation is given by

$$A = \frac{P}{p_a} = \frac{145}{120} = 1.21 \text{ m}^2$$

Select the width of the foundation as B = 1.3 m

Bearing pressure is $\qquad p = \dfrac{P}{BL} = \dfrac{145}{1.3} = 112$ kN/m^2

This value is less than the allowable value.

Design for bending. Cantilever span of the footing for the maximum BM is

$$L_m = \frac{B}{2} - \frac{t}{2} = 0.65 - 0.17 = 0.48 \text{ m}$$

The net bearing load on the footing is equal to the total load minus the self-weight. The load intensity on the footing is

$$q = \frac{P_s}{A} = \frac{140}{1.3} = 107.7 \ kN/m^2$$

Maximum bending moment on the footing slab

$$M = \frac{qL_m^2}{2} = \frac{107.7(0.48)^2}{2} = 12.407 \ kNm/m = 12407 \ Nm/m$$

The value of d is therefore $d^2 \geq \dfrac{M}{Kb\sigma_{cb}} = \dfrac{12407}{0.13(10^6)(1)(7)}$

or $\qquad\qquad\qquad\qquad d \geq 0.12 \ m$

Let $\qquad\qquad\qquad\qquad d = 0.14 \ m = 140 \ mm$

$$A_{st} = \frac{M}{jd\sigma_{st}} = \frac{12407}{0.904(0.14)(230)} = 430 \ mm^2/m$$

Let 12 mm dia bars at 260 mm c/c be provided

$$A_{st} \ (provided) = 435 \ mm^2/m$$

$$\% \ reinforcement = \frac{435(100)}{(1000)(140)} = 0.31$$

Design for shear. The critical section for diagonal tension failure is at a distance d from the face of the wall. The shear span is

$$L_s = L_m - d = 0.48 - 0.14 = 0.34 \ m$$

Shear force $\qquad\qquad V = qL_s = 107.7(0.34) = 36.62 \ kN/m$

Nominal shear stress is $\qquad \dfrac{V}{bd} = \dfrac{36620}{1(0.14)} = 261571 \ N/m^2 = 0.26 \ MPa$

The allowable shear stress for 0.3% of steel is 0.24 MPa. Since the nominal shear stress exceeds the permissible value, either the sectional depth is modified or the shear reinforcement should be provided. It is not desirable to introduce the shear reinforcement in the footing slabs, hence increase the thickness of the footing slab to meet the demand of shear force. The depth of the slab based on shear force consideration is

$$d \geq \frac{V}{b\tau_c}$$

$$\geq \frac{36620}{1(0.24)} = 15 \ mm$$

Let $\qquad\qquad\qquad\qquad d = 0.15 \ m$

Provide a cover of about 50 mm to reinforcement, then the overall depth of the slab is

$$h = d + cover + \frac{1}{2} \ (dia \ of \ bars)$$

$$= 0.15 + 0.05 + 0.01 = 0.21 \ m = 210 \ mm$$

Self-weight. Volume of the footing is

$$V_1 = LB \ h = 1(1.3)(0.21) \quad = 0.273 \ m^3$$

Volume of the stem of the wall over the footing is

$$V_2 = Lt(D - h) = 1(0.34)(0.79) \quad = 0.269 \ m^3$$

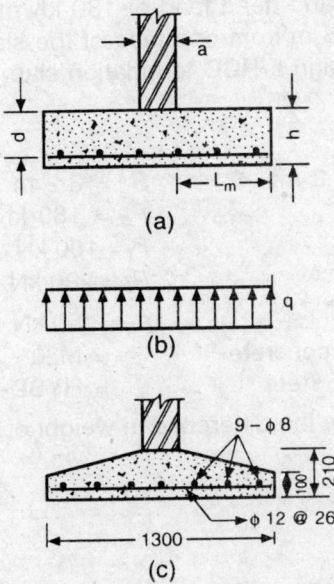

Fig. 13.5 Structural Details of Strip Footing, Example 13.1.

The density of concrete is taken as 25 kN/m³, that of brickwork as 20 kN/m³ and that of the soil as 17 kN/m³.

The additional difference in load on the foundation soil is

$$W = V_1(24) + V_2(20) - (V_1 + V_2)(17)$$
$$= 0.273(25) + 0.269(20) - 0.542(17) = 3.26 \text{ kN/m}$$

The weight difference assumed was 5 kN/m compared to the actual value of 3.26 kN/m. Therefore, the bearing pressure decreases marginally. The size of the footing does not require any modification.

The area of tension reinforcement is

$$A_{st} = \frac{M}{jd\sigma_{st}} = \frac{12407}{0.896(0.15)(230)} = 401 \text{ mm}^2/\text{m}$$

Provide 12 mm *bars at* 260 mm *spacing*

$$A_{st} \text{ (provided)} = \frac{113(1000)}{275} = 411 \text{ mm}^2/\text{m}$$

0.15% of distribution reinforcement should be provided. The area of distribution steel needed is

$$A_s = \frac{0.15Bh}{100} = \frac{0.15(1300)(210)}{100} = 410 \text{ mm}^2$$

Provide 9 numbers of 8 mm *dia* bars as distribution reinforcement. Figure 13.5 illustrates the reinforcement details. The thickness of the slab at the free ends can be reduced to 100 mm.

EXAMPLE 13.2 *Design of eccentrically loaded footing of a wall.* 200 mm thick RCC wall carries a live load of 100 kN/m and dead load of 180 kN/m. The wall is resting on a slab of 3.2 m width at a distance of 1.25 m from one face of the slab. This makes the wall eccentric with respect to the slab CG. Design a RCC foundation slab. Safe bearing capacity of the soil $= p_a = 150$ kN/m^2.

Design data

Wall thickness 0.2 m;	$B = 3.2$ m
Deal load	$P_d = 180$ kN/m
Live load	$P_l = 100$ kN/m
Total load	$P_s = 280$ kN/m
Bearing capacity	$p_a = 150$ kN/m^2
Let the grade of concrete	$= $ M20
and the grade of steel	$= $ HYSD-Fe415

Design of foundation. Assume the difference in weight of the foundation concrete and that of equivalent soil as

$$W' = 0.06\, P_s = 16 \text{ kN/m}$$

Eccentricity of the load is

$$e = \frac{B}{2} - 1.25 = 1.6 - 1.25 = 0.35 \text{ m}$$

Total load on the foundation is

$$P = P_s + W' = 296 \text{ kN/m}$$

The moment caused by the ecentric load on the foundation is

$$M_f = P_s e = 280(0.35) = 98 \text{ kNm/m}$$

The area and moment of inertia of the base are:

$$A = LB = (1)(3.2) = 3.2 \text{ m}^2$$

$$I = \frac{LB^3}{12} = \frac{3.2^3}{12} = 2.731 \text{ m}^4$$

The maximum and minimum pressure on the foundation are:

$$p_1 = \frac{P}{A} + \frac{MB}{2I} = \frac{296}{3.2} + \frac{98(3.2)}{2(2.732)}$$

$$= 92.5 + 57.4 = 149.9 \text{ kN/m}^2$$

$$p_2 = 92.5 - 57.4 = 35.1 \text{ kN/m}^2$$

Since $p_1 < p_a$, the size of the footing is acceptable.

Structural design. The intensities of the pressure loads at the extreme points and also under the two faces of the wall are (see Fig. 13.6)

$$q_1 = \frac{P_3}{A} + \frac{M_f y_1}{I} = \frac{280}{3.2} + \frac{98}{2.732}(y_1)$$

$$= 87.5 + 35.87\,(1.6) = 144.9 \text{ kN/m}^2$$

$$q_2 = 87.5 - 35.87(1.6) = 30.12 \text{ kN/m}^2$$

$$q_5 = 87.5 + 35.87 \, (e + 0.5a)$$

$$q_5 = 87.5 + 35.87(0.35 + 0.1) = 103.6 \text{ kN/m}^2$$

$$q_4 = 87.5 + 35.87 \, (0.35 - 0.1) = 96.5 \text{ kN/m}^2$$

The cantilever span on the outer side is

$$L_{m1} = 1.25 + 0.1 = 1.35 \text{ m}$$

The bending moment at the outer face of the wall on the cantilever slab is

$$M_1 = \frac{L_{m1}^2}{6} (2q_1 + q_3) = \frac{1.35^2}{6} (2(144.9) + (103.6))$$

$$= 119.5 \text{ kNm/m}$$

The cantilever span on the inside face is

$$L_{m2} = 0.5 \, B + e - 0.5a = 1.6 + 0.35 - 0.1 = 1.85 \text{ m}$$

The bending moment on the inside face of the cantilever is

$$M_2 = \frac{L_{m2}^2}{6} (2q_2 + q_4) = \frac{1.85^2}{6} (2(30.1) + 96.5)$$

$$= 89.38 \text{ kNm/m}$$

The design moment should be equal to the higher of the two values

$$M = M_1 = 119.5 \text{ kNm/m}$$

Equating the resisting moment to the working moment, we have

$$d > \sqrt{\frac{119500}{0.910(1)(10^6)}} = 0.37 \text{ m}$$

Let

$$d = 0.38 \text{ m}$$

The diagonal tension is critical at a distance d from the face, so the shear force span is

$$L_s = L_m - d = L_{m1} - d = 1.35 - 0.38 = 0.97 \text{ m}$$

The intensity of force on the slab at a distance $L_s = 0.97$ is the same as that for distance y_5 from the CG of the foundation. The distance of the point from CG is

$$y_5 = \frac{B}{2} - L_s = 1.6 - 0.97 = 0.63 \text{ m}$$

The intensity of the load at y_5 is

$$q_5 = \frac{P_s}{A} + \frac{M_f}{I} y_5 = \frac{280}{3.2} + \frac{98(0.63)}{2.732}$$

$$= 87.5 + 22.6 = 110.1 \text{ kN/m}^2$$

The maximum shear force at the critical face is:

$$V = \frac{L L_s (q_1 + q_5)}{2} = 1(0.97) \frac{(144.9 + 110.1)}{2} = 123.68 \text{ kN}$$

The normal shear stress on the section is

$$\tau = \frac{V}{bd} = \frac{123680}{1(0.38)} = 325473 \text{ N/m}^2 = 0.325 \text{ MPa}$$

The approximate per cent of balanced steel = 0.71

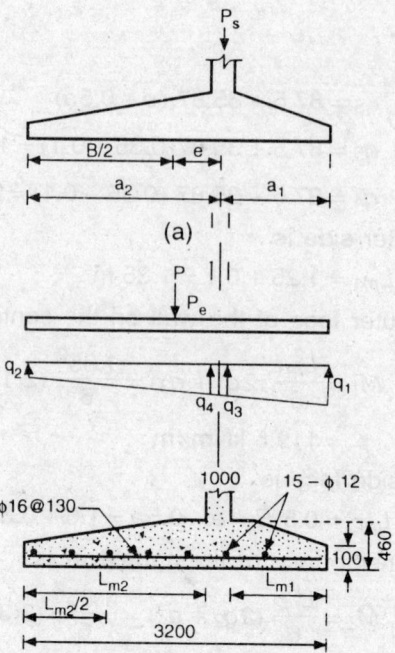

Fig. 13.6 Reinforcement Details in Foundation Slab with Eccentric Wall, Example 13.2.

The allowable shear stress is 0.33 MPa for 0.7% reinforcement. Shear stress is within the allowable range, therefore the depth adopted is adequate. The overall depth of the section at the outer face of the wall is

$$h = d + \text{cover} + \phi/2 = 380 + 50 + 10 = 440 \text{ mm} = 0.44 \text{ m}$$

Let the overall depth of the slab at the free ends be 100 mm. The volume of the concrete in the foundation and that of the wall step below the ground level is approximately given by

$$V_{0l} = \frac{1 (0.1 + 0.44)}{2} (3.2) + (1 - 0.44)(0.2) = 0.976 \text{ m}^3$$

The difference in the weight of concrete and that of the equivalent volume of earth is

$$W' = 0.976 (25 - 17) = 7.808 \text{ kN}$$

The actual weight difference that was assumed for the purpose of computing the bearing pressure was 14 kN which is higher than the actual value. Therefore, the design is slightly on the safer side. The difference of (14 − 7.8) kN in 296 kN is not going to be of much consequence to the pressure on the soil.

Design of reinforcement. The area of tension reinforcement is

$$A_{st} = \frac{M}{jd\sigma_{st}} = \frac{119500}{0.904(0.38) 230 (10^6)} = 1512 \text{ mm}^2$$

Provide $\phi 16$ at 130 mm spacing. The allowable shear stress for such a percentage of reinforcement is 0.31 MPa; however, the nominal shear stress was computed to be 0.325 MPa. This calls for either provision of shear stirrups or increase in the effective depth. If the effective depth is proportionately increased, then

$$d = 0.38 \left(\frac{0.325}{0.310} \right) = 0.398 \text{ m}$$

Use
$$d = 0.4 \text{ m} = 400 \text{ mm}$$
$$h = 400 + 50 + 10 = 460 \text{ mm}$$

Minimum distribution reinforcement is 0.15% and it is equal to

$$A_s = \frac{0.15}{100} A_c = 0.0015 \ A_c$$

The area of cross-section of the concrete in foundation is

$$A_c = 0.1 \ (3.2) + \left(\frac{(0.2 + 0.4 + 0.4) + 3.2}{2} \right) (0.46 - 0.1)$$

$$= 1.076 \text{ m}^2$$

(See Fig. 13.7 for the dimensions)

$$A_s = 0.0015 \ (1.076) \ (10^6) = 1614 \text{ mm}^2$$

Provide 15 numbers of 12 mm bars over the width.

13.6 Isolated Footing with Tapering Width

Isolated footing or even some combined footing are subjected to maximum bending moment at the face of the column. Therefore, the depth of the footing at the face of the column should be maximum and can be decreased to minimum towards the free edge as shown in Fig. 13.7a. The moment capacity of such a section is not same as that of a rectangular section as the width of the section is variable. Figure 13.7c illustrates the variable width and the neutral axis of the section.

Let nd = neutral axis distance from the top

b_n = width at NA

The width at neutral axis can be expressed as

$$b_n = b_1 + (b_2 - b_1) \ \frac{nd}{d_1} \tag{13.32}$$

If the allowable stresses are reached in the concrete and steel simultaneously, then the strain compatibility condition gives

$$\frac{\in_c}{nd} = \frac{\in_s}{d - nd} \tag{13.33}$$

Expressing the strains in terms of stresses in the above equation, we have

$$\frac{\sigma_{cb}}{n} = \frac{m\sigma_{st}}{(1 - n)} \tag{13.34}$$

in which $\in_c$ and $\in_s$ are the strains in the exteme fibre concrete and steel respectively, and m is modular ratio.

The value of n can be obtained by rearranging the above equation as

$$n = \frac{1}{1 + \dfrac{m\sigma_{st}}{\sigma_{cb}}} \tag{13.35}$$

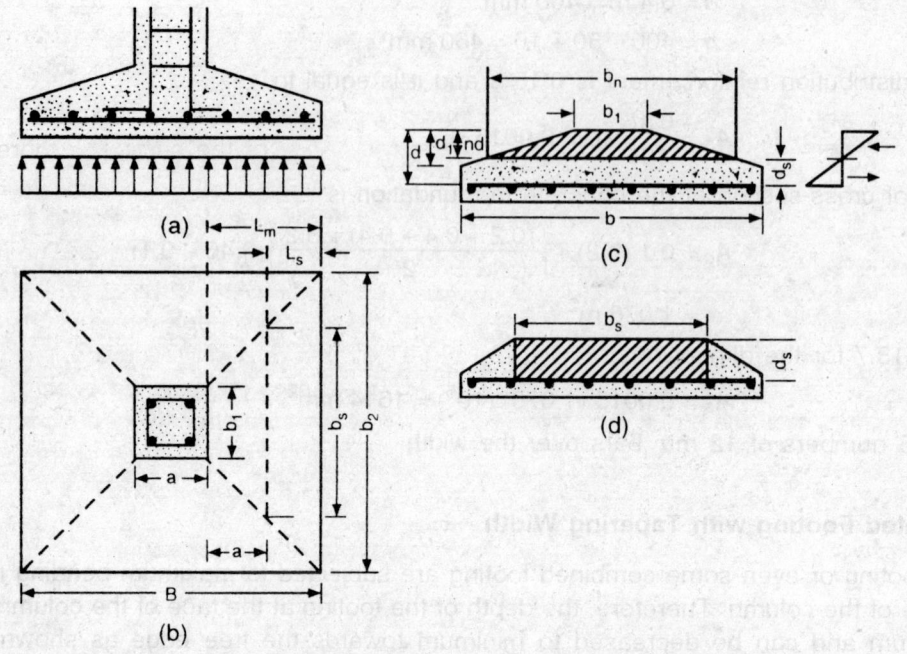

Fig. 13.7 Notations of Isolated Footing with Tapering Top.

Total tension force in the steel is

$$T = A_{st} \sigma_{st} \tag{13.36}$$

The total compression force on the concrete can be calculated by considering the trapezoidal compression area as rectangular-cum-triangular. The stress variation across the depth is taken linear. The total compression forces on the rectangular and triangular portions are:

$$C_1 = \frac{b_1 nd}{2} \sigma_{cb} \tag{13.37}$$

$$C_2 = n^2 \left(\frac{b_2 - b_1}{6} \right) (d\sigma_{cb}) \tag{13.38}$$

The equilibrium of forces on the section gives

$$C_1 + C_2 = A_{st} \sigma_{st}$$

or $$A_{st} = \frac{C_1 + C_2}{\sigma_{st}}$$

$$= \frac{nd}{2} \left(b_1 + n \frac{(b_2 - b_1)}{3} \right) \frac{\sigma_{cb}}{\sigma_{st}} \tag{13.39}$$

The value of b_n can be obtained by substituting the value of n in Eq. (13.32). The moment capacity of the section can be obtained by taking moments about the centre of the reinforcement and it is

$$M_r = R b_1 d^2 + R_1 (b_2 - b_1) d^2 \tag{13.40}$$

where
$$R = \frac{n}{2}\left(1 - \frac{n}{3}\right)\sigma_{cb}$$

$$R_1 = \frac{n^2(2-n)\,\sigma_{cb}}{12}$$

The design criterion should ensure that the bending resistance of the section is more than the moment acting on the section. The values of b_1 and b_2 are obtained by other considerations. If the value of $(d - d_1)$ which corresponds to the least depth at the free end is also selected, then one can solve for d and A_{st} from the above set of equations. The overall depth of the section at the free edge is usually taken in the range of 100 to 150 mm.

In case the thickness of footing is taken uniform, the effective depth can be calculated from the standard formula. The method of design is illustrated through examples.

Design against shear force in tapered sections becomes more complicated even to select the location of the critical diagonal tension zone. This is because the effective depth decreases along the span. As a practical design, the section at distance d from the face of the column section is usually checked for shear stress.

EXAMPLE 13.3 *Isolated footing with axial load and uniform thickness.* A column is subjected to dead and live loads of 340 kN and 260 kN respectively. The size of the column is 400 mm by 400 mm. Design a RCC footing if the allowable bearing capacity of the soil is 100 kN/m^2.

Dead load	$P_d = 340$ kN
Live load	$P_l = 260$ kN
Superimposed load on the footing	$P_s = \overline{600\ \text{kN}}$
Bearing capacity	$p_a = 100$ kN/m^2

Select a square footing as the column is square and axially loaded. Also select the following materials:

Grade of concrete   = M20, $f_{ck} = 20$ N/mm^2

Grade of steel   = mild steel

The working stress design coefficients taken from Table 13.1 are:
$$j = 0.904,\ R = 0.91$$

Design of foundation. Since the soil is clay with low bearing capacity, select the depth of foundation as 1.5 m. The approximate size of the footing be 2.5 by 2.5 by 0.7 m; the volume of the concrete is 2.5 (2.5)(0.7)

The difference in weight of the conrete and the equivalent weight of the soil is:
$$W' = 2.5(2.5)(0.7)(25-17) = 35\ \text{kN}$$

Let $W' = 40$ kN

Total design load on the foundation soil is
$$P = P_s + W' = 640\ \text{kN}$$

Area of the foundation should be such that the bearing pressure on the soil should be less than the allowable bearing capacity of the soil. This condition leads to

$$\frac{P}{A} = \frac{640}{B^2} \le p_a = 100$$

or

$$B^2 \ge \frac{640}{100} = 6.4 \text{ m}^2$$

Select

$$B = 2.55$$

Bearing pressure is

$$p = \frac{P}{A} = \frac{640}{B^2} = \frac{640}{6.5025} = 98.42 \text{ kN/m}^2$$

Intensity of the load on the footing for design is

$$q = \frac{P_s}{A} = \frac{600}{6.5025} = 92.27 \text{ kN/m}^2$$

Structural design. The cantilever span for maximum bending moment is the distance from the free end to the face of the column and it is (see Fig. 13.8)

$$L_m = \frac{B}{2} - \frac{a}{2} = 1.275 - 0.2 = 1.075 \text{ m}$$

Maximum bending moment is

$$M = \frac{BL_m^2 q}{2} = \frac{2.55(1.075)m}{2}(92.27)$$

$$= 135.953 \text{ kNm} = 135953 \text{ Nm}$$

Equating the moment capacity to the working moment, we have

$$d = \sqrt{\frac{135953}{0.91(10^6)(2.55)}} = 0.240 \text{ m}$$

Let

$$d = 0.25 \text{ m}$$

The overall depth of the footing is

$$h = d + \phi + \phi/2 + \text{cover}$$

$$= 0.25 + 0.02 + 0.01 + 0.05 = 0.33 \text{ m}$$

The section can be designed similar to the flat slab against shear force. The critical section is at a distance of (d/2) from the face of the column and the length of the periphery of critical section is

$$b_0 = 4(b_1 + d) = 4(0.4 + 0.25) = 2.6 \text{ m}$$

The side of the square where shear force is maximum is

$$b_s = b_1 + d$$

The shear force on the periphery is

$$V = q(B^2 - b_s^2) = 92.27(2.55^2 - 0.65^2) = 561 \text{ kN}$$

The nominal shear stress at the critical section is

$$\tau = \frac{V}{b_0 d} = \frac{561000}{2.6(0.25)} = 0.863(10^6) \text{ N/m}^2$$

The permissible shear stress is given by

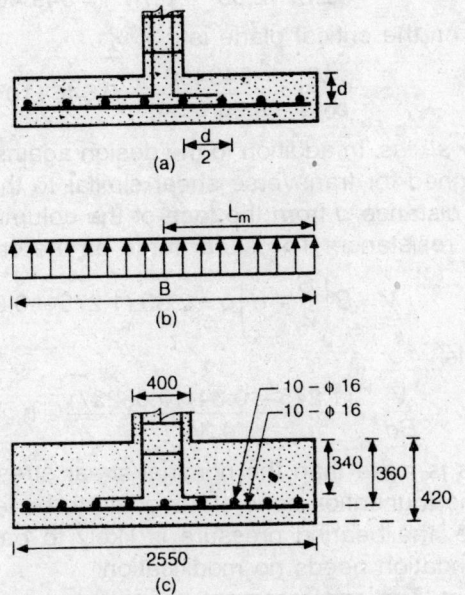

Fig 13.8 Reinforcement Details of Footing of Example 13.3.

$$\tau_a = k_s \tau_c$$

where
$$k_s = (0.5 + \beta_c) \not> 1$$

$$\tau_c = 0.16 \sqrt{f_{ck}}$$

$$\beta_c = \frac{\text{short side of shear periphery}}{\text{long side}} = 1$$

Since $k_s \not> 1$, the value of $k_s = 1$

The substitution of various quantities gives

$$\tau_a = k_s \tau_c = 1 \, (0.16) \sqrt{f_{ck}} = 0.710 \, \text{N/mm}^2$$

The allowable shear stress is less than the nominal shear stress, therefore the section is not safe against shear force. Either shear reinforcement or depth is to be increased. In falt slabs and footings it is better to increase the effective depth. Let the effective depth be increased approximately in the following ratio

$$d = \left(\frac{\tau}{\tau_a}\right) d = \frac{0.863}{0.71}(0.25) = 0.31 \, \text{m}$$

Use
$$d = 0.34 \, \text{m and } h = 0.41 \, \text{m}$$

The peripheral distance at critical shear zone is

$$b_0 = 4 \, (b_1 + d) = 4 b_s = 1.6 + 1.36 = 2.96 \, \text{m}$$

The shear force on the periphery is

$$V = q \, (B^2 - b_s^2)$$

$$= 92.27 \, (2.55^2 - 0.74^2) = 549.46 \text{ kN}$$

The nominal shear stress on the critical plane is

$$\tau = \frac{V}{b_0 \, d} = \frac{549460}{2.96(0.34)} = 0.546 \, (10^6) \text{ N/m}^2$$

Check for transverse shear stress. In addition to the design against punching diagonal tension, the footing has to be designed for transverse shear similar to the one done in wall footings. The critical section is at a distance *d* from the face of the column and the entire width of the footing is accounted for in resistance. The shear force on the critical plane is

$$V = B\left(\frac{B}{2} - d\right) q = 2.55 \, (1.275 - 0.34) \, (92.27)$$

The nominal shear stress is

$$\frac{V}{Bd} = \frac{(1.275 - 0.34) \, (0.09227)}{0.34} = 0.25 \text{ MPa}$$

The allowable shear stress is more than the nominal shear stress.

The actual weight of the foundation concrete is going to be less than what was assumed by about 20 kN. Therefore, the bearing pressure is likely to be slightly less than that was computed. The size of foundation needs no modification.

Design of reinforcement. The reinforcement should be provided in both directions. The effective depth of the steel in one direction is slightly less than that in the other direction.

The depth or steel in one direction = 0.34 m. Area of steel in that direction is

$$A_{st} = \frac{M}{jd\,\sigma_{st}} = \frac{135953}{0.904(0.34)(230)} = 1923 \text{ mm}^2$$

Provide 10 numbers of 16 mm bars. A_s (provided) is 10(201) = 2010 mm^2

The effective depth of steel in another direction is

$$d_x = d_y + \phi = 0.34 + 0.02 = 0.36 \text{ m}$$

Area of steel required is

$$A_{st} = \frac{M}{jd\,\sigma_{st}} = \frac{135953}{0.904(0.36)(230)} = 1816 \text{ mm}^2$$

Provide 10 numbers of 16 mm bars, then A_s (provided) is 2010 mm^2.

EXAMPLE 13.4 *Design of an isolated rectangular footing.* A square column of 400 mm size is carrying a live load of 260 kN and dead load of 340 kN including the self-weight. The allowable bearing capacity of the soil is 100 kN/m^2 and allowable width of the footing is 2 m. Design an isolated footing with M20 concrete and HYSD-Fe415 reinforcement.

Design of foundation

Dead load $P_d = 340$ kN

Live load $P_l = 260$ kN

Superimposed load $P_s = 600$ kN

(on the footing)

Bearing capacity $p_a = 100$ kN/m^2

Let the difference in weight of the concrete in the footing and equivalent volume of the soil be
$$= 0.06(600) = 36 \text{ kN. Therefore,}$$
Total net load on the soil $P = 600 + 36 = 636$ kN

The area of the footing $\quad A_f = \dfrac{P}{p_a} = \dfrac{636}{100} = 6.36 \text{ m}^2$

The width of the footing is restricted to 2 m, therefore the length of the footing is
$$L = \frac{A_f}{B} = \frac{6.36}{2} = 3.18 \text{ m}$$

Use the footing size as
$$B = 2 \text{ m and } L = 3.2 \text{ m}$$
i.e.,
$$A_f = 2 \, (3.2) = 6.4 \text{ m}^2$$

The net reaction acting on the footing is
$$q = \frac{P_s}{A_f} = \frac{600}{6.4} = 93.75 \text{ kN/m}^2$$

The footing slab is designed with tapering slopes.

Design of the section. The cantilever projection of the footing in one direction is (2 – 0.4)/2 = 0.8 m and in another direction it is (3.2 – 0.4)/2 = 1.4 m. The maximum bending moment on the slab occurs at the face of the column along the long span and the corresponding moment arm is
$$L_m = \frac{3.2 - 0.4}{2} = 1.4 \text{ m}$$

The maximum bending moment is
$$M = \frac{qBL_m^2}{2} = 93.75 \, \frac{(2) \, (1.4)^2}{2} = 183.75 \text{ kNm}$$

The working stress moment capacity of a trapezoidal section is given by
$$M_r = Rb_1 d^2 + R_1(b_2 - b_1)d^2$$
where
$$b_1 = 0.4 \text{ m and } b_2 = B = 2 \text{ m}$$
$$R = 0.91 \text{ N/mm}^2$$
$$R_1 = \frac{n^2}{2}\left(1 - \frac{n}{3}\right)\sigma_{cb} = 0.107 \text{ N/mm}^2$$

The substitution of different quantities in the moment capacity gives
$$M_r = (0.91(0.4) + 0.107 \, (2 - 0.4)) \, 10^6 \, d^2$$
$$= 0.54 \, (10^6) \, d^2$$

Equating the moment capacity to the external moment gives
$$0.54(10^6)d^2 \geq 183.75 \, (10^3) \text{ Nm}$$
or
$$d \geq 0.590 \text{ m}$$

The overall depth of the section is

$$h = 0.590 + 0.01 + 0.05 = 0.653$$

Use $h = 0.660$ and $h_m = 0.15$ m .

Volume of the concrete in the footing is

$$2(3.2)(0.15) + \left(\frac{2(3.2) + 0.4(0.4)}{2}\right)(0.66 - 0.15) = 2.6328 \text{ m}^3$$

The difference in weight of the concrete and that of the equivalent soil of the footing is

$$W_g = 2.6328(25 - 17) = 21.06 \text{ kN}$$

This value is less than the assumed one for the purpose of calculation of net pressure on the soil. Therefore no revision is needed.

Check for shear capacity. The critical section where shear failure is likely to occur is at a distance $d/2$ from the face of the column. The periphery of the shear boundary is

$$b_0 = 4(a + d) = 4(0.4 + 0.590) = 3.770 \text{ m}$$

The net shear force acting on the critical shear periphery is

$$V = q(BL - (a + d)^2)$$

$$= 93.75(2(3.2) - 0.993^2) = 507.558 \text{ kN}$$

The effective depth of the steel at a distance $d/2$ from the face of the column is

$$d_s = h - \frac{(h - h_0)d}{2L_m} - \text{cover}$$

$$= 0.66 - \frac{(0.66 - 0.15)(0.590)}{2(1.4)} - 0.05 = 0.47 \text{ m}$$

Use $d_s = 0.5$ m

The nominal shear stress at the critical zone is

$$\tau = \frac{V}{b_0 d_s} = \frac{507558}{3.770(0.5)} = 0.269(10^6) \text{ N/m}^2 = 0.269 \text{ MPa}$$

The allowable shear stress in M20 concrete for flat slabs is

$$\tau_a = 0.71 \text{ MPa}$$

The section is safe against shear failure including transverse shear.

Design of reinforcement. The area of reinforcement for a balanced trapezoidal section is

$$A_{st} = \frac{nd}{2}\left(b_1 + \frac{n(b_2 - b_1)}{3}\right)\frac{\sigma_{cb}}{\sigma_{st}}$$

$$= \frac{0.289(0.590)}{2}\left(0.4 + \frac{0.289}{3}(3.2 - 0.4)\right)\frac{7(10^6)}{230}$$

$$= 1738 \text{ mm}^2$$

Use 9 numbers of 16 mm bars in the long span

$$A_s \text{ (provided)} = 1735 \text{ mm}^2$$

Design of reinforcement in the short span direction. The reinforcement required in the short span direction is proportioned with respect to the bending moment and the effective depth in comparision to the long span. The maximum bending moment is again proportional to the square of the cantilever span which is $(2 - 0.4)/2 = 0.8$ m. The effective depth of the steel in this direction could be taken equal to

$$d_y = d - \phi = 0.593 - 0.02 = 0.573 \text{ m}$$

The area of the reinforcement required in the short span is

$$A_{sty} = A_{st} \left(\frac{L_{my}}{L_m}\right)^2 \frac{d}{d_y}$$

$$= 1738 \left(\frac{0.8}{1.4}\right)^2 \left(\frac{0.590}{0.570}\right) = 588 \text{ mm}^2$$

Provide 6 numbers of 12 mm bars A_s (provided) is 678 mm^2

Figure 13.9 illustrates the design details.

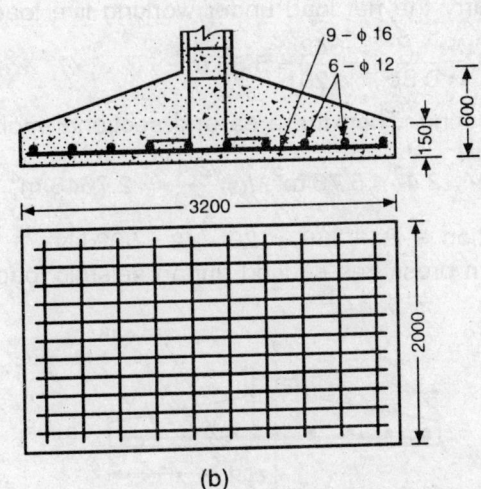

Fig. 13.9 Reinforcement Details of Footing of Example 13.4.

EXAMPLE 13.5 *Design footing subjected to axial force and Seismic moment.* A column is transferring a live load of 240 kN and dead load of 360 kN to the foundation. In addition a bending moment of 250 kNm is applied on the foundation during earthquake load condition for 50% of the live load. Design an isolated footing for a net allowable bearing capacity of the clayey soil is 120 kN/m^2.

Dead load	$P_d = 360$ kN
Live load	$P_l = 240$ kN
Earthquake moment	$M_f = 250$ kNm
Live load during earthquake	$P_{l1} = 120$ kN
Allowable bearing pressure	$p_a = 150$ kN/m^2

The size of the column (square) $a = 0.6$ m

Use M20 concrete with mild steel.

The allowable bearing pressure on the soil during seismic load condition is same as that during normal live load condition.

Let the additional load on the soil due to replacement of soil by concrete be about 7% of the total load, and it.

$$W'_g = 0.07\,(P_d + P_l) = 0.07\,(360 + 240) = 42\ \text{kN}$$

The net load on the soil is

$$P = P_d + P_l + W'_g = 642\ \text{kN}$$

The net load during seismic load conditions is

$$P_e = P_d + 0.5\,P_l + W'_g = 360 + 120 + 42 = 522\ \text{kN}$$

Design of foundation. Net working load on the foundation

$$P = 642\ \text{kN}$$

Bearing area required to carry the net load under working live load condition is

$$A_f = \frac{P}{0.8 p_a} = \frac{642}{120} = 5.35\ \text{m}^2$$

Select 2.4 by 2.4 m size footing. Then the area and sectional moment of inertia are:

$$A_f = 2.4^2 = 5.76\ \text{m}^2,\ I_f = \frac{2.4^4}{12} = 2.7648\ \text{m}^4$$

Axial load at seismic condition $= P_e = 360 + 120 + 42 = 522\ \text{kN}.$

The maximum and minimum pressures caused during seismic load condition are:

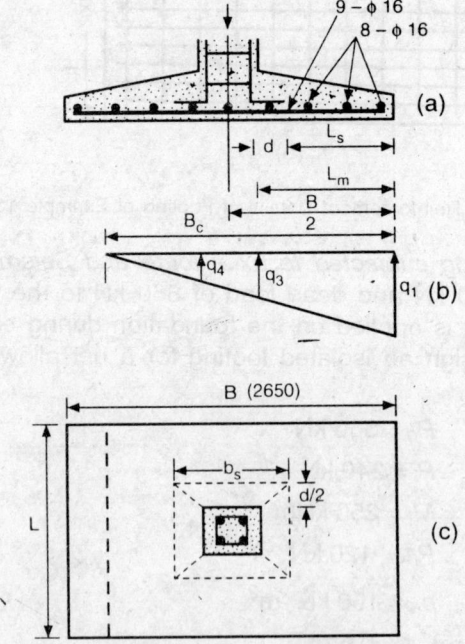

Fig. 13.10 Notations of Footing Subjected to Combined Axial Force and Bending.

$$p_1 = \frac{P_e}{A_f} + \frac{M_f y}{I_f} = \frac{522}{5.76} + \frac{250(1.2)}{2.7648}$$

$$= 90.625 + 108.507 = 199.132 \text{ kN/m}^2$$

$$p_2 = 90.625 - 108.507 = -17.882 \text{ kN/m}^2$$

The maximum pressure under seismic load condition is far higher than the allowable value and in addition tension occurs on the other face. Therefore, the size of the footing has to be increased so as to limit the maximum pressure within the limits. Since the bending moment acting on the soil is large and the soil cannot resist the tension, the total area of the footing will not be effective.

Let B = width of the footing

     B_c = actual width in contact with the soil (Fig. 13.10)

     L = length of the footing $= 2.8$ m

The equilibrium of forces in seismic load condition on the footing gives

$$\frac{L B_c p_1}{2} = P_e = 522 \tag{1}$$

$$\frac{L B_c p_1}{2} \left(\frac{B}{2} - \frac{B_c}{3} \right) = M_f = 250 \tag{2}$$

in which p_1 = maximum pressure on the soil. Let p_1 be set equal to the allowable bearing pressure under seismic load condition, then

$$p_1 = 1.25 p_a = 150 \text{ kN/m}^2$$

Form Eq. (1), we have

$$B_c = \frac{2P_e}{LP_1} = \frac{2(522)}{(2.8)(150)} = 2.486 \text{ m}$$

From Eq. (2), we have

$$\frac{B}{2} = \frac{2M_f}{L B_c p_1} + \frac{B_c}{3}$$

$$= \frac{2(250)}{2.8(2.486)(150)} + \frac{2.486}{3} = 0.479 + 0.829$$

or $B = 2.616$

Adopt the following sizes:

     $L = 2.80$ m and $B = 2.65$ m

and check for the pressures on the soil. From Eqs. (1) and (2), we have

$$\frac{B}{2} - \frac{B_c}{3} = \frac{M_f}{P_e} = \frac{250}{522}$$

or $B_c = 3 \left(\frac{B}{2} - \frac{250}{522} \right) = 3 \left(\frac{2.65}{2} - \frac{250}{522} \right) = 2.538 \text{ m}$

From Eq. (1), we have

$$p_1 = \frac{2P_e}{L B_c} = \frac{2(522)}{2.8(2.538)} = 146.91 \text{ kN/m}^2$$

Since the soil is clayey, let the depth of the foundation be = 1 m. The seismic moment can occur on either way, so the footing need to be symmetric.

Design of the section. The pressure that causes the bending moment on the slab is the effective pressure which is free from the weight of the soil or the self-weight of the slab. The effective load for this purpose is the superimposed load on the footing. The effective pressure is computed using P_s instead of P_e.

$$P_s = P_d + 0.5P_{l1} \quad , \text{(under seismic load)}$$
$$= 360 + 120 = 480 \text{ kN}$$

Let q_1 = maximum pressure as shown in Fig. 13.10. The equilibrium of forces gives

$$\frac{LB_c q_1}{2} = P_s = 480 \tag{3}$$

$$\frac{LB_c q_1}{2}\left(\frac{B}{2} - \frac{B_c}{3}\right) = M_f = 250 \tag{4}$$

The value of B_c in the above expressions is not necessarily the same as the one in the previous computations of net pressures.

From Eqs. (3) and (4), we have

$$\frac{B}{2} - \frac{B_c}{3} = \frac{M_f}{P_s} = \frac{250}{480}$$

or

$$B_c = 3\left(\frac{B}{2} - \frac{250}{480}\right) = 2.4125 \text{ m}$$

Then from Eq. (3), we have

$$q_1 = \frac{2P_s}{LB_c} = \frac{2(480)}{2.8(2.4125)} = 142 \text{ kN/m}^2$$

The intensity of the earth pressure at the face of the column has to be computed so as to calculate the maximum bending moment on the slab. This pressure can be computed from the triangular pressure distribution as shown in Fig. 13.10b:

$$q_3 = \frac{q_1}{B_1}(B_c - L_m) = q_1\left(1 - \frac{L_m}{B_c}\right)$$

where $L_m = $ cantilever span

$$= \frac{B}{2} - \frac{a}{2} = 1.325 - 0.3 = 1.025 \text{ m}$$

then

$$q_3 = 142\left(1 - \frac{1.025}{2.4125}\right) = 81.7 \text{ kN/m}^2$$

Since there are two load conditions and the allowable stresses are different in different load conditions, one must calculate the bending moment in both the load conditions and then select the governing moment. The bending moment on the slab at the face of the column in the seismic load condition is

$$M_1 = \frac{LL_m^2}{6}(2q_1 + q_3)$$

$$= \frac{2.8(1.025)^2}{6}(284 + 81.7) = 179.3 \text{ kNm}$$

Bending moment at the critical section under normal live load condition is computed here.

Axial force $\qquad P_s = 360 + 240 = 600$ kN

Pressure $\qquad q_0 = \dfrac{P_s}{LB} = \dfrac{600}{2.8(2.65)} = 80.86$ kN/m^2

The cantilever span in lengthwise direction is more than that in widthwise, therefore the maximum bending moment is caused on the lengthwise span during the normal live load condition.

The cantilever span $L_0 = \dfrac{L}{2} - \dfrac{a}{2} = 1.4 - 0.3 = 1.1$ m

Bending moment in the live load condition is

$$M_2 = \frac{Bq_0 L_0^2}{2} = \frac{2.65(80.86)(1.1)^2}{2} = 129.64 \text{ kNm}$$

The bending moment caused under seismic load condition is more than 1.33 times than that caused in the live load condition. Therefore, the seismic load condition governs the structural design.

The footing can be designed as a tapering section. Let the size of the footing at the top surface be 1.2 by 1.2 m, then the size of the trapezoidal section on which the maximum bending moment is acting has the following sides

$$b_1 = 1.2 \text{ m} \quad \text{and} \quad b_2 = L = 2.8 \text{ m}$$

The resisting moment capacity of a balanced section with M20 concrete and mild steel reinforcement under seismic load condition is

$$M_r = (R_1 b_1 d^2 + R_2(b_2 - b_1)d^2)1.33$$
$$= 0.91(1.2) + 0.107(1.6)(10^6)d^2 \times (1.33)$$
$$= 1.26(1.33)(10^6)d^2 \text{ Nm}$$

Equating the moment capacity to the external moment acting on the section, we have

$$1.26(1.33)(10^6)d^2 = 179300 \text{ Nm}$$
or $\qquad\qquad d = 0.330$ m

The critical section for shear failure under seismic load condition will be at a distance of $d/2$ from the face of the column. The intensity of pressure at this point is

$$q_5 = \frac{q_1}{B_c}(B_c - (L_m - 0.5d)) = \frac{q_1}{B_c}(B_c - L_s)$$
$$= \frac{142}{2.4125}(2.4125 - (1.025 - 0.166)) = 91.44 \text{ kN/m}^2$$

Maximum shear force is

$$V = LL_s \frac{(q_1 + q_5)}{2}$$
$$= 2.8(1.025 - 0.166)\frac{(142 + 91.44)}{2} = 280.73 \text{ kN}$$

The section is trapezoidal with top width as b_1 and bottom width as b_2. The nominal shear stress on the section is

$$\tau_v = \frac{2V}{(b_1 + b_2)d}$$

$$= \frac{2(280730)}{(1.2 + 2.8)(0.332)(10^6)} = 0.42 \text{ MPa}$$

The percentage of reinforcement in the footing is in the range of 0.5% and the allowable shear stress for such a per cent of reinforcement is 0.3(1.33) = 0.4 MPa. The depth of the section is not adequate against shear force. Therefore, revise the depth of the section to $d = 0.36$ m

The nominal shear stress is

$$\tau_v = \frac{2V}{(b_1 + b_2)d} = \frac{2(280730)}{4(0.36)(10^6)} = 0.39 \text{ MPa}$$

Design of reinforcement. The approximate lever arm between the centres of compression and tension is 0.9d

The actual lever arm is likely to be less than the above value, therefore use

$$jd = 0.9d = 0.324 \text{ m}$$

Allowable stress in steel = 230(1.33) MPa

Area of the tension reinforcement is

$$A_{st} = \frac{M}{jd\sigma_{st}(1.33)} = \frac{179300}{0.324(230)(1.33)} = 1809 \text{ mm}^2$$

Provide 9 numbers of 16 mm bars at the bottom spanning 2.65 m.

$$A_s \text{ (provided)} = 1809 \text{ mm}^2$$

Effective area of the concrete for computation of percentage of steel is

$$A_c = \frac{(1.2 + 2.8)(0.36)}{2} = 0.72 \text{ m}^2$$

Percentage of effective reinforcement is

$$p = \frac{100 A_s}{A_c} = \frac{180900}{0.72(10^6)} = 0.25$$

Allowable shear stress for 0.25% of reinforcement under seimic load condition is

$$\tau_a = 0.28(1.33) = 0.29 \text{ MPa}$$

This value is less than the nominal shear stress, therefore, increase the effective depth to 0.48 from 0.36 m. The area of steel will be reduced by about 25%.

Design of reinforcement in the width direction. The maximum bending moment caused on the widthwise cantilever during live load condition governs the design. This moment is

$$M_2 = 129.62 \text{ kNm}$$

The reinforcement in this direction is placed over the main reinforcement, therefore the depth of the steel in this direction is

$$d = 0.48 - 0.02 = 0.46 \text{ m}$$

Area of the reinforcement required is

$$A_{st} = \frac{M_2}{jd(\sigma_{st})} = \frac{129620}{0.9(0.46)(230)} = 1361 \text{ mm}^2$$

Provide 8 numbers of 16 mm bars.

The reinforcement details are shown in Fig. 13.10. The overall depth of the slab at column face is

$$h = 0.48 + 0.01 + 0.05 = 0.54 \text{ m}$$

The depth at free end is $h_0 = 0.15$ m

EXAMPLE 13.6 *Design circular footing.* A circular cross-sectioned column of 600 mm diameter is transferring a load of 1200 kN to the footing. The allowable net bearing pressure of the clayey soil is 80 kN/m². Design an isolated circular footing.

Axial load $\qquad = P_s = 1200$ kN

Bearing capacity $\quad = p_a = 80$ kN/m²

Concrete is M20 and reinforcement is of HYSD-Fe415 bars

Diameter of the column $= a = 0.6$ m

Design of foundation

Depth of foundation $\quad = 1$ m

Superimposed load $P_s = 1200$ kN

Difference in weight of concrete and soil in foundation

$$W'_g = 0.06 \, P_s = 72 \text{ kN}$$

Net load on the soil $= P = 1272$ kN

Bearing area needed for the foundation is

$$A_f = \frac{P}{p_a} = \frac{1272}{80} = 15.9 \text{ m}^2$$

Let $R =$ radius of the circular slab of the footing, then

$$\pi R^2 \geq 15.9 \text{ m}^2$$

or $\qquad\qquad R \geq 2.25$ m

Use $\qquad\qquad R = 2.3$ m and

$$A_f = \pi R^2 = 16.62 \text{ m}^2$$

Design of the section. Effective soil reaction on the RCC slab is

$$q = \frac{P_s}{A_f} = \frac{1200}{16.62} = 72.21 \text{ kN/m}^2$$

The net effect of a circular footing slab loaded with a central load can be obtained by superposition of two separate cases of circular slabs for which moments are available and they are: a slab simply supported along the periphery with central load and a simply supported slab with uniformly distributed reaction. The radial and circumferential bending moments on the slab due to the combined simply supported slabs in which the reaction from the periphery

644

works out to be zero are given by:

For $r > a$

$$M_r = \frac{P_a}{4\pi}\left(\left(\frac{a^2}{4}\left(\frac{1}{R^2} - \frac{1}{r^2}\right) - \ln\frac{r}{R}\right) - \frac{3}{4}\left(1 - \frac{r^2}{R^2}\right)\right) \tag{1}$$

$$M_\theta = \frac{P_s}{4\pi}\left(\left(1 + \frac{a^2}{4}\left(\frac{1}{R^2} + \frac{1}{r^2}\right) - \ln\frac{r}{R}\right) - \frac{1}{4}\left(3 - \frac{r^2}{R^2}\right)\right) \tag{2}$$

$$V = \frac{P_s}{2\pi r} \tag{3}$$

in which r = radial distance of the point under consideration

R = radius of the slab

a = radius of the column

The moments on the slab are critical at $r = a$ and $r = R$ and they are computed and given below.

At $r = a$

$$M_r = \frac{P_s}{4\pi}\left(\left(\frac{a^2 - R^2}{4R^2} - \ln\frac{a}{R}\right) - \frac{3}{4}\left(\frac{R^2 - a^2}{R^2}\right)\right) \tag{4}$$

$$M_\theta = \frac{P_s}{4\pi}\left(\left(1 + \frac{a^2 + R^2}{4R^2} - \ln\frac{a}{R}\right) - \frac{3R^2 - a^2}{4R^2}\right) \tag{5}$$

At $r = R$

$$M_r = 0$$

$$M_\theta = \frac{P_s}{4\pi}\left(\frac{1}{2} + \frac{a^2}{2R^2}\right) \tag{6}$$

Substitution of the parameters in the above expressions results:

At $r = a$ (for $a = 0.3$ and $R = 2.3$)

$$M_r = 95.5(1.791 - 0.737) = 100.7 \text{ kNm/m}$$
$$M_\theta = 95.5(3.291 - 2.601) = 65.9 \text{ kNm/m}$$

At $r = R$

$$M_\theta = 95.5(0.5 + 0.008) = 48.6 \text{ kNm/m}$$

The radial moment at the face of the column is maximum and the size of the section is governed by this value. Equating the resisting moment to the maximum moment, we have

$$0.91\, bd^2 (10^6) = 100.7 (1000) \text{ Nm}$$

in which $b = 1$,

The above equation therefore gives:

$$d = \sqrt{\frac{0.1007}{0.91}} = 0.34 \text{ m}$$

The thickness of the slab at the free end is governed by the circumferential moment and the corresponding effective depth is given by

$$0.91\, bd_f^2\, (10^6) = 48.6(1000) \text{ Nm}$$

in which $b = 1$ m and then d_f is

$$d_f = \sqrt{\frac{0.0486}{0.91}} = 0.230 \text{ m}$$

Use $\qquad\qquad d_f = 0.24$ m

Check for shear stress. Critical diagonal failure section is at distance $0.5d$ from the face of the column. Therefore

$$r = a + 0.5d = 0.3 + 0.17 = 0.47 \text{ m}$$

The effective shear force on the circle is

$$V = \frac{P_s}{2\pi r} = \frac{1200}{2\pi(0.47)} = 406 \text{ kN}$$

Nominal shear stress at the critical shear force circle is

$$\tau_v = \frac{V}{d} = \frac{406000(10^{-6})}{0.34} = 1.2 \text{ MPa}$$

This value is far higher than the allowable shear stress in the flat slabs which is equal to $0.16\sqrt{f_{ck}} = 0.71$ MPa. So increase the effective thickness of the slab at the support or increase the diameter of column at the footing.

Let $\qquad\qquad d = 0.54$

then $\qquad\qquad r = 0.3 + 0.27 = 0.57$ m

and $\qquad\qquad \tau = \dfrac{P_s}{2\pi r} = \dfrac{1200000(10^{-6})}{2\pi(0.57)(0.54)} = 0.62$ MPa

Design of reinforcement

Radial reinforcement: The radial reinforcement required at the face of the column is

$$A_{st} = \frac{M}{jd\sigma_{st}} = \frac{100700}{0.904(0.54)(230)} = 897 \text{ mm}^2/\text{m}$$

Provide 16 mm bars at 220 mm spacing at the bottom surface in radial direction.

Circumferential reinforcement at $r = 0.3$ m

$$A_{st} = \frac{M}{jd\sigma_{st}} = \frac{65900}{0.904(0.52)(230)} = 610 \text{ mm}^2/\text{m}$$

Provide 16 mm bars at 320 mm spacing at the bottom as circumferential reinforcement near the support.

Circumferential reinforcement at free end is

$$A_{st} = \frac{48600}{0.904(0.22)(230)} = 1062 \text{ mm}^2/\text{m}$$

Provide 16 mm bars at 190 mm spacing.
The overall depth at support is $h = d + 0.06 = 0.60$ m.
The details of the reinforcement are shown in Fig. 13.11.

EXAMPLE 13.17 *Design of a circular footing supporting a shaft.* A circular cylindrical shaft of

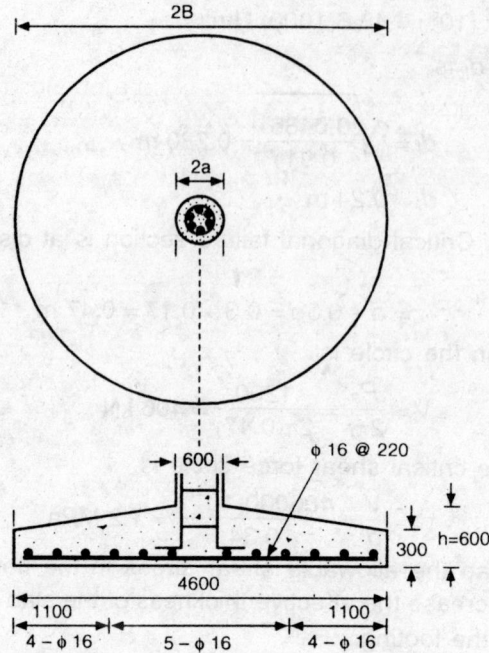

Fig. 13.11 Structural Details of RCC Circular Footing of Example 13.16.

radius 4.5 m with thickness of wall = 300 mm is transferring a load of 13000 kN to the footing. The safe bearing capacity of the soil is 100 kN/m². Design a RCC raft foundation.

Axial load	$= P_s = 13000$ kN
Radius of the shaft	$= c = 4.5$ m
Thickness of the wall	$= t = 0.3$ m
Net bearing capacity of the soil	$= p_a = 100$ kN/m²

Concrete is M20 and reinforcement is HYSD-Fe415.

Design of foundation

Superimposed load $P_s = 13000$ kN

Difference in the weight of the concrete in the raft and the equivalent soil is about 0.07(10640) $W_g = 1000$ kN

Net load $P = 14000$ kN

Area of the bearing surface required for the foundation

$$A_f = \frac{P}{p_a} = \frac{14000}{100} = 140 \text{ m}^2$$

Let R = outer radius of the raft then

$$\pi R^2 = A_f$$

or $R = \sqrt{\dfrac{140}{\pi}} = 6.68$ m

Use $\qquad R = 6.8$ m

Force analysis. The raft slab is a solid circular one and is subjected to an axial load through concentric shaft. The moments on the slab can be obtained by superposition of two separate problems: a circular slab simply supported on the edges and loaded through a concentric circle, and an inverted simply supported slab with uniformly distributed soil reaction. The moments caused by the combined action of the two problems with equal loads in both the cases are:

For $r < c = 4.5$ m

$$M_r = \frac{P_s}{16\pi}\left(2\left(1 + 2\ln\frac{R}{c} - \left(\frac{c}{R}\right)^2\right) - 3\left(1 - \left(\frac{r}{R}\right)^2\right)\right) \tag{1}$$

$$M_\theta = \frac{P_s}{16\pi}\left(2\left(1 + 2\ln\frac{R}{c} - \left(\frac{c}{R}\right)^2\right) - \left(3 - \left(\frac{r}{R}\right)^2\right)\right) \tag{2}$$

For $r > c$

$$M_r = \frac{P_s}{16\pi}\left(2\left(2\ln\frac{R}{r} + \left(\frac{c}{r}\right)^2 - \left(\frac{c}{R}\right)^2\right) - 3\left(1 - \left(\frac{r}{R}\right)^2\right)\right) \tag{3}$$

$$M_\theta = \frac{P_s}{16\pi}\left(2\left(2 + 2\ln\frac{R}{r} - \left(\frac{c}{r}\right)^2 - \left(\frac{c}{R}\right)^2\right) - \left(3 - \left(\frac{r}{R}\right)^2\right)\right) \tag{4}$$

$$V = \frac{P}{2\pi r} \tag{5}$$

The bending moments on the raft are computed for typical values of r and shown in the Table 13.2.

Table 13.2 BM on the Raft Slab (kNm/m)

r	0	2.25	4.65	6.80
M_r	− 58	27	238	0.0
M_θ	− 58	− 29	62	65

Design of section. The depth of the slab at the face of the shaft wall is govened by the radial moment and that at the free end is governed by the circumferential moment. The working stress balanced moment capacity of a rectangular section is

$$M_b = 0.91bd^2 (10^6) \text{ Nm where } b = 1 \text{ m.}$$

Equating the moment capacity to the radial moment gives:

$$d = \sqrt{\frac{M_r}{0.91(10^6)}} = \sqrt{\frac{0.238}{0.91}} = 0.52 \text{ m}$$

The failure of shear plane is at $\pm 0.5d$ from the face of the wall. Therefore, radius of the outer critical shear plane circle is

$$r = c + 0.5t + 0.5d = 4.911 \text{ m}$$

The shear force on the plane is

$$V = \frac{P_s}{2\pi r}$$

The nominal shear stress on the plane is

$$\tau = \frac{P_s}{2\pi rd} = \frac{13000\,000(10^{-6})}{2\pi(4.911)(0.52)} = 0.8 \text{ MPa}$$

The shear stress on the outer circle exceeds the permissible value 0.71 MPa. Therefore, the thickness needs revision.

The radius of the critical shear plane inside the shaft is

$$r = c - 0.5t - 0.5d = 4.5 - 0.15 - 0.261 = 4.089 \text{ m}$$

The shear force on the circle is

$$V_s = P_s\left(1 - \frac{r^2}{R^2}\right)$$

$$= 13000\left(1 - \frac{4.089^2}{6.8^2}\right) = 3800 \text{ kN}$$

Nominal shear stress on the plane

$$\tau = \frac{V}{2\pi rd} = \frac{8300000(10^{-6})}{2\pi(4.089)(0.52)} = 0.6 \text{ MPa}$$

The thickness of the slab at the free end is governed by the circumferential moment and the effective depth (d_f) at the free end is

$$d_f = \sqrt{\frac{65000}{0.91(10^6)}} = 0.270 \text{ m}$$

Increase the thickness of the slab at the face of the wall to 0.68 and check for shear stress at the outer face plane.

$$r = 4.5 + 0.15 + 0.34 = 4.99 \text{ m}$$

The nominal shear stress is

$$\tau_v = \frac{13000(10^{-6})}{2\pi(4.99)(0.68)} = 0.61 \text{ MPa}$$

Use

$$h = 0.68 + 0.06 = 0.74 \text{ m}$$

at free end

$$h = 0.272 + 0.06 = 0.34 \text{ m}$$

Design of reinforcement

Radial reinforcement. The area of reinforcement needed at the centre is

$$A_{st} = \frac{58000}{0.904(0.270)(230)} = 1033 \text{ mm}^2/\text{m}$$

Provide φ16 bars at 190 mm c/c radially at top

At support ($r = 4.5$ m), the tension reinforcement is

$$\frac{238000}{0.904(0.68)(230)} = 1683 \text{ mm}^2/\text{m}$$

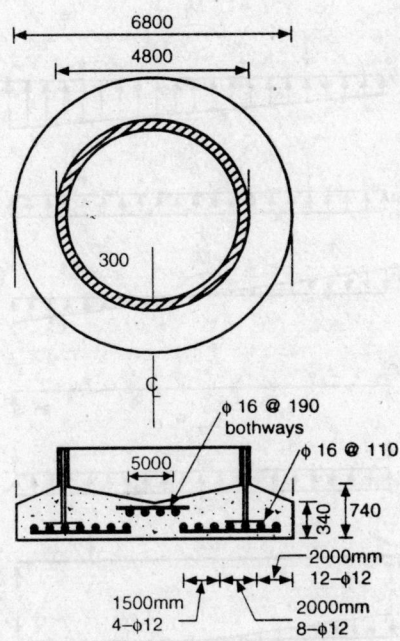

Fig. 13.12 Structural Details of Raft Foundation Supporting of Circular Shaft of Example 13.17.

Provide φ16 bars at 110 mm c/c at bottom as radial bars.

At $r = 0$, provide φ16 at 180 mm c/c at top as circumferential bars, same as the radial.

At $r = 4.5$ m, the circumferential reinforcement is

$$A_{st} = \frac{62000}{0.904(0.66)(230)} = 451 \text{ mm}^2/\text{m}$$

Provide φ12 bars at 250 mm c/c at bottom as circumferential bars,

At $r = 6.8$ m the circumferential reinforcement is

$$A_{st} = \frac{65000}{0.904(0.27)(230)} = 1158 \text{ mm}^2/\text{m}$$

Provide φ16 bars at 170 mm c/c at bottom.

EXAMPLE 13.8 Design of RCC circular solid raft subjected to axial force and bending moment from a shaft.

Total loads on the raft during seismic load condition are:

Axial force	$P_s = 10640$ kN
Bending moment	$M_f = 13500$ kNm

Radius of the shaft through which the load is transferred $c = 4.5$ m and

Thickness of shaft in contact with the raft $\quad t = 0.3$ m

Outer radius of the raft $\quad R = 6.8$ m

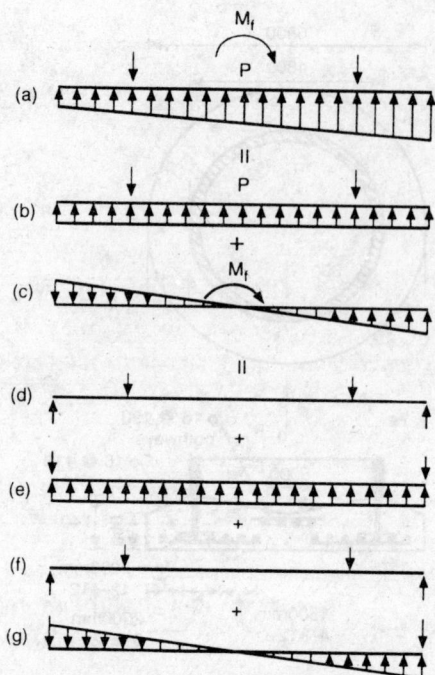

Fig. 13.13 Superposition Principle in Circular Raft Foundation of Example 13.8

Concrete used is M20 and reinforcement used is HYSD-Fe415.

Force analysis. The axial force and moment acting on the raft as shown in Fig. 13.13a can be replaced by two separate axial and moment forces for the purpose of simpler analysis. These two cases are shown in Figs. 13.13b and c respectively. The circular slab is subjected to force through a concentric circular shaft and supported by the soil. Such an effective force can be obtained by combination of two simply supported slabs around the periphery and subjected to the forces and the compatible earth pressure. Figures. 13.13d and e refers to the equivalent of a circular slab subjected to axial force and Figs. 13.13f and g refer to the slab subjected to a moment. The moments acting on the slab are analysed in two separate cases and then superimposed for the final result.

Case 1: *Slab with axial force.* The radial and circumferential moments on the slab caused by axial force are given below.

Let r = radial distance of the point under consideration

c = radius of the shaft

For r < c

$$M_r = \frac{P}{16\pi} [3((r/R)^2 - 1) + 2(2\ln(R/c) + 1 - (c/R)^2)] \tag{1}$$

$$M_0 = \frac{P}{16\pi}\left[((r/R)^2 - 3) + 2\{2\ln(R/c) + 1 - (c/R)^2\}\right] \tag{2}$$

For $r > c$

$$M_r = \frac{P}{16\pi}\left[(3(r/R)^2 - 1) + 2\{2\ln(R/r) - (c/R)^2 + (c/r)^2\}\right] \tag{3}$$

$$M_0 = \frac{P}{16\pi}\left[((r/R)^2 - 3) + 2\{2\ln(R/r) - (c/R)^2 - (c/r)^2 + 2\}\right] \tag{4}$$

$$V = \frac{P}{2\pi r} \tag{5}$$

The moments in the slab are computed at critical locations and are given in Table 13.3.

Table 13.3 BM (kNm) on Slab with Axial Force (kN)

$r =$	0	2.25	4.65	6.8 m
M_r	−48	22	195	0
M_0	−48	−25	51	53
V	0	0	364	0

Case 2: *Slab with moment.* The maximum radial and circumferential moments caused by a moment of M_f acting on the slab through a shaft are given by:
For $r < c$

$$M_r = \frac{M_f r}{4\pi R^2 c^2}(3R^2 - 2c^2 - c^4/R^2) - \frac{5M_f r}{12\pi R^4}(R^2 - r^2) \tag{6}$$

$$M_\theta = \frac{M_f r}{24\pi c^2 R^2}(3R^2 - 2c^2 - c^4/R^2) - \frac{M_f r}{36\pi R^4}(5Rr - 3r^2) \tag{7}$$

For $r > c$

$$M_r = \frac{M_f}{8\pi R^2 r}(2(R^2 - r^2) + c^2(R^4 - r^4)/r^2 R^2) - \frac{5M_f r}{12\pi R^4}(R^2 - r^2) \tag{8}$$

$$M_\theta = \frac{M_f}{24\pi R^2 r}(2(3R^2 - r^2) - c^2(3R^4 + r^4)/R^2 r^2) - \frac{M_f r}{36\pi R^4}(5Rr - 3r^2) \tag{9}$$

The values of M_r and M_θ are given in Table 13.4 for different values of r for the problem.

The resultant moment on the raft of the foundation caused by the combined action of the axial force and moment on the raft are given in Table 13.5.

Design of the size. The bending moment is critical at the face of the shaft wall. The depth of the slab should be designed at the centre, at face of the shaft wall and also at the free end. Since the bending moment at the centre is same with or without seismic force, this

Table 13.4 BM (kNm) in Slab Due to External Moment

r(m)	0	2.25	4.65	6.8 m
M_r	0	153	112	0
M_0	0	31	55	25

Table 13.5 Critical BM (kNm) on the Raft Slab

r(m)	0	2.25	4.65	6.8
M_r	-48	175	307	00
M_0	-48	6	106	78

section should be designed for normal live load condition and for seismic load situation.

Design of section at face of the shaft wall. Equating the moment capacity to the external moment we have

$$0.91 \ (1.33) \ (10^6) \ bd^2 = 307000 \ \text{Nm/m}$$

$$b = 1 \ \text{m then from the above relation, we have}$$

$$d = \sqrt{\frac{0.307}{0.91(1.33)}} = 0.510 \ \text{m}$$

Let

$$d = 0.52 \ \text{m}$$

The shear stress is critical on a circle at a distance $\pm \ 0.5d$ from the face of the shaft wall. The radius of this plane is

$$r = c + 0.5 \ (t + d)$$

$$= 4.5 + 0.5 \ (0.3 + 0.52) = 4.91 \ \text{m}$$

The nominal shear stress on the plane is

$$\tau_v = \frac{P_s}{2\pi rd} = \frac{10640 \ (10^{-3})}{2\pi(4.91)(0.52)} = 0.66 \ \text{MPa}$$

The allowable shear stress under seismic load condition is

$$\tau_c = 1.33 \ (0.71) = 0.94 \ \text{MPa}$$

The thickness of the slab at the free end is governed by the circumferential bending moment, and the corresponding effective depth of steel is

$$d_f = \sqrt{\left(\frac{78000}{0.91(1.33)(10^6)}\right)} = 0.26 \ \text{m}$$

Provide the overall thickness of the raft at the support as

$$h = 0.52 + 0.06 = 0.58 \ \text{m}$$

at the free edge as $h_f = 0.26 + 0.06 = 0.34 \ \text{m}$

Also at the centre $h_f = 0.34 \ \text{m}$

Design of reinforcement

Radial reinforcement. The area of the reinforcement at the mid point is

$$A_{st} = \frac{48000}{0.904(0.26)(230)(1.33)} = 668 \ \text{mm}^2/\text{m}$$

Provide $\phi16$ bars at 300 mm. c/c at top in both ways. The area of the reinforcement required at the face of the wall is

$$A_{st} = \frac{307000}{904(0.52)(230)(1.33)} = 2135 \text{ mm}^2/\text{m}$$

Provide φ16 bars at 90 mm/cc at bottom.

Circumferential reinforcement
  At free edge

$$A_{st} = \frac{78000}{0.904(0.26)(230)(1.33)} = 737 \text{ mm}^2/\text{m}$$

Provide φ16 bars at 180 mm c/c at bottom.
  At the face of the wall

$$A_{st} = \frac{106000}{0.904(0.52)(230)(1.33)} = 737 \text{ mm}^2/\text{m}$$

Provide φ16 bars at 270 mm c/c at bottom.

The reinforcement details are shown in Fig. 13.14.

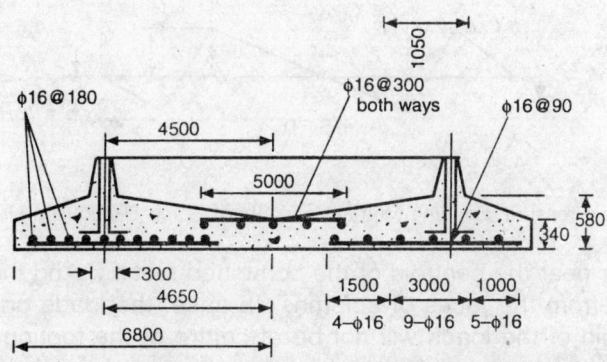

Fig. 13.14 Structural Details of Circular Raft Foundation of Example 13.8.

13.7. Combined Footings

Sometimes, columns are closely spaced because of high loading, constraints and considerations in building overall plan. At times, even if the columns are reasonably well spaced, the bearing capacity of the soil may be small and will not allow separate footings to each of the columns. Practical considerations and economic considerations may force a combined footing for two or more columns. Even though that the design of combined footings is normally discussed for two columns, it is applicable to multiple columns. When a footing is designed for a row of columns, it can be considered as a combined strip footing and designed as a continuous beam. Similarly, a footing designed for a set of columns is usually called a raft or mat foundation. This section presents design of footings combined for two columns only. Typical footings are shown in Figs. 13.14 to 13.17. Figure 13.14a gives the elevation of the footing and Fig. 13.14b gives the distribution of the bearing pressure in which the centroid of the load coincides with the controid of the footing. Typical bending and shear force diagrams are shown in Figs. 13.14c and d. The critical bending moments occur at the

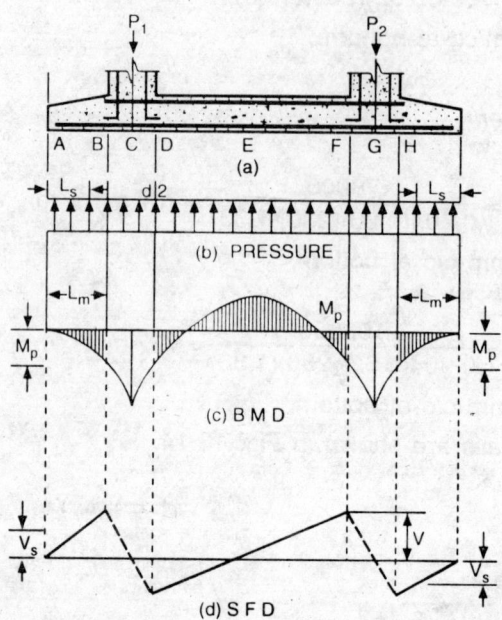

Fig. 13.14a-d Notations, Pressure, Bending Moment and Shear Force Distribution Under Combined Footing.

face of the columns or near the centroid of the combined footing. And the critical shear forces occur at a distance *d* from the faces of columns. In case, the loads on the columns are not same, then the centroid of the loads will not be at centre of the footing. A footing of uniform width would generate uniform or uniformly varying pressure along the length of the footing. However, a uniform pressure under the footing can be achieved by adjusting the width of the footing so as to have centroids of the forces and reactions in a line as shown in Figures 13.15 and 13.16. It is not essential that the bearing pressure should be made uniform to minimize the foundation area. Sometimes, it may not be possible to have a uniform bearing reaction because of space restrictions. The method of design of the combined footing is similar

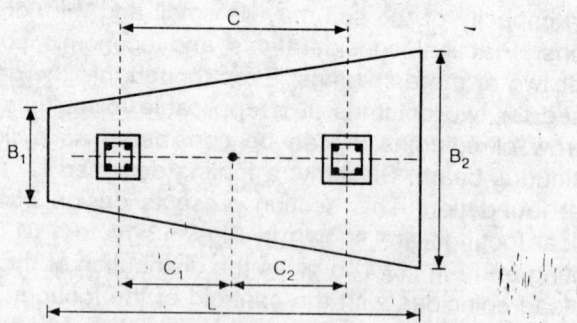

Fig. 13.15 Notations and Foundation of Trapezoidal Footing.

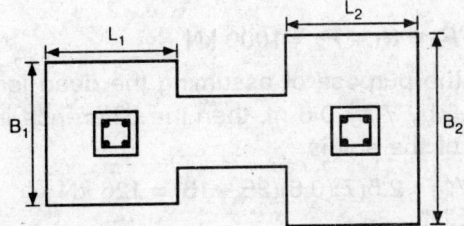

Fig. 13.16. Notations for Combined Footing.

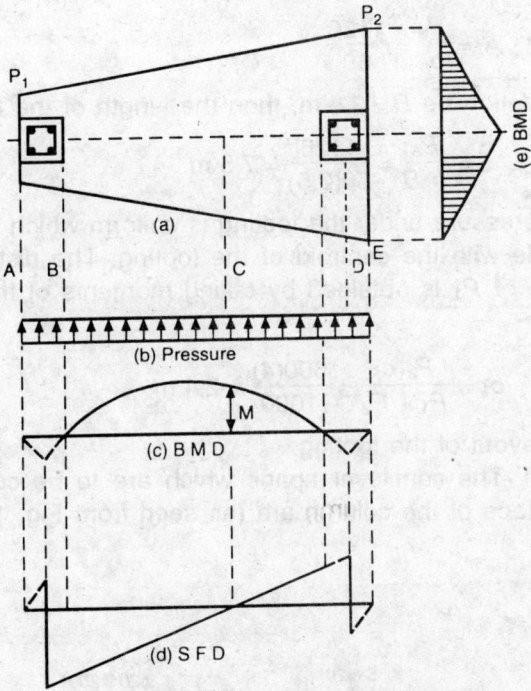

Fig. 13.17 Notations of Bending Moment and Shear Force Distributions under Combined Footing.

to that already applied to the isolated footings. The design is illustrated through a set of examples.

EXAMPLE 13.9 *Combined footing for unequally loaded two columns and uniform in width.* Two columns carrying 400 kN and 600 kN are spaced 4 m apart and they have to be provided with a foundation on soil having a net safe bearing capacity of 60 kN/m². The width of the footing must be restricted to 2.5 m. The columns size is 500 by 500 mm.

The loads on the columns are

$$P_1 = 400 \text{ kN}, P_2 = 500 \text{ kN}$$

Net safe bearing capacity $p_a = 60 \text{ kN/m}^2$

Use $f_{ck} = 20 \text{ MPa}, f_y = 415 \text{ MPa}$

Total superimposed load $= P_s = P_1 + P_2 = 1000$ kN

Design of foundation : For the purpose of assuming the dead load of the foundation, let the size of the foundation be 2.5 by 7 by 0.8 m, then the difference in the weight of the concrete and the equivalent volume of the soil is

$$W'_g = 2.5(7)(0.8)(25 - 16) = 126 \text{ kN}$$

The net load on the foundation soil is

$$P = P_s + W_g' = 1126 \text{ kN}$$

Let the depth of foundation be 1 m below GL. The area of the foundation needed is

$$A_f = \frac{P_s}{p_a} = \frac{1126}{60}$$

Let the width of the foundation be $B = 2.5$ m, then the length of the foundation is

$$L = \frac{P_s}{p_a B} = \frac{1126}{60(2.5)} = 7.5 \text{ m}$$

It was assumed that the pressure under the footing is uniform which means that the centroid of the loads must coincide with the centroid of the footing. The distance of the centroid of the loads from the centre of P_1 is obtained by taking moments of the forces about the line of action of P_1 and it is

$$c_1 = \frac{P_2(c)}{P_1 + P_2} = \frac{600(4)}{1000} = 2.4 \text{ m}$$

Figure 13.18 shows the layout of the footing.

Design of the section. The cantilever spans which are to be considered for maximum bending moments at the face of the column are (as seen from Fig. 13.18):

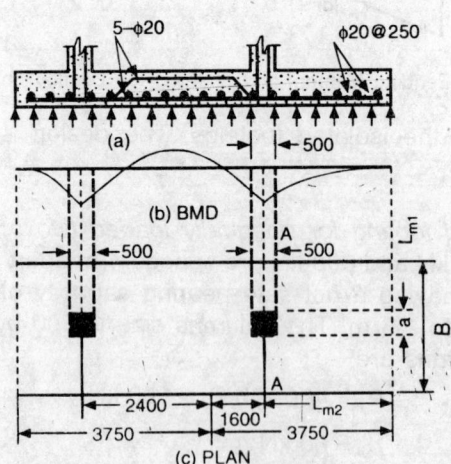

Fig. 13.18 Structural Details of Combined Footing of Example 13.9.

$$L_{m1} = 0.5(B - a) = 0.5(2.5 - 0.5) = 1 \text{ m}$$

$$L_{m2} = \frac{L}{2} - (4 - 9) - \frac{a}{2} = 3.75 - 1.6 - 0.25 = 1.9 \text{ m}$$

Since L_{m2} is larger than L_{m1}, the critical bending moment span is L_{m2} and the maximum positive BM is

$$M_1 = \frac{BqL_{m2}^2}{2}$$

where

$$q = \frac{P_1 + P_2}{LB} = \frac{1000}{7.5(2.5)} = 53.333 \text{ kN/m}^2$$

$$M_1 = \frac{B(P_1 + P_2)L_{m2}^2}{2BL} = \frac{1(1.9)^2}{2(7.5)} = 0.24067 \text{ MNm}$$

The maximum negative bending moment will occur at the centroid of the footing and is

$$M_2 = -\left(\frac{qBL^2}{8} - P_1 c_1\right)$$

$$= -(qBL)\frac{L}{8} + P_1 c_1 = -(P_1 + P_2)\frac{L}{8} + P_1 c_1$$

$$= -\frac{(7.5)}{8} + 0.4(2.4) = 0.0225 \text{ MNm}$$

The positive bending moment is far higher than the magnitude of the negative BM, therefore the depth of the footing is governed by M_1. Equating the resisting moment to the external moment, we have

$$Rbd^2 \geq M_1 = 0.24067 \text{ MNm}$$

or

$$d^2 \geq \frac{0.24067}{0.91(2.5)}$$

or

$$d \geq 0.34 \text{ m}$$

Invariably the depth of the footing is governed by the shear rather than the bending moment. It is, therefore, recommended to select the depth as 10 to 30% more than that required for bending resistance.

Use $\qquad$ $d = 0.4$ m

Design for transverse shear. The critical section is at a distance d from the face of the column and the shear span is

$$L_s = L_{m2} - d = 1.9 - 0.4 = 1.5 \text{ m}$$

The shear force on the plane is

$$V = BL_s q$$

and the nominal shear stress is

$$\tau_v = \frac{V}{bd} = \frac{BL_s q}{bd}, \text{ where } b = B = 2.5 \text{ m}$$

$$= \frac{1.5(0.05333)}{0.4} = 0.2 \text{ MPa}$$

The value is less than 0.24 MPa which is the allowable value for minimum reinforcement.

Check for diagonal tension. The critical shear plane of punching through diagonal tension is at 0.5d from the face of this column. The peripheral length of the plane is

$$b_0 = 4(a + d) = 3.6 \text{ m}$$

The maximum shear force occurs around the column 2 and is

$$V = q\left(\frac{LB}{2} - (a + d)^2\right)$$
$$= \frac{(P_1 + P_2)}{LB}\left(\frac{LB}{2} - (a + d)^2\right)$$
$$= \frac{1(3.75(2.5) - 0.81)}{7.5(2.5)} = 0.4568 \text{ MN}$$

The nominal shear stress is

$$\tau_v = \frac{V}{b_0 d} = \frac{0.4568}{3.6(0.4)} = 0.32 \text{ MPa}$$

The allowable shear stress for diagonal failure is

$$\tau_a = 0.16 \sqrt{f_{ck}} = 0.16 \sqrt{20} = 0.71 \text{ MPa}$$

The section is safe against shear force.

Reinforcement design. The area of the tensile reinforcement at the bottom in the longitudinal direction

$$A_{st} = \frac{M_1}{j d \sigma_{st}} = \frac{240670}{0.904(0.4)(230)} = 3017 \text{ mm}^2$$

Minimum reinforcement

$$\frac{0.3 \, bd}{100} = 0.003(2500)(400) = 3000 \text{ mm}^2$$

Provide 10 numbers of 20 mm bars in the longitudinal direction and crank up half of the bars (spacing is 250 mm c/c).

The bending moment in the widthwise direction is far lesser than lengthwise, however the minimum reinforcement governs the design. Therefore, provide $\phi 20$ bars at 250 mm c/c. Figure 13.18 illustrates the reinforcement details.

EXAMPLE 13.10 *Combined footing for unequally loaded columns and varying in width.* Two columns carrying 400 kN and 600 kN are spaced at 4 m apart and have been provided with a combined footing on soil having a net safe bearing capacity of 60 kN/m². The length of the footing is to be restricted to 6.0 m, with a maximum projection of 1 m beyond the centre of each column. Size of column is 500 × 500 mm².

Size of the column $a = 0.5$ m

Loads on the columns are

$$P = 400 \text{ kN}, P_2 = 600 \text{ kN}, P_s = P_1 + P_2 = 1000 \text{ kN}$$

Net safe bearing capacity of the soil $p_a = 60$ kN/m²

$$f_{ck} = 20 \text{ MPa}, f_y = 415 \text{ MPa}$$
$$\sigma_{st} = 230 \text{ MPa}, \sigma_{cb} = 7 \text{ MPa}$$

Spacing of the loads $c = 4$ m

Design of foundation. Let the difference between the weight of the foundation and the corresponding weight of the soil be taken as

$$W_g' = 0.1P = 100 \text{ kN}$$

The net load on the soil is

$$P = P_s + W_g' = 1100 \text{ kN}$$

The area of the foundation needed without eccentricity is

$$A_f = \frac{P}{p_a} = \frac{1100}{60} = 18.33 \text{ m}^2$$

Since the projection of the footing on either side is restricted to 1 m, the width of the foundation has to be varying.

Let $B_1 =$ width of the foundation near P_1

$B_2 =$ width of the foundation near P_2

One should adjust the following two quantities:

1. Area of the foundation equal to or greater than that is needed, which means

$$(B_1 + B_2)\frac{L}{2} \geq A_f = 18.33 \tag{1}$$

2. The centroid of the loads should coincide with the centroid of the foundation. The centroid of the load from line of action of P_1 is

$$c_1 = \frac{P_2(c)}{P} = \frac{600\,(4)}{1000} = 2.4 \text{ m}$$

Its distance from end B_1 is

$$c_3 = c_1 + 1 = 3.4 \text{ m} \tag{2}$$

The centroid of the footing from end B_1 is given by the moments of the area about that end and it is

$$B_1 L\left(\frac{L}{6}\right) + B_2 L\left(\frac{2L}{6}\right) = c_3\,(B_1 + B_2)\frac{L}{2}$$

or $\qquad\qquad c_3 = \dfrac{(B_1 + 2B_2)}{3}$

or $\qquad\qquad B_1 + 2B_2 = 3c_3 \tag{3}$

From Eqs. (2) and (3), we have

$$B_1 + 2B_2 = 3(3.4) = 10.2 \text{ m} \tag{4}$$

The length of the footing $L = 6$ m, so Eq. (1) reduces to

$$B_1 + B_2 = 6.11 \text{ m} \tag{5}$$

Equations (4) and (5) result in

$$B_2 = 10.2 - 6.11 = 4.09 \text{ m}$$

$$B_1 = 6.11 - 4.09 = 2.02 \text{ m}$$

Use $B_1 = 2.1 \text{ m and } B_2 = 4.1 \text{ m}$

Depth of foundation $= 1 \text{ m}.$

Since the centroid of the foundation slab coincides with that of the column loads, the pressure under the foundation would be uniform.

The foundation area is

$$A_f = \frac{(B_1 + B_2) L}{2} = 3.1(6) = 18.6 \text{ m}^2$$

Design of the foundation slab. The net reaction of the soil on the slab

$$q = \frac{P_s}{A_f} = \frac{1000}{18.6} = 53.763 \text{ kN/m}^2$$

There are three sections of critical bending moments. It may be observed that the maximum cantilever is widthwise and it is near the end close to the column 2. The section *AA* as shown in Fig. 13.19 will be the critical one and the bending moment on this plane is calculated. Since the cantilever span is not constant over the entire length, there will be an extra torison on this section. Since the width of the section is large (*L* = 6 m) the asymmetry in the load on the cantilever projection is not going to be of a serious consequence. The asymmetry produces torsion which in turn produces equal and opposite shear forces. The other critical section is *BB* having a cantilever span of 0.75 m. This will not be of much consequence. The third critical section is the centroidal sections along the width on which a

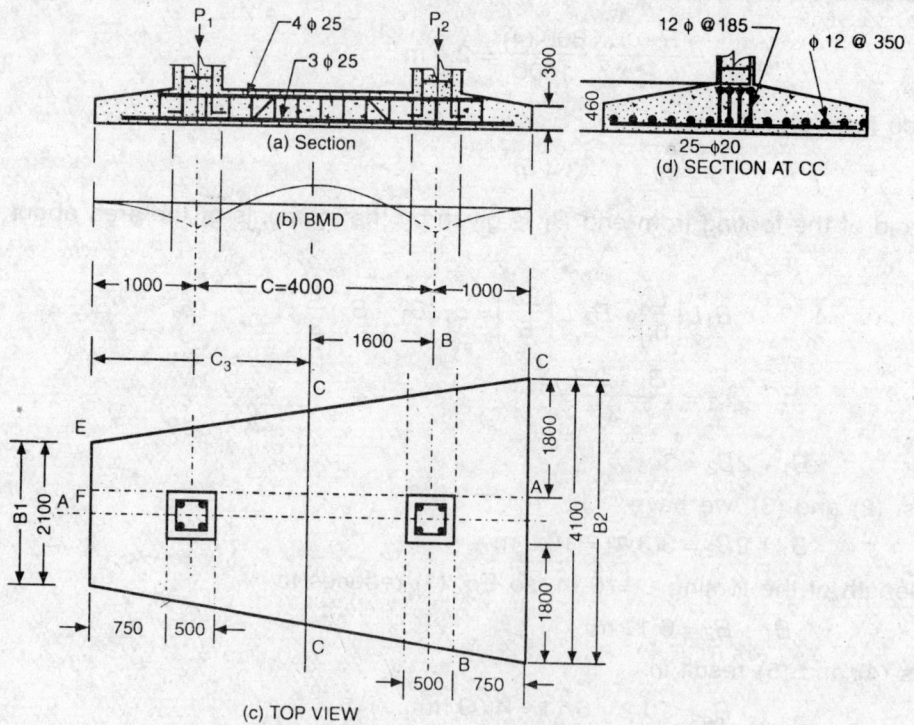

Fig. 13.19 Structural Details of Combined Footing of Example 13.10

negative bending moment is likely to act. The bending moment on each of the sections are computed first.

Bending moment on section AA. Consider the trapezoidal segment *ADEF*, and the bending moment about line *FA*.

The total reaction on the segment is

$$Q = qL \frac{(B_1 - a + B_2 - a)}{4}$$

$$= \frac{qL(B_1 + B_2 - 2a)}{2}$$

$$= 53.763(6)\frac{(2.1 + 4.1 - 1)}{4} = 419.35 \text{ kN}$$

Its line of action from the line *FA* is

$$a_1 = \frac{(B_1 - a + B_2 - a)}{4} = \frac{5.2}{4} = 1.3 \text{ m}$$

The line of action from the middle point of the section is

$$e = \frac{L}{3}\frac{(B_1 + 2B_2)}{(B_1 + B_2)} - \frac{L}{2}$$

$$= \frac{L(B_2 - B_1)}{6(B_1 + B_2)} = \frac{6(2)}{6(6.2)} = 0.323 \text{ m}$$

The bending moment on the plane is

$$M = Qa_1 = 419.35(1.3) = 545.155 \text{ kNm} = 0.545155 \text{ MNm}$$

The torsional moment on the section is

$$T = Qe = 419.35(0.323) = 135.45 \text{ kNm}$$

The section should be designed for the equivalent bending moment and it is equal to

$$M_t = \frac{T(1 + D/b)}{1.7}$$

where $b = L$ and the value of D has to be assumed

Let D be assumed in the range of $B/7$

Let $D = 0.6$ m, therefore

$$M_t = \frac{T(1 + 0.6/6)}{1.7} = 0.647T$$

The effective bending moment about the section *AA* is

$$M_e = M + M_t$$

$$= 0.545155 + 0.647(0.13545) = 0.63279$$

The bending moment per unit width is

$$M_A = \frac{M_e}{L} = \frac{0.63279}{6} = 0.10546 \text{ MNm/m}$$

Bending moment about section CC. The centroid of the foundation is at 1.6 m from the column P_2. Its distance from the free right end is

$$c_2 = 1.6 + 1 = 2.6 \text{ m}$$

The width of the section at CC is

$$b_c = B_2 - \frac{(B_2 - B_1)(c_2)}{L}$$

$$= 4.1 - \frac{(2)(2.6)}{6} = 3.233 \text{ m}$$

The bending moment about the section is

$$M = q\left(\frac{B_2 c_2}{2}\left(\frac{2}{3}c_2\right) + b_2 c_2\frac{(c_2)}{3}\right) - P_2 \ (1.6)$$

$$= \frac{53.763}{6}\left((4.1)(2) + 3.233\right)2.6^2 - 600(1.6)$$

$$= -267.469 \text{ kNm} = -0.267469 \text{ MNm}$$

The bending moment per unit width of the slab is

$$M_c = \frac{M}{b_c} = -\frac{0.267469}{3.233} = -0.082731 \text{ MNm/m}$$

The bending moment is negative on this plane and its magnitude is less than that on the plane BB. Therefore, the depth of the section is governed by moment on the section BB. The working moment capacity of the section is

$$M_r = Rbd^2 = 0.91(1)d^2 = 0.91d^2 \text{ MNm/m}$$

Equating the above to the effective moment M_B, we have

$$M_r = M_A$$

or

$$d = \sqrt{\frac{0.10546}{0.91}} = 0.34 \text{ m}$$

It has been observed that normally the shear force governs the depth of the section, we can assume the effective depth as:

Try $d = 0.5$ m

Design for shear force

   The transverse shear force. The transverse shear force is critical on a section at a distance d from the line B and this shear force is

$$V = \frac{qL}{2}\left(\frac{B_2 - a}{2} - d + \frac{B_2 + a}{2} - d\right)$$

$$= 53.763(3)(1.8 - 0.5 + 0.8 - 0.5) = 258.062 \text{ kN}$$

Let the total depth of the section at section BB be 0.56 m and that at the free edge be 0.15 m. Then the total depth at a distance d from the section BB is

$$d_r = d - \frac{(0.5 - 0.09)(d)}{1.8}$$

$$= 0.5 - \frac{(0.41)(0.5)}{1.8} = 0.386 \text{ m}$$

The torsion on this plane can be prorated with the mean cantilever span and is given by

$$T_s = T\left(\frac{(B_1 + B_2 + 2a)}{2} - d\right)\Big/\left(\frac{B_1 + B_2 - 2a}{2}\right) = 109.45 \text{ kNm}$$

The equivalent shear force is

$$V_c = V + 1.6\frac{T_s}{b} = 0.258062 + 1.6(0.10945)/6 = 0.28725$$

The nominal shear stress on the critical plane is

$$\tau_v = \frac{V_c}{bd_s} = \frac{0.28725}{06(0.386)} = 0.12 \text{ MPa}$$

This value is far less than the allowable value whch is 0.22 MPa for 0.25% reinforcement. It appears that the effective depth can be left at 0.40 as required for the moment.

Check for diagonal tension. The diagonal tension or punching type of failure takes place around the column P_2 at a distance 0.5d from the column face. The size of the square of critical shear force is

$$b_s = a + d = 1 \text{ m}$$

The net punching shear force at column 2 is

$$V = P_2 - q\frac{(B_2 + b_c)c_2}{2}$$

$$= 600 - 53.763(4.1 + 3.233)(1.3) = 87.483 \text{ kN}$$

The nominal shear stress on the plane is

$$\tau_v = \frac{V}{4b_sd} = \frac{0.087483}{4(1)(0.5)} = 0.04374 \text{ MPa}$$

Since the shear stresses in both the cases are far below the allowable values, the depth need not be increased.

Use $\qquad$ $d = 0.4$ m.

The reinforcement required widthwise is

$$A_{st} = \frac{M_B}{jd\sigma_{st}} = \frac{0.10546(10^6)}{0.904(0.4)(230)} = 1290 \text{ mm}^2/\text{m}$$

(average 20 mm bars at 240 mm spacing are needed)

$$\frac{100A_{st}}{bd} = + \frac{31400(1000)}{240(1000)(400)} = 0.32$$

This value is more than the minimum. As a safety precaution and as the bending moment near the column two is more than that near column one.

Provide 20 mm bars at 200 mm spacing near column 2 and then increase the spacing to 280 near the column one.

Total number of bars are $N = \dfrac{6000}{240} = 25$

Reinforcement in the longitudinal direction. The magnitude of negative bending moment at the section CC was computed and it is

$$M = 0.267\ 469\ \text{MNm}$$

It can be seen from Fig. 13.19d that the cross section of the beam is trapezoidal and the moment resistance of such a section is

$$M_r = Rb_1d^2 + R_2(b_2 - b_1)\ d^2$$

where

$$R = 0.91\ \text{MPa}$$

$$R_1 = \frac{n^2(2-n)}{12}\ \sigma_{cb} = 0.080$$

$$b_1 = 3.233\ \text{m},\ b_2 = 0.5\ \text{m}$$

The effective depth of the section can be taken equal to that for positive BM assuming almost equal covers at the top and bottom of the slab

$$d = 0.4\ \text{m}$$

Then

$$M_r = 0.91\ (3.233)(0.16) + 0.080\ (-2.733)(0.16) = 0.435\ \text{MNm}$$

The balanced moment resistance of the section is higher than the bending moment acting on the section. Therefore, the size of the section is adequate. The reinforcement area is given by

$$A_{st} = \frac{M}{jd\sigma_{st}}$$

Use $jd = 0.9\ d$ in this case as it is similar to a T-beam

$$A_{st} = \frac{267\ 469}{0.904(0.4)(230)} = 3230\ \text{mm}^2$$

Provide 7 numbers of $\phi 25$ at top in the longitudinal direction and crank three bars as shown in Figure 13.19a. Also provide a nominal reinforcement at the bottom in the longitudinal direction. 12 mm bars at 350 mm spacing.

$$A_{st}\ (\text{provided}) = 7(490) \qquad = 3430\ \text{mm}^2$$

The overall depth $= 0.4 + 0.01 + 0.05 = 0.46\ \text{m}$

Shear force on the interior of the span at distance d from the column will also be critical and this should be checked with shear capacity.

The width of the footing at distance d from the face of the column 2 is

$$b = B_2 - \frac{(B_2 - B_1)\ (1 + d)}{2} = 4.1 - \frac{1.4}{3} = 3.633\ \text{m}$$

The shear force on the plane is

$$V = P_2 - q\ \frac{(B_2 + b)(1 + d)}{2}$$

$$= 600 - 53.563\ \frac{(7.73)(1.4)}{2} = 310\ \text{kN}$$

Area of cross-section of the slab at the point is

$$A_c = d\ (a + b)/2 = 0.4\ (0.5 + 3.633)(0.5) = 0.8266\ \text{m}^2$$

The nominal shear stress on the plane is

$$\tau_v = \frac{V}{A_c} = \frac{0.310}{0.4(0.5 + 3.633)(0.5)} = 0.375 \text{ MPa}$$

$$\frac{100 A_{st}}{A_c} = \frac{343000}{400(4133)(0.5)} = 0.4$$

Allowable shear stress is

$$\tau_c = 0.26 \text{ MPa}$$

Since the shear stress is more than the allowable value, either the depth can be revised or a shear reinforcement be provided. If shear reinforcement is provided as stirrups in the middle 0.5 m width, only the shear capacity of that width is to be taken. In the present case it is economical to provide shear reinforcement. Considering the middle 0.5 m width as a beam, the percentage of reinforcement is

$$\frac{100 A_{st}}{b_w d} = \frac{343\,000}{(500)(400)} = 1.7$$

where $b_w = a = 0.5$ m

And the allowable shear stress for the above percentage of reinforcement is

$$\tau_c = 0.44 \text{ MPa}$$

The shear capacity of the beam portion is

$$V_a = \tau_c b_w d = 0.44(0.5)(0.4) = 0.088 \text{ MN}$$

Select 4-legged 12 mm bars as stirrups, then area of the stirrup reinforcement is

$$A_{sv} = 4\,(113) = 452 \text{ mm}^2$$

The spacing of the stirrups is

$$s_v = \frac{A_{sv}\sigma_{sv}d}{V - V_a} = \frac{452\,(230)\,(40)}{310000 - 88000} = 187 \text{ mm}$$

Maximum spacing of the stirrups is

$$s_{vm} = \frac{A_{sy}f_v}{(0.4)b} = \frac{456(415)}{0.4(500)} = 946 \text{ mm}$$

Provide 4-legged $\phi12$ stirrups at 185 mm spacing near the column face and increase the spacing to 300 mm after 1 m from the face of the second column.

EXAMPLE 13.11 *Design of a combined footing with beam-slab construction.* Two columns spaced at 4 m are subjected to 400 and 600 kN superimposed loads respectively. The size of each column is 500 mm square. The total length of the foundations is to be restricted to 4.5 m, that is outer to outer face of the columns. The net safe bearing capacity of the soil is 60 kN/m².

Spacing of the columns $c = 4$ m

Size of the columns $a = 0.5$ m

Loads on the columns are

$$P_1 = 400 \text{ kN}, \ P_2 = 600 \text{ kN}$$

$$P_s = P_1 + P_2 = 1000 \text{ kN} = 1 \text{ MN}$$

Net safe bearing capacity of the soil is $p_a = 60 \text{ kN/m}^2$

Use $\qquad f_{ck} = 20 \text{ MPa}, f_y = 415 \text{ MPa}$

Design of foundation. Let the difference between the weight of the foundation concrete and the equivalent soil be

$$W'_g = 0.1 \, P_s = 100 \text{ kN}$$

The centroid of the loads of the columns from the centre of the first column is

$$c_1 = \frac{P_2(c)}{P_s} = \frac{600(4)}{1000} = 2.4 \text{ m}$$

Since the length of the foundation is restricted to 4.5 m, the foundation is desgined as a beam slab construction. The two columns are supported on a beam which in turn is supported by a slab. The centroid of the foundation is at middle of the columns whereas the centroid of the column loads is at 2.4 m from column 1. This causes an eccentricity of the total load and that is

$$e = c_1 - \frac{c}{2} = 2.4 - 2 = 0.4 \text{ m}$$

The bending moment on the foundation soil is

$$M_f = P_s e = 1 \, (0.4) = 0.4 \text{ MNm}$$

The foundation width must selected so as to withstand the axial force and the moment. The maximum pressure on the soil is

$$p_2 = \frac{P}{A_f} + \frac{M_f}{Z_f}$$

where $\qquad A_f = LB = 4.5 \, B \text{ m}^2$

$$Z_f = \frac{L^2 B}{6} = 3.375 \, B \text{ m}^3$$

The maximum pressure should be restricted to the allowable pressure and it is given by

$$p_2 = \frac{P}{A_f} + \frac{M_f}{Z_f} \leq p_a$$

$$= \frac{1100}{4.5B} + \frac{400}{3.375 \, B} \leq 60 \text{ kN/m}^2$$

or $\qquad B = 6.05 \text{ m}$

$W'_g =$ assumed W as large so that the width can be rounded
to the lower side.

Let $\qquad B = 6 \text{ m}$

The minimum pressure on the soil is near the column 1 end and it is

$$p_1 = \frac{P}{A_f} - \frac{M_f}{Z_f}$$

$$= \frac{1100}{4.5(6)} - \frac{400(6)}{(4.5)(4.5)(6)}$$

$$= 40.75 - 19.75 = 21 \text{ kN/m}^2$$

Therefore, there is no tension on the foundation slab.

Design of the foundation slab. The foundation is designed as a beam slab construction. Let the width of the beam be equal to that of the column, then the cantilever span for BM is

$$L_m = \frac{B}{2} - \frac{a}{2} = 3 - 0.25 = 2.75 \text{ m}$$

The soil pressure is varying with maximum near the column 2 so the base slab be designed under the column 2 and the same can be adopted throughout for convenience of construction. The net soil reactions at 2.25 m and 1.25 m (that is, 1 m width under column 2) are:

$$q_2 = \frac{P_s}{A_f} + \frac{M_f y}{I_f}$$

$$= \frac{1100}{27} + \frac{400(2.25)}{45.5625} = 56.8 \text{ kN/m}^2$$

$$q_3 = \frac{P_s}{A_f} + \frac{M_f}{I}(1.25) = 48 \text{ kN/m}^2$$

The average intensity of the pressure is

$$q = q_3 + q_2 = (48 + 56.8)/2 = 52.4 \text{ kN/m}^2$$

(see Figs. 13.20c and d.)

The design bending moment on the slab is

$$M = \frac{q L_m^2}{2} = \frac{52.4(2.75)^2}{2} = 198.1375 \text{ kNm/m}$$

The resisting moment of a balanced section is

$$M_r = Rbd^2 = 0.91 d^2 \text{ MNm/m}$$

Equating the bending moment to the moment resistance, we have

$$0.91 \, d^2 = M = 0.1981375 \text{ MNm/m}$$

or
$$d = 0.48 \text{ m}$$

Use
$$d = 0.56 \text{ m}$$

The critical shear force occurs at a distance d from the face of the column and the shear span is

$$L_s = L_m - d = 2.75 - 0.56 = 2.19 \text{ m}$$

The shear force on the section is

$$V = q L_s = 52.4(2.19) = 114.756 \text{ kN}$$

The nominal shear stress on the section is

$$\tau_v = \frac{V}{bd} = \frac{0.114\,756}{1(0.56)} = 0.2 \text{ MPa}$$

The allowable shear stress for minimum percentage of the reinforcement is 0.24 MPa, therefore, on need to revise the depth. The area of the tension reinforcement is

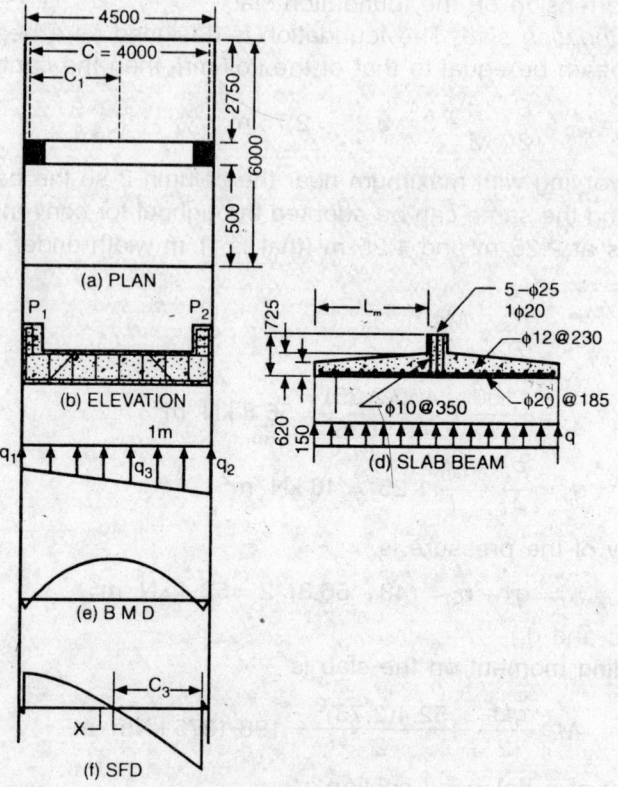

Fig. 13.20 Structural Details of Combined Footing with Beam and Slab Construction.

$$A_{st} = \frac{M}{jd\sigma_{st}} = \frac{198137.5}{0.904(0.56)(230)} = 1702 \text{ mm}^2$$

Provide 20 mm bars at 185 mm spacing at the bottom of the slab.

Then the percentage of the reinforcement is

$$p_t = \frac{100\, A_{st}}{bd} = \frac{169800}{1000(560)} = 0.3\%$$

The overall thickness of the foundation slab is

$$t = d + 0.06 = 0.62 \text{ m at support}$$

Let the thickness of the slab near the free edge be reduced to 150 mm. This will reduce the effective depth at the critical shear zone and consequently increase the nominal shear stress. Even then the increased shear stress will be less than the allowable value.

Provide a minimum of 0.12% distribution reinforcement. The area of the concrete cross-section on one side of the beam is

$$A_c = \frac{(0.15 + 0.62)(2.75)}{2} = 1.05875 \text{ m}^2$$

Minimum reinforcement is

$$A_s = 0.0012 \, (A_c)$$

$$= 0.0012 \, (1.05875)(10^6) = 1270 \text{ mm}^2$$

Provide 12 numbers of $\phi 12$ bars in the 2.75 m width on either side of the beam ($\phi 12$ @ 230 mm).

Design of beam

$$\text{Total length of the foundation} = L_0 = 4.5 \text{ m}$$

$$\text{The effective span of the beam} = L = 4\text{m}$$

The bending moment and shear force distributions are shown in Figs. 13.20e and f. The intensity of the load on the beam is same as the soil pressure on the slab and its value at a distance x from the centre of the beam is

$$w = \frac{P_s B}{A_f} + \frac{M_f B x}{I_f} = \frac{1(6)}{27} + \frac{0.4(6)x}{45.5625}$$

$$= 0.222 + 0.0527 \, x \text{ MN/m}$$

Let c_3 be the distance from the free end of column 2 where the SF is zero. Then setting shear force at distance x, where x is less than 2 m is

$$V = \int_{x_0}^{0.5 \, L_0} wdx - P_2$$

$$= [(0.222 \, x + 0.02635 \, x^2)]_{x_0}^{0.5 L_0} - 0.6 \text{ MN}$$

Let $x_1 = x_0$ where the shear force is zero, then from the above equation, we have

$$(0.222 \, x_0 + 0.2635 \, x_0^2) - (0.222(2.25)$$

$$+ 0.02635(2.25)^2) + 0.6 = 0$$

or

$$0.02635 \, x_0^2 + 0.222 \, x_0 - 0.0333 = 0$$

or

$$x_0^2 + 8.425 \, x_0 - 1.266 = 0$$

or

$$x_0 = 0.148 \text{ m}$$

Then

$$c_3 = 2 - x_0 = 1.852 \text{ m}$$

The bending moment at the critical section is

$$M = \int_{x_0}^{0.5 L_0} wx \, dx - P_2 \, c_3$$

$$= \left[\left(\frac{0.222 \, x^2}{2} + \frac{0.0527 \, x^3}{3} \right) \right]_{0.148}^{2.25} - 0.6 \, (1.852)$$

The negative sign for the bending moment indicates that the tension occurs on the top fibre. The beam is an inverted T-beam so there is an additional advantage. Let the thickness of the slab at the free end be 0.15 m, then the average thickness of the slab which act as a flange is

$$t = \frac{0.62 + 0.15}{2} = 0.385 \text{ m}$$

The effective flange width would be

$$b_f = \frac{L}{3} = 1.333 \text{ m}$$

Since the thickness of the flange is large for the span, one can except the neutral axis to fall in the flange itself. The moment resistance of the beam would be same as that of a rectangular section with width equal to the effective flange width

$$b = b_f$$

Equating the moment resistance to the external moment gives

$$Rbd^2 = 0.35163 \text{ MNm}$$

or

$$d = \sqrt{(0.35163/0.91(1.333))} = 0.54 \text{ m}$$

Let

$$d = 0.65 \text{ m}$$

The web thickness $b_w = 0.5$ m same as that of the column size. The area of the tension reinforcement is

$$A_{st} = \frac{M}{jd\sigma_{st}} = \frac{351\,630}{0.904(0.65)(230)} = 2602 \text{ mm}^2$$

Use 5 nos. of $\phi25$ and 1 no. of $\phi20$ bars

$$A_{st} \text{ (provided)} = 2764 \text{ mm}^2$$

All the six bars can be arranged in one row within 500 mm web. The overall depth of the beam is

$$D = d + 0.013 + 0.05$$

Use

$$D = 0.725 \text{ m}$$

Design for shear. The critical section for shear force would be at a distance d from the face of the column. This section from the mid-point is

$$x_s = \frac{L_0}{2} - \frac{a}{2} - d$$

$$= 2 - 0.25 - 0.65 = 1.1 \text{ m}$$

The shear force at this section can be computed from the previously derived value and is

$$V = -\left[0.222x + 0.02635x^2 \right]_{1.1}^{2.25} + 0.6$$

$$= 0.2432 \text{ MN}$$

Nominal shear stress is

$$\tau_v = \frac{V}{b_w d} = \frac{0.2432}{0.5(0.65)} = 0.748 \text{ MPa}$$

The percentage of reinforcement assuming all the bars are continued right through is

$$p = \frac{100\,A_{st}}{b_w d} = \frac{276\,400}{500(650)} = 0.85\%$$

If two bars are cranked, then the percentage of reinforcement will be about 0.6 and the allowable shear stress is

$$\tau_c = 0.31 \text{ MPa and } \tau_{cmax} = 1.6 \text{ MPa}$$

Let the transverse reinforcement be designed to resist the excess shear force. The shear resistance of the concrete is

$$V_c = b_w d \tau_c = 0.5(0.65)(0.31) = 0.1105 \text{ MN}$$

Net shear force to be resisted by the stirrups is

$$V_s = 0.2432 - 0.1105 = 0.1327 \text{ MN}$$

Select four-legged 10 mm bars for stirrups, then the area of the stirrups reinforcement is

$$A_{sv} = 4(78) = 312 \text{ mm}^2$$

The spacing of the stirrups is

$$s_v = \frac{A_{sv}\,\sigma_{sv}\,d}{V_s}$$

$$= \frac{312(230)(65)}{132700} = 351 \text{ mm}$$

Maximum spacing of the stirrups is

$$s_{vm} = \frac{A_{sv} f_y}{0.4 b_w} = \frac{312(415)}{0.4(500)} = 647 \text{ mm, or}$$

or

$$= 0.75 d = 488 \text{ mm}$$

Provide four-legged 10 mm bar stirrups at 350 mm spacing.

13.8 Design of Raft Foundations

A raft foundation is basically a shallow foundation in which the load variation on the foundation is a function of 2 directions. It is a plate type of structure spread over a large area, supporting a number of columns or the entire superstructure as a single unit. The bending moments on the raft and the soil pressure distribution are functions of the two directions. A raft foundation is also called as *mat* foundation or *spread* foundation. Such foundations are used when the columns of a structure are closely spaced, the loads on each column are large and in the foundations of multistorey buildings, overhead water takns, chimneys, etc. A raft foundation might become unavoidable in some structures such as submerged conditions or in some multistorey structures where basement is to be provided and soil bearing capacity is inadequate. The mat or raft foundation is designed as a flat slab. The reader is advised to read Chapter 4 on flat slabs. Section 13.4 which discusses the design of the circular footings has already dealt the design of the raft foundation. The method of design is illustrated with examples.

EXAMPLE 13.12 *Design of rectangular raft foundation.* A large building is provided with columns spaced at 6 m apart in two perpendicular directions and the load from each column on the foundation is 2880 kN. The size of each of the columns is 500 mm square. The soil is silty clay, with safe net bearing capacity of 90 kN/m². Design the foundation using M25 concrete and HYSD-Fe415 bars.

Spacing of the column $\quad L_1 = L_2 = 6$ m
Size of the column $\quad\quad a = 0.5$ m
Height of the column above foundation $= 5$ m
Load from each column $\quad W_s = 2880$ kN
$$f_{ck} = 25 \text{ MPa}, f_y = 415 \text{ MPa}$$

The working stress design coefficients are:

$$n = 0.289, j = 0.904 \text{ and } K = 0.13.$$

Net bearing capacity of the soil $p_a = 90$ kN/m^2.

Design of the foundation. Let the average thickness of the raft be assumed as 0.6 m for the purpose of calculating the self-weight of the slab. The difference in the weight of the concrete slab and the soil can be assumed as 10 kN/m^2. The gross load on the foundation per one panel size consists of the load from one column, plus the weight of the slab plus the weight of the soil overburden. Since the net bearing capacity is given, only the net load on the soil need to be computed for the purpose of bearing pressure. Assume a mat foundation.

Bearing area per panel $\quad A_f = 6(6) = 36$ m^2
Load from column $\quad\quad P_s = 2880$ kN

Difference in weight of the slab and the soil is

$$6(6)(0.6)(10) = 218 \text{ kN}$$

Total net load per panel $\quad = \overline{3098 \text{ kN}}$

The net bearing pressure on the soil is

$$p = \frac{3098}{36} = 86.06 \text{ kN/m}^2$$

The bearing pressure is about 96% of the net safe bearing capacity of the soil, therefore the mat foundation is just the right type of foundation. The depth of foundation can be 1 m below the ground level.

Design of the section. The mat slab acts as an inverted flat slab with 6 m square panels. Let the load be transferred through pedestals which are considered similar to drops in flat slabs. Such pedestals minimize the failure of the slab by diagonal tension or punching shear. The size of the pedestal can be selected in the range of $L/4$ with a depth in the range of $L/8$.

Let the pedestal (drop) size $\quad D_e = 1.5$ m
Clear spacing between the drops $\quad = L_0 = L - D_e = 4.5$ m

Active load intensity $\quad = \dfrac{W_g}{A_f} = w_t = \dfrac{2880}{36} = 80$ kN/m^2

Total load on the clear panel is $\quad W = w_t L L_0 = 80(6)(4.5) = 2160$ kN

The sum of the magnitudes of the positive and negative BM in a panel is

$$M_0 = \frac{WL_0}{8} = \frac{2160(4.5)}{8} = 1215 \text{ kNm}$$

As the mat foundation slab is an inverted floor slab with the load intensity is the reaction of the soil acting upward. The *positive* bending moment is the bending moment which causes compression on the face of the slab on which the active load is acting. The active load is

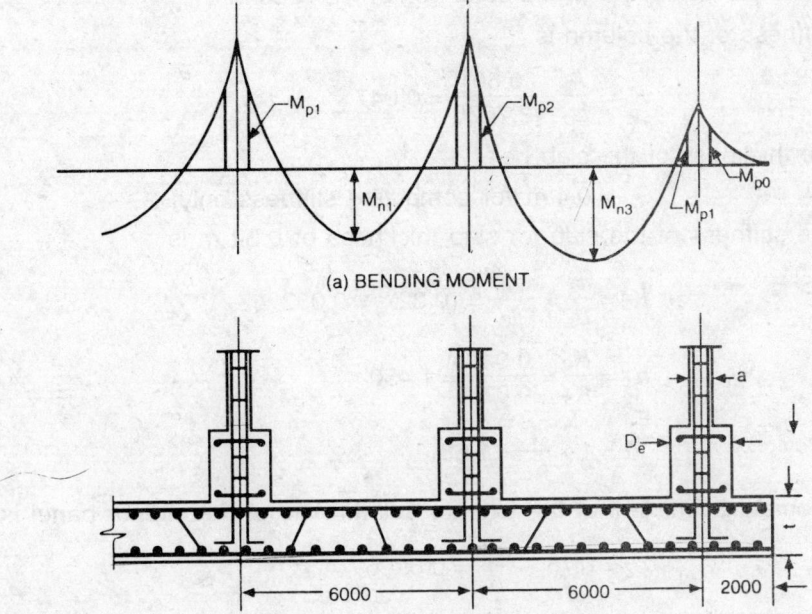

(a) BENDING MOMENT

Fig. 13.21 Structural Details of Mat Foundations.

from the columns and reaction is from the soil. Hence the bending moment which causes compression on the top surface is considered positive. With this notation, the positive bending moment is occurring along the face of the drops and the negative BM is at the middle span. The positive and negative is exactly reverse of the notation that is normally followed in the flat slab design. What are the negative BM coefficients in the flat slab, should be considered as positive moment coefficients in the mat slab.

Let the mat slab project as a cantilever beyond the face of the exterior column pedestals. This overhang produces a positive bending moment at the outer columns (see Fig. 13.21 for notations). The projection of the slab beyond the centre line of the outer column line = 2 m.

The cantilever span beyond the pedestals is

$$L_m - 2 - 0.5 \, D_e = 2 - 0.75 = 1.25 \text{ m}$$

The positive BM at the face of the pedestals is

$$M_{p0} = \frac{w_t LL_m^2}{2} = \frac{80(6)(1.25)^2}{2} = 375 \text{ kNm}$$

The bending moment coefficients of the flat slab are associated with the relative stiffness of the slab and the columns; so these relative stiffnesses are computed first.

Let the depth of the drop   = 0.75 m

Clear height of the column = (5 − 0.75) m

The column can be assumed as fixed at both ends as it is connected to flat slabs with drops and each floor is not allowed to sway by means of infilled walls.

674

Effective height of the column is $L_e = 0.65(5 - 0.75) = 2.7625$ m

The relative stiffness of the column is

$$K_c = \frac{l_c}{L_c} = \frac{0.60^4}{2.7625} = 0.047$$

Let the average thickness of the slab be

$$= 0.4 \text{ m (for computing stiffness only)}$$

then the relative stiffness of the slab for slab thickness of 0.32 m is

$$K_s = \frac{l_s}{L_s} = \frac{L_0 t^3}{L_0} = (0.32)^3 = 0.033$$

$$\alpha_c = \frac{K_c}{K_s} = \frac{0.047}{0.033} = 1.450$$

$$\alpha = 1 + \frac{1}{\alpha_c} = 1.68$$

The bending moment coefficient at the exterior column line of the interior panel is

$$c_2 = 0.75 - \frac{0.1}{\alpha} = 0.6906$$

The bending moment at the inside face of the exterior pedestals is

$$M_{p1} = \frac{w_t L (L_m + D_e)^2}{2} - \frac{R_0 \, D_e}{2}$$

where
$$R_0 = \text{reaction of the external column}$$
$$= w_t L(0.5L + \text{cantilever span})$$

$$M_{p1} = w_t L \left(\frac{3.25^2}{2} - 5 \right) = 130 \text{ kNm}$$

The bending moments at the face of the pedestals and at mid-span of any of the internal panels are

$$M_{pi} = 0.65 \, M_0 = 0.7898 \text{ MNm}$$

$$M_{ni} = 0.35 \, M_0 = 0.4252 \text{ MNm}$$

The positive and negative bending moments in the exterior panel are (see Fig. 13.21):

$$M_{p2} = c_2 M_0 = 0.6906 \, (1.215) = 0.8391 \text{ MNm}$$

$$M_{n2} = M_0 - 0.5 \, (M_{n1} + M_{n2}) = 0.7551 \text{ MNm.}$$

The bending moment on the column and middle strips are calculated using the distribution coefficients suggested by the code. These coefficients are:

For column strip

Moment coefficient at support $= 0.75$
Moment coefficient at mid-span $= 0.60$

The depth of the slab is controlled by the maximum bending moment which occurs at the pedestal face of the exterior columns of the outer panel. This will also be governed by the shear force which will tend to punch thorugh the slab around the pedestal. The maximum bending moment carried by the column strip is

$$M_{pc2} = 0.75\ M_{p2} = 0.75\ (0.839) = 0.6293\ \text{MNm}$$

width of the column strip $b = 0.5\ L = 3$ m

The effective depth of the slab is given by

$$d = \sqrt{\frac{M_{pc2}}{Kb\sigma_{cb}}} = \sqrt{\frac{0.6293}{0.13(3)(8.5)}} = 0.435\ \text{m}$$

The minimum cover recommended for a mat foundation is 75 mm, so the overall depth of the slab can be taken as

$$t = d + 0.020 + 0.075 = 0.53\ \text{m}$$

The self-weight of the slab is

$$w_g = 0.53\ (25)$$

The net load on the soil due to self-weight of the slab is

$$w_g = \text{weight of equal volume of soil}$$

$$= 0.53(10) = 5.3\ \text{kN/m}^2$$

Add extra 1.7 kN/m^2 for weight difference of pedestal, etc. even then the net pressure on the soil is $80 + 1.7 + 5.3 = 87$ kN/m^2. This is less than the safe net bearing capacity of the soil.

Design for shear. The allowable shear stress in the diagonal tension is

$$\tau_c = 0.16\ \sqrt{f_{ck}} = 0.16\ \sqrt{25} = 0.8\ \text{MPa}$$

The critical shear plane is the peripheral plane which is at a distance of 0.5 d from the face of the pedestal. The total length of this shear plane is

$$b_0 = 4(D_e + d) = 4\ (1.5 + 0.435) = 7.74\ \text{m}$$

The total shear force on this plane is

$$V = w_t\ (L_1 L_2 - (D_e + d)^2)$$

$$= 80(36 - 1.935^2) = 2580.5\ \text{kN}$$

The nominal shear stress on the concrete is

$$\tau_v = \frac{V}{b_0 d} = \frac{2.5805}{7.74(0.435)} = 0.78\ \text{MPa} < 0.8$$

The nominal shear stress is less than the allowable shear stress, so the depth selected is safe against shear.

Design of reinforcement. The design bending moments on the column and middle strips are to be obtained from the total bending moments computed at the critical section after multiplying the appropriate coefficients (0.75 or 0.60 for the column strip at support and midspans.) Then the area of the reinforcement required is computed from

$$A_{si} = \frac{M_i}{jd\sigma_{st}}$$

The bending moments and the corresponding reinforcement required at different sections are computed in Table 13.6.

However, the minimum percentage of reinforcement of 0.12 is to be maintained in the

Table 13.6 Critical Bending Moments and the Reinforcements

| S.N. | Section (Type of BM) | Bending moments (MNm) | | | Reinforcement | | | |
| | | Panel Strip | | Needed (mm^2) | | Provided | |
		Column	Middle	Column	Middle	Column	Middle
	Exterior panels						
1	0 (+ ve)	0.3694	0.3694	0	4085	0	13,ϕ20	6,ϕ20
2	1 (+ve)	0.1300	0.1300	0	1437	0	6,ϕ20	6,ϕ20
3	3 (−ve)	0.7551	0.4531	0.3019	5010	3339	16,ϕ20	11,ϕ20
4	2 (+ve)	0.8391	0.6293	0.2098	6958	2320	16,ϕ20 and 4,ϕ25	6,ϕ20 and 1,ϕ25
	Interior panels						
5	i (+ve)	0.7898	0.5444	0.2454	6019	2712	16,ϕ20 and 2,ϕ25	6,ϕ20 and 2,ϕ25
6	i (−ve)	0.4252	0.2551	0.1701	2820	1880	2,ϕ20	6,ϕ20

slab in each direction, and the minimum reinforcement is

$$A_{sm} = \frac{0.12}{100}(530)(3000) = 1908 \text{ mm}^2/3 \text{ m}$$

The minimum reinforcement to be provided is 6 numbers of 20 mm bars in each column or middle strips.

The anchorage length of the bars for overlap or cut off points is

$$L_d = \frac{\phi \sigma_{st}}{4 \tau_{bd}} = \frac{\phi (230)}{4(1.4)} = 41 \phi$$

PROBLEMS

Use IS 456 and sketch reinforcement details in each of the following problems:

13.1 A masonry wall of 400 mm thickness is transmitting a total load of 20 kN/m to a soil having an allowable bearing capacity of 100 kN/m^2. Design a RCC strip foundation using M20 concrete and high yield steel bars.

13.2 A reinforcement concrete wall of 300 mm thickness is subjected to an axial load of 400 kN/m and bending moment of 120 kNm/m. Design a RCC foundation using M20 concrete and HYSD-Fe415 bars. The allowable bearing capacity of the soil is 120 kN/m^2.

13.3 A RCC wall of 300 mm thickness transmits an axial load of 400 kN/m to a soil having an allowable bearing capacity of 120 kN/m^2. The property line is such that it would not permit projection of the foundation beyond 1.2 m on one side of the central line of the wall. Design a RCC foundation using M20 concrete and HYSD-Fe415 bars.

13.4 A column of size 500 mm by 500 mm is transmitting a load of 800 kN to a soil having an allowable bearing capacity of 120 kN/m^2. Design a square RCC footing using M20 concrete and mild steel reinforcement. Assume an uniform thick footing.

13.5 A column 500 mm by 500 mm is transmitting an axial load of 800 kN to a soil having an allowable bearing capacity of 120 kN/m^2. Design a RCC square footing using M20 concrete and HYSD-Fe415 bars. The thickness of the footing at the free end should be taken as 150 mm.

13.6 A square column of 300 mm is transmitting a load of 500 kN to a soil having an allowable bearing capacity of 90 kN/m^2. The width of the foundation cannot exceed 1.8 m. Design an isolated footing made of M20 concrete and mild steel reinforcement.

13.7 A square column of 500 mm size is transmitting an axial load of 800 kN and bending moment of 200 kNm to a soil having an allowable bearing capacity of 100 kN/m^2. Design a RCC square footing using M20 concrete and HYSD-Fe415 bars.

13.8 A square column of 500 mm size is transmitting an axial load of 800 kN and bending moment of 200 kNm to a soil having an allowable bearing capacity of 100 kN/m^2. The bending moment and axial load are generated in a seismic load condition. Design a square RCC footing using M20 concrete and HYSD-Fe415 bars.

13.9 A column of 500 mm diameter is transmitting a load of 900 kN to a soil having an allowable bearing capacity of 90 kN/m^2. Design an isolated circular footing using M20 concrete and HYSD-Fe415 bars. The bending moment coefficients for the problem can be taken from the text of this chapter.

13.10 A circular cylindrical shaft of radius 5.5 m with its thickness of 400 mm is transmitting an axial load of 20,000 kN to a soil having an allowable bearing capacity of 90 kN/m^2. Design a RCC circular raft foundation using M20 concrete and HYSD-Fe415 bars. The bending moment coefficients for the circular raft can be taken from the text of this chapter.

13.11 Two square columns each transmitting 500 kN are spaced at 3 m apart and size of each column is 350 mm. The allowable bearing capacity of the soil is 90 kN/m^2. Design a combined RCC footing using M20 concrete and HYSD-Fe415 bars with its width not exceeding 2 metres.

13.12 Two columns transmitting 600 and 800 kN are spaced at 4 metres apart on a soil having allowable bearing capacity of 70 kN/m^2. Design a combined footing having a width of 3 m using M20 concrete and HYSD-Fe415 bars. The size of each column can be taken as 200 mm square.

13.13 Two columns spaced at 5 m apart are transmitting 600 kN each to a soil having an allowable bearing capacity of 60 kN/m^2. Design a combined beam and slab footing with a width not exceeding 2.6 m using M25 concrete and HYSD-Fe415 bars. The size of each of the columns is 500 mm and square width of the beam should also be taken as 500 mm. The footing is made of M20 concrete and HYSD-Fe415 bars.

13.14 A multistorey building having its columns spaced at 5 m apart in two perpendicular directions is resting on a soil having an allowable bearing capacity of each 100 kN/m^2 and transmits a load of 2000 kN from each column. The size of the column is 500 mm. Design a raft foundation using M20 concrete and HYSD-Fe415 bars.

13.15 A multistorey building having its column spaced at 5 m apart in two perpendicular directions is resting on a soil having an allowable bearing capacity of 100 kN/m^2. Each column of 500 mm square is transmitting an axial load of 1500 kN and a bending moment of 200 kNm. The building has 4 rows of columns and 5 columns in each row. The bending moment in each column is created by a seismic load and it can be in any direction. Design a raft foundation using M20 concrete and HYSD-Fe415 bars.

Chapter 14

Reinforced Concrete Retaining Walls

14.1 Introduction and Classification

Soil in embankment or in cutting has a tendency to slide down and repose in a particular inclination. When such an embankment or cutting is to be kept in vertical position, there should be a supporting structure to keep the soil from falling into an inclined repose formation. Different types of structures and structural forms are used for this purpose. When the height of the soil to be protected from falling is in the order of 3 m to 10 m, a reinforced concrete construction is recommended. Such a construction is defined as a reinforced concrete retaining wall.

A broad classification of retaining walls is : (i) gravity, (ii) cantilever, (iii) counterfort, (iv) buttressed and (v) guyed.

A gravity retaining wall holds the earth by the mass of the structure. The horizontal thrust of the earth has a tendency to overturn the structure. This overturning moment is resisted by the weight of the wall which introduces the stabilizing moment. Figure 14.1 illustrates the gravity type of retaining wall in which the earth pressure is indicated by *F* and the weight of the wall as *W*. The main design considerations of the wall are:

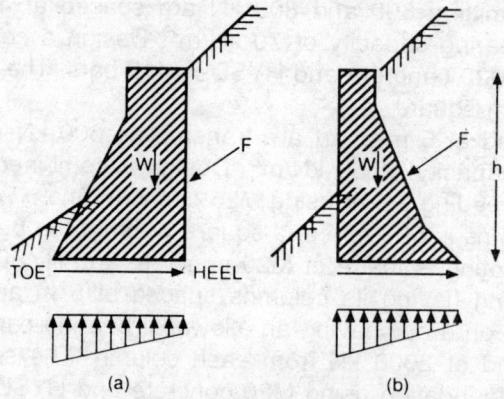

Fig. 14.1 Gravity Type of Retaining Wall.

a) The stabilizing moment due to the self-weight must be more than the overturning moment. There must be a factor of safety of 1.5 to 2 against overturning.
b) The horizontal resistance at the base of the wall must be more than the net horizontal force caused by the earth pressure.
c) The bearing pressure caused on the soil by the active forces must be less than the allowable pressure of the soil.
d) The stresses produced in the masonry must be within the permissible limits.

In most retaining walls, the design criteria (a) and (c) dominate the design.

RCC cantilever type: Figure 14.2 illustrates the typical cantilever type in which there are three structural components of the system. Each of the three elements of the system acts as a cantilever. The stem which retains the earth is supported by the base (or footing) of the wall. The base is further sub-divided into *heel* and the *toe* as shown in the figure. The base of the stem should be designed such that the structure is safe against overturning, sliding and bearing pressure. The cantilever type is usually used upto height of 5 m.

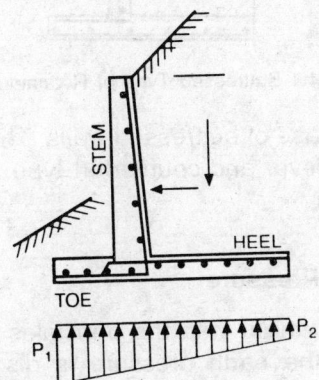

Fig. 14.2 Cantilever Type of Retaining Wall.

RCC counterfort and Buttressed types: As the height of the retaining wall increases, the moment due to the earth pressure increases rapidly and consequently the size of the stem increases if it is cantilevered from the base. Therefore, the stem is further supported by the counterforts as shown in the Fig. 14.3. The support to the stem can be given either from the embankment side or from the other side. In case the support is on the filling (same as embankment) side, then it is called *counterfort*, otherwise *buttressed*. Figure 14.4 illustrates the buttressed type support. The width of the heel is usually large in counterfort retaining

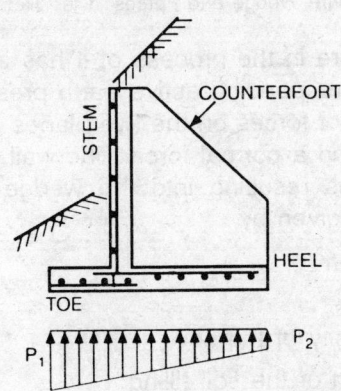

Fig. 14.3 Counterfort Type Retaining Wall.

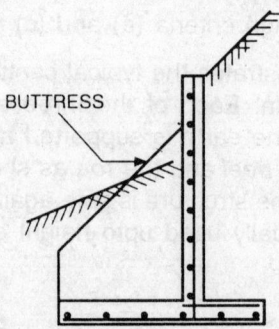

Fig. 14.4 Buttressed Type of Retaining Wall.

walls, whereas the toe is large in case of buttressed walls. The most commonly used reinforced concrete retaining walls are cantilever and counterfort type. The design of these two types is discussed in detail in this chapter.

14.2 Coulomb Theory of Earth Pressure

It is assumed that the reader is familiar with the principles of soil mechanics; therefore only a brief and relevant review on the earth pressure is discussed in this section. Coulomb assumed that a wedge of earth mass has a tendency to slide and this mass of the wedge

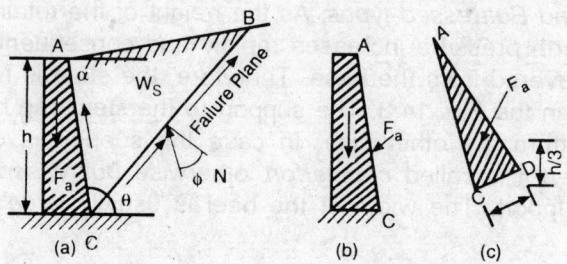

Fig. 14.5 Failure Wedge and Forces in the Retaining Wall.

generates an active earth pressure in the process of it has a tendency to slide. Figure 14.5 illustrates the wedge and the corresponding active earth pressure. The weight of the wedge is held in equilibrium by two sets of forces on the two planes AC and BC as shown. On each plane there is a frictional force and a normal force. The wall is assumed to undergo a small deformation or displacement thus resulting into the wedge action. The total active earth pressure on the retaining wall is given by

$$F_a = c_a \frac{\gamma_a h^2}{2}$$

(14.1)

where γ_s = density of the soil

h = depth of the soil filling

c_a = coefficient of active earth pressure and it is given by

$$c_a = \frac{\sin^2(\theta - \phi)}{\sin^2\theta \sin(\phi + \beta)\left[1 + \sqrt{\left(\dfrac{\sin(\phi + \beta)\sin(\phi - \alpha)}{\sin(\theta + \beta)\sin(\theta - \alpha)}\right)}\right]^2} \qquad (14.2)$$

in which
α = slope of the embankment or backfill

ϕ = angle of internal friction of the soil

β = friction angle between the soil and the wall surface where

$$\tan\beta = \mu \qquad (14.3)$$

μ = coefficient of friction between the soil and the surface of the wall

θ = inclination of the wall face

The ranges of the angles for some soils are given in Table 14.1. In case of retaining walls with upright face, the active pressure coefficient can be obtained from Eq. (14.2) for $\theta = 90°$ and it is

$$c_a = \frac{[\cos\alpha - \sqrt{(\cos^2\alpha - \cos^2\phi)}]\cos\alpha}{\cos\alpha + \sqrt{(\cos^2\alpha - \cos^2\phi)}} \qquad (14.4)$$

In many cases, the slope of the bakcfill will be zero, so substitution of $\alpha = 0$ in Eq. 14.4 gives

$$c_a = \frac{1 - \sin\phi}{1 + \sin\phi} \qquad (14.5)$$

As the retaining wall tends to move, the soil present on the downward side of the wall will resist the movement. Such a resistance of the soil is called passive earth pressure. The passive earth pressure can be expressed as

$$F_p = c_p \frac{\gamma_s h^2}{2} \qquad (14.6)$$

where
$$c_p = \frac{[\cos\alpha + \sqrt{(\cos^2\alpha - \cos^2\phi)}]\cos\alpha}{\cos\alpha - \sqrt{(\cos^2\alpha - \cos^2\phi)}}, \text{ for } \theta = 0 \text{ and} \qquad (14.7)$$

$$= \frac{1 + \sin\phi}{1 - \sin\phi}, \text{ for } \alpha = 0° \qquad (14.8)$$

Table 14.1 gives the approximate characteristic values for some soils. The earth pressure is very sensitive to the angle of the internal friction, therefore it is desirable to investigate for actual value for the soil.

There are situations where the backfill is further loaded by an external load which is called surcharge. The intensity of the surcharge is to be treated uniform over the entire depth as

Table 14.1 Some Earth Pressure Parameters

Soil	γ_s (kN/m^2)	ϕ (mean)	μ	β	c_a	c_p
Sand	17–20	25–35 (30°)	0.55	29°	0.35	2.85
Medium clay	16–18	14–28 (21°)	0.40	22°	0.45	2.22
Soft clay	15–17	4–16 (10°)	0.33	18°	0.70	1.43

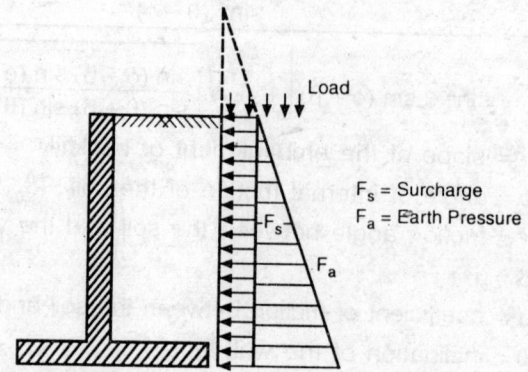

Fig. 14.6 Surcharge and Earth Pressure on Retaining Wall.

shown in Fig. 14.6. The centre of pressure of the surcharge will be at the mid-height.

14.3 Main Forces on the Walls

The cantilever retaining wall acts as a simple cantilever beam, fixed at the base and subjected to earth pressure on one side of the wall. The counterfort retaining wall consists of vertical slab supported by upright cantilever beams. The upright beams are called counterforts. Both types of retaining walls are supported on footings. The footing is divided into two parts, namely toe and heel. The face of the vertical wall in contact with the soil is defined as the rear face of the retaining wall, then the portion of the footing in the rear side of the walls is called the heel and the portion in the front of the wall is called the toe. Because of some practical considerations, the footing may have to be curtailed flush with the front face of the vertical wall. In such a case the footing is called L-type; otherwise it is inverted T-type.

The forces acting on the retaining wall may be grouped as:

- i) Self-weight of the retaining wall.
- ii) Weight of the soil on the heel.
- iii) Vertical component of the soil pressure.
- iv) Horizontal component of the soil pressure.
- v) Surcharge (the force due to loads on the soil).
- vi) Soil reactions on the footing.
- vii) Frictional force on the footing against sliding.
- viii) Guy force if a guy is provided.

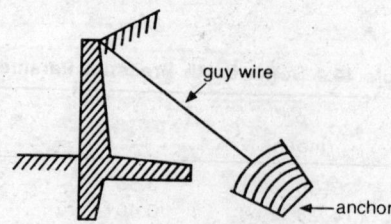

Fig. 14.7 Anchored Type of Retaining Wall.

14.4 Design Criteria

The design of a retaining wall is divided into three parts, namely:

  i)   Stability of the total retaining wall
 ii)   Permissible bearing capacity of the soil
iii)   Structural design of the elements of the retaining wall.

The horizontal soil pressure has the tendency of overturning about the tip of the toe, and sliding forward of the retaining wall. Both these forces are resisted by other forces. The two conditions of the overall equilibrium of the system are called stability conditions. The factor of safety against overturning and sliding should be at least 1.5. Similarly, the pressure caused on the foundation soil by the system of forces must be within the permissible limits.

14.5 Design Criteria for the Footing and Preliminary Size

Selection of the width of the footing (B) and the relative widths of the toe and heel can be based on the criteria of the stability and safe bearing capacity of the soil.

Let

W_f = weight of the footing

W_w = weight of the vertical wall

W_s = weight of the soil filling over the heel

B = width of the footing

kB = width of the heel

Taking moments of the forces about the tip of the toe and using a factor of safety of 2, we get

$$W_s B (1 - 0.5k) + \frac{W_f B}{2} + W_w B (1 - k)$$

$$> (2)F_a(h/3) = 2c_a\gamma_s h^3/6 \qquad (14.9)$$

Relative values of the various weights may be assumed for the purpose of approximate determination of width B

$$\left. \begin{array}{l} W_f = 0.16\ W_s \\ W_w = 0.3\ W_s \\ W_s = kB\gamma_s h \end{array} \right\} \qquad (14.10)$$

The substitution of Eq. (14.10) in Eq. (14.9) yields

$$kB^2(1.38 - 0.8k) \geq c_a h^2/3$$

or

$$B \geq 0.49h \sqrt{\frac{c_a}{k(1 - 0.58k)}} \qquad (14.11)$$

The value of B derived above is based on the assumption of the relative weights W_s, W_f and W_w. These relative weights are likely to vary with the height of the retaining wall. Therefore, the value of B is suggested in the range of

$$B = 0.5h \sqrt{\frac{c_a}{k(1 - 0.6k)}} \qquad (14.12)$$

Determination of width of the footing based on sliding criterion is found to be usually uneconomical for RCC retaining walls. However, the stability against sliding is attained by provision of a key in the footing. The value of k could be obtained from the condition of permissible bearing capacity of the soil. The two equilibrium conditions of the vertical forces and the bending moments of the forces along with soil reaction give.

$$\frac{B}{2}(p_1 + p_2) = W_s + W_f + W_w \tag{14.13}$$

and

$$\frac{B^2}{6}(p_1 + 2p_2) = [W_s(1 - 0.5k) + 0.5W_f + W_w(1 - k)]B - c_a\gamma_sh^3/6 \tag{14.14}$$

in which p_1 and p_2 are the intensities of the soil reaction at the front and rear tips of the footing. Substitution of the approximate values of W's in the above two equations gives

$$p_1 + p_2 = 2.92\,k\gamma_sh \tag{14.15}$$

and

$$p_1 + 2p_2 = \gamma_sh\left[6k(1.38 - 0.8k) - \frac{k(1 - 0.58k)}{0.245}\right] \tag{14.16}$$

Solution of the two equations yields

$$p_1 = (1.66 + 2.43k)k\gamma_sh \tag{14.17}$$

If p_a is the safe bearing capacity of the soil, then the design criterion is

$$p_1 \leq p_a \tag{14.18}$$

Equations (14.17) and (14.18) give

$$k \geq 0.355\left[\sqrt{1 + \frac{3.4p_a}{\gamma_sh}} - 1\right] \tag{14.19}$$

For purpose of design the value of k can be used in the range of

$$k = 0.36\left(\sqrt{\left(1 + \frac{3.4p_a}{\gamma_sh}\right)} - 1\right) \tag{14.20}$$

to

$$k = 0.45\left(\sqrt{\left(1 + 3\frac{p_a}{\gamma_sh}\right)} - 1\right) \tag{14.21}$$

The values of B and k, as suggested above, are to be taken as guidelines for the design of the retaining walls. The design of the retaining walls is illustrated through examples.

14.6 Cantilever Retaining Wall

The cantilever retaining wall consists of three elements : (i) the toe, (ii) the heel and (iii) the vertical slab. These three elements all act as cantilevers and therefore should be designed as cantilevers.

EXAMPLE 14.1 *Design of cantilever retaining wall.* Design of a reinforced concrete retaining wall to retain soil of height 4 m above the ground level. The design data is:

Safe bearing capacity $p_a = 110 \text{ kN/m}^2$

$f_{ck} = 20 \text{ MPa}, f_y = 415 \text{ MPa}$

Angle of internal friction $\phi = 30°$

Unit weight of the soil $\gamma_s = 20 \text{ kN/m}^3$

Coefficient of friction between the soil and the concrete = 0.55.

$$c_a = \frac{1 - \sin \phi}{1 + \sin \phi} = \frac{1}{3}$$

Design of the footing

Minimum depth of foundation $= \dfrac{p_a}{\gamma_s} \left(\dfrac{1 - \sin \phi}{1 + \sin \phi} \right)^2 = 0.62 \text{ m}$

Adopt the depth of foundation as 0.7 m, the total height of the retaining wall will then be

$$h = 4 + 0.7 = 4.7 \text{ m}$$

From Eq. (14.20), we obtain

$$k = 0.36 \left(\sqrt{1 + \frac{3.4\, p_a}{\gamma_s h}} - 1 \right) = 0.36(\sqrt{1 + 3.9} - 1) = 0.43$$

From Eq. (14.12), we have

$$B = 0.5\, h \sqrt{\frac{c_a}{k\,(1 - 0.55k)}} = 0.505h = 2.38 \text{ m}$$

Use $B = 2.50 \text{ m}$ and $kB = 1.10$

The soil reaction can be obtained only with the known values of the weights of the retaining wall. The thickness of the slab and the footing may be assumed as suggested below to start the computation of the forces. Because the weight of the retaining wall is only about 30% of the weight of the soil filling, any error in the assumption of the weight to the extent of about 25% does not effect the final results.

Let

Average thickness of the wall = (about $h/15$) = 300 mm

Average thickness of the footing = (about $h/20$) = 240 mm

The maximum thickness of the wall at the base is taken as 400 mm.

Determination of the reaction of the soil. The reaction of the soil can be obtained from the simple equilibrium criteria of the forces acting on the wall. The forces and the moments are computed through Table 14.2. This type of tabulation of forces systematizes the data and minimizes the possible numerical errors.

The sum of the vertical forces and moments about the tip of the toe can be taken from Table 14.2.

They are

$$1.25(p_1 + p_2) = 146.57 \text{ kN}$$

686

$$1.04(p_1 + 2p_2) = 250.22 - 115.35$$

which reduce to

$$p_1 + p_2 = 117.3$$
$$p_1 + 2p_2 = 129.7$$

Therefore $p_2 = 12.4 \ kN/m^2$ and $p_1 = 104.9 \ kN/m^2$

The maximum pressure is 104.9 kN/m^2 as against a permissible value of 110 kN/m^2. Therefore, the design of the footing is just right.

Table 14.2 Forces on the Retaining Wall 1m Width

No.	Item	Force (in kN)	Distance from toe	Moment about toe tip (kNm)
1	Footing	0.24(2.5)(25) = 15.00	1.25	18.75
2	Wall	0.3(4.46)(25) = 33.45	1.20	40.14
3	Soil filling over the heel	(1.1)(4.46)(20) = 98.12	1.95	191.33
		Total (W) = 146.57	(M_s) =	250.22
4	Horizontal soil pressure	$\dfrac{c_a \gamma_s h^2}{2}$ = 73.63	$-$ 4.7/3	$-$ 115.35
5	Soil reaction	1.25 ($p_1 + p_2$);	$B_i^2(p_1 + 2p_2)/6 = -$ 1.04($p_1 + 2p_3$)	

Check for stability. Table 14.2 gives all the forces from which the stability of the wall can be verified.

Available factor of safety against overturning

$$= \text{stabilizing moment/overturning moment}$$
$$= 250.22/115.35 = 2.17$$

Frictional resistance between the base and the soil is

$$F_y = \mu W = 0.55(146.57) = 80.6 \ kN$$

Horizontal active soil pressure $F_a = 73.63$ kN

The frictional resistance is slightly more than the active force.

Using a factor of safety of 1.5, the horizontal force to be resisted by the key

$$= 1.5(73.63) = 110.45$$

$$\text{Thickness of the key} = \frac{110.45}{b\tau_c} = \frac{110.45}{1(300)} = 0.37 \ m$$

(assuming shear capacity of the concrete as 0.3 MPa).

Let the depth of the key below the ground level be h_0. The active earth pressure is to be resisted by the passive earth pressure upto the bottom of the key. The passive earth pressure on key is then equated to the active pressure:

$$F_p = \frac{c_p \gamma_s h_0^2}{2} \geq 1.5 \left(\frac{c_a \gamma_s h_t^2}{2} \right) - \mu W. \text{ solving we have}$$

or
$$h_0 \geq 1.9 \text{ where } h_t = h_l + h_a; \ h_1 = 4 \text{ m. } c_{pi} = 2.85$$

Depth of key $\qquad = 1.9 - 0.7 = 1.2$ m

Provide a key of 400 mm thick and 1.22 m deep. Otherwise the depth of the footing can be taken to about 1.5 m below the ground level.

Design of the footing (toe and heel). The toe has to be designed as a cantilever beam subjected to the soil reaction. Sometimes, the weight of the soil over the toe be neglected for the purpose of computations. (This assumption is on the safer side.) The forces acting on the toe slab are taken from Fig. 14.8 and shown in Fig. 14.9. Similarly, the forces acting on the heel slab are also shown in Fig. 14.9. The weight of the filling over the heel slab should be considered for the purpose of design.

Assume the thickness of the stem at base about $h/12$.

Let $\qquad t_s = 400$ mm

Cantilever span of the toe $= L_t = B - kB - t_s = 2.5 - 1.1 - 0.4 = 1$ m

The bearing pressure at the base of the toe is

$$p_3 = p_1 - (p_1 - p_2) \frac{L_t}{B} = 104.9 - (104.9 - 12.4)(1)/2.5 = 67.9$$

The reaction from the soil includes the effect of the weight of the footing. Therefore, while calculating the bending moment, the self-weight is to be included.

Self-weight $= 0.24(25) = 6$ kN/m$^2 = w_g$

The bending moment at base of the toe is

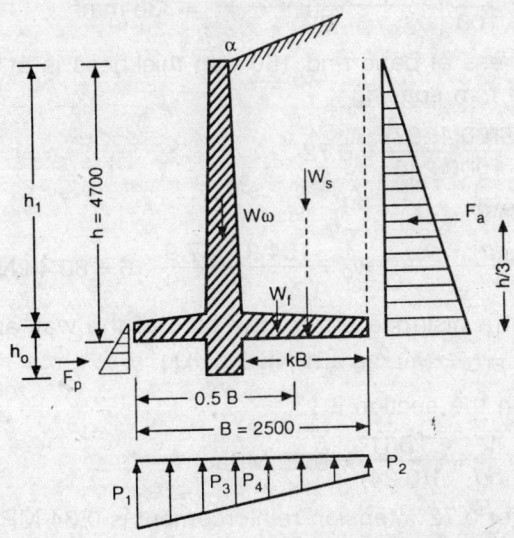

Fig. 14.8 Notations, Forces and Reactions on Cantilever Retaining Wall.

$$M = -w_g \frac{L_t^2}{2} + (p_3 + 2p_1)\frac{L_t^2}{6}$$

$$= -6(1)/2 + (67.9 + (104.9)(2))/6 = 49.28 \text{ kNm/m}$$

Equate the bending moment to the moment capacity of the section, we have

$$Kbd^2\sigma_{cb} = M$$

$$d = \sqrt{\frac{M}{Kb\sigma_{cb}}} = \sqrt{\frac{0.04928}{0.173(1)(5)}} = 0.239 \text{ m}$$

Use $d = 0.24$ m

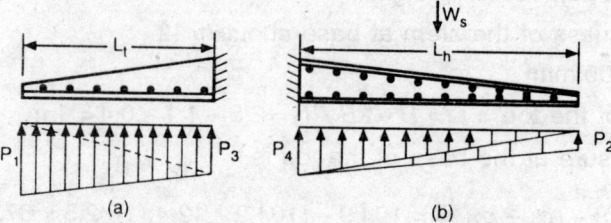

Fig. 14.9 Soil Reaction on Toe and Heel of Retaining Wall.

$$A_{st} = \frac{M}{jd\sigma_{st}} = \frac{49280}{0.904(0.24)(230)} = 988 \text{ mm}^2/\text{m}$$

Provide 16 mm bars at 200 mm spacing
The distribution reinforcement is to be provided at 0.15%, therefore it is equal to

$$A_{sy} = \frac{0.15}{100}\left(\frac{(300 + 150)(1000)}{2}\right) = 338 \text{ mm}^2$$

where 300 mm is the thickness at base and 150 mm thickness is at the free end.
Provide 10 mm bars at 220 mm spacing

$$\frac{100\,A_{st}}{bd} = \frac{100(1697)}{1000(240)} = 0.72$$

Average net load on the toe is

$$p = \frac{(p_1 + p_3)}{2} - w_g = \frac{104.9 + 67.9}{2} - 6 = 80.4 \text{ kN/m}$$

The critical shear force is at a distance d from the face of the wall and it is equal to

$$V = p(L_t - d) = 80.4(0.76) = 61 \text{ kN}$$

The nominal shear stress on the section is

$$\tau_v = \frac{V}{bd} = \frac{0.061}{1(0.24)} = 0.25 \text{ MPa}$$

The allowable shear stress for 0.72% tension reinforcement is 0.34 MPa, therefore the section is safe in shear. The embedment length is given by

$$L_d = \frac{\phi \, \sigma_{st}}{4\tau_{bd}} = \frac{20(230)}{4(1.12)} = 1026 \text{ mm}$$

Design of the heel

Cantilever span of the heel $L_h = 1.1$ m

Pressure at the tip $p_2 = 12.4 \text{ kN/m}^2$

Pressure at the base is

$$p_4 = p_2 + (p_1 - p_2)\frac{L_h}{B} = 12.4 + \frac{(104.9 - 12.4)(1.1)}{2.5}$$

$$= 53.1 \text{ kN/m}^2$$

Weight of the soil over the heel is

$$W_s = \gamma_s(h - 0.24)L_h = 20(4.46)(1.1) = 98.12 \text{ kN/m}$$

Let the average thickness of the heel be 0.24 m, then the self-weight of the heel is

$$w_g = 0.24(25) = 6 \text{ kN/m}^2$$

The net bending moment at the base of the heel is (tension causing on the top face)

$$M = \frac{W_s L_h}{2} + \frac{w_g L_h^2}{2} - (2p_a + p_4)\frac{L_h^2}{6}$$

$$= 98.12\frac{(1.1)}{2} + \frac{6(1.1)^2}{2} - (2(12.4) + 53.1)\frac{1.1^2}{6} = 41.88 \text{ kNm/m}$$

The effective depth of the heel required at the base is

$$d = \sqrt{\frac{M}{Kb\sigma_{cb}}} = \sqrt{\frac{0.04188}{0.13(1)(7)}} = 0.22 \text{ m}$$

The area of the tension reinforcement

$$A_{st} = \frac{M}{jd\sigma_{st}} = \frac{41\,880}{0.904(0.22)(230)} = 915 \text{ mm}^2/\text{m}$$

A_{st} (provided) $= 1570 \text{ mm}^2/\text{m}$

Provide 16 mm bars at 210 mm spacing.

Provide the distribution reinforcement of 10 mm bars at 220 mm spacing.

The criticial section against shear is at the face of the wall. The shear force at this **section** is

$$V = W_s - (p_2 - p_4)\frac{L_h}{2} = 98.12 - (65.5)0.55 = 62.1 \text{ kN/m}$$

The nominal shear stress at the critical section is

$$\tau_v = \frac{V}{bd} = \frac{0.0621}{1(0.22)} = 0.28 \text{ MPa}$$

$$\frac{100\,A_{st}}{bd} = \frac{157000}{1000(220)} = 0.71 \text{ MPa}$$

The allowable shear stress in the concrete for the above percentage of the tension reinforcement is 0.34 MPa; therefore, the section is safe against shear. The development length of the 20 mm bars was computed to be 1026 mm and it should be provided in this case also.

The overall thickness of the slab is $t_h = d + 0.06 = 0.28$ m.

Design of stem (vertical wall). The maximum bending on the stem occurs at the base just above the heel and it is given by

$$M = \frac{c_a \gamma_s}{6}(h - t_h)^3 = \frac{20}{18}(4.7 - 0.28)^3 = 96 \text{ kNm/m}$$

The effective depth and the area of the tension reinforcement at the base of the stem are:

$$d = \sqrt{\frac{M}{Kb\sigma_{cb}}} = \sqrt{\frac{0.096}{0.13(1)(7)}} = 0.33 \text{ m}$$

$$A_{st} = \frac{M}{jd\sigma_{st}} = \frac{96\,000}{0.904(0.333)(230)} = 1400 \text{ mm}^2/\text{m}$$

Provide 16 mm bars at 140 mm spacing, then
Provide the overall thickness of the stem at the base as $t_w = 0.4$ m which checks well with the assumed value. The shear force is critical at the face of the heel or the toe. The depth of filling upto the top face of the toe is the critical depth for calculation of the shear force. The total shear force at the section is

$$V = \frac{c_a \gamma_s}{2}(h - t_l)^2 = \frac{20}{6}(4.7 - 0.3)^2 = 64.53 \text{ kN}$$

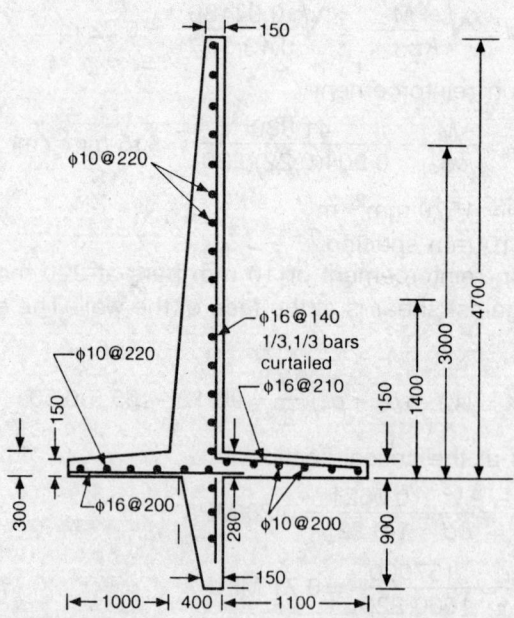

Fig. 14.10 Reinforcement Details of Cantilever Type Retaining Wall of Example 14.1.

The nominal shear stress at the section is

$$\tau_v = \frac{V}{bd} = \frac{0.06453}{1(0.333)} = 0.19 \text{ MPa}$$

The nominal shear stress is less than the allowable value even for the nominal percentage of the reinforcement.

   The bending moment on the wall is proportional to the cube of the depth. Therefore, the thickness of the wall can be minimum at the top and one-third of the reinforcement bars can be curtailed each at one-third and two-thirds of the depth. Provide 10 mm bars at 220 mm spacing as the distribution reinforcement.

14.7 Curtailment of Reinforcement in the Vertical Wall

If there is no surcharge or additional practical criterion for the thickness of the vertical wall at top, then the thickness of the wall at top may be assumed about 150 mm as a minimum. Let the effective depth at the top of the wall be d_0; then effective thickness at any depth h_x from the top surface is given by

$$d_x = d_0 + (d - d_0)\frac{h_x}{h} \tag{14.22}$$

The bending moment at this depth is

$$M_x = \frac{\gamma_s c_a}{6} h_x^3 \tag{14.23}$$

and the area of the reinforcement (A_{sx}) required at this depth is

$$A_{sx} = \frac{M_x}{jd_x f_s} \tag{14.24}$$

The ratio of the steel requirement at any depth h_x with respect to that at the base is obtained from Eqs. (14.23) and (14.24):

$$\frac{A_{sx}}{A_s} = \frac{M_x b}{M d_x} = \left(\frac{d}{d_x}\right)\left(\frac{h_x}{h}\right)^3 \tag{14.25}$$

Let

$$\frac{A_{sx}}{A_s} = r$$

Then the substitution of the Eq. (14.22) in Eq. (14.25) gives

$$\left(\frac{h_x}{h}\right)^3 - r\left(\frac{d - d_0}{d}\right)\frac{h_x}{h} - r\frac{d_0}{d} = 0 \tag{14.26}$$

Let $$x = \frac{h_x}{h},$$ then Eq. (14.26) gives

$$x^3 - r\left(1 - \frac{d_0}{d}\right)x - r\frac{d_0}{d} = 0 \tag{14.27}$$

The value of x could be obtained for any given value of r. The curtailment of the reinforcement in small walls may be done in one or two stages. The value of r will be known for any stage of curtailment. For example, if every third bar is curtailed, then the value of r is 2/3. So the value of r is always a simple fraction.

EXAMPLE 14.2 *Curtailment of bars of the wall in Example 14.1.* Let the top thickness of the stem of Example 14.1 be 150 mm, then d_0 may be taken as 120 mm. If one-third bars are curtailed, then the height at which they should be curtailed is given by the Eq. (14.27). The appropriate values are:

$$r = \frac{2}{3}, \quad \frac{d_0}{d} = \frac{120}{333}$$

$$r\left(1 - \frac{d_0}{d}\right) = 0.425$$

$$\frac{2d_0}{3d} = 0.242$$

Equation (14.27) reduces to $x^3 - 0.425x - 0.242 = 0$

and the solution is $x = 0.85$

or $\qquad h_{x1} = 0.85h = 3.79$ m

Height from the bottom $\qquad = h - h_{x1} = 0.67$ m

Point where the bars can be curtailed

$$= 670 + L_d$$

$$= 670 + 1026 = 1696 \text{ mm.}$$

Similarly, in the second stage of th curtailment, if another one-third of the main reinforcement is curtailed, $r = 1/3$ and Eq. (14.27) reduces to

$$x^3 - 0.213x - 0.121 = 0$$

and the solution is

$$x = 0.64 \text{ and } \quad h_{x2} = 0.64h = 2.86 \text{ m}$$

Minimum development length of the $\phi20$ tensile bar which should project beyond the point it is required for tension is $L_d = 1026$ mm.

14.8 Counterfort Retaining Walls

As the height of the retaining wall increases, the bending moment due to the horizontal thrust of the soil increases rapidly. The bending moment on the vertical cantilever slab, is proportional to the cube of the height of the wall. It is desirable to change the structural action of the vertical wall in such a way that the wall does not act as a cantilever but acts as a slab with closer supports. If a series of upright beams which support the wall are provided along the length of tne wall, then the vertical wall acts as a continuous slab in one direction. Each panel of this vertical wall may be treated as an idealized plate fixed at three edges, free at the top and subjected to the horizontal earth pressure. The upright beams are generally provided at the rear of the vertical wall, and are called *counterforts* If the beams are provided in the front of the wall, they are called *buttresses*. The counterforts which rest on the heel make the heel act as a continuous slab. An idealization of each panel may be taken as a slab fixed at three edges and free at one edge. In many cases, the toe which acts as a cantilever subjected to the soil reaction calls for enormous thickness. Therefore, it is desirable to have front counterforts or buttresses so that the toe also acts like a slab with supports.

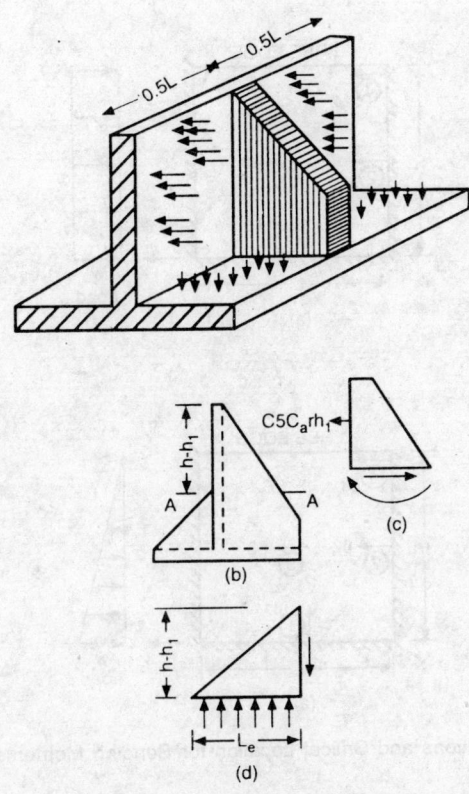

Fig. 14.11 Free Body Diagrams of Elements of Counterforts Retaining Wall.

Design criteria for footing and vertical slab: A typical counterfort retaining wall is shown in Fig. 14.11a. The idealized structural action of a particular vertical panel is shown in Fig. 14.11b. Bending moment coefficients of plates subjected to uniformly distributed load and triangularly distributed load are available from plate theory. Critical moment coefficients of these are given in Table 14.3 for the purpose of design. There are certain locations where the bending moments are critical. Magnitudes of the bending moments are maximum at the middle of the fixed edges. Positive bending moment is generally maximum at the middle zone of the slab. For the purpose of design, the middle of the plate is considered to be under maximum positive bending moment. The location of the critical bending moments indicated by their coefficients are shown in Fig. 14.12. The loads acting on the footing (toe and heel) could always be divided into uniformly and triangularly distributed loads.

The bending moments at the four critical locations on the slab caused by UDL can be expressed as

$$M_i = \alpha_i w L^2 \qquad (14.28)$$

where w = UDL, L = Clear span between the fixed edges.

          i = 1, 2, 3 and 4 are the locations and the directions as marked

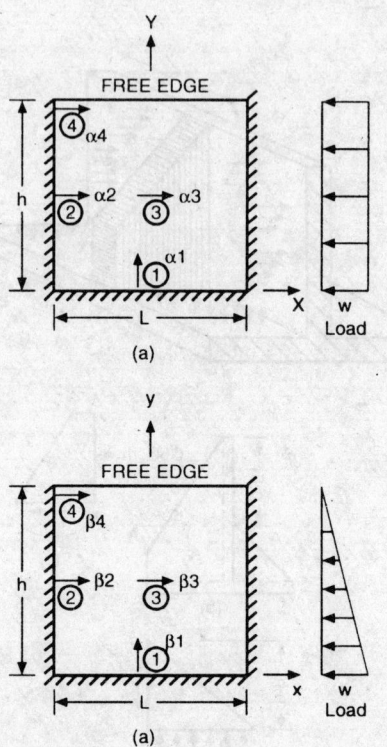

Fig. 14.12 Notations and Critical Location for Bending Moments in Plates.

in Fig. 14.12. The bending moments at the four critical locations on the slab caused by a triangular load can be expressed as

$$M_i = \beta_i w L^2 \qquad (14.29)$$

where $\qquad w$ = load intensity at the base of the triangular load.

The values of α_i and β_i are functions of h/L and are given in Table 14.3.

14.9 Criteria of Design of Counterforts

The spacing of the counterforts have considerable effect on the bending moments. In a vertical slab bending moment is proportional to the height of the wall and the square of the spacing of the counterforts. Therefore the spacing of the counterforts be less than the height of the slab. The following guideline may be adopted

$$L = 0.8 \sqrt{h} \text{ to } 1.2 \sqrt{h} \qquad (14.30)$$

The rear counterforts are subjected to horizontal force due to soil pressure which is transmitted through the vertical wall. The critical section of the counterfort which should be designed is at the point where the front counterfort ends. The soil pressure induces bending moment and axial force which will tend to separate the vertical slab from the counterfort. Similarly, the forces acting on the heel will have a tendency to separate the heel slab from the counterfort.

Table 14.3 Bending Moment Coefficients of a Plate with Three Fixed Edges and One Free Edge

(a) Uniformly distributed load intensity

h/L	$y = 0$ $x = 0$ α_1	$y = h/2$ $x = \pm L/2$ α_2	$y = h/2$ $x = 0$ α_3	$y = h$ $x = \pm L/2$ α_4
0.6	− 0.055	− 0.036	0.017	− 0.074
0.7	− 0.054	− 0.044	0.021	− 0.078
0.8	− 0.053	− 0.051	0.025	− 0.081
0.9	− 0.052	− 0.056	0.029	− 0.083
1.0	− 0.051	− 0.061	0.032	− 0.083
1.25	− 0.047	− 0.071	0.037	− 0.083
1.50	− 0.042	− 0.075	0.040	− 0.083
2.0	− 0.040	− 0.083	0.041	− 0.083

(b) Triangularly Distributed Load with Maximum Intensity at Base (Table 14.3 cont.)

h/L	β_1	β_2	β_3	β_4
0.6	− 0.024	− 0.013	0.006	0.0
0.7	− 0.026	− 0.017	0.008	0.0
0.8	− 0.02	− 0.021	0.010	0.0
0.9	− 0.029	− 0.024	0.012	0.0
1.0	− 0.030	− 0.027	0.013	0.0
1.25	− 0.031	− 0.033	0.017	0.0
1.50	− 0.029	− 0.034	0.019	0.0
2.0	− 0.029	− 0.040	0.021	0.0

Reinforcement is to be provided to resist bending and axial forces. The forces acting on the front counterfort are earth pressure through the wall and the soil reaction transmitted through the toe slab. Both the front and rear counterforts should be designed as beams with variable depth.

The width of the base and that of the heel could be taken approximately equal to those derived in Section 14.5.

EXAMPLE 14.3 *Design of counterfort retaining wall.* Design a counterfort retaining wall to retain soil of 7.5 m above the ground level. The unit weight and the bearing capacity of the soil are 20 kN/m^3 and 200 kN/m^2 respectively. The angle of internal friction of the soil and coefficient of friction are 30° and 0.55 respectively.

Design data and assumptions. Let M20 concrete and HYSD steel be used for the construction. $f_{ck} = 20$ MPa, $f_y = 415$ MPa. The saturated weight of the soil is taken as $\gamma_s = 20$ kN/m^3 and $p_a = 200$ kN/m^2, $\phi = 30°$.

Design of the footing.

$$\text{Minimum depth of foundation} = \frac{p_a}{\gamma_s} \left(\frac{1 - \sin \phi}{1 + \sin \phi} \right)^2 = 1 \text{ m}$$

Overall height $= 7.5 + 1 = 8.5$ m

The width of the footing and heel may be taken as approximately equal to the values given in Eqs. (14.12) and (14.21):

$$k = 0.45 \left[\sqrt{\left(1 + \frac{3p_a}{\gamma_s h}\right)} - 1 \right] = 0.45 \left[\sqrt{\left(1 + \frac{3(20)}{20(8.5)}\right)} - 1 \right] = 0.47$$

$$B = 0.5H \sqrt{\frac{c_a}{k(1 - 0.55k)}} = 0.5h \sqrt{\frac{1}{3(0.47)(0.725)}} = 4.2 \text{ m}$$

Provide width of footing $B = 4.4$ m and width of heel $L_h = 2.1$ m. The weights of various elements of the retaining wall are needed for the purpose of calculation of the pressure on the soil. So the following approximate average thicknesses of various elements are assumed.

Average thickness of:

i) vertical wall (about $h/40$) = 0.21 m

ii) base slab (about $h/30$) = 0.3 m

Clear spacing of counterforts $L = 3$ m

Thickness of counterfort ($L/10$) = 0.3 m, c/c of counterforts = 3.3 m

Height of the front counterfort = 3 m

Soil reaction. The reaction of the soil is obtained from the equilibrium criterion. The forces and the moments on the retaining wall for one counterfort spacing are computed in Table 14.4 The stability of the retaining wall is also computed from the same table.

The equilibrium of the forces obtained from the sub-totals of columns (4) and (5) of Table 14.4:

$$7.26(p_1 + p_2) = 1373$$
$$10.65(p_1 + 2p_2) = 1993$$

which reduces to

$$p_1 + p_2 = 189$$
$$p_1 + 2p_2 = 187$$

and the solution of the equations is

$$p_2 = -2 \text{ kN/m}^2$$
$$p_1 = 191 \text{ kN/m}^2$$

The value of p_2 works out to be negative indicating tension on the soil at the tip of the heel. The soil cannot resist tension, so separation takes place between the soil and the tip of the heel thus resulting into only a partial contact. Since the net tension is only 2 kN/m² and that too is based on the assumed sizes, one can consider it as zero when compared with the maximum value of 191 kN/m². Therefore, assume a pressure distribution as

$$p_2 = 0 \text{ and } p_1 = 190 \text{ kN/m}^2$$

The soil reactions at the base of the toe and the heel slabs are (Fig. 14.13):

$$p_3 = \frac{p_1(2.3)}{4.4} = \frac{190(2.3)}{4.4} = 99.3 \text{ kN/m}^2$$

$$p_4 = \frac{p_1(2.1)}{4.4} = 90.7 \text{ kN/m}^2$$

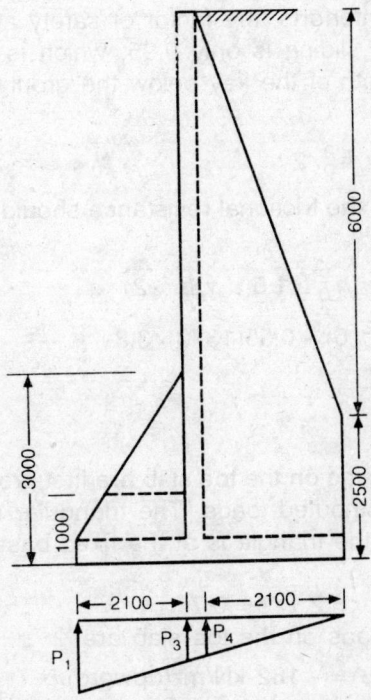

Fig. 14.13 Counterforts Retaining Wall of Example 14.3.

Table 14.4 Forces Acting on the Retaining Wall for One Counterfort Spacing

No	Element		Force (kN)	Distance from tip of toe (m)	Moment about tip of toe (kNm)
1	Footing	0.3(4.4)(3.3)(25)	= 109	2.2	240
2	Wall	0.21(8.2)(3.3)25)	= 142	2.3	327
3	Rear counterfort	0.3(8.2)(2.1)(25)/2	= 65	3.1	200
4	Front counterfort	0.3(3)(2.1)(25)/2	= 24	0.73	17
5	Rear soil	2.1(8.2)(3)(20)	= 1033	3.35	3461
6	Sub totals		W = 1373		M = 4255
7	Active pressure	20(8.5)(8.5)(3.3)/6	= 795	− 8.5/3	− 2252
8	Net Moment about toe		= M_t		1993
9	Soil reaction	4.4(3.3)p_1/2	= − 7.26p_1	4.4/3	− 10.65p_1
		4.4(3.3)p_2/2	= − 7.26p_2	4.4(2)/3	− 21.3p_2
10	Net soil reaction		− 7.26($p_1 + p_2$)		− 10.65($p_1 + 2p_2$)

Check for the stability of the retaining wall. The factor of safety against overturning and sliding can be computed by using the forces listed in Table 14.4. Factor of safety against overturning is = 4255/2255 = 1.9.

Factor of safety against sliding without key is $0.55(1373)/795 = 0.95$.

The factor of safety against overturning is 1.9. By considering the front counterfort and the projections in the rear counterforts, the factor of safety against overturning is adequate. As the factor of safety against sliding is only 0.95, which is less than 1.5, therefore a key has to be provided. Let the depth of the key below the ground level be h_0, then the passive earth pressure is

$$F_p = c_p \gamma_s h_0^2 / 2$$

The passive earth pressure and the frictional resistance should be at least 1.5 times the active earth pressure. This results

$$\mu W/3.3 + 0.5 c_p \gamma_s h_0^2 \geq 1.5 (c_a \gamma_s h^2 / 2)$$

$$0.5(3)(20) h_0^2 > (1.5)(20)(8.5)^2 / 6) - 0.55(1373)/3.3$$

$$h_0 > 2.1$$

Depth of key $= h_0 - 1 = 1.1$ m

Design of toe. The forces acting on the toe slab are first grouped to become a combination of uniformly and triangularly distributed loads. The triangular distribution should be selected in such a way that the base of the triangle is at the fixed base of the toe slab. These forces are shown in Fig. 14.14.

Aspect ratio of the toe slab $r = 2.1/3 = 0.7$

The net intensities of the loads on the toe slab are

$$\text{UDL} = -(190 - 8) = -182 \text{ kN/m}^2 \text{(upward)}$$

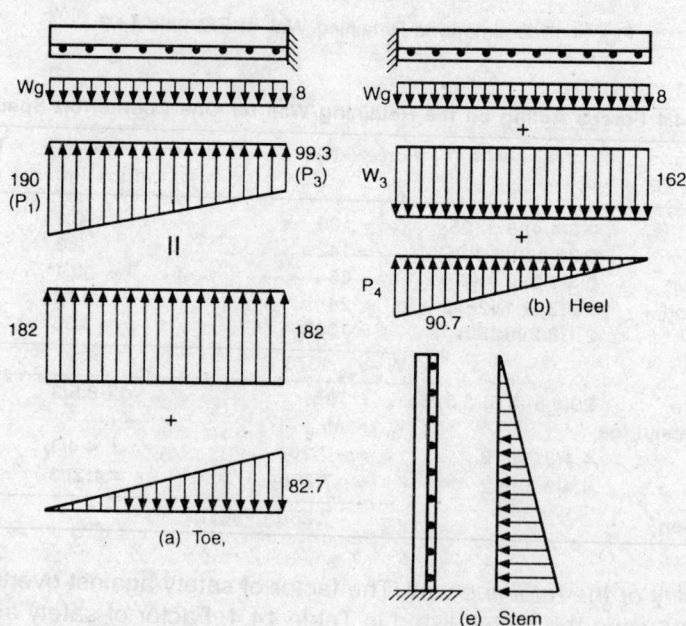

Fig. 14.14 Loads on Toe, Heel and Stem of Retaining Wall of Example 14.3.

Triangular $= 190 - (99.3 - 8) = 82.7$ kN/m^2 (downwards)

The bending moments on the toe slab are computed in Table 14.5 using the moment coefficients of Table 14.3. Keeing the thickness as constant, the amount of reinforcement may be varied according to the bending moment.

The absolute maximum bending moment occurs at the end face of the counterfort and it is $M = M_4 = 133.72$ kNm/m. The effective thickness of the slab is

$$d = \sqrt{\frac{M}{Kb\,\sigma_{cb}}} = \sqrt{\frac{0.13372}{0.13(1)(7)}} = 0.383 \text{ m}$$

Use $d = 0.4$ m and $t = 0.46$ m.

The area of the tension steel required at top face is

$$A_{st4} = \frac{M}{jd\,\sigma_{st}} = \frac{133\,720}{0.904(0.4)(230)} = 1608 \text{ mm}^2$$

Similarly, the reinforcement details of the toe slab are computed and listed in Table 14.6 using the bending moment given in Table 14.5.

The reinforcement requirements in a slab change from location to location since the transverse bending moment along a section is maximum at the middle, but decreases gradually.

Design of the heel slab. The clear span of the slab between the counterforts and the cantilever span are

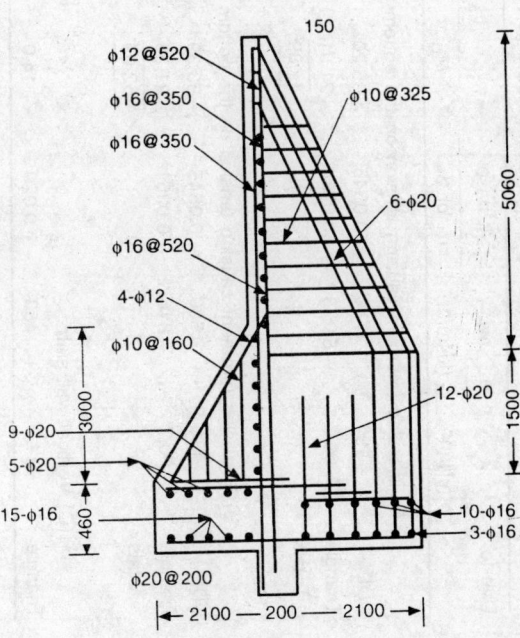

Fig. 14.15 Reinforcement Details in the Counterforts of Example 14.3.

Table 14.5 Critical Moments in Toe, Heel and The Wall Slabs (BM in kNm/m)

Due to	w (kN/m²)	wL^2 (kNm/m)	At base α_1 or β_1	M_1	Middle height α_2 or β_2	M_2	Middle point α_3 or β_3	M_3	Top α_4 or β_4	M_4
(A) Toe slab (moment causing tension on the bottom face is treated positive)										
UDL	−190	1710	−0.0545	93.20	−0.0439	75.07	0.0212	−36.25	−0.0782	133.7
Triangle	90.7	−816.3	−0.0264	−21.31	−0.0171	−13.96	0.0080	6.50	−0.0	0
Totals				71.89		61.11		−29.75		133.7
(B) Heel slab (moment causing tension on the bottom face is positive)										
UDL	170	1530	−0.0545	−83.39	−0.0439	−67.17	0.0212	32.44	−0.0782	119.65
Triangle	−90.7	−816	−0.0261	21.30	−0.0171	13.95	0.008	−6.53	0.0	0
Totals				−62.09		−53.22		25.91		−119.65
(C) Vertical wall slab										
Triangle	53.67	483	−0.029	−14.0	−0.0402	19.42	0.021	10.14	0	0

$$L = 3 \text{ m, and } h = 2.1 \text{ m}$$

The aspect ratio of the slab is $2.1/3 = 0.7$ and the moment coefficients are selected and shown in Table 14.5. The load intensity on the heel consists of self-weight, the soil weight and the soil reaction which are shown in Fig. 14.14b.

The self-weight and the soil weight of the filling acting downwards are

$$w_g = 8 \text{ kN/m}^2$$

and

$$w_s = 8.2(20) = 162 \text{ kN/m}^2$$

that is total downward UDL $= 170 \text{ kN/m}^2$.

Upward triangular intensity is $p_4 = 90.7 \text{ kN/m}^2$ and it is zero at the free end and maximum at the base. THe maximum bending moments caused by the UDL and triangular loads are computed separately using the moment coefficients. The load acting downwards is taken as positive and the corresponding bending moments are shown in Table 14.5. The absolute maximum bending moment occurs near the free edge span and it is

$$M = M_4 = 119.65 \text{ kNm/m}$$

The effective depth required is

$$d = \sqrt{\frac{M}{Kb\sigma_{cb}}} = \sqrt{\frac{0.11965}{0.13(1)(7)}} = 0.362 \text{ m}$$

Use $d = 0.4$ m and $t = 0.46$ m (same as the thickness of the toe). The area of the reinforcement required at different locations are computed in Table 14.6. The critical shear force is at free edge near the counterfort support and it is

$$V = 170(3/2) = 255 \text{ kN, and}$$

$$\tau_v = \frac{V}{bd} = \frac{0.255}{1(0.4)} = 0.63 \text{ MPa}$$

The shear stress is more than the allowable. Therefore, provide 4-legged 10 mm stirrups at

$$s_v = \frac{4(78.5)(140)}{(0.63 - 0.3)1000} = 135 \text{ mm}$$

Stem or vertical wall design. The stem is fixed at three faces and free at top, and it is subjected to lateral earth pressure.

The cantilever height $\qquad h = 8.5 - 0.46 = 8.04$ m

Continuous clear span $\qquad L = 3$ m

Aspect ratio $\qquad h/L = 8.04/3 = 2.68$

The load is linearly varying with maximum at the base and it is

$$w = c_a \gamma_s h = \frac{20}{3}(8.04) = 53.67 \text{ kN/m}^2$$

The bending moment coefficients are taken from Table 14.3, and the bending moments computed have been given in Table 14.5. The absolute maximum bending moment in the stem occurs at the mid-height and it is

$$M = |M_2| = 19.42 \text{ kNm/m}$$

The effective depth required is

$$d = \sqrt{\frac{0.01942}{0.13(1)(7)}} = 0.15 \text{ m}$$

Use $d = 0.15$ m and $t = 0.2$ m

The reinforcement requirements in the wall are computed using the moments in Table 14.5 and are listed in Table 14.6 subject to the minimum requirements.

Note: The assumed thickness of the base slab was 0.3 m whereas the actual one is 0.46 m and the assumed wall thickness was 0.21 m while the actual one is 0.20 m. The increase in the thickness of the base slab increases the pressure on the soil marginally.

Design of the front counterfort. Let H_f = height of the front conterfort.

The counterfort acts as a reaction beam to the toe. Therefore, the soil reaction on the toe acts as a load on the counterfort as shown in Fig. 14.17. The net intensity of load acting on the toe slab are shown in Fig. 14.14. The total forces acting on the counterfort are

$$w_1 = 182(L) = 182(3.3) = 600.6 \text{ kN/m}$$

$$w_3 = -82.7(3.3) = -273 \text{ kN/m}$$

The bending moment caused by these loads is

$$M = w_1 \frac{(L_e)^2}{2} + w_3 \frac{L_e^2}{6} = (3w_1 + w_3) \frac{L_e^2}{6}$$

$$= (3(600.6) - 273) \frac{(2.1)^2}{6} = 1223.67 \text{ kNm}$$

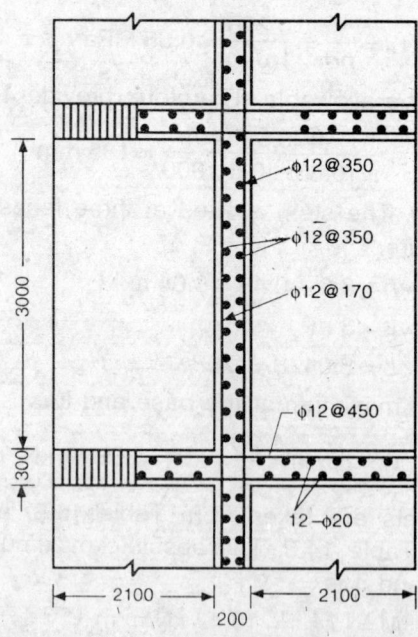

Fig. 14.16 Reinforcement Details in the Wall and Counterforts of Example 14.3.

Table 14.6 Reinforcement Details

Location (face)		Area of the reinforcement required
Toe		
1. Cantilevers span	(b)*	865
2. Under counterfort (middle)	(b)	735
3. Mid-span	(t)	358
4. Under counterfort (edge)	(b)	1608
Heel		
1. Cantilever span	(t)	747
2. Under the counterfort (middle)	(t)	640
3. Mid-span	(b)	312
4. Under counterfort (edge)	(t)	1343
Stem		
1. Cantilever span, bottom-rear face		768
2. At counterforts (middle-rear face)		1067
3. Mid-span (horizontal-front face)		565
4. At counterforts (top-rear face)		0

(b) refers to bottom face, (t) refers to top face.

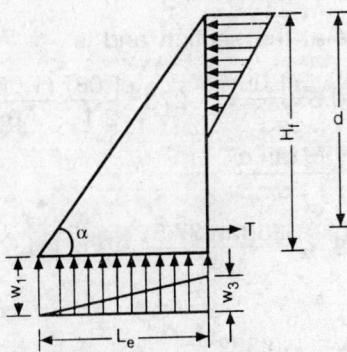

Fig. 14.17 Forces on the Front Counteforts of Example 14.3.

The depth of the counterfort is varying so the moment capacity of a beam with variable depth is given by

$$M_r = K \, bd^2 \cos^2 \alpha$$

where
$$\alpha = \text{slope of the beam}$$

Equating the moment capacity to the external moment, we have

$$Kb \, (d \cos \alpha)^2 \, \sigma_{cb} = M = 1.2267 \text{ MNm}$$

$$d \cos \alpha = \sqrt{\frac{1.22367}{0.173(0.3)(5)}} = 2.17 \text{ m}$$

It may be noted that the beam is really a short beam in which the span is 2.1 m and the depth appears to be more than the span. From the geometry, we have

$$\cot \alpha = \frac{\text{span}}{\text{depth}} = \frac{2.1}{d}$$

whereas

$$\cos \alpha = \frac{2.17}{d}$$

Therefore

$$\mathrm{cosec}\, \alpha = \frac{2.10}{2.17} = 0.97$$

This means that the element cannot be idealised as a beam. Its depth is more than the span and hence the strut action is dominant when compared to the beam. Hence provide nominal reinforcement.

The shear force at this point is

$$V = w_1 L (2.1 - 1.02) + w_3 L(2.1 - 1.02)/2.1$$
$$= 600.6(1.08) - 273 (1.08)/2.1 = 508 \text{ kN}$$

The effective depth of the counterfort at this section is

$$d_s = 1.08/\tan \alpha = 1.41 \text{ m}$$

The design shear force at the section is

$$V_e = V - \frac{M \tan \alpha}{d_e}$$

where

$M = \text{BM}$ at the section and is

$$M = 600.6 \frac{(1.08)^2}{2} - 273 \frac{(1.08)}{2.1} \frac{(1.08)^2}{6} = 323 \text{ kNm}$$

$$V_e = V - \frac{M \tan \alpha}{d_r}$$

$$= 508 - \frac{323(\tan 37.5)}{1.41} = 332 \text{ kN}$$

The nominal shear stress is

$$\tau_v = \frac{V_e}{bd_s} = \frac{0.332}{0.3(1.41)} = 0.79 \text{ MPa}$$

The maximum allowable shear stresses with and without reinforcement are

$$\tau_{cmax} = 1.6 \text{ MPa} \quad \text{and} \quad \tau_c = 0.34 \text{ MPa}$$

Transverse reinforcement be provided. Hence select two-legged 10 mm bars, then

$$A_{sv} = 2(78.5) = 157 \text{ mm}^2$$

The spacing of the stirrups is

$$s_v = \frac{A_{sv}\, \sigma_{sv}}{(\tau_v - \tau_c)b} = \frac{157(230)}{0.45(300)} = 267 \text{ mm}$$

Design of the rear counterfort. The main functions of the rear counterfort is to interlink the heel and the vertical slab and to support the vertical wall. The total downward load on the heel slab is to be carried to the counterforts through tie reinforcement.

The net downward load on the heel is

$$T = 170(3)(2.1) - 90.7(3)(2.1)/2 = 785.3 \text{ kN}$$

$$A_s = \frac{T}{\sigma_{st}} = \frac{785\ 300}{230} = 3413 \text{ mm}^2$$

Provide 12 numbers of 20 mm bars tied into the heel from the counterfort.

Up to the top of the front counterfort from the base the effective depth against bending includes the front and the rear counterforts. The criticial section of bending of the rear counterfort is at the top of the front counterfort.

The depth of the top of the front counterfort is

$$h_1 = 8.5 - 0.46 - 3 = 5.04 \text{ m}$$

The bending moment at this section is

$$M = \gamma_s c_a \frac{h_1^3 L_0}{6} = 20 \left(\frac{1}{3}\right) \frac{(5.04)^3 (3.3)}{6} = 470 \text{ kNm}$$

The counterfort along with the vertical wall will act as a T-beam. The thickness of the slab at this depth is 180 mm, therefore the effective depth needed is

$$d = \frac{t}{2n} + \frac{M}{dt\,j\,\sigma_{cb}}$$

$$= \frac{0.18}{2(0.289)} + \frac{0.475}{3.3(0.3)(0.904)(7)} = 0.338 \text{ m}$$

Let the rear counterfort go straight up to 1.5 m height from base and then reduce to zero at the top of the wall. Then the overall depth of the counterfort at 5.04 m from top is

$$D = \frac{5.04\ (2.1)}{(8.04 - 1.5)} = 1.62 \text{ m}$$

Therefore use

$$d = 1.62 - 0.12 = 1.50 \text{ mm}$$

$$A_{st} = \frac{M}{jd\,\sigma_{st}} = \frac{475000}{0.904\ (1.5)\ (230)} = 1523 \text{ mm}^2$$

Continue 6 of the tie bars into the counterfort.

Design for shear of the rear counterfort. The shear force at the critical moment zone is

$$V = c_a \gamma_s \frac{L_0 h_1^2}{2} = 20\ (3.3) \frac{(5.04)^2}{6} = 281 \text{ kN}$$

The slope of the beam is

$$\tan \alpha = \frac{D}{h_1} = \frac{1.62}{5.04} = 0.32, \text{ or } \alpha = 17.75°$$

The effective shear force is

$$V_e = V - \frac{M \tan \alpha}{d} = 281 - \frac{475\ (0.32)}{1.5} = 180 \text{ kN}$$

The nominal shear stress is

$$\tau_v = \frac{V_e}{bd} = \frac{0.180}{0.3(1.5)} = 0.4 \text{ MPa}$$

This value is only marginally more than the allowable, therefore the nominal shear reinforcement controls the design. Provide two-legged 10 mm stirrups, then

$$A_{sv} = 157 \text{ mm}^2$$

and
$$s_{max} = \frac{A_{sv}f_y}{0.4(b)} = \frac{157\,(415)}{0.4\,(300)} = 543 \text{ mm}$$

Provide two-legged 10 mm stirrups at 500 mm spacing.

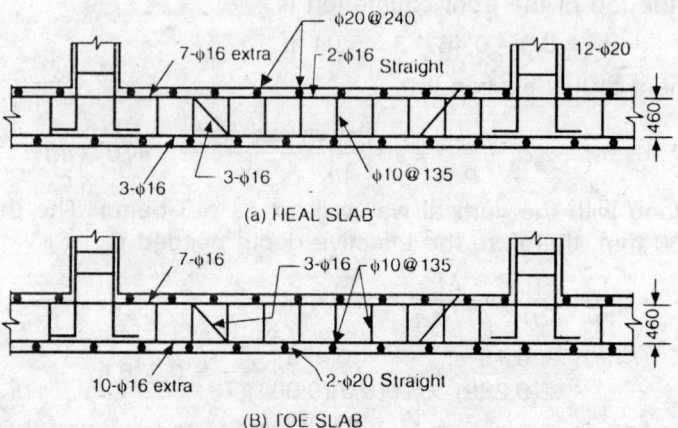

Fig. 14.18 Reinforcement Details in Heel and Toe. Example 14.3.

PROBLEMS
[Use IS 456 and sketch reinforcement details in each
of the following problems]

14.1 A RCC retaining wall is to be designed to retain a soil of 3 m above the ground level. The unit weight of the soil is 18 kN/m³ and its angle of internal friction is 26°. The footing of the retaining wall is to be taken 1 m below the ground level on to a soil having an allowable bearing capacity of 90 kN/m². Design a RCC retaining wall using M20 concrete and mild steel reinforcement. Ensure a factor of safety of 2 against overturning as well as against sliding.

14.2 A counterfort retaining wall is to hold an embankment of 5 m height. The unit weight of the soil is 18 kN/m³ and its angle of internal friction is 20°. Design a counterfort retaining wall with counterforts spaced at 3 m apart using M20 concrete and HYSD-Fe415 bars. The allowable bearing capacity of the soil on which retaining wall rests can be taken as 90 kN/m².

14.3 A buttressed retaining wall holds an embankment of 5 m height and resting on soil having an allowable bearing capacity of 90 kN/m². The unit weight of the soil is 17 kN/m³ and angle of internal friction is 20°. Design a buttressed reinforced concrete retaining wall using M20 concrete and HYSD-Fe415 bars.

Chapter 15

Limit State Design of Footings

15.1 Introduction

The design of footing consists of two parts: namely, the determination of size and depth of the footing which depend mainly on the soil characteristics, and the structural design of the footing slab. An introduction to the design of size and depth of the footing was given in Section 13.1 to 13.3 of Chapter 13. Similary, an introduction to the limit state design was given in Chapter 7. The critical sections of failure are also given in Chapter 13. The moment capacity of a balanced rectangular cross-section is given by

$$M_{rb} = Kbd^2 f_{ck} \tag{15.1}$$

where $\qquad\qquad\qquad\qquad K$ = constant for a given reinforcement material.

The area of the reinforcement is obtained as

$$A_{st} = \frac{1.15M}{jdf_y} \tag{15.2}$$

where $\qquad\qquad\qquad\qquad j = 1 - 0.42\,k_3 \tag{15.3}$

The values of K and k_3 are reproduced in Table 15.1. For a balanced section the area of the tension reinforcement can also be obtained from

$$A_{st} = p_0 bd \frac{f_{ck}}{f_y} \tag{15.4}$$

where $\qquad\qquad\qquad\qquad p_0$ is given in Table 15.1.

In case of triangular sections in which the compression is at the apex of the triangle, the moment capacity and the area of reinforcement are given by

$$M_{rb} = K_2 bd^2 f_{ck} \tag{15.5}$$

$$A_{st} = p_{02} \frac{f_{ck}}{f_y} \tag{15.6}$$

where $\qquad\qquad\qquad\qquad K_2$ and p_{02} are constants and are given in Table 15.1;

$\qquad\qquad\qquad\qquad\qquad b$ = width of the section at the level of the reinforcement.

Table 15.1 Design Coefficients

f_y	$k_3 = \dfrac{x_u}{d}$	j	K	K_2	p_0	p_{02}
250	0.531	0.78	0.149	0.030	0.220	0.066
360	0.495	0.79	0.141	0.026	0.205	0.057
415	0.479	0.80	0.138	0.025	0.198	0.054
500	0.456	0.80	0.133	0.023	0.189	0.049

In case of trapezoidal section, the moment capacity and the area of the reinforcement are given by

$$M_{rb} = ((K_2 - K)b_1 + K_2 b_2)\, d^2 f_{ck} \tag{15.7}$$

$$A_{st} = ((p_0 - p_{02})b_1 + p_{02}b_2)\, d\frac{f_{ck}}{f_y} \tag{15.8}$$

in which b_1 and b_2 are the widths of the section at compression face and at the level of the tension steel.

The design condition is that the strength of the section at any section must be more than the moment acting on the section.

The *critical shear* is computed at a distance equal to the effective depth from the face of the wall or column and it is assumed to act on the entire width at that plane. In case of footings resting on piles, the critical section is at half the effective depth from the face of the wall or column. In case of footing where two-way load action exists, the footing should be designed for *diagonal tension shear*. The critical shear plane is at a distance half the effective depth of the slab from the face of the column. The total effective width is equal to the total perimeter of a rectangle at a distance $d/2$ from the face of the column.

Design of the section. The total width of the footing is designed as one unit and the effective depth of the section is computed taking the total width. The reinforcement should be evenly distributed over the entire width. However, in case of two-way reinforced rectangular footings, the reinforcement in the longitudinal direction is evenly distributed and the reinforcement in short span shall be distributed as follows:

Reinforcement in the central band width

$$= \frac{2}{1+\beta}\ (\text{Total reinforcement in the short direction})$$

where the central band is equal to the width of the footing ($= B$). Therefore

$$\beta = \frac{L}{B}$$

L = long side of the footing

B = short side of the footing

Admissible strengths

a) Bending compressive strength in concrete in footings is same as that in other RCC superstructures ($= 0.67\, f_{ck}$).

b) Transverse shear strengths in concrete in footing is same as that in the superstructure (repeated in Table 15.2 for convenience).

c) The shear strength of concrete against failure through conical separation of the slab (or diagonal tension) is

$$\tau_{cc} = 0.25\,(0.5 + k_c)\,\tau_c \tag{15.9}$$

in which $\quad \tau_c = 0.25\,\sqrt{f_{ck}}$

$$k_c = \frac{\text{short side of the column}}{\text{long side of the column}}$$

and limited to 2 or 1 in case of square columns or capitals.

If the nominal shear on the section exceeds the strength but less than $1.5\ \tau_c$, then shear reinforcement should provided. In footings over soils, the shear reinforcement may be avoided as far as possible. The transverse reinforcement shall be designed for the difference in the shear force and half of the shear capacity of the section. In no case the shear stress on the slab shall exceed $1.5\ \tau_c$.

d) The bearing stress on the area between the column and footing be limited to the bearing strength. That is

$$\sigma_{br} \leq f_{br} \qquad (15.10)$$

in which

σ_{br} = bearing stress on the footing

$$f_{br} = 0.45\ f_{ck}\ \sqrt{\frac{A_1}{A_2}} \qquad (15.11)$$

where

A_1 = actual supporting area of the footing

A_2 = loaded area of the column base

e) The minimum thickness of the footing should be 150 mm in case of footings on soils and 300 mm in case of pile caps. In case of wall footings it can be even 100 mm.

Table 15.2 Design Shear Strength τ_c in MPa

$\dfrac{100\ A_s}{bd}$	Concrete grades		
	M15	M20	M25
0.25	0.35	0.36	0.36
0.50	0.45	0.47	0.49
0.75	0.54	0.56	0.57
1.00	0.60	0.62	0.64
1.25	0.64	0.67	0.70
1.50	0.68	0.72	0.74
2.0	0.71	0.79	0.82
2.5	0.71	0.82	0.88

15.2 Design of Foundations to Wall

The following are some of the typical RCC foundations for wall and isolated columns:

1. A strip foundation
   a) Axially loaded wall.
   b) Wall subjected to eccentric load or subjected to combined axial force and bending moment.

2. Isolated footing with constant thickness
   a) Axially loaded coulmn.
   b) Column subjected to either axial force and moment or eccentric load with symmetrical foundation,
   c) Same as in case (b) with unsymmetrical load.

3. Isolated footings with variable depth.

710

In all of the above cases there are two critical sections in the foundation slab, a section where the bending moment is maximum and another section where the shear force is maximum. The foundation slab cantilevers from the face of the wall or the column. The bending moment is maximum at the face of the column as shown in Fig. 15.1 as section A. However, the critical section where the maximum diagonal tension occurs is at 0.5d from the face of the column. The section which is subjected to critical shear stress, corresponds to a section

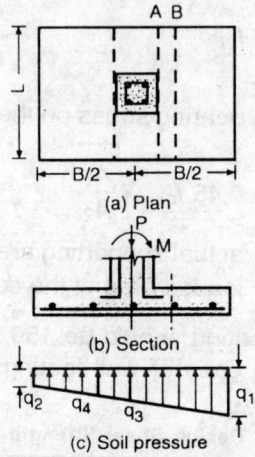

Fig. 15.1 Critical Bending Moment and Shear Force Sections.

at a distance d from the face of the column, where d is the effective depth of the reinforcement (Fig. 15.1). Let

L_m = cantilever span for critical bending moment

L_s = cantilever span for critical shear force

Let the intensity of pressure at the face of the column be equal to q_3 which can be computed as

$$q_3 = q_1 - \frac{(q_1 - q_2) L_m}{L} \qquad (15.12)$$

The bending moment developed at the face of the column or wall is

$$M = \frac{BL_m}{6} (q_3 + 2q_1) \qquad (15.13)$$

where q_i = net intensity of load on the footing from the soil. It is computed from the superimposed load on the footing slab. The weight of the footing and the soil over the footing or the surcharge over the footing are balanced by the bearing pressure corresponding to the loads. Therefore, only net pressure be taken in the moment or the shear force computations. Similarly, the critical maximum shear force for design is given by

$$V = \frac{BL_2}{2} (q_5 + q_1) \qquad (15.14)$$

where

$$q_5 = q_1 - \frac{(q_1 - q_2) L_s}{L} \qquad (15.15)$$

The values of q_1 and q_2 are computed as

$$q_1 = \frac{P_s}{A} + \frac{MB}{2I} \tag{15.16}$$

$$q_2 = \frac{P_s}{A} - \frac{MB}{2I} \tag{15.17}$$

where P_s = superimposed load on the footing and I = moment of inertia.

EXAMPLE 15.1 *Strip foundation to wall under axial force.* A brick wall carrying a live load of 60 kN/m and dead load of 80 kN/m (including the weight of the wall) is to be provided with a reinforced concrete slab foundation. The thickness of the wall next to the foundation is 340 mm. The allowable net bearing capacity of the soil is 115 kN/m².

Dead load $P_d = 80$ kN/m, Live load $P_l = 60$ kN/m
Bearing capacity $p_a = 115$ kN/m², Thickness of wall $a = 0.34$ m
                            $f_{ck} = 20$ MPa, and $f_y = 415$ MPa

For a balanced sectional design, the design coefficients taken from Table 15.1 are $K = 0.138$, and $j = 0.8$.

    Foundation design. Let the depth of foundation be 1 m. The net load on the soil is equal to the superimposed load on the footing plus the difference in weight of the footing and that of the equivalent volume of the soil.

Difference in densities = 8 kN/m³

    Estimate the volume of the concrete in the footing and in the pedestal or stem and add the corresponding difference in weight. This weight can be estimated in the range of 5 to 15% of the superimposed load to start with and then verified. The higher value of 15% is selected for heavy raft foundations. Superimposed load on the footing is

$$P_s = P_d + P_l = 80 + 60 = 140 \text{ kN/m}^2$$

The extra load due to the difference in weight of the concrete and soil is taken as $W' = 5$ kN.

Net load on the foundation = $P = P_s + W' = 145$ kN/m²

    As the foundation is a strip foundation for a wall, it is designed for 1 m length of the wall. So, the length of the foundation $L = 1$ m. The area of the foundation is given by

$$A = \frac{P}{p_a} = \frac{145}{115} = 1.26 \text{ m}^2$$

Select the width of the foundation as $B = 1.3$ m.
Net bearing pressure is

$$p = \frac{P}{BL} = \frac{145}{1.3} = 112 \text{ kN/m}^2$$

This value is less than the allowable pressure.

    Design for bending. Cantilever span of the footing for the maximum BM is

$$L_m = \frac{B}{2} - \frac{t}{2} = 0.65 - 0.17 = 0.48 \text{ m}$$

The net bearing load on the footing is equal to the total load minus the self-weight. The load intensity on the footing is

$$q = \frac{P_s}{A} = \frac{140}{1.3} = 107.7 \text{ kN/m}$$

Maximum collapse bending moment on the footing slab

$$M_c = \frac{\gamma_f q L_m^2}{2} = (1.5) \frac{107.70(0.48)^2}{2} = 18.61 \text{ kNm/m}$$

The design criterion for bending is

$$M_{rb} = Kbd^2 f_{ck} \geq M_c$$

Then the value of d is given by

$$d \geq \sqrt{\frac{0.01861}{0.138(20)(1)}} = 0.082 \text{ m}$$

The effective depth required to resist shear is usually higher than that to resist the bending moment. Therefore, the effective depth is rounded off to a higher figure when compared with that obtained from moment computation.

Use $d = 0.14$, then the area of the reinforcement is

$$A_{st} = \frac{1.15 M_c}{jdf_y} = \frac{1.15(18610)}{0.8(0.14)(415)} = 460 \text{ mm}^2/\text{m}$$

$$\% \text{ reinforcement} = \frac{460(100)}{(1000)(140)} = 0.33$$

Design for shear. The critical section for the diagonal tension failure is at a distance d from the face of the wall. The shear span is

$$L_s = L_m - d = 0.48 - 0.14 = 0.34 \text{ m}$$

The design shear force is

$$V = \gamma_f q L_s = (1.5) \, 107.7 \, (0.34) = 54.93 \text{ kN/m} = 0.05493 \text{ MN/m}$$

Nominal shear stress is $\tau_v = 0.05\,493/1\,(0.14) = 0.39$ MPa

The shear capacity for 0.33% of steel is 0.38 MPa.

The nominal shear stress exceeds the permissible value marginally. Therefore, either the sectional depth be modified or shear reinforcement provided. It is not desirable to introduce shear reinforcement in footing slabs, therefore increase the thickness of the footing slab to meet the demand of shear force. The depth of the slab based on shear consideration is

$$d \geq \frac{V}{b\tau_c} = \frac{0.05493}{1(0.38)} = 0.141 \text{ m}$$

Let $d = 0.15$ m and provide a cover of about 50 mm to the reinforcement, then the overall depth of the slab is $h = d + \text{cover} + \frac{1}{2}$ (dia of bars) $= 0.15 + 0.06 = 0.21$ m

Volume of the footing is $v_1 = LBh = 0.273 \text{ m}^3$

Volume of the stem of the wall over the footing is

$$v_2 = Lt(D - h) = 1(0.34)(0.79) = 0.269 \text{ m}^3$$

The density of concrete is taken as 25 kN/m³, of brickwork as 20 kN/m³ and of the soil as 17 kN/m³. The additional difference in load on the foundation soil is

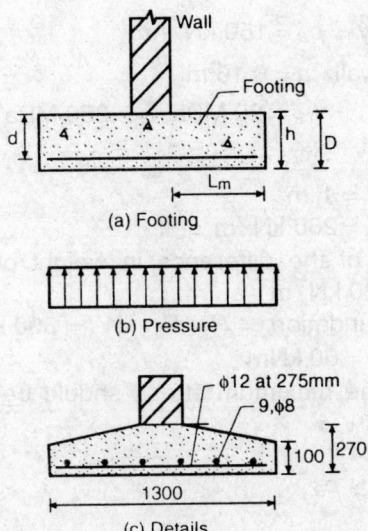

(a) Footing

(b) Pressure

ϕ12 at 275mm
9,ϕ8
100 | 270
1300

(c) Details

Fig. 15.2 Reinforcement Details for Foundation of a Wall Footing, Example 15.1.

$$W' = v_1(25) + v_2(20) - (v_1 + v_2)(17) = 3 \text{ kN/m}$$

The weight difference assumed was 5 kN/m compared to the actual value of 3 kN/m. Therefore, the net bearing pressure is within the permissible limits.

The area of tension reinforcement is

$$A_{st} = \frac{1.15 \, M}{jdf_y} = \frac{1.15(18\,610)}{0.8(0.15)(415)} = 430 \text{ mm}^2/\text{m}$$

Provide 12 mm bars at 260 mm spacing

$$A_{st} \text{ (provided)} = 113 \frac{(1000)}{260} = 435 \text{ mm}^2/\text{m}$$

A minimum of 0.15% distribution reinforcement should be provided so the area of distribution steel needed is

$$A_s = \frac{0.15 \, Bh}{100} = \frac{0.15}{100} (1300)(210) = 410 \text{ mm}^2$$

Provide 9 numbers of 8 mm dia bars as distribution reinforcement. Figure 15.2 illustrates the reinforcement details. The thickness of the slab at the free ends can be reduced to 100 mm.

EXAMPLE 15.2 *Strip foundation subjected to axial force and bending moment.* A reinforced concrete wall of thickness 160 mm is subjected to a live load of 100 kN/m and a dead load of 180 kN/m. A bending moment of 60 kNm/m caused by the live load is also acting on the wall. The dead load indicated above includes the self-weight of the wall upto the ground level. Design a RCC foundation to the wall if the net allowable bearing capacity of the soil is 150 kN/m^2 at 1 m below G.L.

Dead load $P_d = 180 \text{ kN/m}$

Live load $P_l = 100 \text{ kN/m}$

Total load $P_s = 280 \text{ kN/m}$

$$\text{Bearing capacity} \quad p_a = 150 \text{ kN/m}^2$$

$$\text{Thickness of the wall} \quad a = 0.16 \text{ m}$$

$$f_{ck} = 20 \text{ MPa}, \, f_y = 250 \text{ MPa (mild steel)}$$

Foundation design

Let the depth of foundation = 1 m

Total superimposed load $P_s = 280$ kN/m

Add extra load to account of the difference in weight of the concrete and soil in the foundation (about 6%) = $W' = 20$ kN/m

Total design load on the foundation = $P = P_s + W' = 300$ kN/m

Total bending moment = $M_f = 60$ kNm

The design condition that the maximum stress should be less than the allowable value gives

$$\frac{P}{B} + \frac{6M_f}{B^2} \le p_a$$

or

$$B^2 - \frac{PB}{p_a} - \frac{6M_f}{p_a} \ge 0$$

The substitution of the values of P, p_a and M in the above equation results into

$$B^2 - 2B - 2.4 \ge 0$$

or

$$B \ge 2 + \sqrt{\frac{4 + 9.6}{2}} = 2.844 \text{ m}$$

Use

$$B = 2.9 \text{ m},$$

then

$$A_f = 2.9 \text{ m}^2, \, I_f = 2.9^3/12 = 2.032 \text{ m}^4, \, Z_f = 2.9^2/12 = 0.7008 \text{ m}^3$$

where A_f and I_f are the area the 2nd moment of area of the foundation. Actual maximum and minimum pressures under the footing are

$$p_1 = \frac{P}{A_f} + \frac{M_f}{Z_f} = 103.45 + 42.81 = 146.26 \text{ kN/m}^2 < p_a$$

$$p_2 = 103.45 - 42.81 = 60.64 \text{ kN/m}^2 > 0 \; but \; < p_a$$

Structural design. The net reacting intensities of soil pressures on the foundation slab at extreme point and at the face of the wall are (Fig. 15.3):

$$q_1 = \frac{P_s}{A_f} + \frac{M_f y_1}{I_f} = 96.55 + 42.81 = 139.36 \text{ kN/m}^2$$

$$q_3 = \frac{P_s}{A_f} + \frac{M_f y_3}{I_f} = 96.55 + 2.36 = 98.91 \text{ kN/m}^2$$

The maximum bending moment on the footing caused by the net reacting pressure of the soil occurs at the face of the wall and its value is

$$M = \frac{\gamma_f LL_m^2}{6}(2q_1 + q_3) = \frac{(1.5)(1)(1.37)^2(2(139.36) + 98.91)}{6} = 177.19 \text{ kNm/m}$$

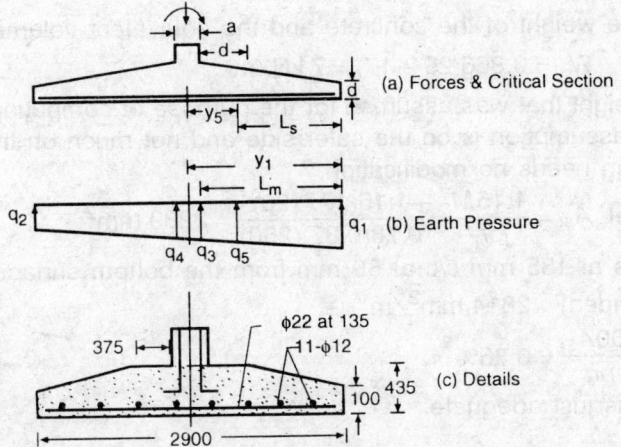

Fig. 15.3 Soil Pressure and Reinforcement Details of Wall Footing, Example 15.2.

The resisting moment capacity of the section must be more than the active moment and this can be expressed as

$$M_r \geq M \quad \text{or} \quad Kbdf_{ck} \geq 0.17719 \text{ MNm/m}$$

or

$$d = \sqrt{\frac{0.11719}{0.149(20)}} = 0.20 \text{ m}$$

Use $d = 0.37$ so as to provide adequate resistance against shear.

Shear span $= L_s = L_m - d = 1.37 - 0.37 = 1 \text{ m}$

The distance of this point from the centre $= y_5 = 1.45 - 1 = 0.45 \text{ m}$

The intensity of the pressure at the section is

$$q_5 = \frac{P_s}{A_f} + \frac{M_f y_5}{I_f} = 108.83 \text{ kN/m}^2$$

The shear force on the cantilever is

$$V = \frac{\gamma_f L L_s (q_1 + q_5)}{2} = \frac{1.5(1)(139.36 + 108.83)}{2}$$

$$= 186.1 \text{ kN/m}$$

The nominal shear stress $= \tau_v = \dfrac{V}{bd} = 0.5 \text{ MPa}$

The shear capacity for 0.7% of steel is 0.5 MPa.
The overall depth of the slab at the face of the wall is equal to

$$h = d + \text{cover} + \phi/2$$

$$= 375 + 50 + 15 = 435 \text{ mm}$$

Let the overall depth of the slab at the free end be equal to 100 mm (= a nominal thickness). The volume of the concrete in the foundation and the wall stem below the ground level is approximately equal to

$$\frac{1(0.1 + 0.435)(2.9)}{2} + (1 - 0.435)(0.16) = 0.866 \text{ m}^3$$

The difference in the weight of the concrete and the equivalent volume of the soil is
$$W' = 0.866(25 - 17) = 7 \text{ kN/m}$$
The difference of weight that was assumed for the purpose of computing the bearing capacity was 20 kN/m. The assumption is on the safer side and not much on the uneconomical side. Therefore, the design needs no modification.

Area of tension steel $A_{st} = \dfrac{1.15M}{jdf_y} = \dfrac{1.15(177190)}{0.78(0.37)(250)} = 2289 \text{ mm}^2$

Provide 22 mm bars at 135 mm c/c at 50 mm from the bottom surface
$$A_{st} \text{ (provided)} = 2814 \text{ mm}^2/\text{m}$$
$$\frac{100A_{st}}{bd} = 0.76\%$$
The shear strength is just adequate.

Provide 11 numbers of 12 mm bars as distribution reinforcement.
Figure 15.3 illustrates the reinforcement details.

EXAMPLE 15.3 *Design of eccentrically loaded footing of a wall.* A RCC wall 200 mm thick carries a live load of 100 kN/m and dead load of 180 kN/m. The wall is resting on a slab of 3.2 m width at a distance of 1.25 m from one face of the slab. This makes the wall eccentric with respect to the slab CG. Design a RCC foundation slab. Bearing capacity of the soil is

$$p_a = 150 \text{ kN/m}^2.$$

Width of foundation	$B = 3.2$ m; thickness of wall = $a = 0.2$ m
Dead load	$P_d = 180$ kN/m
Live dead	$P_l = 100$ kN/m
Total load	$P_s = P_d + P_l = 280$ kN/m

$$p_a = 150 \text{ kN/m}^2, \ f_y = 415 \text{ MPa}, \ f_{ck} = 20 \text{ MPa}$$

Design of the foundation. Assume the difference in weight of the foundation concrete and that of equivalent soil as 0.06 $P_s = 16$ kN

Eccentricity of the load $e = B/2 - 1.25 = 0.35$ m

Total load on the foundation
$$P = P_s + 16 = 296 \text{ kN/m}$$

The moment caused by the eccentric load on the foundation
$$M = M_f = P_s e = 280(0.35) = 98 \text{ kNm/m}$$

The area and the moment of inertia of the base are
$$A_f = LB = (1)(3.2) = 3.2 \text{ m}^2$$
$$I_f = \frac{LB^3}{12} = \frac{3.2^3}{12} = 2.731 \text{ m}^4$$

The maximum and minimum pressures on the foundation soil are
$$p_1 = \frac{P}{A_f} + \frac{MB}{2I_f} = 92.5 + 57.4 - 149.9 \text{ kN/m}^2$$
$$p_2 = 92.5 - 57.4 = 35.1 \text{ kN/m}^2$$

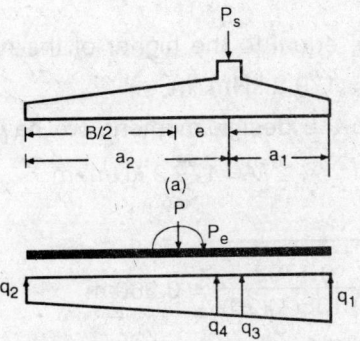

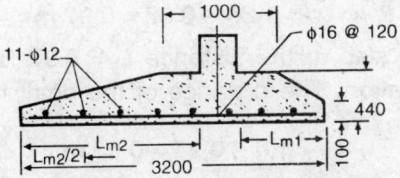

Fig. 15.4 Soil Pressure and Reinforcement Details of Eccentrically Load Wall Footing, Example 15.3.

p_1 is less than p_a, therefore the width of the foundation is adequate.

Structural design. The net pressures on the soil at the extreme points and also at the two faces of the wall are (see Fig. 15.4):

$$q_1 = \frac{P_s}{A_f} + \frac{M_f y_1}{I_f} = 87.5 + 35.87(1.6) = 144.9 \text{ kN/m}^2$$

$$q_2 = 87.5 - 35.87(1.6) = 30.12 \text{ kN/m}^3$$
$$q_3 = 87.5 + 35.87(e + 0.5a)$$
$$\quad = 87.5 + 35.87(0.35 + 0.1) = 103.6 \text{ kN/m}^2$$

$$q_4 = 87.5 + 35.87(0.35 - 0.1) = 96.5 \text{ kN/m}^2$$

The outer cantilever span on the footing slab is

$$L_{m1} = 1.25 + 0.1 = 1.35 \text{ m}$$

The design bending moment at the outer face of the wall on the cantilever slab is

$$M_1 = \gamma_f \frac{L_{m1}^2}{6} (2q_1 + q_3) = \frac{(1.5)\, 1.35^2}{6} (2(144.9) + 103.6) = 179.2 \text{ kNm/m}$$

The cantilever span on the inside face is

$$L_{m2} + 0.5B + e - 0.5a = 1.6 + 0.35 - 0.1 = 1.85 \text{ m}$$

The bending moment on the inside face of the cantilever is

$$M_2 = \gamma_f \frac{L_{m2}^2}{6} (2q_2 + q_4)$$

$$\quad = \frac{1.5(1.85)^2}{6} (60.2 + 96.5) = 134.1 \text{ kNm/m}$$

The design moment is, therefore, equal to the higher of the two values:

$$M = M_1 = 179.2 \text{ kNm/m}$$

Equating the resisting moment to the design moment, we have

$$M_{rb} = Kbd^2 f_{ck} \geq M = 179.2 \text{ kNm/m}$$

or

$$d \geq \sqrt{\frac{0.1792}{0.138(1)(20)}} = 0.260 \text{ m}$$

However, use $d = 0.38$ m to provide adequate resistance to shear. The diagonal tension is critical at a distance d from the face, so the shear force span is

$$L_s = L_{m1} - d = 1.35 - 0.38 = 0.97 \text{ m}$$

The intensity of force on the slab at the distance $L_s = 0.97$ is the same as that for distance y_5 from the CG of the foundation. The distance of the point from the CG is

$$y_5 = \frac{B}{2} - L_s = 1.6 - 0.97 = 0.63 \text{ m}$$

The intensity of the pressure distance y_5 is

$$q_5 = \frac{P_s}{A_f} + \frac{M_f y_5}{I_f} = 110.1 \text{ kN/m}^2$$

The maximum design shear force at the critical face is

$$V = \frac{\gamma_{fb} L_3 (q_1 + q_5)}{2}$$

$$= (1.5)1(0.97)\frac{144.9 + 110.11}{2} = 185.5 \text{ kN/m}$$

The nominal shear stress on the section is

$$\tau_v = \frac{V}{bd} = \frac{0.1855}{1(0.38)} = 0.49 \text{ MPa}$$

$$h = d + \text{cover} + \phi/2 = 380 + 50 + 10 = 440 \text{ mm}$$

Let the overall depth of the slab at the free ends be 100 mm. The volume of the concrete plus that of the wall below the ground level is approximately given by

$$\frac{1(0.1 + 0.44)(3.2)}{2} + (1 - 0.44)(0.2) = 0.976 \text{ m}^3$$

The difference in weight of the concrete and that of the equivalent volume of the earth masonry is $W' = 0.976 (25 - 17) = 7.8$ kN. The actual weight difference that was assumed for the purpose of computing the bearing pressure was 16 kN which is higher than the actual value. Therefore, the design is slightly on the safer side. The difference (16–7.8) kN in 296 kN will not be of much consequence to the pressure on soil.

Design of the reinforcement. The area of tension reinforcement is

$$A_{st} = \frac{1.15 \, M}{jdf_y} = \frac{1.15(179200)}{0.80(0.38)(415)} = 1633 \text{ mm}^2 \text{ m}$$

Provide 16 mm diameter bars at 120 mm c/c

Let this reinforcement be provided on both faces and half of it be curtailed at half the distance from the toe.

$$A_s \text{ (provided)} = \frac{201(1000)}{120} = 1675 \text{ mm}^2/\text{m}$$

Percentage of the reinforcement at the face of the wall

$$P = \frac{A_s(100)}{bd} = \frac{1675}{380000} = 0.44\%$$

The shear strength for such a percentage of reinforcement is 0.48 MPa against the nominal shear stress of 0.49 MPa. The area of the cross-section of the concrete in foundation is

$$A_c = 0.1(3.2) + \frac{((0.2 + 0.4 + 0.4) + 3.2)}{2} (0.46 + 0.1) = 1.076 \text{ m}^2$$

The distribution reinforcement is

$$A_s = 0.0015 \, (1.076)10^6) = 1614 \text{ mm}^2$$

Provide 15 numbers of 12 mm bars over the width (Fig. 15.4).

15.3 Isolated Footing

Isolated footing or even some combined footings are subjected to maximum bending moment at the face of the column. Therefore, a maximum depth of the footing at the face of the column is provided and it is decreased to minimum towards the free edge as shown in Fig. 15.5a. The moment resisting capacity of such a section is not as that of a rectangular section as the width of the section is variable. Figure 15.5c illustrates the variable width and the

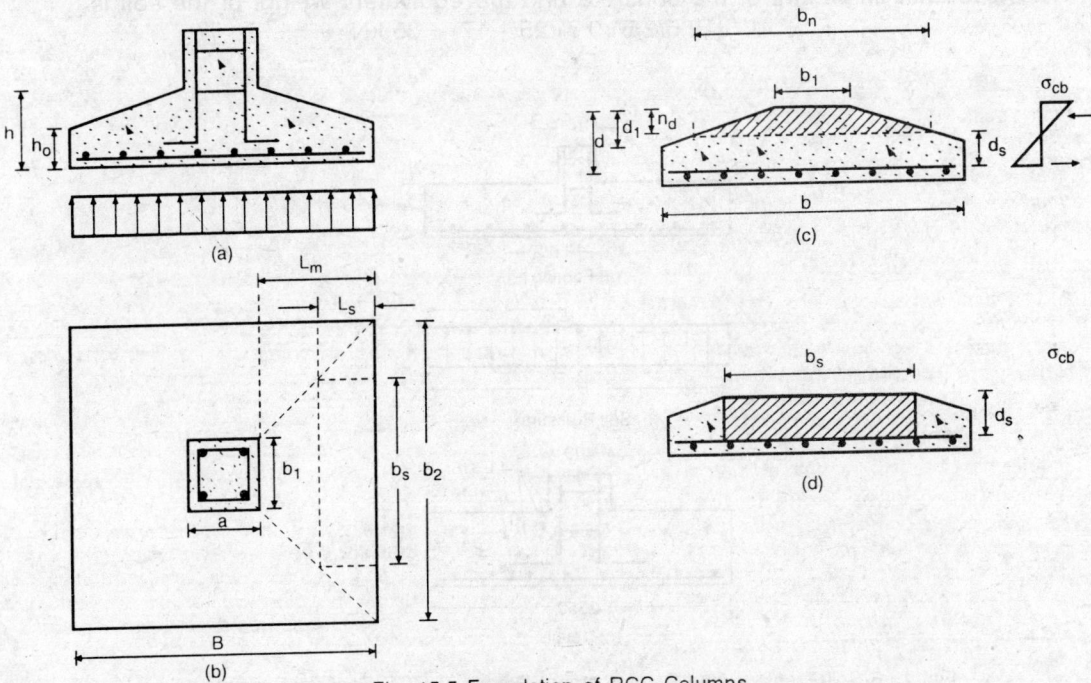

Fig. 15.5 Foundation of RCC Columns.

neutral axis of the section. The balanced moment capacity of a trapezoidal section was given in Eq. (15.7). The design criterion would be to ensure that the bending resistance of the section is more than the design moment on the section. If the value of $(d - d_1)$ which corresponds to the least depth at the free end is selected, then one can solve for d and A_{st} from Eqs. (15.7) and (15.8). The overall depth of the section at the free edge is usually taken in the range of 100 to 150 mm. In case, the footing thickness is taken as uniform, then the effective depth can be calculated from the standard formula. The method of design is illustrated through examples.

Design against shear force in tapered sections becomes more complicated even to select the location of the critical diagonal tension zone. This is because the effective depth decreases along the span. The depth of the section upto a distance d from the face of the column is usually maintained constant.

EXAMPLE 15.4 *Isolated footing with axial load and uniform thickness.* A column is subjected to dead and live loads of 340 kN and 260 kN respectively. The size of the column is 400 mm by 400 mm. Design a RCC footing if the allowable bearing capacity of the soil is 100 kN/m² and the soil is clay.

$$\text{Dead load} \quad P_d = 340 \text{ kN}$$
$$\text{Live load} \quad P_l = 260 \text{ kN}$$
$$\text{Superimposed load} \quad P_s = \overline{600 \text{ kN}}$$

Bearing capacity $\quad p_a = 100 \text{ kN/m}^2, f_{ck} = 20 \text{ MPa}, f_y = 415 \text{ MPa}.$

Design of the foundation. Since the soil is clay with low bearing capacity, select the depth of foundation as 1.0 m. Let the approximate size of the footing be $2.5 \times 2.5 \times 0.7$, then the volume of the concrete is $2.5(2.5)(0.7)$ m³.

The difference in weight of the concrete and the equivalent weight of the soil is
$$W' = 2.5(2.5)(0.7)(25 - 17) = 35 \text{ kN}$$

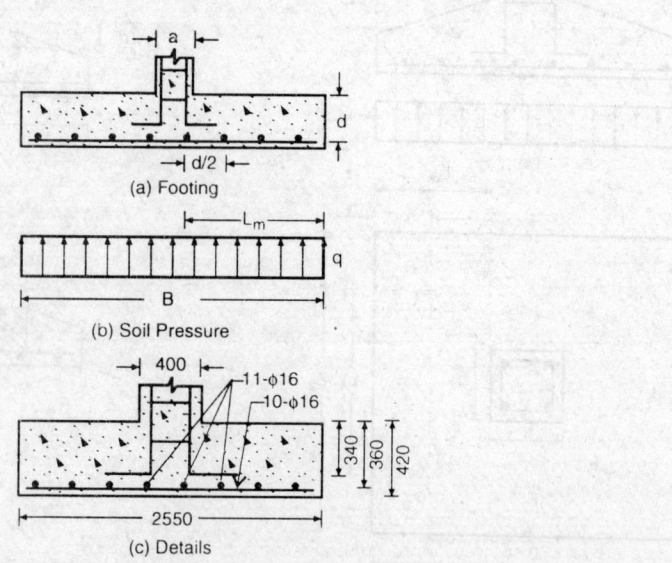

(a) Footing

(b) Soil Pressure

(c) Details

Fig. 15.6 Soil Pressure and Reinforcement Details of RCC Footing, Example 15.4.

Use $\qquad W' = 40$ kN

Total load on the foundation soil is

$$P = P_s + W' = 640 \text{ kN}$$

Area of the foundation should be such that the bearing pressure on the soil should be less than the net allowable bearing capacity of the soil. This condition leads to

$$\frac{P}{A} = \frac{640}{B^2} \le p_a = 100$$

or

$$B^2 \ge \frac{640}{100} = 6.4 \text{ m}^2$$

Select $\qquad B = 2.55$

Bearing pressure is

$$p = \frac{P}{A} = \frac{640}{B^2} = 98.42 \text{ kN/m}^2$$

Intensity of the load on the footing for design is

$$q = \frac{P_s}{A} = \frac{600}{6.5025} = 92.27 \text{ kN/m}^2$$

Structural design:

$$L_m = \frac{B}{2} - \frac{a}{2} = 1.275 - 0.2 = 1.075 \text{ m}$$

Maximum bending moment is

$$M = \gamma_f \frac{B L_m^2 q}{2}$$

$$= \frac{(1.5)2.55(1.075)^2}{2}(92.27) = 204 \text{ kNm}$$

Equating the moment capacity to the design moment, we have

$$Kbd^2 f_{ck} = M$$

or

$$d = \sqrt{\frac{0.204}{0.138(2.55)(20)}} = 0.17 \text{ m}$$

Let $\qquad d = 0.25$ m

The overall depth of the footing is

$$h = d + \phi + \phi/2 + \text{cover} = 0.25 + 0.02 + 0.1 + 0.05 = 0.33 \text{ m}$$

The section can be designed against shear force similar to the flat slab. The critical section is at a distance of ($d/2$) from the face of the column and the length of the periphery of critical section is

$$b_0 = 4(b_1 + d) = 4(0.4 + 0.25) = 2.6 \text{ m}$$

The size of the square where shear force is maximum is

$$b_s = b_1 + d$$

The shear force on the periphery is

$$V = \gamma_f q (B^2 - b_s^2) = (1.5)92.27(2.55^2 - 0.65^2) = 841 \text{ kN}$$

The nominal shear stress at the critical section is

$$\tau_v = \frac{V}{b_0 d} = \frac{0.841}{2.6(0.25)} = 1.29 \text{ MPa}$$

The permissible shear stress is given by

$$\tau_c = k_s \tau_{cc}$$

where $k_s = (0.5 + \beta_c) \not> 1$

$$\beta_c = \frac{\text{short side of shear periphery}}{\text{long side}} = 1$$

Since k_s should be equal or greater than 1, the value of k_s works out to be equal to 1.

$$\tau_{cc} = 0.25 \sqrt{f_{ck}}$$

Substitution of various quantities gives

$$\tau_c = k_s \tau_{cc} = 0.25 \sqrt{20} = 1.1 \text{ MPa}$$

Since the shear strength is less than the nominal shear stress, the section is not safe against shear force. Either shear reinforcement is to be provided or depth is to be increased. In flat slabs and footings it is better to increase the effective depth. Let the effective depth be increased approximately in the following ratio:

$$d = \left(\frac{\tau_v}{\tau_c}\right) d = \frac{1.29}{1.1}(0.25) = 0.3 \text{ m}$$

Use
$$d = 0.34 \text{ m}$$

The peripheral distance at the critical shear zone is

$$b_0 = 4(b_1 + d) = 4b_s = 1.6 + 1.36 = 2.96 \text{ m}$$

The shear force on the periphery is

$$V = \gamma_{fq} (B^2 - b_s^2) = 1.5(92.27)(2.55^2 - 0.74^2) = 824 \text{ kN}$$

The nominal shear stress on the critical plane is

$$\tau_v = \frac{V}{b_0 d} = \frac{0.824}{2.96(0.34)} = 0.82 \text{ MPa}$$

Check for transverse shear stress. In addition to the design against (punching shear) diagonal tension, the footing has to be designed for transverse shear similar to the one done in wall footings. The critical section is at a distance d from the face of the column and the entire width of the footing is accounted in resistance. The shear force on the critical plane is

$$V = \gamma_f B \left(\frac{B}{2} - d\right) q = (1.5)2.55(1.275 - 0.34)(92.27)$$

The nominal shear stress is

$$\tau_v = \frac{V}{bd} = (1.5)\frac{(1.275 - 0.34)(0.09227)}{0.34} = 0.38 \text{ MPa}$$

The shear strength is 0.38 MPa even for 0.3% reinforcement. Therefore, the nominal shear stress will be within the limits.

Design of reinforcement. The reinforcement should be provided in two orthogonal directions. The depth of the steel in one direction is slightly different from that in another direction.

Depth of steel in one direction = 0.34 m.
Area of steel in that direction is

$$A_{st} = \frac{1.15\ M}{jdf_y} = \frac{(1.15)204000}{0.8(0.34)(415)} = 2078\ \text{mm}^2$$

Provide 11 numbers of 16 mm bars.
The effective depth of steel in another direction is

$$d_x = d_y + \phi = 0.34 + 0.02 = 0.36\ \text{m}$$

Area of the steel required

$$A_{st} = \frac{1.15\ M}{jdf_y} = \frac{1.15\ (204\ 000)}{0.8(0.36)(415)} = 1963\ \text{mm}^2$$

Provide 10 numbers of 16 mm bars in another direction. Let the overall depth = 0.42 m

EXAMPLE 15.5 *Design of an isolated rectangular footing.* A square column of 400 mm size is carrying a live load of 260 kN and dead load of 340 kN including the self-weight. The allowable net bearing capacity of the soil is 100 kN/m^2 and allowable width of the footing is 2 m. Design an isolated footing with M20 concrete and HYSD reinforcement.

$$\text{Dead load} \qquad P_d = 340\ \text{kN}$$

$$\text{Live load} \qquad P_l = 260\ \text{kN}$$

$$\text{Superimposed load (on the footing)}\ P_s = \overline{600\ \text{kN}}$$

$$\text{Bearing capacity} \qquad p_a = 100\ \text{kN/m}^2$$

Let the difference in weight of the concrete in the footing and equivalent volume of the soil = 0.06 (600) = 36 kN.

Total net load on the soil $P = 600 + 36 = 636$ kN
Area of the footing needed is

$$A_f = \frac{P}{p_a} = \frac{636}{100} = 6.36\ \text{m}^2$$

The width of the footing is restricted to 2 m, therefore the length of the footing

$$L = \frac{A_f}{B} = \frac{6.36}{2} = 3.18\ \text{m}$$

Use the footing size as

$$B = 2\ \text{m},\ L = 3.2\ \text{m and } A_f = 2(3.2) = 6.4\ \text{m}^2$$

The net reaction acting on the footing is

$$q = \frac{P_s}{A_f} = \frac{600}{6.4} = 93.75\ \text{kN/m}^2$$

The footing slab is designed with tapering top surface.

Design of the section. The cantilever projection of the footing in one direction is (2 – 0.4)/2 = 0.8 m and in the another direction is (3.2 – 0.4)/2 = 1.4 m. The maximum bending moment on the slab occurs at the face of the column corresponding to the longer arm.

The maximum bending moment is

$$M = \frac{\gamma_f q\, BL_m^2}{2} = (1.5)93.75\,\frac{(2)(1.4)^2}{2} = 275.63 \text{ kNm}$$

The moment capacity of a trapezoidal section is given by

$$M_r = ((K - K_2)\, b_1 + K_2 b_2)\, d^2 f_{ck}$$

where $b_1 = 0.4$ m, $b_2 = B = 2$ m, $K = 0.138$, $K_2 = 0.025$

The substitution of the different quantities we have

$$M_r = (0.113(0.4) + 0.025(2))(20)\, d^2 = 1.9\, d^2$$

Equating the moment capacity to the external moment gives

$$d = \sqrt{\frac{0.27563}{1.9}} = 0.38 \text{ m}$$

Select $d = 0.58$ m, $h = 0.66$ m, and $h_0 = 0.15$ m

Volume of the concrete in the footing is

$$2(3.2)(0.15) + \frac{(2(3.2) + 0.4(0.4))(0.66 - 0.15)}{2} = 2.6328 \text{ m}^3$$

The difference in weight of the concrete and that of the equivalent soil of the footing is

$$W'_g = 2.6328(8) = 21.06 \text{ kN}$$

This value is less than the assumed one for the purpose of calculation of net pressure on the soil. Therefore no revision is needed

Check for shear capacity. The critical section where shear failure is likely to occur is at a distance $d/2$ from the face of the column. The periphery of the shear boundary is

$$b_0 = 4(a + d) = 4(0.4 + 0.58) = 3.92 \text{ m}$$

The net shear force acting on the critical shear periphery is

$$V = \gamma_f q(BL - (a + d)^2) = 1.5(93.75)(2(3.2) - 0.98^2) = 765 \text{ kN}$$

The effective depth of the steel at a distance $d/2$ from the face of the column is

$$d_s = h - \frac{(h - h_0)d}{2L_m} - \text{cover}$$

$$= 0.66 - \frac{(0.66 - 0.15)(0.593)}{2(1.4)} - 0.05 = 0.47 \text{ m}$$

The nominal shear stress at the critical zone is

$$\tau_v = \frac{V}{b_0 d_s} = \frac{0.765}{3.22(0.47)} = 0.42 \text{ MPa}$$

The shear strength of M20 concrete for slabs is

$$\tau_c = 0.25\,\sqrt{20} = 1.1 \text{ MPa}$$

The section is safe against shear failure including transverse shear.

Design of the reinforcement. Area of reinforcement for a balanced trapezoidal section is

$$A_{st} = \frac{1.15\, M}{jd f_y} = \frac{1.15(275\,630)}{0.8(0.58)(415)} = 1646 \text{ mm}^2$$

Use 9 numbers of 16 mm bars in the long span

Design of reinforcement in the short span direction. The reinforcement required in the short span direction is proportioned with respect to the bending moment and the effective depth in comparison to the long span. The maximum bending moment is proportional to the square of the cantilever span. The effective depth of the steel in this direction could be taken equal to

$$d_y = d - \phi = 0.58 - 0.02 = 0.56 \text{ m}$$

The area of the reinforcement required in the short span is

$$A_{sty} = A_{st}\left(\frac{L_{my}}{L_m}\right)^2 \frac{d}{d_y} = 1646\left(\frac{0.8}{1.4}\right)^2\left(\frac{0.58}{0.56}\right) = 369 \text{ mm}^2$$

Provide 5 numbers of 10 mm bars
Figure. 15.7 illustrates the design details.

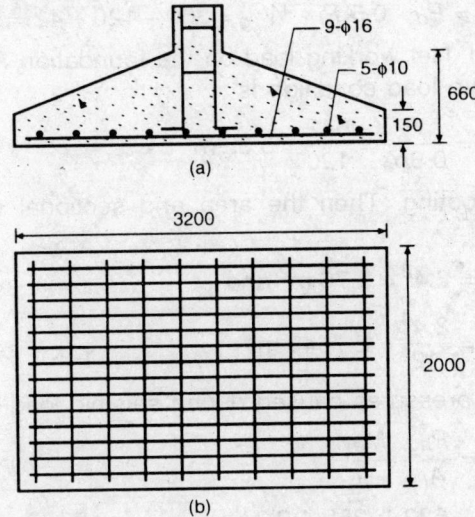

Fig. 15.7 Reinforcement Details of Tapered Footing, Example 15.5.

The same example was solved by WSD in which the overall depth was found to be 0.7 m as against 0.58 m in this method. However, the reinforcement in the WSD method is about 2/3 of that of this method.

EXAMPLE 15.6 *Design of footing subjected to axial force and seismic moment.* A column is transferring a live load of 240 kN and dead load of 360 kN to the foundation. In addition, a bending moment of 250 kNm is applied on the foundation during earthquake load condition for 50% of the live load. Design an isolated footing for an allowable bearing capacity of the soil of 120 kN/m^2.

    Dead load $P_d = 360$ kN
    Live load $P_l = 240$ kN
    Earthquake moment $M_e = 250$ kNm

    Live load during earthquake

$$P_{l1} = 120 \text{ kN}$$

Allowable bearing $\qquad p_a = 150 \text{ kN/m}^2$

Size of the column $\qquad a = 0.6 \text{ m}$

Use M20 concrete with HYSD bars

The allowable bearing pressure on the soil during seismic load condition is same as that during normal live load condition for isolating footings.

Let the additional load on the soil due to replacement of the soil by the concrete be about 7% of the total load, and is given by

$$W'_g = 0.07(P_d + P_l) = 0.07(360 + 240) = 42 \text{ kN}$$

The net load on the soil is

$$P = P_d + P_l + W'_g = 642 \text{ kN}$$

Net load during seismic load condition is

$$P_e = P_d + 0.5 \, P_l + W'_g = 360 + 120 + 42 = 522 \text{ kN}$$

Design of the foundation. Net working load on the foundation $P = 642$ kN. Bearing area required based on the net live load condition is

$$A_f = \frac{P}{0.8p_a} = \frac{642}{120} = 5.35 \text{ m}^2$$

Select 2.4 by 2.4 m size footing. Then the area and sectional moment of inertia of the foundation are

$$A_f = 2.4^2 = 5.76 \text{ m}^2 \text{ and}$$

$$I_f = \frac{2.4^4}{12} = 2.7648 \text{ m}^4$$

The maximum and mininum pressures caused during seismic load condition are :

$$p_1 = \frac{P_e}{A_f} + \frac{M_e y}{I_f}$$

$$= \frac{522}{5.76} + \frac{250(1.2)}{2.7648} = 90.625 + 108.507 = 199.132 \text{ kN/m}^2$$

$$p_2 = 90.625 - 108.507 = -17.882 \text{ kN/m}^2$$

The maximum pressure under seismic load condition is far higher than the allowable value and in addition tension occurs on the soil. Therefore, the size of the footing has to be increased so as to limit maximum pressure within the limits. Since the bending moment acting on the soil is large and the soil cannot resist the tension, the total area of the footing will not be effective.

Let $\qquad B$ = width of the footing (Fig. 15.8)

$\qquad B_c$ = actual width in contact with the soil

$\qquad L$ = length of the footing = 2.8 m

The equilibrium of forces in the seismic load condition on the footing gives

$$\frac{LB_c \, p_1}{2} = P_e = 522 \text{ kN} \tag{1}$$

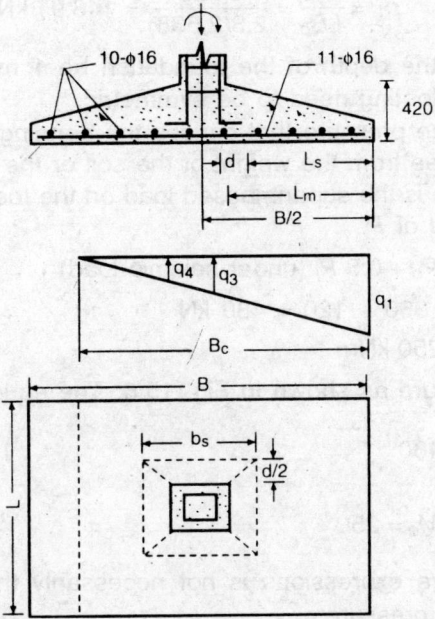

Fig. 15.8 Reinforcement Details and Pressure Distribution of Footing Subject to Combined Load and Bending of Example 15.6.

$$\frac{LB_c p_1}{2}\left(\frac{B}{2} - \frac{B_c}{3}\right) = M_f = 250 \qquad (2)$$

in which p_1 = maximum pressure on the soil. Let p_1 be set equal to the allowable pressure under seismic load condition, then

$$p_1 = 150 \text{ kN/m}^2$$

From Eq. (1), we have

$$B_c = \frac{2P_e}{Lp_1} = \frac{2(522)}{(2.8)(150)} = 2.486 \text{ m}$$

From Eq. (2), we have

$$\frac{B}{2} = \frac{2M_f}{LB_c p_1} + \frac{B_c}{3} = \frac{2(250)}{2.8(2.486)(150)} + \frac{2.486}{3} = 2.616 \text{ m}$$

Use $L = 2.80$ m and $B = 2.65$ m and check for the pressure on the soil. From Eqs. (1) and (2), we have

$$\frac{B}{2} - \frac{B_c}{3} = \frac{M_e}{P_e} = \frac{250}{522}$$

or

$$B_c = 3\left(\frac{B}{2} - \frac{250}{522}\right) = 3\left(\frac{2.65}{2} - \frac{250}{522}\right) = 2.538 \text{ m}$$

From Eq. (1), we have

$$p_1 = \frac{2P_e}{LB_c} = \frac{2(522)}{2.8(2.538)} = 146.91 \text{ kN/m}^2$$

Since the soil is clayey, let the depth of the foundation be 1 m. The seismic moment can occur on either side, so the footing need to be symmetric.

Design of the section. The pressure that causes the bending moment on the slab is the effective pressure which is free from the weight of the soil or the self-weight of the slab. The effective load for this purpose is the superimposed load on the footing. The effective pressure is computed using P_s instead of P:

$$P_s = P_d + 0.5 \, P_l \text{ (under seismic load)}$$

$$= 360 + 120 = 480 \text{ kN}$$

$$M_e = 250 \text{ kNm}$$

Let q_1 = maximum pressure as shown in Fig. 15.6. The equilibrium of forces gives

$$\frac{LB_c q_1}{2} = P_s = 480 \tag{3}$$

$$\frac{LB_c q_1}{2}\left(\frac{B}{2} - \frac{B_c}{3}\right) = M_e = 250 \tag{4}$$

The value of B_c in the above expressions is not necessarily the same as the one in the previous computations of net pressures.

From Eqs. (3) and (4), we have

$$\frac{B}{2} - \frac{B_c}{3} = \frac{M_e}{P_s} = \frac{250}{480}$$

or

$$B_c = 3\left(\frac{B}{2} - \frac{250}{480}\right) = 2.4125 \text{ m}$$

Then from Eq. (3), we have

$$q_1 = \frac{2P_s}{LB_c} = \frac{2(480)}{2.8(2.4125)} = 142 \text{ kN/m}^2$$

The intensity of the earth pressure at the face of the column has to be computed so as to calculate the maximum bending moment on the slab. This pressure can be computed from the triangular pressure distribution as shown in Fig. 15.6b.

$$q_3 = \frac{q_1}{B_c}(B_c - L_m) = q_1\left(1 - \frac{L_m}{B_c}\right)$$

where

$$L_m = \text{cantilever span}$$

$$= \frac{B}{2} - \frac{a}{2} = 1.325 - 0.3 = 1.025 \text{ m}$$

$$q_3 = 142\left(1 - \frac{1.025}{2.4125}\right) = 81.7 \text{ kN/m}^2$$

Since there are two load conditions and the allowable stresses are different in different load conditions, one must calculate the bending moment in both the load conditions and then select the governing moment. The limit bending moment on the slab at the face of the column in seismic load condition is:

$$M_1 = \gamma_f \frac{L L_m^2}{6} (2q_1 + q_3)$$

$$= (1.5) \frac{2.8(1.025)^2}{6} (284 + 81.7) = 269 \text{ kNm}$$

(An alternate partial safety factor of 1.2 for dead, full live and seismic loads is recomended by the code.)

Bending moment at the critical section under the normal live load condition is computed below.

Axial force $\qquad P_s = 360 + 240 = 600$ kN

Pressure $\qquad q_0 = \dfrac{P_s}{LB} = \dfrac{600}{2.8(2.65)} = 80.86 \text{ kN/m}^2$

Since cantilever span lengthwise is more than that widthwise, the maximum bending moment is caused on the lengthwise span during the normal live and load condition.

The cantilever span $L_0 = \dfrac{L}{2} - \dfrac{a}{2} = 1.4 - 0.3 = 1.1$ m

Bending moment in the live load condition is

$$M_2 = \gamma_f \frac{B q_0 L_0^2}{2} = (1.5) \frac{(2.65)(80.86)(1.1)^2}{2} = 194.5 \text{ kNm}$$

The bending moment caused under the seismic load condition is more than 1.33 times that caused in the live load condition, therefore the seismic load condition governs the design.

The footing can be designed as a tapering section. Let the size of the footing at the top surface be 1.2 m by 1.2 m, then the size of the trapezoidal section on which the maximum bending moment is acting has the follwoing sides:

$$b_1 = 1.2 \text{ m} \quad \text{and} \quad b_2 = L = 2.8 \text{ m}$$

The resisting moment capacity under seismic load of a balanced section with M20 concrete and mild steel reinforcement is

$$M_r = ((K - K_1) b_1 + b_2 K_2)) d^2 f_{ck}$$

$$= (0.113(1.2) + 2.8 (0.25))(20)d^2 = 4.01 \, d^2$$

Equating the moment capacity to the external moment acting on the section, we have

$$d = \sqrt{\frac{0.269}{4.01}} = 0.26 \text{ m}$$

Use $\qquad d = 0.3$ m

The critical section for the shear failure under seismic load condition will be at distance of $d/2$ from the face of the column. The intensity of pressure at this point is

$$q_5 = \frac{q_1}{B_c} (B_c - (L_m - 0.5 \, d)) = \frac{q_1}{B_c} (B_c - L_s)$$

$$= \frac{142}{2.4125} (2.4125 - (1.025 - 0.15)) = 90.5 \text{ kN/m}^2$$

Maximum shear force is

$$V = \gamma_f L L_2 \frac{(q_1 + q_2)}{2} = 1.5\,(2.8)(1.025 - 0.15)\,\frac{(142 + 90.5)}{2}$$

$$= 427.2 \text{ kN}$$

The section is trapezoidal with top width as b_1 and bottom width as b_2. The nominal shear stress on the section is

$$\tau_v = \frac{2V}{(b_1 + b_2)d} = \frac{2(1.4272)}{(1.2 + 2.8)(0.3)} = 0.71 \text{ MPa}$$

The percentage of reinforcement in the footing is in the range or 0.5% and the shear capacity for such a percent of reinforcement is 0.48 (1.33) = 0.64 MPa. The depth of the section is slightly inadequate against shear force. Therefore, revise the depth of the section.

Let $\qquad d = 0.36$ m.

Design of the reinforcement. The area of the tension reinforcement is

$$A_{st} = \frac{1.15\,M}{jd(1.33)\,f_y} = \frac{1.15\,(269\,000)}{0.8(0.36)(1.33)(415)} = 1950 \text{ mm}^2$$

Provide 11 numbers of 16 mm bars at the bottom spanning 2.65 m.

$$A_s \text{ (provided)} = 2211 \text{ mm}^2$$

Effective area of the concrete for computation of percentage of steel:

$$A_c = \frac{(1.2 + 2.8)(0.32)}{2} = 0.64 \text{ m}^2$$

Percentage of effective reinforcement is

$$p = \frac{100\,A_s}{A_c} = \frac{221100}{640\,000} = 0.35\%$$

The shear capacity for 0.35% of reinforcement under seismic load condition is

$$\tau_c = 0.42\,(1.33) = 0.56 \text{ MPa}$$

Design of the reinforcement in the width direction. The maximum bending moment caused on the width cantilever during live load condition governs the design. This moment was computed and it is

$$M_2 = 195.4 \text{ kNm}$$

The reinforcement in this direction is placed over the main reinforcement, therefore the depth of the steel in this direction is:

$$d = 0.36 - 0.02 = 0.34 \text{ m}$$

Area of the reinforcement required is

$$A_{st} = \frac{1.15 M_2}{jdf_y} = \frac{195\,400\,(1.15)}{0.8(0.34)(415)} = 1919 \text{ mm}^2$$

Provide 10 numbers of 16 mm bars.

The reinforcement details are shown in Fig. 15.6. The overall depth of the slab at column face is

$$h = 0.36 + 0.01 + 0.05 = 0.42 \text{ m}$$

EXAMPLE 15.7 *Design circular footing.* A circular cross-sectioned column of 600 mm diameter is transferring a load of 1200 kN to the footing. The allowable net bearing capacity of the clayey soil is 80 kN/m². Design an isolated circular footing.

Axial load $\qquad P_s = 1200$ kN

Bearing capacity $\qquad p_a = 80$ kN/m²

M20 concrete and HYSD-Fe415 bars are used.

Diameter of the column $a = 0.6$ m

Design of the foundation. Let the depth of foundation = 1 m.

Let the net load on the soil be $= P = P_s + 0.07\, P_s = 1272$ kN

Bearing area of the foundation needed is

$$A_f = \frac{P}{p_a} = \frac{1272}{80} = 15.9 \text{ m}^2$$

Let R = radius of the circular slab of the footing, then

$$\pi R^2 \geq 15.9 \text{ m}^2 \text{ or } R \geq 2.25 \text{ m}$$

Use $R = 2.3$ m, then $A_f = \pi R^2 = 16.62$ m².

Design of the section. Effective soil reaction on the RCC slab is

$$q = \frac{P_s}{A_f} = \frac{1200}{16.62} = 72.21 \text{ kN/m}^2$$

The limit bending moments for collapse of the footing are

$$M_r = \frac{\gamma_f P_s}{4\pi} \text{ (at the centre)} \tag{1}$$

$$M_\theta = \frac{\gamma_f P_s}{4\pi} \text{ (constant throughout)} \tag{2}$$

The numerical values are

$$M_r = \frac{1.5\,(1200)}{4\,\pi} = 143.24 \text{ kNm/m}$$

$$M_\theta = \frac{1.5\,(1200)}{8\,\pi} = 71.62 \text{ kNm/m}$$

The radial moment at the face of the column is maximum and the size of the section is governed by this value. Equating the resisting moment to the maximum moment, we have

$$d = \sqrt{\frac{M_r}{K f_{ck}}} = \sqrt{\frac{0.14324}{0.138(20)}} = 0.230 \text{ m}$$

Use $\qquad d = 0.3$ m.

The thickness of the slab at the free end is governed by the circumferential moment and the corresponding effective depth is

$$d_0 = \sqrt{\frac{M_\theta}{K f_{ck}}} = \sqrt{\frac{0.07162}{0.138(20)}} = 0.16 \text{ m}$$

Check for the shear stress. Critical diagonal failure section is at distance 0.5 d from the face of the column, that is

$$r_s = a + 0.5\ d = 0.3 + 0.15 = 0.45\ \text{m}$$

The effective shear force on the circle is

$$V = \frac{\gamma_f P_s(1 - (r_s/R)^2)}{2\pi r_s} = 612\ \text{k/N}$$

Nominal shear stress at the critical shear force circle is

$$\tau = \frac{V}{d} = 0.612/0.30 = 2.04\ \text{MPa}$$

This value is far higher than the allowable shear stress in the flat slabs which is equal to $0.25\ \sqrt{f_{ck}} = 1.1$ MPa. Increase the effective thickness of the slab at the support or increase the diameter of the column at the footing.

Let $\qquad d = 0.50$, then $r_s = 0.3 + 0.25 = 0.55$ m

$$\tau_v = \frac{\gamma_f P_s(1 - (r_s/R)^2)}{2\pi r_s d} = 0.97\ \text{MPa}$$

Design of the reinforcement

    Radial reinforcement. The radial reinforcement required at the face of the column is

$$A_{st} = \frac{1.15 M_r}{jdf_y} = \frac{143240\ (1.15)}{0.8(0.5)(415)} = 993\ \text{mm}^2/\text{m}$$

Provide 16 mm bars at 200 mm c/c at the bottom surface in the radial direction.

    Circumferential reinforcement

$$A_{st} = \frac{1.15\ M_\theta}{jdf_y} = (1.15)\ \frac{71620}{0.8(0.48)(415)} = 516\ \text{mm}^2/\text{m}$$

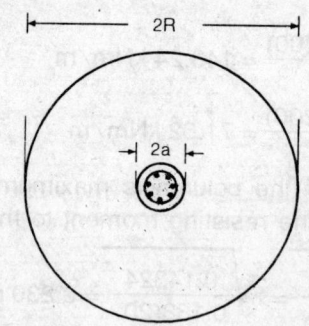

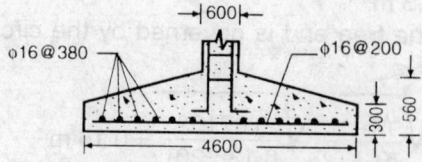

Fig. 15.9 Reinforcement Details of Circular Footing, Example 15.7.

Provide 16 mm bars at 380 mm c/c at the bottom as circumferential reinforcement near the support.

The overall depth at support is $h = 0.5 + 0.06 = 0.56$ m.

The details of the reinforcement are shown in Fig. 15.7.

15.4 Combined Footings

Sometimes, the columns of a structure are closely spaced because of high loading and other building constraints. At times, even if the columns are reasonably well spaced, the bearing capacity of the soil may be small and will not allow separate footings to each of the columns. A footing provided for more than two columns is normally called *combined footing*. A footing combined for a row of columns is called a combined *strip footing* and designed as a continuous beam. Similarly, a footing designed for a set of columns of two or more rows is usually called a *raft or mat foundation*. Figure 15.10a gives the elevation of the footing and Fig. 15.10b gives the distribution of the bearing pressure when the centroid of the load coincides the centroid of the footing. Typical bending and shear force diagrams are shown in Figs. 15.10c and d. The critical bending moments occur at the face of the columns or near the centroid of the combined footing. And the critical shear forces occur at a distance d from the faces of the columns. A footing of uniform width would generate uniform or uniformly varying pressure along the length of the footing. However, a uniform pressure under the footing can be achieved by adjusting the width of the footing so as to have centroids in line as shown in Fig. 15.11. Sometimes it may not be possible to have a uniform bearing reaction because of the bending

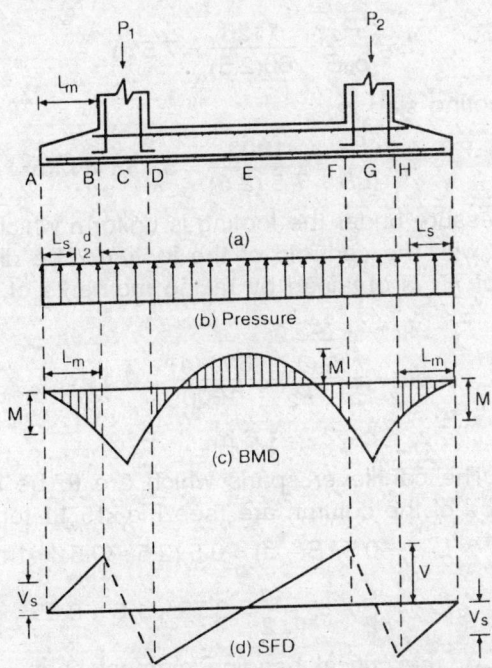

Fig. 15.10 Pressure and Force Distribution of Combined Footing.

moments or space restrictions. The method of design of the combined footing is similar to that already applied to the isolated footings. Figure 15.12 shows a typical bending moment and shear force for a typical combined footing.

EXAMPLE 15.8 *Combined footing for unequally loaded two columns and uniform in width.* Two columns carrying 400 kN and 600 kN are spaced 4 m apart and they have to be provided with a foundation on soil having a net safe bearing capacity of 60 kN/m². The footing must be restricted to 2.5 m width. The columns size is 500 by 500 mm.
Loads on the columns are

$$P_1 = 400 \text{ kN}, \ P_2 = 600 \text{ kN} \ \text{and} \ P_s = P_1 + P_2 = 1000 \text{ kN}$$

Net safe bearing capacity $p_a = 60 \text{ kN/m}^2$

Distance between the loads $c = 4 \text{ m}$.

Use
$$f_{ck} = 20 \text{ MPa}, \ f_y = 415 \text{ MPa}.$$

Design of the foundation. For the purpose of assuming the dead load of the foundation, let the size of the foundation be 2.5 m by 7 m by 0.8 m. Then the difference in the weight of the concrete and the equivalent volume of soil is

$$W_g^1 = 2.5 \ (7) \ (0.8) \ (25 - 16) = 126 \text{ kN}$$

The net load on the foundation soil is

$$P = P_s + W_g^1 = 1126 \text{ kN}$$

Let the depth of foundation = 1 m below GL and the width of the foundation $B = 2.5$ m, then the length of the foundation is

$$L = \frac{P_s}{p_a B} = \frac{1126}{60(2.5)} = 7.5 \text{ m}$$

The net pressure on the footing slab is

$$q = \frac{P_s}{LB} = \frac{1000}{7.5 \ (2.5)} = 53.33 \cdot \text{kN/m}^2$$

It was assumed that the pressure under the footing is uniform which means that the centroid of the loads must coincide with the centroid of the footing. The distance of the centroid of the loads from the centre of P_1 is obtained by taking moments of the forces about the line of action of P_1 and it is

$$c_1 = \frac{P_2 \ (c)}{P_1 + P_2} = \frac{600 \ (4)}{1000} = 2.4 \text{ m}$$

and the other distance $c_2 = c - c_1 = 1.6 \text{ m}$

Design of the section. The cantilever spans which are to be considered for maximum bending moments at the face of the column are (see Fig. 15.11 for notations):

$$L_{m1} = 0.5 \ (B - a) = 0.5 \ (2.5 - 0.5) = 1 \text{ m}$$

$$L_{m2} = \frac{L}{2} - c_2 - \frac{a}{2} = 3.75 - 1.6 - 0.25 = 1.9 \text{ m}$$

Since L_{m2} is larger than L_{m1}, the critical bending moment span is L_{m2} and the maximum positive BM is

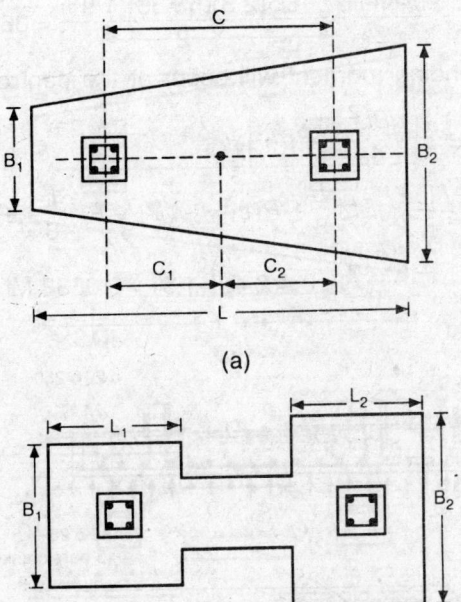

(a)

Fig. 15.11 Typical Combined Footing.

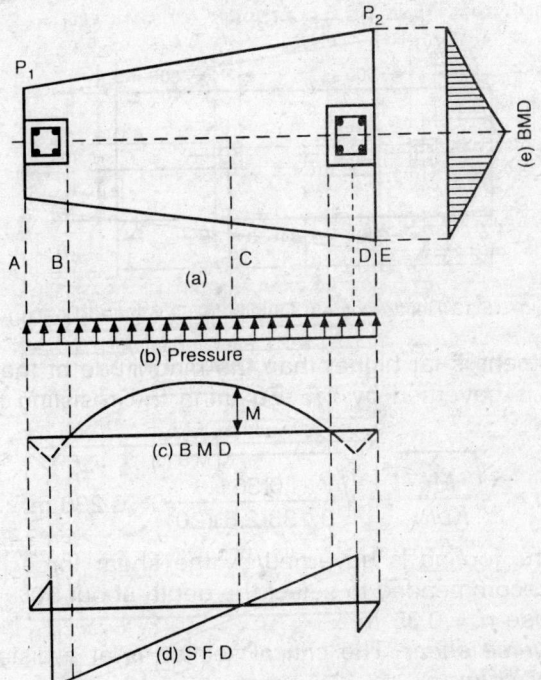

Fig. 15.12 Pressure and Force Distribution of Eccentric Combined Footing.

$$M_1 = \frac{\gamma_f B q L_{m2}^2}{2} = \frac{1.5(2.5)(53.33)(1.9^2)}{2} = 361 \text{ kNm}$$

The maximum negative bending moment will occur at the centroid of the footing and it is

$$M_2 = -\left(\frac{qBL^2}{8} - P_1 c_1\right)\gamma_f$$

$$= \left(-(qBL)\frac{L}{8} + P_1 c_1 = -(P_1 + P_2)\frac{L}{8} + P_1 c_1\right)\gamma_f$$

$$= \left(-\frac{(7.5)}{8} + 0.4(2.4)\right)(1.5) = 0.0338 \text{ MNm}$$

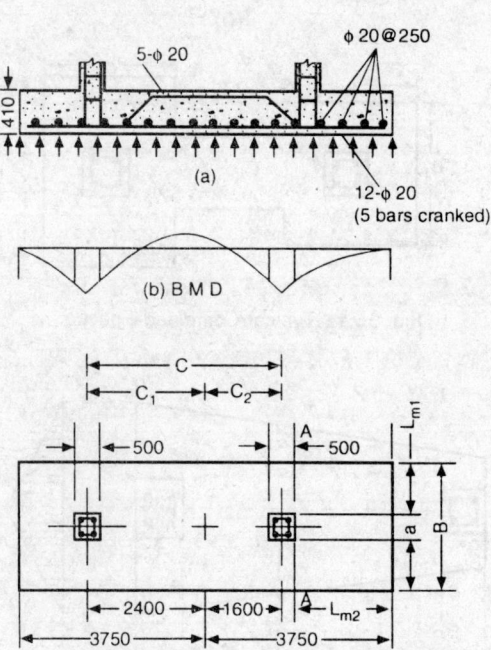

Fig. 15.13 Reinforcement Details for Combined Footing.

The positive bending moment is far higher than the magnitude of the negative BM, therefore the depth of the footing is governed by M_1. Equating the resisting moment to the collapse moment, we have

$$d = \sqrt{\frac{M_1}{K b f_{ck}}} = \sqrt{\frac{0.361}{0.138(2.5)(20)}} = 0.236 \text{ m}$$

Invariably the depth of the footing is governed by the shear force rather than the bending moment. It is, therefore, recommended to select the depth about 50% more than that required for bending resistance. Use $d = 0.35$ m.

Design for the transverse shear. The critical section is at a distance d from the face of the column and the shear span is

$$L_s = L_{m2} - d = 1.9 - 0.4 = 1.5 \text{ m}$$

The nominal shear stress is

$$\tau_v = \frac{V}{bd} = \frac{\gamma_f BL_sq}{bd} = \frac{(1.5)(1.5)(0.05333)}{(1)0.35} = 0.34 \text{ MPa}$$

The value is just less than 0.39 MPa which is the shear capacity for the minimum reinforcement.

Check for diagonal tension (*punching type shear stress*). The critical shear plane of punching through diagonal tension is at $0.5d$ from the face of this column. The peripheral length of the plane is

$$b_0 = 4 (a + d) = 3.4 \text{ m}$$

The maximum shear force occurs around column 2 and it is

$$V = q \left(\frac{LB}{2} - (a + d)^2 \right) \gamma_f = \frac{(P_1 + P_2)}{LB} \left(\frac{LB}{2} - (a + d)^2 \right) \gamma_f$$

$$= \frac{1 \, (3.75(2.5) - 0.723)(1.5)}{7.5(2.5)} = 0.692 \text{ MN}$$

The nominal shear stress is

$$\tau_v = \frac{V}{b_0 d} = \frac{0.692}{(3.4)(0.35)} = 0.58 \text{ MPa}$$

The shear strength for the diagonal failure is

$$\tau_a = 0.25 \sqrt{f_{ck}} = 1.1 \text{ MPa}$$

The section is safe against shear force.

Reinforcement design. The area of the tensile reinforcement at the bottom in the longitudinal direction is

$$A_{st} = \frac{1.15 \, M_1}{jdf_y} = \frac{361 \, 000 \, (1.15)}{(0.8)(0.35)(415)} = 3572 \text{ mm}^2$$

Provide 12 numbers of 20 mm bars in the longitudinal direction and crank up half of the bars.

$$A_s \text{ (provided)} = 3768 \text{ mm}^2$$

The bending moment in the widthwise direction is much lesser than that in the length. But since the minimum reinforcement governs the design, therefore provide $\phi20$ bars at 250 mm c/c. Figure 15.13 illustrates the reinforcement details.

Note: In the isolated or combined footings, there is no question of redistribution of the bending moment. Therefore, bending moments in the limit state design can be obtained by multiplying those in the working stress design by the partial safety factors.

15.5 Design of Raft Foundation

A raft foundation is basically a shallow foundation in which the load on the foundation is function of orthogonal directions. It is plate type of structure, spared over a large area and supporting a number of columns or the entire superstructure as a single unit. The bending moments on the footing. and the soil pressure distribution are functions of the two directions.

A raft foundation is also called as *mat* or spread foundation. Such foundations are used when the columns of a structure are closely spaced, or the loads on the columns are large and they are usually provided for multistorey buildings, overhead water tanks, chimneys, etc. A raft foundation might become unavoidable in submerged structures in some multistorey structures where basement is to be provided and in retaining walls, etc. The mat or raft foundation is designed as a flat slab. The reader is advised to read the chapter on flat slabs.

EXAMPLE 15.9 *Design of rectangular raft foundation.* A large building is provided with columns spaced at 6 m apart in two perpendicular directions and the load from each column on the foundation is 2880 kN. The size of each of the columns is 500 mm square. The soil is silty clay and has safe net bearing capacity of 90 kN/m^2. Design the foundation using M25 concrete and HYSD-Fe415 bars.

Spacing of the column $L_1 = L_2 = L = 6$ m

Size of the column $a = 0.5$ m

Height of the column above foundation = 5 m

Load from each column $W_s = 2880$ kN

$$f_{ck} = 25 \text{ MPa}, f_y = 415 \text{ MPa}$$

Design of the foundation. Let the average thickness of the raft be assumed as 0.6 m for the purpose of calculating the self-weight of the slab. The difference in the weight of the concrete slab and the soil can be assumed as 10 kN/m^3. The gross load on the foundation per panel size consists of the load from the one column plus the weight of the slab plus the weight of the soil overburden. Since the net bearing capacity is given, only the net load on the soil need to be computed for the purpose of bearing pressure.

Bearing area available per panel

$$A_f = 6(6) = 36 \text{ m}^2$$

Load from each column $P_s = 2880$ kN

Difference in the weight of the slab and the soil is assumed as

$$6(6)(0.6)(10) = 218 \text{ kN}$$

Total net load per panel $= 2880 + 218 = 3098$ kN

The net bearing pressure on the soil is

$$p = \frac{3098}{36} = 86.06 \text{ kN/m}^2$$

The bearing pressure is about 96% of the net safe bearing capacity of the soil, therefore the mat foundation is just the right type. The depth of the foundation can be 1 m below the ground level.

Design of the section. The mat slab acts as an inverted flat slab with 6 m square panels. Let the load be transferred through pedestals which are considered similar to the drops in the flat slabs. Such pedestals minimize the failure of the slab by diagonal tension or punching shear. The size of the pedestal can be selected in the range of $L/4$ with a depth in the range of $L/8$.

Let the pedestal size= $D_e = 1.5$ m

Clear spacing between the pedestal

$$L_0 = L - D_e = 4.5 \text{ m}$$

Active load intensity

$$= \frac{W_s}{A_f} = q = \frac{2880}{36} = 80 \text{ kN/m}^2$$

Total load on the clear panel is $W = qLL_0 = 80(6)(4.5) = 2160 \text{ kN}$

The sum of the magnitudes of the positive and negative BM in a panel with a partial safety factor of 1.5 is

$$M_0 = \frac{\gamma_f W L_0}{8} = (1.5)\frac{2160(4.5)}{8}$$

$$= 1822.5 \text{ kNm} = 1.8225 \text{ MNm}$$

As the mat foundation slab is an inverted flat slab with the load intensity which is the reaction of the soil acting upwards. The *positive* bending moment is the bending moment which causes compression on the face of the slab on which the active load is acting. The active load is from the columns and reaction is from the soil. Hence the bending moment which causes compression on the top surface is considered positive. With this notation, the positive bending moment is occurring along the face of the drops and the negative BM is at the middle spans.

Let the mat slab projects as a cantilever beyond the face of the exterior column pedestals. This overhang produces a positive bending moment at the outer columns (see Fig. 13.21 for notations). The projection of the slab beyond the centre line of the outer column line is 2 m. The cantilever span beyond the pedestals is

$$L_m = 2 - 0.5 \, D_e = 2 - 0.75 = 1.25 \text{ m}$$

The positive BM at the face of the pedestal is

$$M_{p0} = \frac{\gamma_f q L L_m^2}{2} = (1.5)\frac{80(6)(1.25)^2}{2} = 562.5 \text{ kNm}$$

The bending moment at the inside face of the exterior pedestals is

$$M_{p1} = \frac{\gamma_f q L (L_m + D_e)^2}{2} - \frac{R_0 D_e}{2}$$

where

$$R_0 = \text{reaction of the external column and it is}$$

$$\gamma_f q L(0.5L + \text{cantilever span}) = \gamma_f (q)L(3 + 2)$$

$$M_{p1} = \gamma_f q L\left(\frac{3.25^2}{2} - \frac{5(1.5)}{2}\right) = 1102.5 \text{ kNm}$$

The resisting moment capacities can be selected and the collapse moment can be appropriately adjusted. Assumed a distribution similar to that used in the working stress design. The bending moments at the face of the pedestals and at mid-span of any of the internal penels be proportioned as

$$M_{pt} = 0.65 \, M_0 = 1.1847 \text{ MNm}$$

$$M_{ni} = 0.35 \, M_0 = 0.6378 \text{ MNm}$$

The positive and negative bending moments in the exterior panel be proportioned as

$$M_{p2} = c_2 M_0 = 0.7(1.8225) = 1.2758 \text{ MNm}$$

$$M_{p2} = M_0 - 0.5(M_{p1} + M_{p2}) = 0.6334 \text{ MNm}$$

The bending moments on the column and the middle strips are calculated using the distribution coefficients suggests by the code. These coefficients are

For column strip
   Moment coefficient at the support = 0.75
   Moment coefficient at the mid span = 0.60

The depth of the slab is controlled by the maximum bending moment which occurs at the face of the pedestal at the exterior columns of the outer panel. This will also govern the shear force which will tend to punch through the slab around the pedestal. The maximum bending moment carried by the column strip is

$$M_{pc2} = 0.75\ M_{p2} = 0.75(1.2758) = 0.9569\ \text{MNm}$$

Width of the column strip $b = 0.5L = 3\ \text{m}$

The effective depth of the slab is given by

$$d = \sqrt{\frac{M_{pc2}}{Kb\,\sigma_{cb}}} = \sqrt{\frac{9569}{0.138(3)(25)}} = 0.305\ \text{m}$$

The minimum cover recommended for a mat foundation is 75 mm, so the overall depth of the slab can be taken as

$$t = d + 0.020 + 0.075 = 0.4\ \text{m}$$

Select $t = 0.5$ m and $d = 0.41$ m to ensure safety against shear force.

$$w_g = \text{Weight of equal vol. soil}$$

$$= 0.50(10) = 5\ \text{kN/m}^2$$

Add extra 1.0 kN/m^2 for weight different of pedestal, etc. Even then the net pressure on the soil is 80 + 1.0 + 5.0 = 86 kN/m^2. This is less than the safe net bearing capacity of the soil.

Design for shear. The shear strength in diagonal tension is

$$\tau_c = 0.25\sqrt{f_{ck}} = 0.25\sqrt{25} = 1.25\ \text{MPa}$$

The critical shear plane is the peripheral plane which is at a distance of 0.5d from the face of the pedestal. The total length of this shear plane is

$$b_0 = 4(D_e + d) = 4(1.5 + 0.41) = 7.64\ \text{m}$$

Table 15.3 Critical Bending Moments and the Reinforcements

Sl. No.	Section (type of BM)	Bending moments (MNm) Panel	Strip Column	Middle	The reinforcement Needed (mm²) Column	Middle	Provided Column	Middle
		Exterior panels						
1	0(+ve)	0.5625	0.5625	0	4874	0	10—φ25	1—φ25
2	1(+ve)	1.1025	1.1025	0	9553	0	19—φ25	5—φ25
3	3(−ve)	0.6334	0.3800	0.2534	3293	2195	7—φ25	5—φ25
4	2(+ve)	1.2758	0.9568	0.3190	8290	2764	17—φ25	6—φ25
		Interior panels						
5	i(+ve)	1.1847	0.8855	0.2962	7899	2567	16—φ25	6—φ25
6	i(-ve)	0.6378	0.3826	0.2552	3318	2211	7—φ25	5—φ25

The total shear force on this plane is

$$V = \gamma_f q (L_1 L_3 - (D_e + d)^2) = 120(36 - 1.91^2) = 3886 \text{ kN}$$

The nominal shear stress on the concrete is

$$\tau_v = \frac{V}{b_o d} = \frac{3.886}{7.64(0.41)} = 1.24 \text{ MPa} < \tau_c = 1.25$$

The nominal shear stress is less than the shear strength so the depth selected is safe against shear.

Design of reinforcement. The design bending moments on the column and middle strips are to be obtained from the total bending moments computed at the critical section after multiplying the appropriate coefficients (0.75 or 0.60 for the column strip at support and mid spans). Then the area of the reinforcement required is computed from

$$A_{st} = \frac{1.15 \, M_t}{jdf_y}$$

The bending moments and the corresponding reinforcements required at different sections are computed and given in Table 15.3. However, the minimum percentage of reinforcement of 0.12% is to be maintained in the slab in each direction. The minimum reinforcement is

$$A_{sm} = \frac{0.12}{100}(500)(3000) = 1800 \text{ mm}^2/3 \text{ m}$$

The minimum reinforcement to be provided is 6 numbers of 20 mm bars in each column and middle strips.

PROBLEMS

Use IS 456 and sketch reinforcement details in each of the following problems

15.1 A masonry wall of 400 mm thickness is transmitting a total load of 200 kN/m to a soil having an allowable bearing capacity of 100 kN/m². Design a RCC strip foundation using M20 concrete and high yield steel bars.

15.2 A reinforced concrete wall of 300 mm thickness is subjected to an axial load of 400 kN/m and bending moment of 129 kNm/m. Design a RCC foundation using M20 concrete and HYSD-Fe415 bars. The allowable bearing capacity of the soil is 120 kN/m².

15.3 A RCC wall of 300 mm thickness transmits an axial load of 400 kN/m to a soil having an allowable bearing capacity of 120 kN/m². The property line is such that it would not permit projection of the foundation beyond 1.2 m on one side of the central line of the wall. Design a RCC foundation using M20 concrete and HYSD-Fe415 bars.

15.4 A column of size 500 mm by 500 mm is transmitting a load of 800 kN to a soil having an allowable bearing capacity of 120 kN/m². Design a square RCC footing using M20 concrete and mild steel reinforcement. Assume an uniform thick footing.

15.5 A column of 500 mm by 500 mm is transmitting an axial load of 800 kN to a soil having an allowable bearing capacity of 120 kN/m². Design a RCC square footing using M20 concrete and HYSD-Fe415 bars. The thickness of the footing at the free end should be taken as 150 mm.

15.6 A square column of 300 mm is transmitting a load of 500 kN to a soil having an allowable bearing capacity of 90 kN/m^2. The width of the foundation cannot exceed 1.8 m. Design an isolated footing made of M20 concrete and mild steel reinforcement.

15.7 A square column of 500 mm size is transmitting an axial load of 800 kN and bending moment of 200 kNm to a soil having an allowable bearing capacity of 100 kN/m^2. Design a RCC square footing using M20 concrete and HYSD-Fe415 bars.

15.8 A square column of 500 mm size is transmitting an axail load of 800 kN and bending moment of 200 kNm to a soil having an allowable bearing capacity of 100 kN/m^2. The bending moment and axial load are generated in a seismic load condition. Design a square RCC footing using M25 concrete and HYSD-Fe415 bars.

15.9 Square cross section column of 500 mm diameter is transmitting a load of 900 kN to a soil having an allowable bearing capacity of 90 kN/m^2. Design an isolated circular footing using M20 concrete and HYSD-Fe415 bars. The bending moment coefficients for the problem can be taken from the text of this chapter.

15.10 A circular cylindrical shaft of radius 5.5 m with its thickness of 400 mm is transmitting an axial load of 20,000 kN to a soil having an allowable bearing capacity of 90 kN/m^2 Design a RCC circular raft foundation using M20 concrete and HYSD-Fe415 bars. The bending moment coefficients for the circular raft can be taken from the text of this chapter.

15.11 Two columns transmit 500 kN each are spaced at 3 m apart and size of the each square column is 350 mm. The allowable bearing capacity of the soil is 90 kN/m^2. Design a combined RCC footing using M25 concrete and HYSD-Fe415 bars with its width not exceeding 2 m.

15.12 Two columns transmitting 600 and 800 kN are spaced at 4 m apart on a soil having allowable bearing capacity of 70 kN/m^2. Design a combined footing having a width of 3 m using M20 concrete and HYSD-Fe415 bars. The size of each square column can be taken as 300 mm.

15.13 Two columns spaced at 5 m apart are transmitting 600 kN each to a soil having an allowable bearing capacity of 60 kN/m^2. Design a combined beam and slab footing with a width not exceeding 2.6 m using M20 concrete and HYSD-Fe415 bars. The size of each of the columns is 500 mm square and width of the beam should also be taken as 500 mm. The footing is made of M20 concrete and HYSD-Fe415 bars.

15.14 A multistorey building having its columns spaced at 5 m apart in two perpendicular directions is resting on a soil having an allowable bearing capacity of each 100 kN/m^2 and transmits a load of 2000 kN from each column. The size of the square column is 500 mm. Design a raft foundation using M20 concrete and HYSD-Fe415 bars.

15.15 A multistorey building having its column spaced at 5 m apart in two perpendicular directions is resting on a soil having an allowable bearing capacity of 100 kN/m^2. Each column of 500 mm is transmitted an axial load of 1500 kN and a bending moment of 200 kNm. The building has 4 rows of columns and 5 columns in each row. The bending moment in each column is created by a seismic load and it can be in any direction. Design a raft foundation using M20 concrete and HYSD-Fe415 bars.

Chapter 16

Design of Deep Foundations

16.1 Introduction

A foundation which derives its main strength and stability from the property of the depth of foundation is called a *deep foundation*. Even the shallow foundation does derive strength from its depth; however, influence of the depth of foundation plays a vital role in deep foundations. The design of foundation has two distinct parts. The design of depth and size of foundation controlled by soil characteristics. A brief outline without going into the theoretical basis is presented as the theory on mechanics of soils is outside the scope of the book. The student is supposed to have studied the mechanics of soils and foundation design in another subject. The structural design which forms the second part of the foundation is discussed in detail. The deep foundations can be classified into two main groups:

1. Pile foundation
2. Well foundations

The pile foundations can further be classified into three parts based on the behaviour, and they are:

a) Bearing piles;
b) Friction piles; and
c) Compaction piles.

Based on the construction method, the piles can be classified as

a) Precast piles; and
b) Cast in-situ piles (bored piles).

The bored piles can be divided into bored compaction piles, driven cast in-situ piles and bored cast in-situ piles.

16.2 Bearing Piles

A pile which transmits the load to the soil through bearing at the tip of the pile is called a bearing pile. A hard strata or rockey strata should be available at a reasonable depth below the ground level and the soil above the rockey soil is relatively soft or stiff variety, for providing bearing piles. An investigation on the soil profile and its properties need to be carried out to determine the possibility of establishing the bearing piles. If the hard strata is too far below the ground level, the relative economics of friction and bearing piles need to be investigated. A pile when rests on a hard strata, its tendency to settle is minimized, consequently friction between the soil and the pile along its depth will not be generated. A combination action of bearing-cum-frictional reaction in case of piles resting on hard rock is normally not possible. The bearing capacity of the rock and the bearing area of the pile decides the capacity of the pile. The ultimate capacity of the pile is given by

$$Q_u = A_p \, p_0 \tag{16.1}$$

where A_p = bearing area of the pile at tip

                p_0 = ultimate bearing capacity of the rock

The piles must be spaced at at least 2.5 times the bulb diameter of the pile or 750 mm so as to derive the full benefit of the bearing capacity of the soil. In such a case the capacity of a group of piles can be taken equal to the direct sum of the capacities of the piles in the group. Point bearing piles can either be precast or cast in-situ. Piles bearing on gravel or stiff clay have a tendency to slide down thus mobilizing the frictional resistance from the soil. The load carrying capacity of such piles is discussed under the heading of friction piles.

16.3 Friction Piles

A pile which transmits the load to the soil primarily by skin friction between the pile and the soil is called a friction pile. The load carrying capacity of the pile can be determined either by *static formula* or *dynamic formula*. The ultimate load capacity by static formulae as per Indian Code of Practice are listed for ready reference.

Pile capacity in granular soil

$$Q_u = \sum_{i=1}^{n} (SK\, q_d \tan \delta)_i + A_p \left(\frac{\gamma}{2} DN_r + q_{dn}N_q \right) \tag{16.2}$$

where Q_u = ultimate resisting capacity

                 A_p = area of cross-section at tip

                 q_{di} = effective over burden pressure at the ith layer

                 D = size of the pile at tip

                 γ = unit weight of soil at toe

     N_r and N_q = bearing capacity factors of the soil at toe

                 K_i = coefficient of lateral earth pressure at ith layer

                      (it varies from 1 to 3 in loose to medium sands)

                 S_i = surface area of the pile in the ith layer,

                 δ_i = angle of wall friction between pile and soil in the ith layer,

                      it may be taken equal to ϕ

                 ϕ = angle of internal friction of the soil

                 n = number of different layers through which the pile rests

The first part of the expression gives the contribution from friction and the second part gives that from the bearing at the tip of the pile. The load carrying capacity can be approximated for a type of soil as

$$Q_u = S(q_0 + \gamma H/2)\, K \tan \phi + \gamma A_p(0.5\, DN_r + HN_q) \tag{16.3}$$

where S = surface area of the pile

           q_0 = permanent surcharge

           H = embedded depth of the pile

Table 16.1 Value of N Factors

ϕ	N_r	N_q	N_c
10	–	–	9
20	5	8	18
25	10	15	26
30	20	24	37
35	50	40	55
40	120	70	–
44	260	–	–

The values of N_r and N_q suggested by Terzaghi are given in Table 16.1.
Pile capacity in cohesive soils is

$$Q_u = A_p N_c C_p + \alpha C A_s \tag{4.4}$$

where N_c = bearing capacity factor and it is taken as 9 for clays or soft soils

C_p = cohesion of soil at the tip of the pile

C = average cohesion of the soil

α = reduction coefficient

A_s = surface area of the pile shaft

Table 16.2 Cohesion Values

Soil	N-value	α Bored	α Driven	$C(kN/m^2)$
Soft	4 or less	0.7	1.0	0 to 35
Medium	4 to 8	0.7 to 0.5	1.0 to 0.7	35 to 70
Stiff	8 to 15	0.5 to 0.4	0.7 to 0.4	70 to 150
Hard	15 or more	0.4 to 0.3	0.4 to 0.3	150

Approximate values of α and cohesion values are given in Table 16.2.

Dynamic formula: The ultimate bearing capacity of a pile based on dynamic formula of Indian Code of Practice is given by

$$Q_u = \frac{\eta W h}{s + 0.5c} \tag{16.5}$$

where W = mass of the hammer or ram

h = height of free fall of the hammer

= full free fall for trigger-operated hammers

= 80 per cent of fall for winch-operated hammers

= 90 per cent of fall of single acting hammers

η = efficiency of the blow

$$= \left(1 - \frac{\text{energy absorbed by the pile, etc.}}{\text{energy of strike}}\right)$$

s = final set or penetration per blow

c = sum of elastic compressions in pile, dolly, packing and ground

The efficiency of the blow can be computed from

$$\eta = \frac{W + W_g^2 e^2}{W + W_g} \tag{16.6}$$

for $W > W_g e$

where W_g = weight of the pile, anvil, helmet and follower, if any

e = coefficient of restitution of the materials under impact

= 0.5 for steel double acting ram on steel anvil

= 0.4 for single acting drop hammer on RCC head

= 0.25 for single acting drop hammer on hard wood dolly and

= 0 for deteriorated condition of head

$$= \frac{W + W_g^2 e}{W + W_g} - \left(\frac{W - W_g^2 e}{W + W_g}\right)^2 \tag{16.7}$$

for $W < W_g e$

The value of the temporary compression is given by

$$c = c_1 + c_2 + c_3 \tag{16.8}$$

where c_1 = temporary compression in dolly and packing

$$= 0.175(10^{-9}) \frac{Q_u}{A_p} \tag{16.9}$$

(c_1 in metres. A_p = area of pile in m^2 and Q_u in N)

c_2 = temporary compression in pile and it is

$$= 0.9 (10^{-9}) \frac{Q_u L}{A_p} \tag{16.10}$$

(c_2 in metres, L = length of pile in metres)

$$c_3 = 0.35(10^{-9}) \frac{Q_u}{A_p} \tag{16.11}$$

(c_3 in metres)

16.4 Uplift Capacity

Some piles are subjected to uplift in which case they must be designed to resist the upward force. The uplift capacity of a pile is given by

$$Q_f = W_g + S(C + p \tan \phi) \tag{16.12}$$

where $\quad W_g$ = weight of the pile plus that portion of the pile cap and overburden

S = surface area of the pile

C = cohesion on the soil

p = normal pressure on the pile

ϕ = angle of internal friction

16.5 Structural Design

Foundation piles are normally subjection axial compression, bending moment and shear force of which axial force dominates. They are designed as short columns in most cases. The bending moment on the pile cap causes axial force and a secondary bending moment on the piles. The lateral force on the pile cap cause transverse shear and bending moment on the piles. The lateral load carrying capacity of a pile depends not only on the structural strength but also on the subgrade modulus. As the lateral deflection of the pile takes place, the top portion of the pile will act as a cantilever. The cantilever span of the pile is a function of the horizontal subgrade property. As per the suggestion of the Indian Code of Practice, the equivalent cantilever can be obtained with the help of graphs. First estimate the value of horizontal subgrade reaction k_s or the modulus of subgrade reaction k_c from Tables 16.3 or 16.4 respectively.

Table 16.3 Approximate Values of k_s^-

Sand	k_s(N/mm^2)	
	Dry	Submerged
Loose	0.026	0.015
Medium	0.078	0.053
Dense	0.210	0.125

Table 16.4 Approximate Subgrade Constant for Clays (k_c) in N/mm^2.

Unconfined compressive strength	Range	Approximate
0.02 to 0.04	0.7 to 4.2	0.77
0.1 to 0.2	3.2 to 6.5	4.90
0.2 to 0.4	6.5 to 13	10.00
above 0.4	–	20.00

The equivalent cantilever span for a set of values of k_s and k_c are given in Table 16.5.
    The bending moment caused by the lateral force can be computed as

$$M = LF_h \tag{16.13}$$

where $\qquad F_h$ = lateral force on the pile

L = cantilever span taken from Table 16.5

Table 16.5 Equivalent Cantilever in Sandy and Clayey Soils

k_s (MPa)	Sand		k_s (MPa)	Clay	
	L/D			L/D	
	Fixed head	Free head		Fixed head	Free head
0.005	8	11	0.7	6.5	7.5
0.01	6	10	2.0	5.5	6.5
0.05	5	7	5.0	4.0	5.2
0.10	3.5	5.5	10.0	3.4	5.0
0.20	3.0	5.0	20.0	3.0	4.0

L = cantilever span, D = size of pile.

Whenever there are more than two piles in the direction of the latral force and they are connected by a pile cap, one can assume a fixed head and obtain the equivalent cantilever.

The design of the pile by limit states design can be done using the procedures given in Chapter 9 of Part I. The method is illustrated with a set of examples.

16.6 Precast Concrete Piles

Pile of square or hexagonal or octagonal in cross section can be precast with M20 to M30 concrete with adequate reinforcement details. These piles must be designed for handling and impact loads in addition to the other working loads.

Handling condition: The precast pile can be picked up and moved around with two or three pick-up points. The distance of pick-up piont from end in two and three pick-up points for minimum bending moments along with the design moments are listed in Table 16.6. While hoisting with single pick-up point from head, the recommended distance and the corresponding design bending moment are given in Table 16.6.

Table 16.6 Pick-up Points and Bending Moments

Number of pick-up points	Location of pick-up points	Design BM
Three	0.145L from each end and at mid-span	0.011 WL
Two	0.207L from each end	0.0215 WL
One	0.293L from head	0.043 WL

Note: L = Length of pile, W = Weight of pile

The cross-section should be designed to withstand the impact or driving force.

The stress induced in the material of the pile as recommended by the Indian Code of Practice is

$$\sigma = \frac{F(2/\sqrt{\eta - 1})}{A_t}$$

(16.14)

where η = efficiency of the blow

F = resistance against driving

A_t = transformed area of the cross-section

$$= A_c + 1.5 \, m \, A_s$$

A_c = area of concrete

A_s = area of the reinforcement

m = modular ratio

Minimum reinforcement: A set of minimum reinforcement requirements as recommended by the Indian Code of Practice is listed below.

a) Minimum main reinforcement shall be as follows:
1.25 per cent for pile whose length is less than 30 times the least size
(that is $L/b \leq 30$, where b is the least dimension)
1.50 per cent for piles of $30 < L/b \leq 40$,
2.0 per cent for piles of $L/b > 40$,

b) Minimum lateral reinforcement (ties) is controlled by:
0.6 per cent at each ends of the pile of about 3 times depth and 0.2 per cent in the body of the pile.

The ties must be as close as possible at the driving end. The minimum cover to the ties should be 40 mm in ordinary conditions of exposure and 50 mm to 70 mm in corrosive exposure condition.

16.7 Bored Piles

A pile formed in the ground through bored excavation can be called a bored cast in-situ pile. If the soil is stiff enough to stand on its own then no casing need to be provided while boring. In some cases the bore hole is stabilized by filling it with mud slurry or a temporary casing which is withdrawn while laying the concrete. Such piles are called *uncased* piles. In some marine or underwater situations, a steel tube is driven and left as it is even after placing the concrete, then such pile is called *encased* pile. The cohesion between the soil and the pile in bored piles is less when compared with that in the driven piles. The reduction factors for cohesion are given in Table 16.2. The minimum spacing of the piles is 2.5 times the diameter and the minimum percentage of the main reinforcement in piles is 0.4.

16.8 Under-reamed Piles

A bored cast in-situ pile with one or more enlarged bulbs are called under-reamed piles. The ultimate load carrying capacities of the piles based on soil characteristics as suggested by Indian code are listed here.

For sandy soil

$$Q_u = \frac{\pi D^2 \gamma}{4}(0.50 N_r + h N_q)$$

$$+ \frac{\pi(D_u^2 - D^2)\gamma}{4}\left(0.5 \, D_u n N_r + N_q \sum_{i=1}^{n} h_i\right) + \frac{\pi D \gamma}{2}(K \tan \delta)(h_1^2 + h^2 - h_n^2) \quad (16.15)$$

where D = diameter of the stem

D_u = diameter of the bulb = (2 to 3)D

h = total depth of the pile

γ = unit weight of the soil

n = number of the bulbs

K = earth pressure coefficient (about 1.75)

δ = angle of wall friction (= ϕ)

ϕ = angle of internal friction

h_i = depth of centre of the ith bulb

h_n = depth of centre of last bulb from top

N_r and N_q = bearing capacity factors.

For clayey soils

$$Q_u = \frac{\pi D^2}{4} N_c\, C_p + \frac{\pi(D_u^2 - D^2)}{4} N_c C_u + \alpha\, CS + C_u S_u \qquad (16.16)$$

where C_p = cohesion of soil around the toe

C_u = cohesion of soil around the bulb

C = average cohesion of the soil

S = surface area of the stem

S_u = surface area of the cylinder circumscribing the bulb

α = reduction factor (about 0.5)

N_c = bearing capacity factor (about 9)

Typical single and double under-reamed piles are shown in Fig. 16.1. There are some limiting values which one should apply while designing the piles, these are listed here.

a) Minimum diameter of the stem = 250 mm

b) $D_u = (2\ to\ 3)\ D$

c) Maximum spacing of the bulbs = 1.5 D_u

d) Minimum depth of the top most bulb = 2D_u

e) Location of bottom most bulb from toe = bucket length from the toe or 500 mm

f) Minimum distance between the underside of the pile cap and the bulb = 1.5 D_u

g) Normally recommended maximum number of bulbs = 2

h) Minimum spacing of the piles = 2 D_u

i) Minimum spacing of the piles can be 1.5 D_u, then a reduction of 10 per cent in the pile capacity be allowed. No reduction need to be given in compacted piles.

j) Maximum spacing of the piles = 3 m

k) Slump of concrete for use in water free unlined bore holes is about 100 to 150 mm.

l) Minimum cement content upto 10 m depth piles should be 350 kg/m³.

m) Minimum longitudinal mild steel reinforcement is 0.4 per cent or 0.3 per cent in case of HYSD-Fe415 bars are used, subject to a minimum of six numbers of 12 mm bars.

n) Spacing of tie be less than 300 mm or the diameter of the stem.

o) Minimum clear cover to the reinforcement shall be 40 mm.

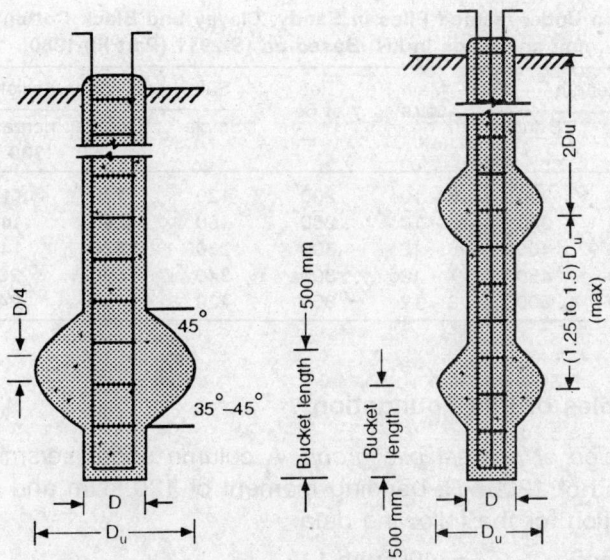

Fig. 16.1 Typical Under-Reamed Piles.

Indian Code of Practice recommends safe loads on piles for different details. Some of them are given in Table 16.7 for ready reference. From the experience of field tests of the author, the load capacities given in the code are on the higher side. A reduction as much as 25 per cent of the recommended values is suggested in case of piles placed in subsoil water condition and loose silty or clayey soils.

16.9 Design of Pile Cap

A reinforced concrete slab or block which interconnects a group of piles and acts as a medium to transmit the load from wall or column to the piles is called a *pile cap*. The dispersion of load from the column or reaction from the pile cap be taken at 45° angle. The critical sections of bending moment and shear force depend on the general arrangement of the piles and pile cap. The following minimum requirements be met while designing the pile cap.

a) Minimum thickness is governed by
   i) Minimum anchorage length of the main reinforcement of the column and that of the pile.
   ii) Rigid enough to distribute the load uniformly to the piles. Span to thickness ratio of the cap should be less than five.

iii) 300 mm thickness at the free edges and 500 mm thickness in the body of the cap.

b) Clear overhang of the pile cap beyond the outermost pile should be 150 mm.

c) A levelling course of 75 mm thick lean concrete be provided underneath.

d) Clear cover to the main reinforcement shall be 60 mm.

e) Pile should project at least 50 mm into the pile cap.

Table 16.7 Safe Loads on Under-reamed Piles in Sandy, Clayey and Black Cotton Soils (all dimensions in mm and loads in kN) Based on [S-2911 (Part III)-1980].

D	D_u	Length		Main bars	Ties of 6ϕ	Safe Load (compression)			Lateral load	
		Single	Double			Single	Double	Increase for 300 mm	Single	Double
250	625	3500	3500	$6 - 10\phi$	200	120	180	11.5	15	18
300	750	3500	3500	$6 - 10\phi$	250	160	240	14	20	24
400	1000	3500	4000	$6 - 12\phi$	300	230	420	14	34	40
450	1100	3500	4500	$7 - 12\phi$	300	340	500	20	40	48
500	1250	3500	5000	$9 - 12\phi$	300	420	630	24	45	54

16.10 Design Examples of Pile Foundation

EXAMPLE 16.1. *Design of precast pile group.* A column under seismic load condition is to transfer a vertical load of 1200 kN, bending moment of 120 kNm and shear force of 40 kN. Design a pile foundation for the following data:

Size of the column $\qquad = 400$ mm

Soil is sandy with $\qquad \phi = 25°$

$\gamma = 18$ kN/m^3, $\qquad K$ (for soil) $= 1.75$

surcharge $\qquad = q_0 = 20$ kN/m^2

M20 concrete with HYSD-Fe415 bars.

Assume the thickness of the pile cap as 800 mm for the purpose of estimating self-weight. Similarly, let the pile size (D) be 300 mm to estimate some basic properties (see Fig. 16.2 for arrangement of piles).

Let,

Size of the pile cap $\qquad = 5\,D = 1500$ mm

Weight of the pile cap $\qquad = (1.5)\,(1.5)\,(0.8)\,(25) = 45$ kN

Spacing of the piles $\qquad = s = 3D = 900$ mm and

Number of piles $\qquad = 4$

Superimposed loads on the pile cap are:

$W_s = 1200$ kN, $M_s = 120$ kNm, $V_s = 40$ kN

Total load on the 4 pile $= 1200 + 45 = 1245$ kN

Load on each pile $P_1 = 1245/4 = 312$ kN

BM on the pile cap $M_1 = 120$ kNm

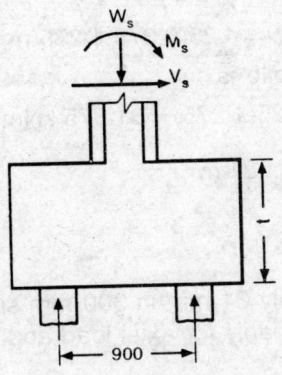

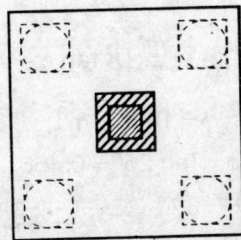

Fig. 16.2 Arrangement of Piles in Example 16.1.

The shear force at the top of the pile cap will cause a moment at the bottom face of the pile cap and it is equal to

$$M_1 = V_s t = 40(0.8) = 32 \text{ kNm}$$

Total $\qquad$ BM $= M_2 = M_s + M_1 = 120 + 32 = 152$ kNm

The axial load caused on a pair of piles by the bending moment M_2 is

$$2P_2 = \frac{M_2}{s} = \frac{152}{0.9} = 170 \text{ kN}$$

Maximum working load on each pile is

$$P_w = P_1 + P_2 = 312 + 170/2 = 397 \text{ kN}$$

use $P_w = 410$ kN and shear force $V_w = V_s/4 = 10$ kN

Applying a factor of safety of 2.5 to the piles, the ultimate loads acting at the head of the pile are:

$$P = 2.5\, P_w = 2.5(410) = 1025 \text{ kN}$$
$$V = 2.5\, V_w = 2.5(10) = 25 \text{ kN}$$

Calculation bending moment on pile

Let the medium sand subgrade reaction (taken from Table 16.3) be $k_s = 0.072$.
The equivalent cantilever span (taken from Table 16.5) is

$$L/D = (5 + 3.5)/2 = 4.25$$

$$L = 4.25 \, D = 4.25(0.3) = 1.275 \text{ m}$$

Design bending moment on the pile is

$$M = VL = 25(1.275) = 31.875 \text{ kNm}$$

$$e = \frac{M}{P} = 0.031 \text{ m}$$

$$\frac{e}{D} = 0.1$$

Since equivalent eccentricity is only 31 mm in 300 mm size, the section will be an uncracked one. The design can be done primarily for axial load and checked for combined BM and axial force.

Length of the pile: The total depth of the pile can be determined from the static formula (Eq. 16.3) as :

$$Q_u = \frac{P}{1.25} = S\left(q_0 + \frac{\gamma H}{2}\right) K \tan \phi + A_p (0.5 D \gamma N_r + \gamma H N_q)$$

$$q_0 = 20 \text{ kN/m}^2, \, \gamma = 18 \text{ kN/m}^3$$

$$K = 1.75, \tan \phi = \tan 25° = 0.466$$

$$N_r = 10, \, N_q = 15$$

$$S = 4DH = 1.2 \, H$$

$$A_p = D^2 = 0.09 \text{ m}^2$$

Substitution of the various quantities in the equation gives

$$\frac{1025}{1.25} = 1.2H (20 + 9H) \, 1.75 \, (0.466)$$

$$+ \, 0.09 \, (0.15(18)(10) + 18(15) \, H)$$

$$= 19.572H + 8.81H^2 + 2.43 - 24.3H$$

or $\quad\quad 8.81H^2 + 43.872H - 820 = 0$

The solution gives

$$H = 7.46 \text{ m}$$

Try $\quad\quad H = 7.5 \text{ m} \quad (= \text{overall depth})$

Structural design. Limit state axial load capacity of the a short concrete column is

$$P_w = 0.4f_{ck}A_c + 0.67f_yA_s \geq Q_u$$

Slenderness factor $\quad = H/D = 7.5/0.3 = 25$

Minimum reinforcement for precast concrete pile of H/D equal to 30 is about 1.5 per cent

$$A_s = \frac{1.5}{100}(300)(300) = 1350 \text{ mm}^2$$

Use 4 numbers of 16 mm and 12 mm bars each, then

$$A_s \text{ (provided)} = 4(201 + 113) = 1256 \text{ mm}^2$$

The axial load capacity of the pile is

$$P_u = 0.4 f_{ck} A_c + 0.67 f_y A_s$$

$$= 0.4(20)(90000 - 1256) + 0.67(415)(1256)$$

$$= 709\,952 + 349\,230 = 1059\,182 \text{ N}$$

P_u is more than the design load, so the minimum reinforcement provided is adequate. However, because of the combined bending and axial force select the area of the reinforcement more than the minimum desired.

Provide 8 numbers of 18 mm bars, then

$$A_s \text{ (provided)} = 8(254) = 2032 \text{ mm}^2$$

Design for combined action

$$P_{u0} = 0.45\, f_{ck} A_c + 0.75 f_y A_s$$

$$= 0.45(20)(90000 - 2032) + 0.75(415)(2032) = 1449\,470 \text{ N} = 1449.5 \text{ kN}$$

$$M_{u0} = K b d^2 f_{ck} = 0.138(300)(300 - 50)^2(20)$$

$$= 51.75(10^6) \text{ Nmm} = 51.75 \text{ kNm}$$

Note: K here refers to that associated with the moment capacity of concrete section

$$\frac{P}{P_{u0}} + \frac{M}{M_{u0}} = \frac{1025}{1449.5} + \frac{31.875}{51.75} = 0.707 + 0.616 = 1.323$$

This value is more than 1.25 which is the allowed value under seismic load condition. The section, therefore, needs to be modified marginally. Recheck with the following detail:

Try $D = 325$ mm with same reinforcement details

$$P_{u0} = 0.45(20)(325^2 - 2032) + 0.75(415)(2032) = 1564800 \text{ N}$$

M_{u0} based on tension reinforcement is

$$M_{u0} = A_{st} f_y j d / 1.15$$

$$= 3(254)(415)(0.8)(325 - 50)/1.15$$

$$= 60\,496\,170 \text{ Nmm}$$

Use $M_{u0} = 60.5$ kNm

$$\frac{P}{P_{u0}} + \frac{M}{M_{u0}} = \frac{1025}{1564} + \frac{31.875}{60.5} = 1.18$$

The depth of the pile need to be redesigned for the revised size of the pile as $D = 325$ mm.

$$S = 4(0.325)\ H = 1.3\ H \text{ m}^2$$

$$A_p = 0.325^2 = 0.1056 \text{ m}^2$$

From the static formula, we have

$$\frac{1025}{1.25} = 1.3 H (20 + 9H)(1.75)(0.466)$$

$$+ 0.1056(0.15(18)(10) + 18(15)\ H)$$

$$= 21.2H + 9.42\ H^2 + 2.85 + 28.5\ H$$

The solution of the above equation gives $H = 7.02$ m. Provide 7.1 m depth piles.

Design for handling. Assuming a two point pick up, the bending moment on the pile while handling is taken from Table 16.6 (partial safety factor for load is taken 1.5):

$$M = 0.022\ WL$$

$$= 0.022(7.1)(0.325)^2(25)(1.5)(7.1) = 4.4\ \text{kNm}$$

This bending moment is far less than that caused by the seismic load; the pile is therefore safe in handling.

Design of ties. Provide 2 sets of 8 mm ties. The minimum area of ties per metre length of the pile is 0.2% and is

$$A_{sv} = \frac{0.20}{100}(325)(1000) = 650\ \text{mm}^2/\text{m}$$

Spacing of two sets of 8 mm ties is

$$s_v = \frac{650}{2(2)(50)} = 325\ \text{mm}$$

Provide two sets of 8 mm ties at 300 mm spacing in the body of the pile and at 100 mm spacing at each end of the pile for a distance $2(325) = 650$ mm. Figure 16.3 gives the reinforcement details.

Design of pile cap.

Spacing of the piles = 0.900 m

Bending moment is critical at the face of the column and shear force is critical at a distance half the effective depth of the pile cap from the face of the column.

$$\text{BM span} = \left(\frac{0.9}{2} - \frac{0.4}{2}\right) = 0.25\ \text{m}$$

$$\text{Shear span} = 0.25 - \frac{1}{2}d = 0$$

For an assumed effective depth of pile cap equal to 500 mm.

Design BM on the pile cap is

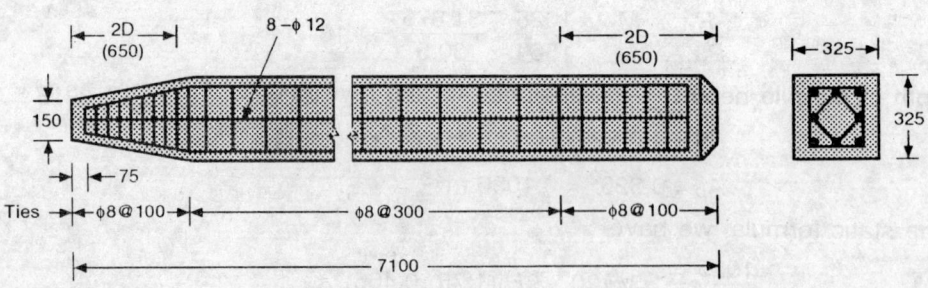

Fig. 16.3 Details of Precast Pile.

$$M = \text{(load from a pair of piles) (BM span)}$$

$$= 2\left(\frac{1025}{1.25}\right)(0.25) = \frac{512.5}{1.25} \text{ kNm}$$

(A reduction partial safety factor of 1.25 is applied for seismic load condition.)

Minimum width of the pile cap is

$$b = 2(150) + 325 + 900 = 1525 \text{ mm}$$

Provide 1525 by 1525 mm pile cap. The effective depth required is

$$d \text{ (required)} = \sqrt{\frac{M}{Kbf_{ck}}}$$

$$= \sqrt{\frac{512.5(10^6)}{0.138(1525)(20)(1.25)}} = 312 \text{ mm}$$

To accommodate the development lengths, etc., a minimum thickness of 600 mm is required. Then

$$d = 600 - \text{covers at top and bottom}$$

$$= 600 - 60 - 60 = 480 \text{ mm}$$

$$A_{st} = \frac{1.15\,M}{jdf_y} = \frac{1.15(512.5)(10^6)}{0.8(480)(415)(1.25)} = 2960 \text{ mm}^2$$

Provide 10 numbers of 20 mm bars in each direction.

$$A_{st} \text{ (provided)} = 3140 \text{ mm}^2$$

Development length $= \dfrac{\phi f_y}{4\tau_{bd}} = \dfrac{20(415)}{4(1.2)(1.6)} = 1080 \text{ mm}$

Distance of the free edge of the pile cap from the face of the column is $(1525 - 400)/2 = 562$ mm.

Length of the bar to be bent up to generate full development length is

$$= 1080 - 502 - 4(20) = 498 \text{ mm}$$

The reinforcement details of the pile cap are shown in Fig. 16.4.

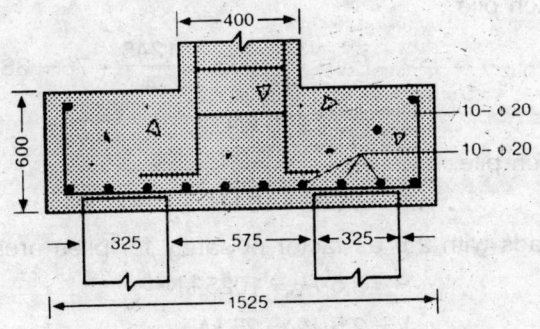

Fig. 16.4 Pile Cap Details.

Note: The assumed thickness of the pile cap was 800 mm as against 600 mm provided. The bending moment at the top level of the pile cap will be only 600/800 of that used in the calculations. Axial force caused by this moment on the piles will be less than the value used in the computations.

EXAMPLE 16.2 *Design of bored pile group in clayey soil.* A column under seismic load condition is to transmit 1200 kN axial load, 120 kNm bending moment and 40 kN shear force. Design a pile foundation for the following data:

Size of the column	= 400 mm square
Clayey soil (medium) with	$N = 7$
Cohesion value	$= C = 70 kN/m^2$
Unconfined compressive strength	$= 0.2 N/mm^2$

$\gamma = 16$ kN/m^3, surcharge $= q_0 = 20$ kN/m^2

concrete = M20, reinforcement = HYSD-Fe415 bars

(Problem is similar to Example 16.1 except the piles are bored type.)

Let pile cap size	$= 1500 \times 1500 \times 800$ mm
Let diameter of pile	$= D = 400$ mm

Superimposed loads on pile cap are

$$W_s = 1200 \text{ kN}, \ M_s = 120 \text{ kNm}, \ V_s = 40 \text{ kN}$$

Weight of the pile cap	$= 1.5(1.5)(0.8)25 = 45$ kN
Total load on piles	$= W_1 = W_s + 45 = 1245$ kN
Spacing of the pile	$= s = 2.5 D = 1000$ mm

(Refer Fig. 16.2 for the general arrangement.)

BM at the top level of the piles is

$$M_2 = M_s + V_s t = 120 + 40(0.8) = 152 \text{ kNm}$$

The axial load caused by the BM is

$$P_2 = \frac{M_2}{2s} = 76 \text{ kN}$$

Total axial force on each pile

$$P_w = \frac{W_1}{4} + P_2 = \frac{1245}{4} + 76 = 388 \text{ kN}$$

Use $\quad P_w = 410$ kN

The shear force on each pile is

$$V_w = V_s/4 = 10 \text{ kN}$$

The ultimate design loads with 2.5 as factor of safety for piles are:

$$P = 2.5 \ P_2 = 1025 \text{ kN}$$
$$V = 2.5 \ V_w = 25 \text{ kN}$$

Bending moment on the pile. The value of the horizontal subgrade modulus from Table 16.3 for 0.2 N/mm^2 unconfined compressive strength is:

$$k_c = 4.9 \text{ N/mm}^2$$

Equivalent cantilever span from Table 16.5 is

$$L/D = 4$$

or

$$L = 4D = 4(0.40) = 1.60 \text{ m}$$

The bending moment on the pile is

$$M = VL = 25(1.60) = 40 \text{ kNm}$$

$$e = \frac{M}{P} = \frac{40}{1025} = 0.04 \text{ m}$$

Length of pile. The load carrying of pile from Eq. (4.4) is

$$Q_u = A_p \, N_c \, C_p + \alpha \, CS \geq P/1.25$$

where

$$A_p = \frac{\pi D^2}{4} = 0.1257 \text{ m}^2$$

$$N_c = 9$$

$$\alpha = 0.7$$

$$C = C_p = 70 \text{ kN/m}^2$$

$$S = \pi DH = 1.257 \, H$$

Reduction of safety under seismic load is 1.25. Substituting different quantities in the load capacity criterion, we get

$$0.1257(9)(70) + 0.7 \, (70)(1.257H) \geq 820$$

or

$$H \geq 12.03 \text{ m}$$

Try

$$H = 12.1 \text{ m (overall, depth)}$$

Structural design. It could be seen from the pervious example that about 8 numbers of 18 mm bars were needed for a pile so the capacity of the pile with 8 numbers of 18 mm bars is

$$P_{u0} = 0.45 \, f_{ck} A_c + 0.75 \, f_y A_s$$

$$= 0.45(20)(0.97)(0.1257) + 0.75(415)(0.002032) = 1.73 \text{ MN}$$

Vide Chapter 9 of Part I for design of column under combined action and use the following assumptions and coefficients from Table 9.12:

$$D_s = D - 0.10 = 0.3 \text{ m}$$

$$u = 0.2, \ c = D_s/D = 0.75$$

$$c_{ab} = 0.28, \ c_{xb} = 0.22$$

$$\beta = \frac{f_y}{f_{ck}} = \frac{415}{20} = 20.8$$

$$p = \frac{A_{st}}{A_c} = \frac{2032}{125700} = 0.016$$

Let

$$p_c = p_t = p/2$$

$$P_{ab} = 0.45(1 - p)\, c_{ab} D^2 f_{ck}$$

$$= 0.45(0.984)(0.28)(0.16)(20) = 0.4 \text{ MN}$$

$$\frac{P_{a0}}{P_{ab}} = \frac{1.73}{6.4} = 4.325$$

$$e_s = \frac{(0.45(1 - p)(1 - u - e_{xb})c_{ab} + 0.68 p_c \beta c)D}{0.45(1 - p)c_{ab} + 0.68\,(p_c - p_t)\beta}$$

$$e_s = \frac{(0.45(0.984)(1 - 0.2 - 0.22)(0.28) + 0.68(0.008)(20.8)(0.55))(0.4)}{0.45(0.984)(0.28)}$$

$$= 0.78 \text{ m}$$

$$e_b = e_s - 0.5\, D_s = 0.78 - 0.5(0.3) = 0.63 \text{ m}$$

$$\frac{e}{e_b} = \frac{0.04}{0.63} = 0.063$$

The load carrying capacity of the column is

$$P_a = \frac{P_{ab}}{1 + (P_{u0}/P_{ab} - 1)\, e/e_b}$$

$$= \frac{1.73}{1 + (3.325)(0.063)} = 1.43 \text{ MN}$$

This value is more than 1.025 MN. The size of the pile is mainly governed by the depth requirement rather than the structural strength requirement.

Design of pile cap. Design of pile cap is exactly similar to that of the previous example with minor modification in the bending moment. The reinforcement details of the pile and pile cap are shown in Figs. 16.5 and 16.6.

16.11 Design of Well Foundations

Bridge piers, quay walls, and pump houses constructed on river beds or sea shores are subjected to very heavy loads. The foundations in such cases should be taken to large depths so as to avoid failure against scoure and instability. The well foundations must be designed to resist failures or loads listed below:

a) Scoure of the soil (stability).

b) Lateral earth pressures (stability).

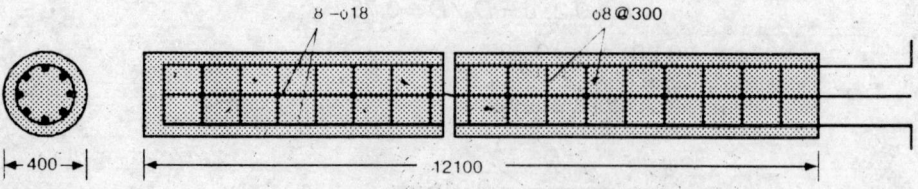

Fig. 16.5 Cast in-situ Pile. Example 16.2.

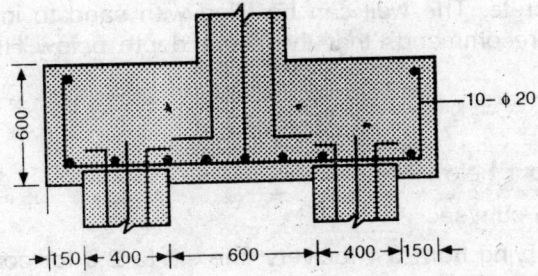

Fig. 16.6 Details of Pile Cap, Example 16.2.

c) Bearing pressure on the soil.

d) Structural strength to resist forces on the well.

During high flood level condition, the soil on the downstream side of the well will be scoured out and some deposit on the upstream may also take place. The depth of the foundation must be taken below scoure depth so as to have enough anchorage to provide stability against earth pressures and external loads. The load of the well is transferred primarily through bearing resistance of the base of the well. Open cylindrical or similar types of shafts are first sunk into the soil slowly by adding loading the well. The rate of sinking of the well can be increased by dredging of the soil inside the shaft and jetting of water on the outer surface of the shell. The bottom-most portion of the shell is provided with a cutting edge so as to facilitate cutting into the soil and settlement during sinking. Typical cutting edge is shown in Fig. 16.7. After sinking the well to the desired depth; the bottom of the well is plugged with

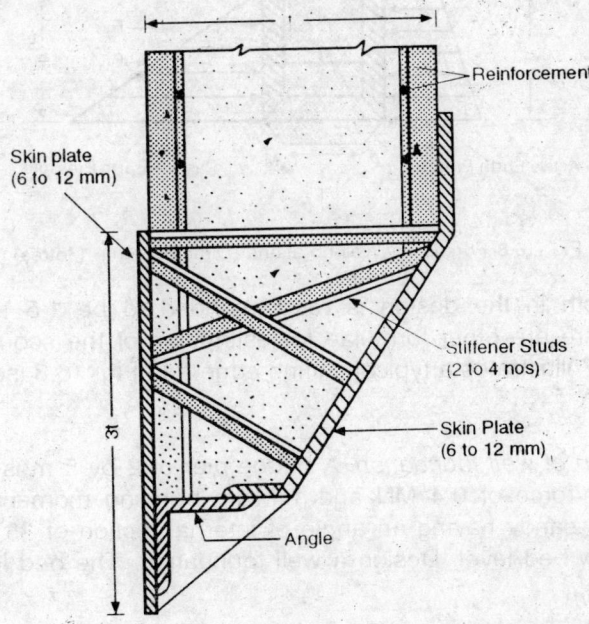

Fig. 16.7 Typical Cutting Edge in Wells.

plain or reinforced concrete. The well can be filled with sand to increase the stability. The Indian Road Congress recommends that the scour depth below HFL can be computed by Lacey formula as under:

$$h_s = 0.473(Q/f)^{1/3} \qquad (16.17)$$

where h_s = depth of scour below HFL

Q = discharge in cum/sec

f = silt factor varying from 0.4 for very fine silt to 2.0 for coarse sand

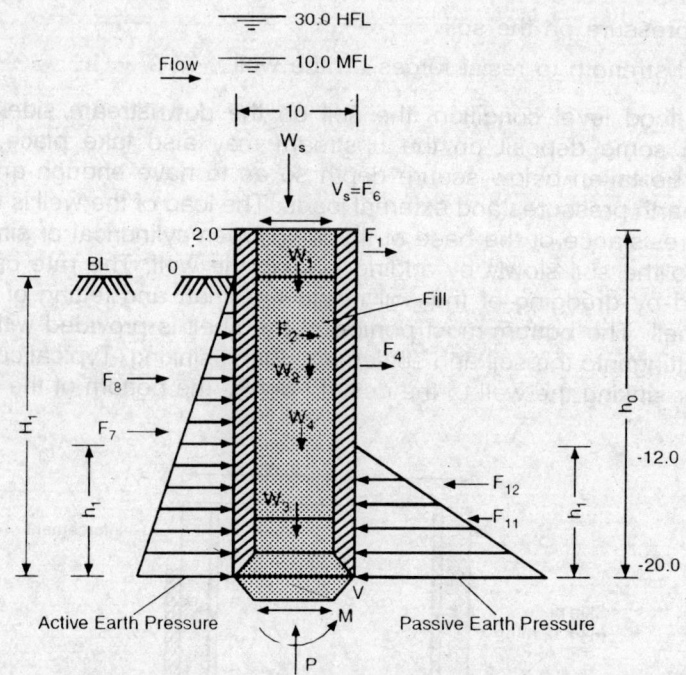

Fig. 16.8 Forces on Well-notations (Dimensions in Metres).

The actual scour depth in the design is recommended to be 1.5 to 2 times the values suggested above. There are other formulae for estimation of the scour depth which are not listed here. Figure 16.7 illustrates a typical cutting edge and Fig. 16.8 indicates different forces acting on the well.

EXAMPLE 16.3 *Design of well foundation.* A bridge pier of 2 by 5 m is transmitting a vertical load of 10 MN, shear force of 0.4 MN and 1 MNm bending moment to the top of a well foundation. The soil is sandy having an angle of internal friction of 25° at bed level and 40° at –16 to –30 m below bed level. Design a well foundation. The bed level (BL) is = 0.

Data of the problem

$$W_s = 10 \text{ MN}, \ V_s = 0.4 \text{ MN (seismic)}$$

M20 concrete with HYSD-Fe415 bars.

Depth of scour	= 12 m for 10 m dia shaft
Bearing capacity	= 2000 kN/m^2
Top of the well	= + 2 m above BL
Depth of well below scour	= 8 m
MFL	= 10 m above BL
LWL	= 2 m above BL
HFL	= 30 m above BL
Density of water	$\gamma = 10$ kN/m^3
Density of soil	$\gamma_s = 18$ kN/m^3
Density of concrete	$\gamma_c = 25$ kN/m^3

The active and passive earth pressure coefficients under normal condition are:
$$c_a = 0.3, \quad c_p = 3.6$$

and in seismic load condition are: $\quad c_{ac} = 0.35, \quad c_{pe} = 3.2$

Note: For notations of W, F etc. vide Fig. 16.8.

Seismic coefficient $\quad\quad \alpha_h = 0.05$

The *design is carried* out by trial and correction

Head of active earth pressure	$H_1 = 20$ m
Head of passive earth press	$H_2 = 8$ m

The following assumptions are made:

Outer diameter of the well	$D = 10$ m
Thickness of the well =	$t = 0.6$ m
Thickness of top plug =	$t_1 = 2$ m
Thickness of bottom plug =	$t_2 = 3$ m

Inside diameter of the well is
$$D_1 = D - 2t = 10 - 1.2 = 8.8 \text{ m}$$

Weight of top plug
$$W_1 = \frac{\pi D_1^2}{4} t_1 \gamma_c = \frac{\pi}{4}(8.8)^2 (2.0)(25) = 3041 \text{ kN}$$

The corresponding seismic force F_1 and its height above base and the overturning moment are:
$$F_1 = \alpha W_1 = 152 \text{ kN}$$
$$h_1 = (22 - 1) = 21 \text{ m}$$
$$M_1 = F_2 h_1 = 152(21) = 3192 \text{ kNm}$$

Weight of sand filling

$$W_2 = \frac{\pi D_1^2}{4} (H_1 + 2 - t_1 - t_2) \gamma_s$$

$$= \frac{\pi}{4} (8.8)^2 (22 - 5) (18) = 18610 \text{ kN}$$

The corresponding seismic force and its height of action above case, and the moment are:

$$F_2 = \alpha W_2 = 930 \text{ kN}$$

$$h_2 = \frac{(22 - 5)}{2} + 3 = 11.5 \text{ m}$$

$$M_2 = F_2 h_2 = 930 (11.5) = 10695 \text{ kNm}$$

Weight of bottom plug is

$$W_3 = \frac{\pi D_1^2}{4} t_2 \gamma_c = \frac{\pi}{4} (8.8)^2 (3) (25) = 4562 \text{ kN}$$

The corresponding seismic force and its height above the base and the moment are

$$F_3 = \alpha W_3 = 0.05 (4562) = 228 \text{ kN}$$

$$h_3 = t_2/2 = 1.5 \text{ m}$$

$$M_3 = F_2 h_3 = 228 (1.5) = 342 \text{ kNm}$$

The weight of shell of the well is

$$W_4 = \frac{\pi}{4} (D^2 - D_1^2) (H_1 + 2)\gamma_c$$

$$= \frac{\pi}{4} (10^2 - 8.8^2) (22) (25) = 9745 \text{ kN}$$

The corresponding seismic force and its height of action above the base, and the moment are:

$$F_4 = \alpha W_4 = 0.05 (9745) = 487 \text{ kN}$$

$$h_4 = H_0/2 = 22/2 = 11 \text{ m}$$

$$M_4 = F_4 h_4 = 487 (11) = 5357 \text{ kNm}$$

Bouyancy force:

$$W_5 = -\frac{\pi D^2}{4} H_0 \gamma = -\frac{\pi}{4} (100) (22) (10) = -17279 \text{ kN}$$

Its height of line of action from toe

$$h_5 = D/2 = 5 \text{ m}$$

And the corresponding moment is

$$M_5 = -W_5 h_5 = 17279 (5) = 86395 \text{ kNm}$$

Overturning moment caused by the superimposed seismic shear force from the superstructure is

$$M_6 = F_6 H_0 = 400(22) = 8800 \text{ kNm}$$

where

$$F_6 = V_5$$

The active and dynamic earth pressures, lines of action above the base and the corresponding overturning moments are:

$$F_7 = 0.5\ c_a\,(\gamma_s - \gamma)\ DH_1^2$$

$$= 0.5\ (0.3)\ (8)\ (10)\ (20^2) = 4800\ \text{kN}$$

$$h_7 = h_1/3 = 6.67\ \text{m}$$

$$M_7 = F_7 H_7 = 4800\ (6.67) = 32016\ \text{kNm}$$

$$F_8 = 0.5\ (c_{ae} - c_a)\,(\gamma_s - \gamma)\ DH_1^2$$

$$= 0.5(0.35 - 0.3)(8)(10)(20)^2 = 800\ \text{kN}$$

$$h_8 = H_1/2 = 10\ \text{m}$$

$$M_8 = F_8 h_8 = 8000\ \text{kNm}$$

The total overturning moment about the toe of the base is

$$M_0 = \sum_{t=1}^{8} M_4 = 164542\ \text{kNm}$$

The restoring forces and the moment caused by the top plug, sand filling, bottom plug, the shell and the load is

$$M_{10} = (W_1 + W_2 + W_3 + W_4 + W_s)\ D/2$$

$$= (3041 + 18610 + 4562 + 9745 + 10000)\ 5$$

$$= (45958)(5) = 229\ 790\ \text{kNm}$$

Static and dynamic passive earth pressures and the corresponding moments are:

$$F_{11} = 0.5\ c_p\,(\gamma_s - \gamma)\ DH_2^2$$

$$= 0.5(3.6)(8)(10)(64) = 9216\ \text{kN}$$

$$h_{11} = H_2/3 = 8/3 = 2.67\ \text{m}$$

$$M_{11} = F_{11} h_{11} = 24606\ \text{kN}$$

$$F_{12} = 0.5\ (c_{pe} - c_p)\,(\gamma_s - \gamma)DH_2^2$$

$$= 0.5(3.2 - 3.6)(8)(10)(64) = -1024\ \text{kN}$$

$$h_{12} = 2H_2/3 = 16/3 = 5.33\ \text{m}$$

$$M_{12} = F_{12}\ h_{12} = -5458\ \text{kNm}$$

The net restoring (passive) moment is

$$M_p = M_{10} + M_{11} + M_{12}$$

$$= 229\ 790 + 24\ 606 - 5458 = 248\ 938\ \text{kNm}$$

Stability analysis. Active horizontal force is

$$F_a = (F_1 + F_2 + F_3 + F_4 + F_6 + F_7 + F_8) = 7797\ \text{kN}$$

Passive earth pressure

$$F_p = F_{11} + F_{12} = 8192\ \text{kN}$$

The frictional force between the soil and the base plug is

$$F_f = (W_1 + W_2 + W_3 + W_4 + W_5 + W_s)\ \mu$$

$$= 0.4(45968 - 17279) = 11475\ \text{kN}$$

net passive resistance is

$$F_h = F_q + F_f = 8192 + 11475 = 19667 \text{ kN}$$

Factor of safety against sliding is

$$\frac{F_h}{F_a} = \frac{19667}{7797} = 2.52$$

Factor of safety against overturning is

$$= \frac{M_p}{M_0} = \frac{248938}{164542} = 1.51$$

Factor of safety for stability under seismic load condition can be 1.5. Therefore, the well is stable under seismic load condition.

Bearing pressure. Downward force at the base is

$$W_0 = W_s + \sum_{i=1}^{5} W_i$$

$$= 10000 + 3041 + 18610 + 4562 + 9745 - 17279$$

$$= 45958 - 17279 = 28\,779 \text{ kN}$$

Net bending moment about the centre of the base is

$$M_a = M_6 + M_7 + M_8 - M_{11} - M_{12} + \sum_{i=1}^{4} M_i$$

$$= 8800 + 32016 + 8000 - 24606 + 5458 = 49254 \text{ kNm}$$

Maximum and minimum bearing pressures on the soil are:

$$p_t = \frac{W_0}{A_0} \pm \frac{M_a}{Z_0}$$

where

$$A_0 = \frac{\pi D^2}{4} = 78.54 \text{ m}^2 \text{ and}$$

$$Z_0 = \frac{\pi D^3}{32} = 98.17 \text{ m}^2$$

$$p_1 = \frac{28779}{78.54} + \frac{49254}{98.17}$$

$$= 366.42 + 501.71 = 868.13 \text{ kN/m}^2$$

$$p_2 = 366.42 - 501.71 = -135.29 \text{ kN/m}^2$$

Factor of safety against bearing pressure on the soil (neglecting the sking friction) is

$$\beta_b = \frac{q_u}{p_1} = \frac{2000}{668.13} = 2.3$$

A factor of safety of 2.5/1.25 to 3/1.25 is adequate under seismic load condition. We can see that the factors of safety are marginally higher than those minimum desired. The size of the well could be decreased by about ten per cent to achieve some more economy. However, such a reduction is not carried out at this state as the structural design is yet to be finalized. (Pressure distribution to be modified as tension in soil not permitted)

Structural design. The structural components of the well consists of:

a) Thickness of steining (shell thickness).

b) Reinforcement details.

c) Bottom plug (usually plain concrete).

d) Top plug.

(Shell wall or steining are used interchangeably.)

The well is subjected to the following external forces for which it should be designed:

a) Radial pressure just above the bottom plug.

b) Longitudinal stress caused by axial load and bending moment on the well.

c) Local bending moment on the wall.

One can adapt a limit states design of the structural components of the well. However, because of the combined overall and local actions and biaxial state of stress, only working stress design is used in this example.

Design of steining. The thickness of the wall is assumed to be 600 mm and for such a value the ratio of the radius to the thickness is 8.33. A cylindrical shell with radius to thickness ratio more than 15 can be designed by thin shell theory. The thin shell theory with minor corrections can be applied to this shell as an initial guide. The critical section is just above the bottom plug and the total radial pressure at this level is

$$p = \gamma \, (\text{HFL} - (\text{base level} + t_1))$$
$$+ \, c_a \, (\gamma_s - \gamma) \, (\text{BL} - (\text{base level} + t_1)$$
$$= 10 \, (30 + 20 - 3) + 0.3(8) \, (0 + 20 - 3)$$
$$= 470 + 40.8 = 510.8 \text{ kN/m}^2$$

Hoop compressive thrust on the shell is

$$T = \frac{pD}{2} = \frac{510.8(10)}{2} = 2554 \text{ kN/m} = 2554\,000 \text{ N/m}$$

(use M20 concrete)

The working design condition is

$$P = A_e \, \sigma_{ac} + A_{sc} \, \sigma_{sc}$$
$$2554\,000 = 600 \, (1000) \, (5) + A_{sc} \, (190)$$

or
$$A_{sc} = \frac{2554000 - 300000}{190} = \text{negative}$$

Provide a minimum of 0.2% reinforcement, then

$$A_{sc} = 0.002 \, (500) \, (1000) = 1000 \text{ mm}^2/\text{m}$$

We can also decrease the thickness of the shell to 500 mm.

Provide 20 mm bars at 500 mm at each face of the shell.

Longitudinal stress. The area and the modulus of the shell are

$$A = \pi \, (D - t) \, t = \pi \, (9.5) \, (0.5) = 14.77 \text{ m}^2$$

$$Z = \frac{\pi}{32} \frac{(D^4 - (D - 2t)^4)}{D} = 33.76 \text{ m}^3$$

The maximum compressive stress on the concrete neglecting the reinforcement is

$$f = \frac{W_0}{A} + \frac{M_0}{Z} = \frac{28779}{14.77} + \frac{49254}{33.76}$$

$$= 1948 + 1459 = 3407 \text{ kN/m}^2 = 3.5 \text{ MPa}$$

Provide only nominal reinforcement and that can be 20 mm bars at 500 mm on each face.

Check for local bending stress. The local bending moment which is critical at the bottom plug can be approximate as:

$$M = \frac{pRt}{\sqrt{12}} \left(1 - \frac{1}{H\beta}\right)$$

where

p = radial thrust = 510.8 kN/m^2

t = thickness = 0.5m

R = radius = 5 m

H = height of the cylinder = H_0 = 22 m

$$\beta = \sqrt[4]{\frac{3}{R^3 t^3}} = \sqrt[4]{\frac{3}{25(0.25)}} = 0.83$$

Substituting various quantities, we obtain

$$M = \frac{510.8(5)(0.5)}{3.464} \left(1 - \frac{1}{22(0.83)}\right) = 348 \text{ kNm/m}$$

This bending moment is in addition to the axial longitudinal force. The balanced moment capacity of the concrete section is about 120 kNm/m which means the section should be designed as a doubly-reinforced one. The area of the tension reinforcement needed is

$$A_{st} = \frac{M}{jd\sigma_{st}} = \frac{348000000}{0.86(430)(230)} = 4091.5 \text{ mm}^2/\text{m}$$

Provide 25 mm bars at 110 mm spacing at both faces in the bottom 10 m depth. The radial thrust at top of the well is about 50 per cent of that at the bottom plug, therefore provide 25 mm bars at 200 mm spacing at top portion as extra bars.

The radial thrust from the bottom plug will cause hoop tension and bending moment in the wall of the well. These values are:

$$T_1 = \frac{pD_c^2}{16 \, r_0}$$

where

D_c = mean diameter of the curb (cutting edge)

r_0 = vertical height of an inverted arch

p = net bearing pressure on the soil.

Let the depth of the cutting edge be

$$h_c = 2t = 1 \text{ m}$$

sag of the arch of the bulb $= \dfrac{D}{4} = 2.5$

then

$$r_0 - h_c + 2.5 = 3.5$$
$$D_c = D - t = 0.5 \text{ m}$$

The net bearing pressure is computed by considering the critical condition which will adversely affect this part of the design. The net pressure also refers to the force transmitted to the steining and it is critical at low water lever when the buoyancy effect is least.

Net load on the bottom plug is equal to superimposed load from pier, weight of top plug, weight of steining and 50% of buoyancy

$$W_{p0} = W_s + W_1 + W_4 + 0.5 \; W_5$$
$$= 10000 + 3041 + 9745 - 0.5 \,(17279) = 14147 \text{ kN}$$

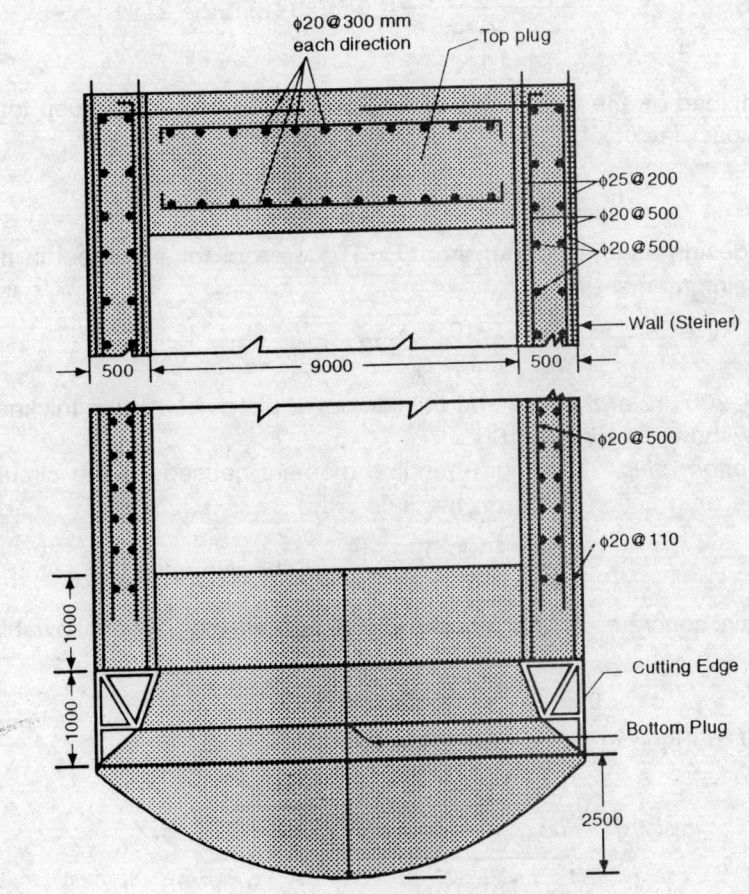

Fig. 16.9 Details of Well.

$$p = \frac{W_{10}}{A_0} = \frac{14147}{78.54} = 180 \text{ kN/m}^2$$

$$Q_1 = \frac{pD_c^2}{8r_0} = \frac{180(9.5)^2}{8(3.5)} = 580 \text{ kN/m}$$

Bending moment on the shell is

$$M = Q_1 \frac{h_c}{2} = 580(0.5) = 290 \text{ kNm/m}$$

This value is less than the BM caused by the radius inward force so it does not govern the deisgn. Reinforcement details in the well are given in Fig. 16.9.

Design of cutting edge (curb). Net hoop tension caused on the cutting edge is given by

$$T_2 = \frac{D_0}{2} \left(\frac{pD_c^2}{8r_0} - (0.5c_a(\gamma_s - \gamma)(H_0 + (H - h_c))h_c) \right)$$

$$= \frac{9.5}{2} \left(\frac{180(9.5)^2}{8(3.5)} - 0.5(0.3)(8)(22 + 21)1 \right)$$

$$= 1084 \text{ kN}$$

Assume the total load on the well during sinking is 5000 kN, then the hoop tension on cutting edge with an impact factor of 1.2 is

$$T_3 = \left(\frac{5000}{D_c} \right) \left(\frac{h_c}{t} \right) \left(\frac{D_c}{2} \right)(1.2) = 1910 \text{ kN}$$

T_3 controls the design as it is greater than T_2. The area of tension steel in the cutting edge along with the reinforcement is

$$A_s = \frac{T_3}{\sigma_{st}} = \frac{1910}{0.15} = 12733 \text{ mm}^2$$

Provide ISA 200, 200, 12 and inside and outside cover plates of 12 mm thickness. The cutting edge details are shown in Fig. 16.10.

Design of bottom plug. Maximum bending moment caused on the circular bottom plug is (neglecting the arch action to be on the safe side)

$$M = \frac{3p D_c^2}{16} = \frac{3(180)(9.5)^2}{16} = 3046 \text{ kNm/m}$$

Provide only plain concrete of M20 quality in the bottom plug. The allowable stress in the plain concrete is

$$\sigma_{at} = 0.35 \sqrt{f_{ck}} = 1.56 \text{ N/mm}^2$$

The thickness of the plug at the middle zone is

$$t_{p2} = \sqrt{\frac{6M}{b\sigma_{at}}}$$

$$= \sqrt{\frac{6(2046)(10^6)}{1000(1.56)}} = 2806 \text{ mm}$$

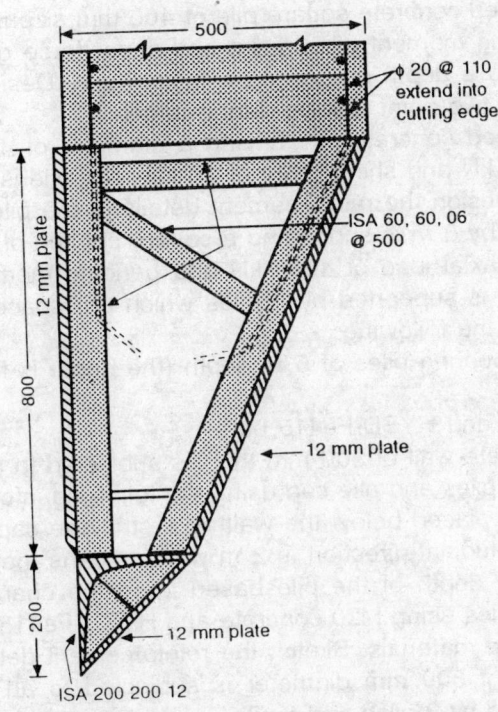

500

φ 20 @ 110
extend into
cutting edge

12 mm plate

ISA 60, 60, 06
@ 500

800

12 mm plate

200

12 mm plate

ISA 200 200 12

Fig. 16.10. Cutting Edge of Well.

The actual bending moment on the bottom plug will be less than that computed as the plug acts partly as a dome. Provide two metre thick bottom plug plus 2.5 m bulb. The overall thickness provided at middle of the plug is

$$t_{p2} = 2 + 2.5 = 4.5 \text{ m}$$

An average value of 3 m was assumed for the purpose of stability calculations.

Design of top plug. Well is provided with well cap which rests on the steining, therefore the load carried by the top plug is only nominal. Only nominal thickness with nominal reinforcement is provided.

Use $t = 1000$ mm

$$A_{st} = 0.002(1000)(1000) = 2000 \text{ mm}^2/\text{m}.$$

Provide 20 mm bars at 300 mm both at top and bottom each and in radial and circumferential directions.

PROBLEMS

16.1   A precast reinforced concrete square pile of 300 mm side is subjected to an axial load of 250 kN and 20 kNm bending moment. Design reinforcement details of the pile such that it can be handled and transported with two pick-up points. Use M20 concrete and

HYSD-Fe415 bars.

16.2 A precast reinforced concrete square pile of 400 mm size is subjected to an axial load of 400 kN, bending moment of 50 kNm and shear force of 30 kN. The pile is driven in clayey soil upto a depth of 7 m as a friction pile. Design reinforcement details of the pile assuming two point pick-up during handling.

16.3 A circular reinforced concrete pile having a diameter of 400 mm is subjected to an axial force of 300 kN and shear force of 20 kN. The pile is driven into a sandy soil to a depth of 6 m. Design the reinforcement details of the pile.

16.4 A pile cap of 3 m by 3 m is supporting a square column of 650 mm size. The column is transferring an axial load of 4000 kN and bending moment of 80 kNm about one axis. The pile cap is supported by 4 piles which are spaced 2.5 m apart in both the directions. Design the following:
 (a) The piles as bearing piles of 5 m depth (the depth is fixed by soil condition).
 (b) The pile cap.
 Use M20 concrete and HYSD-Fe415 bars.

16.5 A reinforced concrete wall of 300 mm thick is subjected to an axial load of 600 kN/m. Design foundation piles and pile cap using the following information. Two rows of piles at 1.5 m apart are placed below the wall as a strip pile cap. The spacing of the piles alongwith the longitudinal direction is 2 m which means that each pile is supposed to carry 600 kN. The depth of the pile based upon the characteristics is 6 m. Design precast concrete piles using M20 concrete and HYSD-Fe415 bars. Also design the pile cap using the same materials. Sketch the reinforcement details.

16.6 A square column of 600 mm diameter is subjected to an axial load of 2 MN. The column is supported by 3 piles and a pile cap of equilateral triangular in shape in plan. Locate the piles such that each pile carries 1/3rd of the total load of the column. Design bored compaction reinforced concrete pile using M20 concrete and mild steel reinforcement. Depth of the pile based on the soil characteristics is 8 m. The minimum spacing between the piles should be restricted to 2.5 times diameter of the pile. The minimum projection of the pile cap beyond the face of the pile should be 500 mm. Sketch the layout of the piles and the reinforcement details of the piles. Assume thickness of the cap as 800 mm.

16.7 A pile cap is supporting a wall of 250 mm thickness and is subjected to 500 kN/m. Assuming that the pile cap is resting on 2 rows of the piles spaced 1.4 m symmetrically about the wall, design the pile cap using M20 concrete and mild steel reinforcement.

16.8 A square column of 600 mm size is subjected to an axial load of 3 MN and bending moment of 0.1 MNm about one axis. Assume that the pile cap is square in shape and resting on 5 piles, 4 placed, one at each of the 4 corners and 1 at the centre. Calculate the maximum load on the piles assuming a rigid cap and the piles are spaced 2.5 m apart. Design the pile cap with M20 concrete and HYSD-Fe415 bars.

16.9 A reinforced concrete column of 650 mm square in shape is subjected to a concentrated load of 4000 kN and a bending moment of 100 kNm of about one axis. Design a foudation consisting of 4 bearing piles and a pile cap. The depth of the strong bearing strata is 6 m below the ground level and its bearing capacity is 800 kN/m^2. Design precast reinforced concrete circular piles.

16.10 Two reinforced concrete columns of each 500 mm diameter are placed at 4 m apart. Each of the columns is subjected to an axial load of 800 kN and bending moment of

150 kNm. Design the combined pile cap to support the two columns using 350 mm diameter board piles. Capacity of each of the piles is 300 kN. Calculate the minimum number of piles and locate them appropriately under the combined piles cap. Assume a maximum of 9 piles symmetrically arranged under the pile cap. Design the pile cap using M20 concrete and HYSD-Fe415 bars.

16.11 Four reinforced concrete circular columns of 600 mm size are provided with common spread pile cap. The spacing of the columns is 5 m in both the directions and each column is subjected to an axial load of 2 MN. The size of the pile cap is 6.5 by 6.5 m symmetrically placed with respect to the 4 columns. Assume the columns to be rigid to transfer the loads of the columns to the piles, and 36 piles are provided under the pile cap having equal load carrying capacity. Design the pile cap as a flat slab, using M20 concrete and HYSD-Fe415 bars.

16.12 Design a circular well foundation to carry an axial load of 10 MN in sandy soil. The depth and diameter of the well based soil characteristics works to be 20 and 10 m respectively. The soil is sandy having an active and passive earth coefficients as 0.3 and 3.5 respectively. The heights of the soil on the up and down streams above the base level of the well are 15 m and 7 m respectively.

Design the following:

a) well steining with M20 concrete and mild steel reinforcement;
b) bottom plug with M20 concrete.

16.13 A bridge pier of 1×5 m is transmitting a vertical load of 10 MN, shear force is 0.2 MN and bending moment of 0.6 MNm, to the well foundation. The well foundation is to be provided in a sandy soil having an angle of internal friction of 30°. Design the well foundation using M20 concrete and mild steel reinforcement. The height of soil above the bottom level of the well on the up and down streams are 15 and 8 respectively. The total depth of the well is 20 m and safe bearing capacity of the soil at the bottom of the soil is 700 kN/m^2.

Chapter 17

Introduction to Machine Foundations

17.1 Introduction

Machine foundationing is rather a specialized subject requiring knowledge of dynamics and soil structure interactions. Most machines have many moving parts having forces, accelerations and frequencies of motions which are partly transmitted to the foundations. There are variety of machines having different combination of forces and motions. This chapter presents an introduction to machine foundations and illustrates the design of typical ones. Every machine is designed for certain operating frequency to give optimal performance. The foundation of the machine must be designed so as to transmit the forces to the soil with minimum disturbances to the machine. Machines are anchored to the foundation through foundation bolts, so the movement of the machine bases are partly damped and partly transmitted to the foundations. The foundation which may be made of a block or a number of blocks of concrete or piles, transmits the forces and part of the vibrations to the soil. The response of the soil to the vibrations depends on the soil characteristics, mass and inertia properties of the foundation system and characteristics of the machine. Therefore, there is a considerable soil structure interaction in machine foundations. It is difficult to classify the machines into a fixed set of groups as there are hosts of machinery already developed and still being developed. However, one can classify the machines into a number of groups depending on the types of motions and forces transmitted to the foundation. Every point in three-dimensional space has six degrees of freedom. A machine may have only one degree of forced movement when isolated from the foundation but when mounted on a foundation, the degrees of motion of the machine and foundation system can be more or at least equal to that of the machine depending on the type of foundation, and soil and anchorage system. The machines may be classified into following three groups based on the dynamics of the foundation:

  a) Rotary machines.
  b) Reciprocating machines.
  c) Impact machines.

17.2 Machine Vibrations

Machines operate at certain frequencies and excite the foundations to vibrate. The vibration of the machine-foundation system depends on several factors such as masses, base area, inertia properties of the foundation, location and anchoring of the machine with relation to the foundation and soil characteristics. The main aim of the designer is to limit the amplitudes of vibrations and to see that the natural frequency of the system is different from that of the operating frequency of the machine. As the natural frequency tends towards the operating value, the amplitudes of vibrations increase—which sometimes may increase without bounds

The present section gives some formulae which help in the computation of the natural frequency without going into theoretical development. The student is referred to the list of reference given at the end of the book for further study on the dynamics of foundations. The study of the dynamics of systems is outside the scope of this book.

Rotary machines. Rotary machines are classified into three groups based on their speeds:

a) Low frequency machines with speeds less than 1500 rpm.
b) Medium frequency machines with speeds between 1500 and 3000 rpm.
c) High speed machines with speeds exceeding 3000 rpm.

Some of the typical machines that fall under the rotary type are: pumps, motors, generators, lathes, drilling machines, boring machines, crushing mill, rolling mill and turbines.

If the foundation is more or less symmetrical and the machine is mounted concentric with the foundation, the foundation will have a maximum tendency of *vertical vibration*. The natural frequency of the system is given by

$$f_z = \sqrt{\frac{K}{m}} \qquad (17.1)$$

where
$m =$ total mass of the foundation and the machine.

$K =$ equivalent of soil spring constant also called the vertical stiffness of the soil base and it is

$$K = k_e A_f \qquad (17.2)$$

$k_e =$ coefficient of elastic uniform compression of the soil

$\quad =$ force per unit area per unit of settlement.

$A_f =$ area of the foundation

The frequency given by Eq. (17.1) is an angular rotation and measured in radians. It can be converted into number of oscillations per second (same as hertz) as

$$n_z = \frac{f_z}{2\pi} \qquad (17.3)$$

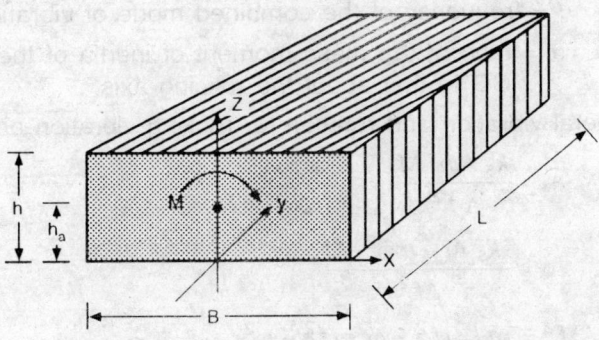

Fig. 17.1 Rectangular Block and Axis of Rotation.

where n_z = number of oscillations per second.

The amplitude of the vibration of the system can be expressed as

$$a_z = \frac{F_z}{m(\omega^2 - f_z^2)}$$

(17.4)

where F_z = superimposed force of excitation

ω = angular velocity of the rotation of the machine and is

$$= 2\pi n$$

(17.5)

n = cycle per second of oscillations of the machine

a_z = vertical amplitude of vibration (see Fig. 17.1 for notations)

Rotary machines can also develop rocking vibration in case of unbalanced systems. The *rocking frequency* is given by

$$f_0 = B \sqrt{\frac{A_f k_g}{12 \, I_{m0}}}$$

(17.6)

where f_0 = rocking frequency of the system (rotation about y- or x-axis)

B = width of the foundation and it is the dimension perpendicular to the axis of rotation

k_g = coefficient of non-uniform compression (it is about $2k_e$)

I_{m0} = mass moment of inertia of the machine and foundation about the horizontal axis of rocking.

The *amplitude of vertical* vibration of the edge is given by

$$a_z = \frac{MB}{2I_{m0} (f_0^2 - \omega^2)}$$

(17.7)

where M = excitation moment of the system

A machine which has a *combined mode* of vibration both vertical and rocking, its frequency can be obtained from the following equation as suggested by Barkan (vide reference given at the end of the book)

$$\alpha f^4 - (f_0^2 + f_z^2) f^2 + f_0^2 f_z^2 = 0$$

(17.8)

where f = frequency of the combined mode of vibration

α = ratio of the mass moment of inertia of the system about is own CG to that about the rokcing axis.

The amplitudes of lateral vibration and rotation for rocking vibration are approximated as

$$a_x = \frac{k_s A_f h_c M}{\Delta f^2}$$

(17.9)

$$\phi = \frac{(k_s A_f - m\omega^2) M}{\Delta f^2}$$

(17.10)

where $\Delta f^2 = m \, I_m (f_1^2 - \omega^2) (f_2^2 - \omega^2)$

(17.11)

f_1 and f_2 = solutions of the quadratic equation (17.8)

$\quad a_x$ = horizontal translational amplitude

$\quad \phi$ = amplitude of rotation

$\quad k_s$ = shear modulus of the subgrade which is approximately equal to 0.5 k_e

The total horizontal displacement of a working point above CG of the system is

$$a_{xc} = a_x + \phi\, h_c \tag{17.12}$$

where $\qquad\qquad h_c$ = working height of the system from the CG

The mass moment of inertia of a rectangular parallelepiped can be written as

$$I_m = \frac{m}{12}(B^2 + H^2) \tag{17.13}$$

where $\qquad\qquad m$ = total mass

$\qquad\qquad B$ = width which is perpendicular to axis of rotation

$\qquad\qquad H$ = height perpendicular to the axis of vibration

The mass moment of inertia about another axis at a distance h_0 from the centroid on z-axis is

$$I_{m0} = I_m + mh_0^2 \tag{17.14}$$

$$= \frac{m(B^2 + H^2) + 12h_0^2)}{12}$$

$$\alpha = \frac{I_m}{I_{m0}} \tag{17.15}$$

$$= \frac{B^2 + H^2}{B^2 + H^2 + 12h^2}$$

$$= \frac{1}{1 + 12h^2/(B^2 + H^2)} \tag{17.16}$$

Figure 17.1 illustrates typical dimensions of a rectangular block. The properties of a foundation block are expressed here for clarity and notation. Properties about one axis of importance are given. Foundation base area is

$$A_f = BL \tag{17.17}$$

Second moment of area (moment of inertia) of the base

$$I_f = \frac{LB^3}{12} \tag{17.18}$$

Mass moment of inertia about CG axis

$$I_m = \frac{m}{12}(H^2 + B^2) \tag{17.19}$$

Mass moment of inertia about base

$$I_{m0} = \frac{m}{12}(H^2 + B^2) + mh_0^2$$

778

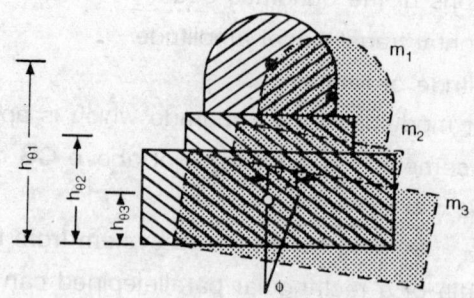

Fig. 17.2 Rocking Vibration.

Normally, a foundation may consist of more than one rectangular block in addition to the machine itself. A typical illustration is shown in Fig. 17.2. In such a case the mass moment of inertia about the base can be computed as

$$I_{m0} = I_{m1} + I_{m2} + I_{m3} + m_1 h_{01}^2 + m_2 h_{02}^2 + m_3 h_{03}^2 \qquad (17.20)$$

$$= \Sigma I_{mi} + \Sigma m_i h_{0i}^2 \qquad (17.21)$$

where
I_{mi} = mass moment of inertia of ith mass about its own CG

h_{0i} = height of the ith mass about the base

Depending on the type of force and the foundation system, the foundation can also have a *lateral shift vibration* which is called as *shear vibration*. This mode of vibration is not common in heavy machinery and heavy foundations. The natural frequency fo vibration of shear can be expressed as

$$f_x = \sqrt{\frac{k_s A_f}{m}} \qquad (17.22)$$

where
f_x = natural frequency of shear vibration.

Some machines might generate a rotation of the foundation about z-axis depending on the excitation force. The natural frequency of *twisting or torsional vibration* is given by

$$f_t = \sqrt{\frac{k_t J}{I_{mz}}} \qquad (17.23)$$

where
f_t = natural frequency of torsional vibration
k_t = coefficient of nonuniform shear (it is equal to $2 k_s = k_e$)
I_{mz} = mass moment of inertia about z-axis

$$= \frac{m}{12} (L^2 + B^2) \qquad (17.24)$$

J = polar moment of inertia of the base = $I_x + I_y$

$$= \frac{A_f}{12} (L^2 + B^2) \qquad (17.25)$$

The amplitude of torsional vibration is given by

$$\psi_z = \frac{T}{I_{ms}\,(f_t^2 - \omega^2)} \tag{17.26}$$

where $\qquad\qquad T$ = torsional excitation

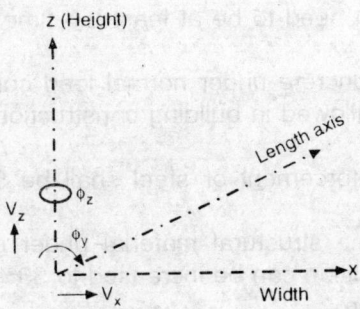

Fig. 17.3 Typical Degrees of Freedom.

Figure 17.3 shows the degrees of freedom of a rotating machine.

Reciprocating machinery. Generators, turbines, compressors, steam engines, etc. are normally coupled with another rotary type of machine. The forces of excitation in such a system are multi-dimensional and the system is likely to have a rocking vibration. The natural frequency and amplitudes of vibration are governed by the expressions already discussed. The forces of excitations in reciprocating engines are likely to be relatively higher than those in the rotary type of machines.

17.3 Structural Design of Foundation to Rotary Machines

The following criteria can be used in the design of rotary machine foundations.

1. The vibrations developed should not cause disturbances to the foundations of the building and other machinery. The *amplitudes* of vibrations at bearing levels be limited to the following:

    a) For low frequency machines:
        Vertical vibrations: 0.06 to 0.12 mm
        Horizontal vibrations: 0.10 to 0.15 mm

    b) For medium frequency machines:
        Vertical vibrations: 0.04 to 0.06 mm
        Horizontal vibrations: 0.07 to 0.09 mm

    c) For high frequency machines
        Vertical vibrations: 0.02 to 0.03 mm
        Horizontal vibrations: 0.04 to 0.05 mm

As the frequency of the machine increases, the allowable amplitude of vibration should be reduced. The amplitudes are also controlled by the service requirements of the machines.

d) The amplitudes of vibration of the foundation or its components be limited to 50% of the values specified in (a), (b) and (c).

2. The *natural frequency* of the system should not be within 20% of the operating frequency of the machine. Preferably 40% off from the operating speeds.

3. The bearing pressure on the soil should be within 60 to 70% of the allowable value used for building foundations.

4. The weight of the foundation need to be at least 2.5 times the weight of the machine to minimize the vibrations.

5. The allowable stresses in concrete under normal load condition in machine foundations shall be only 40% of those allowed in building construction. This is to account for fatigue effects.

6. The allowable stress in reinforcement or steel shall be 55% of the allowable value in building construction.

7. The allowable stresses in the structural material under combined action of shrinkage, creep or in seismic load condition can be increased to 33% of the allowable values under normal working load condition.

8. Minimum reinforcement in each direction should be as follows:
   0.20% for low frequency machines
   0.30% for medium frequency machines
   0.40% for high frequency machines
   and the total reinforcement for one cubic metre of concrete shall be 50 kg for low frequency, 75 kg for medium and 100 kg for high frequency machines.

9. Minimum cover to the reinforcement should be:
   75 to 100 mm at bottom
   50 to 60 mm at sides
   40 to 50 mm at top faces

10. Minimum diameter of the reinforcement bars be 12 mm and maximum spacing be 200 mm.

EXAMPLE 17.1 *Simple rotary machine*. A lathe has the following characteristics:

Weight W_m $\quad\quad\quad\quad$ = 150 kN

Base area 1.2 × 5 m^2

Height of CG = 0.9 m

Speed $\quad\quad\quad\quad\quad$ = 1200 m

Mass inertia $\quad\quad\quad$ = I_{mm} = 7500 kg m^4

Vertical excitation force $\quad F_z = 50$ kN

Soil characteristics are:

Safe allowable pressure $\quad p_a = 110$ kN/m^2

(coefficient of elastic uniform compression) = $k_e = 50$ MN/m^3

$$k_s = 0.5 k_e = 25 \text{ MN/m}^3$$

The foundation is made of M20 concrete and HYSD-Fe415 bars.

Allowable amplitude   = 0.10 mm

The machine is supported on three pairs of foundation bolts spaced 2.5 m intervals.

Solution. Operating frequency of the machine is

$$\omega = \frac{2\pi n}{60} = \frac{2\pi\,(1200)}{60} = 126 \text{ rad/sec}$$

Base area of the machine frame is 1.2×5 m^2; therefore assume the base area of the foundation as about 2×6 m^2

$$B = 2 \text{ m}, L = 6 \text{ m}$$
$$A_f = BL = 12 \text{ m}^2$$

Safe load carrying capacity of the soil is

$$P_a = 0.6 p_a A_f = 0.6(110)(12) = 792 \text{ kN}$$

The total load on the foundation is likely to be less than 792 kN as the machine, weight is 150 kN only. For the purpose of computing the natural frequency of the system, one must assume the weight of the foundation. Let the weight of foundation block is approximately 3 times the weight of the machine, that is

Try :

$$W_f = 3\,W_m = 3(150)\ 450 \text{ kN}$$
$$W = W_f + W_m = 600 \text{ kN}$$
$$m = \frac{W}{g} = \frac{600000}{9.81} = 61162 \text{ kg}$$

The natural frequency of the system is

$$f_z = \sqrt{\frac{k_e\,A_f}{m}} = \sqrt{\frac{50(10^6)(12)}{61162}} = 99 \text{ rad/sec}$$

$$\frac{\omega}{f_z} = \frac{126}{99} = 1.27$$

The natural frequency is about 27% off from the operating speed of the machine, therefore it is within the reasonable limits. The amplitude of vibration is

$$a_z = \frac{F_z}{m\,(\omega^2 - f_z^2)}$$

$$= \frac{50000}{61162(126^2 - 99^2)} = 0.00013 \text{ m} = 0.13 \text{ mm}$$

The amplitude of vibration is 0.13 mm which is 30% more than the allowable value of 0.10 mm. Therefore, increase the mass of the system. Increase the weight of the foundation about 40%.

Try:

$$W_f = 1.4(3)\ W_g = 4.2(150) = 630 \text{ kN}$$

Then the depth of the block is

$$D = \frac{W_f}{A_f} = \frac{630}{25(12)} = 2.1 \text{ m}$$

$$W = W_f + W_m = 630 + 150 = 780 \text{ kN}$$

$$m = \frac{W}{g} = \frac{780000}{9.81} = 79510 \text{ kg}$$

$$f_z = \sqrt{\frac{k_e A_f}{m}} = \sqrt{\frac{50(10^6)(12)}{79510}} = 87$$

$$a_g = \frac{F_z}{m\,(\omega^2 - f_z^2)}$$

$$= \frac{50000}{79510(126^2 - 87^2)} = 7.6\,(10^{-5}) \text{ m}$$

$$= 0.076 \text{ mm} \le 0.1 \text{ mm}$$

Since the amplitude is within the limits, the mass of the foundation is acceptable. The depth of the foundation block is 2.1 m and the machine frame is fitted with three pairs of anchor bolts. The base can be treated as two span beam with two equal spans of 2.5 m each. The maximum bending moment on the foundation block is

$$M = \frac{W_m L}{9} = \frac{150(2.5)}{9} = 41.7 \text{ kNm}$$

The area of the tension reinforcement required for an effective depth of the block as (2.1 − 0.1) = 2 m is

$$A_{st} = \frac{41.7(10^6)}{jd\sigma_{st}(0.50)}$$

$$= \frac{41700000 \times 2}{0.87(2000)(230)} = 208 \text{ mm}^2$$

This value is very small when compared with the nominal reinforcement. The nominal reinforcement required in each direction is

$$A_s = \frac{0.2D(1000)}{100} = \frac{0.2(2100)(1000)}{100}$$

$$= 4200 \text{ mm}^2/\text{m}$$

Provide 20 mm bars at 200 mm at top and bottom faces in each direction and along the vertical faces and extra reinforcement in between. The vertical reinforcement required is

$$A_{sv} = \frac{0.2(b)(1000)}{100} = 4000 \text{ mm}^2/\text{m}$$

Provide 20 mm bars at 150 mm spacing on each vertical side. Figure 17.4 illustrates the details.

EXAMPLE 17.2 *Generator Foundation*. A generator having a base size 7.5 × 4 m² has a set of nine loads at points marked in Fig. 17.5. The load and other characteristics of the machine are:

$$W_1 = W_2 = 50 \text{ kN},\ W_5 = 300 \text{ kN}$$

$$W_3 = W_4 = 100 \text{ kN},\ W_6 = 300 \text{ kN}$$

$$W_7 = W_8 = 50 \text{ kN},\ W_9 = 50 \text{ kN}$$

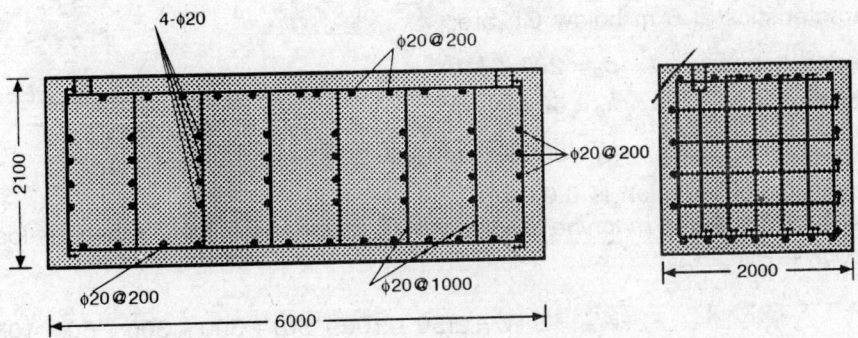

Fig. 17.4 Reinforcement Details of Rotary Machine Foundation.

$N = 1500$ rpm; $I_{mm} =$ mass inertia $= 150000$ kgm^4

CG of the machine above its base is 1.5 m. The vertical excitation force and excitation moment about an axis parallel to the width of the machine are:

$$F_z = 200 \text{ kN}$$
$$M_x = 400 \text{ kNm}$$

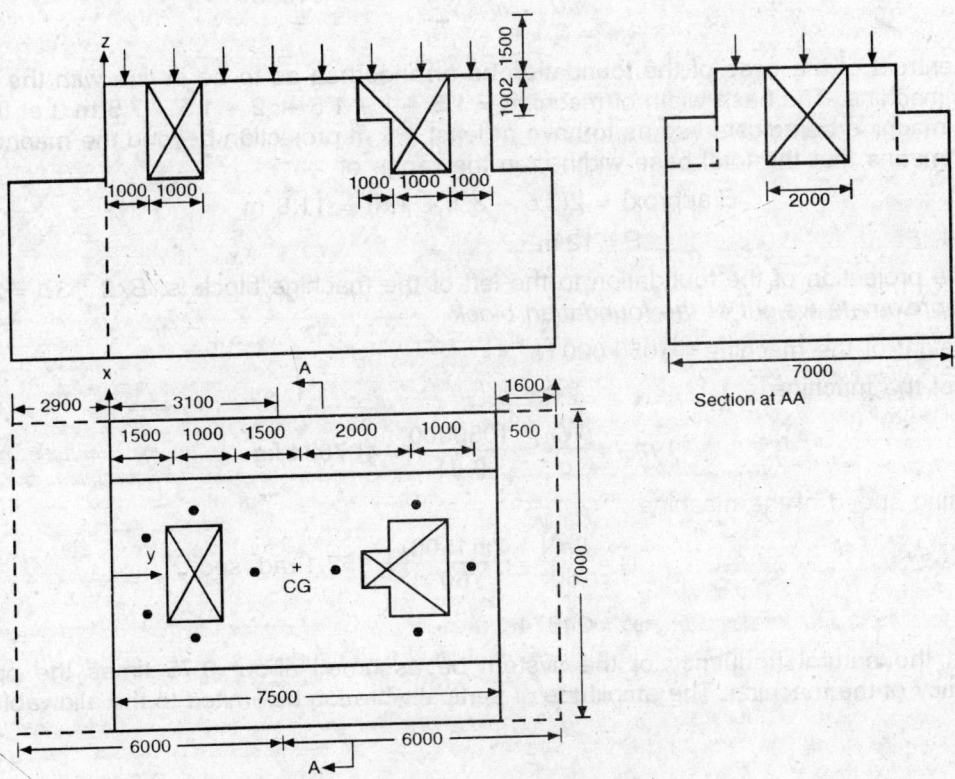

Section at AA

Fig. 17.5 Locations of Loads, Example 17.2.

The soil characteristics at 5 m below GL are:

Safe bearing capacity $\qquad p_a = 200 \text{ kN/m}^2$

Subgrade modulus $\qquad k_e = 50 \text{ MN/m}^3$

$$k_s = 30 \text{ MN/m}^3, \ k_q = 90 \text{ MN/m}^3$$

Allowable amplitude of variation is 0.04 mm.

Calculate of CG of the machine and width of foundation. Total machine load on the foundation block is

$$W_m = \overset{9}{\underset{i=1}{\Sigma}} W_i = 2(50 + 100 + 50) + 300 + 300 + 50 = 1050 \text{ kN}$$

where W_1, W_2, ..., refer to loads at points 1, 2, respectively.

For convenience, a reference axes system is assumed as shown in Fig. 17.5. The forces are symmertic about the middle horizontal y-axis; therefore the centroid of the machine is on the middle line. The algebraic sum of the moments of machine forces about x-axis gives

$$x_c = \frac{\Sigma W_1 Y_i}{W_m}$$

$$= \frac{2(50(0.5) + 100(1.5) + 50(6)) + 300(2.5 + 4) + 50(7)}{1050}$$

$$= 3.1 \text{ m}$$

The centroid of the base of the foundation be arranged so as to be in line with the centroid of the machine. The base width of machine = 1.5 + 1 + 1.5 + 2 + 1.5 = 7.5 m. Let the width of the machine be adjusted so as to have at least 1.5 m projection beyond the machine base which means that the total base width is in the range of

$$B(\text{approx}) = 2(7.5 - 3.1 + 1.5) = 11.8 \text{ m}$$

Let $\qquad B = 12 \text{ m}.$

The projection of the foundation to the left of the machine block is: $B/2 - 3.1 = 2.9$ m. *Approximate weight of the foundation block*

Weight of the machine = 1050 000 N

Mass of the machine

$$m_m = \frac{W_m}{g} = \frac{1050000}{9.81} = 107034 \text{ kg}$$

Operating speed of the machine

$$\omega = \frac{2\pi N}{60} = \frac{2\pi(1500)}{60} = 157.1 \text{ rad/sec}$$

$$\omega^2 = 24674$$

Let the natural frequency of the system be assumed about 0.75 times the operating frequency of the machine. The amplitude of vertical vibration be limited to the allowable value, then

$$d_a = \frac{F_z}{m(\omega^2 - f_z^2)} \le a_{az}$$

or
$$m \geq \frac{F_g}{a_{az}(\omega^2 - f_z^2)} = \frac{F_g}{a_{az}\,\omega^2(1 - f_z^2/\omega^2)}$$

$$= \frac{200000}{0.00004(24675)(1 - 0.75^2)} = 463,182 \text{ kg}$$

The mass of the block is to be about 4.3 times that of the machine. One can choose about 4 to 10 times the machine weight for the foundation block. The foundation block can be divided into two units, the lower rectangular base and the upper anchor block. Opening of depth 2.5 m are to be provided, and this portion of the block is idealized as a solid one.

Assume the depths of blocks, etc. as shown in the Fig. 17.5. Weight of lower and upper blocks are:

$$W_{f1} = 12(7)(3)(25) = 6300 \text{ kN}$$
$$W_{f2} = 7.5(7)(1.5)(25) = 1970 \text{ kN}$$
$$W_f = W_{f1} + W_{f2} = 8270 \text{ kN}$$

(neglecting topmost 1 m portion to compensate for the openings)

$$m_f = \frac{W_f}{g} = \frac{8270(1000)}{9.81} = 843016 \text{ kg}$$

$$(m_{f1} = 642\,200 \text{ kg}, \; m_{f2} = 200\,816 \text{ kg})$$

$$m = m_m + m_f = 950{,}050 \text{ kg}$$

The axis of rocking is parallel to 7 m; therefore $B = 12$ m, $L = 7$m. Base area and second moment of area are:

$$A_f = BL = 12(7) = 84 \text{ m}^2$$

$$I_f = \frac{LB^3}{12} = 7(144) = 1008 \text{ m}^4$$

The CG's of the two blocks from base are:

$$z_1 = 1.5 \text{ m}$$
$$z_2 = 3 + 0.75 = 3.75 \text{ m}$$

CG of the machine above base

$$z = 1.5 + 1 + 1.5 + 3 = 7 \text{ m}$$

The combined CG of the foundation and machine from the base is

$$z_0 = \frac{m_{f1}z_1 + m_{f2}z_2 + m_m z_m}{m}$$

$$= \frac{642200(1.5) + 200816(3.75) + 107034(7)}{950050} = 2.6 \text{ m}$$

The mass moment of inertia of the foundation blocks and the combined one are:

$$I_{mf1} = m_{f1}(3^2 + 12^2)/12 = 8188\,050 \text{ kg m}^2$$

$$I_{mf2} = m_{f2}(1.5^2 + 7.5^2)/12 = 978\,978 \text{ kg m}^2$$

Mass moment of inertia about the base is

$$I_{m0} = I_{mf1} + m_{f1}z_1^2 + I_{mf2} + m_{f2}z_2^2 + I_{mm} + m_m z_m^2$$

$$= 8,188,050 + 1,444,950 + 978,978 + 2,823,975 + 150,000 + 5,244,666$$

$$= 18,830,619 \text{ kg m}^2$$

Mass moment of inertia about its CG is

$$I_m = I_{mf1} + m_{f1}(z_1 - z_0)^2 + I_{mf2} + m_{f2}(z_2 - z_0)^2 + I_{mm} + m_m(z_m - z_0)^2$$

$$I_m = 8188050 + 642200(1.5 - 2.6)^2 + 978978$$

$$+ 200816(3.7 - 2.6)^2 + 150000 + 107034(7 - 2.6)^2$$

$$= 12,409,255 \text{ kg m}^2$$

$$\alpha = \frac{I_m}{I_{m0}} = \frac{12,409,255}{18,830,619} = 0.66$$

Bearing pressure and natural frequency. Bearing pressure on the soil is

$$p = \frac{W}{A_f} = \frac{mg}{A_f} = \frac{950050(9.81)}{84(1000)} = 111 \text{ kN/m}^2$$

Allowable bearing pressure under the dynamic load condition is about

$$p_{ad} = 0.6\, p_a = 120 \text{ kN/m}^2$$

The pressure on the soil is within the limits. The natural frequency of vertical vibration is

$$f_z = \sqrt{\frac{k_z A_f}{m}}$$

$$= \sqrt{\frac{50(10^6)\,(84)}{950050}} = 66.5 \text{ rad/sec}$$

$$\frac{f_z}{\omega} = \frac{66.5}{157.1} = 0.423$$

The assumed ratio was 0.75 as against the actual 0.423, therefore the design is on the safer side. The amplitude of vertical vibration is

$$u_z = \frac{F_z}{m(\omega^2 - f_z^2)}$$

$$= \frac{200000(1000)}{950050(157.1^2 - 66.5^2)}$$

$$= 0.01 \text{ mm} \leq 0.04 \text{ mm}$$

The rocking frequency of the system is

$$f_\phi = B\sqrt{\frac{A_f k_q}{12 I_m}}$$

$$= 12\sqrt{\frac{84(90)(10^6)}{12(950050)}} = 309 \text{ rad/sec}$$

Amplitude vertical vibration is

$$a_z = \frac{MB}{21_m(f_\phi^2 - \omega^2)}$$

$$= \frac{400000(12)(1000)}{2(950050)(309^2 - 157.1^2)}$$

$$= 0.036 \text{ mm} < 0.04 \text{ mm}$$

The combined vibration frequency expression is

$$\alpha f^4 - (f_\phi^2 + f_z^2)f^2 + f_\phi^2 f_z^2 = 0$$

or

$$0.66f^4 - (309^2 + 66.5^2)f^2 + 309^2(66.5)^2 = 0$$

$$f^4 - (151368)f^2 + 6.3975(10^8) = 0$$

which gives

$$f_1^2 = 147016; \text{ and } f_2^2 = 4352$$

or

$$f_1 = 383, \text{ and } f_2 = 66$$

$$\Delta f^2 = ml_m(f_1^2 - \omega^2)(\omega^2 - f_2^2)$$

$$= 950,050(12,409,255)(147,016 - 24,674)(24674 - 4352)$$

$$= 2.945(10^{22})$$

Amplitude of vibration in

$$\alpha_x = \frac{k_s A_f h_c M}{\Delta f^2}$$

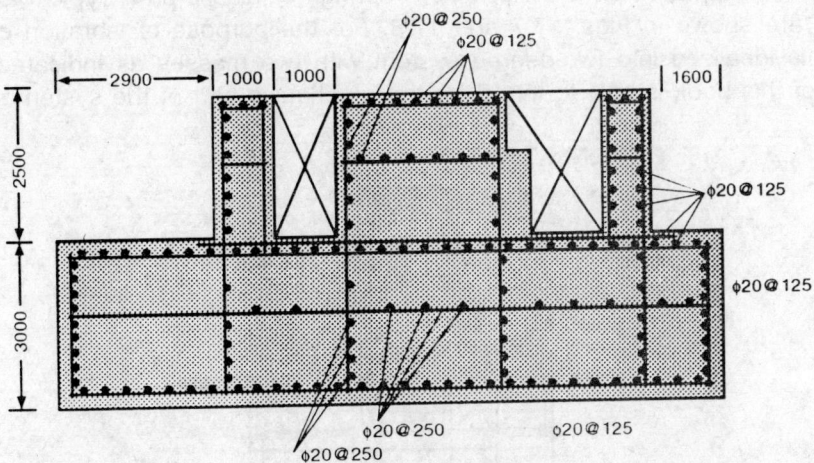

Reinforcement is in Each direction

Fig. 17.6 Reinforcement Details of Machine Foundation, Example 17.2.

$$= \frac{30(10^6)(84)(7 - 2.6(400000)}{2.945(10^{22})} \text{ m}$$

$$= \frac{3(84)(4.4)(4)}{2.945(10^7)} = 0.0002 \text{ mm}$$

$$\phi = \frac{(k_sA_f - m\omega^2)M}{\Delta f^2}$$

$$= \frac{(30(10^6)(84) - 950050(157.1)^2)400000}{2.945(10^{22})} = 0.3(10^{-6})$$

The amplitudes of vibrations are within the permissible limits. Some of them are far below allowable values.

Design of reinforcement. The bending moment and shear forces on the foundation block are only nominal when compared with the size of the block. Only nominal reinforcement of 0.3% in each direction is to be provided. The reinforcement detail as shown in Fig. 17.6.

17.4 Impact Machines

Forge hammer consists of three main parts: anvil, frame and ram or drop hammer called tup. The anvil rests on good quality timber packing to receive the impact loads. The impact load could be imparted either by a drop hammer or by a hydraulic ram. The anvil and the frame are be provided solid concrete foundations. The impact force on the anvil must be partly absorbed by a semi-elastic padding so as to minimize the damage to concrete. The impact force of the anvil will generate vibration in the foundation block. The magnitude of vibrations and the frequency of the system depend on the masses and the elastic properties of the soil. The foundation block can rest directly on the soil in case of small hammers. If the amplitude of vibration of a machine resting on soil exceeds the allowable limits, then the foundation block should be supported on a set of elastic springs or elastic pad. Typical arrangement of foundations are shown in Figs. 17.7 and 17.8. For the purpose of vibration characteristics, the system is idealized into two-degree system with two masses as indicated in Fig. 17.9. The scope of this book is only to introduce the machanics part of the system so no detailed

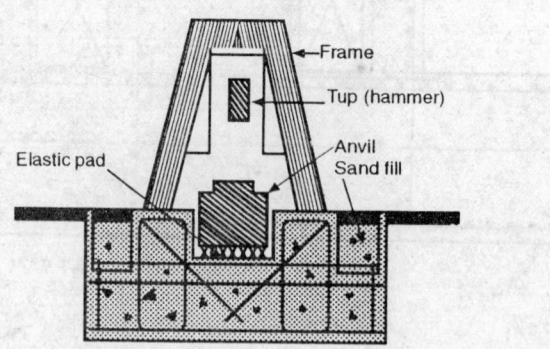

Fig. 17.7 Impact Machine Foundation Block on Soil.

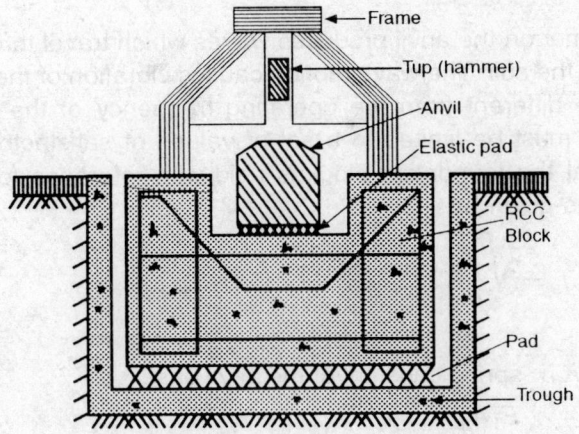

Fig. 17.8 Impact Machine Foundation Block.

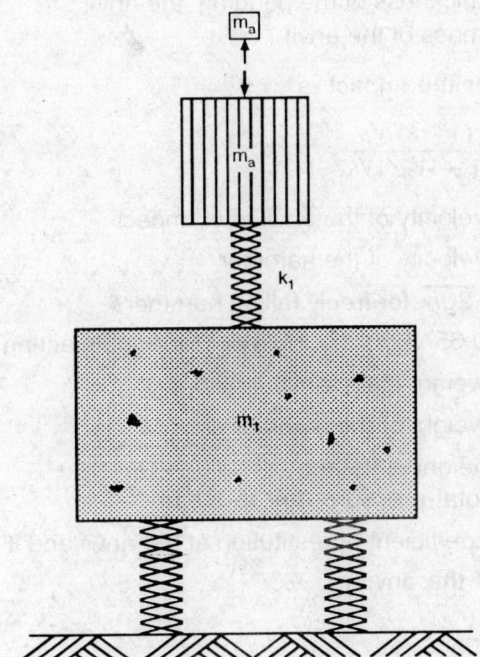

Fig. 17.9 Idealised Vibrating of Impact Machine Foundation.

discussion on the behaviour or derivation of the results is presented. The important aspects of the design based on the Indian Code of Practice of Foundations for impact type machines are presented here.

17.5 Vibration Characteristics

The impact of the hammer on the anvil produces waves which travel through the anvil padding and block foundation to the soil. The wave motion causes vibration of the system. The vibration of the system must be different from the operating frequency of the machine. Further, the amplitudes of vibration must be limited to a set of values of satisfactory performance of the machine. Assuming that the foundation block is rigid with reference to the anvil, the natural frequency of the anvil is given by

$$f_a = \sqrt{\frac{K_a}{m_a}} \tag{17.27}$$

where

$$K_a = \text{spring constant of the anvil base}$$

$$= \frac{E_1 A_a}{t_1} \tag{17.28}$$

$E_1 = $ Young's modulus of elasticity of the anvil base material
$A_a = $ area of the anvil base
$t_1 = $ thickness of the padding the anvil
$m_a = $ mass of the anvil

The velocity of the anvil after the impact is

$$V_a = \frac{(1 + k) V_h}{1 + W_a/W_h} \tag{17.29}$$

where

$V_a = $ velocity of the anvil after impact

$V_h = $ velocity of the hammer

$= \sqrt{2gh}$, for freely falling hammers

$= 0.65 \sqrt{2gh (1 + P_s/W_h)}$, for double acting steam hammer

$W_a = $ weight of the anvil

$W_h = $ weight of the hammer

$h = $ height of the tup
$P_s = $ total pressure due to steam

$k = $ coefficient of restitution at the anvil and it is usually taken as 0.6

The amplitude of vibration of the anvil is

$$a_a = \frac{V_a}{f_a} \tag{17.30}$$

and the corresponding dynamic force on the anvil pad (also called cushion) is

$$F_1 = K_a a_a \tag{17.31}$$

The elastic and total compressions of the anvil cushion are:

$$v_{e1} = \frac{W_a}{K_a} \tag{17.32}$$

$$v_1 = v_{e1} \pm a_a \tag{17.33}$$

$$= \frac{W_a}{K_a} \pm \frac{V_a}{f_a} \tag{17.34}$$

The natural frequency of the foundation block is given by

$$f_f = \sqrt{\frac{K_f}{m_a + m_r + m_f}} \tag{17.35}$$

where

f_f = natural frequency of the foundation block

m_f = mass of the foundation block

m_r = mass of the frame of the machine (to be considered if it is on the foundation block)

K_f = spring constant of the foundation. It is either equal to the total modulus of the soil subgrade if the foundation is resting on soil or it is the total spring constant if the block is on springs or on cushion.

17.6 Design Considerations of Foundations to Impact Machines

The mass of the system must be large so as to absorb the impact energy and provide minimum disturbances. Therefore, the mass of the foundation is usually in the range of 30 to 200 times the weight of the tup or the hammer. The smaller value is chosen for small hammers founded on stiff clays or sandy soils. The depth of the foundation increase with increase in the impact energy. The thickness of the foundation varies from 1.5 to 6 m. The base area of the foundation or the trough plate should be such that the maximum pressure on the soil is within the limits. Further, the combined effect of the base and mass of the foundation is such that the maximum amplitude is within the allowable limits. The maximum allowable amplitudes are:

1 mm for 10 kN hammer,

2 mm for 20 kN hammer,

3 mm to 4 mm for 30 kN hammer and above.

The natural frequency of the system must be about two and half times the operating frequency of the machine. In case the natural frequency falls less than the operating value, then it must be below 70% of the operating frequency.

The thickness of the cushion below the anvil must be able to absorb the energy and withstand a large number of repeated blows. The stress in the cushion must be exceed 50% of the allowable value under static load condition. The thickness of the cushion increase with increase in the weight of the hammer. It varies from 200 mm to 1200 mm, the higher value is for heavy hammer. A minimum reinforcement of 25 to 50 kg/m^3 of concrete is recommended.

EXAMPLE 17.3 *Design of a Hammer Foundation.* A drop hammer has the following data:

Weight of hammer	$W_h = 25$ kN
Weight of anvil	$W_a = 75$ kN
Weight of frame	$W_r = 60$ kN

Anvil base area $\qquad A_a = 1.75 \times 2.75$ m $= 4.81$ m^2
Drop height of tup $\qquad h = 1.4$ m
Coefficient of restitution $\quad k = 0.6$
Allowable bearing pressure $p_a = 150$ kN/m^2
Modulus of subgrade $\qquad k_e = 50$ MN/m$^3 =$ coeff. uni. compression

The velocity of the foundation block after impact is:

$$V_f = \frac{(1 + k)V_a}{1 + (m_f + m_r)/m_a} \qquad (17.36)$$

The amplitude of vibration of the foundation block is

$$a_f = \frac{V_f}{f_f} \qquad (17.37)$$

and the corresponding dynamic force of the foundation on its base is

$$F_2 = K_f a_f \qquad (17.38)$$

The combined system consisting of the anvil and the foundation will vibrate at a natural frequency given by the solution of the following equation:

$$f^4 - (1 + \beta)(f_a^2 + f_f^2)f^2 + (1 + \beta)f_a^2 f_f^2 = 0 \qquad (17.39)$$

where $\qquad f =$ natural frequency of the system

$$\beta = \frac{m_a}{m_f + m_r} \qquad (17.40)$$

The maximum amplitudes of the anvil and foundation are:

$$a_{am} = \frac{V_a(f_a^2 - f_1^2)}{(f_1^2 - f_2^2)f_2} \qquad (17.41)$$

$$a_{fm} = \frac{V_a(f_a^2 - f_1^2)(f_a^2 - f_2^2)}{f_a^2(f_1^2 - f_2^2)f_2} \qquad (17.42)$$

where f_1^2 and f_2^2 are the natural frequencies obtained from Eq. (17.39); f_1 corresponds to the first mode. The corresponding maximum dynamic forces can be obtained by multiplying the amplitude with the corresponding spring constants, and are

$$F_{2m} = a_{af}K_f \qquad (17.43)$$

$$F_{1m} = (a_{fm} - a_{am})K_a \qquad (17.44)$$

Design the foundation with M20 concrete and HYSD-Fe415 bars.

Solution. Recommended thickness of the cushion under the anvil is from 200 mm to 1200 mm. Select 500 mm thickness as a first trial. The Young's modulus of timber is 2000 MN/m^2.

$$t_1 = 0.5 \text{ m}$$

$$E_1 = 2000 \text{ MN/m}^2$$

Let the weight of the foundation block be 50 times the weight of the tup, that is about 1250 kN. The total weight of the system is

$$W_a = 75 \text{ kN}$$

Let
$$W_f = 1250 \text{ kN}$$

$$W_t = 1325 \text{ kN}$$

Base area of the foundation needed is

$$A_f = \frac{W_t}{0.5 p_a} = \frac{1325}{0.5(150)} = 17.7 \text{ m}^2$$

Provide 3.5 m by 5.5 m base area, and total height 3.75 m with pocket of $3.25 \times 2.25 \times 2.5$ m

Base area
$$A_f = 3.5(5.5) = 19.25 \text{ m}^2$$

Weight of the foundation is

$$W_f = ((3.75)(5.5)(3.5) - 2.5(2.25))25 = 1347 \text{ kN}$$

The vibration characteristics are computed below:

$$m_a = \frac{75000}{9.81} = 7645 \text{ kg}$$

$$m_r = \frac{60000}{9.81} = 6116 \text{ kg}$$

$$m_f = \frac{1347000}{9.81} = 137308 \text{ kg}$$

$$K_a = k_a A_a = \frac{E_1 A_a}{t_1}$$

$$= \frac{2000(4.81)}{0.5} = 19240 \text{ MN/m}$$

$$f_a = \sqrt{\frac{K_a}{m_a}} = \sqrt{\frac{19240(10^6)}{9645}} = 1412 \text{ rad/sec}$$

Velocities of the hammer before impact and the anvil after impact are:

$$V_h = \sqrt{2gh} = \sqrt{2(9.81)(1.4)} = 5.24 \text{ m/sec}$$

$$V_a = \frac{(1+k)V_h}{1 + W_a/W_h} = \frac{1.6(5.24)}{1 + 75/25} = 2.1 \text{ m/sec}$$

$$a_a = \frac{V_a}{f_a} = \frac{2.1(1000)}{1412} = 1.49 \text{ mm}$$

$$K_f = k_e A_f = 50(19.25) = 962.5 \text{ MN/m}$$

$$f_f = \sqrt{\left(\frac{K_f}{m_a + m_r + m_f}\right)}$$

$$= \sqrt{\left(\frac{962.5(10^6)}{7645 + 6116 + 137308}\right)} = 80 \text{ rad/sec}$$

$$V_f = \frac{(1+k)V_a}{1 + (m_f + m_r)/m_a}$$

$$= \frac{1.6(2.1)}{1 + 18.76} = 0.17 \text{ m/sec}$$

$$a_f = \frac{V_f}{f_f} = \frac{0.17(1000)}{80} = 2.13 \text{ mm}$$

$$\beta = \frac{m_e}{m_f + m_r} = 0.0533$$

The quadratic of the natural frequency equation is

$$f^4 - (1 + \beta)(f_a^2 + f_f^2)f^2 + (1 + \beta)f_a^2 f_f^2 = 0$$

or

$$f^4 - (1.0533)(2000144)f^2 + (1.0533)(1.276)(10^{10}) = 0$$

The solution of the equation is

$$f_1^2 = 2190,350$$

and

$$f_2^2 = 6400$$

The maximum amplitudes of vibration of the anvil and the foundation are:

$$a_{am} = \frac{V_a(f_a^2 - f_1^2)}{(f_1^2 - f_2^2)f_2}$$

$$= \frac{(2100(1412^2 - 2100350)}{2100350 - 6400)80} = 1.28 \text{ mm}$$

$$a_{fm} = - \frac{V_a(f_a^2 - f_1^2)(f_a^2 - f_2^2)}{(f_a^2(f_1 - f_2^2)f_2}$$

$$= - \frac{2100(1412^2 - 2100350)(1412^2 - 6400)}{1412^2(2100350 - 6400)80} = 1.33 \text{ mm}$$

The allowable maximum amplitudes of vibration of the anvil and the foundation are in the range of 2 to 3 mm for 25 kN hammer. The actual amplitudes of vibration are within the limits. The maximum dynamic forces on the anvil base and the soil are:

$$F_{1m} = (a_{fm} - a_{am})K_a = \frac{1.33 - 1.28}{1000}(19240)$$

$$= 0.962 \text{ MN}$$

$$F_{2m} = a_{fm}K_f = \frac{1.33}{1000}(962.5) = 1.28 \text{ MN}$$

The dynamic stresses carried on the timber padding, concrete below the anvil and the soil below the foundation are:

$$\sigma_{11} = \frac{F_{1m}}{A_a} = \frac{0.962}{4.81} = 0.2 \text{ MPa}$$

$$\sigma_{12} = \frac{F_{1m}}{3.25(2.25)} = 0.13 \text{ MPa}$$

$$\sigma_2 = \frac{F_{2m}}{A_f} = \frac{1.28}{19.25} = 0.066 \text{ MPa} = 66 \text{ kN/m}^2$$

Allowable stresses in timber padding, concrete and soil under static loading are 2 MPa, 5 MPa and 0.15 MPa. The dynamic stresses on timber and concrete are far below the allowable values. The dynamic stress on soil is also lower than the allowable value and the fatigue factor available is

$$\eta = \frac{p_a}{\sigma_2} = \frac{0.15}{0.066} = 2.27$$

The recommended fatigue factor varies from 2 to 3; therefore the design is safe.

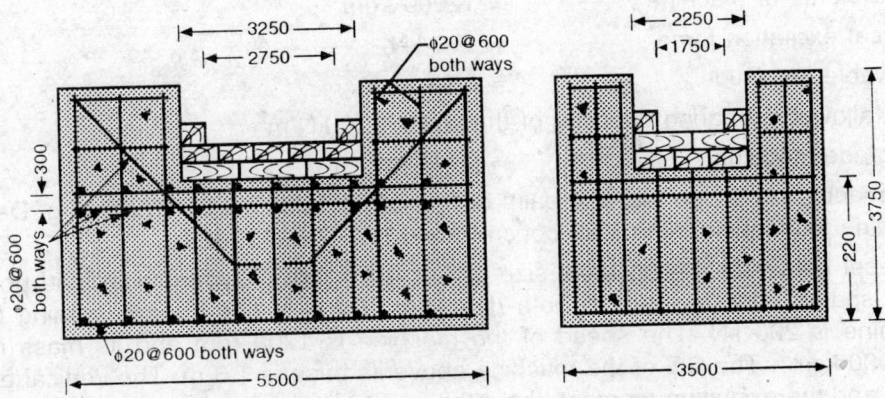

Fig. 17.10 Details of Hammer Foundation, Example 17.3.

Structural design. The smallest depth of concrete in the block is 2.25 m and the cantilever span is $(5.5 - 3.25)/2 = 1.125$ m. The block transfers the load through shear and bearing as the span to depth ratio is less than 0.5. The bearing stress on concrete was found to be nominal. Therefore, only minimum reinforcement be provided in the concrete. The amount of the reinforcement can be taken about 250 N/m^3 of concrete. The reinforcement detailing is very important to avoid cracking and spalling. The details are shown in Fig. 17.10.

PROBLEMS

17.1 A rotary machine has the following properties:

Weight of the machine	= 200 kN
Base area	= 2×5 m^2
Height of CG of above the base	= 1 m
Speed of the machine	= 900 rpm
Mass inertia	= 9000 kgm^4
Vertical excitation	= 60 kN
Safe allowable bearing pressure	= 140 kN/m^2
Subgrade modulus	= 70 MN/m^3

The machine is supported by 2 rails spaced at 1.5 m apart which are parallel to the longitudinal section of the machine. Design the machine foundation using M20 concrete and HYSD-Fe415 bars. The allowable amplitude of the machine is 0.2 mm.

17.2 A medium speed of horizontal boring machine has the following properties:

Weight of the machine = 30 kN

Base area of the machine = 1.5×5 m^2

Height of CG above the base = 1.5 m

Speed of the machine = 1500 rpm

Mass inertia of machine = 12000 kgm^4

Vertical excitation force = 100 kN

Allowable amplitude = 0.15 mm

Safe allowable bearing capacity of the soil = 120 kN/m^2

Subgrade modulus = 60 MN/m^3

The machine is supported at 4 pairs of foundation bolts on 1.5 m spacing. Design the machine foundation using M25 concrete and HYSD-Fe415 bars.

17.3 A diesel generator having base size of 6×4 m^2 with 6 symmetrical load points in 2 rows spaced at 2 m apart in both directions. Each of the 6 loads coming from the machine is 200 kN. The speed of the machine is 1200 rpm and its mass inertia is 100,000 kgm^4. The CG of the machine above its base is 1.6 m. The vertical excitation force and the excitation moment about the longitudinal axis of the machine are 150 kN and 200 kNm respectively. The soil characteristic at about 4 m below ground level are:

Safe net allowable bearing capacity = 150 kN/m^2

Subgrade modulus = 50 MN/m^3

Shear modulus = 25 MN/m^3

Coefficient of non-uniform compression = 100 MN/m^3

Design the machine foundation for a permissible amplitude of 0.06 mm using M20 concrete and HYSD-Fe415 bars.

17.4 A power hammer has the following characteristics:

Weight of hammer = 15 kN

Weight of anvil = 60 kN

Weight of frame = 40 kN

Anvil base area = 1.2 by 2.2 m

Drop height of tup = 1.6 m

coefficient of restitution = 0.6

The soil characteristic on which foundation is supposed to rest are:

Safe allowable bearing pressure = 120 kN/m^2

Subgrade modulus = 50 MN/m^3

Design a RCC foundation using M20 concrete and HYSD-Fe415 bars with 300 mm thick wooden cushion to the anvil. assume Young's modulus of timber as 1600 MN/m^2.

1 Codes of Practice Useful in Design of RCC Structures

(IS Code Reference is Bureau of Indian Standards (BIS) New Delhi 110002)

AGGREGATES IN CONCRETE

1.1. Aggregates, coarse and fine from natural sources for concrete, IS 383–1970
1.2. Glossary of terms relating to cement: Part I Concrete agregates, IS: 6461 (Part I)–1972
1.3. Method of test for determining aggregates impact value of soft coarse aggregates, IS: 5640–1970
1.4. Methods of test for aggregates for concrete:

   (a) Part I particle size and shape, IS: 2386 (Part I)–1963
   (b) Part II Estimation of deleterious materials and organic impurities, IS: 2386 (Part II)–1963
   (c) Part III Specific gravity, density, voids, absorption and bulking, IS: 2386 (Part III)–1963
   (d) Part IV Mechanical properties, IS: 2386 (Part IV)–1963
   (e) Part V Soundness, IS: 2386 (Part V)–1963
   (f) Part VI Measuring mortar making properties of fine aggregates, IS: 2386 (Part VI)–1963
   (g) Part VII Alkali aggregate reactivity, IS: 2386 (Part VII)–1963
   (h) Part VIII petrographic examination, IS: 2386 (Part VIII)–1963.

CEMENT

1.5. Definitions and terminology relating to hydraulic cement, IS: 4845–1968
1.6. High alumina cement for structural use, IS: 6452–1972
1.7. High strength ordinary Portland cement, IS: 8112–1976
1.8. Hydraulic cements, methods of sampling, IS: 3535–1966
1.9. Hydraulic cement, physical tests for methods of: IS: 4031–1968
1.10. Method of chemical analysis of hydraulic cement, IS: 4032–1968
1.11. Ordinary and low heat Portland cement, IS: 269–1989
1.12. Portland–Pozzolana cement, IS: 1489–1991

   (a) Coment or concrete admixtures IS: 9103
   (b) 43 Grade OPC; IS: 8112–1989
   (c) 53 Grade OPC; IS: 12264–1987
   (d) Rapid hardening PC, IS: 8041–1989
   (e) Portland slag cement, IS: 455–1989
   (f) Sulphate resisting PC IS: 12330–1988
   (g) Hydrophobic cement, IS: 8043–1991

CONCRETE DESIGN AND CONSTRUCTION

1.13. Code of practice for plain and reinforced concrete, IS: 456–1999 (final draft)

1.14. Code of practice on prestressed concrete, IS: 1343–1981
1.15. Ready-mixed concrete, IS: 4926–1976

TESTING OF CONCRETE

1.16. Abrasion resistance of concrete, method of tests, IS: 9284–1979
1.17. Autoclaved cellular concrete products, methods of test for:

    (a) Determination of unit weight or bulk density and moisture content, IS: 6441 (Part I)–1972
    (b) Determination of drying shrinkage, IS: 6441 (Part II)–1972
    (c) Corrosion protection of steel reinforcement in autoclaved cellular concrete, IS: 6441 (Part IV)–1972
    (d) Determination of compressive strength, IS: 6441 (Part V)–1972
    (e) Strength, deformation and cracking of flexural members subject to bending short duration loading test, IS: 6441 (Part VI)–1973
    (f) Strength, deformation and cracking of flexural members subject to bending-substained loading test, IS: 6441 (Part VII)–1973
    (g) Loading tests for flexural members in diagonal tension, IS: 6441 (Part VIII)–1973
    (h) Jointing of autoclaved cellular concrete elements, IS: 6441 (Part IX)–1973

1.18. Determining setting time of concrete by penetration resistance, IS: 8142–1976
1.19. Method of making, curing and determining compressive strength of accelerated cured concrete test specimens, IS: 9013–1978
1.20. Method of test for permeability of cement mortar and concrete, IS: 3085–1965
1.21. Sampling and analysis of concrete, IS: 1199–1959
1.22. Method of test for splitting tensile strength of concrete cylinders, IS: 5816–1970
1.23. Strength of concrete, IS: 516–1959
1.24. Methods of Pull-out test for testing bond in reinforced concrete, IS: 2770 (Part I)–1977
1.25. Methods of test for determination of water soluble chlorides in concrete admixtures, IS: 6925–1973

FOUNDATION ENGINEERING

1.26. Code of practice for calculation of settlements of foundations, shallow foundations subjected to symmetrical static vertical loads, IS: 8009 (Part I)–1976
1.27. Code of practice for design and construction of conical and hyperbolic paraboloidal types of shell foundations IS: 9456–1980
1.28. Code of practice for Design and Construction of Diaphragm walls, IS: 9556–1980
1.29. Code of practice for design and construction of foundations for transmission line towers and poles, IS: 4091–1979
1.30 Code of practice for design and construction of machines foundation:

    (a) Foundations for reciprocating type machines, IS: 2974 (Part I)–1969
    (b) Foundations for impact type machines (hammer foundations), IS: 2974 (Part II)–1980
    (c) Foundations for rotary type machines (medium and high frequency), IS: 2974 (Part III)–1975
    (d) Foundations for rotary type machines of low frequency, IS: 2974 (Part IV)–1979
    (e) Foundations for impact type machines other than hammers (forgning and stamping press, pig breaker elevator and hoist tower), IS: 2974 (Part V)–1970

1.31. Code of practice for design and construction of pile foundation:

    (a) Concrete piles, driven cast in-situ concrete piles, IS: 2911 (Part I/Sec. I)–1979
    (b) Concrete piles, bored cast in-situ piles, IS: 2911 (Part I/Sec. 2)–1979
    (c) Concrete piles, driven precast concrete piles, IS: 2911 (Part I/Sec. 3)–1979

(d) Timber piles, IS: 2911 (Part II)–1980

(e) Under-reamed piles, IS: 2911 (Part III)–1980

(f) Load test on piles, IS: 2911 (Part IV)–1979

1.32. Code of practice for design and construction of raft foundations, IS: 2950 (Part I)–1974

1.33. Code of practice for design and construction of simple spread of foundations, IS: 1080–1980

1.34. Code of practice for design and construction of well foundations, IS: 3955–1967

1.35. Design of generator foundation for hydel power stations, IS: 7207–1974

1.36. Code of practice for determination of allowable bearing pressure on shallow foundations, IS: 6403–1971

1.37. Code of practice for structural safety of buildings: shallow foundations, IS: 1904–1978

LOADING STANDARDS, STRUCTURAL SAFETY

1.38. Code of practice for structural safety of buildings: loading standards, IS: 875–1987

1.39. Methods of test for concrete poles for overhead power and telecommunication lines, IS: 2905–1966

REINFORCEMENT IN CONCRETE

1.40. Code of practice for bending and fixing of bars for concrete reinforcement, IS: 2502–1963

1.41. Cold-worked steel high strength deformed bars for concrete reinforcement, IS: 1786–1979

1.42. Deformed Bars for concrete reinforcement hot rolled mild steel and medium tensile steel, IS: 1139–1966

1.43. Fabric for concrete reinforcement, hard drawn steel wire, IS: 1566–1967

1.44. Glossary of terms relating to cement concrete: Concrete Reinforcement, IS: 6461 (Part III)–1972

1.45. High tensile steel bars used in prestressed concrete IS: 2090–1962

1.46. Indented wire for prestressed concrete: IS: 6003–1970

1.47. Mild Steel and Medium Tensile steel bars and hard-drawn steel wire for concrete reinforcement:

(a) Mild steel and medium tensile steel bars, IS: 432 (Part I)–1966

(b) Hard-drawn steel wire, IS: 432 (Part II)–1966

1.48. Plain hard-drawn steel wire for prestressed concrete:

(a) Cold drawn stress-relieved wire, IS: 1785 (Part I)–1966

(b) As-drawn wire, IS: 1785 (Part II)–1967

1.49. Recommendations for detailing of reinforcement in reinforced concrete works, IS: 5525–1969

1.50. Uncoated stress relieved strand for prestressed concrete, IS: 6006–1970

STRUCTURAL DESIGN

1.51. Code of practice for concrete structures for the storage of liquids:

(a) General, IS: 3370 (Part I)–1965

(b) Reinforced Concrete Structure, IS: 3370 (Part II)–1965

(c) Prestressed Concrete Structures, IS: 3370 (Part III)–1967

(d) Design tables, IS: 3370 (Part IV)–1967

1.52. Criteria for design and construction of precast concrete trusses, IS: 3201–1965

1.53. Code of practice for design and construction of stone slab over joint floor IS: 2792–1964

1.54. Criteria for design of reinforced concrete bins for the storage of granular and powdery materials:

(a) General requirements and assessment of bin loads, IS: 4995 (Part I)–1974

(b) Design criteria, IS: 4995 (Part II)–1974

1.55. Criteria for design of reinforced concrete chimneys:

(a) Design criteria, IS: 4998 (Part II)–1975

1.56. Criteria for design of reinforced Concrete Shell Structures and Folded Plates, IS: 2210–1962
1.57. Code of Practice for Earthquake Resistant Design and Construction of Buildings, IS: 4326–1976
1.58. Criteria for Earthquake Resistant Design of Structures, IS: 1893–1984

OTHER CODES OF PRACTICES

1.59. CP110–1972 Code of Practice of Structural Use at Concrete, BSI, U.K.
1.60. Handbook on Unified Code for Structural Concrete (CP110–1972) 1974, CCA, London.
1.61. ACI 318–77, ACI Standard Building Code Requirements for Reinforced Concrete, ACI, Detroit, USA.
1.62. BS 5337–1976, BS Code of Practice for Structural Use of Concrete for Retaining Aqueous Liquids, BSI, U.K.
1.63. ACI Manual of Concrete Inspection, ACI, SP-2, Detroit, 1974.
1.64. CRSI Handbook, 2nd ed., CRSI, Chicago, 1975.
1.65. Manual of Standard Practice for Detailing Reinforced Concrete Structures, (ACI 315–75), ACI, Detroit, USA.
1.66. NSC-Code for Design of Concrete Structures for Buildings. Canadian Standards Association, Ontario, Canada, 1977.
1.67. Causes, Mechanism and Control of Cracking in Concrete, ACI: SP-20, ACI, Detroit, 1968.
1.68. Standard Specification for Highway Bridge, AASHO, Washington, DC.

(a) Test on water quality, IS: 3025 (Several parts).

2 SELECTED BOOKS

2.1. C.H. Whitney; Earth Pressures and Retaining Walls, John, Wiley & Sons Inc., 1957.
2.2. C.D. Barkan; Dynamics of Bases and Foundations, McGraw-Hill Book Co. NY, 1960.
2.3. R.D. Chellio; Pile Foundations, McGraw-Hill Book Co. NY, 1961
2.4. J.A. Blume; N.M. Newmark and L.H. Corning; Design of Multistorey Reinforced Concrete Buildings for Earthquake Motions, PCA, Skokie, 1961.
2.5. R.H. Wood; Plastic and Elastic Design of Slabs and Plates, Ronald Press, NY, 1961.
2.6. E. Sigalov and S. Strongin, Reinforced Concrete, Foreign Languages Publishing House, Moscow, 1962.
2.7. K.W. Johanson; Yield Line Theory, Cement and Concrete Association, London, 1962.
2.8. L.L. Jones; Ultimate Load Analysis of Reinforced Concrete Structures, Interscience Publishers, NY, 1962.
2.9. Frame Analysis Applied to Flat Slab Bridges, PCA, Skokie, USA.
2.10. C.W. Dunham; Advanced Reinforced Concrete, McGraw-Hill Book Co., NY, 1964.
2.11. B. Balwanta Rao and C. Muthuswamy; Considerations in the Design and Sinking of well Foundations for Bridge Piers, IRC, Paper No. 238, IRC, N. Delhi 1966.
2.12. V. Murashev; E. Signalov and V. Baikov; Design of Reinforced Concrete Structures, MIR Publishers, Moskow.
2.13. K.H. Gerstle; Basic Structural Design, McGraw-Hill Book Co., NY, 1967.
2.14. K. Terzaghi and R.B. Pekck; Soil Mechanics in Engineering Practices, John Wiley and Sons, NY, 1968.
2.15. G.S. Ramaswamy; Design and Construction of Concrete Shell Roofs, McGraw-Hill Co., NY, 1969

2.16. N. Khachaturian and G. Gurfinkel; Prestressed Concrete, McGraw-Hill Co., NY, 1969.
2.17. T. Whitaker; The Design of Pile Foundations, Pergamon Press, 1970.
2.18. M. Trichy and M. Volicek; Statical Theory of Concrete Structures, Irish Univ. Press Shannon, Academia, Prague, 1972.
2.19. R.B. Peck, W.H. Hanson and T.H. Thornburn; Foundation Engineering John Wiley and Sons, NY, 1974.
2.20. R. Park and T. Paulay; Reinforced Concrete Structures, John Wiley and Sons, NY, 1975.
2.21. J.B. Kennedy and A. Neville; Basic Statistical Methods for Engineers, and Scientists, Harper Int. Edition, Harper & Row publishers, NY, 1976.
2.22. P. Sriramulu and C.V. Vaidyanathan; Handbook of Machine Foundations, Tata McGraw-Hill Publishing Co., New Delhi, 1976.
2.23. P. Dayaratnam; Prestressed Concrete Structures, Oxford & IBH Publishing Co., New Delhi, 1982.
2.24. Jai Krishna and O.P. Jain; Plain and Reinforced Concrete, Nemchand Bro., Roorkee, 1981.
2.25. T.V. Lin; Design of Prestressed Concrete Structures, John Wiley and Sons NY USA.
2.26. P. Kumar Mehta and Paulo. J.M. Monteiro; Concrete, Microstructure, Properties and Materials, Indian Edition, Indian Concrete Institute, Chennai, 1997
2.27. A.M. Neville, Properties of Concrete, Longman Group Limited, England, 1995

3 SELECTED PAPERS

3.1. H.M. Westergaard and M.A. Slater; Moments and Stresses in Slabs, ACI Proc. 17, 1921.
3.2. C.S. Whitney; Plastic Theory of Reinforced Concrete Design, Trans. of ASCE, V. 68, 1942.
3.3. C.P. Siess and N.W. Newmark; Rational Analysis and Design of Two-way Concrete Slab, J. of ACI, Dc. 1948.
3.4. E. Hognestad; Yield Line Theory for the Ultimate Flexural Strength of Reinforced Concrete Slabs, J. of ACI, March 1953.
3.5. G.C. Ernst; Ultimate Slopes and Deflections—A Brief for Limit Design, ASCE Trans. 121, 1956.
3.6. A.L.L. Baker; Report of Institution Research Committee on USD, Proc. of Inst. Civil Engineers (London), 1962.
3.7. P. Dayaratnam; Design of Two-way Slab as an Orthotropic Plate, The Indian Concrete Journal, March 1964.
3.8. R.L. Crawford; Limit Design of Reinforced Concrete Slabs, ASCE-EM5, Oct. 1964.
3.9. D.L.N. Rao, P. Dayaratnam and K.K. Charyulu; Ultimate Load of RCC Columns with Initial Curvature in Reinforcement, The Indian Concrete Journal, Vol. 39. No. 9, 1965.
3.10. Phil. M. Ferguson and J.N. Thomson; Development Length of Large High Strength Reinforcing Bars, J. of ACI, Jan., 1965.
3.11. M.Z. Cohn; Rotational Compatibility in the Limit Design of Reinforced Concrete, SP-12, ACI/ASECE 1965.
3.12. ACI Committee 435, Deflections of Reinforced Concrete Flexural Members, J. of ACI, Jan. 1966.
3.13. ACI Committee 436; Suggested Design Procedure for Combined Footings and Mats, J. of ACI, Oct. 1966.
3.14. ACI Committee 435, Allowable Deflections, J. of ACI, June 1968.
3.15. E.G. Nawy and G.S. Orenstein, Crack width Control in Reinforced Concrete Two-way Slabs, ASCE ST3, Mar. 1970.
3.16. W.G. Corley and J.O. Jirsa; Equivalent Frame Analysis for Slab Design, J. of ACI, Nov. 1970.
3.17. J.G. MacGregor, J.E. Breen and E.O. Pfrang; Design of Slender Columns, J. of ACI, Jan. 1970.
3.18. R.W. Furlong, Column Slenderness and Charts for Design, J. of ACI. Jan. 1971.
3.19. P. Gergely; Distribution of Reinforcement for Crack Control, J. of ACI, May, 1972.
3.20. ACI: Reinforced Concrete Columns, SP-50, ACI, Detrot, 1975.

3.21. F.P. Wiesinger; Yield-Line Method-Strip Method-Segment Equilibrium Method, ASCE National Convention, April 1975.

3.22. D.I. Fraser; Equivalent Frame Method of Beam—Slab Structures, J. of ACI, May, 1977.